XIX International Symposium on Very High Energy Cosmic Ray Interactions (ISVHECRI 2016)

EPJ Web of Conferences Volume 145 (2017)

Moscow, Russia
22 - 27 August 2016

Editors:

B. Pattison

ISBN: 978-1-5108-4326-4

Printed from e-media with permission by:

Curran Associates, Inc.
57 Morehouse Lane
Red Hook, NY 12571

Some format issues inherent in the e-media version may also appear in this print version.

Printed by Curran Associates, Inc. (2017)

For additional information, please contact EDP Sciences – Web of Conferences at the address below.

EDP Sciences – Web of Conferences
17, Avenue du Hoggar
Parc d'Activité de Courtabœuf
BP 112
F-91944 Les Ulis Cedex A
France

Phone: +33 (0) 1 69 18 75 75
Fax: +33 (0) 1 69 28 84 91

contact-edps@webofconferences.org

Additional copies of this publication are available from:

Curran Associates, Inc.
57 Morehouse Lane
Red Hook, NY 12571 USA
Phone: 845-758-0400
Fax: 845-758-2633
Email: curran@proceedings.com
Web: www.proceedings.com

TABLE OF CONTENTS

edp sciences

EPJ Web of Conferences

Statement of Peer review

In submitting conference proceedings to EPJ Web of Conferences, I certify to the Publisher that I adhere to the **Policy on Publishing Integrity** of the journal in order to safeguard good scientific practice in publishing.

1. All articles have been subjected to peer review administered by the proceedings editors.
2. Reviews have been conducted by expert referees, who have been requested to provide unbiased and constructive comments aimed, whenever possible, at improving the work.
3. Proceedings editors have taken all reasonable steps to ensure the quality of the materials they publish and their decision to accept or reject a paper for publication has been based only on the merits of the work and the relevance to the journal.

Title, date and place of the conference

ISVHECRI 2016

XIX International Symposium on Very High Energy Cosmic Ray Interactions

P.N. Lebedev Physical Institute of the Russian Academy of Sciences, Moscow (LPI RAS), 22-27 August 2016

Proceedings editor(s): B. Pattison

Date and editor's signature

30 March 2017

 DOI: 10.1051/epjconf/201714500001

Preface

The 19th International Symposium on Very High Energy Cosmic Ray Interactions (ISVHECRI 2016), held at the P.N. Lebedev Physical Institute of the Russian Academy of Sciences, Moscow (LPI RAS) from 22 to 28 August 2016, attracted more than 120 participants. The Symposium was carried out under the auspices of the International Union of Pure and Applied Physics (IUPAP) with financial support from the Federal Agency for Scientific Organizations and the Russian Foundation for Basic Research.

ISVHECRI is a biennial conference series launched in 1980 at Nakhodka (Russia) dedicated to very high energy cosmic ray research. A specific aspect of the ISVHECRI series is to focus on high energy hadron collisions of cosmic ray particles and related topics including all directions of very, and extremely, high energy cosmic ray studies with air showers. These Symposia highlight the accelerator studies of hadronic interactions at very high energies and the consequent interpretation of the cosmic ray data as well as an understanding of hadronic interactions from cosmic ray data.

Since 2002 this series has brought together cosmic ray physicists and scientists involved with high energy accelerators and colliding beam facilities, including the LHC and RHIC. The previous ISVHECRI meeting at CERN in 2014 was an ideal opportunity for cosmic-ray and particle physicists engaged in accelerator experiments to directly discuss problems of mutual interest following the successful start-up of the Large Hadron Collider.

Topics covered in this Symposium:

- Recent accelerator data and results
- Cosmic Rays and Gamma Rays up to 10^{18} eV
- Extensive air showers
- Hadronic interaction models and cross sections
- Exotic phenomena
- Neutrinos and Inclusive Muons

The Symposium consisted of plenary sessions, including invited review talks, as well as highlight talks and posters selected from the submitted abstracts. A Public Lecture highlighted the history and origins of the Symposium.

The participants had the opportunity to visit the Joint Institute for Nuclear Research with a one-day excursion to Dubna, Moscow Region and learned about the on-going studies at the famous accelerator and nuclear centre as well as to get an idea of new projects under construction such as NICA. Another highlight of the week was the Symposium Dinner held on an evening river cruise on the Moscow River.

The complete Timetable and Speakers' slides can be found on the Symposium website `http://isvhecri-2016.lebedev.ru/en/main`.

We wish to thank the International Advisory Committee, the Scientific Program Committee, the Local Organizing Committee and all the speakers who made this a very successful meeting. Thanks are also due to the administration of the LPI RAS for their support for the Symposium.

Bryan Pattison
Alexander Borisov

Tunka Advanced Instrument for cosmic rays and Gamma Astronomy (TAIGA): Status, results and perspectives

L. Kuzmichev[1,2,a], I. Astapov[9], P. Bezyazeekov[2], V. Boreyko[10], A. Borodin[10], M. Brückner[8], N. Budnev[2], A. Chiavassa[4], O. Gress[2], T. Gress[2], O. Grishin[2], A. Dyachok[2], S. Epimakhov[1], O. Fedorov[2], A. Gafarov[2], V. Grebenyuk[10], A. Grinyuk[10], A. Haungs[14], D. Horns[6], T. Huege[14], A. Ivanova[2], D. Jurov[2], N. Kalmykov[1], Y. Kazarina[2], V. Kindin[9], V. Kiryuhin[2], R. Kokoulin[9], K. Kompaniets[9], E. Korosteleva[1], D. Kostunin[14], V. Kozhin[1], E. Kravchenko[11,12], M. Kunnas[6], V. Lenok[2], B. Lubsandorzhiev[3], N. Lubsandorzhiev[1], R. Mirgazov[2], R. Mirzoyan[5,2], R. Monkhoev[2], R. Nachtigal[6], E. Osipova[1], A. Pakharukov[2], M. Panasyuk[1], L. Pankov[2], A. Petrukhin[9], V. Poleschuk[2], M. Popesku[13], E. Popova[1], A. Porelli[8], E. Postnikov[1], V. Prosin[1], V. Ptuskin[7], A. Pushnin[2], G. Rubtsov[3], E. Ryabov[2], Y. Sagan[10], V. Samoliga[2], F.G. Schröder[14], Yu. Semeney[2], A. Silaev[1], A. Silaev(junior)[1], A. Sidorenko[3], A. Skurikhin[1], V. Slunecka[10], A. Sokolov[11,12], C. Spiering[8], L. Sveshnikova[1], V. Sulakov[1], V. Tabolenko[2], B. Tarashansky[2], A. Tkachenko[10], L. Tkachev[10], M. Tluczykont[6], R. Wischnewski[8], A. Zagorodnikov[2], V. Zurbanov[2], and I. Yashin[9]

[1] Skobeltsyn Institute of Nuclear Physics MSU, Moscow, Russia
[2] Institute of Applied Physics ISU, Irkutsk, Russia
[3] Institute for Nuclear Research of RAN, Moscow, Russia
[4] Dipartimento di Fisica Generale Universiteta di Torino and INFN, Torino, Italy
[5] Max-Planck-Institute for Physics, Munich, Germany
[6] Institut fur Experimentalphysik, University of Hamburg, Germany
[7] IZMIRAN, Moscow, Russia
[8] DESY, Zeuthen, Germany
[9] NRNU MEPhI, Moscow, Russia
[10] JINR, Dubna, Russia
[11] Novosibirsk State Univesity, NSU, Novosibirsk, Russia
[12] Budker Institute of Nuclear Physics SB RAS, Novosibirsk, Russia
[13] ISS, Bucharest, Romania
[14] Institut für Kernphysik, Karlsruhe Institute of Technology, KIT, Karlsruhe, Germany

Abstract. We present the current status of high-energy cosmic-ray physics and gamma-ray astronomy at the Tunka Astrophysical Center (AC). This complex is located in the Tunka Valley, about 50 km from Lake Baikal. Present efforts are focused on the construction of the first stage of the gamma-ray observatory TAIGA - the TAIGA prototype. TAIGA (Tunka Advanced Instrument for cosmic ray physics and Gamma Astronomy) is designed for the study of gamma rays and charged cosmic rays in the energy range 10^{13} eV–10^{18} eV. The array includes a network of wide angle timing Cherenkov stations (TAIGA-HiSCORE), each with a FOV = 0.6 sr, plus up to 16 IACTs (FOV - 10° × 10°). This part covers an area of 5 km^2. Additional muon detectors (TAIGA-Muon), with a total coverage of 2000 m^2, are distributed over an area of 1 km^2.

1. Introduction

The progress in understanding the nature of sources of high-energy cosmic rays from our Galaxy and from the Metagalaxy is going along in three directions:

1. The study of secondary gamma quanta, produced by cosmic rays (CR) in the vicinity of the source, where particles are accelerated (experiments H.E.S.S. [1], VERITAS [2], MAGIC [3], MILAGRO [4], HAWC [5], etc.). About 10 sources with gamma-ray spectra extending up to several tens of TeV have been discovered. This implies that the parent protons or electrons must have been accelerated to several hundreds of TeV. The measured energy spectra, in general, support the standard model of supernova remnants to be the main sources of Galactic TeV radiation. However, not a single source of gamma rays with energies above 100 TeV has been detected so far.
2. High-energy neutrinos produced by hadronic interactions are studied by neutrino telescopes. Several dozen neutrinos with energies up to a few PeV have been registered by IceCube [6]. This "breaking the wall" [7] was achieved after four decades of effort and required an array of one cubic kilometer size. Detection of individual astrophysical sources will likely have to wait for future neutrino telescopes (Baikal-GVD [8], KM3NeT [9] and IceCube-Gen2 [10]) which are going to be completed in the next 5–15 years.

[a] e-mail: kuz@dec1.sinp.msu.ru

 DOI: 10.1051/epjconf/201714501001

3. The precise determination of the energy spectrum, mass composition and anisotropy of CRs by detailed measurements of all EAS parameters. In the past, the Tunka Astrophysical Center in the Tunka Valley (50 km from Lake Baikal) was solely devoted to this third approach. We will continue the study of CRs with the Cherenkov air-shower array Tunka-133 ([11]). This array permits a detailed study of the energy spectrum and mass composition in the energy range from $5 \cdot 10^{15}$ to 10^{18} eV. Complementary opportunities to study the mass composition at $10^{17} - 10^{18}$ eV have been opened by the Tunka Radio Extension array (Tunka-REX) [12] and the scintillation array Tunka-Grande [13].

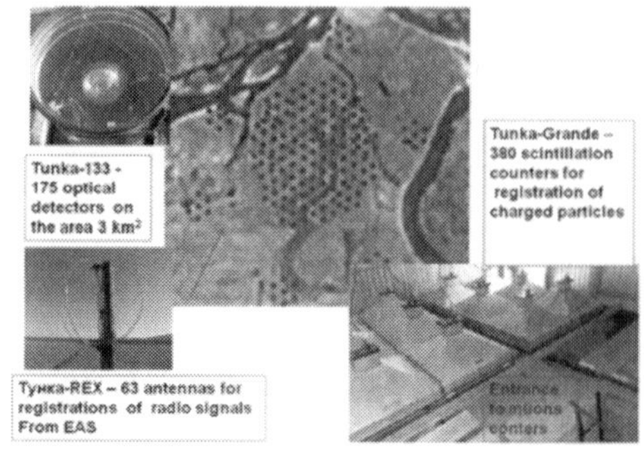

Figure 1. The EAS array complex for study of cosmic rays with $E \geq 10^{15}$ eV at Tunka Valley.

Recently we have also addressed gamma-ray astronomy (i.e. the hitherto most successful way to find individual sources of CRs) and started to deploy a gamma-ray observatory at the Tunka site. The name of this observatory is TAIGA (Tunka Advanced Instrument for cosmic rays and Gamma Astronomy).

In this paper we describe briefly all components of the Tunka Center, present physical results and discuss plans for future extensions.

2. The EAS array complex for CR studies at $E \geq 10^{15}$ eV

At present, three arrays to study charged CRs operate at the Tunka site: Tunka-133, Tunka-REX and Tunka-Grande (Fig. 1). Their measurement of the energy spectrum and mass composition is important in order to understand the acceleration limit of the Galactic CR sources and the transition from Galactic to extragalactic CR.

2.1. Tunka-133

Tunka-133 consists of 175 optical detectors on the basis of PMTs of type EMI 9350, with a hemispherical photocathode of 20 cm diameter.

The detectors are grouped into 25 clusters, each with 7 detectors–six hexagonally arranged detectors and one in the center. The distance between the detectors in the cluster is 85 m. 19 clusters are installed in an inner circle of 500 m radius ("inside" clusters), 6 clusters are placed at a distance of 700–1000 m from the center ("outside" clusters).

An optical detector consists of a 50 cm diameter metallic cylinder, containing the PMT. The PMT output pulses are sent via 95 m coaxial cable RG58 to the center of a cluster and digitized there. The dynamic range of the amplitude measurement is about $3 \cdot 10^4$. This is achieved by means of two channels for each detector extracting the signals from the anode and an intermediate dynode of the PMT with different additional amplification factors.

The cluster electronics includes the cluster controller, four 4-channel FADC boards, an adapter unit for connection with the optical detectors and a special temperature controller. The 12 bit and 200 MHz sampling FADC board is based on AD9430 fast ADCs. Station synchronization and data transmission between DAQ-center and the clusters is realized via optical cables.

2.2. Tunka-REX

Detection of air shower radio emission is another technique that can be used to measure EAS parameters. The radio signal is sensitive to the shower energy and the depth of the shower maximum X_{max}. It is mainly due to the geomagnetic deflection of relativistic electrons and positrons in the shower inducing a time-variable current.

Tunka-REX consists of 63 radio antennae spread over an area of 3 km^2, 57 antennae are placed in an inner circle of 500 m radius. Antennae are connected to the FADC boards in the Tunka-133 clusters and in the Tunka-Grande stations. Readout of the radio data is initialized by the local trigger of the Tunka-133 cluster or scintillation stations and performed via their DAQ systems.

There is a clear correlation between the energy reconstructed from Tunka-REX data and the energy measured by Tunka-133. The Tunka-REX energy precision is about 15%.

A more detailed description of Tunka-REX construction and results is given in [14].

2.3. Tunka-Grande

The Tunka-Grande array consists of 19 scintillation stations, with each of the stations having a surface and an underground part. The surface part includes 12 scintillation counters with size $80 \times 80 \times 4$ cm^3 which formerly operated as part of the EAS-TOP and KASCADE-Grande arrays. There are 8 counters of the same type located underground and operated as muon detectors. The Tunka-Grande DAQ and its synchronization and control systems are practically the same as those of the Tunka-133 array. Each station is operated independently under the station local trigger. Counters of the ground placed counters of the station are divided into two groups, with 6 counters in each. The condition for a station local trigger formation is the presence of a signal from one relativistic particle in each of the two groups in a 100 ns interval.

A description of MC simulation and data processing is presented in [15]. The EAS parameters reconstructed by the Tunka-Grande data are in good agreement with those reconstructed by the Tunka-133 data (Fig. 2).

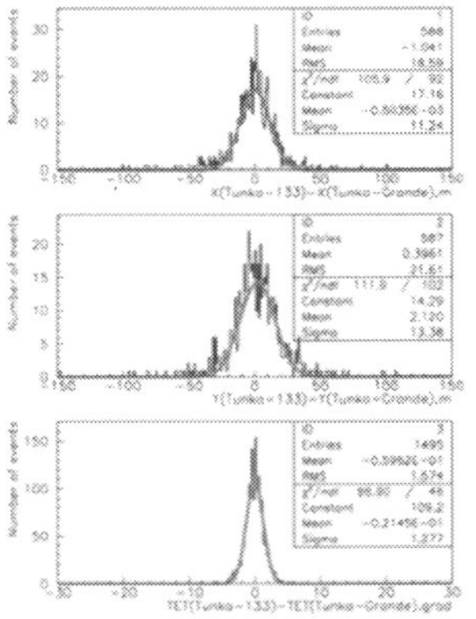

Figure 2. Comparison of the EAS parameters (core position and zenith angle) reconstructed from the data of Tunka-133 and Tunka-Grande.

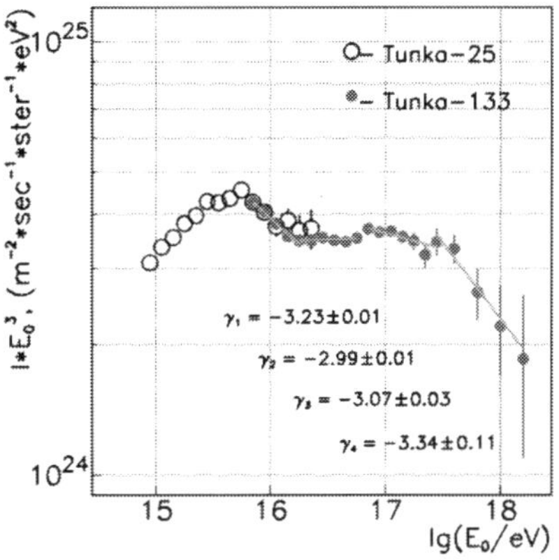

Figure 3. Differential primary cosmic-ray energy spectrum.

3. Short review of the main results

3.1. Methods of EAS parameter reconstruction with Cherenkov light

Our methods of EAS parameter reconstruction have been presented in several papers ([11, 16]). The reconstruction of the EAS core position is performed by fitting measured amplitudes A_i with an amplitude distance function (ADF):

$$A(R) = A(200) \cdot f(R) \quad (1)$$

The function $f(R)$ is a fit to four different parametrizations according to the distance R (in meters) to the shower core [17]. Four variables in this parametrization (R_0, R_{kn}, a and b_2), describing the ADF shape in the different ranges of core distance, are related to a single parameter of the ADF shape–the steepness b_A. The parameter b_A is treated as an independent variable during the minimisation procedure. But if two other independent variables define the core position far from the dense parts of the array, b_A is treated as a fixed parameter. Its value is derived from the value of X_{max}, obtained from the mean width τ_{eff} of the Cherenkov light pulses at a distance of 400 m to the core. The connection τ_{eff} vs. X_{max} and b_A vs. X_{max} has been obtained and is discussed in [17].

As a measure of energy we use the Cherenkov light flux density at a core distance of 200 m–$Q(200)$. Reconstruction of $Q(200)$ is made by fitting the measured values of Q_i with the lateral distribution function (LDF) [17]. The connection between the EAS energy E_0 and $Q(200)$ can be expressed by the following formula:

$$E_0 = C \cdot Q(200)^g \quad (2)$$

For an energy range of 10^{16}–10^{18} eV, a zenith angle range of 0°–45° and a complex composition, consisting of equal contribution of protons and iron nuclei, CORSIKA simulations provide an index $g = 0.94$.

To reconstruct the EAS energy from the Cherenkov light flux one needs to know the absolute sensitivities of the Cherenkov detectors and the atmosphere transparency. To avoid these problems, the method of normalization of the integral energy spectrum to a reference spectrum is used. The reference energy spectrum was measured by the QUEST experiment [18, 19]. The integral energy spectrum obtained for each night of the Tunka-133 operation is normalised to that reference spectrum.

3.2. All particle energy spectrum

The Cherenkov light array Tunka-133 operates on clear moonless nights between October and early April. During the rest of the year, the nights are too short and the weather conditions are mostly unsatisfactory. Tunka-133 took data during 5 winter seasons 2009–2010 to 2013–2014. Here we present the data of these 5 seasons. The total time of data acquisition is 1540 hrs.

To reconstruct the Tunka-133 spectrum we selected events with zenith angles $\theta \leq 45°$ and the core position inside a circle of radius $R_c \leq 450$ m for energy $E_0 < 5 \cdot 10^{16}$ eV and a circle of radius $R_c < 800$ m for showers with energy $E_0 \geq 5 \cdot 10^{16}$ eV. Comparison of the spectra for these two effective areas showed that starting from the above mentioned energy, the spectra within the error bars are the same, but the event statistics in the second case is three times larger which is essential for energies $E_0 \geq 10^{17}$ eV.

The efficiency of shower selection inside the circle with $R_c \leq 450$ m reaches 100% for energies $E_0 \geq 6 \cdot 10^{15}$ eV. The total number of events above this energy is 270000. About 3000 events, selected in the circle with $R_c < 800$ m have $E_0 \geq 10^{17}$ eV.

The resulting differential energy spectrum is shown in Fig. 3 along with the previous spectrum of Tunka-25 [20]. The spectrum of Tunka-133 shows a number of features which indicate deviations from the power law. One can interpret the picture as a much more complicated behaviour than a power law (or multiply broken power law). The power law description can be used only for small parts of the spectrum–not more than half an order of magnitude. At an energy of about $2 \cdot 10^{16}$ eV the power law index changes from $\gamma = 3.23 \pm 0.01$ to $\gamma = 2.99 \pm 0.01$. This feature was first observed in the KASCADE-Grande experiment [21]. Points of the spectrum are consistent with such an index till $E_0 = 5 \cdot 10^{16}$ eV. Above this energy one can notice a single point deflecting from the power law description by about 2 standard deviations. The energy of this point ($6.3 \cdot 10^{16}$ eV) coincides with that for a "peak", observed in the GAMMA experiment [22].

DOI: 10.1051/epjconf/201714501001

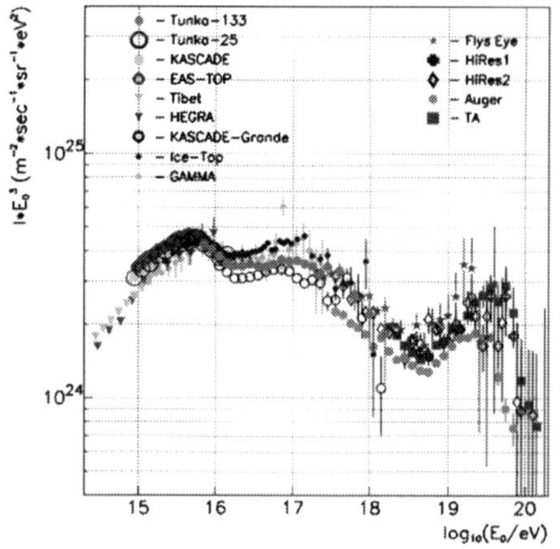

Figure 4. Comparison of energy spectra obtained at Tunka with some other experiments.

Between this point and $E_0 = 3 \cdot 10^{17}$ the index is $\gamma = 3.07 \pm 0.03$. The spectrum becomes much steeper with $\gamma = 3.34 \pm 0.11$ above the last point (the second "knee").

In Fig. 4 the spectrum is compared with a number of other experimental data. The spectra of all the experiments shown in Fig. 4: KASCADE [23], EAS-TOP [24], Tibet [25], HEGRA [26], are practically indistinguishable at the energy of the first (classical) knee.

In the intermediate energy range $10^{16}-10^{17}$ eV, the Tunka-133 spectrum agrees with those measured by GAMMA [22], KASCADE-Grande [21] and Ice-TOP [27]. Noticeably, in Fig. 4, the difference among the spectra at E_0 about 10^{17} eV can be eliminated by an energy correction of only 3%. Such an energy shift is much smaller than the absolute accuracy of these experiments.

For the high energy edge, the Tunka-133 spectrum is in agreement with Fly's Eye [28], HiRes [29] and Telescope Array (TA) [30] data, and also (to a lesser extent) with Auger [31].

The fine structure of the spectrum in the range of $10^{15}-10^{17}$ eV does not contradict the so called rigidity-dependent model of the knee origin [32]: any single galactic source accelerates particles up to a limited maximal energy depending on the charge of nucleus Z, $E_{max}(Z) = Z \cdot E_{max}(Z = 1)$ [33]. In this interpretation the position of the first (classical) knee $3 \cdot 10^{15}$ eV corresponds to the proton cutoff, and the energy $8 \cdot 10^{16}$ eV indicates the cutoff of iron nuclei ($Z = 26$). The existence of more energetic particles can demand the discussion of new galactic sources of unknown nature or (less probable) extragalactic sources.

4. From high-energy cosmic rays to multi-TeV gamma-ray astronomy

For the energy range of gamma quanta above 30 TeV there are a number of fundamental questions which presently have no answers. First of all, there is the question on the sources of Galactic cosmic rays with energies around 1 PeV, the energy region approximately adjoining the classical knee in the all-particle energy spectrum. It should be noted that no single photon with an energy of more than 80 TeV has been detected till now.

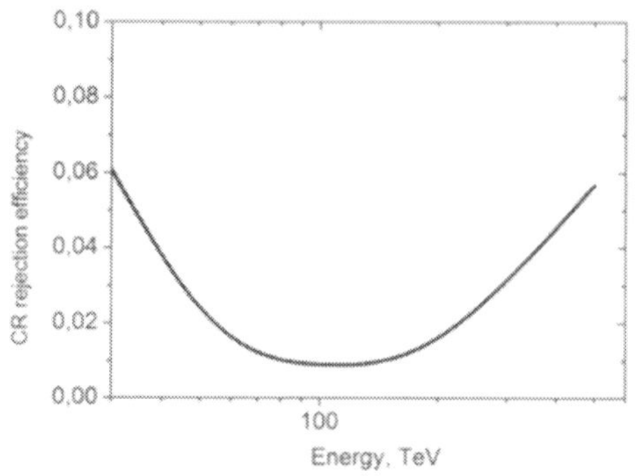

Figure 5. CR rejection efficiency for point-like sources (preliminary).

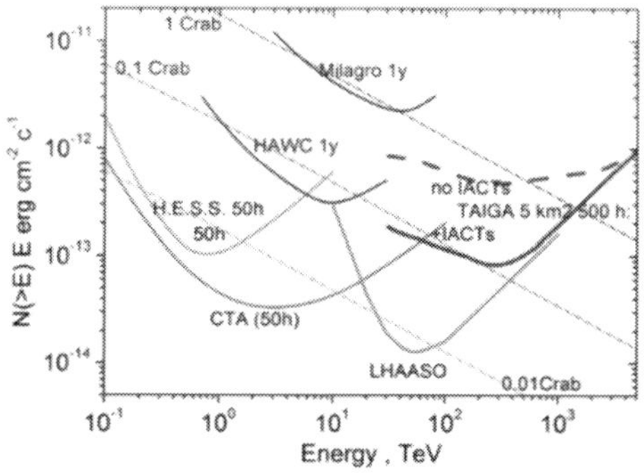

Figure 6. Integral sensitivity for detection of local sources.

4.1. Concept of the TAIGA experiment

The gamma-ray Observatory TAIGA is designed to study gamma radiation and the charged cosmic rays in the energy range of $10^{13}-10^{18}$ eV [34]. The observatory will include a network of wide field of view (0.6 sr) timing Cherenkov light stations, named TAIGA-HiSCORE [35,36] (High Sensitivity Cosmic Origin Explorer), and up to 16 imaging atmospheric Cherenkov telescopes (FOV 10×10 degrees), covering an area of $5\,km^2$. The capabilities of these Cherenkov arrays are enhanced by muon detectors (TAIGA-Muon) with a total coverage of $2000\,m^2$, distributed over an area of $1\,km^2$.

The advantage of a few IACTs added to the wide-angle timing array is their better gamma/hadron separation by image parameter information, while core position, direction and energy can be better reconstructed by the timing array. Combination of information from both components gives a superior result compared to the performance of each single component.

This allows to reduce the CR showers by about 100 times at an energy of 100 TeV for point-like sources, even when the distance between the IACTs is 600 m (Fig. 5). The detection sensitivity for local sources of a 5 km^2 observatory in the energy range of 30–200 TeV is expected to be 10^{-13} erg cm^{-2} sec^{-1} for 500 h of observation or 10 detected events (Fig. 6).

4.2. The TAIGA prototype-2017

By the end of 2017, the prototype of TAIGA (Fig. 7) will start operation. The number of detectors (60 stations

DOI: 10.1051/epjconf/201714501001

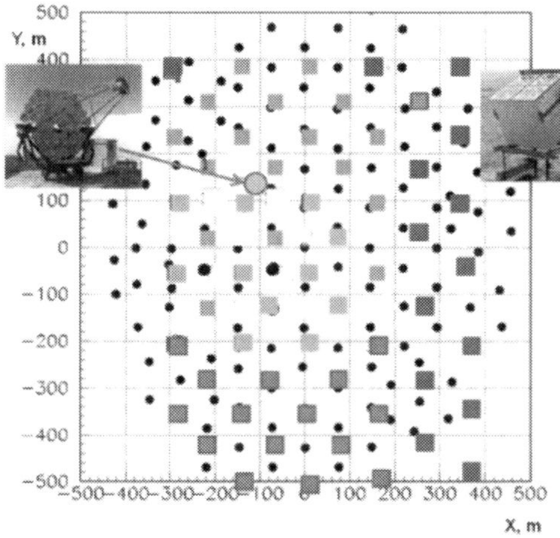

Figure 7. The TAIGA-prototype-2017. Green squares – optical stations of TAIGA-HiSCORE installed in 2014, blue squares - optical stations to be installed in 2017. Yellow circle – position of the first TAIGA-IACT. Small black circles - optical detectors of Tunka-133.

Figure 8. Left figure: Cluster of 28 PMTs. Right figure: The first TAIGA-IACT.

of TAIGA-HiSCORE and one IACT) and the area of the prototype (0.6 km^2) are approximately 1/10 of the full observatory. The expected integral sensitivity for 200 hours of observation (about 2 seasons of operation) in the range 30–200 TeV is about 10^{-12} erg cm^{-2} sec^{-1}.

The imaging atmospheric Cherenkov telescope is of Davis-Cotton type with 34 mirrors, 60 cm diameter each, the focal length is 4.75 m, the camera comprises 560 PMTs of XP1911 type with 2 cm photocathode diameter. The FOV of the camera is $10° \times 10°$ [34].

The camera consists of identical clusters, each with 28 PMTs (Fig. 8, left figure). The basis of the cluster electronics is a 64-channel ASIC MAROC-3. Each channel includes a preamplifier with adjustable gain, a charge sensitive amplifier and a comparator with adjustable threshold. This chip has a multiplexed analog output signal which is proportional to the input charge. The chip is connected to a 12-bit external ADC. Signals from each PMT go to 2 channels with gains different by a factor 30. This results in a full dynamic range of 3000 photoelectrons.

In December 2016 the first TAIGA-IACT was put into operation (Fig. 8, right figure).

Currently, the TAIGA-HiSCORE array is composed of 28 optical stations distributed in a regular grid over a surface area of 0.25 km^2 with an inter-station spacing of 106 m (Fig. 6) [37]. All stations are tilted into the southern direction by 25° to increase the time for study gamma-quanta fluxes from the first test object–the Crab Nebula. Each optical station contains four large area PMTs with 20 or 25 cm diameter, namely EMI ET9352KB, or Hamamatsu R5912 and R7081. Each PMT has a Winston cone (made of ten segments of ALANOD 4300UP foil with reflectivity 80%) with 0.4 m diameter and a 30° viewing angle (field of view is ~0.6 sr).

The anode signals of all 4 PMTs of the station are summed up. It leads to additional lowering of the energy threshold by a factor of 2. The minimal distance between stations is 106 m. Each station is connected with the DAQ center by a fiber optic cable for data transfer and synchronization. The synchronization stability of the optical stations reaches about 0.2 ns. Precision calibration is achieved by external light sources. Each station operates independently from the others. The condition for the formation of the local trigger signal is an excess in the sum of anode signals over a threshold level. The threshold is approximately 200 p.e, which corresponds to a Cherenkov light flux of 0.3 photon cm^{-2} [38]. The counting rate of a single station trigger is about 10-15 Hz. Signals from anode and intermediate dynode are digitized with a step of 0.5 ns by a special board based on the DRS-4 chip. A more detailed description of DAQ and synchronization systems is given in [39].

The energy of primary particles is reconstructed by the Cherenkov light flux, Q200, at a distance of 200 m from the EAS core [16].

Reconstruction of shower parameters was performed using algorithms developed for the Tunka-133 array [11, 16]. Arrival directions of showers are determined by the relative delay of Cherenkov light at each station. In a first step, the arrival direction is reconstructed with a plane wave model of the front. This reconstructed direction is used in the reconstruction of the EAS core. The pulse amplitude is fitted by a parametrization of the amplitude distance function (ADF) [16]. The final EAS arrival direction is reconstructed for the found core position assuming a curved front of the shower.

The accuracy of the reconstruction procedure was checked by MC simulation as well as with experimental data. The latter are presented in [40,41] (chess-board method).

We used two samples of events with zenith angles 28–39° corresponding to the Crab Nebula when it was in the FOV of TAIGA-HISCORE, and for zenith angles 0–25°. For both samples energies were distributed with a power law energy spectrum with index $\gamma = -2.7$ for protons and helium primaries and index $\gamma = -2.3$ for gamma primaries and energy threshold 30 TeV. In simulations the reconstruction method reproduced all the steps of data processing. The accuracy of the arrival direction reconstruction strongly depends on the number of hit stations. The angular resolution is equal to 0.4–0.5°

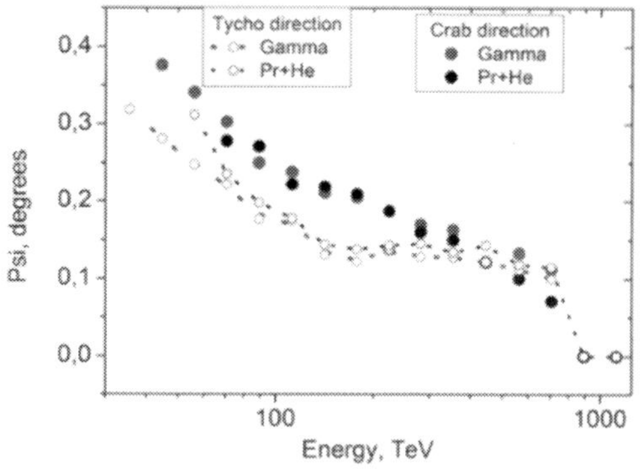

Figure 9. Arrival direction resolution for two samples of events, observed at different zenith angles at the Tunka site: with zenith angles 28–39° (for the Crab Nebula) and with zenith angles 0–25° (for the Tycho Nebula). The slope of the energy spectra is equal to $\gamma = -2.3$ for primary gamma rays and $\gamma = -2.7$ for background events (P+He).

for events with 4 – 5 hit stations and about 0.10° for events with more than 10 hit stations (Fig. 9).

4.3. First results from TAIGA-HiSCORE

About 10 million EAS with simultaneous hits in 4 or more stations were recorded during the 35 clean moonless nights of the 2015–2016 winter season. The total time of data acquisition was 210 hours. The energy spectrum obtained by above described procedure is shown in Fig. 10. The energy threshold of almost 100% efficiency of registration is about 250 TeV. Events with zenith angles less than 15° are used to reconstruct the spectrum below 10^{15} eV and events with zenith angles less than 40° are used for higher energies.

Our preliminary spectrum is compared with the results of previous experiments in the Tunka Valley, as well as with the results of the direct balloon experiment ATIC-2 and the satellite experiment NUCLEON in Fig. 10. Data taking of the latter experiment is ongoing and one can expect increased statistics by at least a factor of 5 by the end of its operation. Presently the extrapolation of our spectrum to the lower energies does not contradict the results of the direct experiments within their statistical errors.

The Crab Nebula with the pulsar at its center is the most prominent and most studied source in TeV gamma-ray astronomy. Therefore it is used as a standard candle for calibration of gamma - ray telescopes. Moreover it is one of a few sources which radiates in the energy range of 10–80 TeV, although data spread is still very high. The source can be observed in the field of view of the array from October till March for up to 3.5 hours per night. The maximum full observation time of the source for the described array is about 230 hours per year. The *typical* observation time is twice less due to bad weather. In the winter of 2015–2016 the achieved observation time was about 60 hours.

The most promising energy range for the gamma-ray search is the range close to the array threshold. The shower core position for such events is reconstructed as the center of gravity of the pulse amplitudes. The energy is estimated from the average Cherenkov light

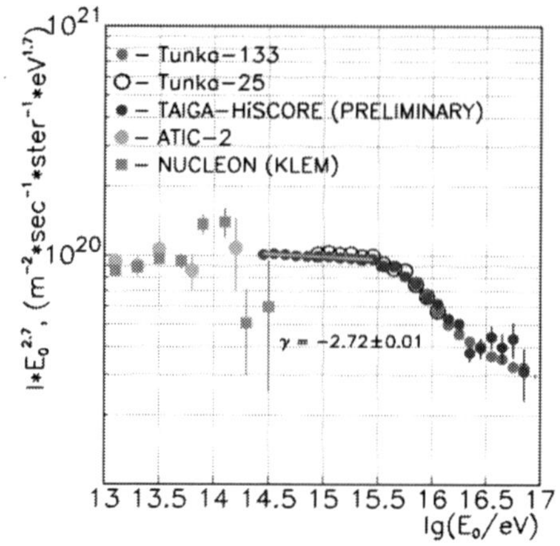

Figure 10. The energy spectrum of primary cosmic rays from the data of the TAIGA-HiSCORE array in comparison with the results of other experiments.

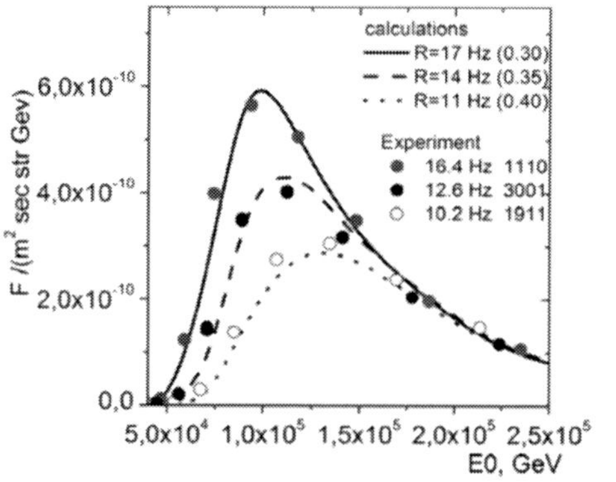

Figure 11. The energy distribution of events in the threshold region.

flux at the four stations closest to the core. The energy distribution for the events close to the array threshold is shown in Fig. 11. The distributions measured in October 2015 and January 2016 are shown separately because the counting rate of the array was different, namely 17 Hz in October and 10 Hz in January. The difference of the counting rate is due to different triggering thresholds of the stations. The CORSIKA simulated energy distribution for the different rates is shown in Fig. 11. The peak energies of these distributions change from 100 to 130 TeV in good agreement with the experimentally measured ones. The gamma-ray shower energy is 1.8 times less than that of the charged particle for the same amount of Cherenkov light. So the peak energy of the gamma-ray flux will be in the range of 50–70 TeV. At this peak energy the expected number of gamma quanta from the Crab is 10–25 events for 100 hours of observation, depending on the used extrapolation of the spectrum from low energies. The experimental event excess within a 0.4° bin around the Crab direction is compatible with those expectations.

A first, fast moving pointlike source with TAIGA-HiSCORE was detected, while observing the night sky during the 2015/2016 winter-season: a LIDAR operating onboard the International Space Satation (ISS). We found several episodes of ∼1 s duration with extremely high array trigger rate (few kHz as compared to the usual 15 Hz array rate), with data being compatible with a 4 kHz periodic light source. The reconstructed event directions are coincident with the trajectory of the International Space Station (ISS), when it passes the zenith over the TAIGA site [40]. The strong light source is the CATS-LIDAR [42] onboard the ISS, emitting $\sim$ 1 mJ at a wavelength of 532 nm at 4 kHz, and pointing almost vertically downwards. This ISS light source is an interesting object for TAIGA calibration, a detailed analysis is underway.

4.4. TAIGA-Muon

We plan to develop a rather cheap and effective detector of muons for the future TAIGA-Muon array [43]. The prototype detector consists of a polystyrene based scintillator in the form of an isosceles triangle. The length of its hypotenuse is 98 cm, the thickness of the plate is 2 cm. Along the legs of the triangle two wavelength shifting bars are arranged. The bars are made from PMMA plastic doped with BBQ dye. The cross section of the bar is 3×17 mm^2. The light from the bars is detected by a PMT of type FEU-84 (QE = 12% at 500 nm). The reflectors made from PTFE film are placed on the far ends of the bars. The prototype was tested with cosmic ray muons in several points. The mean number of detected photoelectrons from a single muon varies from 9 to 15. Based on these results we estimate the number of photoelectrons in the future counter to be larger than 20. This will be achieved by using PMTs with higher quantum efficiency than that of FEU-84 and by optimizing the amount of BBQ dye in the WLS bar with larger cross-section 5×20 mm^2. We assume that 20 detected photoelectrons will be enough to ensure 100% detection efficiency of a cosmic muon and a long lifetime of the detector.

5. Conclusion

A unique complex of installations for the study of high-energy cosmic rays is constructed at the Tunka Astrophysical Center. The complex includes the Tunka-133 array registering EAS Cherenkov light, Tunka-REX for registration of EAS radio emission and the Tunka-Grande array for registration of charged particles (electrons and muons). The joint operation of these arrays opens new opportunities for studying the cosmic radiation with energies more than 10^{16} eV.

The first significant steps towards the gamma-ray observatory TAIGA have been done. The first IACT of TAIGA is now in the commissioning stage.

The first season of operation of the TAIGA-HiSCORE prototype as part of the TAIGA prototype with its 28 optical stations was rather successful and demonstrated good performance and high reliability of equipment.

32 additional stations will be deployed in 2017. Before the winter season 2017–2018 the TAIGA configuration will include 60 wide angle stations arranged over an area of 0.6 km^2, and one single IACT. The expected integral sensitivity for 200 hours of a source observation (about 2 seasons of operation) in the range 30–200 TeV is about 10^{-12} erg cm^2 sec^{-1}.

This work was supported by the grant 15-12-20022 of the Russian Science Foundation, the Russian Federation Ministry of Education and Science (agreement 14.B25.31.0010, zadanie 3.9678.2017/BCh, 3.904.2017/PCh) (part 4), by the Deutsche Forschungsgemeinschaft (DFG) and by the German Helmholtz Association.

References

[1] A. Abramowski et al. (HESS Colaboration), Nature **531**, 476 (2016)
[2] V. Acciari et al. (Veritas Collaboration), Astrophys. J. **730**, L20 (2011)
[3] S. Ansoldi et al. (MAGIC Collaboration), Astrn. Astrophys. **585**, A133 (2016)
[4] A. A. Abdo et al., Astrophys. J. **700**, L127 (2009)
[5] A. Abeysekara et al. (HAWC Collaboration), arXiv:1701.01778 (2017)
[6] M. Aartsen et al. (IceCube Collaboration), Phys. Rev. Lett. **115**, 081102 (2015)
[7] C. Spiering, Proc. 16th Lomonosov Conference on Elementary Particle Physics, Moscow (2013) 105
[8] A. Avrorin et al. (Baikal Collaboration), EPJ Web Conf. **121**, 05003 (2016)
[9] S. Adrian-Martinez et al. (KM3NeT Collaboration), J. Phys. G: Nucl. Part. Phys. **43**, 084001 (2016)
[10] M. Aartsen et al. (IceCube Collaboration), arXiv:1510.05228
[11] S. Berezhnev et al. (Tunka Collaboration), NIM A **692**, 98 (2012)
[12] P.A. Bezyazeekov et al. (Tunka-REX Collaboration), JCAP **01**, 052 (2016)
[13] R. Monhoev et al. (Tunka Collaboration), Bull. Russian Academy of Sciences: Physics **81**, 504 (2017)
[14] D. Kostiunin et al. (Tunka-REX Collaboration), these proceedings
[15] N. Budnev et al., Moscow University Physics Bulletin **4**, 357 (2014)
[16] S. Berezhnev et al (Tunka Collaboration), NIM A **756**, 94 (2014)
[17] Prosin V.V., et al., NIM A **756**, 94 (2014)
[18] E. Korosteleva et al., Int.J.Mod Phys. A **20**, 6837–6839 (2005)
[19] E.E. Korosteleva et al., Nuclear Physics B (Proc. Supp.) **165**, 74–80 (2007)
[20] Budnev N., et al., Astropart. Phys. **50**, 18–25 (2013)
[21] Apel W.D., et al. (KASCADE-Gr. Collaboration), Astropart. Phys. **36**, 183 (2012)
[22] Garyaka A.P., et al., Journal of Physics G: Nuclear and Particle Physics **ID=35**, 115201 (2008)
[23] Antony T., et al., Nucl. Instr. and Methods in Physics Research A **513**, 490 (2003)
[24] Aglietta M., et al., EAS-TOP Collaboration, Astropart. Phys. **10**, 1 (1999)
[25] Amenomori M., et al., (Tibet Col.), Astrophys. J. **678**, 1165 (2008)
[26] Arqueros F., et al., The HEGRA Collaboration, Astron. Astrophys. **359**, 682 (2000)
[27] Aartsen M.G., et al., Phys. Rev. Lett. D **88**, 042004 (2013)
[28] Abu-Zayyad T., et al., Phys. Rev. Lett. **84**, 4276 (2000)
[29] Sokolsky P. for the HiRes Collaboration, Nuclear Physics B (Proc. Suppl.) **212-213**, 74 (2011)
[30] Abu-Zayyad T., et al., Astropart. Phys. **48**, 16 (2013)

[31] Schulz A. for the Pierre Auger Collaboration, Proc. 33rd ICRC Rio De Janeiro **ID=769** (2013)

[32] Sveshnikova L., et al., Nuclear Physics B (Proc. Suppl.) **218-224**, 256 (2014)

[33] Erlykin A., Wolfendale A., Martirosov R., CERN-Courier **21**, 201 (2011)

[34] N. Budnev et al. (TAIGA Collaboration), Journal of Physics: Conference Series **718**, 052006 (2016)

[35] M. Tluczykont et al., Adv. Space Res. **48**, 1935–1941 (2011)

[36] M. Tluczykont et al., Astropaticle Phys. **56**, 42 (2014)

[37] I.Yashin et al (TAIGA Collaboration), ICPPA-2015, J. of Physics: Conference Series **675**, 032037 (2015)

[38] S.F. Berezhnev et al. (TAIGA Collaboration), Bull. Russ. Acad. Sci. Phys. **79**, 38 (2015)

[39] Gress O. et al. (TAIGA Collaboration), NIM A **845**, 367 (2017)

[40] M.Tluczykont et al (TAIGA Collaboration), Proceedings of RICAP16 6th Roma International Conference on Astroparticle Physics. EPJ Web Conferences

[41] A.Porelli et al. (TAIGA Collaboration) Proceeding of ECRS 2014, J.of Phys: Conf.Ser. **632**, 012041 (2015)

[42] `http://cats.gsfc.nasa.gov/`

[43] Budnev N. et al. (TAIGA Collaboration), NIM A **845**, 384 (2017)

The ALPACA Project

Masato Takita[a] for the ALPACA Collaboration

Institute for Cosmic Ray Research, the University of Tokyo, 5-1-5, Kashiwanoha, Kashiwa-shi, Chiba 277-8582, Japan

Abstract. We have started up the ALPACA (Andes Large area PArticle detector for Cosmic ray physics and Astronomy) project. The ALPACA experiment is composed of an 83,000 m^2 air shower array and a 5,400 m^2 underground muon detector array to make wide field-of-view high-sensitivity observations of high-energy cosmic rays/cosmic gamma rays on the Cerro Estuqueria highland, 4,740 m above sea level around Mount Chacaltaya, Bolivia. We briefly report on the design concept of the new project and its physics targets.

1. Introduction

In the southern hemisphere, the HESS group reported the discovery of 14 new gamma-ray sources by the galactic plane survey [1]. Surprisingly, most of them were UNIDentified (UNID) sources and faint in X-rays or other wavelengths. As the HESS survey was limited within the galactic plane in the southern hemisphere due to its narrow field of view, the importance of a wide field-of-view unbiased survey is emphasized. Furthermore, many of the 14 sources have a harder energy spectrum at TeV energies (indices: −1.8 to −2.8) than the standard Crab candle (index; −2.6). Although, the energy spectra were mostly measured below approximately 10 TeV due to the sensitivity limit of the HESS air Cherenkov telescopes, they suggest that they extend up to the 100 TeV region. The number of TeV sources now amounts to approximately 200. However, the origin of the cosmic rays has not yet been clarified.

Cosmic rays are supposed to be accelerated up to the knee energy (PeV) region at supernova remnants (SNRs) in our galaxy. Therefore, we naturally expect gamma rays at 100 TeV energies, which originate in π^0 decays produced by the accelerated cosmic rays interacting with matter surrounding the SNRs. However, on-going experiments focus on measuring gamma rays in the 1–10 TeV region. The gamma-ray emission of electron origin might be highly suppressed above 10 TeV due to the rapid decrease of the inverse-Compton cross section by the Klein-Nishina effect as well as synchrotron radiation energy losses in the strong magnetic field around SNRs.

The detection and spectral measurement of gamma rays in the 100 TeV region from their celestial sources, together with multi-wavelength (radio, X-ray, gamma-ray) observations, will be an important experiment enabling us to discriminate between the two processes (cosmic-ray/electron origins), to locate the acceleration site of cosmic rays (PeVatrons which accelerate cosmic rays up to PeV energies) and to verify the standard acceleration model. Some of the HESS 14 sources are very promising candidates for the cosmic-ray PeVatron. Furthermore, diffuse gamma rays from Fermi bubbles recently reported by the Fermi-LAT group and sub-PeV neutrino events [2] detected by IceCube suggests that Fermi bubbles are PeVatron candidates. Similarly, the energy spectrum of diffuse gamma rays from the extended region around the galactic center marginally measured by HESS up to approximately 10 TeV also strongly indicates the existence of PeVatrons [3], from which we expect to detect gamma rays at 100 TeV energies. Thus, a wide field-of-view gamma-ray imaging at 100 TeV energies in the southern sky, where the HESS sources, the Fermi bubbles and the galactic center are located within the field of view, will be a key experiment.

2. Experiment

The experimental site (approximately 500 m × 500 m ∼250,000 m^2 in total area) is located on a flat high land called Cerro Estuqueria (4,740 m above sea level, 16°23′ S, 68°08′ W), as shown in Fig. 1, around Mount Chacaltaya, near La Paz, Bolivia. Our detectors will be set up in some part of this area.

We plan to set up a 5,400 m^2 underground (approximately one to a few meters) muon detector array (MD) and an 83,000 m^2 air shower array (AS), shown in Fig. 2. The MD of the water Cherenkov type is composed of eight pools with each pool (approximately 1 m deep) containing twelve 56 m^2 unit detectors. AS is made up of 401 1 m^2 plastic scintillation counters at 15 m spacing.

The AS field of view is roughly 2 steradians. The expected angular resolution of AS is approximately 1 degree at 5 TeV and 0.2 degrees around 100 TeV for gamma rays. For 100 TeV gamma rays, the AS energy resolution is estimated to be ∼25%. The hadron rejection power of MD is more than 99.9% at 100 TeV, while keeping most of gamma-ray events. The long-term detector stability, angular resolution, pointing accuracy and energy scale can be calibrated by the cosmic ray shadow of the Moon as well as by some of the bright stable TeV gamma ray sources in the southern sky.

[a] e-mail: takita@icrr.u-tokyo.ac.jp

Figure 1. Experimental site for the ALPACA experiment, Cerro Estuqueria (4,740 m above see level, 16°23′ S, 68°08′ W), near Mount Chacaltaya, in Bolivia.

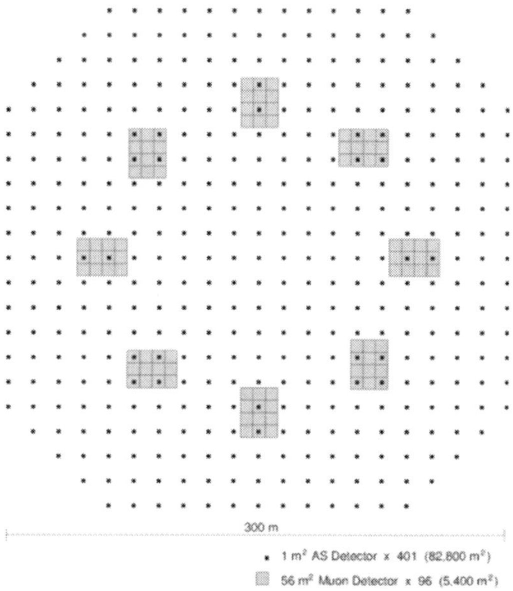

Figure 2. Schematic view of the ALPACA experiment. The small black squares indicate 4011 m^2 plastic scintillator detectors, forming an air shower array with 83,000 m^2 in area. The grey rectangles indicate eight underground muon detector pools, each of which contains twelve 56 m^2 muon detector units. The total area of the underground muon detector array is 5,400 m^2.

3. Physics targets

We have four physics targets in our new project:

1. Measurement of high-energy (5 TeV–1 PeV) cosmic gamma rays.
2. Measurement of cosmic ray energy spectra around the Knee energy region (100 TeV–100 PeV).
3. Measurement of cosmic ray anisotropy > 5 TeV in the sidereal time frame.
4. Measurement of the Sun shadow in cosmic rays >5 TeV.

We aim at low-background detection of celestial gamma rays in the 100 TeV region with a world-best sensitivity

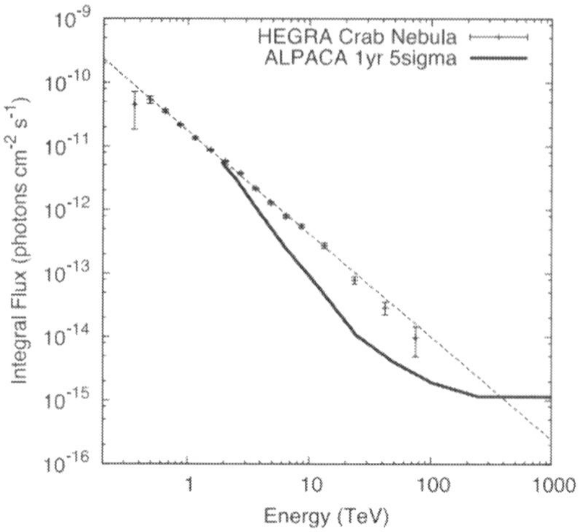

Figure 3. Sensitivity of ALPACA to high-energy gamma-ray point sources. The experimental points are from Ref. [4]. The ALPACA sensitivity is evaluated from Ref. [5].

(an order of magnitude better than any previous/existing experiments) and at locating the origins of cosmic rays accelerated up to the knee energy region in the southern sky. Around 100 TeV, the angular resolution of this air shower array is 0.2 degrees with a wide field-of-view of 2 steradians, while the energy resolution is estimated to be 25%. The hadron rejection power of MD is more than 99.9% at 100 TeV, while keeping most of gamma-ray events. Assuming the presence of a Crab-like gamma-ray source in the southern sky, with a power-law differential energy spectrum proportional to $E^{-2.6}$ extending up to above 100 TeV where E is the energy of a gamma ray, the ALPACA experiment is sensitive to the source with $\sim$15% Crab intensity during one calendar year, as demonstrated in Fig. 3.

This southern-hemisphere AS + MD will be a unique experiment complementary to on-going experiments (FERMI, HESS, VERITAS, MAGIC, CALET, Tibet ASγ, HAWC) and future projects (LHAASO, CTA) in this field, which are either located in the northern hemisphere or aiming at gamma-ray astronomy below 10 TeV region, or having a narrow field-of-view. Thus, the new energy window in the 100 TeV region observing gamma rays with a wide field-of-view will be opened first in the southern sky by the ALPACA experiment. We expect to detect more than a dozen established sources, i.e., young SNRs (SN1006, RX J1713.7-3946, RX J0852.0-4622), Pulsar Wind Nebulae, the galactic center, etc in the 100 TeV region, some of which may be cosmic-ray PeVatron candidates. Furthermore, our wide field-of-view sensitivity to diffuse gamma rays allows us to study extremely diffuse gamma-ray sources which are difficult to detect by IACTs. The diffuse gamma rays from the Fermi bubbles recently reported by the Fermi-LAT group may be clearly detected, if they extend up to the 100 TeV region. Similarly, the detection of diffuse gamma rays above 100 TeV from an extended region from the galactic center is promising, where the gamma-ray energy spectrum strongly suggests the existence of PeVatrons. Detection and spectral measurement of gamma rays in the 100 TeV region from these celestial sources, together with multi-wavelength observations (radio, X-ray, gamma-ray), are key points enabling us to discriminate between the two processes (cosmic-ray/electron origins), to locate the acceleration site of cosmic rays and to examine the standard acceleration model of cosmic rays.

From the astronomical point of view, we pioneer ultra-high energy gamma-astronomy (above 100 TeV) in the southern sky. Besides, gamma-ray emission from near-by extragalactic sources, e.g., M87, Cen A, gamma rays of dark matter origin, those from the Sun disk recently obseved by Fermi [6] may be interesting subjects.

We also aim to measure the energy spectra of proton, helium and iron components separately around the knee energy region with the new AS + MD. The standard cosmic-ray acceleration model at SNRs predicts the knee energy of each nucleus component being proportional to Z (atomic number). We can discriminate proton and iron components by MD, as an iron nucleus produces approximately twice as many muons as a proton with the same energy. Thus, the cosmic-ray acceleration scenario (SNR shock acceleration) will be verified by observing the linearly Z(atomic number)-dependent knee(= bent) positions of proton, helium, iron components around the knee energy region.

Precise cosmic-ray anisotropy measurements in the sidereal time frame in the TeV energy region in the southern sky provide unique data for the community to understand the magnetic field structure in the heliosphere. The ALPACA experiment gives complementary data in the TeV region to those from IceCube above a few tens of TeV.

Furthermore, measurements of the Sun shadow in cosmic rays above the TeV energy region in the southern hemisphere also helps understand the modeling of the magnetic fields between the Sun and the Earth, complementary to the observations in the northern hemisphere.

4. ALPAQUITA

As a pilot experiment, the ALPAQUITA AS without MD, having ~10% of the ALPACA AS in area will be constructed at the experimental site, Cerro Estuqueria, in 2017.

5. Summary

We have started up the ALPACA (Andes Large area PArticle detector for Cosmic ray physics and Astronomy) project which aims at imaging the southern sky by means of gamma rays in the 100 TeV region, composed of a southern-hemisphere air shower array (83,000 m^2) with a large (5,400 m^2) underground water Cherenkov muon detector array in the Andes highland (4,740 m above sea level) in Bolivia. The muon detector array enables us to discriminate the muon-poor gamma-ray component from the muon-rich hadronic component, allowing us to make low-background observations of gamma rays in the 100 TeV region. This will be the first important attempt in the southern sky in this field. It should be stressed that, until now, there has not existed any high-sensitivity experiment capable of probing the gamma-ray southern sky around 100 TeV with a wide field-of-view and high duty cycle.

The ALPACA experiment covers a wide range of cosmic ray physics, e.g., measurement of the chemical composition and energy spectra around the knee energy region, measurement of the cosmic ray anisotropy above the TeV energy region in the southern hemisphere, measurement of the Sun shadow in cosmic rays in the southern hemisphere, etc.

Thus, the expected impacts upon the field are promising, as the project gives an important clue to the problem which has remained over 100 years since the discovery of cosmic rays: the origin, acceleration and propagation mechanism of cosmic rays.

In 2017, we will construct the ALPAQUITA air shower array with ~ one-tenth of the ALPACA air shower array in Cerro Estuqueria, as a pilot expriment.

References

[1] F. Aharonian et al., Science **307**, 1938–1942 (2005)
[2] C. Lunardini et al., PRD **92**, 021301-1–5 (2015)
[3] A. Abramoswski et al., Nature **531**, 476 (2016)
[4] F. Aharonian et al., ApJ **614**, 897 (2004)
[5] T.K. Sako et al., Astroparticle Physics **32**, 177 (2009)
[6] C.Y. Kenny et al., `arXiv:1508.06276v1`

Investigating cosmic rays and air shower physics with IceCube/IceTop

Hans Dembinski[1,2,a] for the IceCube collaboration[b]

[1] Bartol Institute, University of Delaware, 217 Sharp Lab, 19716 Newark, USA
[2] Max-Planck-Institute for Nuclear Physics, PO Box 103980, 69029 Heidelberg, Germany

Abstract. IceCube is a cubic-kilometer detector in the deep ice at South Pole. Its square-kilometer surface array, IceTop, is located at 2800 m altitude. IceTop is large and dense enough to cover the cosmic-ray energy spectrum from PeV to EeV energies with a remarkably small systematic uncertainty, thanks to being close to the shower maximum. The experiment offers new insights into hadronic physics of air showers by observing three components: the electromagnetic signal at the surface, GeV muons in the periphery of the showers, and TeV muons in the deep ice. The cosmic-ray flux is measured with the surface signal. The mass composition is extracted from the energy loss of TeV muons observed in the deep ice in coincidence with signals at the surface. The muon lateral distribution is obtained from GeV muons identified in surface signals in the periphery of the shower. The energy spectrum of the most energetic TeV muons is also under study, as well as special events with laterally separated TeV muon tracks which originate from high-p_T TeV muons. A combination of all these measurements opens the possibility to perform powerful new tests of hadronic interaction models used to simulate air showers. The latest results will be reviewed from this perspective.

1. Introduction

Cosmic-ray research is a conundrum. Despite high-quality data and sophisticated theoretical models, many open questions remain. We can identifiy two reasons why this is so, one concerns the propagation of cosmic rays in space, the other the indirect detection of high-energy cosmic rays. The second issue is connected to theoretical uncertainties in the description of soft quantum chromodynamics, and therefore to an active field of fundamental physics research.

The first obstacle is that cosmic rays are electrically charged. They are bent onto complex paths by the various magnetic fields present in outer space [1]. The arrival direction of a cosmic ray at Earth can be accurately measured, but it cannot be traced back to its source. Therefore, we cannot select between several plausible theories of origin and propagation of cosmic rays [2]. Searches for sources are reduced to searches for anisotropies in the cosmic ray arrival directions. Several experiments have observed such anisotropies (a recent summary can be found in Ref. [3]).

The two remaining identifying properties of a cosmic ray are its energy and mass. Measurements of the flux of cosmic rays combined with the mass composition are able to discriminate between different origin and propagation scenarios [2], but here we face the second obstacle.

Cosmic rays above about 0.1 PeV are too rare and too energetic to be directly measured in detectors mounted on balloons or satellites. One retreats to indirect observation through air showers initiated in Earth's atmosphere. The downside of this approach is that the inference of energy and mass from air shower data is now model-dependent and requires accurate simulation of several cascades of hadronic particle interactions in the regime of soft quantum chromodynamics.

Since these simulations carry uncertainties themselves, they have to be tested in experiments. This requires both dedicated analyses of data from collider experiments which measure particles going into the forward direction [4,5], and complementary analyses of data from air shower experiments themselves. The latter can test models by measuring several observables for the same set of air showers simultaneously. A model passes the test if it is able to describe all observables consistently. To avoid degeneracy, at least three observables are needed; two to determine the primary energy and mass, and a third to test the hadronic model.

Muons in air showers play an important role here. They carry information about the last hadronic interaction that created their parent pion. Due to the near light-like propagation of muons, this information is reflected in the lateral density profile and in the arrival times of muons when an air shower reaches the ground. Several experiments have measured the lateral density profile and found a consistent deficit in the simulated numbers of muons in the lateral profile compared to experimental data [6–8]. This discrepancy is called the muon puzzle.

The IceCube Neutrino Observatory is an excellent detector for measuring cosmic rays and muons in air showers. It has unique capability to shine new light on the muon puzzle: the ability to separately measure the GeV and TeV muon component in air showers (see Fig. 1). Measurements further benefit from the high

[a] e-mail: `hdembins@mpi-hd.mpg.de`
[b] `http://icecube.wisc.edu`

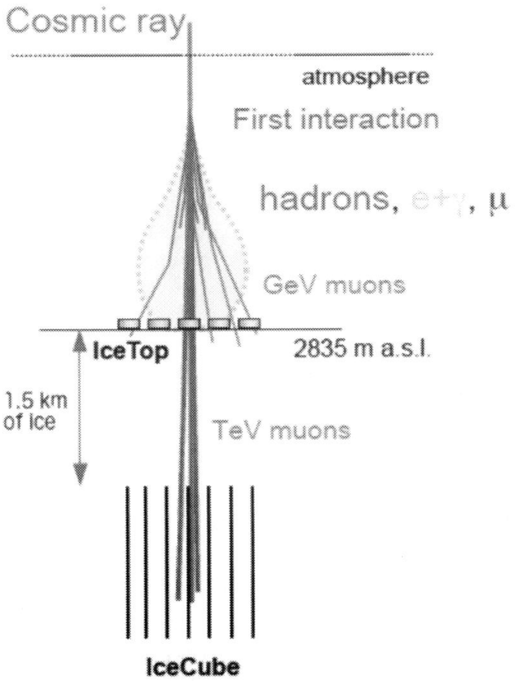

Figure 1. Sketch of an air shower which develops over the IceCube Neutrino Observatory. The IceTop array is close to the shower maximum. Signals near the shower axis are dominated by contributions from electrons and photons, which allows us to infer the shower energy with comparably small systematic uncertainty. Signals far from the shower axis are dominated by hits of GeV muons. TeV muons form collimated particle bundles, and are the only air shower particles apart from atmospheric neutrinos which penetrate the ice shield and reach the deep detector.

altitude of IceTop, its surface detector array. The proximity of an array to the shower maximum significantly reduces the systematic uncertainties in the inferred cosmic-ray energy.

Predictions for muon densities vary in the two energy regimes for different hadronic interaction models, as shown in Fig. 2. For determining the cosmic-ray mass, measurements of TeV muons are better suited, because the model-uncertainty is smaller compared to the effect from the primary mass. A simultaneous measurement of the GeV and TeV muon abundance allows one to test hadronic interaction models.

Although not discussed in this proceeding, we also emphasize that the absence of TeV-scale muons in an otherwise normal-looking air shower is strong evidence for a shower produced by a PeV-scale gamma ray. Limits on the flux of PeV gamma rays have been published [14], and studies with more recent data are ongoing.

In the following we will summarize results obtained with IceCube on the flux of cosmic rays, their mass composition, and on air shower physics.

2. IceCube Neutrino Observatory

The IceCube Neutrino Observatory, shown in Fig. 3, is a cubic-kilometer Cherenkov detector located at the South

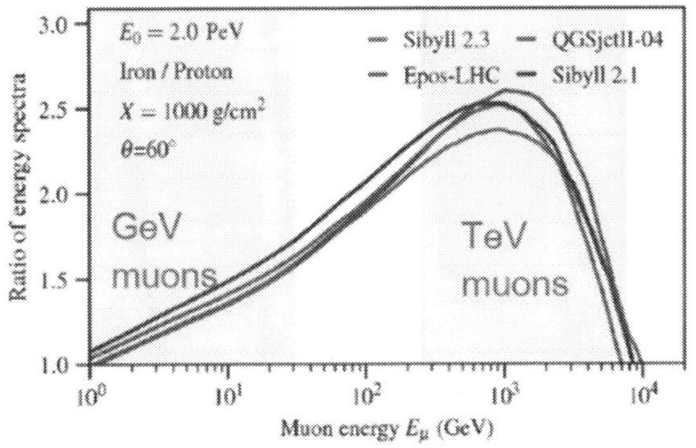

Figure 2. Average ratio of muon spectra in air showers initiated by proton and iron showers as a function of the muon energy. Shown are 2 PeV air showers with a zenith angle of 60° simulated using CORSIKA [9] with several different hadronic interaction models [10–13] to a slant depth of 1000 g.cm^{-2}. The plot was kindly contributed by F. Riehn.

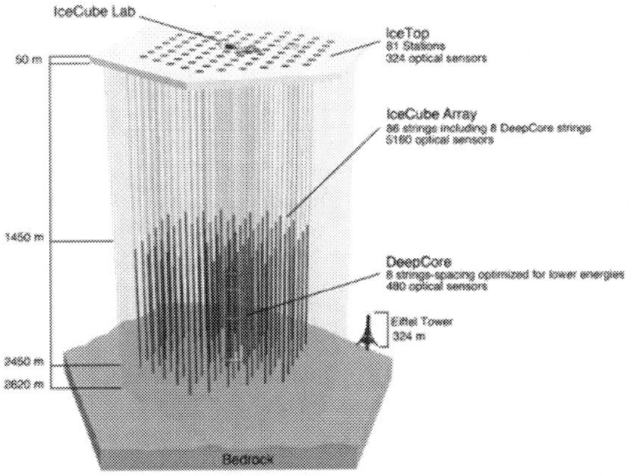

Figure 3. The IceCube Neutrino Observatory. Charged particles produce Cherenkov light in the clear deep ice at the South Pole. Cherenkov light is detected by Digital Optical Modules (DOMs) placed on 86 strings between roughly 1450 m to 2450 m below the surface. The strings in the center are more densely instrumented (DeepCore). The IceTop array is formed by 81 stations with two ice-filled tanks each.

Pole [15]. The main detector, IceCube, is embedded in the Antarctic ice shelf. Its surface component, IceTop [16], is the primary detector for cosmic-ray induced air showers. IceCube detects charged particles via Cherenkov light generated in the ice, which is collected and digitized by Digital Optical Modules (DOMs). The bulk of the DOMs are attached to 86 strings in the deep ice, with a horizontal spacing of 125 m and a vertical spacing of 17 m. The DeepCore sub-detector has a denser instrumentation [17]. IceTop uses the same DOMs embedded in light-tight tanks filled with ice.

IceCube's buried array is protected from low-energy cosmic radiation by a 1.5 km-thick ice shield. The protective layer imposes an energy threshold of about 0.3 TeV for particles with vertical incidence. At the surface, 81 strings are equipped with an IceTop station. Each station consists of two IceTop tanks separated by 11 m. Each tank contains two DOMs with different gain settings to cover a large dynamic signal range.

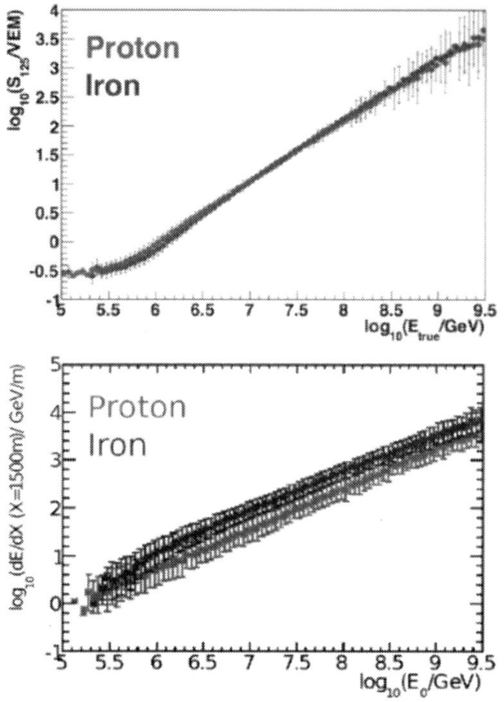

Figure 4. Top: S_{125} as a function of primary energy in simulated air showers, which has minimal sensitivity to composition. Bottom: Muon energy loss at a path length of 1500 m in the ice for the same air showers, which is highly composition-sensitive [18].

IceCube is large and dense enough to observe cosmic rays over three orders of magnitude in energy, from PeV to EeV. The high altitude of the IceTop array of 2835 m above sea level corresponds to an average atmospheric slant depth of 692 g.cm^{-2}. This places IceTop closer to the maximum of shower development than most other air shower experiments.

3. Cosmic-ray energy spectrum and mass composition

Air shower properties are primarily reconstructed from the size and time of signals collected by IceTop. The air shower direction is reconstructed from the arrival times of signals in at least three stations. The achieved angular resolution is better than one degree above a PeV [16].

The energy of an air shower is inferred from the shower size parameter S_{125}, the equivalent signal at a lateral distance of 125 m from the shower axis. The equivalent signal is obtained by fitting a model of the lateral density to the measured signals. The energy resolution obtained is better than 25% above 2 PeV, improving with energy.

Thanks to the high altitude of the IceTop array, the conversion of S_{125} to energy has a very small systematic uncertainty of only 10%, which is illustrated by the overlap of the proton and iron showers in Fig. 4. Another important systematic effect are the varying snow layers on top of the IceTop tanks. While there is very limited snow fall at the South Pole, there is considerable snow drift, which leads to snow buildup of up to 1 m per year. The snow buildup is precisely monitored by shifters at the South

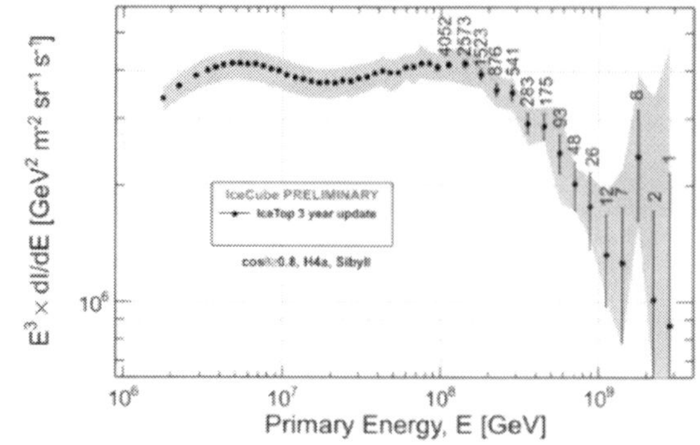

Figure 5. Unfolded all-particle flux of cosmic rays obtained from 3 years of IceTop data, derived using the H4a mass composition assumption. The gray band indicates the systematic uncertainty of the flux [18].

Pole and is corrected for using an exponential attenuation model [16]. The snow-corrected signals show a negligible remaining dependency on the snow height, except for sub-PeV showers near the trigger threshold of the array, which are not used in standard analyses.

The small systematic uncertainty of the energy scale leads to a corresponding small systematic uncertainty in the flux of cosmic rays, shown in Fig. 5. The latest spectrum was derived from three years of data and shows two peaks, the so-called knee around 5 PeV and a second "knee" around 100 PeV. The discovery of the second knee structure is relatively recent, but it is now established by at least three different experiments [19, 20]. Another estimate of the total flux was derived from coincident events seen in IceTop and IceCube, which does not depend on the H4a model and is in perfect agreement with the main result [18].

IceCube was not built with intrinsic particle identification capabilities, but it turns out that muons can nevertheless be separated from other charged particles. GeV muons are detected in IceTop on a statistical basis at large lateral distance from the shower axis, where they leave a characteristic signal (see next section). TeV muons are the only charged particles able to penetrate the 1.5 km-thick ice shield to reach IceCube.

TeV muons are primarily used to estimate the cosmic-ray mass. The number of TeV muons can be roughly deduced from the reconstructed energy loss per path length dE/dX in the ice. For the mass composition analysis, IceCube uses the median energy loss for a reference path length of 1500 m, shown in Fig. 4. The median energy loss is the most important mass-sensitive observable. Since TeV muons are not minimum-ionizing, they also undergo stochastic losses. These show up as spikes in the energy loss profile. The number of such stochastic losses above two different thresholds is used to further refine the discrimination of different masses.

The shower size S_{125}, the zenith angle of the shower, the median energy loss, and the two counts of stochastic energy losses are fed into a neutral network trained on simulations to produce estimates of the energy and mass of the cosmic rays.

The output of the network is binned in the energy and mass estimator for proton, helium, nitrogen, and iron showers simulated with CORSIKA and SIBYLL2.1 [9, 10], and likewise for real events. Binning the output for simulated known masses produces templates, which are fitted to the binned histograms of the mass estimator obtained for real events. This allows us to determine which fraction of the showers in each energy bin are similar in mass to proton, helium, and so on.

With this method, the component spectra shown in Fig. 6 are produced. We observe an initially light composition that becomes increasingly heavy as the energy increases, then stabilizes around 100 PeV [18]. Our measurement indicates a heavier composition around 1 EeV than measurements from the Pierre Auger Observatory based on the depth of shower maximum [21]. This apparent inconsistency could originate from a deteriorating description of TeV muons in SIBYLL2.1 at such high energies (hints for such a deterioration are found in GeV muons and discussed in the next section). We emphasize, however, that the results were cross-checked using QGSJet-II.03 [22], which gave a consistent interpretation in terms of cosmic-ray mass.

A sudden drop in the helium and iron spectra around 6 PeV is observed, with corresponding elevated levels of proton and oxygen. This feature is under intense study. So far the most likely explanation is a statistical fluctuation. We note that the plotted standard error bars cannot be interpreted in the usual way, because they do not show the correlations. The four component spectra are not statistically independent, since they sum up to the smooth total flux by construction.

4. GeV muons in IceTop

IceTop tanks are ice-Cherenkov detectors. A minimum-ionizing particle which passes vertically through the detector produces a characteristic signal, which produces a stable peak in a histogram of photo-electron counts. IceTop uses this very feature for its continuous signal calibration. Signals expressed in units of Vertical Equivalent Muons (VEMs) are comparable from tank to tank and between real and simulated events, while the number of photo-electrons collected per vertical muon hit may vary from tank to tank and over time.

It was determined by IceTop that this effect can be used to statistically separate muon hits from hits of other charged particles in tanks far from the shower axis. If the tank is sufficiently far from the shower axis, the probability of an individual tank to be hit by a particle is much smaller than one. Then, the typical charge deposited by a single muon (for vertical showers it is around 1 VEM by definition of the unit) stands out statistically over the continuum of charges from electrons and photons.

By counting how many tanks in a narrow ring around the shower axis have a muon hit, the average muon density for air showers within a narrow interval in cosmic-ray energy can be derived using Poisson statistics [23, 24] and the known cross-sectional area of an IceTop tank. For near vertical showers, the average lateral muon density profiles derived in this way are shown in Fig. 7. These profiles are raw results, not corrected for biases in the method.

A detailed study of simulated proton and iron showers showed that the muon density obtained in this way is slightly biased. The bias is corrected using these simulations, by taking the average effect for proton

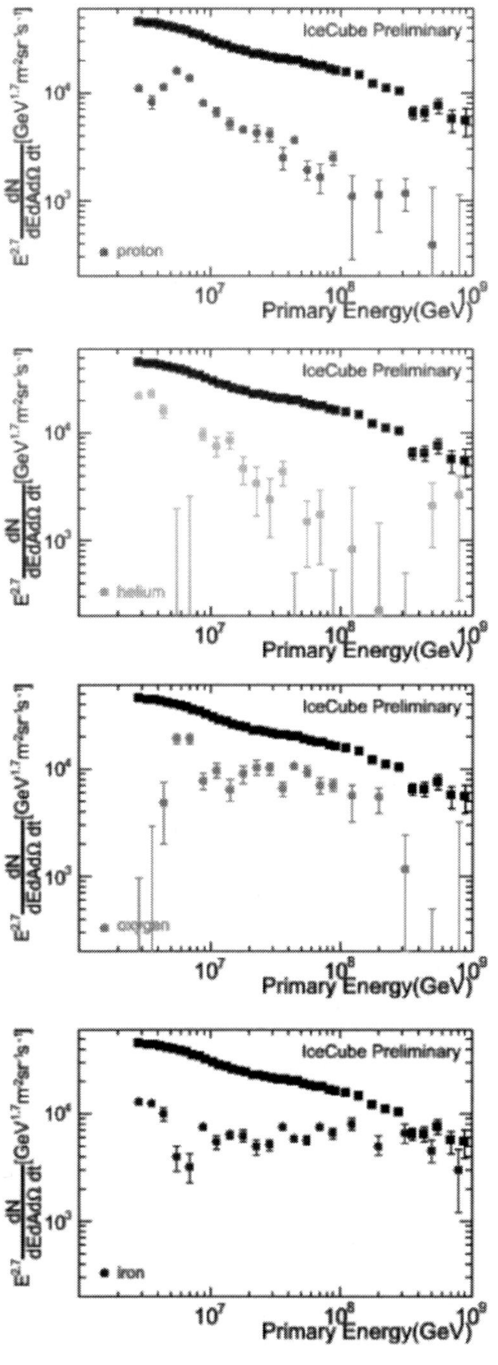

Figure 6. Flux of cosmic rays split into components (colored dots), based on similarity to simulated proton, helium, oxygen, and iron showers. The sum of the components is shown in each plot for reference (black dots) [18].

and iron showers into account. This leads to a small systematic uncertainty due to the mass composition. Another systematic uncertainty arises from a variation of the empirical statistical model of the non-muon hits.

For two lateral distances at 600 m and 800 m the muon density is shown as a function of the air shower energy in Fig. 8, and compared to simulations for proton and iron

DOI: 10.1051/epjconf/201714501003

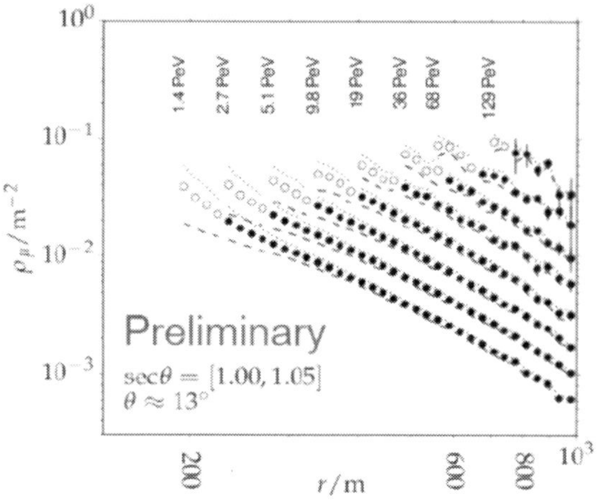

Figure 7. Raw lateral profiles of muon density observed in IceTop for near vertical showers (average zenith angle is 13°) obtained for air showers in several energy bins (average energy is shown in rotated text). Dotted and dashed lines indicate estimates based on alternative models used to fit the background of non-muon signals. The impact of non-muon background on the analysis is larger for bins closer to the shower axis up to a point where the systematic uncertainty becomes too large and results are discarded (open dots). The profiles are raw results obtained from fits of signal distributions, before applying a composition-dependent correction derived from air shower simulations.

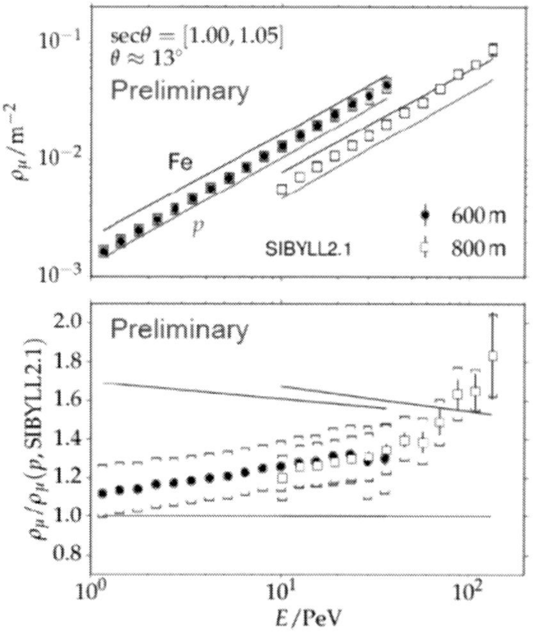

Figure 8. Top: muon density at 600 m and 800 m as a function of primary energy for near vertical showers. Lines show expected values for proton and iron showers simulated with CORSIKA and SIBYLL2.1. Bottom: same as above, but with all densities normalized to the expected density for proton showers. Brackets indicate systematic uncertainties. The densities shown here were corrected for a small composition-dependent bias found when applying the method to air shower simulations.

showers. The measurement is dominated by the systematic uncertainty, which however remains below 15%.

For the hadronic interaction model SIBYLL2.1, we find a similar behavior in GeV muons and in TeV muons. The results are compatible with a light mass composition around 1 PeV and then indicate an increase in the average mass. However, around 100 PeV the muon density rises even beyond the value for iron showers, which is astrophysically not plausible. The more likely explanation is a deterioration of the description of muons in SIBYLL2.1 as the air shower energy increases. This interpretation is consistent with another measurement by the HiRes-Mia group [6], which measured the muon density in air showers from 100 PeV to 1 EeV and found a similar deficit of muons in simulations compared to real showers.

5. High-energy muon flux

As previously mentioned, TeV muons in the deep ice occasionally have stochastic energy losses which generate large localized amounts of light. This experimental signature can be used to identify air showers events with high-energy muons (HE-muons). HE-muons are typically produced by prompt decays of charmed hadrons, which also give rise to prompt neutrinos.

The IceCube collaboration is very interested in reliable predictions of the flux of PeV neutrinos that originate from air showers, because they form the main background to astrophysical neutrinos. Measuring the flux of HE-muons allows one to constrain the flux of atmospheric PeV neutrinos very well.

In a recent study [25], machine learning tools were applied to discriminate air showers with a HE-muon from other events, which was defined as a single muon that carries more than 50% of the energy of the whole muon bundle that reaches the deep detector. The classifier was trained on simulated air showers, using 30 variables which quantify various aspects of the air shower event in IceCube. After training, a signal-to-background ratio of better than 10 is achieved, with about 40% of the signal surviving.

Estimating the energy of the HE-muon based on its energy loss in the deep ice is still difficult, so the energy resolution is not very high. Unfolding methods are used to compensate distortions in the estimated flux due to low energy resolution, and also correct for the detector acceptance and selection losses. The final result is the unfolded flux of HE-muons at the South Pole surface, shown in Fig. 9. The measurement is very promising, but so far still compatible with both a flux predicted by SIBYLL2.1, in which charmed hadrons are not simulated, and an alternative flux where decays of such charmed hadrons have been added.

The situation will improve in the near future. The current measurement is limited by the statistical accuracy of the simulated response matrices which are used in the flux unfolding. More simulations are in preparation to significantly reduce the flux uncertainty between 0.1 and 1 PeV.

6. Laterally-separated TeV muons

The last muon analysis presented here targets a special class of air shower events discovered by IceCube. These

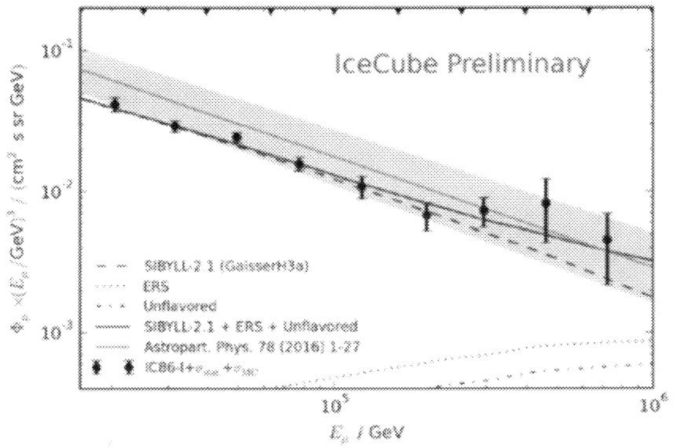

Figure 9. Unfolded spectrum of TeV muons measured with IceCube (black dots). Shown for comparison are a previous result (gray line and error band), a prediction computed with SIBYLL2.1 without prompt hadrons (dashed line), the same with charmed hadrons (ERS) added (solid line) [25].

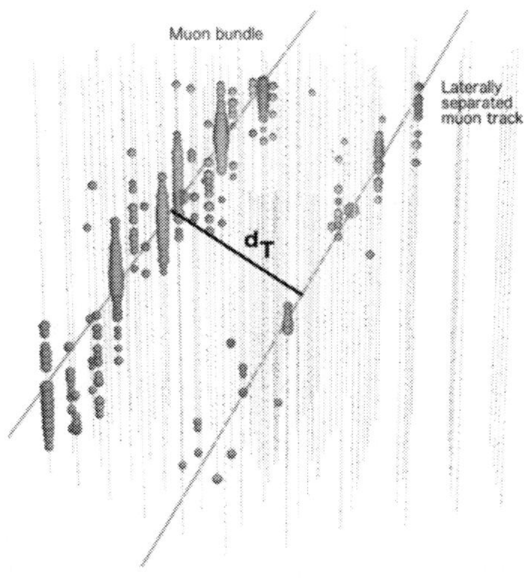

Figure 10. Event in IceCube with a muon bundle and a laterally-separated muon track that arrives in coincidence and with near-parallel direction. The experimental observable for these type of events is lateral distance d_T between bundle and track.

events feature a coincidence of two parallel tracks in the deep ice, which are interpreted as a muon bundle and a single laterally separated muon (LS-muon), see Fig. 10. An early measurement of the distribution of the lateral separation d_T, shown in Fig. 11, has been already published already in 2013 [26].

Because of the light-like propagation of muons, the lateral separation d_T of the LS-muon is geometrically connected to the transverse momentum p_T of the meson which decayed into the LS-muon, roughly described as

$$d_T \simeq \frac{p_T \, H_{\text{int}}}{E_\mu \, \cos\theta}, \tag{1}$$

where H_{int} is the height of the primary interaction of the cosmic ray, E_μ is the energy of the LS-muon and θ is the zenith angle of the air shower.

Using simulations, it was shown that hadrons which produce LS-muons are themselves produced predominantly in the primary interaction of the cosmic ray. LS-muons are therefore direct probes of the primary interaction, which is very sensitive to the mass of the cosmic ray. Simulated distributions of transverse momentum of hadrons produced in the first interaction vary greatly from model to model and with the mass of the cosmic ray, as shown in Fig. 12.

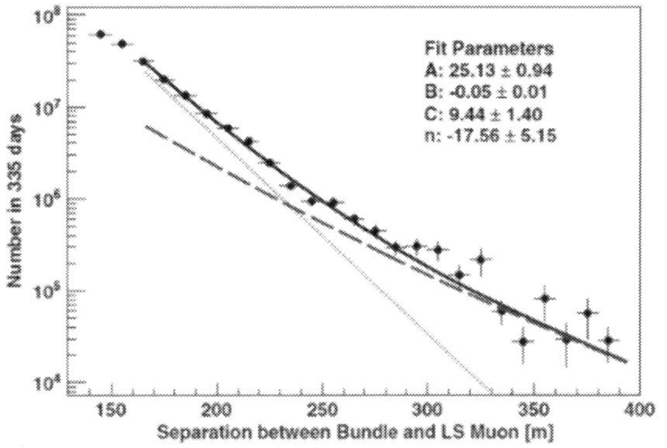

Figure 11. Distribution of lateral separation muon bundle and isolated muon, after subtracting a background contributed by random coincidences. Also shown are the best-fit parameters for the empirical model $N = \exp(A + Bx) + 10^C(1 + x/400)^n$, which describes a transition from an exponential, expected for a soft process, and a power law, expected for a hard process. The P-value for this fit is 4%. For a purely exponential function it is 0.001% [26].

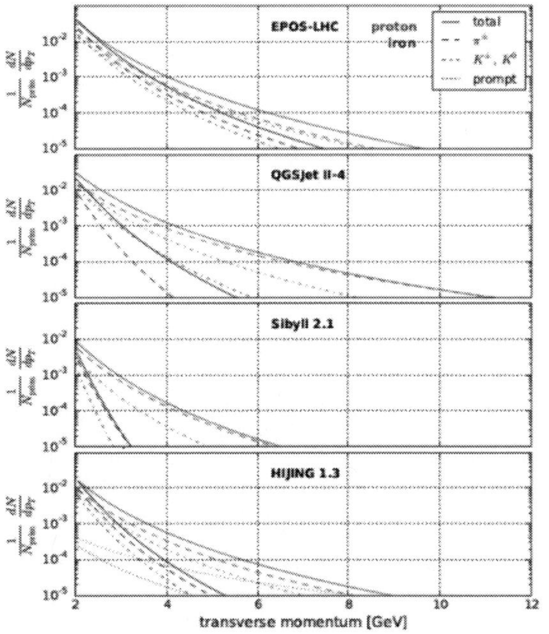

Figure 12. Distributions of transverse momentum for hadrons produced in the first interaction of a 0.1 PeV proton or iron cosmic ray with nitrogen, as predicted by several hadronic interaction models [27].

A new study is underway, which benefits from a larger data set and a new sophisticated simulation approach for laterally separated muons. Muons with large transverse momentum are separately generated and then embedded in regular CORSIKA air showers [27,28]. In the new

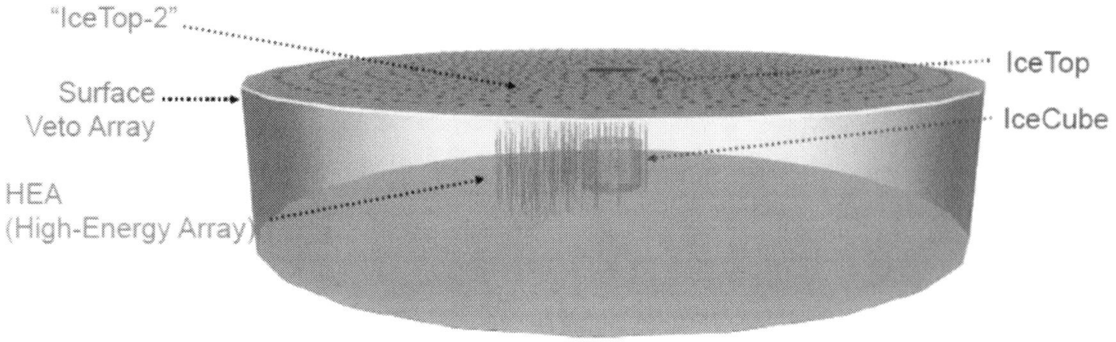

Figure 13. Proposed design for the Generation-Two IceCube Neutrino Observatory. IceCube is extended by the High-Energy Array to $10\,\text{km}^3$, similarly IceTop is extended to $10\,\text{km}^2$. A larger surface of about $75\,\text{km}^2$ is covered by the Surface Veto Array, which consists of simpler detectors and used to veto air shower events for neutrino searches.

approach, all details in the production of LS-muons can be controlled. Also, the Monte-Carlo production of such special events is greatly accelerated, which rarely occur in standard CORSIKA simulations.

The analysis of LS-muons is particularly interesting. If the high-p_T tail of the LS-muon distribution is dominated by hard processes, the tail could be predicted to necessary accuracy based on pertubative quantum chromodynamics. This opens a possibility to avoid the theoretical uncertainties which are always present when conventional hadronic interaction models are used to infer the cosmic-ray mass [29]. The change of slope in Fig. 11 is a hint that the tail is indeed dominated by hard processes. If this approach turns out to be feasible, it would allow us to infer the mass composition of PeV cosmic rays to unprecedented theoretical accuracy.

7. Conclusion and outlook

The IceCube Neutrino Observatory is a powerful and versatile cosmic-ray detector. It has the unique ability to detect TeV muons, which along with neutrinos are the only air shower particles able to reach the deep detector under the ice shield. For the first time, predictions from hadronic interaction models for TeV muons can be experimentally tested. New measurements, which have not been possible before, include the flux of muons with hundreds of TeV and the lateral distribution of high-p_T muons.

Thanks to the high altitude of the IceTop array, the energy of air showers can be measured with systematic uncertainties smaller than 10%, which results in a very precise measurement of the cosmic-ray flux. IceTop is among the first experiments which observed the so-called second knee structure around 100 PeV.

Finally, the ice-Cherenkov tanks of the IceTop array allow us to statistically identify GeV muon hits based on the deposited signal alone, and correspondingly measure the lateral density of GeV muons far from the shower axis without additional detector equipment. The preliminary results indicate a deficit of GeV muons in air showers simulated with SIBYLL2.1 around 100 PeV, with is consistent with similar measurements done by other experiments at higher cosmic-ray energies.

Studies are currently ongoing for a proposed extension of the IceCube Neutrino Observatory, shown schematically in Fig. 13. The goal is to increase IceCube to $10\,\text{km}^3$ and likewise IceTop to $10\,\text{km}^2$. This would extend the accessible range of cosmic-ray energies to 3 EeV and the rate of coincident events observed in both IceTop and IceCube by a factor of 50, due to the increased acceptance in the zenith angle. Also planned is an even larger veto array of about $75\,\text{km}^2$ made of simpler detectors to veto air showers for neutrino studies [30].

Both the near and far future of cosmic-ray research with IceCube/IceTop are very promising, with many interesting results in preparation.

References

[1] R. Jansson, G.R. Farrar, Astrophys. J. **761**, L11 (2012), 1210.7820
[2] K.H. Kampert, M. Unger, Astropart. Phys. **35**, 660 (2012), 1201.0018
[3] M.G. Aartsen et al. (IceCube) (2017), 1701.03731
[4] A. Van Spilbeeck, PoS **DIS2016**, 192 (2016)
[5] M. del Prete et al., EPJ Web Conf. **126**, 04014 (2016)
[6] T. Abu-Zayyad et al. (MIA, HiRes), Phys. Rev. Lett. **84**, 4276 (2000), astro-ph/9911144
[7] A. Aab et al. (Pierre Auger), Phys. Rev. **D90**, 012012 (2014), 1407.5919
[8] R.P. Kokoulin et al., Nucl. Phys. Proc. Suppl. **196**, 106 (2009)
[9] D. Heck, G. Schatz, T. Thouw, J. Knapp, J. N. Capdevielle, Tech. Rep. FZKA-6019, Karlsruhe, Germany (1998)
[10] R.S. Fletcher, T. Gaisser, P. Lipari, T. Stanev, Phys. Rev. **D50**, 5710 (1994)
[11] F. Riehn, R. Engel, A. Fedynitch, T. Gaisser, T. Stanev (2015), 1510.00568
[12] S. Ostapchenko, Phys. Rev. **D83**, 014018 (2011), 1010.1869
[13] T. Pierog, I. Karpenko, J.M. Katzy, E. Yatsenko, K. Werner, Phys. Rev. **C92**, 034906 (2015), 1306.0121
[14] M.G. Aartsen et al. (IceCube), Phys. Rev. **D87**, 062002 (2013), 1210.7992
[15] M.G. Aartsen et al. (IceCube), (2016), 1612.05093
[16] R. Abbasi et al. (IceCube), Nucl. Instrum. Meth. **A700**, 188 (2013), 1207.6326
[17] T. DeYoung (IceCube), Nucl. Instrum. Meth. **A692**, 180 (2012), 1112.1053
[18] K. Rawlins, T. Feusels (IceCube), PoS **ICRC2015**, 334 (2016)

[19] Z. Zundel (Telescope Array), PoS ICRC2015, 445 (2016)
[20] V.V. Prosin et al., EPJ Web Conf. **99**, 04002 (2015)
[21] M. Settimo (Pierre Auger), J. Phys. Conf. Ser. **718**, 052037 (2016)
[22] S. Ostapchenko, Nucl. Phys. Proc. Suppl. **151**, 143 (2006), hep-ph/0412332
[23] H.P. Dembinski, J. Gonzalez (IceCube), PoS **ICRC2015**, 267 (2016)
[24] J.G. Gonzalez (IceCube), J. Phys. Conf. Ser. **718**, 052017 (2016)
[25] T. Fuchs (IceCube), Development of a Machine Learning Based Analysis Chain for the Measurement of Atmospheric Muon Spectra with IceCube, in 25th European Cosmic Ray Symposium (ECRS 2016) Turin, Italy, September 04-09, 2016 (2017) 1701.04067
[26] R. Abbasi et al. (IceCube), Phys. Rev. **D87**, 012005 (2013), 1208.2979
[27] D. Soldin (IceCube), EPJ Web Conf. **99**, 06001 (2015), 1411.4448
[28] D. Soldin , PoS **ICRC2015**, 256 (2016)
[29] S.R. Klein, Nucl. Phys. Proc. Suppl. **175-176**, 346 (2008), astro-ph/0612051
[30] M.G. Aartsen et al. (IceCube) (2014), 1412.5106

DOI: 10.1051/epjconf/201714502001

Astrophysical neutrinos and atmospheric leptons

T.K. Gaisser[a]

Bartol Research Institute and Department of Physics and Astronomy, University of Delaware, Newark, DE 19716, USA

Abstract. IceCube measurements of the neutrino flux from TeV to PeV show the signal of astrophysical neutrinos standing out at high energy well above the steeply falling foreground of atmospheric neutrinos. The astrophysical signal appears both in measurements of neutrino-induced muons and in the starting event sample, which responds preferentially to electron and tau neutrinos, but which also includes muon neutrinos. Searches for point sources of astrophysical neutrinos have, however, not yet identified a single source or class of sources for the astrophysical component. Some constraints on astrophysical sources implied by the current observations will be described in this talk. Uncertainties in the fluxes of atmospheric leptons resulting from an incomplete knowledge of the primary cosmic-ray spectrum and from a limited understanding of meson production, including charm will also be reviewed. The ultimate goal is to improve the understanding of the astrophysical spectrum in the transition to lower energy where atmospheric neutrinos dominate.

The main aspects of this presentation will be included in the author's Review Talk at the end of the Symposium.

[a] e-mail: `gaisser@bartol.udel.edu`

Cosmic Ray Astrophysics using The High Altitude Water Cherenkov (HAWC) Observatory in México

Eduardo de la Fuente[1,a], Juan Carlos Díaz–Vélez[2,3,b], Alberto Hernández Almada[3,c], Alberto Nigoche–Netro[1,d], and THE HAWC Collaboration[e]

[1] Instituto de Astronomía y Meteorología, Departamento. de Física, Centro Universitario de Ciencias Exactas e Ingenierías (CUCEI), Universidad de Guadalajara, AV. Vallarta 2602, CP. 44130, Guadalajara, Jalisco, México
[2] Centro Universitario de los Valles (CUVALLES), Universidad de Guadalajara, Ameca Km. 45.5, CP. 46600, Ameca, Jalisco, México, and WIPAC & Department of Physics, University of Wisconsin–Madison, Madison, WI 53706, USA
[3] Instituto de Física, Universidad Nacional Autónoma de México, CU, Av. Universidad 3000, CP. 04510, Ciudad de México, México

Abstract. The High-Altitude Water Cherenkov (HAWC) TeV gamma–ray Observatory in México is ready to search and study gamma-ray emission regions, extremely high-energy cosmic-ray sources, and to identify transient phenomena. With a better Gamma/Hadron rejection method than other similar experiments, it will play a key role in triggering multi–wavelength and multi–messenger studies of active galaxies (AGN), gamma-ray bursts (GRB), supernova remnants (SNR), pulsar wind nebulae (PWN), Galactic Plane Sources, and Cosmic Ray Anisotropies. It has an instantaneous field-of-view of ~2 str, equivalent to 15% of the whole sky and continuous operation (24 hours per day). The results obtained by HAWC–111 (111 detectors in operation) were presented on the proceedings of the International Cosmic Ray Conference 2015 and in [1]. The results obtained by HAWC–300 (full operation) are now under analysis and will be published in forthcoming papers starting in 2017 (see preliminary results on `http://www.hawc-observatory.org/news/`). Here we present the HAWC contributions on cosmic ray astrophysics via anisotropies studies, summarizing the HAWC detector and its upgrading by the installation of "outriggers".

1. Introduction

When high-energy particles or photons, referred to as primaries, with $E > E_c$ (where E_c ~600 MeV /Z E_c is the critical energy and Z the atomic number) enter the Earth's atmosphere (Z=7) at an altitude of ~30 km, an extensive air shower (EAS) is induced. EAS are produced by the interaction of primaries with atmospheric nuclei. The theory of EAS is fully described by Bhabha & Heitler (1937) [2] and the pioneer works of Carlson & Oppenheimer (1937) [3] and Zatsepin (1949) [4] as a complement.

A primary cosmic ray (usually high energy proton or iron nucleus) produces an EAS with three components:

1. The weak, soft component, or electromagnetic, produces gamma rays through neutral pion decay,
2. the hard component (muons and hadrons), where charged and neutral kaons and pions decay into gamma rays, muons, neutrinos, antineutrinos, and nuclear fragments like protons and neutrons. For example, approximately 20% of 1 GeV muons produced at 10 km will reach sea level before decay, and
3. a nucleonic component, where high-energy nucleons, disintegration-product nucleons, and nuclear disintegration are present.

On the other hand, if a gamma ray primary hits a nucleus in the Earth's atmosphere, an electron-positron pair is produced by Coulomb interaction. These electrically charged particles, interact with other atmospheric nucleus producing secondary gamma rays via bremsstrahlung radiation ($e^{\pm} + N \longrightarrow e^{\pm} + N + \gamma$). The length scale for the energy loss of e^- and photons when they interact with matter (in gr cm^{-2}) is known as the radiation length. When the $e^{\pm}$ from the primary produce a secondary gamma via bremsstrahlung, we obtain one radiation length. When this secondary gamma pair-produce another $e^{\pm}$, and the resulting $e^{\pm}$ gives way to other secondary gammas via bremsstrahlung a second radiation length is obtained, and so on. The radiation length, X_0, is the distance over which a high energy electron decreases its energy through bremsstrahlung to $1/e$. The average distance that a high energy photon travels before converting to an electron pair is $\sim \frac{9}{7}$ X_0, where $X_0 \sim 180\frac{A}{Z^2} \sim 36.6\frac{g}{cm^2}$, and A is the atomic mass.

When a charged particle (e^- or muons) from secondaries interacts with a polarized medium (e.g., air or water), it can produce "Vavilov–Cherenkov" radiation as a result of the particle travelling faster than light in the medium [5–8]. While IACT's are excellent telescopes for high resolution observations, Water Cerenkov Detectrs

[a] e-mail: `edfuente@astro.iam.udg.mx`
[b] e-mail: `juan.diazvelez@alumnos.udg.mx`
[c] e-mail: `albertby@hotmail.com`
[d] e-mail: `anigoche@gmail.com`
[e] `http://www.hawc-observatory.org/`

(WCD) are excellent as monitoring instruments for transients, AGNs, and large scale structures.

2. The HAWC detector

2.1. The main detector

The High-Altitude Water Cherenkov (HAWC) Observatory is a unique water Cherenkov detector, studying the Universe through gamma rays in an energy range from 100 GeV to 100 TeV. HAWC was constructed on Sierra Negra, Puebla, México, at 4100 meters above sea level. At this elevation, ~10% of the energy from a 1 TeV gamma-ray shower will reach the detector level, which is about 5× the energy at the Milagro elevation (2600 m) and more than 25× the energy that reaches sea level.

HAWC consists of 300 water Cherenkov detectors, each made of a corrugated metallic cylindrical structure, of 7.3 m diameter and 4.5 m high, bracing a bladder with 180,000 liters of ultra clean water and 4 photomultiplier tubes (PMT) in its interior that work as sensors of the Cherenkov radiation produced by charged particles from extended air showers; in the center of the detector is a high quantum efficiency 10″ Hamamatsu R7081-HQE PMT, surrounded by three 8″ Hamamatsu R5912 PMTs recovered from Milagro. HAWC is an ideal survey instrument with a big synergy with other observatories like AUGER[1], IceCube[2], and Air Cherenkov Detectors like Veritas [3], HESS[4], and Magic[5]. The main parameters are presented in Table 1. Figure 1 shows the layout of the full HAWC WCD (top), the diagram of a single WCD (middle), and a true picture of the detector as it is now (bottom).

All the signals from the PMTs are recorded using Time over Threshold (ToT) as the method to determine both the arrival time and the amplitude of the signals without dead time. The analog PMT pulses are connected to shaping electronics in the counting house through a 175 m cable path. The pulses are shaped and discriminated with 2 thresholds, at $\frac{1}{4}$ PE and 5 PE. Threshold crossing times are digitized using CAEN VX1190A multi-hit time-to-digital converters (TDC). All hits are recorded to memory and triggering is entirely done using software. The total data collection rate is ~400 MB/s and drops to 15—20 MB/s with the application of a trigger. Events are reconstructed in real–time (few second latency) using an online analysis computing farm so that HAWC can rapidly alert other experiments of transient sources. The shower front is first fitted with a plane followed by a fit with the curvature corrections and shower time spread given by the actual propagation of particles in the atmosphere. The position of the shower core is determined by fitting a Nishimura-Kamata-Greisen (NKG) distribution [9,10] to the amplitude of the detected signals. This can be reliably done if the core falls inside the array (22,000 m^2 effective area) but becomes more ambiguous if the core falls outside of the array [11].

HAWC began full operation at the end of March 2015, and is expected to be ~15 times more sensitive (Crab) than its predecessor, the Milagro observatory. After the operation of the engineering prototype "The Verification And Measuring of the Observatory System (VAMOS)" [12] built in 2011, 4 stages were considered during the modular construction of HAWC. In September 2012, the first 30 WCD were completed and operated as an engineering prototype. Then, the array was expanded to 77 WCD, then to 95 WCD, and in August 2013, the operation of a 111–tank array (HAWC–111) began. This array was operated until July 2014 and it was about 3–5 times more

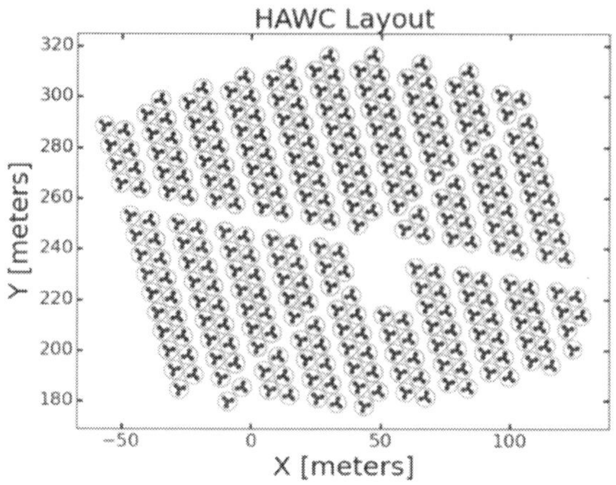

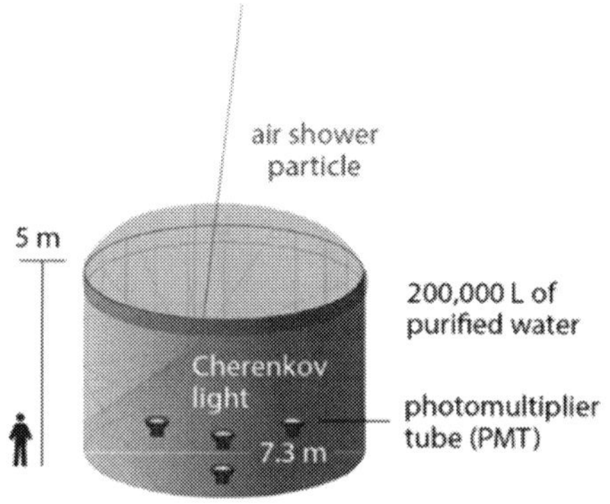

Figure 1. Layout of the HAWC array (top), a diagram of a single HAWC tank (middle), a real picture of the full HAWC observatory.

[1] https://www.auger.org/
[2] https://icecube.wisc.edu/
[3] http://veritas.sao.arizona.edu/
[4] https://www.mpi-hd.mpg.de/hfm/HESS/
[5] https://magic.mppmu.mpg.de/

Table 1. HAWC Main Parameters.

Parameter	Value
Hemisphere	Northern
Latitude	19°
Altitude	4100 m a.s.l. (630 gm/cm^2)
Effective Area	22,000 m^2
Area of each WCD	42 m^2
Field of View	$\sim$2 str (15% of the whole sky)
Trigger rate	16 kHz
Detection method	cascades produced by CR and γ
Median primary energy	2 TeV
Approx. angular resolution	$0.3^\circ - 1.5^\circ$

sensitive than Milagro. On November 26, 2014, data taking began with 250 tanks. The detector was expanded through the winter of 2014–2015 and full HAWC operations officially commenced on March 20, 2015. More details about the HAWC construction, performance and operation is presented by Smith et al. (2015) [13].

2.2. HAWC near future: The outriggers era

Although the gamma/hadron discrimination method on HAWC is working well (e.g., [11,13–15]), the method performs better as the energy of the gamma increases and at the highest energies the events are almost background free. Nevertheless, there is a large fraction of showers where the core falls outside of the HAWC array, and although they leave enough information in the WCDs to make a gamma/hadron discrimination, there is ambiguity in the core position, the direction of the shower, and the shower size. Also, the HAWC trigger for the highest energy gammas reaches an effective area of 10^5 m^2 but again, many of them are poorly reconstructed because the shower core falls outside the array.

To recover a large fraction of these showers and increase the present fraction of well reconstructed showers above 10 TeV, especially at the highest energies where we are limited by the statistics, an upgrade for the HAWC observatory is mandatory.

One way to do this is by the installation of a sparse array of small water Cherenkov detectors around HAWC-300. These "outriggers" (OR's) can help pinpoint the core position and thus improve the angular resolution of reconstructed showers. This same approach was utilized by the Milagro observatory. In 2004, a sparse outrigger array to Milagro was added and this dramatically increased the sensitivity of the observatory by being able to determine the shower core position over an area much bigger than the main detector and thereby correctly reconstructing partially detected showers.

As a result of these considerations, an effort was started to install outriggers around HAWC in 2016. The goal of this is to:

1. enhance sensitivity above 10 TeV, accurately determine core position for showers off the main WCD array (HAWC 300),
2. increase effective area above 10 TeV by 3–4 times, and
3. determine the position of the shower core for showers falling outside the HAWC WCD array and that still leave enough information in HAWC to reconstruct the shower front and discriminate between gamma and cosmic-ray initiated showers. This improvement is naturally limited only to high energy showers above 1 TeV.

The HAWC outrigger array consists of 350 cylindrical tanks of diameter 1.55 m and height 1.65 m, 1/80th size of one HAWC WCD (see Fig. 2 top). Each tank with 2500 liters of ultra-purified water, will include one Hamamatsu R5912 8″ PMT at the bottom of the tank. The outrigger array will be deployed in a circular symmetric way around the main HAWC array with an inter-tank separation of 12 m to 18 m (see Fig. 2 down). To optimize the layout, detailed simulations are being performed using the Monte Carlo method. The FlashCAM readout electronics (see [16] and references therein) is used for trigger and readout of the system. Preliminary results are presented in [16] and [11]. Details and final results will be presented in a forthcoming paper.

3. HAWC cosmic ray astrophysics results

3.1. Anisotropy in the arrival direction distribution of TeV cosmic rays

A number of experiments have observed an anisotropy with an amplitude of 10^{-4} at energies of order 1 TeV including the Tibet ASγ [17], Super-Kamiokande [18], Milagro [19], EAS-TOP [20], MINOS [21], ARGO-YBJ [22] experiments in the Northern Hemisphere, and IceCube [23–25] and its surface air shower array IceTop [26] in the Southern Hemisphere.

There are two significant features in the observed anisotropy. The first is a large-scale structure with an amplitude of about 10^{-3}, and the second consists of several localized regions of cosmic-ray excesses with an amplitude of 10^{-4} with and deficits of angular size 10° to 30°. The origin of this anisotropy is not yet well understood since it is expected that cosmic rays should lose any correlation with their original direction due to diffusion as they traverse interstellar magnetic fields.

The anisotropy of cosmic rays is measured by computing the relative intensity as a function of equatorial coordinates (α, δ) in a sky grid of equal-area pixels using the HEALPix library. A binned data map $N(\alpha, \delta)_i$ with a resolution of 0.2° per bin is used to store the arrival directions of air showers recorded by the detector for each angular bin, i.

A reference map $\langle N\rangle\ (\alpha, \delta)_i$ that gives a description of the arrival direction distribution of an isotropic flux at Earth is produced by collecting all events recorded during a time period, Δt, and then integrating the local arrival direction distribution against the detector event rate. This is done using the direct integration technique described in Ref. [27].

This procedure effectively smooths out the true arrival direction distribution in right ascension on angular scales $\sim \Delta t \cdot 15^\circ$ hr^{-1} and also compensates for variations in the detector rate. We calculate the deviations from isotropy using the reference by computing the relative intensity given in Eq. (1)

$$\delta I(\alpha, \delta)_i = \frac{\Delta N}{\langle N\rangle} = \frac{N(\alpha, \delta)_i - \langle N\rangle(\alpha, \delta)_i}{\langle N\rangle(\alpha, \delta)_i}, \qquad (1)$$

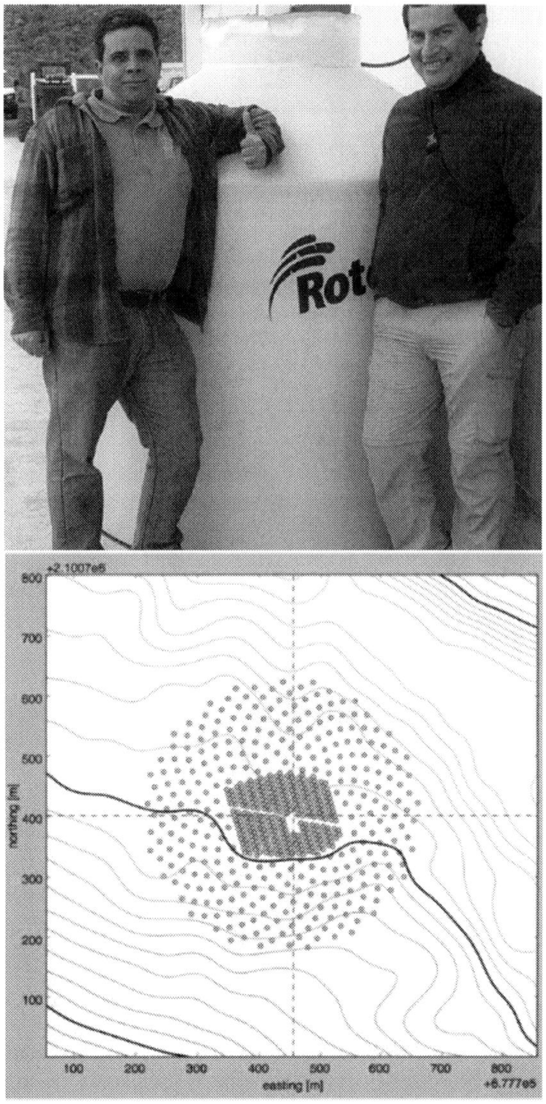

Figure 2. Top: image of the first Outrigger installed on HAWC for upgrade. Each OR is 1/80th size of a main HAWC detector. Bottom: layout showing the position of the 350 outriggers to be installed around HAWC main detector improving the effective area by a factor of 4.

which gives the amplitude of deviations from the isotropic expectation in each angular bin, i. The significance of the deviation can be calculated using the method of Li and Ma [28]. A choice of $\Delta t = 24$ is needed to ensure a uniform angular scale as a function of declination.

The analysis originally published in [29] was carried out on HAWC-111 data using $\Delta t = 24\,\mathrm{hr}$ to obtain sensitivity to all angular features equally over the sky. The resulting large scale relative intensity is a combination of sidereal anisotropy and the Solar dipole effect which causes an excess of cosmic rays in the direction of the Earth's motion around the Sun. However, the two signals are difficult to disentangle unless the data cover an integer number of years of exposure such that each signal makes a complete transit in the other reference frame.

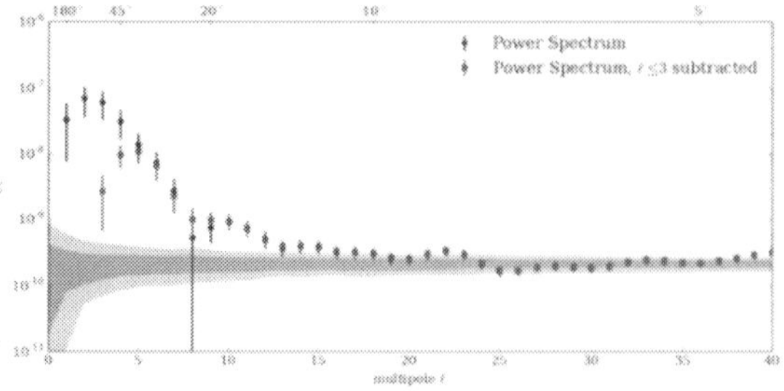

Figure 3. Angular power spectra of the unsmoothed relative intensity map before (blue) and after (red) subtracting the large-scale structure ($\ell \leq 3$). Gray error bands show the 68% and 95% spread of the C_ℓ for isotropic data sets. Comparing the band to the data shows which *ell*-modes significantly contribute to the sky map. The error bars on the C_ℓ are the square root of the variances returned by a fit using a power spectrum estimator (PolSpice).

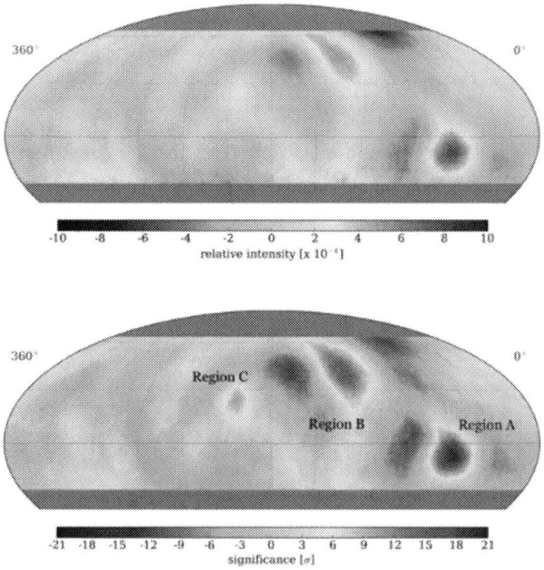

Figure 4. Relative intensity (*top*) and pre-trial significance (*bottom*) of the cosmic-ray flux after fit and subtraction of the dipole, quadrupole, and octupole terms from the map shown in above. The map is shown with 10° smoothing applied.

Figure 3 shows the angular power spectrum of the resulting map. The large-scale signals can be subtracted from the resulting map using a multipole fit to all multipole moments $\ell \leq \ell_{max} = 3$, so we are left with the small-scale structure in the sidereal frame. The residual structure will contain power at angular scales less than $180°/\ell_{max} = 60°$. The resulting small-scale anisotropy map is shown in Fig. 4 along with the statistical significance map using the Li & Ma method. A 10° smoothing has been applied.

Three significant features remain, the strong region of excess flux which are labeled as regions A, B and C in Fig. 4. These excesses coincide with the 10° – 20° regions of cosmic-ray excess first observed by Milagro [30] (Regions A and B) as well as a third region observed only by ARGO-YBJ [31] (Region C).

Connecting the northern and southern measurements would eliminate biases from partial sky coverage. No clear connection of the small-scale anisotropy that is present

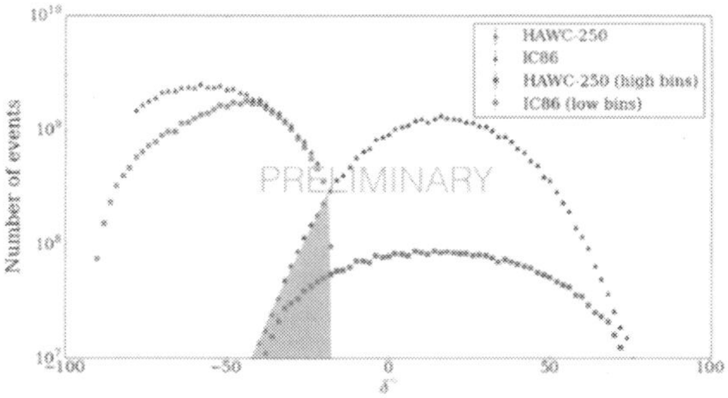

Figure 5. Distribution of events as as a function of declination for IceCube (IC) and HAWC. The shaded area corresponds to the overlapping region for both experiments [32].

in both hemispheres has been made yet. The published IceCube maps are of higher energy (20 TeV median) HAWC and other northern measurements. Unlike it's predecessor Milagro (35.88° N latitude), HAWC (19.0°N latitude) has a field of view that overlaps with that of IceCube (see Fig. 5). Because of this, there is some work to combine HAWC and IceCube data using cuts to bring their median energies closer to a central value [32].

References

[1] Abeysekara, et al., The Astrophysical Journal **817**, 3–13 (2016)

[2] H.J. Bhabha and W. Heitler, Proceedings of the Royal Society of London A **159**, 432–458 (1937)

[3] J.F. Carlson and R. Oppenheimer, Physics Review **61**, 220–231 (1937)

[4] G.T. Zatsepin, Doklady Akademii Nauk SSSR **67**, 993–1001 (1949)

[5] P.A. Cerenkov, Doklady Akad. Nauk SSSR **2**, 451–455 (1934)

[6] S.I. Vavilov, Doklady Akad. Nauk SSSR **2**, 457–459 (1934)

[7] P.A. Cerenkov, Phys. Rev. **52** 378–381 (1937)

[8] Tamm, I.E., and Frank, I.M., Doklady Akad. Nauk SSSR **14**, 107–112 (1937)

[9] Kamata, K., & Nishimura, J. 1958, Progress of Theoretical Physics Supplement **6**, 93–98

[10] Greisen, K. 1960, Annual Review of Nuclear and Particle Science **10**, 63–68

[11] A. Sandoval et al., Proceeding of Science **ICRC2015**, 1–6 (2015)

[12] Abeysekara et al., Astroparticle Physics **62**, 125–133 (2015)

[13] A. Smith et al., Proceeding of Science **ICRC2015**, 1–8 (2015)

[14] Zigfried Hampel–Arias, & Stefan Westerhoff, Proceedings of Science **ICRC2015**, 1–8 (2015)

[15] J. Pretz et al., Proceeding of Science **ICRC2015**, 1–16 (2015)

[16] V. Joshi, Proceedings of the 6th Roma International Conference on Astroparticle Physics, in press, (2017)

[17] M. Amenomori et al., Astrophys. J. Lett. **626**, L29-L32 (2005)

[18] G. Guillian et al., Phys. Rev. D **75**, 062003 (2007)

[19] A. A. Abdo et al., Phys. Rev. Lett. **101**, 221101 (2008)

[20] M. Aglietta et al., Astrophys. J. Lett. **692**, L130–L133 (2009)

[21] J. De Jong, International Cosmic Ray Conference **4**, 46 (2011)

[22] G. Di Sciascio, Euro. Phys. J. Web of Conf. **52**, 4004 (2013)

[23] R. Abbasi et al., Astrophys. J. Lett. **718**, L194–L198 (2010)

[24] R. Abbasi et al., Astrophys. J. **740**, 16 (2011)

[25] R. Abbasi et al., Astrophys. J. **746** 33, (2012)

[26] M. G. Aartsen et al., Astrophys. J. **765**, 55 (2013)

[27] R.W. Atkins et al., Astrophys. J. **595**, 803–811 (2003)

[28] T.-P. Li and Y.-Q. Ma, Astrophys. J. **272**, 317–324 (1983)

[29] A. Abeysekara et al., The Astrophysical Journal **796**, 108 (2014)

[30] A.A. Abdo et al.,, Astrophys. J. **698**, 2121–2130 (2009)

[31] B. Bartoli et al., Phys. Rev. **D88**, 082001 (2013)

[32] J. Díaz-Vélez and D. Fiorino, Proc. 34th ICRC, in press (2015)

Origin of Cosmic Rays: Modern status

Vladimir Ptuskin[a]

IZMIRAN, Troitsk, Moscow 108840, Russia

Abstract. Brief discussion on the origin of cosmic rays at energies from 10^6 to 10^{20} eV.

1. Introduction

The directly measured cosmic-ray energy spectrum extends from about 1 MeV to more than 10^{20} eV. The radio-astronomical, X-ray, gamma-ray observations and the first detection of very high energy astrophysical neutrinos shed light on the physical processes with cosmic rays. The majority of the observed cosmic rays are of Galactic origin and are accelerated in supernova remnants and pulsars. The Fermi Bubble is a prominent galactic source of energetic particles in the central part of the Galaxy. The origin of observed cosmic rays with energies above $\sim 10^{18}$ eV is associated with extragalactic sources: Active Galactic Nuclei, progenitors of Gamma-Ray Bursts, fast spinning newborn pulsars, large-scale structure formation shocks and some other objects.

2. Last decade progress in cosmic ray research

The last decade can be rightfully considered as a "golden age" of new cosmic-ray measurements. The list of remarkable experiments includes amongst others the Voyagers 1, 2 traveling throughout the outer heliosphere and the interstellar medium; ACE, PAMELA and AMS spacecraft missions; long-duration balloon experiments ATIC, BESS, CREAM and TRACER; IceCube Neutrino Observatory at the South Pole; the ground based Extensive Air Shower (EAS) detectors KASCADE-Grande, MILAGRO, ARGO-YBJ and Tunka; Auger and Telescope Array for studies of EAS produced by the highest-energy cosmic rays; the Fermi gamma-ray space observatory and the ground based HESS, MAGIC and VERITAS telescopes for gamma-ray observations of cosmic ray sources. The references to the experimental results of cosmic ray research can be found in [1].

The detection of astrophysical neutrinos with PeV and sub-PeV energies in the IceCube experiment [2] has opened a new way to study the sources of very high energy cosmic rays. The observed neutrino events are scattered over the sky without evident correlation with any Galactic or extragalactic classes of astronomical objects. The discussion about potential sources of very high energy neutrinos can be found in [3]. The measured neutrino flux is close to the Waxman-Bahcall upper bound [4] calculated under the assumptions that neutrinos are produced by ultra-high energy cosmic rays in a process of their exit from the sources. However, the observed neutrino energy spectrum has a cutoff at ~ 2 Pev that corresponds to the maximum energy of primary protons $\sim 10^{17}$ eV, much smaller than the maximum energy of cosmic rays observed at the Earth. It was suggested [5] that extragalactic Type IIn supernova remnants are sources of observed diffuse high energy neutrinos. Based on the well developed theory of cosmic ray shock acceleration in supernova remnants it was shown that the large kinetic energy of explosion and very high gas density in the acceleration region (in the wind of a progenitor star with large mass loss rate) typical for Type IIn supernovae lead to the required energy of accelerated protons and efficiency of neutrino production in pp interactions.

Cosmic rays of the highest energies continue to hold the great interest of physicists. The HiRes, Auger and Telescope Array installations experimentally established the suppression of cosmic ray flux starting at energy about 5×10^{19} eV as expected due to the Greisen-Zatsepin-Kuzmin effect, the energy loss of the highest energy cosmic rays interacting with the extragalactic microwave and optic background radiation (see review [6]). The development of detailed models of acceleration and propagation of the highest energy cosmic rays is hampered by the uncertainty of the mass composition. The interpretation of HiRes and Telescope Array data favours the proton composition at energies 10^{18} to 5×10^{19} eV, whereas the Auger data indicate that cosmic ray composition is becoming heavier for energies above 2×10^{18} eV. The interpretation of recent results of the Auger experiment [7] favors a very hard source spectrum $\sim E^{-1}$ with mixed composition. The maximum particle magnetic rigidity at the sources $R_{\max} \approx 5 \times 10^{18}$ V corresponds to the maximum energy $\approx 1.3 \times 10^{20}$ eV for Iron nuclei. This has the very important consequence that the shape of the all particle spectrum is likely due to the concurrence of two effects: maximum energy reached at the sources and energy losses during propagation.

The Voyger 1 and 2 spacecrafts provide direct measurements of cosmic ray spectrum and composition on opposite side of the energy spectrum. Launched in 1977 and expected to operate till 2020–2025, Voyager 1 is 138 a.u. from the Sun at present and Voyager 2 is at 114 a.u.. Voyager 1 is in "interstellar space" and Voyager 2 is currently in the "heliosheath" – the outermost layer

[a] e-mail: vptuskin@izmiran.ru

of the heliosphere where the solar wind is slowed by the pressure of interstellar gas. The Voyager 1 spacecraft is measuring the cosmic ray intensity not affected by the solar wind modulation. This experiment extends our knowledge of the local interstellar cosmic ray spectrum and composition down to MeV energies [8]. The spectra of protons, nuclei and electrons look as a natural continuation of spectra at energies above a few hundred MeV and no additional low energy component was revealed. The intensity of the so called Anomalous Cosmic Rays (ACR) is diminished to zero and it confirms that these particles are accelerated somewhere in the inner part of the heliosheath (probably, at the hot spots of the solar wind termination shock).

The remarkable result of the PAMELA experiment was the detection of an unexpectedly large flux of positrons with energies above 10 GeV [9]. It was confirmed with higher statistics and extended to higher energies by the AMS-02 experiment [10,11]. The measured flux of positrons is larger than the calculated flux of secondary positrons produced in cosmic ray interactions with interstellar gas atoms. The AMS-02 positron measurements are extended now up to 600 GeV and show the rapid turn down of the spectrum above 250 GeV. The high content of positrons can be explained by the contribution of high-energy electron-positron pairs generated by pulsars [12,13] or by the acceleration of positrons in the reverse shocks moving through the radioactive ejecta in supernova remnants [14]. The conclusive elucidation of the nature of positron flux in cosmic rays is very important since an alternative explanation suggests that these positrons are the products of dark matter decay.

Recent high-accuracy ATIC-2, CREAM and PAMELA measurements revealed deviations of the cosmic ray spectrum from a plain power law at energies from 10 to 10^5 GeV/n [15,16]. These deviations include the spectral hardening in proton, helium and nuclear spectra at magnetic rigidity $\sim$240 GV and an increase of the He/p ratio with energy. One of the explanations of these features implies cosmic ray acceleration by reverse shocks in supernova remnant [17]. New experimental results require the improvement of existing models of cosmic ray acceleration and propagation in the Galaxy although the basic model with the cosmic ray acceleration in supernova remnants (including pulsars) followed by the propagation in galactic magnetic fields with a slow leakage from the Galaxy remains immutable.

The fine structure of the all particle cosmic ray spectrum at 10^5 GeV to 10^9 GeV and the data on elemental composition were clarified in the KASCADE-Grande and Tunka experiments [18,19]. This structure corresponds to the superposition of fluxes of different nuclei with similar spectra on magnetic rigidity and the breaks (knees) at $3 \times 10^6 Z$ GV (Z is the particle charge). It results in the all-particle spectrum with a proton-helium knee at 3×10^6 GeV and the steepening of a power law energy spectrum after this energy. The second knee at energy just below 10^{17} eV where composition is dominated by heavy (iron) nuclei was found. The hard light component emerges at about 10^{17} eV. This component may be of Galactic origin or reflect the transition to extragalactic cosmic rays.

Efforts of theorists in the last few decades were mainly focused on the problem of particle acceleration by shocks propagating in the turbulent interstellar plasma [20,21]. The modeling of nonlinear shock acceleration in the different types of supernova remnants demonstrated that the calculated energy spectrum of cosmic rays after propagation through the Galaxy is in good agreement with the measured spectrum up to approximately 3×10^{18} eV [22].

3. Dark matter problem and cosmic rays

An important aspect of cosmic ray research is associated with a fundamental problem of physics the clarification of the nature of dark matter in the Universe. It is established that the energy budget of the Universe consists of only 4% of baryonic matter, 73% is made of the dark energy and the 23% of dark matter. Dark matter particles are not described by the Standard Model of elementary particle physics. The promising candidates for dark matter are the WIMPs, weakly interacting massive particles with a mass in the range between tens of GeV and TeV. A direct search for weak interactions of these particles with large mass detectors is conducted in underground laboratories. The indirect search in space and at the top of the atmosphere is based on the assumption that WIMPs can annihilate each other and produce standard elementary particles as the final state. Using gamma-astronomical methods, one can search for specific signals in continuum and line radiation from galaxies, satellite galaxies, galaxy clusters, Galactic Center and other regions (clumps) where dark matter may be accumulated. The gamma-ray space telescope Fermi is used for this purpose [23]. The cosmic ray antiparticles, positrons and antiprotons, are convenient to use for the dark matter search. They are produced only in small amounts via interactions of CRs with interstellar matter and it is easier to disentangle dark matter signal from the astrophysical background (see e.g. discussion in [24]).

As was said above, the anomalously high flux of positrons in cosmic rays can, in principle, be explained by the contribution from pulsars or the production of positrons via decays of dark matter.

Antiprotons in cosmic rays are produced as secondary particles at a level $\sim 10^{-4}$ relative to protons. Accurate antiproton measurements done at 70 MeV to 200 GeV by the PAMELA instrument [25] and up to 400 GeV by AMS-02 [11] did not reveal any unambiguous evidence of excess above the predicted secondary antiproton flux although some preference for a flatter component probably starts to emerge above 100 GeV [26,27].

No firm evidence of dark matter signal was found so far, but the investigations of cosmic rays play an important role in constraining dark-matter models for understanding the nature of dark matter (see [28,29]).

The present work was partly supported by RFBR grant 16-02-00255.

References

[1] K.A. Olive et al. (Particle Data Group), Chinese Phys. C **38**, 090001 (2014)
[2] IceCube Collaboration, Science **342**, 1242856 (2013)
[3] T. Gaisser, F. Halzen, Annu. Rev. Nucl. Part. Sci. **64**, 101 (2014)
[4] E. Waxman, J. Bahcall, Phys. Rev. **59**, 023002 (1999)

[5] V.N. Zirakashvili, V.S. Ptuskin, Astropart. Physics **78**, 28 (2016)
[6] K. Kotera, A. V. Olinto, Annu. Rev. Astron. Astrophys. **49**, 119 (2011)
[7] A. Aab et al., `arXiv:16107155` (2016)
[8] E. Stone et al., Science **341**, 150 (2013)
[9] O. Adriani et al., Nature **458**, 607 (2009)
[10] M. Aguilar et al., Phys. Rev. Lett. **113**, 121102 (2014)
[11] S. Ting, AMS days at CERN https://indico.cern.ch/event/381134 (2015)
[12] A. Harding, R. Ramaty, Proc. 20th Intern. Cosmic Ray Conf. Moscow **2**, 92 (1987)
[13] M. Di Mauro et al., JCAP **4**, 006 (2014)
[14] V.N. Zirakashvili, F.A. Aharonian, Phys. Rev. D **84**, 083010 (2011)
[15] O. Adriani et al., Science **332**, 69 (2011)
[16] H.S. Ahn et al., ApJ Lett. **714**, L89 (2010)
[17] V.S. Ptuskin, V.N. Zirakashvili, E.S. Seo, ApJ **763**, 47 (2013)
[18] A. Haungs et al., Proc. 33rd Intern. Cosmic Ray Conf. Rio de Janeiro, `arXiv:1308.1485v1` (2013)
[19] V.V. Prosin et al., Nucl. Instrum. Methods in Physics Research A **756**, 94 (2014)
[20] A.R. Bell, Astropart. Physics **43**, 56 (2013)
[21] D. Caprioli, Nucl. Phys. B (Proc. Suppl. **256**, 48 (2014)
[22] V.S. Ptuskin, V.N. Zirakashvili, E.S. Seo, ApJ **718**, 31 (2010)
[23] W. B. Atwood et al., ApJ **697**, 1071 (2009)
[24] P. Picozza, R. Sparvoli, NIMPA **729**, 45 (2013)
[25] O. Adriani et al., Phys.Rev. Lett. **111**, 1102 (2013)
[26] G. Giesen et al., JCAP **1509**, 023 (2015)
[27] R. Kappl, A. Reinert, M.W. Winkler, JCAP **10**, 034 (2015)
[28] G. Beck, S. Colafrancesco, `arXiv150801386` (2015)
[29] K. Hamaguchi, T. Mori, K. Nakayama, Phys. Lett. B **747**, 523 (2015)

New look on the origin of cosmic rays

Ya.N. Istomin[1,2,a]

[1] P.N. Lebedev Physical Institute, Leninsky Prospect 53, Moscow 119991, Russia
[2] Moscow Institute Physics and Technology, Institutskii per. 9, Dolgoprudnyi, Moscow region 141700, Russia

Abstract. From the analysis of the flux of high energy particles, $E > 3 \cdot 10^{18}$ eV, it is shown that the distribution of the power density of extragalactic rays over energy is of the power law, $\bar{q}(E) \propto E^{-2.7}$, with the same index of 2.7 that has the distribution of Galactic cosmic rays before the so called 'knee', $E < 3 \cdot 10^{15}$ eV. However, the average power of extragalactic sources, which is of $\mathcal{E} \simeq 10^{43}$ erg s^{-1}, exceeds by at least two orders the power emitted by the Galaxy in cosmic rays, assuming that the density of galaxies is estimated as $N_g \simeq 1$ Mpc^{-3}. Considering that such power can be provided by relativistic jets from active galactic nuclei with the power $\mathcal{E} \simeq 10^{45} - 10^{46}$ erg s^{-1}, we estimate the density of extragalactic sources of cosmic rays as $N_g \simeq 10^{-2} - 10^{-3}$ Mpc^{-3}. Assuming the same nature of Galactic and extragalactic rays, we conclude that the Galactic rays were produced by a relativistic jet emitted from the Galactic center during the period of its activity in the past. The remnants of a bipolar jet are now observed in the form of bubbles of relativistic gas above and below the Galactic plane. The break, observed in the spectrum of Galactic rays ('knee'), is explained by fast escape of energetic particles, $E > 3 \cdot 10^{15}$ eV, from the Galaxy because of the dependence of the coefficient of diffusion of cosmic rays on energy, $D \propto E^{0.7}$. The obtained index of the density distribution of particles over energy, $N(E) \propto E^{-2.7-0.7/2} = E^{-3.05}$, for $E > 3 \cdot 10^{15}$ eV agrees well with the observed one, $N(E) \propto E^{-3.1}$. The estimated time of the termination of the jet in the Galaxy is $4.2 \cdot 10^4$ years ago.

1. Introduction

The main hypothesis of the origin of cosmic rays in the Galaxy is the acceleration of charged particles to high energies on the fronts of shock waves formed by supernova explosions. To ensure the power of cosmic rays, observed in the Galaxy, which equals about 10^{41}erg s^{-1}, it is necessary to transfer to accelerated particles about 15% of the kinetic energy of expanding shock waves. These strong shocks produce the universal spectrum of particle energy distribution, $N(E) \propto E^{-2}$. This allows to put together cosmic rays originating from different supernova, and forms the unique particle spectrum, extending from energies in a few units of GeV to energies $\simeq 10^{18}$ eV. The observed spectrum $N(E)$ significantly differs from the Universal spectrum produced via acceleration at shock fronts, that is explained by the propagation of cosmic rays in the Galaxy. It has the character of diffusion in space because of the scattering of charged particles by magnetic field inhomogeneities. The diffusion coefficient increases with particle energy, i.e. the lifetime of the fast particles in the Galaxy decreases. Therefore, the observed spectrum over energy differs from the spectrum given by sources, remaining a power law, $N(E) \propto E^{-\beta}$. But the value of β is not constant, it changes from $\beta_1 = 2.7$ for $E < 3 \cdot 10^{15}$ eV to $\beta_2 = 3.1$ for $E > 3 \cdot 10^{15}$ eV. The important circumstance is the fact that the spectrum at high energies becomes softer, not harder. This suggests that the source of cosmic rays at energies $E < 3 \cdot 10^{15}$ eV and at energies $E > 3 \cdot 10^{15}$ eV is single. It is possible that it is the superposition of two independent sources, if it would be vice versa, $\beta_2 < \beta_1$. But it is unlikely that the spectrum produced by one source at $E < 3 \cdot 10^{15}$ eV would cut off at higher energies, whereas another independent source at energies $E > 3 \cdot 10^{15}$ eV would cut off at lower energies, and they were joined at the same energy $E \simeq 3 \cdot 10^{15}$ eV.

The fact that the source of Galactic cosmic rays is single in the whole energy range, $E_{min} < E < 3 \cdot 10^{18}$ eV, suggests that if there are many in our Galaxy, they have the same universal properties. For the hypothesis of the origin of cosmic rays by supernova explosions this suggests that the shock wave is both strong enough and slightly loaded by cosmic rays. Otherwise, the spectral index of the accelerated particles will be different for different shock waves. This makes the hypothesis of the origin of cosmic rays by supernova explosions less attractive than previously assumed ([1,2]). In this article we do not discuss the paradigm of acceleration of Galactic cosmic rays by shock waves from supernova explosions. The last review made by Blasi [3] gives a good representation of this paradigm in many aspects. Observational data certainly show the acceleration of particles on the fronts of supernova shock waves. Is that enough to consider a hypothesis of the origin of Galactic cosmic rays by shock wave acceleration is proven? Here you can quote from Blasi's review: "Despite all this circumstantial evidence, no proof found yet that SNRs can accelerate CRs up to the knee energy."

If the source is single, it would also produce cosmic rays in other galaxies with characteristics similar to those observed in the Galaxy. The aim of our work is to analyze the properties of cosmic rays of superhigh energies $E > 3 \cdot 10^{18}$ eV observed on the Earth, coming outside, from

[a] e-mail: istomin@lpi.ru

other galaxies, and compare them with the properties of Galactic cosmic rays.

2. Ultrahigh energy cosmic rays

The energy distribution of the flux of cosmic rays $I(E)$ observed on the Earth at energies $E > 3 \cdot 10^{18}$ eV noticeably deviates from the power law distribution, which is typical for lower energies, $3 \cdot 10^{15}$ eV $< E <$ $3 \cdot 10^{18}$ eV, where $I(E) \propto E^{-3.1}$. This region is called the 'ankle', here cosmic rays of Galactic origin are replaced by particles, the origin of which has extragalactic nature. They come from active galaxies located at large distances. In particular, there is identification of particles with energies $E \simeq 10^{19}$ eV coming from the galaxy Cen A, located at a distance of 4.6 Mpc (Auger Collaboration [4]), and correlation with active galactic nuclei (AGN) within $\simeq$ 100 Mpc (Auger Collaboration [5,6]). Charged particles (protons or nuclei) of such an energy, $E \simeq 10^{19}$ eV, have a large cyclotron radius, r_L, in the intergalactic magnetic field $B \simeq 10^{-9}$ G,

$$r_L = \frac{E}{ZeB} = 11Z^{-1}\left(\frac{E}{10^{19}\text{eV}}\right)\left(\frac{B}{1n\text{G}}\right)^{-1} \text{Mpc}.$$

The magnetic field changes its direction on the scale of the order of 1 Mpc, i.e. the correlation length of the magnetic field equals $l_c \simeq 1$ Mpc. As a result, a charged particle moving to our Galaxy, walking distance, r, deviates from the initial direction by the angle $\Delta\theta$,

$$\Delta\theta = \left(\frac{l_c r}{r_L^2}\right)^{1/2} = 0.09Z\left(\frac{E}{10^{19}\text{eV}}\right)^{-1}\left(\frac{B}{1n\text{G}}\right) \times \left(\frac{l_c}{1\text{Mpc}}\right)^{1/2}\left(\frac{r}{1\text{Mpc}}\right)^{1/2}.$$

Therefore, the propagation of cosmic rays of high energy from sources, that are closer than $r < 100$ Mpc, can be considered as rectilinear to a good approximation, while from more distant sources as diffusive. Let us find the distribution function, $N(t, \mathbf{r}, E)$, of cosmic rays of high energy, $E \simeq 10^{19}$ eV, over the space, $\mathbf{r}$, and the energy, E, considering that charged particles come to the given point from many sources in space located at points of $\mathbf{r} = \mathbf{r}_i$ and having powers $Q_i(E)$. Considering the distribution of sources in space as homogeneous on average, the particle distribution function of N can be considered as isotropic, depending only on the distance, r. In addition, due to the large number of sources, the particle distribution can also be considered stationary, $N = N(r, E)$. We discuss two cases: rectilinear and diffusive propagation of particles.

2.1. Rectilinear motion

In this case, the equation for the distribution function of cosmic rays has the form

$$-c\frac{\partial N}{\partial r} + \frac{\partial}{\partial E}\left(\frac{dE}{dt}N\right) = \sum_i \frac{Q_i(E)}{4\pi r_i^2}\delta(r - r_i). \quad (1)$$

Here c is the velocity of the light, $\delta(x)$ is the Dirac delta function. The value of dE/dt is energy losses of particles. They consist of two parts: losses connected with the interaction of energetic particles with the relic radiation and losses associated with the expansion of the Universe. To the first, it is the so called GZK effect (Greisen [7]; Zatsepin & Kuz'min [8]) - pion production in the reaction $p + \gamma \to N + \pi$. The characteristic particle energy of this process is $E_\pi = \mu c^2(1 + \mu/m_p)m_p c^2/2T \simeq 4 \cdot 10^{20}$ eV (m_p is the mass of the proton, μ is the mass of the muon, T the temperature of the relic radiation). The interaction of a proton with a photon produces a neutral π-meson, the interaction of a photon with a nucleus can also produce charged π-mesons. At lower energies the generation of electron-positron pairs, $p + \gamma = p + e^+ + e^-$ becomes significant. The characteristic particle energy of this reaction is $E_e = m_e c^2 m_p c^2/T \simeq 2.1 \cdot 10^{18}$ eV (m_e is the electron mass). Energy losses in photo-pion reactions are well described by the expression (Stanev et al. [9])

$$\left(\frac{d\epsilon}{dt'}\right)_\pi = -(1+\epsilon)\exp(-\epsilon^{-1}). \quad (2)$$

Here we introduced the dimensionless energy $\epsilon = E/E_\pi = E/4 \cdot 10^{20}$ eV and the dimensionless time $t' = ct/L$, where L is the characteristic distance, beyond which a particle loses the energy of the order of its initial value, $L = 13.7$ Mpc. These are convenient distances to measure in the same units, $r' = r/L$. The energy losses due to the generation of electron-positron pairs (see the paper by Berezinsky, Gazizov & Grigorieva [10]) can be approximated by the expression, which well describes that near the maximum of its relative losses,

$$\left(\frac{d\epsilon}{dt'}\right)_e = -a\epsilon(\epsilon_e^s + b\epsilon_e^p)^{-1}\exp(-\epsilon_e^{-1}),$$
$$\epsilon_e = E/E_e = 190\epsilon. \quad (3)$$

Here indices s and p are $s = -1.5$, $p = 0.6$, and constants a and b are $a = 4.5 \cdot 10^{-4}$, $b = 8.4 \cdot 10^{-3}$ respectively. Adiabatic particle losses associated with the general expansion of the Universe are described by the Hubble law, $dE/dt = -HE$, where H is the Hubble constant, $H = 72\,\text{kms}^{-1}/$ Mpc. In dimensionless variables adiabatic losses are

$$\left(\frac{d\epsilon}{dt'}\right)_a = -\alpha\epsilon,\ \alpha = 3.3 \cdot 10^{-3}. \quad (4)$$

The total losses of particle energy are equal to

$$\frac{d\epsilon}{dt'} = -(1+\epsilon)\exp(-\epsilon^{-1}) - a\epsilon(\epsilon_e^s + b\epsilon_e^p)^{-1} \times \exp(-\epsilon_e^{-1}) - \alpha\epsilon. \quad (5)$$

Although the value of α is small, the adiabatic energy losses become important for particles of not very high energies, $\epsilon_e < (\alpha/a)^{-1/s} \simeq 3$, $E < 6 \cdot 10^{18}$ eV, when interaction with relict photons becomes insignificant.

It should also be mentioned that in addition to relic photons in the intergalactic environment, there also exists the light from galaxies, which contributes to the energy losses of cosmic rays. But, as shown by calculations, their contribution is small compared with that of the relic radiation (Aloisio, Berezinsky & Grigorieva [11]).

Let us introduce the dimensionless values of the power of the sources and the distribution function,

$Q_i' = Q_i L/c$, $N' = NL^3$. The dimensionless variables (with index $'$, which we omit further) will appear in Eq. (1). We also introduce the variable τ instead of ϵ,

$$\tau = \int_{\epsilon}^{\epsilon'} \frac{dx}{\mid d\epsilon/dt \mid_{\epsilon=x}}.$$

The value of τ is the time during which a particle loses its energy from the initial value ϵ' to the current one ϵ. As a result, Eq. (1) will take the form

$$\frac{\partial}{\partial r}\left(\mid \frac{d\epsilon}{dt} \mid N\right) - \frac{\partial}{\partial \tau}\left(\mid \frac{d\epsilon}{dt} \mid N\right) = -\sum_i \frac{Q_i(\epsilon)\mid\frac{d\epsilon}{dt}\mid}{4\pi r_i^2}\delta(r - r_i). \quad (6)$$

Integrating Eq. (6) over r from 0 to r_m with the constant sum $r + \tau = const$, we get

$$N(r = 0, \epsilon) = \frac{1}{|d\epsilon/dt|}\sum_i \frac{Q_i(\epsilon_i)\mid \frac{d\epsilon}{dt} \mid_{\epsilon=\epsilon_i}}{4\pi r_i^2} + \frac{\mid d\epsilon/dt \mid_{\epsilon=\epsilon_m}}{\mid d\epsilon/dt \mid} N(\epsilon_m, r_m). \quad (7)$$

The value of r_m is the distance to which the approximation of rectilinear propagation is valid, $r_m \simeq 100\,\text{Mpc}/13.7\,\text{Mpc} = 7.3$. The energy ϵ_m is determined by the ratio

$$r_m = \int_{\epsilon}^{\epsilon_m} \frac{dx}{\mid d\epsilon/dt \mid_{\epsilon=x}}.$$

In Eq. (7) the first term is the contribution of the sources of cosmic rays of superhigh energies, located at distances $r < r_m$ from the Galaxy, whose particles are observed in the Galaxy ($r = 0$). Because, according to the analysis by Takami & Sato [12], the density of such sources is of $10^{-2} - 10^{-4}\text{Mpc}^{-3}$, and the number of these sources in the sphere of radius r_m is significantly greater than unity, it is possible to go from the summation to the integration over volume, $4\pi \int_0^{r_m} r_i^2 dr_i$, introducing the average density of sources $\bar{q}(\epsilon)$. The value of energy, on which the quantity $\bar{q}(\epsilon)$ depends, is the initial energy of a particle, which overcomes the distance r_i from the source to the Galaxy, $\epsilon = \epsilon_i$,

$$r_i = \int_{\epsilon}^{\epsilon_i} \frac{dx}{\mid d\epsilon/dt \mid_{\epsilon=x}}.$$

However, at constant energy ϵ the distance r_i and the energy ϵ_i are connected, so the integration over r_i can be replaced by the integration over ϵ_i. The result is

$$\frac{1}{|d\epsilon/dt|}\sum_i \frac{Q_i(\epsilon_i)\mid \frac{d\epsilon}{dt} \mid_{\epsilon=\epsilon_i}}{4\pi r_i^2} = \frac{\int_{\epsilon}^{\epsilon_m} \bar{q}(\epsilon')d\epsilon'}{|d\epsilon/dt|}.$$

Thus, the distribution of cosmic rays of high energy, $E > 3 \cdot 10^{18}$ eV, observed in the Galaxy, is largely determined by the loss function $|d\epsilon/dt|$ (Eq. (5)) in the intergalactic space, but not by the sources, the distribution of which over the energy $\bar{q}(\epsilon)$ is in average form.

2.2. Diffusive motion

In order to determine the contribution of distant ($r > r_m$) sources into the distribution of cosmic rays observed in our Galaxy (the second term in Eq. (7)) we have to consider the diffusion region $r > r_m$. The equation for the distribution function $N(r, \epsilon)$ is

$$\frac{\partial}{\partial E}\left(\frac{dE}{dt}N\right) - \frac{D}{r^2}\frac{\partial}{\partial r}\left(r^2\frac{\partial N}{\partial r}\right) = \sum_i \frac{Q_i(E)}{4\pi r_i^2}\delta(r - r_i). \quad (8)$$

Here D is the coefficient of diffusion of cosmic rays in intergalactic space. As well as in the Galaxy, it can depend on the particle energy. Let us suppose that this dependence is a power law with some index κ, $D = D_0 E^{\kappa}$. Also as before, we introduce the dimensionless variables: the distances are measured in L, time - in L/c, the energy - in $4 \cdot 10^{20}$ eV, the diffusion coefficient - in units of $D_0' = D_0(4 \cdot 10^{20}\,\text{eV})^{\kappa}/Lc$. We introduce also the effective time τ_D,

$$\tau_D = \int_{\epsilon}^{\epsilon'} \frac{x^{\kappa}dx}{\mid d\epsilon/dt \mid_{\epsilon=x}}.$$

As a result, Eq. (8) becomes

$$\frac{\partial}{\partial \tau_D}\left(\left|\frac{d\epsilon}{dt}\right|N\right) - \frac{D_0}{r^2}\frac{\partial}{\partial r}\left(r^2\frac{\partial}{\partial r}\left(\left|\frac{d\epsilon}{dt}\right|N\right)\right) = \epsilon^{-\kappa}\left|\frac{d\epsilon}{dt}\right|\sum_i \frac{Q_i(\epsilon)}{4\pi r_i^2}\delta(r - r_i). \quad (9)$$

Knowing the Green function of the equation of diffusion in a spherical region $0 < r < \infty$ (the diffusion approximation includes also the region $0 < r < r_m$ of the scale of the free path length),

$$G(r, r', t) = \frac{1}{(4\pi D_0 t)^{1/2}}\frac{r'}{r}\left\{\exp[-\frac{(r - r')^2}{4D_0 t}] - \exp[-\frac{(r + r')^2}{4D_0 t}]\right\}. \quad (10)$$

Integrating this with the right hand part of equation (9) over r' and τ_D, then, as before, going from the summation over i to the integration over r_i from r_m to infinity, introducing again the density of sources $\bar{q}(\epsilon)$, and transforming the integration over τ_D to the integration over ϵ', we get

$$N(r_m, \epsilon) = \frac{1}{|d\epsilon/dt|}\int_{\epsilon}^{\epsilon_0} \bar{q}(\epsilon')\Phi\left[\left(\frac{r_m^2}{4D_0\tau_D(\epsilon')}\right)^{1/2}\right]d\epsilon'. \quad (11)$$

Here the function $\Phi(y)$ is equal to

$$\Phi(y) = 2 - \frac{2}{\pi^{1/2}}\int_0^{2y} \exp(-z^2)dz + \frac{1}{\pi^{1/2}y}[1 - \exp(-4y^2)].$$

It varies monotonically from the value $\Phi = 2$ at $y = 0$ to $\Phi = 1$ for $y >> 1$. And the fast transition of Φ to the value $\Phi \simeq 1$ occurs already at $y \simeq 2$. The typical value of y is the ratio of the length of the mean free path length of particles r_m to the path L, travelling along which a particle loses a significant part of its energy, $y \simeq r_m/L = 7.3$. In addition, the large time τ_D, small y, corresponds to high initial energies of particle ϵ'. Because the source power $\bar{q}(\epsilon)$ decreases with energy, and, as we will see, rather fast, $\bar{q}(\epsilon) \propto \epsilon^{-2.7}$, then the contribution of the function $\Phi(y)$ at small values of its argument into the integral (Eq. (11)) is small. Thus, we can consider $\Phi \simeq 1$ in the expression (Eq. (11)). Altogether, the required distribution function of particles on the boundary of $r = r_m$ at $\epsilon = \epsilon_m$ equals

$$N(r_m, \epsilon_m) = \frac{1}{\mid d\epsilon/dt \mid_{\epsilon=\epsilon_m}} \int_{\epsilon_m}^{\epsilon_0} \bar{q}(\epsilon')d\epsilon'. \qquad (12)$$

The value of ϵ_0 is the maximum energy of particles in sources.

3. Energy distribution of extragalactic particles observed in the Galaxy

Substituting the resulting distribution (Eq. (12)) into the expression (Eq. (7)), we obtain the sought distribution function of extragalactic cosmic rays, observed in the Galaxy,

$$N(r = 0, \epsilon) = \frac{\int_{\epsilon}^{\epsilon_0} \bar{q}(\epsilon')d\epsilon'}{|d\epsilon/dt|}. \qquad (13)$$

Equation (13) has a simple physical meaning: the particle flux in the energy space, $N|d\epsilon/dt|$, is equal to the total flux produced by sources Q_i. Moreover, since we have many sources, their fluxes are summing, ranging from close sources, providing a full range of energies from $\epsilon = \epsilon_{min}$ up to the maximum one $\epsilon = \epsilon_0$, to distant sources from which we observe only large initial energy $\epsilon \simeq \epsilon_0$. Suppose that the distribution $\bar{q}(\epsilon)$ is the power law function, $\bar{q}(\epsilon) = q_0\epsilon^{-\beta}$. Then the distribution function $N(\epsilon)$ is equal to

$$N(\epsilon) = \frac{q_0}{\beta - 1} \frac{\epsilon^{-\beta+1} - \epsilon_0^{-\beta+1}}{F(\epsilon)}, \qquad (14)$$

$$F(\epsilon) = (1 + \epsilon)\exp(-1/\epsilon) + a\epsilon(\epsilon_e^s + b\epsilon_e^p)^{-1}\exp(-1/\epsilon_e) + \alpha\epsilon.$$

The graph of the function $\epsilon^3 N(\epsilon)$ for $\epsilon_0 = 25$, $E_0 = 10^{22}$ eV, and $\beta = 2.7$ is shown on Fig. 1. We also show here the experimental data (points) obtained by Telescope Array (see the Fig. 2). One can see that at energies $E < 10^{18}$ eV and $E > 4 \cdot 10^{20}$ eV the distribution $N(\epsilon)$ is the power law, $N(\epsilon) \propto \epsilon^{-\beta}$. The same slope is observed at the intermediate energies, in the region of the maximum relative losses of energy, $(dE/dt)/E$, on the birth of electron-positron pairs, $E \simeq 10^{19}$ eV. In the region $0.2 < \epsilon < 1$, N grows exponentially with decreasing energy, $N(\epsilon) = q_0\epsilon^{-\beta+1}\exp(1/\epsilon)/(\beta - 1)$. According to observations summarized by Berezinsky [13], the flux of particles $I(\epsilon) = cN(\epsilon)/4\pi$ in the energy range $6 \cdot 10^{18}$ eV $< E < 6 \cdot 10^{19}$ eV, $1.5 \cdot 10^{-2} < \epsilon < 0.15$, is indeed the power law function of the energy with the index $\beta = 2.7$. As an illustration of this we show here observations of the Telescope Array (Tinyakov [14]). One can see in Fig. 2 that the slope is indeed equal to 0.33 ± 0.03 in the range $6 \cdot 10^{18} eV < E < 6 \cdot 10^{19}$ eV.

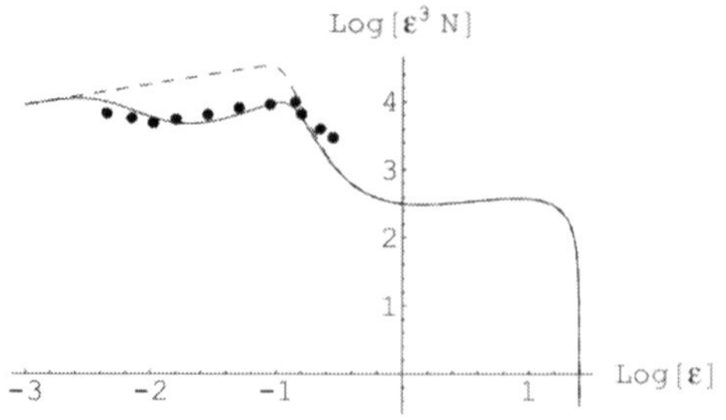

Figure 1. The function $\epsilon^3 N(\epsilon)$, $N(\epsilon)$ is given by the formula (14). Here we choose the value of maximum energy $E_0 = 10^{22}$ eV, $\epsilon_0 = 25$, and the index $\beta = 2.7$. The dashed line represents this function when electron-positron pair production is absent. Points are observational data by the Telescope Array (see Fig. 2).

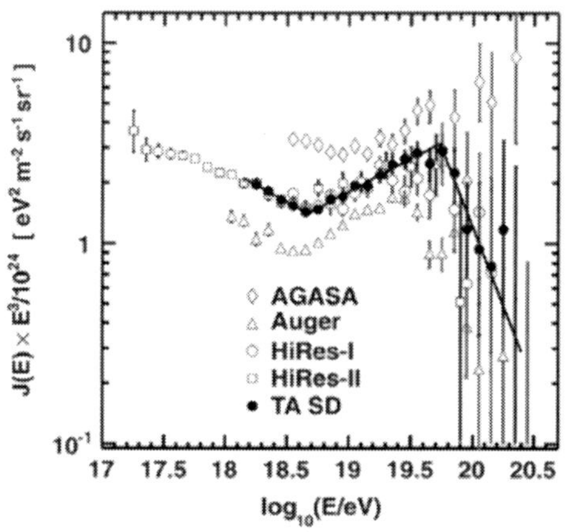

Figure 2. The spectrum of high energy cosmic rays observed by Telescope Array (black points) (Tinyakov [14]). In the range $6 \cdot 10^{18}$ eV $< E < 6 \cdot 10^{19}$ eV the spectrum is the power law function of the energy with the index $\beta = 2.67 \pm 0.03$. Another points present also observations by different instruments.

When analyzing the data of observations it is convenient to use the function $\epsilon^3 N(\epsilon)$, which clearly describes the transition from the Galactic distribution ($E^3 I(E) \propto E^{-0.1}$) for $E < 3 \cdot 10^{18}$ eV to the extragalactic one ($E^3 I(E) \propto E^{0.3}$) for $6 \cdot 10^{18}$ eV $< E < 6 \cdot 10^{19}$ eV. The function $\epsilon^3 N(\epsilon)$ has a maximum at $\epsilon = \epsilon_1 = 0.1$, $E_1 = 4 \cdot 10^{19}$ eV (see Fig. 2). At energies $\epsilon > \epsilon_1$ the function $\epsilon^3 N(\epsilon)$ exponentially decreases not to zero, but to the minimum value at ϵ_2, $\epsilon_2 \simeq 1$, $E_2 \simeq 4 \cdot 10^{20}$ eV. After that the distribution $N(\epsilon)$ continues to fall down by a power law manner with the same index β, $N(\epsilon) \propto \epsilon^{-\beta}$. The fall down of the function $\epsilon^3 N(\epsilon)$ at energy ϵ_1 to its value at ϵ_2 is approximately one and half orders. The strong growth of the density of particles for $\epsilon < \epsilon_2$ is explained by the sharp decrease of the rate of energy losses of particles when braking by relict photons becomes small. Here particles are accumulated. The distribution of $N(E)$ for energies $E > 4 \cdot 10^{20}$ eV does not go exponentially to small values that can seem as the result of the GZK effect. Here we observe that the distribution of $N(E)$ repeats the distribution of extragalactic sources $\bar{q}(E)$, unless the maximum energy E_0 is not close to the energy $4 \cdot 10^{20}$ eV. However, at energies $E > 4 \cdot 10^{20}$ eV the observational data have great uncertainty, which does not allow to make a conclusion about the growth of the function $E^3N(E)$ at high energies $E > 4 \cdot 10^{20}$ eV under the condition $E_0 > 4 \cdot 10^{20}$ eV. The energy range $E < 6 \cdot 10^{18}$ eV (see Fig. 1), where the distribution $N(E)$ also reproduces the distribution of extragalactic sources $\bar{q}(E)$, is hidden by Galactic cosmic rays.

It should be noted that the solution of the kinetic equation in the form (13) is not only valid for protons, but also for arbitrary nuclei with atomic number A, $N = N_A$. One only needs to take into account the fact that sources of the nucleus A are not only point sources q^s, in which nuclei are accelerated, but also the intergalactic space, in which photo-disintegration of nuclei occurs when they collide with photons,

$$A' + \gamma = A + N, \quad A' > A,$$

$$q_A(E) = q^s_A - N_A(E) \sum_{A'<A} \nu_{A,A'}(E) + \sum_{A'>A} N_{A'}(m_{A'}E/m_A)\nu_{A,A'}(m_{A'}E/m_A).$$

Here $\nu_{A,A'}$ is the frequency of the disintegration of the nucleus A' into the nucleus A,

$$\nu_{A,A'}(E) = c \int_{E_{th}}^{\infty} n_{ph}\sigma_{A,A'}(E, E_{ph})dE_{ph}.$$

The value of n_{ph} is the spectral density of photons in intergalactic space, $\sigma_{A,A'}$ is the cross-section of photo-disintegration, E_{th} is the threshold energy of a photon. In the expression (13) for the energy loss dE/dt we should also add the energy loss of the nucleus during photo-disintegration, which is maximum at the energy of the nucleus $E_A/A \simeq E_b m_p c^2/2T \simeq 10^{19}$ eV (E_b is the binding energy of the nucleon N in the nucleus A, $E_b \simeq$ 10 MeV). Thus, the distribution of nuclei over energy is determined by the system of equations

$$\left|\frac{dE}{dt}\right|_A N_A = \int_E^{E_0} q^s_A(E')dE' - \int_E^{E_0} N_A(E')\nu^t(E')dE' + \sum_{A'>A} \int_E^{E_0} \nu(m'_A E'/m_A)N_{A'}(m'_A E'/m_A)dE'. \tag{15}$$

The value of $\nu^t(E)$ is a total frequency of destruction of the nucleus A with formation of all nuclei $A' < A$, $\nu^t(E) = \sum_{A'<A} \nu_{A,A'}(E)$. The system of Eq. (15) can be simply solved for two limiting cases:

a) In sources at $E > 10^{18}$ eV protons are the main component. Then for protons the second and the third terms in the right hand side of (15) are equal to zero. The solution for $N_p(E)$ looks as described above, Eq. (14).
b) In sources at $E > 10^{18}$ eV iron is the main component. Then for $N_{Fe}(E)$ we get

$$\left|\frac{dE}{dt}\right|_{Fe} N_{Fe}(E) + \int_E^{E_0} \nu^t(E')N_{Fe}(E')dE' = \int_E^{E_0} q^s_{Fe}(E')dE'. \tag{16}$$

The solution of (16) is

$$N_{Fe}\left|\frac{dE}{dt}\right|_{Fe} = \int_E^{E_0} q^s(E')\exp\left\{-\int_E^{E'} \frac{\nu^t(E'')}{|dE/dt|''}dE''\right\}dE'. \tag{17}$$

It is seen that the factor $\exp\{-\int_E^{E'}(\nu^t(E")/|dE/dt|")dE"\}$ suppresses sources at high energies due to the destruction of iron in interstellar space into light nuclei. Near the maximum of relative losses, where $|dE/dt| \propto E$, the index β in sources ($q^s \propto E^{-\beta}$) must be less than the observed index β^{obs}, $\beta < \beta^{obs} = 2.7$. Generally speaking, the factor $\exp\{-\int_E^{E'}(\nu^t(E")/|dE/dt|")dE"\}$ is not an exponential function of the energy E', and if sources (galaxies) produce a power-law spectrum, the observed distribution of $N_{Fe}(E)$ is not a power-law function. However, the average value of β, obtained by Monte Carlo simulation by Taylor, Ahlers & Aharonian [15], is of the value of $\beta \simeq 1.8$. If we consider the effect of propagation of nuclei in the host galaxy, this suggests that in the accelerating machine the index $\beta \simeq 1$. Accelerating mechanisms, in which the main energy is released at maximum energies $E \simeq E_0$, are difficult to find in nature. As for other nuclei, the same suppressing factor $\exp\{-\int_E^{E'}(\nu^t(E")/|dE/dt|")dE"\}$ associated with their disintegration, but to a less degree than for iron, will lead to an effective value of β less than β^{obs}, $2.7 < \beta < 1.8$. This index will be different for different nuclei $A < A_{Fe}$.

The statement that in the energy range where $|dE/dt| \propto E$ the observed spectrum of extragalactic cosmic rays $N(E)$ repeats the source spectrum $q(E) \propto E^{-\beta}$, see (13), generally speaking, is true if the maximum energy E_0 for sources is larger, on average, than that in the range considered $6 \cdot 10^{18}$ eV $< E < 6 \cdot 10^{19}$ eV. However, if E_0 is in this range, and is different for different sources, the index β can differ from the observed value of β^{obs}. In order to find a connection of β with β^{obs} it is necessary to average Eq. (13) over different realizations of E_0.

$$N(E) \propto \langle E^{-1}(E^{-\beta+1} - E_0^{-\beta+1})\rangle.$$

In this expression the contribution only gives sources that have E_0 larger than E, $E_0 > E$. We introduce the probability $P(E_0)$ to have a given value of E_0,

$$\langle\ldots\rangle = \int_E^{\infty} \ldots P(E_0)dE_0, \quad \int_{E^*}^{\infty} P(E_0)dE_0 = 1.$$

For the distribution $N(E)$ to be an exponential function of E, $P(E_0)$ has also to be a power-law function $P(E_0) = (r-1)(E^*/E_0)^r/E^*$. The index r is not only larger than unit, but larger than two, $r > 2$, in order for the average value of E_0 to be a finite value. The result is

$$\langle E^{-1}(E^{-\beta+1} - E_0^{-\beta+1})\rangle = \frac{\beta - 1}{\beta + r - 2}E^{-\beta-r+1}E^{*r-1}.$$

Thus, $\beta = \beta^{obs} - r + 1 < \beta^{obs} - 1 < 1.7$. To explain the observed power-spectrum of extragalactic cosmic rays at $E > 6 \cdot 10^{18}$ eV as a result of the superposition of different sources with different values of maximum energy E_0, we must have hard sources with $\beta < 1.7$, that is probably small.

Let us now estimate the average power density of sources $\bar{q}$. Observations give the value of the extragalactic particle flux, where there is a maximum value of the product $E^3 I(E)$ at the energy $E_1 = 4 \cdot 10^{19}$ eV,

$$I(E_1) = 4.7 \cdot 10^{-39}\,\text{cm}^{-2}\text{s}^{-1}\text{sr}^{-1}\,\text{eV}^{-1}.$$

Accordingly, the density of particles $N(E)$, $N(E) = 4\pi I(E)/c$, is

$$N(E_1) = 5.9 \cdot 10^{25}\text{Mpc}^{-3}\,\text{eV}^{-1}.$$

On the other hand, the dimensionless density of particles $N(\epsilon_1)$ and average power of sources $\bar{q}(\epsilon_1)$ are connected by the relation (14),

$$\bar{q}(\epsilon_1) = (\beta - 1)\left[(\epsilon_1^{-1} + 1)\exp(-1/\epsilon_1) + a(\epsilon_{e1}^{-2} + b\epsilon_{e1}^{0.6})^{-1}\right.$$

$$\left.\exp(-1/\epsilon_{e1}) + \alpha\right] N(\epsilon_1) = 2 \cdot 10^{-2} N(\epsilon_1).$$

Bearing in mind that the dimensional quantity $\bar{q}$ is $c/L = 7.1 \cdot 10^{-16} s^{-1}$ times less than the dimensionless one, we get

$$\bar{q}(E_1) = 8.4 \cdot 10^8\,\text{eV}^{-1}\,\text{s}^{-1}\text{Mpc}^{-3}.$$

Considering that, as well as in our Galaxy, the power law spectrum with index $\beta = 2.7$ continues up to small energies $E_{min} \simeq 5$ GeV, we get

$$\bar{q}(E) = 8.4 \cdot 10^8 \left(\frac{E}{4 \cdot 10^{19}\text{eV}}\right)^{-2.7} \text{eV}^{-1}\text{s}^{-1}\text{Mpc}^{-3}. \tag{18}$$

The total average power density of extragalactic sources $\mathcal{E} = \int_{E_{min}}^{E_0} \bar{q}(E) E dE$ is equal to

$$\mathcal{E} = 1.8 \cdot 10^{43} \left(\frac{E_{min}}{5GeV}\right)^{-0.7} \text{erg}\,\text{s}^{-1}\text{Mpc}^{-3}. \tag{19}$$

This value divided by the density of galaxies, $N_g \simeq 1\text{Mpc}^{-3}$, exceeds the power of the Galaxy in cosmic rays, $10^{41}\text{erg}\,\text{s}^{-1}$, by at least two orders of magnitude. If we assume that the sources of ultrahigh energy particles are active galactic nuclei with relativistic jets generated inside, the power of which is $(10^{45} - 10^{46})\text{erg}\,\text{s}^{-1}$ (Mao-Li, et al. [16]), then their density in the Universe is of $N_g \simeq (10^{-2} - 10^{-3})\text{Mpc}^{-3}$. The same estimation of the density of extragalactic sources of cosmic rays follows also from the conditions of arrival isotropy of particles in the range of energies $E \simeq 10^{19}$ eV (Abbasi et al. [17]).

4. Galactic cosmic rays

The most surprising fact, following from the previous consideration, is that the spectrum of the power density of extragalactic cosmic rays (18) has the the same slope as the Galactic cosmic ray density before the 'knee', $E < 3 \cdot 10^{15}$ eV, $\beta = 2.7$. Let us note here that understanding the cosmic ray power density $\bar{q}(E)$ as the average power radiated by a galaxy as a whole, is proportional to the density of cosmic rays in a galaxy $N(E)$, $\bar{q}(E) \propto N(E)Sc$, where the value of S is the galactic surface radiated cosmic rays. This indicates that the nature of the origin of cosmic rays in the Galaxy and in active galactic nuclei is the same. The formation of the spectrum with index $\beta = 2.7$ in the Galaxy is explained by the fact that the index of the power law energy spectrum of source is of universal value $\beta = 2$. This is valid as for the acceleration of particles on fronts of strong shock waves ([1, 2]), as well as for the acceleration at the base of the jets emitted near massive black holes in centers of galaxies (Istomin [18]).

Next, accelerated particles, spreading over a galaxy, are scattered by inhomogeneities of a magnetic field. Their motion becomes diffusive. Moreover, the coefficient of diffusion is larger for particles with larger energies, $D \propto E^{\delta}$, $\delta \simeq 0.6$ (Ptuskin [19]). It should be noted that the index δ is the same for protons and for different nuclei. This follows from the general selfsimilarity considerations. The diffusion of charged particles by fluctuations of the magnetic field depends, in the general case, on three dimensional quantities: a particle velocity v, its Larmor radius r_L and the correlation length of the random field l_c. Thus, $D = cl_c f(r_L/l_c)$, where c (the speed of the light) is the speed of motion of a relativistic particle, $v = c$, f is a dimensionless function of the ratio of a particle Larmor radius, $r_L = E/eZB$, to the correlation length l_c. This function depends on the spectrum and amplitude of the random magnetic field. Because the various charged particles are scattered by the same magnetic field, the power-law index δ is the same for different nuclei. Only the diffusion coefficients can slightly vary at the same energy E. But the diffusion is the same for particles of the same rigidity $\mathcal{R}$ (Blasi & Amato [20]), $r_L = \mathcal{R}/B$. The observed spectrum of Galactic cosmic rays under the assumption about the universality of the source gives the value of δ of 0.7. We are taking this value as following from the observations, regardless of other values obtained from theoretical and numerical calculations of the scattering of charged particles and of the paradigm of particle acceleration by shock waves from supernova explosions.

The density of particles is equal to the product of the power of a source, Q, to the lifetime of particles τ, $N = Q\tau$. The lifetime is the escape time of particles from a galaxy, $\tau = R^2/D$, R is the radius of a galaxy. Thus, $N(E) \propto E^{-2-0.7}$. There arises the question: why in our Galaxy does the spectral index of cosmic rays at energies $E > 3 \cdot 10^{15}$ eV deviate from the value of $\beta = 2.7$? The spectrum becomes softer, $\beta = 3.1$. If the source of cosmic rays in a galaxy is through strong shock waves from supernova explosions, it is not clear why the spectrum of high energy particles from active galaxies remains with index $\beta = 2.7$ up to energies $\sim 10^{20}$ eV. It is more natural to assume that the source of cosmic rays is a jet emitted from an active galactic nucleus and whose power significantly exceeds the power transformed to particles by supernova explosions. Istomin [18] suggested that Galactic cosmic rays were produced by the jet, emitted from the center of the Galaxy. Giant bubbles of relativistic gas, observed above and below the Galactic plane, are remnants of this bipolar jet that existed previously. From the size of the bubbles it follows that the jet switched on $2.4 \cdot 10^7$ years ago, and worked for at least 10^7 years.

We consider that cosmic rays at sufficiently large times of filling of the Galaxy spread uniformly over the Galactic plane. This follows from the observations. However, the diffusion model of propagation of cosmic rays in the Galaxy and the halo can not provide such uniform distribution with any distribution of sources $Q(\rho)$ over ρ in the Galactic plane, where ρ is the cylindrical radius. This is due to the fact that the second derivative $\partial^2 N/\partial z^2 \simeq N/H^2$ is not equal to zero. The value of H is the vertical scale of the halo. Thus, in the steady state $\partial N/\partial \rho \neq 0$. Even the uniform distribution of sources in the Galactic plane $Q(\rho) = const$, $\rho < R$, leads to the dependence $N(\rho) = (Q/\pi^2 DR)E(\rho/R)$, where $E(x)$ is the complete elliptic integral of the second kind, which varies from $\pi/2$ at $x = 0$ to 1 at $x = 1$ and has singularities of derivatives at $x = 1$. For one source located in the center of the Galaxy, the distribution of $N(\rho)$ is also not uniform, $N(\rho) \propto 1/\rho$. One way to overcome this contradiction is the assumption that the convective motion of cosmic rays, not only the diffusion motion, is very important in the Galaxy (Istomin [18]). The motion along the spiral arms, where there is a regular magnetic field, can occur with a speed close to the speed of light. Then in the steady state $\mathbf{v}\nabla N \simeq 0$, $|v| \simeq c$. If the source is located in the center of the Galaxy, the time to fill the Galaxy by cosmic rays is determined by two processes: the fast filling of arms during the time $\tau \simeq R/c \simeq 5 \cdot 10^4$ years; and slow diffusive overcoming of the central region of the Galaxy of the scale $L \simeq (1-2)$ kpc, and also of the interarm space of the same size approximately, $\tau \simeq L^2/D \simeq 10^7$ years. The total time to fill the Galaxy is determined by the slower process. However, the evacuation of cosmic rays from the Galaxy, if the source of cosmic rays stops working, differs from the filling, because the escape of particles is determined by the output into the outer regions of the Galaxy and the halo of the size of R, and has mainly a diffusive character. The time of the evacuation is of $\tau \simeq R^2/D \simeq 10^9$ years. The given estimates refer to particles of small energy $E \simeq 1$ GeV. For energetic particles the times of filling and escape are less due to an increase of the diffusion coefficient with increasing energy.

After the source is turned off particles continue to flow out from the Galaxy, and their density begins to decrease with time. Consider how this may happen. Suppose that at time $t = 0$ cosmic rays with density $N_0(E)$ uniformly filled the spherical region (Galaxy and halo) of radius R. Then, assuming that the motion of particles is of diffusion character, using the Green function (equation (10)), we find the density of cosmic rays $N(r, E, t)$ at a point located at distance r from the center of the Galaxy and at time t, $N(r, E, t) = N_0(E)F(p, r)$,

$$F(p,r) = \frac{1}{\pi^{1/2}}\left\{\frac{R}{2rp}\left[\exp[-p^2(1-r/R)^2] - \exp[-p^2(1+r/R)^2]\right] + \int_{-p(1-r/R)}^{p(1+r/R)} \exp(-y^2)dy\right\} \tag{20}$$

where the parameter p is equal to $p(E,t) = [R^2/4D_g(E)t]^{1/2}$. For small values of p, $p << 1$, the function F is equal to $F = 4p/\pi^{1/2}$, for large values $p >> 1$, $F = 1$. The transition of density of cosmic rays from the initial distribution N_0 $(p > 1)$ to the decrease with time, $N = N_0(4R^2/\pi D_g t)^{1/2}$, takes place at $p \simeq 1$. At this point there is a transition of the spectrum of cosmic rays from the original, $N \propto E^{-2.7}$, with energies $E < E_k$ to the distribution of $N \propto E^{-2.7-0.7/2} = E^{-3.05}$ for $E > E_k$. The value of energy E_k is defined by the equality $p = 1$, $D_g(E_k) = R^2/4t$. In the center of the Galaxy $(r = 0)$ the transition occurs exactly at the point $p = 1$, for us $(r \simeq R/2)$, this point is also very close to unity. When $p = 1$, there is a local increase of density of cosmic rays,

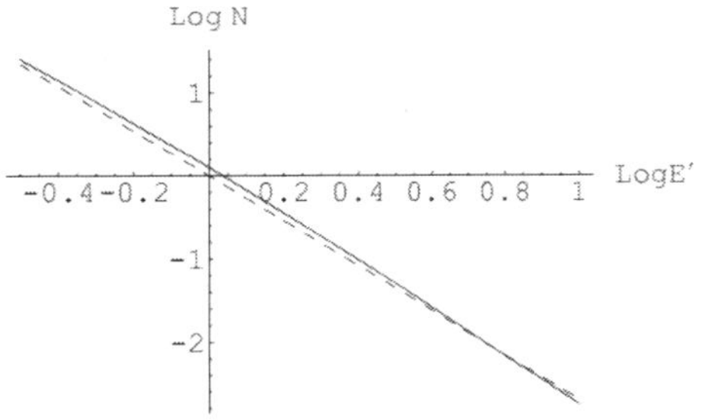

Figure 3. The distribution of Galactic cosmic rays $N(E) \propto E^{-2.7}F(p)$ over energy E near the "knee" at the point $r = R/2$, $E' = E/E_k$. The function $F(p)$ is defined by the expression (20). The dashed line is the dependence $N(E) \propto E^{-2.7}$.

i.e. near the energy $E \simeq E_k$ the distribution $N(E)$ is really observed as the 'knee' (see Fig. 3).

It should be noted that results obtained do not depend on our assumption of a spherically symmetric distribution of Galactic cosmic rays in the disk and the halo. The distribution may have different scales in the plane of the Galaxy R and across H. In diffusion models different scales R and H are due to the different diffusion coefficients of cosmic rays in the galactic plane, D_g, and in the direction above the Galactic plane, D_z. By that $D_g/R^2 = D_z/H^2$. If we introduce a new coordinate above the Galactic plane $z' = z(D_g/D_z)^{1/2}$, then in the coordinates (r, z') the distribution of rays will be spherically symmetric. Because we are interested in the distribution in the Galactic plane, $z = 0$, it will be the same as at $z' = 0$.

The position of the 'knee' depends on the time t, passed after turning off the source. Knowing the position of the 'knee' at the present time $t = t_0$, we find

$$t_0 = \frac{R^2}{4D_g(E_k)} = 4.2 \cdot 10^4 \left(\frac{R}{5 \cdot 10^{22}\,\text{cm}}\right)^2 \times \left(\frac{D_g(1GeV)}{2.2 \cdot 10^{28}\,\text{cm}^2\,\text{s}^{-1}}\right)^{-1}\left(\frac{E_k}{3 \cdot 10^{15}\,\text{eV}}\right)^{-0.7} \text{year}.$$

It should be noted that the 'knee' is moving with time, $E_k = 3 \cdot 10^{15}(t/t_0)^{-1.43}$ eV. The speed is now equal to $dE_k/dt = -1.43E_k/t_0 = -10^{11}$eV/year.

5. Conclusions

We showed that the distribution of power of sources of extragalactic cosmic rays in the energy range of $3 \cdot 10^{18}$ eV $< E < 10^{21}$ eV is a power law, $\bar{q}(E) \propto E^{-2.7}$. Thus, it is the same as in our Galaxy in the energy range below the 'knee', $E < 3 \cdot 10^{15}$ eV. This indicates the common nature of the origin of cosmic rays in the Galaxy and in other galaxies. However, the power of extragalactic sources exceeds by at least two orders of magnitude the capacity of the Galaxy (see formula (19)). The conclusion from this is that a 'normal' galaxy, to which our Galaxy belongs, is not the source of cosmic rays. The estimation of the density of extragalactic cosmic rays sources, $N_g \simeq 10^{-2} - 10^{-3}$ Mpc^{-3}, indicates galaxies with active nuclei. The possible source of energetic charged particles is relativistic jets emitted from the surroundings of massive black holes. Thus, acceleration by shock waves from supernova explosions is not possible to explain the origin of cosmic rays up to energies $\simeq 10^{18}$ eV. Otherwise, all galaxies in more or less degree would be sources of cosmic rays, because in all of them there are supernova explosions. Why, nevertheless, we observe cosmic rays in the Galaxy is explained that once in the past the Galaxy was also active. From the center of the Galaxy there emitted a relativistic bipolar jet, the remnants of which is observed now above and below the Galactic plane as bubbles of relativistic gas with the power law energy spectrum E^{-2} (Su, Slatyer & Finkbeiner [21]).

Due to the fact that this source of cosmic rays in the Galaxy is now not working, it appears the deflection of the spectrum of Galactic cosmic rays from the dependence $N(E) \propto E^{-2.7}$ in the whole energy range. For $E > 3 \cdot 10^{15}$ eV, the spectrum becomes softer, $N(E) \propto E^{-3.1}$. This is because particles, that once filled the Galaxy, leave it faster the larger their energy, because their diffusion coefficient increases with energy, $D \propto E^{0.7}$. Particles of larger energies leave the Galaxy quicker than particles of lesser energies. For the diffusive motion of particles, $N(E) \propto N_0(E)(Dt)^{-1/2} \propto E^{-2.7-0.7/2}$ for energies $E > E_k$ and $N(E) \simeq N_0(E)$ for energies $E < E_k$, where $N_0(E)$ is the initial distribution of particles. Thus, the 'knee' formation reflects the escape of particles from the Galaxy. Knowing the current position of the 'knee' one can estimate the time when the source of cosmic rays in the Galaxy stopped, it occurred $4.2 \cdot 10^4$ years ago. The position of the 'knee' is not constant in time, it must move, $E_k \propto t^{-1.43}$. This motion can be noticed if we have a sufficient accuracy of the measurement of the 'knee' position. During the last 50 years the 'knee' has moved by $\Delta E_k = -5 \cdot 10^{12}$ eV.

The release of cosmic rays from the Galaxy also explains the appearance of heavy nuclei in the energy range 10^{16} eV $< E < 10^{18}$ eV, just between the knee and extragalactic particles. Indeed, the particle Larmor radius is inversely proportional to the nucleus charge Z. For a fixed energy particle diffusion is less for heavy nuclei than for protons. Protons leave the Galaxy faster.

The recent discovery of the source of PeV particles in the Galactic center with the power of 10^{37-38} erg/s, made by the HESS group [22], confirms the hypothesis that centers of galaxies with massive black holes are powerful sources of ultrahigh energy particles. If we take into account the fact that the luminosity of Sgr A* is now 10^{-9} of the Eddington luminosity, one can imagine what will be the luminosity of the Galactic center in cosmic rays during its active phase.

Author thanks V. Berezinsky for fruitful discussions. This work was done under support of the Russian Foundation for Fundamental Research (grant numbers 14-02-00831 and 15-02-03063).

References

[1] Krymskii, G.F., Soviet Physics-Doklady **22**, 327 (1977)
[2] Bell, A.R., MNRAS **182**, 147 (1978)
[3] Blasi, P., The Astronomy and Astropart. Review **21**, 70 (2013)
[4] Pierre Auger Collaboration, 2010, Astropart. Phys. **34**, 314 (2010)
[5] Pierre Auger Collaboration, Science **318**, 939 (2007)
[6] Pierre Auger Collaboration, Astropart. Phys. **29**, 188 (2008)
[7] Greisen, K., Phys. Rev. Lett. **16**, 748 (1966)
[8] Zatsepin, G.T., Kuz'min, V.A., JETP Letters **4**, 78 (1966)
[9] Stanev, T., Engle, R., Mucke A., Protheroe, R.J., Rachen, J.P., Phys. Rev. D **62**, 093005 (2000)
[10] Berezinsky, V., Gazizov, A., Grigorieva S., Phys. Rev. D **74**, 043005 (2006)
[11] Aloisio, R., Berezinsky V., Grigorieva, S., Astropart. Phys. **41**, 94 (2013)
[12] Takami, H., Sato, K., Astropart. Phys. **30**, 306 (2009)
[13] Berezinsky, V., EPJ Web of Conference **53**, 01003 (2013)
[14] Tinyakov, P. for the Telescope Array Collaboration, Nuclear Instruments and Methods in Physics Reseach A **724**, 29 (2014)
[15] Taylor, A.M., Ahlers, M., Aharonian F.A., Phys. Rev. D **84**, 10500 (2011); `ArXiv:astro-ph/1107.2055`
[16] Mao-Li, M., Xin-Wu, C., Dong-Rong, J., Min-Feng, G., Chin. J., Astron. Astrophys. **8**, 39 (2008)
[17] Abbasi, R.U., et al., ApJ **610**, L73 (2004)
[18] Istomin, Ya.N., New Astronomy **27**, 13 (2014)
[19] Ptuskin, V.S., Phys. Uspechi **50**, 534 (2007)
[20] Blasi, P., Amato, E., JCaP Issue 1, id. 011 (2012)
[21] Su, M., Slatyer, T.R., Finkbeiner, D.P., ApJ **724**, 1044 (2010)
[22] HESS Collaboration, Nature **531**, Issue 7595, 476 (2016)

The search of the anisotropy of the primary cosmic radiation by the difference method

Victor Pavlyuchenko[1,a], Romen Martirosov[2], Natalia Nikolskaya[1], and Anatoly Erlykin[1]

[1] P.N.Lebedev Physical Institute, Moscow 119991, Russia
[2] A Alikhanian National Laboratory, Yerevan, Armenia

Abstract. On the basis of experimental data obtained in the knee energy region with the GAMMA array an anomaly has been found in the mass composition of primary cosmic rays coming from the region of the VELA cluster. We used an original difference method which has high sensitivity, stability against accidental experimental errors and the possibility to separate anomalies connected with the laboratory coordinate system from anomalies observed in the celestial coordinates. The multiple scattering of the charged particles in the galactic magnetic fields makes it possible to study regions of the sky outside the direct visibility of the array.

1. Introduction

The origin of the knee at energies of 3–4 PeV in the all-particle spectrum of the primary cosmic rays (PCR) has caused interest since its discovery in 1958 [1] because it is important for understanding the mechanism of PCR acceleration in the Galaxy. At the present time there are three main astrophysical models which try to explain the origin of the knee in this energy region. The selection between them can be made studying the anisotropy of arriving PCR particles. These models are the classic diffusion model [2], the model of the limited maximum energy for particle acceleration in the shells of supernova remnants [3] and the model of a nearby source, particles from which overlap with the smooth galactic spectrum, create an excess in the PeV region and imitate the formation of the knee [4]. This source has to be sufficiently close to us in order for its particles to have insufficient time to mix with particles from other sources.

So far there is no reliable experimental confirmation of the validity of these models, since the measured amplitude of the anisotropy in the 0.1–1 PeV region does not exceed 0.1% [5]. This is mainly caused by the mixing of the particle trajectories in chaotic magnetic fields due to which the propagation of PCRs in the Galaxy is similar to the large-scale diffusion transport. However, it is possible to study not only an anisotropy of the PCR intensity, but also the anisotropy of the mass composition since for the same conditions nuclei with different masses scatter differently in the same magnetic fields. In this case the multiple scattering of particles does not prevent, but rather helps to study PCRs from sources outside the direct visibility of the array. To do so we searched for regions of the sky having the maximum difference of the mass composition for particles coming from the opposite directions.

[a] e-mail: pavict@rambler.ru

2. The difference method

The main idea of the method [6] is the following: we take directions (ℓ_0, b_0) in the galactic coordinate system and build a cone around each of them. Its width angle θ (or $H_0 = \cos\theta$) is selected in such a way that it divides all registered particles in two sets with an equal number of particles in both of them - inside and outside the cone: $n^{in} = n^{out}$. In both sets we build the distributions of the parameter depending on the mass composition and then both distributions are subtracted from each other. The difference between both distributions is evaluated by the standard estimate of $\chi^2/J = J^{-1}\Sigma_i(\Delta_i/\sigma_i)^2$. Here J is the number of degrees of freedom, $\Delta_i = m_i^{in} - m_i^{out}$ is the difference between distributions in the interval $'i'$ and its error $\sigma_i = \sqrt{m_i^{in} + m_i^{out} + 1} = \sqrt{n_i + 1}$ is calculated from the Poisson distribution. If $n^{in} = n^{out}$ then there is no need to use the normalisation for the subtraction and the total number of events n_i in the interval $'i'$ does not depend on the galactic coordinates (ℓ_0, b_0). It is crucially important for the comparison of χ^2/J values with each other when searching for their maximum. One can see from the above formulae that χ^2/J values calculated for the (ℓ_0, b_0) and the opposite $(\ell_0 + 180°, -b_0)$ directions are equal. That is why the equation $n^{in} = n^{out}$ allows us to study the entire sky within the sensitivity of the method using the array with the limited fiducial sector. However, it does not allow to distinguish the (ℓ_0, b_0) direction from the opposite one, $(\ell_0 + 180°, -b_0)$.

In order to divide the total number of events in two sets for the studied (ℓ_0, b_0) direction we used the expression $H = \cos\psi = \sin b_0 \cdot \sin b + \cos b_0 \cdot \cos b \cdot \cos(\ell - \ell_0)$. Here H is the cosine of the angle ψ between (ℓ, b) - the arrival direction of the particle and (ℓ_0, b_0) - the studied direction in the spherical coordinate system. Events with $H > H_0$ enter the interiors of the cone and are included in n^{in}. Those with $H < H_0$ enter the outer regions of the cone and are included in n^{out}. The common background and possible methodical errors are equal in

 DOI: 10.1051/epjconf/201714503003

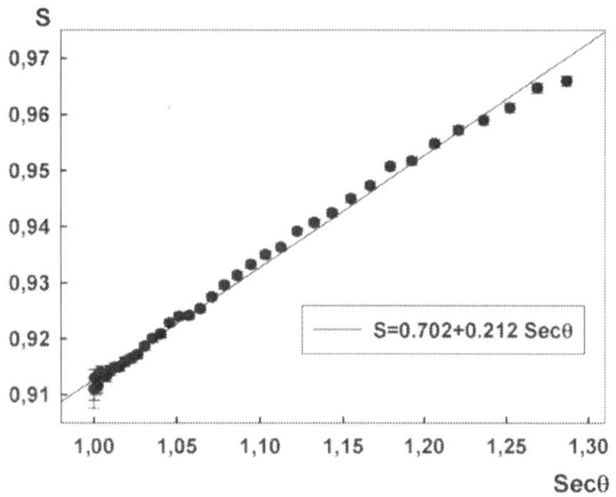

Figure 1. The experimental dependence of the lateral age parameter S from $\sec\theta$, θ being the zenith angle of the EAS arrival direction. The straight line is a linear fit of this dependence by the expression shown in the inset.

both sets of events and if $n^{in} = n^{out}$ they disappear in the process of the subtraction.

The requirements for the experimental array are common: it has to be stable and its registration efficiency should not depend on the azimuth of the particle arrival direction.

3. Results

The blind search of the anisotropy in the PCR mass composition has been made with experimental data ($3.38 \cdot 10^6$events) obtained with the extensive air shower (EAS) array GAMMA, operated at an altitude of 3200 m above sea level on Mount Aragats in Armenia. Its geographical coordinates are 40.28° N and 44.11° E. The array consists of a system of surface and underground detectors. The detailed description of the array, its technical characteristics and main results are given in [7–10].

The PCR mass composition at high energies is determined with large uncertainties by indirect methods studying various EAS characteristics and model dependent simulations. In this study we used the lateral age, S, of EAS where S is the experimental dimensionless parameter connected with the steepness of the lateral distribution function (LDF) of the charged EAS particles approximated by the so called NKG function [11].

The age S depends on the mass composition since PCR nuclei heavier than protons interact higher in the atmosphere for the same conditions. The cascade development of EAS initiated by these nuclei is faster and at the observation level EAS have wider LDF with bigger S. The S parameter is one of the basic shower parameters, which is well determined in experiments without an involvement of models. In Fig. 1 we show the measured experimental dependence of the S parameter on the zenith angle θ. With increasing θ the geometrical path length from the first interaction point to the array also increases and also the thickness of the atmosphere above the array as $X = X_0 \sec\theta$ ($X_0 = 700\,\mathrm{g\,cm^{-2}}$ is the vertical thickness). The near linear growth of S with $\sec\theta$ in the zenith range

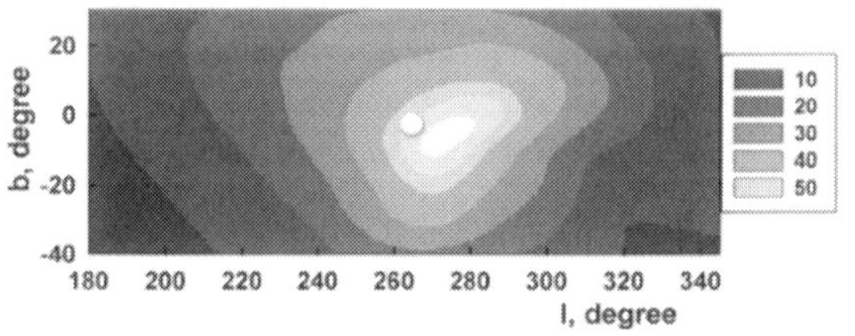

Figure 2. The dependence of χ^2/J value which evaluates the difference between the S-distributions in two sets of EAS coming from the opposite directions on galactic coordinates. The white circle in the center -the position of the Vela cluster.

from 0° to 40° justifies the good sensitivity of S to θ and therefore to the PCR mass composition.

We have made a blind search for anomalies in the vicinity of the galactic plane ($b = 0$) in the range of galactic latitudes $-40° \leq b \leq 40°$ and in the full range of galactic longitudes from $\ell = 0°$ to 360°. A single maximum of $\chi^2/J = 57.3$ for $J = 17$ has been found in the direction of $\ell_m = 277.3°; b_m = -5.3°$ (Fig. 2). Using the normalized fit of the two-dimensional ℓ, b distribution then the coordinates of the maximum are $\ell_m = 270 \pm 09°; b_m = -0.1 \pm 0.8°$.

The value of $\chi^2/J = 57.6$ with $J = 17$ is very high. For random fluctuations χ^2/J has to be close to $1 \pm \sqrt{2/J}$. By dividing the total number of events by 2,3,5 and 10 we ensured that the big value of χ^2/J is due to the large statistics and the high sensitivity of the method, which used the full statistics in each point of the sky. When dividing the statistics the maximum value of χ^2/J decreases from 57.6 to 29.4, 20.3, 12.6 and 7.9 respectively, i.e. with a nearly invariable shape of the distributions the value of $\chi^2/J - 1$ depends linearly on the total number of events in our case.

An additional study [12] showed that showers from the maximum region are initiated by PCR with a lighter mass composition ('young' EAS) compared with those from the opposite direction at PeV energies. Since light nuclei at a given energy are less deviated in magnetic fields than heavier ones it means that the source of the anomaly is in the direction of the maximum. The analysis of the stability of this result [12] showed that accidental errors of the S parameter do not change the position of the anomaly but reduce the hight of the maximum χ^2/J, which depends on the error values. If the errors are asymmetric in azimuth angles of the laboratory coordinate system, which rotates with the Earth, then they induce the appearence of an additional maximum in the direction of the Earth's rotation axis ('World's Pole').

Thus the difference method not only supresses background and methodical uncertainties but allows to separate effects connected with the laboratory coordinate system from those connected with the system of fixed celestial coordinates. The asymmetry in azimuth angles is revealed by the appearence of the peak at the Pole and a stable anomaly in other directions is most likely related to celestial coordinates. In the difference method there are no model assumptions.

Nearby the found anomaly there is the Vela cluster with closely located supernova remnants (SNR): the central Vela SNR and two others - Vela X (263.9°, −3.3°) and

Vela Jr (266.2°, −1.2°), the distances from which are about 0.2 and 0.3 k pc respectively. It is close to the anomaly's coordinates (see Fig. 2) and located at a relatively close distance to the Earth. Evidently this cluster is a good candidate for being a nearby PCR source. However, the difference method does not allow to calculate their PCR intensity and to tell whether it is completely responsible for the formation of the knee.

4. Conclusion

The difference method demonstrated its simplicity, high sensitivity and stability against accidental errors. From the very beginning it accounts for the PCR diffusive transport and the rotation of the Earth. That is why it is able to survey the whole celestial sphere using the experimental array with a limited fiducial sector and to separate anomalies connected with the laboratory coordinate system from anomalies in celestial coordinates.

Using this method we have found an anomaly of the PCR mass composition in the knee energy region with a high statistical security. Closest to this anomaly is the Vela cluster of SNR, which can be regarded as a good candidate for being the nearby single source responsible for the formation of the knee.

This work is done with the financial support from the Program of Fundamental Research of the Presidium of the Russian Academy of Sciences 'High energy physics and neutrino astrophysics'.

Authors are very thankful to Dr.B.Pattison for his help in the preparation of the paper and useful discussions.

References

[1] G.V. Kulikov, G.B. Khristiansen, Journ. Exp. Theor. Phys. **35**, 635 (1958) (in Russian)
[2] V.S. Ptuskin, Usp. Fiz. Nauk **177**, 558 (2007) (in Russian)
[3] E.G. Berezhko, H. Völk, Astron. Astrophys. **451**, 981 (2006)
[4] A.D. Erlykin, A.W. Wolfendale, J. Phys. G: Nucl. Part. Phys. **23**, 979 (1997)
[5] G. Guillian et al., Phys. Rev. D **75**, 062003 (2007)
[6] V.P. Pavlyuchenko, Bull. Lebedev Phys. Inst. **41/3**, 55 (2014)
[7] A.P. Garyaka et al., Astropart. Phys. **28**, 169 (2007)
[8] A.P. Garyaka et al., J. Phys. G:Nucl. Part. Phys. **35**, 115201 (2008)
[9] R.M. Martirosov et al., Proc. 32 Int. Cosm. Ray Conf., Beijing **1**, 178 (2011)
[10] A.P. Garyaka et al., J. Contemp. Phys. Armenian Acad. Sci. **48/2**, 51 (2014)
[11] C. Hayakawa, Cosm. Ray Phys: Nucl. Astrophys. Aspect, Wiley (1969)
[12] V.P. Pavlyuchenko et al., Bull. Lebedev Phys. Inst. **5**, 55 (2014) arxiv1406.0799; 1509.00316

Origin of very high energy emission in galaxy clusters

V.G. Sinitsyna[a] and V.Y. Sinitsyna

P. N. Lebedev Physical Institute, Russian Academy of Science, Russia

Abstract. The Perseus cluster of galaxies with the central galaxy NGC 1275 is ideally suitable for studying both the physics of relativistic jets from Active Galactic Nuclei and for revealing the feedback role of the central galaxy. We present the results of fifteen-year-long observations of the AGN NGC 1275 at energies 800 GeV–40 TeV discovered by the SHALON telescope in 1996. The data obtained at very high energies by SHALON, namely the images of the galaxy and its surroundings, and the flux variability indicate that TeV γ-ray emission is produced by a number of processes: in particular, part of this emission is generated by relativistic jets in the nucleus of NGC 1275 itself. Unique data on GK Per(Nova 1901) and the IC 310 radio galaxy TeV γ-ray emission were obtained with the SHALON experiment for the first time.

1. Introduction

Clusters of galaxies have long been considered as possible candidates for the sources of TeV gamma rays emitted by protons and electrons accelerated at large-scale shocks or by a galactic wind or active galactic nuclei (AGNs) [1–9]. The cluster of galaxies in Perseus is one of the best-studied clusters owing to its relative proximity (its distance ~ 100 Mpc or redshift $z = 0.0179$) and brightness. The dominant galaxy in the Perseus cluster is NGC 1275 (Figs. 1, 2, 3).

NGC 1275 has been classified in various ways, for example, as a Seyfert 1.5 galaxy, because broad emission lines were detected in its spectrum at optical wavelengths [10]. However, within the unified model of active galactic nuclei [11,12], it also belongs to the main class of BL Lacertae objects owing to its large and fast flux variability and polarization [13]. It should be noted that there is evidence that the mentioned AGN unification scheme can be even simplified [14].

NGC 1275 is a powerful source of radio and X-ray emission. In the radio band, the object found in NGC 1275, also known as Perseus A and 3C 84, has a powerful and compact core that has been well studied with VLBI [15–17]. NGC 1275 is extremely bright in the radio band and was classified as an FR I radio galaxy; it has a prominent structure that consists of a compact central source and an extended jet [15,18]. Having a supermassive black hole (with a mass of $3.4 \times 10^8\,M_\odot$) at its center [19], NGC 1275 also exhibits jet precession, which can be interpreted as a possible manifestation of the fact that NGC 1275 is the result of a merger between two galaxies [20]. The radio emission extends to great distances and shows a clear interaction with the gas inside the Perseus cluster of galaxies. ROSAT [8] and, subsequently, Chandra [9] observations revealed cavities in the gas located inside the cluster, whose presence suggests that the jets from 3C 84 sweep up numerous "bubbles" in the atmosphere of the Perseus cluster (Fig. 1).

NGC 1275 surrounded by extended filamentary structures historically aroused great interest owing to both its position at the center of the Perseus cluster and its possible "feedback" role [21]. Evidence for the "feedback" role of NGC 1275 can be obtained from ROSAT and Chandra observations, which reveal shells of hot gas and cavities that spatially coincide with the radio structures (Fig. 3) extending from the central, active part of the AGN. NGC 1275 also arouses interest owing to its close proximity to the Earth at redshift $z = 0.0179$ [22], making it possible to study the physics of relativistic jets.

2. NGC 1275 viewed in TeV gamma-rays

Extragalactic sources of very high energy γ-rays have been searched for in the SHALON experiment from the very beginning of its operation [23,24]. In 1996, observations with the SHALON mirror Cherenkov telescope revealed a new metagalactic source of γ-ray emission at very high energies $E > 800$ GeV [25,26] (Figs. 2, 3). The coordinates of an emission source detected in our experiment coincides with the Seyfert galaxy NGC 1275 [25–33]. NGC 1275 was observed by the SHALON telescope for 271.2 h in different years (from 1996 to 2012) during clear moonless nights at zenith angles from 3° to 33°. Gamma-ray emission from NGC 1275 was detected by the SHALON telescope at energies above 800 GeV at a 31.4σ confidence level determined according to Li & Ma [34]. The average integral flux at energies above 800 GeV for NGC 1275 is $(7.8 \pm 0.5) \times 10^{-13}\,\mathrm{cm^{-2}\,s^{-1}}$. The γ-ray energy spectrum in the observed energy range is well described by a power law $F(E_0 > 0.8\,\mathrm{TeV}) \propto E^{k_\gamma}$, where $k_\gamma = -2.24 \pm 0.09$ (see Fig. 2). Figure 2 shows the source's image at TeV energies observed by the SHALON experiment. The color scale in Fig. 2 right is in units of the excess above the minimum detected signal.

Possible correlations between the emission regions of TeV γ-rays and low-energy (radio and X-ray) photons should be established to elucidate the mechanisms of the generation of very high energy emission in the source and

[a] e-mail: sinits@sci.lebedev.ru

 DOI: 10.1051/epjconf/201714504001

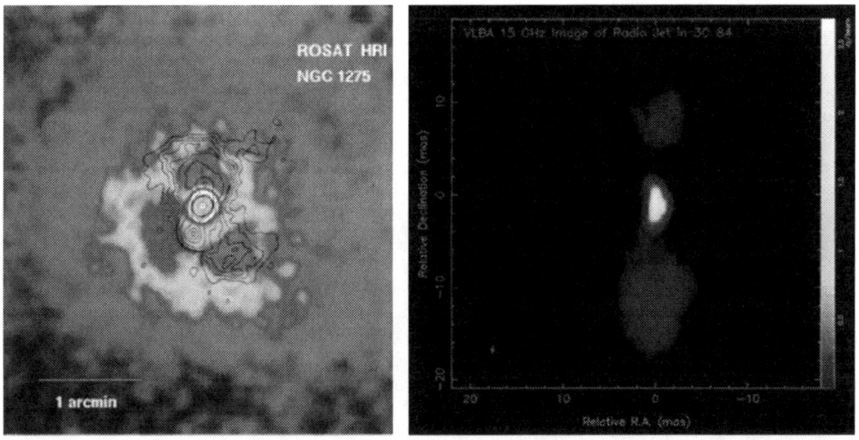

Figure 1. Left: ROSAT X-ray (0.1–2.4 keV) image of NGC 1275 [8]. The contours represent the source's radio structure from VLA radio observations. Right: VLBA 15 GHz image of NGC 1275.

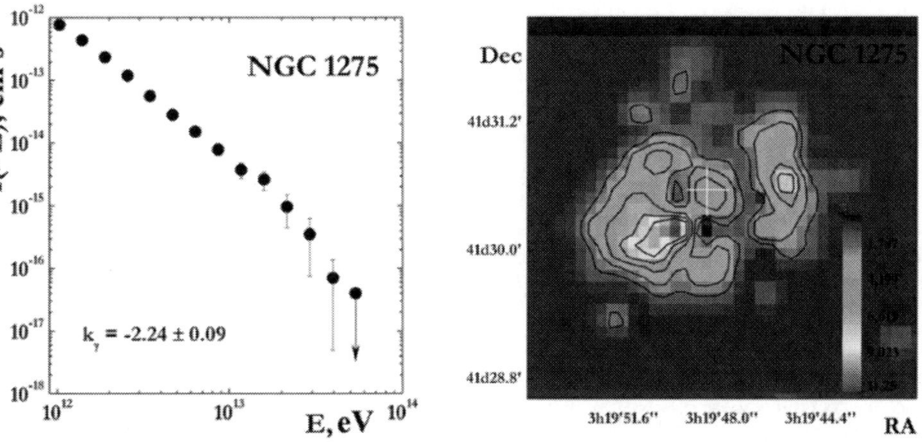

Figure 2. Left: gamma-ray spectrum of NGC 1275 with a power-law index $k_\gamma = -2.24 \pm 0.09$. Right: the image of the gamma-ray source NGC 1275 at energies > 0.8 TeV by SHALON.

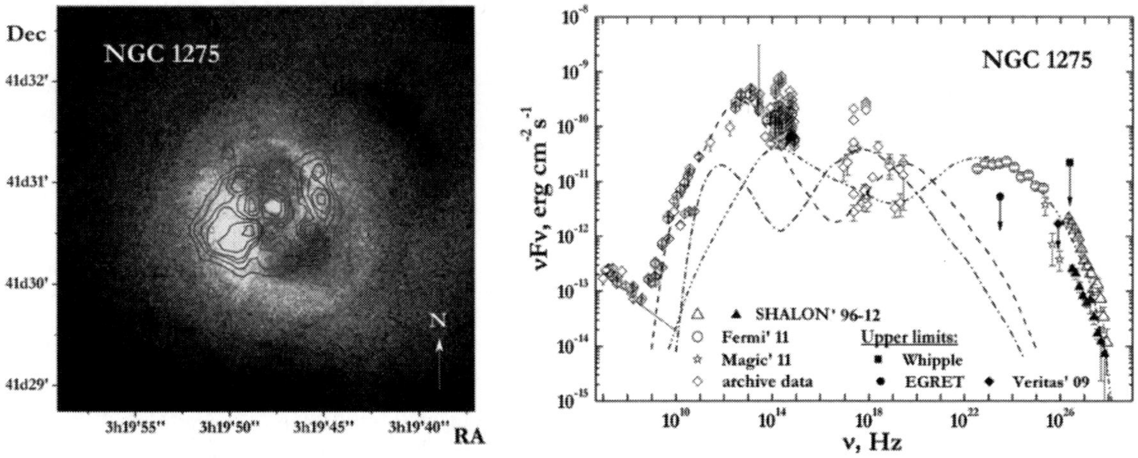

Figure 3. Left: Chandra X-ray (1.5–3.5 keV) image of NGC 1275 [35]; the red contours indicate the SHALON image of NGC 1275 in the energy range 800 GeV–40 TeV. Right: spectral energy distribution of the γ-ray emission from NGC 1275. △ and ▲ represent the data from the SHALON Cherenkov telescope in comparison with experimental data (see [33]). The dashed, dash-dotted, and dash-dotted with two dots curves indicate the spectral energy distributions of NGC 1275 obtained in the CM model [36] and see text below.

to test the models describing them. Figure 1 left shows a ROSAT X-ray image of NGC 1275; the contours represent the source's radio structure from VLA radio observations and the right is the VLBA 15 GHz image of NGC 1275 with jets (taken from MOJAVE[1]). The radio and X-ray emission maxima coincide with AGN NGC 1275 (Fig. 3). But, the X-ray emission disappears almost completely near the bright areas of the radio components located symmetrically relative to the core in the north and the south [8]. We also combined the SHALON-1 (0.8–40 TeV) and Chandra (1.5–3.5 keV X-ray) images. Figure 3 (black-and-white scale) presents a Chandra X-ray (1.5–3.5 keV) image for the central part of the Perseus cluster centered on NGC 1275 with a size of $\sim$5.5 arcmin [35]. In the X-ray energy range, the core of the Perseus cluster, on the whole, appears as a clear circularly symmetric structure with a distinct maximum on NGC 1275 (Fig. 3).

The clearly seen dimming in X-ray flux, along with the dip NW of the center correlates with the components of the extended double radio structure 3C 84 (Fig. 3). These dips are surrounded by bright (at energies 1.5–3.5 keV) arc regions from the north and the south. The simplest interpretation is that the intense emission from these rims comes from the shells surrounding the radio lobes [35]. A bright emission spot is also observed to the east.

The emission regions of very high energy γ-rays observed by SHALON from NGC 1275 have a structure similar to that described above (see [35]) and well correlates with the photon emission regions in the energy range 1.5–3.5 keV (Fig. 3). A correlation of the emission with energies 0.8–40 TeV [28–30] and the X-ray emission in the range 0.3–7 keV [35] was also found. Thus, the TeV γ-ray emission recorded by SHALON from NGC 1275 has an extended structure with a distinct core centered at the source's position.

To analyze the emission related to this core, we additionally identified the emission component corresponding to the central region of NGC 1275 with a size of 32″. The emission from the central region of NGC 1275 was detected at energies above 0.8 TeV at a 13.5σ confidence level determined by the Li & Ma method [34] with an average integral flux $(3.26 \pm 0.30) \times 10^{-13}\,\mathrm{cm^{-2}\,s^{-1}}$. The γ-ray energy spectrum of the central component in the entire energy range from 0.8 to 40 TeV is described by a power law with an exponential cutoff, $I(>E_\gamma) = (2.92 \pm 0.11) \times 10^{-13} \times E^{-1.55\pm0.10} \times \exp(-E_\gamma/10\,\mathrm{TeV})\,\mathrm{cm^{-2}\,s^{-1}}$. The SHALON spectrum corresponding to the emission from the central region of NGC 1275 is represented in Fig. 3 with the black triangles.

NGC 1275 was detected at high energies of 100 MeV–300 GeV, by the Fermi LAT experiment [37]. Recently, the AGN NGC 1275 was also recorded by the MAGIC ground-based mirror Cherenkov telescope at energies above 100 GeV in the 2010–2011 observations [38].

Figure 3 compares the integral γ-ray spectrum of NGC 1275 ($\triangle$) and its central region ($\blacktriangle$) obtained from SHALON data (1996–2012) [33] with the Fermi LAT (2009–2011) [37] and MAGIC (2010–2011) [38] experimental data.

[1] http://www.physics.purdue.edu/astro/MOJAVE/

3. Flaring activity of NGC 1275 in high and very high energies

The revealing correlations between the emissions in different energy ranges, comparing the emission regions, and, in particular, the detection of the flux changes remains necessary, because it makes it possible to judge the nature of the source, its evolution, and the emission generation mechanisms in various objects.

The observed γ-ray flux variations do not, on average, exceed 20% of $(7.8 \pm 0.5) \times 10^{-13}\ \mathrm{cm^{-2}\,s^{-1}}$. The SHALON mirror Cherenkov telescope has detected three short-time (within five days) increases and one decrease of the very high energy γ-ray flux in the entire time of observations of NGC 1275. Given these variations, the flux decrease below the average was recorded in 1999 and the integral flux was $(4.7 \pm 1.3) \times 10^{-13}\ \mathrm{cm^{-2}\,s^{-1}}$. The increases were detected in late Jan. 2001, late Nov. early Dec. 2005, and late Oct. 2009. The fluxes in these periods were $(21.2 \pm 7.5) \times 10^{-13}$, $(35.5 \pm 12.4) \times 10^{-13}$, and $(23.4 \pm 4.5) \times 10^{-13}\ \mathrm{cm^{-2}\,s^{-1}}$, respectively. The duration of the flux increase in October 2009 was 3 days. No duration of intervals of flux increase were found in 2001 and 2005, because the observations were interrupted due to weather conditions in both cases.

To reveal possible correlations of the emissions in various energy ranges, including those at high and very high energies, we compared the NGC 1275 γ-ray fluxes from SHALON in the periods when the observations were simultaneous with the ones by the Fermi LAT experiment. The published Fermi LAT data were obtained from August 4, 2008, to September 30, 2010 [37]. SHALON observations of NGC 1275 were performed in November 2008 with a break for the Moon's time, October 2009, and mid-November-early December 2010. In this time, only one γ-ray flux increase to $(23.4 \pm 4.5) \times 10^{-13}\ \mathrm{cm^{-2}\,s^{-1}}$ was detected in the period of October 18–20, 2009. These periods of SHALON observations do not coincide with the times of the main flares observed at Fermi LAT [37]. A slight local flux increase can be seen in the period of mid-October 2009 [37], which corresponds to the above-mentioned γ-ray flux increase observed by SHALON.

4. Sources during observations of NGC 1275

Additionally an analysis of the region around NGC 1275 was performed in order to search for TeV gamma-ray sources in the SHALON telescope field of view during the observation of the Perseus Cluster.

4.1. SN2006 gy

The flux increase was detected from the region NGC 1275 in autumn 2006. The detailed analysis of γ-shower direction turned out to be the detection of a metagalactic object. This object was identified with the supernova SN2006 gy (see [39]) that is about 10 minutes away from NGC 1275 (see details in [40]).

Observations had been made in 2006 September, October, November, December and then during the winter of 2007. No flux increase was found in the September observations. In the flare, observed on October 22, the flux increased 6 times from the NGC 1275 average flux

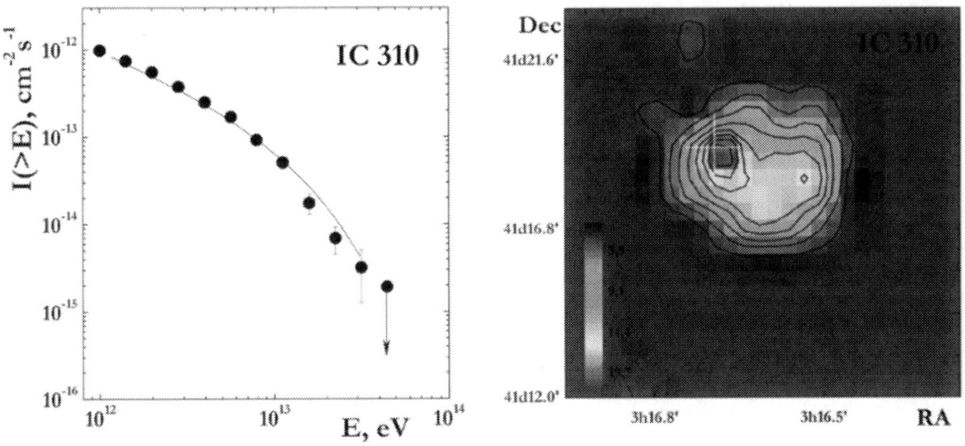

Figure 4. Left: gamma-ray spectrum of IC 310 by SHALON. Right: the image of the gamma-ray source IC 310 at energies > 0.8 TeV by SHALON.

and stayed at this level throughout the October moonless period. The integral γ-ray flux for SN 2006gy was found to be $(3.71 \pm 0.65) \times 10^{-12}$ cm^{-2} s^{-1} [40,41] at energies of > 0.8 TeV. Follow-up observations at the end of November showed that the flux of SN2006 gy had dropped to a level of about $(0.69 \pm 0.17) \times 10^{-12}$ cm^{-2} s^{-1} and was constant during the November, December period. The results of observation analysis of 2007 have not revealed any TeV gamma-ray emission from the region of SN 2006gy. So, the explosion of extragalactic supernova was observed at TeV energies for the first time with the SHALON Cherenkov telescope.

4.2. GK Per

During the observations of NGC 1275 the SHALON field of view contains the source of non-thermal radio and X-ray emission GK Per (Nova 1901) of classical nova type as it is located at $\sim 3°$ SW from NGC 1275. So due to the large telescopic field of view ($\geq 8°$) all observations of NGC 1275 are naturally followed by the tracking of GK Per. These long-term observations of the Perseus Cluster revealed γ-ray emission from a nearby object –GK Per (see details in [42]). The γ-ray source associated with GK Per was detected above 800 GeV with average γ-flux $I_{GK\ Per}(> 0.8\,\mathrm{TeV}) = (2,9 \pm 1,3) \times 10^{-13}$ cm^{-2} s^{-1} with a statistical significance [34] of 9.2σ. The analysis of γ-ray shower arrival direction revealed the main TeV-emission region coinciding with the position of the central source of GK Per and the weak emission of shell, that is also observed in X-ray by Chandra [43].

The TeV γ-ray emission of classical nova GK Per, that could be a shell-type supernova remnant in the early evolution stage, was detected for the first time by SHALON. Also, very high energy γ-rays from the shell of GK Per, visible in X-rays, were detected with SHALON experiment for the first time.

4.3. IC 310

The radio galaxy IC 310 (redshift z = 0.0189) is located in the Perseus Cluster at 0.6° (about 1Mpc) from the cluster's central galaxy, NGC 1275. It is one of the brightest objects of this Cluster of galaxies at radio frequencies and at X-ray energies. It has been detected in the high energy γ-ray band above 30 GeV with Fermi LAT [44] and above 260 GeV with the MAGIC telescopes [45].

The radio morphology of such types of galaxies consists of a bright "head", which is located at the core of the galaxy, and the "tail" of a radio lobe pointing away from the center of the cluster (Fig. 5). The jet direction is determined by the motion of galaxy through the medium of the cluster. In the case of IC 310, the radio lobe with a length of about 400 kpc extends in the direction parallel to the line connecting IC 310 with the center of the Perseus cluster.

A recent investigation of the radio structure of IC 310 by VLBA shows that the "head" contains a parsec-scale core-jet structure oriented in the same direction as the kiloparsec structure of the "tail". That means that there is no evidence of an interaction with the intracluster medium determining the direction of the "tail" [46]. In X-rays IC 310 looks as a point source at the position of the radio "head" [47].

IC 310 as a source accompanying NGC 1275 was observed with SHALON telescope between 1996 and 2012 for a total of 157 hours. The γ-ray source associated with IC 310 was detected above 0.8 TeV by SHALON with a statistical significance 17.2σ determined by the Li & Ma method [34]. The signal significance for this source is less then one for the source with a similar flux and spectrum index obtained in the same observation hours because of smaller collection field of view relative to the standard procedure of the SHALON experiment [33]. Corrections for the effective field of view were made to calculate the source flux and energy spectrum. The average integral flux at energies above 0.8 TeV for IC 310 is $(0.89 \pm 0.09) \times 10^{-13}$ cm^{-2} s^{-1} (Fig. 4). Taking into account the proximity to a nearby source NGC 1275, we made the observation data procession first associated with NGC 1275 and then with IC 310. We found that 5% of showers are common for both sources. After the detailed analysis of the arrival direction of these showers and angular distance [33], less than 2% of NGC 1275 showers were recognized to be IC 310 showers. This didn't change the average flux of NGC 1275 [33]. The energy spectrum

DOI: 10.1051/epjconf/201714504001

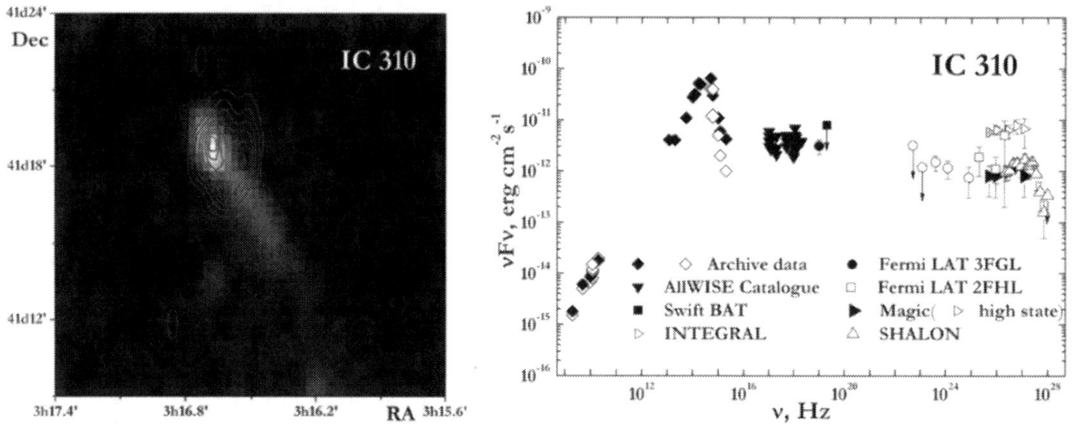

Figure 5. Left: radio image of IC 310 from WENSS sky survey; the red contours indicate the SHALON image of IC 310 in the energy range 800 GeV–50 TeV. Right: Spectral energy distribution of the γ-ray emission from IC310. $\triangle$ represent the data from the SHALON Cherenkov telescope.

of γ-rays in the observed energy region from 0.8 to 40 TeV is well described by a hard power law with an exponential cutoff $I(>E_\gamma) = (0.89 \pm 0.09) \times 10^{-12} \times E^{-0.84\pm0.10} \times \exp(-E_\gamma/12.5\,\mathrm{TeV})\ \mathrm{cm}^{-2}\,\mathrm{s}^{-1}$ (see Fig. 4). The spectral energy distribution and detailed images of IC 310 are shown in Fig. 5.

The main TeV γ-ray emission region was revealed in observations of IC 310 corresponding to the core of the galaxy. It coincides with the position of the "head" of the radio structure of IC 310. Figure 5, left (black-and-white scale) shows a radio image of the IC 310 galaxy from the WENSS survey [48] (SkyView interface[2]), red lines show the TeV structure of IC 310 by SHALON.

5. NGC 1275 as a point and extended source

As has been pointed out above, the Perseus cluster of galaxies with the central galaxy NGC 1275 is ideally suitable for studying both the physics of relativistic jets from AGNs and for revealing the feedback role of the central galaxy. Evidence for the latter was obtained in ROSAT and Chandra observations at low energies, from which shells of hot gas and cavities that spatially coincide with the radio structures originating in the central, active part of the AGN can be seen (Fig. 1 and [8,9,35,49]). The observational data for NGC 1275 at energies 800 GeV–40 TeV, namely the images of the galaxy and its surroundings (Figs. 2, 3), as well as the flux variations suggest that the TeV gamma-ray emission in these regions is produced by a number of processes.

The extended structure around NGC 1275 (Fig. 3) that spatially coincides with the X-ray emission regions (Fig. 3) can be produced by mechanisms related to the generation of an X-ray structure [9,35,49]. The brightness distribution of the X-ray emission and the observed TeV emission shows a sharp increase in intensity right outside the bubbles blown by the central black hole and visible in the radio band. This suggests that the X-ray-generating particles are swept up from the region of the radio lobes under the pressure of cosmic rays and magnetic fields generated in the jets at the center of NGC 1275 [9,49]. The structures visible in TeV γ-rays are formed through the interaction of very high energy cosmic rays with the gas inside the Perseus cluster and interstellar gas heating at the boundary of the bubbles blown by the central black hole in NGC 1275.

The presence of emission in the energy range 1–40 TeV from a central region of $\sim 32''$ in size around the nucleus of NGC 1275 (see Fig. 3, the black triangles) and the short-time flux variability [33] point to the origin of the very high energy emission as a result of the generation of jets ejected by the central supermassive black hole of NGC 1275.

The multifrequency spectral energy distribution for the nucleus of NGC 1275, up to high and very high energies, was described in the CM model [36] and is a composition of the components of inverse Compton scattering of the intrinsic synchrotron radiation from relativistic electrons (synchrotron self-Compton) of three separate plasma blobs ejected from the inner regions of the NGC 1275 nucleus (Fig. 3, the dashed, dash-dotted, and dash-dotted-dotted curves). The available Fermi LAT data at high energies and the SHALON observations at very high energies in a region $< 32''$ around NGC 1275 are described in terms of this model with one of the components producing synchrotron self-Compton emission of the relativistic jets from the nucleus itself (Fig. 3, the dash-dotted-dotted curve).

6. Conclusion

We present the results of fifteen-year-long observations of the AGN NGC 1275 at energies 800 GeV–40 TeV discovered by the SHALON telescope in 1996. Having analyzed the SHALON data, we have determined characteristics of NGC 1275 such as the spectral energy distributions and images at energies >800 GeV for the first time. The emission regions of very high energy gamma-rays observed by SHALON from NGC 1275 well correlates with the photon emission regions viewed in X-rays by Chandra. Thus, the TeV gamma-ray emission recorded by SHALON has an extended structure with a distinct core centered at the source's position. To analyze

[2] `http://skyview.gsfc.nasa.gov/`

the emission related to this core, we additionally identified the emission component corresponding to the central region of NGC 1275, and obtained the spectral energy distribution of this component as a result. Variations of TeV gamma-ray fluxes were also found. The data obtained at very high energies, namely the images of the galaxy and its surroundings, and the flux variability indicate that the TeV gamma-ray emission is generated by a number of processes: in particular, part of this emission is generated by relativistic jets in the nucleus of NGC 1275 itself. The presence of an extended structure around NGC 1275 is evidence of the interaction of cosmic rays and magnetic fields generated in the jets at the galactic center with the gas of the Perseus cluster.

References

[1] B. Dennison, Astrophys. J. **239**, L93 (1980)
[2] B.P. Houston et al., J. Phys. G: Nucl. Phys. **10**, L147 (1984)
[3] S. Colafrancesco et al., Astropart. Phys. **9**, 227 (1998)
[4] C.L. Sarazin Astrophys. J. **520**, 529 (1999)
[5] F.T. Miniati et al., Astrophys. J. **62**, 233 (2001)
[6] A.N. Timokhin et al., Astron. Astrophys. **417**, 391 (2004)
[7] S. Colafrancesco, P. Marchegiani, Astron. Astrophys. **502**, 711 (2009)
[8] H. Böhringer et al., Mon. Not. R. Astron. Soc. **264**, L25 (1993)
[9] A.C. Fabian et al., Mon. Not. R. Astron. Soc. **366**, 417 (2006)
[10] M.P. Veron-Cetty, P. Veron. Quasars and Active Galactic Nuclei, 8th ed., ESO Sci. Rep. No. 18 (ESO, 1998); ADC/CDS Catalogue No.7207 (1998)
[11] B.L. Fanaroff, J.M. Riley, Mon. Not. R. Astron. Soc. **167**, 31P (1974)
[12] C.M. Urry, P. Padovani, Publ. Astron. Soc. Pacif. **54**, 215 (1995)
[13] J.R.P. Angel, H.S. Stockman, Ann. Rev. Astron. Astrophys. **18**, 321 (1980)
[14] P. Kharb et al., Astrophys. J. **710**, 764K (2010)
[15] R.C. Vermeulen, et al., Astrophys.J. **430**, L41 (1994)
[16] G.B. Taylor, R.C. Vermeulen, Astrophys. J. **457**, L69 (1996)
[17] K. Asada et al., Publ. Astron. Soc. Jpn. **58**, 261 (2006)
[18] K. Asada et al., Astron. Soc. Pacif. Conf. **402**, 91A (2009)
[19] R.J. Wilman et al., Mon. Not. R. Astron. Soc. **359**, 755 (2005)
[20] F.K. Liu, X. Chen, Astrophys. J. **671**, 1272 (2007)
[21] J.S. Gallagher, Astron. Nachr. **220**, 1040G (2009)
[22] M. Strauss et al., Astrophys. J. Suppl. Ser. **83**, 29S (1992)
[23] V.G. Sinitsyna, Proc. of the Workshop, Towards a Major Atmospheric Cherenkov Detector IV, Padova, Italy, ed. M. Cresti (Univ. of Padova, Padova, Italy, 1995), p. 133
[24] V.G. Sinitsyna, Nuovo Cim. **19C**, 965 (1996)
[25] V.G. Sinitsyna, Proc. of the Workshop, Towards a Major Atmospheric Cherenkov Detector V, Kruger Park, South Africa, ed O. De Jager (Westprint, Potchefstroom, 1997), p. 136
[26] V.G. Sinitsyna, AIP (Conf. Proc.) **515**, 293 (2000)
[27] V.G. Sinitsyna, Rad. Phys. Chem. **75**, 880 (2006)
[28] V.G. Sinitsyna et al., Nucl. Phys. B (Proc. Suppl.) **122**, 409 (2003)
[29] V.G. Sinitsyna et al., Bull. Russ. Acad. Sci. **71**, 906 (2007)
[30] V.G. Sinitsyna et al., Nucl. Phys. B (Proc. Suppl.) **196**, 442 (2009)
[31] S.I. Nikolsky, V.G. Sinitsyna, Phys. At. Nucl. **67**, 1900 (2004)
[32] V.G. Sinitsyna et al., Nucl. Phys. B (Proc. Suppl.) **75A**, 352 (1999)
[33] V.G. Sinitsyna, V.Y. Sinitsyna, Astron. Lett. **40**(2–3), 91 (2014)
[34] T.-P. Li and Y.-Q. Ma, Astrophys. J. **272**, 317 (1983)
[35] A.C. Fabian et al., Mon. Not. R. Astron. Soc. **318**, L65 (2000)
[36] S. Colafrancesco et al., Astron. Astrophys. **519**, A82 (2010)
[37] A.M. Brown, J. Adams, Mon. Not. R. Astron. Soc. **413**, 2785 (2011)
[38] J. Aléksic et al., Astron. Astrophys. **539**, L2 (2012)
[39] E.O. Ofek et al., `arXiv:astro-ph/0612408v1`
[40] V.G. Sinitsyna, V.Y. Sinitsyna, EPJ Web of Conferences **52**, 10005 (2013)
[41] V.G. Sinitsyna et al., 2011, "Cosmic Rays For Particle And Astroparticle Physics" Proc. 12th ICATPP (Eds. S. Giani, C. Leroy, and P.G. Rancoita; Singapore: World Scientific) **6**, 296 (2011)
[42] V.G. Sinitsyna, V.Y. Sinitsyna, J. Phys.: Conf. Ser. **718**, 052046 (2016)
[43] S. Balman, Astrophys. J. **627**, 933 (2005)
[44] A. Neronov et al., Astron. Astrophys. **519**, L6 (2010)
[45] J. Aleksić et al., Astrophys. J. Lett. **723**(2), L207 (2010)
[46] M. Kadler et al., Astron. Astrophys. **538**, L1 (2012)
[47] K. Sato et al., PASJ **57**, 743 (2005)
[48] R.B. Rengelink et al., A&AS **124**, 259 (1997)
[49] E. Churazov et al., A&A **356**, 788 (2000)

Particle acceleration in binaries

V.G. Sinitsyna[a] and V.Y. Sinitsyna

P. N. Lebedev Physical Institute, Russian Academy of Science, Russia

Abstract. Cygnus X-3 massive binary system is one of the powerful sources of radio and X-ray emission consisting of an accreting compact object, probably a black hole, with a Wolf-Rayet star companion. Based on the detections of ultra high energy gamma-rays by Kiel and Havera Park, Cygnus X-3 has been proposed to be one of the most powerful sources of charged cosmic ray particles in the Galaxy. The results of long-term observations of the Cyg X-3 binary at energies 800 GeV–85 TeV detected by SHALON in 1995 are presented with images, integral spectra and spectral energy distribution. The identification of source with Cygnus X-3 detected by SHALON was secured by the detection of its 4.8 hour orbital period in TeV gamma-rays. During the whole observation period of Cyg X-3 with SHALON significant flux increases were detected at energies above 0.8 TeV. These TeV flux increases are correlated with flaring activity at a lower energy range of X-ray and/or at observations of Fermi LAT as well as with radio emission from the relativistic jets of Cygnus X-3. The variability of very high-energy gamma-radiation and correlation of radiation activity in the wide energy range can provide essential information on particle mechanism production up to very high energies. Whereas, modulation of very high energy emission connected to the orbital motion of the binary system, provides an understanding of the emission processes, nature and location of particle acceleration.

1. Introduction

SHALON is a high-altitude imaging atmospheric Cherenkov telescope for the detection of very high energy γ-rays from 800 GeV to 100 TeV. γ-astronomical research has been carryied out with the SHALON telescope since 1992. During the period 1992–2016 SHALON has been used for observations of different galactic and extragalactic objects (see these Proceedings and [1]). Among them are objects of the Cygnus region.

The Cygnus region contains a number of sources of radio and X-ray emission and some of them were detected at high energies. The most powerful have also been supposed to be GeV-TeV γ-ray emitters. Among them are the Cyg X-3 massive binary system (Figs. 1, 2, 3, 4, 5, 6, 7).

2. SHALON mirror Cherenkov telescopes

The SHALON γ-ray telescopes are located at an altitude of 3340 m above sea level each of which has a composite mirror with an area of 11.2 m^2. The detector array consisting of 144 FEU-85 photomultipliers assembled into a square array and mounted at the mirror focus has characteristics sufficient to record information about the shower structure in the energy range under consideration. The detector has the largest field of view in the world, $> 8°$. This allows one to monitor the background from charged cosmic-ray particles and the atmospheric transparency continuously during observations and expands the area of observation and, hence, the efficiency of observations [2]. The technique for simultaneously obtaining information about the cosmic-ray background and the showers initiated by γ-rays is unique and has been applied in the SHALON experiment from the very beginning of its operation [2]. This technique serves to increase the useful source tracking time and, what is particularly important, such source and background observation conditions as the thickness and state of the atmosphere remain the same. In addition, the wide field of view allows recording the off-center showers arriving at distances of more than 30 m from the telescope axis completely and almost without any distortions; they account for more than 90% of all the showers recorded by the telescope.

Due to criteria used in our experiment the background is rejected with 99.92% efficiency and the amount of background gamma-like events is less than 10%. During a primary analysis, the primary particle arrival direction is determined with an accuracy up to $\leq 0.1°$. The subsequent analysis specially developed for the SHALON telescopes and based on the Tikhonov regularization method [3,4] improves the accuracy to a value of less than $0.01°$ [5].

3. Cygnus X-3

The binary system Cyg X-3 is one of the brightest Galactic X-ray sources, displaying high and low states and rapid variability in X-rays. In addition to being a powerful X-ray source, Cyg X-3 is seen in the infrared and is a strong and variable radio source. It is also the strongest radio source among X-ray binaries (Fig. 1) and shows both huge radio outbursts and relativistic jets. The radio-activity is closely linked with the X-ray emission and the different X-ray states [7,8]. Cygnus X-3 is a high mass X-ray binary and microquasar, with a compact object, which is either a neutron star or may be a black hole, and a companion

[a] e-mail: sinits@sci.lebedev.ru

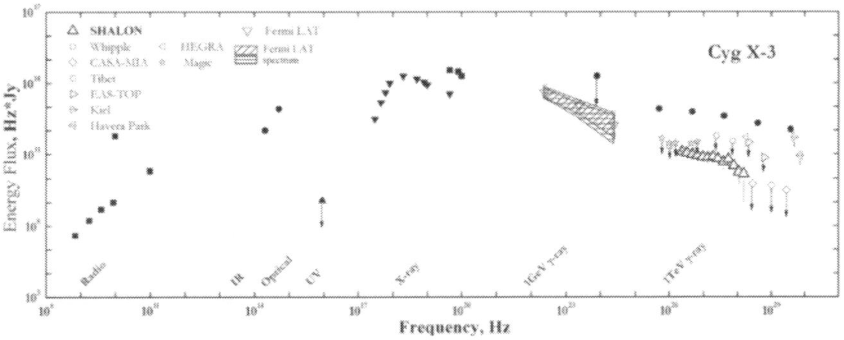

Figure 1. Spectral energy distribution of γ-ray emission from Cyg X-3. △ represent the data from the SHALON Cherenkov telescope. Black points are the archival data from [6].

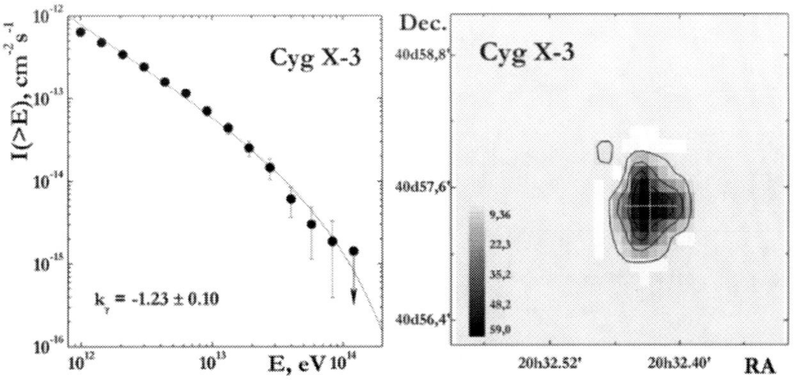

Figure 2. Left: the γ-ray integral spectra of Cyg X-3 averaged over the period 1995–2015. Right: TeV γ-ray emission map of Cyg X-3.

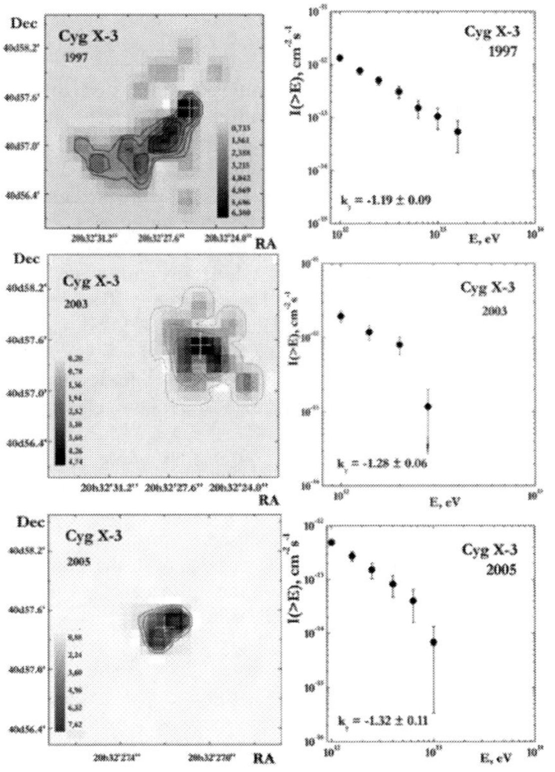

Figure 3. Left: the image of γ-ray emission from Cyg X-3 in 1997, 2003 and 2005 by SHALON; Right: the γ-ray integral spectra with the power index k_γ.

object, which is a Wolf-Rayet star. But Cygnus X-3 shows a short orbital period of 4.8 hours, typical of low mass binaries. It has been inferred from the modulation of both X-ray (Fig. 6) and infrared emissions. The nature of the compact object is still uncertain. Based on the detections of ultra high energy γ-rays [9, 10], Cygnus X-3 has been proposed to be one of the most powerful sources of charged cosmic ray particles in the Galaxy (Fig. 1).

Attempts to detect TeV emission from Cygnus X-3 were first made in the mid 1970s and continued through the mid 1980s. Two observations were particulary important: the Kiel [9] results and contemporaneous observations at Haverah Park [10]. These results indicated a very large UHE flux from Cygnus X-3. So, these results stimulated the construction of many new detectors. The upper limits of the Cygnus X-3 flux are over an order of magnitude lower than detected in the 1980s. Figure 1 also shows upper limits on the steady flux from Cygnus X-3 reported between 1990 and 1995 compared with earlier observations.

4. Cyg X-3 viewed in TeV gamma-rays

Cyg X-3 has been regularly observed since 1995 with the SHALON telescope [2] during a total of 303.2 hours. The observations were performed using the standard (for SHALON) technique of obtaining information about the cosmic-ray background and γ-ray-initiated showers in the same observing session. The SHALON method of selecting γ-ray showers from background cosmic-ray showers allows to reject 99.92% of the background showers [1]. The γ-ray source associated with Cyg

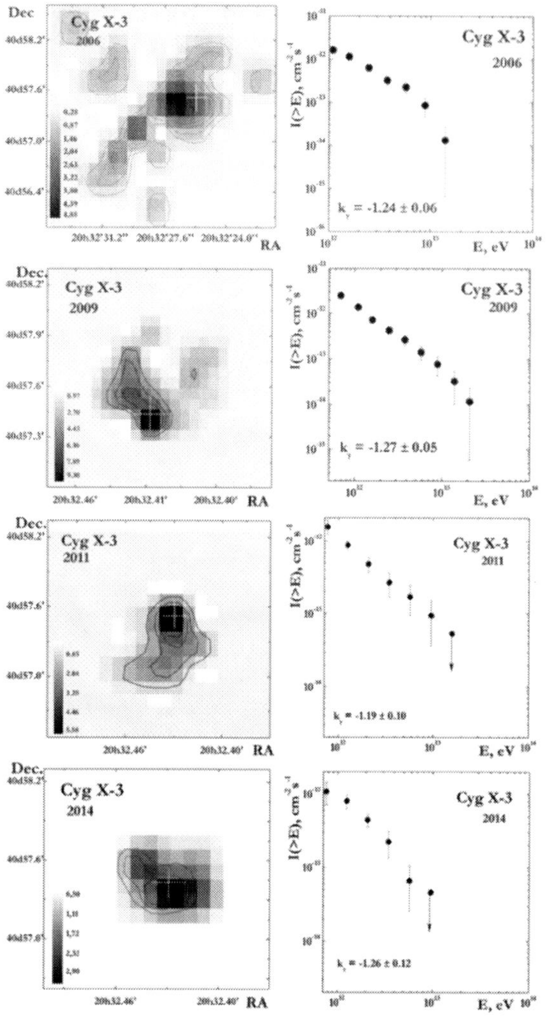

Figure 4. Left: the image of γ-ray emission from Cyg X-3 in 2006, 2009, 2011 and 2014 by SHALON; Right: the γ-ray integral spectra with the power index k_γ.

X-3 was detected above 800 GeV with a statistical significance [11] of 41.2σ with a average integral γ-ray flux above 800 GeV $F(E_\circ > 0.8\,\text{TeV}) = (6.8 \pm 0.4) \times 10^{-13}\,\text{cm}^{-2}\,\text{s}^{-1}$ [12–16]. The energy spectrum of Cyg X-3 at 800 GeV–85 TeV can be approximated by the power law $F(> E_O) \propto E^{-1.23\pm0.10}$. Also, the γ-ray energy spectrum of Cyg X-3 can be fitted by a power law with an exponential cutoff $I(> E_\gamma) = (0.68 \pm 0.04) \times 10^{-12} \times E^{-1.15\pm0.08} \times \exp(-E_\gamma/75\,\text{TeV})\,\text{cm}^{-2}\,\text{s}^{-1}$ (see Fig. 2). In Fig. 2, right TeV emission region in energy range of 0.8–100 TeV during the whole period of SHALON observations of Cyg X-3 is shown.

5. Flaring activity of Cyg X-3

Extreme variability in different wavelengths including VHE γ-rays is the remarkable feature of Cyg X-3 (Fig. 5). A number of high activity periods of Cyg X-3 were detected with SHALON at energies > 800 GeV during the whole period of observation since 1995. In the last 10 years Cyg X-3 has shown activity in all wavelengths. Some results on TeV activity detected by SHALON are presented in (Figs. 3, 4).

The binary Cyg X-3 came to a period of flaring activity at radio-and X-ray energies in 2006 (Fig. 4). In May and July 2006 a significant increase of the Cyg X-3 flux was detected with SHALON at TeV energies with flux values $(4.12 \pm 1.01) \times 10^{-12}\,\text{cm}^{-2}\,\text{s}^{-1}$ and $(1.62 \pm 0.75) \times 10^{-12}\,\text{cm}^{-2}\,\text{s}^{-1}$ respectively. The γ-ray flux in June $(0.46 \pm 0.18) \times 10^{-12}\,\text{cm}^{-2}\,\text{s}^{-1}$ [16].

The average γ-ray flux detected by SHALON in the flaring period from May to July 2006 was estimated as $(1.47 \pm 0.24) \times 10^{-12}\,\text{cm}^{-2}\,\text{s}^{-1}$ (Fig. 4). This intensity increase was also observed by the Crimea Observatory and the integral flux was estimated to be $F(E_\circ > 1\,\text{TeV}) \sim (3\text{–}5) \times 10^{-12}\,\text{cm}^{-2}\,\text{s}^{-1}$. [17].

Cyg X-3 was in a quiet period at TeV energies in 2005. The average γ-quantum flux from Cyg X-3 for $E > 0.8\,\text{TeV}$ is estimated to be $F(E_O > 0.8\,\text{TeV}) = (5.4 \pm 0.73) \times 10^{-13}\,\text{cm}^{-2}\,\text{s}^{-1}$ (Fig. 3). The images and spectra of Cyg X-3 in the 2005 silent period are shown in Fig. 3. No features revealed during flaring periods were found.

The γ-ray flux detected by SHALON in 2003 was estimated as $(1.79 \pm 0.33) \times 10^{-12}\,\text{cm}^{-2}\,\text{s}^{-1}$ (Fig. 3). Earlier, in 1997, a comparable increase of the flux over the average value was also observed and estimated to be $(1.2 \pm 0.5) \pm 10^{-12}\,\text{cm}^{-2}\,\text{s}^{-1}$ (Fig. 3).

The last significant increase of very high energy γ-quantum flux was detected in May 2009: $(3.1 \pm 0.98) \times 10^{-12}\,\text{cm}^{-2}\,\text{s}^{-1}$ (Fig. 4). The increase is correlated with the flaring activity at a lower X-ray energy range and with observations of Fermi LAT [18,19] with average flaring flux of $F = (190 \pm 40) \times 10^{-8}\,\text{phcm}^{-2}\,\text{s}^{-1}$ above 100 MeV (Fig. 5). In the autumn of 2009 Cyg X-3 comes again to a quiet period at TeV energies. The average γ-quantum flux from Cyg X-3 for energies above 0.8 TeV is estimated to be $(0.46 \pm 0.07) \times 10^{-12}\,\text{cm}^{-2}\,\text{s}^{-1}$ with $k_\gamma = -1.10 \pm 0.12$. No significant change of spectrum index was observed in total between high and low states of Cyg X-3.

The last increases of very high energy γ-ray flux were detected in October 2011 and 2014, which were correlated with flaring activity at lower energy range of soft X-ray and/or by observations of Fermi LAT [19]. A high TeV γ-ray flux was detected by SHALON during the X-ray flares of end-September and mid-October 2014 observed by MAXI [20]. Images and spectra of TeV emission are presented in Fig. 4.

Reliably revealing flares and their duration in long-term observations with mirror Cherenkov telescopes is complicated by the fact that the technique makes a continuous tracking of the source impossible, because it requires such conditions as moonless nights, which already creates a gap in the data for more than 10 days; an ideal atmosphere without clouds and haze and, in addition, the source's passage at a distance of no more than 35° from the zenith are needed, because the influence of a change in atmospheric thickness should be minimal.

Nevertheless, revealing correlations between the emissions in different energy ranges, comparing the emission regions, and, in particular, the detection of the flux changes remains necessary, because it makes it possible to judge the nature of the source, namely Cyg X-3, its evolution, and the emission generation mechanisms.

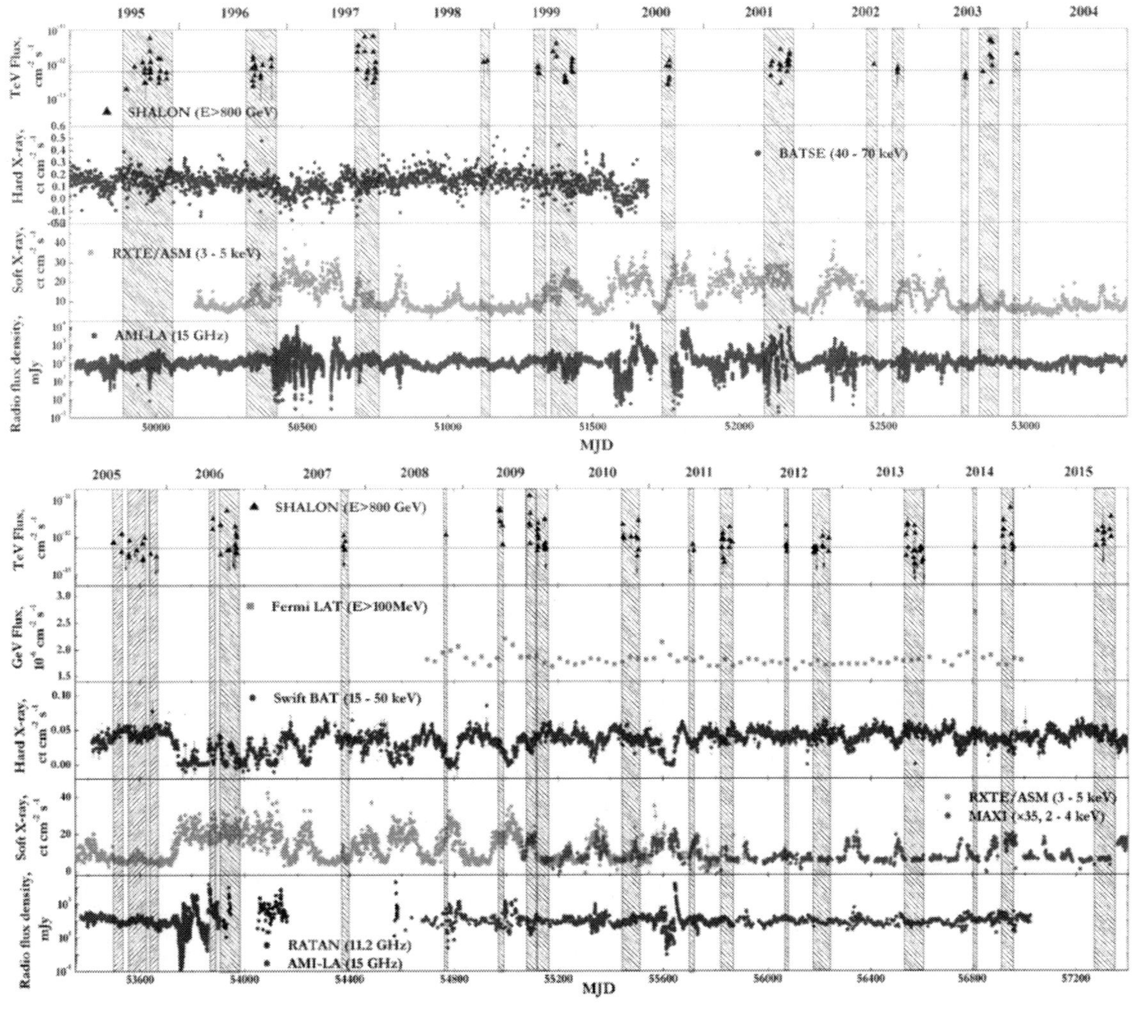

Figure 5. The light curves of Cyg X-3 in a wide energy range.

To reveal possible correlation of periods of activity in the TeV energy range with the flares at the low energies the light curves of Swift/BAT (15–50 keV)[1], MAXI (2–4 keV) [20], RXTE/ASM (3–5 keV)[2], the fluxes at radio-ranges from RATAN (11.2 GHz) (see [21]), AMI-LA (15 GHz)[3] and TeV fluxes from SHALON observations were analyzed (Fig. 5). Analysis of the Cyg X-3 light curve obtained by SHALON shows that the intervals of low flux at TeV energies are at the low activity periods of radio and soft X-rays, but at high flux periods at hard X-rays. There are three type of low energy Cyg X-3 behavior when TeV gamma-ray flares appear.

Cyg X-3 flares at energies >0.8 TeV, detected by SHALON, occur in the case of the intermediate radio flare (< 1000 mJy) at the sharp radio-flux fall before the increase (7–15 days before the radio-flare, as in 1995, 1997, 1999 years). And a very high energy increased flux was detected during the decay period in soft X-rays some 6–8 days after the maximum.

A TeV gamma-ray flux increase appears simultaneously with the main radio-flare (>1000 mJy) at the period of high level flux in soft X-rays (2001).

It was also found that TeV flares occur at the period of high flux of soft X-rays and low activity of hard X-rays within the 5–8 days before the intermediate and main radio flare (2006, 2009) and 7–15 days after the main radio-flare (2006).

The duration of the increase of the detected TeV-flare is from 1 to 4 days andthe duration of fall is 1 day.

A correlation of TeV and soft X-ray fluxes was found in the 1997 observation period. But the flux increase of 2003 didn't obey this scheme, it was in the quiet period of the soft X-rays. In general, the correlation of soft X-ray and TeV energy γ-ray fluxes is traced since 1996.

During the period of observations of Cyg X-3 with SHALON 8, significant flux increases were detected at energies above 0.8 TeV. The significant anti-correlation of the fluxes at TeV and hard X-rays and the correlation of very high energy flux and soft X-ray were found. It is

[1] `http://swift.gsfc.nasa.gov/results/transients/CygX-3/`
[2] `http://xte.mit.edu/asmlc/srcs/cygx3.html`
[3] `http://www.mrao.cam.ac.uk/ guy/cx3/`

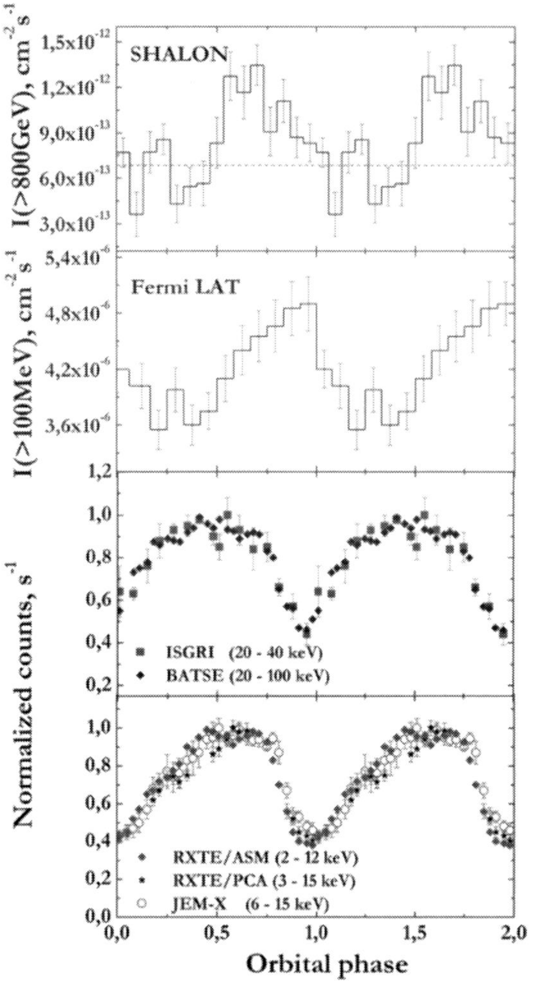

Figure 6. SHALON, Fermi LAT [19] and X-ray light curves of Cyg X-3 folded on the orbital period; The dashed line shows the level of averaged integral TeV γ-ray flux by SHALON.

noted, that TeV flaring activities occur within 5–8 days of strong radio flares (Fig. 5). It is probably, linked with the powerful ejection from the regions close to the centers blackhole. This ejection is accompanied with a relativistic shock where the relativistic electrons and magnetic field are effectively generated.

6. Orbital period

To securely identify the detected emission with Cyg X-3, a timing analysis to search for the 4.8-hour orbital period of Cyg X-3 was performed (see. Fig. 6). We compared the SHALON light curve folded on the orbital period to the folded >100 MeV light curve of Cyg X-3 from Fermi LAT together with X-ray data by RXTE/ASM (2–12 keV), RXTE/PCA (3–15 keV), JEM-X (6–15 keV) and ISGRI (20–40 keV), BATSE (20–100 keV).

The folded X-ray, GeV and TeV light curves have a similar asymmetric shape with a slow rise followed by a faster decay. Also, the SHALON and Fermi LAT light curves are shifted by about 0.15 in phase and its shape have the local maxima. But the phase of the SHALON minimum flux is close to the phase of the minimum of the soft X-ray count rate, shifted by 0.3 in phase (Fig. 6). The observed modulation of high and very high energy gamma-rays suggests that the emission region size and location are bounded by the system size.

7. Phase-resolved energy spectra

The determination of the energy spectrum of very high energy emission is important for understanding the mechanism of radiation production. In the case of the binaries the phase resolved spectra, especially if they vary with phase, can bring information about the generation mechanism of TeV gamma-rays and point out the site of origin.

Spectra of three intervals of orbital phase of Cyg X-3 binary system were extracted.

1. The interval with maximum flux corresponding to inferior conjunction at phase $0.0 \leq \phi < 0.05$; $0.45 \leq \phi < 1.0$; The γ-ray energy spectrum is well described by a relatively soft power law with an exponential cutoff:
   $I(>E_\gamma) = (1.5 \pm 0.4) \times 10^{-12} \times E^{-1.15\pm0.10} \times \exp(-E_\gamma/20\,\mathrm{TeV})\,\mathrm{cm}^{-2}\,\mathrm{s}^{-1}$
   (see black points in Fig. 7)
2. The interval with additional maximum at the epoch of superior conjunction at phase $0.12 \leq \phi < 0.25$; The γ-ray energy spectrum is described by a hard power law:
   $I(>E_\gamma) = (0.67 \pm 0.10) \times 10^{-12} \times E^{-0.93\pm0.09}\,\mathrm{cm}^{-2}\,\mathrm{s}^{-1}$
   (see open triangles in Fig. 7)
3. The interval with minimum flux corresponding to superior conjunction at phase $0.05 \leq \phi < 0.12$; $0.25 \leq \phi \leq 0.45$; Here the γ-ray energy spectrum is described by a hard power law with an exponential cutoff:
   $I(>E_\gamma) = (0.4 \pm 0.08) \times 10^{-12} \times E^{-0.70\pm0.09} \times \exp(-E_\gamma/14.5\,\mathrm{TeV})\,\mathrm{cm}^{-2}\,\mathrm{s}^{-1}$
   (see open squares in Fig. 7).

Inverse Compton (IC) scattering of photons from the WR star on very high energy electrons is a natural candidate to explain the TeV gamma-ray emission. IC scattering directly produces a modulation of the flux because of the orbital motion. The maximum occurs when stellar photons are backscattered towards the observer (inferior conjunction). The observed connection between gamma-ray and radio flares suggests that the high energy electrons are located in the relativistic jet and situated in two symmetric locations in a relativistic jet. The detection of gamma-rays in the phases corresponding to additional maximum at superior conjunction in the large energy range of 800 GeV–100 TeV with hard energy spectrum in whole interval without the presence of a cutoff is argued in favor of the hadronic origin of these energetic photons.

8. Conclusion

The Cygnus Region contains a number of powerful sources of radio and X-rays which are supposed to be potential TeV-emitting objects. The results of 20-year observations of Cyg X-3 at energies 0.8–85 TeV, first detected by

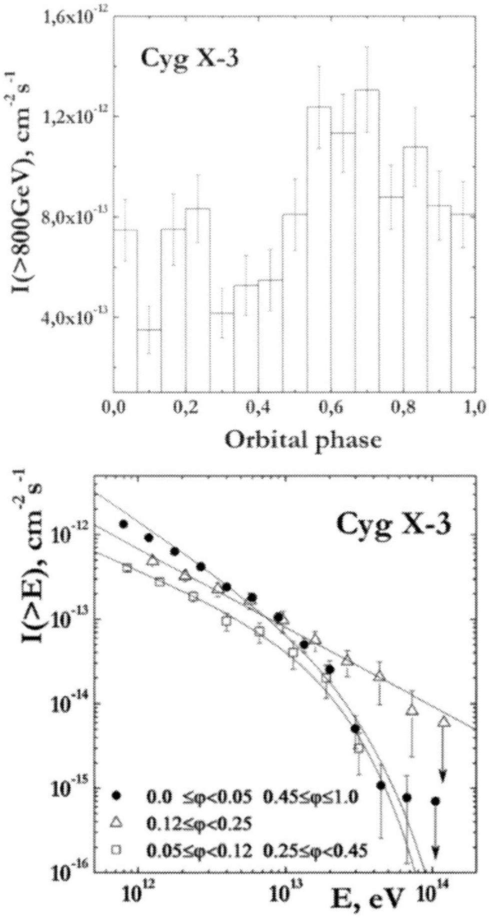

Figure 7. Top: SHALON light curves of Cyg X-3 folded on the orbital period; Bottom: phase-resolved energy spectra by SHALON (see text).

the SHALON telescope in 1995, are presented. The identification of a source associated with Cygnus X-3, detected by SHALON, was secured by the detection of its 4.8 hour orbital period in TeV gamma-rays. A number of high activity periods of Cyg X-3 were detected by SHALON during the whole period of observation. The significant increases of flux are correlated with flaring activity at the lower energy range of X-ray and/or at observations of Fermi LAT. The detected modulation of very high energy emission connected to the orbital motion of the binary system, provides understanding of the emission processes, nature and location of particle acceleration.

References

[1] V.G. Sinitsyna, V.Yu. Sinitsyna, Astron. Lett. **37**(9), 621 (2011)
[2] V.G. Sinitsyna, Nuovo Cim. **19C**, 965 (1996)
[3] A.N. Tikhonov, V.Ya. Arsenin, *Solutions of Ill-Posed Problems* (Nauka, Moscow, 1979; Halsted, New York, 1977)
[4] A.V. Goncharskii et al., *Ill-Posed Problems in Astrophysics* (Nauka, Fizmatlit, Moscow, 1985)
[5] V.G. Sinitsyna, V.Yu. Sinitsyna, Astron. Lett. **40**(2–3), 75 (2014)
[6] F.A.-D. Cordova, Los Alamos Science, Spring, 39 (1986)
[7] M.L. Mccollough et al., Astrophys. J. **517**, 951 (1999)
[8] S.A. Trushkin et al., Proc. of IAU Symposium No. 238, 2006 (ed V. Karas, G. Matt), p. 463
[9] M. Samorscki, W. Stamm, Astrophys. J. **268**, L17 (1983)
[10] J. Lloyd-Evans et al., Nature **305**, 784 (1983)
[11] T.-P. Li and Y.-Q. Ma, Astrophys. J. **272**, 317 (1983) 317
[12] V.G. Sinitsyna, AIP **515**, 205, 293 (2000)
[13] V.G. Sinitsyna et al., Int. J. Mod. Phys. A **29**, 7023 (2005)
[14] V.G. Sinitsyna, Rad. Phys. and Chem. **75**, 880 (2006)
[15] V.G. Sinitsyna et al., J. Phys. Soc. Jpn., Suppl. A **78**, 92 (2009)
[16] V.G. Sinitsyna, V.Yu. Sinitsyna, *"Cosmic Rays For Particle And Astroparticle Physics"* (Proc. 12th ICATPP, 2010) ed S. Giani, C. Leroy, and P.G. Rancoita; (Singapore: World Scientific 2011) vol. 6, p. 132
[17] Yu.T. Neshpor et al., Bull. of RAS; Physics **73**(5), 694 (2009)
[18] M. Tavani et al., Nature Letters **462/3**, 620 (2009)
[19] A.A. Abdo et al., Science **326**, 1512 (2009)
[20] M. Matsuoka et al., PASJ **61**, 999 (2009)
[21] J. Alekšic et al., Astrophys. J. **721**, 843 (2010)

DOI: 10.1051/epjconf/201714504003

Bursts of the Crab Nebula gamma-ray emission at high and ultra-high energies

A.S. Lidvansky[a]

Institute for Nuclear Research, Russian Academy of Sciences, Moscow, Russia

Abstract. Characteristics of the flares of gamma rays detected from the Crab Nebula by the AGILE and Fermi-LAT satellite instruments are compared with those of a gamma ray burst recorded by several air shower arrays on February 23, 1989 and with one recent observation made by the ARGO-YBJ array. It is demonstrated that though pulsar-periodicity and energy spectra of emissions at 100 MeV (satellite gamma ray telescopes) and 100 TeV (EAS arrays) are different, their time structures seem to be similar. Moreover, maybe the difference between "flares" and "waves" recently found in the Crab Nebula emission by the AGILE team also exists at ultra-high energies.

1. Introduction

The Gamma-ray flares from the Crab Nebula were discovered in a few-hundred-MeV energy range by the AGILE [1] and Fermi LAT [2] satellite telescopes. Since then, both telescopes continue recording such flares approximately once a year, the strongest of them (super-flare) having been detected in April 2011 [7,8]. The energy spectra of additional emission during the flares were measured to be different from those of the Nebula, however, "the mechanism driving the flares, their impulsive nature, the 12-month recurrence time, and the location, remain unknown" [12]. For all that, based on multi-wavelength campaign to study the Crab using Keck, the Hubble Space Telescope (HST), and the Chandra X-ray Observatory, the authors of [12] suggested the so-called "inner knot" to be just the emitting region for the flares. In addition, the analysis made by the AGILE collaboration for the September-October 2007 event has found a fine structure in the flare's time behavior [10]. They have demonstrated that there is a difference between shorter "flares" and longer "waves" in the Crab Nebula emission during this flare.

After the sensational discovery of gamma ray flares by AGILE and Fermi LAT it was recalled [3] that one event of this type (though at much higher energies) had been discovered many years ago by several EAS arrays [4–6].

2. Gamma-Ray emission burst on February 23, 1989

The first announcement about this burst was made during the International Workshop on Gamma-Ray Astronomy in the Crimea in 1989 [4] [4]. An increase in the intensity of extensive air showers (EAS) was detected by the Carpet air shower array of the Baksan Neutrino Observatory on February 23, 1989. After this communication, the group at Kolar Gold Fields (KGF) in India confirmed this result of Baksan and published a paper [5] on the simultaneous detection of a gamma-ray burst in the Crab Nebula at ultra high-energies independently by two EAS arrays. The final publication [6] by the Baksan and Durham University teams summarized the data of all arrays that could observe the source on this day. It was demonstrated that with different significance the burst was detected by all air shower arrays located in the longitude range from India to Italy (KGF, Tien Shan, Baksan and EAS TOP). The arrays located to the east and west of this interval (at that time OHYA MINE and Akeno SPICA in Japan and HEGRA at La Palma, Canary Islands, were in operation) showed no excess from the source direction. Thus, the total duration of the observed effect was no longer than about 7 hours. One can speak also about a possible decay of intensity whose maximum probably fell on observations with the KGF array. Figures 1 and 2 taken from paper [6] demonstrate the most remarkable features of the burst as detected by the Baksan Carpet air shower array.

It is clearly seen that the Baksan array data demonstrate an obvious dependence of the effect on the pulsar phase: events from a single phase bin can produce the entire observed excess of intensity (the KGF group also found some phase irregularity, though not so well pronounced). Summarizing, one can state that with a rather high probability (combined probability of random coincidences was estimated in [6] as $1.25\ 10^{-7}$) a gamma-ray burst from the Crab Nebula was detected on February 23, 1989 in the energy range 10^{14}–10^{15} eV, and the excess emission in this burst is somehow connected with the pulsar's activity.

3. Different and similar types of flares at different energies

In [3] attention was drawn to the fact that the three-pulse temporal structure in the event of February 23, 1989 might be reproduced in at least one of the AGILE flare events (September 2007). A bit later it has been discovered [10] that in addition to flares there exists

[a] e-mail: `lidvansk@lebedev.ru`

 DOI: 10.1051/epjconf/201714504003

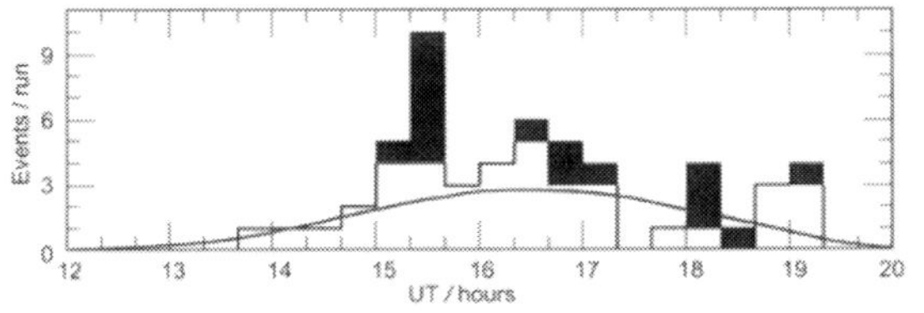

Figure 1. The number of events within the Crab cell per 20 min run for 23 February 1989. The smooth curve represents the expected background. Blacked events are from the 9th bin in Fig. 2.

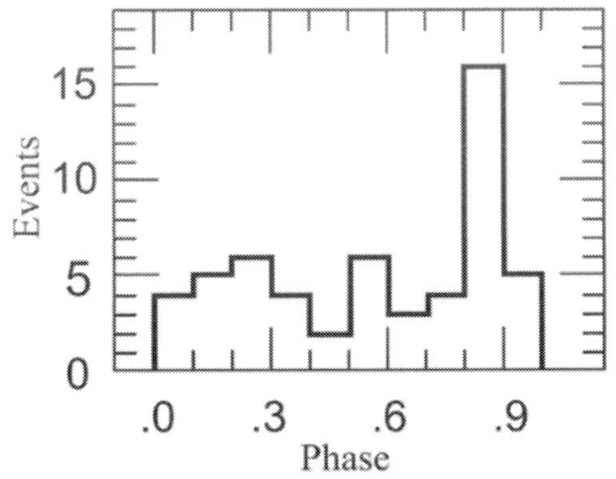

Figure 2. The relative phase distribution of the 55 events within the Crab cell after barycentring the arrival times and applying the values of the pulsar period and its derivative according to the Jodrell Bank Crab ephemeris.

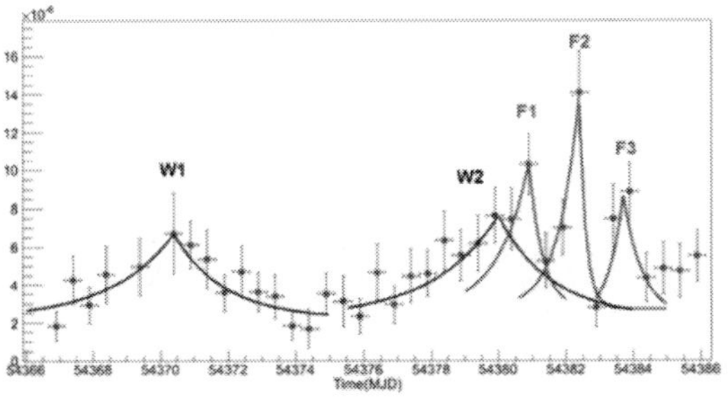

Figure 3. The difference between flares and waves in the Crab Nebula gamma-ray emission according to Striani et al. [10]. The three-flare structure of the event is similar to three bunches of narrow-phase emission constituting almost all excess counts in Fig. 1.

another type of intensity increase called waves. And exactly during September 2007 three short flares (F1, F2, and F3) were identified together with two waves W1 and W2 (see Fig. 3).

The similarity of the temporal structures for the events of Figs. 1 and 2 is obvious. In one case three bunches shorter than one hour are repeated with a period slightly longer than one hour. The other event includes three flares with duration of about one day, and they are repeated with a period a bit longer than one day. So, a sort of a stable pattern with a scale factor is observed at drastically different energies of gamma rays. It should be noted that the fine temporal structure was also reported by the Fermi LAT team for the March 2013 flare: "The light

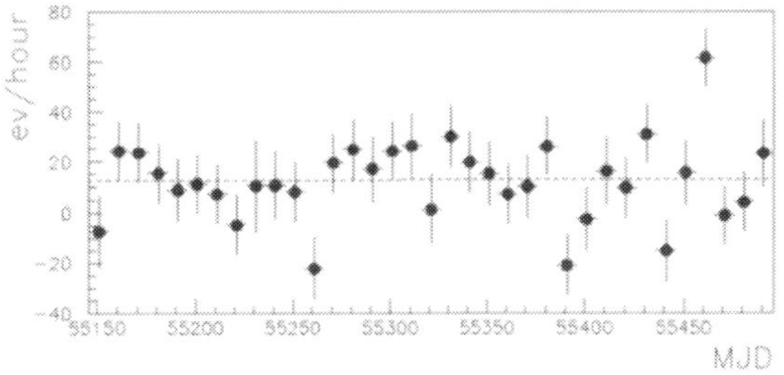

Figure 4. The Crab light curve obtained in September 2010 by the ARGO-YBJ array with 10-day bins. The flare recorded by satellite gamma-ray telescopes took place on September 8–22. The maximum point is for September 19-26, 2010. Its significance is 5.4 standard deviations.

curve shows three sharp spikes (MJD 56357.1, 56357.9, 56360.1) on top of a strongly increased flux level." [15]. Here, the period between the spikes is not so constant, but its duration, nevertheless, is close to that of the Fig. 3 event.

The results of AGILE and Fermi LAT stimulated other groups to search for possible Crab flares, and one of them seemed, at first, to be successful. The ARGO-YBJ array after processing the data for the Crab direction has found [9] an enhancement for the period September 19-26, 2010, which is shown in Fig. 4 for a 10-day period of averaging. This result was obtained at energies (about 1 TeV) intermediate between the two regions considered above. Nevertheless, there is a temptation to think that the ratio of durations (several hours for the 1989 event and about a week in this case) gives some evidence in favor of the existence of the same "flares and waves" dichotomy at ultra-high energies. However, in the next publication [11] of ARGO-YBJ this piece of data was partially disavowed, though another interesting fact was presented: a correlation between Fermi and ARG-YBJ Crab intensities: "Even if the ARGO-YBJ rate variations are consistent with statistical fluctuations, the Fermi and ARGO-YBJ data seems to follow a similar trend. The ARGO-YBJ rate appears higher in the "hot" Fermi periods." [11].

It is commonly believed that the energy spectrum of the Crab Nebula is formed by synchrotron emission below 1 GeV and by inverse Compton radiation at higher energies. As far as flaring radiation is concerned, it is presumed to be associated with the synchrotron component. Spectral behavior of the flares appeared to be rather different: according to Fermi LAT, the flare in February 2009 only exhibited a flux increase with no spectral change. On the contrary, the flares in September 2010 and April 2011 had fluxes strongly correlated with the spectral index. In the March 2013 flare "the spectrum of the synchrotron nebula hardens as the flux increases" [15]. At the same time, according to an analysis made in [15], within the measurement accuracy the spectrum of the pulsar did not change in the entire analysis window with respect to the all-time average, and only the synchrotron component of the Crab Nebula is rapidly variable.

At the moment no information on spectra is available for the ultra-high energy bursts.

The Crab pulsar emission has a very complicated energy dependence. In the radio waveband in addition to the main pulse (MP) and interpulse (IP) there are precursors of the MP, the low-frequency component (LFC) and two high-frequency components (HFC1 and HFC2), as well as giant radio pulses that randomly appear on different phases. At higher energies the Crab light curve becomes more regular and the gamma-ray domain has only two pulses P1 and P2. The amplitudes of these two pulses depend on energy. For example, the P2 pulse is twice larger in amplitude than P1 at an energy less than 10 MeV (COMPTEL), but already at energies >30 MeV (EGRET) the situation is opposite. At energies exceeding 100 GeV (the data of the VERITAS and MAGIC Cherenkov telescopes) pulse P2 becomes dominant again. There are also some indications to the possible appearance of a new pulse P3. So, it would be not surprising if at still higher energies only a single pulse survives, as it takes place in Fig. 2.

4. Discussion and conclusions

Interpretation of gamma-ray flares produced by synchrotron radiation of accelerated electrons in compact regions of an intensified magnetic field (plasma instabilities) requires the energy of such electrons to be as high as 10^{15} eV (see for example, [10]). If these electrons do really exist, one can recall the idea of Atoyan and Aharonian [16] that Crab Nebula wisps probably have sufficient amount of matter to make bremsstrahlung radiation effective. If so, 1000 TeV electrons can easily produce 100 TeV gamma rays. Even without this hypothesis accelerated particles can simultaneously produce multi-TeV emission via inverse Compton effect and hundred-MeV - GeV emission through synchrotron radiation [17]. When explaining the two sets of data discussed in this paper, the problem is that at ultra-high energies pulsar periodicity takes place, while it is not observed in hundred-MeV flares. As a possible explanation one can suggest that the initial acceleration is directly connected with the pulsar, while at lower energies an observer receives gamma rays after randomization of the phase in the pulsar-wind driven shock.

Nevertheless, many details of the mechanisms of flare generation (like, for example, triple scaled temporal structure and distinction between flares and waves) are unclear as before, and the Crab Nebula still remains an enigmatic "astrophysical chimera" [13]. One can hope, however, that the above juxtaposition of data at different energies will be useful for construction of a realistic model of the source internal mechanism.

The work is partially supported by the Russian Foundation for Basic Research, grant no. 16-02-00687.

References

[1] M.Tavani et al., Science **331**, 736–39 (2011)
[2] A.A. Abdo et al., Science **331**, 739-42 (2011)
[3] A.S. Lidvansky, Journal of Physics: Conference Series **409**, 012114 (2013)
[4] V.V. Alexeenko, A.S. Lidvansky and V.A. Tizengauzen V.A., Proc. of Int. Workshop on Very High Energy Gamma Ray Astronomy (Crimea, USSR), 17–21 April, 1989. Ed. by A.A. Stepanyan, D.J. Fegan and M.F. Cawley, p. 187
[5] B.S Acharya, M.V.S. Rao, K. Sivaprasad, B.V. Sreekantan and P.R. Vishwanath, Nature **347**, 364–5 (1990).
[6] V.V. Alexeenko, Yu.M. Andreyev, A.E. Chudakov, Ya.S. Elensky, A.S. Lidvansky, S.Kh. Ozrokov, Yu.V. Stenkin, V.A. Tizengauzen, L.J. Graham, J.L. Osborne and A.W. Wolfendale, J. Phys. G: Nucl. Part. Phys. **18**, L83–L88 (1992)
[7] R. Buehler et al., Astrophys. Journ. **749**, 26 (2012)
[8] E. Striani et al., Astrophys. Journ. Lett. **741**, L5 (2011)
[9] S. Vernetto et al., Proc. 32nd ICRC, Beijing, 2011
[10] E. Striani et al., Astrophys. Journ. **765**, 52 (2013)
[11] B. Bartoli et al. Astrophys. Journ. **798**, 119 (2015)
[12] A. Rudy et al., Astrophys. Journ. **811**, 24 (2015)
[13] J.J. Hester, Annu. Rev. Astron. Astrophys. **46**, 127–55 (2008)
[14] R. Bühler and R. Blandford, Rep. Prog. Phys. **77**, 066901 (2014)
[15] M. Mayer et al., Astrophys. Journ. Lett. **775**, L37 (2013)
[16] A.M. Atoyan and F.A. Aharonian, Mon. Not. Roy. Astron Soc. **278**, 525 (1996)
[17] W. Bednarek and W. Idec, Mon. Not. Roy. Astron Soc. **414**, 2229 (2011)

Fermi bubbles as sources of cosmic rays above 1 PeV

Dmitry Chernyshov[1,2,3,a], Kwong-Sang Cheng[2,b], Vladimir Dogiel[1,2,4,c], and Chung-Ming Ko[3,5,d]

[1] I.E.Tamm Theoretical Physics Division of P.N.Lebedev Institute of Physics, Leninskii pr. 53, 119991 Moscow, Russia
[2] Department of Physics, University of Hong Kong, Pokfulam Road, Hong Kong, China
[3] Institute of Astronomy, National Central University, Jhongli, Taoyuan, Taiwan, R.O.C.
[4] Moscow Institute of Physics and Technology, 141700 Moscow Region, Dolgoprudnii, Russia
[5] Department of Physics and Center for Complex Systems, National Central University, Jhongli, Taoyuan, Taiwan, R.O.C.

Abstract. Fermi bubbles are giant gamma-ray structures extended north and south of the Galactic center with characteristic sizes of the order of 10 kpc discovered by the Fermi Large Area Telescope. Good correlation between radio and gamma-ray emission in the region covered by Fermi bubbles implies the presence of high-energy electrons in this area. Due to high energy losses it is rather problematic to transfer relativistic electrons from the Galactic disk toward the Fermi bubbles. Therefore it is natural to assume that these electrons are accelerated *in-situ*. Additionally this acceleration mechanism should also affect protons. In particular it may re-accelerate Galactic cosmic rays produced by supernova remnants. Unlike electrons, protons have huge lifetimes and therefore re-acceleration should not be a local effect but affect the whole Galaxy. The effect may even be observed near the Earth. In our model we propose that hadronic cosmic rays (CR) below the "knee" of the observed CR spectrum are produced by Galactic supernova remnants distributed in the Galactic disk. Re-acceleration of these particles in Fermi Bubbles produces CRs above the knee. This model provides a natural explanation of the observed CR flux, spectral indices, and matching of spectra at the knee.

1. Introduction

Fermi bubbles were discovered by [1] and [2] almost seven years ago and they still remain as one of the most attractive astrophysical events. The bubbles are presented by giant structures located above and below the Galactic plane visible in gamma-ray [3] and radio [4] energy bands. Despite their nature being still enigmatic, the location of these objects indicates their connection with past or present activity in the center of our Galaxy. Different models relate the bubbles to starburst activity [5], single [2,6–8] or multiple [9] energy release events on a central black hole.

Such objects should not be unique phenomena observed only in our Galaxy. However, Fermi bubbles are too faint and it is impossible to detect exactly the same structures elsewhere. Yet some much more powerful objects with similar properties and probably similar nature are observed in some galaxies with active nuclei. For example even more giant structures are clearly seen in the direction of Cen-A in GHz radio [10,11], GeV [12] and TeV [13] gamma-ray ranges. Giant X-ray and radio lobes (bubbles) were found also in the galaxies NGC 3801 [14], Mrk 6 [15] and Circinus Galaxy [16].

Taking all facts together one can conclude that Fermi bubbles are indeed real structures. Since they enclose huge energies [2] and their volume is comparable to that of the Galactic disk, Fermi bubbles have a potential to affect the distribution of the cosmic rays (CRs) in the Galaxy. In particular in our paper [17] we showed that Fermi bubbles may be responsible for the formation of the CR spectrum above the "knee". Below we will briefly recapitulate the major points of this model and also discuss some consequences.

2. Fermi bubbles as giant re-acceleration sites

The origin of the gamma-ray emission from Fermi bubbles is not yet established. However, is was shown by Cheng et al. [18] they should act like giant accelerators. Indeed even in the case of hadronic origin of the gamma-ray emission some additional primary leptonic component is necessary to reproduce microwave emission. The emission produced by secondary electrons have too steep a spectrum to match the experimental data [4]. These electrons have relatively high energies and should be accelerated *in-situ*, otherwise an unrealistic velocity of outflow from the central part of the Galaxy is required.

Such large accelerators which fill a substantial volume of the Galaxy should also modify the distribution of Galactic cosmic rays. It is generally accepted that supernova (SN) explosions in our Galaxy can provide enough energy to produce the observed total luminosity of the CR [see e.g., reference 19]. Moreover diffusive shock acceleration naturally explains the power-law CR spectrum [20,21], and taking into account propagation effects, their spectral index. However, many fundamental questions related to the assumption of SNRs being the sources of the Galactic CR are still open. One of them is the maximum energy of CR which can be estimated from

[a] e-mail: chernyshov@lpi.ru
[b] e-mail: hrspksc@hku.hk
[c] e-mail: dogiel@lpi.ru
[d] e-mail: cmko@astro.ncu.edu.tw

the age of a typical SNR, T, expanding with a velocity of u_{sh}

$$E_{max} \sim Ze\beta_{sh}BTu_{sh}, \quad (1)$$

where $\beta_{sh} = u_{sh}/c$, B is the magnetic field strength at the shock and the term $\mathcal{E} = \beta_{sh}B$ in this case can be interpreted as an effective electric field.

For the parameters of the standard Galactic SNR and for reasonable values of the Galactic magnetic field the maximum energy of CR protons cannot exceed 10^{13}–10^{14} eV. In more complex models outside quasi-linear approximation it was shown that the magnetic field at the shock can be amplified. As a result for a conservative set of parameters the maximum energy reaches a value of about 10^{15} eV.

The important point of the CR spectrum is a sudden steepening around 3×10^{15} eV which indicates a change of the acceleration or propagation mechanism. Smooth attachment of the spectra above and below this energy and sharpness of the transition indicates that we are dealing with a single spectrum rather than with the sum of two distinct components. The review of different models suggested to explain this phenomenon can be found in the original paper [17].

In summary, it is generally agreed that SN shocks can only accelerate particles to energies $E < 10^{15}$ eV. However, Fermi bubbles with age and size exceeding that of typical SNR by 4 orders of magnitude can easily accelerate particles to much higher energies. Using Eq. (1) one can estimate that the bubbles have a potential to form the spectrum of CR in between 3×10^{15} eV and 10^{18} eV. We do not consider the possibility that Fermi bubbles can form a whole spectrum of CR both below and above 3×10^{15} eV. The reason for this is that the total power needed for the luminosity of CRs in our Galaxy, $L_{CR} \sim 10^{41}$ erg/s [22], and the average energy release toward the Fermi bubbles is also 10^{41} erg/s [9]. To explain the origin of CR by acceleration in Fermi bubbles one requires the acceleration efficiency to be close to 100% which is unlikely. On the other hand the sudden change of the slope of the spectrum around the "knee" can be naturally explained due to a change of the acceleration properties. Thus our model can be described in the following way: SNRs in the disk accelerate particles with power-law distribution up to energies of 3×10^{15} eV and Fermi bubbles further re-accelerate these particles up to 10^{18} eV.

3. Structure of shocks in the Fermi bubble

In the model of Fermi bubbles origin we suggested that the activity responsible for the formation of the bubbles was due to stellar capture and tidal disruption by a central black hole [9,17]. The average time between two successive captures in the Galaxy is between 10^4 years and 10^5 years [23]. Thus the activity is periodic and the characteristic period between two events is shorter than the characteristic lifetime of the Fermi bubbles. Periodic energy releases in the Galactic center should form a series of shocks propagating through the halo. We note that from the numerical simulations it appears that only very powerful events expected in the case of capture of massive stars can form shocks. Thus the amount of shocks inside the Fermi bubbles should not be very large.

The separation between shocks can be estimated from their velocity, u, and the characteristic period of capture, τ_{cap}, as

$$l_{sh} = \tau_{cap}u = 30(\tau_{cap}/3\times10^4\ \mathrm{yr})(u/10^8\ \mathrm{cm/s})\ \mathrm{pc}. \quad (2)$$

However, the exact separation between two consecutive shocks depends on the actual time separation of two consecutive capture events and their energy releases. There is another important spatial scale which characterizes processes of particle acceleration by a single shock, i.e., the acceleration length scale in a single shock, $l_D \sim D/u$, where D is the spatial diffusion coefficient near a shock whose value depends on particle interactions with small scale magnetic fluctuations and u is the shock velocity.

4. Particle acceleration by the bubble shocks

The problem of particle acceleration in conditions of supersonic turbulence (multi-shock structure) was extensively analyzed before. In a series of papers by [24–26] applied to acceleration processes in OB associations, which is quite similar to the structure of the Bubble, they introduce a nondimensional parameter characterizing the acceleration regime as

$$\psi = \frac{l_{sh}}{l_D} = \frac{ul_{sh}}{D} \sim \frac{ul_{sh}}{cr_L}. \quad (3)$$

The corresponding energy, E_1, that separates different regimes can be estimated from the condition $\psi \sim 1$ or $l_D(E_1) \sim l_{sh}$ which for the conditions of the Fermi bubble is

$$\begin{aligned} E_1 &= eBl_{sh}u/c \\ &= 10^{15}\left(\frac{B}{5\,\mu G}\right)\left(\frac{l_{sh}}{30\ \mathrm{pc}}\right)\left(\frac{u}{10^8\ \mathrm{cm/s}}\right)\ \mathrm{eV}. \quad (4)\end{aligned}$$

In the case of $\psi \gg 1$ or $l_{sh} \gg l_D$ analyzed in [24,25], there is a combined effect of a fast particle acceleration by a single shock, which generates the spectrum E^{-2} and relatively slow transformation of this spectrum due to interactions with other shocks (stochastic Fermi acceleration) into a hard E^{-1} spectrum in the intershock medium at relatively low energies. However, it is unclear if such slow transformation can be completed within the life time of the shocks in the Bubble. Furthermore, the hard E^{-1} spectrum requires a significantly higher power to be formed in comparison with a softer spectrum with a slope of E^{-2}. Thus the transformation of the spectrum requires a power which significantly exceeds 10^{41} erg/s which the bubbles, as we mentioned above, can not supply. It is reasonable to assume that Fermi bubbles do not affect the spectrum of CR below E_1. This statement is in agreement with our initial assumption that SNRs are the major contributors for CRs with energies $E < 10^{15}$ eV and the exact particle spectrum generated from the Bubble is unimportant in the energy range of $E < 10^{15}$ eV.

For $\psi \ll 1$ or $l_{sh} \ll l_D$ the acceleration regime changes to a pure stochastic acceleration by a supersonic turbulence. In the stationary case the equation for accelerated CRs can be presented in the form [25]

$$\frac{\partial}{\partial z}D(\rho)\frac{\partial f}{\partial z}+\frac{1}{\rho}\frac{\partial}{\partial \rho}D(\rho)\rho\frac{\partial f}{\partial \rho}+\frac{1}{p^2}\frac{\partial}{\partial p}\kappa(\rho)p^2\frac{\partial f}{\partial p}=0 \quad (5)$$

where ρ and z are the cylindrical spatial coordinates, p is the particle momentum, $D(\rho)$ is the spatial diffusion coefficient and κ is the momentum diffusion coefficient.

Proton acceleration in the Bubble depends sensitively on the acceleration parameters and structure of the Bubble. In the following we present a detailed analysis. We present the bubble region as a cylinder extending above and below the Galactic plane from $z=0$ to $z=\pm H$ with a radius $\rho=\rho_B$. As boundary conditions we put the density of particles f equal to zero at the Galactic halo surface

$$f|_\Sigma = 0 \text{ at } \rho=\rho_G, \text{ and } z=H \quad (6)$$

The diffusion coefficients inside and outside the bubble are supposed to be different

$$D(\rho)=D_B\theta(\rho_B-\rho)+D_G\theta(\rho-\rho_B)$$
$$\kappa=\kappa_B\theta(\rho_B-\rho)\Theta(E-E_1) \quad (7)$$

where $D_B=Lc/3$ is the coefficient inside the bubble due to interactions with a supersonic turbulence and D_G is the average diffusion coefficient in the Galaxy defined for example in [19]. The momentum diffusion coefficient is $\kappa_B=u^2/D_D$. The momentum dependence of f can be presented by a power-law function, $f(p)\propto p^{-\gamma}$, where γ should be determined from Eq. (5) and is approximately

$$\gamma\approx\frac{3}{2}+\sqrt{\frac{9}{4}+\pi^2\frac{D}{\rho_B^2\kappa}}=\frac{3}{2}+\sqrt{\frac{9}{4}+\frac{\pi^2D_B^2}{u^2\rho_B^2}}. \quad (8)$$

Thus a proper choice of D_B allows to form a power-law distribution with a certain value of a spectral index.

To consolidate this idea, we work out a concrete numerical model. Essentially, we solve the stationary state of the CR transport Eq. (5) in our Galaxy with two Fermi Bubbles (one on each side of the Galactic plane). We modeled our Galactic halo as a cylinder of radius $\rho_G=20$ kpc, and the top and bottom at ± 10 kpc from the mid-plane. Each Fermi Bubble is also a cylinder of the same height ± 10 kpc, but with a radius $\rho_B=3$ kpc. The spatial diffusion coefficients are different inside and outside the bubble as described by Eq. (7).

Since we expect the average separation between shocks in the Fermi bubbles to be of the order of 100 pc [17] we consider a constant spatial diffusion coefficient and adopt $D_B=2.08\times10^{30}\,\text{cm}^2\,\text{s}^{-1}$. Outside the bubble, we take into account the energy (or momentum) dependence of the spatial diffusion coefficient and adopt $D_G=D_0(pc/4\,\text{GeV})^{0.6}$, $D_0=6.2\times10^{28}\,\text{cm}^2\,\text{s}^{-1}$ [cf. 27].

We assume that there is little or no stochastic acceleration outside the bubble (see Eq. (7)), and adopt $\kappa_B H^2/D_B=1.9$ (i.e., $\kappa_B=4.4\times10^{-15}\,\text{s}^{-1}$ or the corresponding acceleration time scale is 7.6 Myr).

For Galactic SNRs we adopt the distribution suggested by [28] and modified it with a Gaussian thickness profile

$$Q_{\text{SNR}}(\rho,z)\propto\left(\frac{\rho}{R_\odot}\right)^{1.2}\exp\left(-\frac{3.22\rho}{R_\odot}\right)\exp\left(-\frac{z^2}{h^2}\right), \quad (9)$$

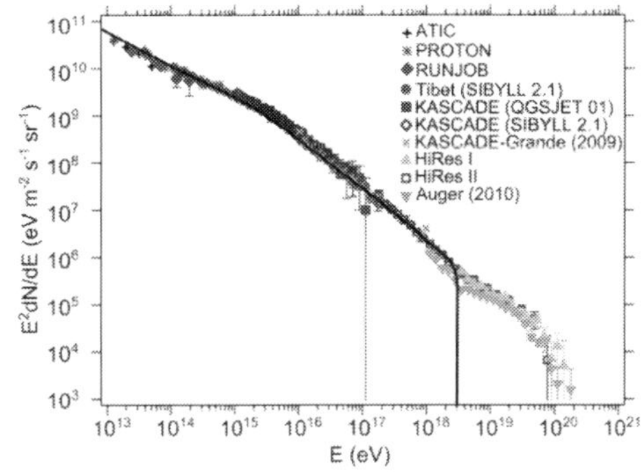

Figure 1. The CR spectrum at the Earth as a combination of the SNR contribution (in the Galactic disk) and the stochastic acceleration in Fermi Bubbles. The black solid line is the spectrum from our numerical model. For references to the experimental data see the original paper [17]. In this model, $D_B=2.08\times10^{30}$ cm^2 s^{-1} inside the bubbles and $D_G=6.2\times10^{28}(pc/4\,\text{GeV})^{0.6}$ cm^2 s^{-1} outside, $\kappa_B H^2/D_B=1.9$, and the injection spectrum from SNR $\mu=4.35$ (see Eq. (10)).

here we take $h=100$ pc, $R_\odot=8$ kpc. We adopt the idea that SNRs inject energetic particles in the form of a power law with a high-energy cutoff at $p_{\max}c\approx3\times10^{15}$ eV. Therefore, together with the SNR distribution (Eq. (9)), the source function is

$$Q(\rho,z,p)=Q_0\left(\frac{p}{p_{\max}}\right)^{-\mu}\exp\left(-\frac{p}{p_{\max}}\right)\left(\frac{\rho}{R_\odot}\right)^{1.2}$$
$$\times\exp\left(-\frac{3.22\rho}{R_\odot}\right)\exp\left(-\frac{z^2}{h^2}\right). \quad (10)$$

Finally, the appropriate boundary conditions for the momentum coordinate are

$$\left.\frac{p}{f}\frac{\partial f}{\partial p}\right|_{p=p_{\text{low}}}=-4.7\,, \quad f|_{p=p_{\text{up}}}=0\,, \quad (11)$$

where the energy of the lower momentum boundary is $p_{\text{low}}c=10^{12}$ eV, and the upper momentum boundary is $p_{\text{up}}c=3\times10^{18}$ eV. The condition at the lower momentum ensures that the spectral index matches that of low-energy CRs (say $E<10^{12}$ eV).

The spatial boundary conditions are

$$\left.\frac{\partial f}{\partial \rho}\right|_{\rho=0}=0\,, \quad f|_{\rho=\rho_G}=0\,, \quad (12)$$

and

$$\left.\frac{\partial f}{\partial z}\right|_{z=0}=0\,, \quad f|_{z=\pm H}=0\,. \quad (13)$$

The spectrum evaluated at the Earth's position is the solid line shown in Fig. 1. The model fits the data reasonably well and it is not coincident that the spectra join smoothly at the knee. The value of the parameter E_1 based on Eq. (4) is not reliable. To check how the result depends on this parameter we estimated the spectrum for different values of E_1 (see Eq. (7)). It turns out that the shape of the spectrum does not depend on E_1 as long as $E_1<3\times10^{15}$ eV.

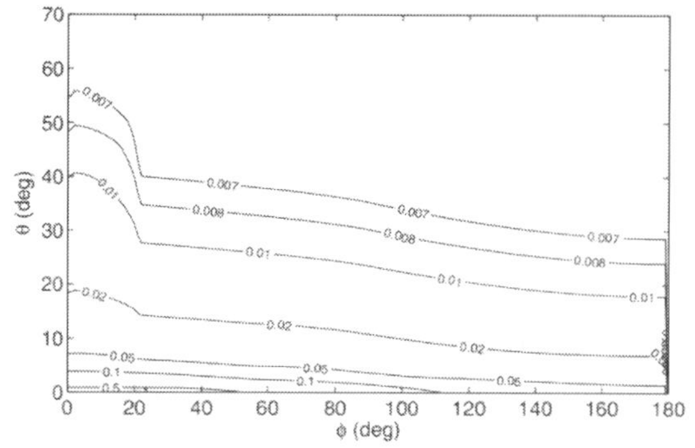

Figure 2. The relative intensity of high-energy neutrino emission across the sky. Here ϕ is the Galactic longitude and θ is the Galactic latitude. The picture is symmetrical across $\theta = 0$ and across $\phi = 0$ so only one corner is shown.

We should note that in the estimations above we assumed that the CR spectrum consists only of protons. This assumption is for illustrative purposes only. To obtain the spectrum of different elements one should note that the propagation and acceleration parameters depend only on rigidity. Thus the spectra of all elements should have the same shape but should be shifted along the energy axis in accordance to their charge Z [see e.g. Fig. 5 in reference 29, for the details]. Due to the same shape of the spectrum one should expect several "knees" in the spectrum corresponding to different elements at energies of $3Z \times 10^{15}$ eV. According to [30] there is a knee-like structure in the heavy component of CR spectrum at about 8×10^{16} eV (apparently corresponding to iron) that strengthen this idea.

In this case the spectrum of particles above the knee should be slightly softer than the one estimated by us for the proton-only case. But as one can see from Eq. (4) the value of the spectral index can be easily adjusted by a proper choice of the acceleration parameters.

The important test of this model can be done via neutrino emission. Indeed we expect that high-energy CRs are concentrated in the area of Fermi bubbles [17]. Therefore one should expect there should be an enhancement of the emission from this region.

It is interesting that *IceCube* has detected 4–5 out of 28 events above 30 TeV, which may originate from cosmic rays with energies $> 10^{15}$ eV from Fermi Bubbles, Lunardini et al. [31]. This discovery seem to strengthen our theory. To test it we estimated expected intensities of the neutrino emission.

Since the statistics of *IceCube* data is very poor we cannot consider numerical values of the intensities as reliable data. Instead we estimated the neutrino flux from the Galactic disk and compared it to the neutrino flux from the Bubble area. The spatial distribution of relative intensity of the high-energy neutrino emission across the sky is shown in Fig. 2. Despite that there is a prominent excess at high latitudes, it turns out that the Bubble to disk ratio is very low: it does not exceed 10% in even the most promising scenarios.

Since in the *IceCube* data there is no prominent signal from the Galactic disk we have to conclude that the observed excess is not related to that expected in our model and is probably due to an external source.

5. Conclusion

We suggested a model of CR origin in the energy range 3×10^{15} eV $< E < 10^{18}$ eV. In the model we assumed that CRs with energies below 3×10^{15} eV are produced by SNRs in the Galactic disk. These particles diffuse out to the Galactic halo and are then re-accelerated in Fermi bubbles. For a proper choice of the acceleration parameters it is possible to reproduce the right value of the spectral index in the energy range above the "knee". The spectrum below the "knee" remains undisturbed. This model also naturally explains the smooth attachment of the spectra near the "knee". It also explains the change of chemical composition of CRs at high energies if the rigidity dependence of the acceleration and propagation is taken into account.

We also estimated the possible enhancement of the high-energy neutrino emission from the central area of the Galaxy. It turns out that the flux from Fermi bubbles should not exceed 10% of the flux expected from the Galactic disk. This value is very low and can not be detected at the present moment.

DOC and VAD acknowledge support from the RFFI grants 15-52-52004 and 15-02-02358. KSC is supported by the GRF Grants of the Government of the Hong Kong SAR under HKU 17310916. DOC, KSC, and VAD acknowledge support from the International Space Science Institute to the International Team "New Approach to Active Processes in Central Regions of Galaxies". CMK is supported, in part, by the Taiwan Ministry of Science and Technology grants MOST 104-2923-M-008-001-MY3 and MOST 105-2112-M-008-011-MY3.

References

[1] Dobler, G., Finkbeiner, D.P., Cholis, I., et al., ApJ **717**, 825 (2010)
[2] Su, M., Slatyer, T.R., & Finkbeiner, D.P., ApJ **724**, 1044 (2010)
[3] Ackermann, M., Albert, A. et al., ApJ **793**, 64 (2014)
[4] Ade, P. et al. (Planck Collaboration), A&A **554**, 139 (2013)
[5] Crocker, R.M., & Aharonian, F., Phys. Rev. Lett. **106**, 101102 (2011)
[6] Fujita, Y., Ohira, Y., Yamazaki, R., ApJ **775**, 20 (2013)
[7] Guo, F., Metheus, W.G., Dobler, G. & Oh, S.P., ApJ **756**, 182 (2012)
[8] Yang, H.-Y.K., Ruszkowski, M., Ricker, P.M., Zweibel, E. & Lee, D., ApJ **761**, 185 (2012)
[9] Cheng, K.-S., Chernyshov, D.O., Dogiel, V.A., Ko, C.-M., & Ip, W.-H., ApJ **731**, L17 (2011)
[10] Feain, I.J., Cornwell, T.J., Ekers, R.D. et al., ApJ **740**, 17 (2011)
[11] Junkes, N., Haynes, R.F., Harnett, J.L., & Jauncey, D.L., A&A **269**, 29 (1993)
[12] Yang, R.-Z., Sahakyan, N., de Ona Wilhelmi, E., et al., A&A **542**, 19 (2012)
[13] Aharonian, F., Akhperjanian, A.G., Anton G. et al., ApJL **695**, 40 (2009)
[14] Croston, J.H., Kraft, R.P. & Hardcastle, M.J., ApJ **660**, 191 (2007)
[15] Mingo, B., Hardcastle, M.J., Croston, J.H. et al., ApJ **731**, 21 (2011)

[16] Mingo, B., Hardcastle, M.J., Croston, J.H. et al., ApJ **758**, 95 (2012)
[17] Cheng, K.-S., Chernyshov, D.O., Dogiel, V.A., Ko, C.-M., Ip, W.-H., Wang, Y., ApJ **746**, 116 (2012)
[18] Cheng, K.-S., Chernyshov, D.O., Dogiel, V.A., Ko, C.-M., ApJ **799**, 112 (2015)
[19] Berezinskii, V.S., Bulanov, S.V., Dogiel, V.A., Ginzburg, V.L., and Ptuskin, V.S. *Astrophysics of Cosmic Rays*, ed. V.L.Ginzburg (Norht-Holland, Amsterdam, 1990)
[20] Bell, A.R., MNRAS **182**, 147 and 443 (1978)
[21] Krymskii, G.F., Akademiia Nauk SSSR, Doklady (DoSSR, in Russian) **234**, 1306 (1997)
[22] Berezhko, E.G., Yelshin, V.K. & Ksenofontov, L.T., Astropart. Phys. **2**, 215 (1994)
[23] Syer, D. & Ulmer, A., MNRAS **306**, 35 (1999)
[24] Bykov, A.M. & Fleishman, G.D., MNRAS **255**, 269 (1992)
[25] Bykov, A.M. & Toptygin, I.N., Physics Uspekhi **36**, 1020 (1993)
[26] Bykov, A.M. & Toptygin, I.N., Astronomy Letters **27**, 625 (2001)
[27] Ackermann, M., et al., ApJ **750**, 3 (2012)
[28] Stecker F.W., & Jones, F.C., ApJ **217**, 843 (1977)
[29] Kampert, K.-H. et al., (KASCADE Collaboration), `arXiv:astro-ph/0102266` (2001)
[30] Apel, W.D. et al., (KASCADE-Grande Collaboration), PRL **107**, 171104 (2011)
[31] Lunardini, C., Razzaque, S., Theodoseau, K.T., Yang, L., Phys. Rev. D **90**, 023016 (2014)

The Pierre Auger Observatory status and the AugerPrime upgrade program

Daniele Martello[1,a] on behalf of the Pierre Auger Collaboration[2,b]

[1] Dipartimento di Matematica e Fisica "E. De Giorgi" and INFN Lecce, via per Arnesano, Lecce, Italy
[2] Observatorio Pierre Auger, Av. San Martin Norte, 304, 5613 Malargue, Argentina

Abstract. The nature and the origin of ultra-high energy cosmic rays (UHECRs), above 10^{17} eV, are still unknown. The Pierre Auger Observatory with its huge exposure provides us with a large set of high quality data. The analysis of these data has led to major breakthroughs in the last decade, but a coherent interpretation is still missing. To answer the open questions the Observatory has started a major upgrade, with an emphasis on improved mass composition determination using the surface detectors. The latest results and the planned detector upgrade will be presented. The expected performance and the improved physics sensitivity of the Observatory will be discussed.

1. Introduction

The nature and the origin of ultra-high energy cosmic rays above 10^{17} are still unknown, even if in the last decade measurements have shed light on these puzzling questions. Understanding the sources, nature and propagation properties of UHECRs is one of the key questions in astroparticle physics. From the experimental point of view, their study can be performed indirectly, by using the extensive air showers (EASs) they produce by interacting with the nuclei of the molecules that compose the atmosphere of the Earth.

Anyway, the all particle spectrum by itself cannot provide sufficient discrimination between the different astrophysical hypotheses, and the determination of the primary composition is mandatory to reach any reliable conclusion.

2. The Pierre Auger Observatory

The Pierre Auger Observatory [1] with its 3000 km^2 of instrumented surface, brings unique capabilities to the UHECR study. The Observatory is located in the province of Mendoza (Argentina) near the city of Malargue in a vast high plain. It is the world's largest cosmic ray observatory. The Observatory consists of an array (SD) of 1600 water-Cherenkov particle detector stations (WCD) overlooked by 24 air fluorescence telescopes (FD). In addition, three high elevation fluorescence telescopes (HEAT) overlook a surface of 23.5 km^2 where 61 additional WCDs (Infill) are installed (see Fig. 1).

Each WCD of the surface detector is a water-Cherenkov detector which samples the particle content of the EAS falling on the array. The WCD consists of a 12,000 liter polyethylene water tank containing a sealed laminated polyethylene liner with a reflective inner surface. Cherenkov light from the passage of charged particles is collected by three 230 mm photomultiplier tubes (PMTs) that look through windows of clear polyethylene into highly purified water. The WCD is self contained. A solar power system provides power for the PMTs and electronics package. The electronics package, consisting of a processor, GPS receiver, radio transceiver and power controller, is mounted on the container. The WCD stations are placed on a triangular grid with 1500 m spacing (SD-1500).

The 24 telescopes of the FD overlook the SD array from four sites: Los Leones, Los Morados, Loma Amarilla and Coihueco [1] (see Fig. 1). Six independent telescopes are located at each FD site in a clean climate-controlled building. A single telescope has a field of view of 30° × 30° in azimuth and elevation, with a minimum elevation of 1.5° above the horizon. The telescopes face towards the interior of the array so that the combination of the six telescopes provides 180° coverage in azimuth.

Three additional fluorescence telescopes (HEAT) with an elevated field of view were built at the FD site at Coihueco. These telescopes are very similar to the original fluorescence telescopes but are tilted by 29° upward. These three telescopes work independently of other FD sites and form the "fifth site" of the Observatory. The HEAT telescopes were designed to cover the elevation range from 30° to 58°, which lies above the field of view of the other FD telescopes. The HEAT telescopes allow a determination of the cosmic ray spectrum and X_{max} distributions in the energy range from below the second knee up to the ankle.

The Infill array consists of a denser WCD array with 750 m spacing (SD-750) nested within the 1500 m array covering a surface of 23.5 km^2. The area is centered 6 km away from the Coihueco fluorescence site. The Infill array is fully efficient from 3×10^{17} eV onwards for air showers with zenith angle $\leq 55°$.

[a] e-mail: daniele.martello@le.infn.it
[b] e-mail: auger_spokespersons@fnal.gov
http://www.auger.org/archive/authors_2016_09.html

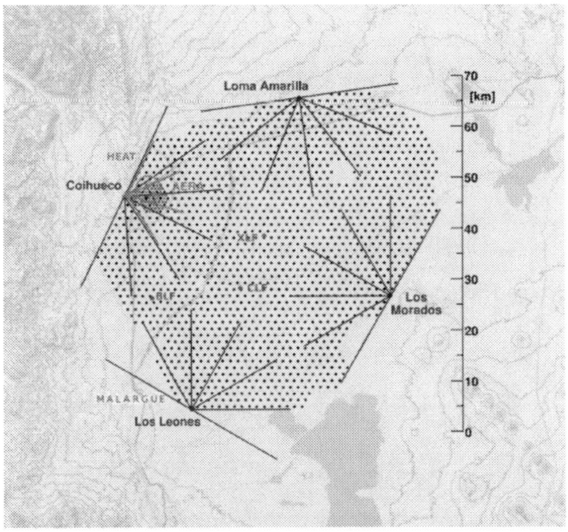

Figure 1. The Auger Observatory layout. Each dot corresponds to one of the 1660 surface detector stations. The four fluorescence detector sites are shown, each with the field of view of its six telescopes. The Coihueco site hosts three extra high elevation (HEAT) telescopes. The 750 m array is located a few kilometers from Coihueco.

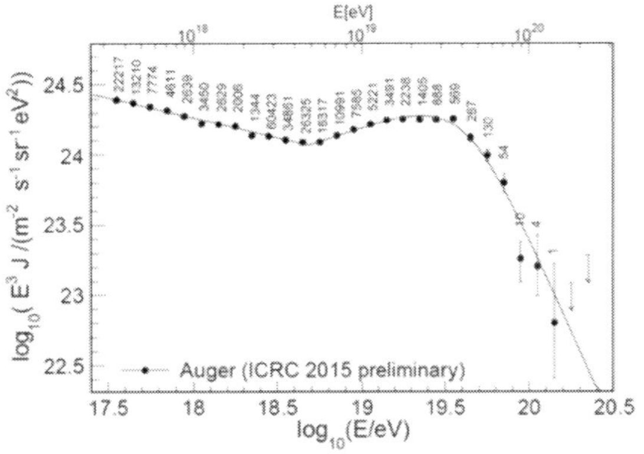

Figure 2. The combined energy spectrum measured by the Auger Observatory, fitted with a flux model. The data points include only the statistical uncertainties. For each data point is reported the number of events.

On-line and long-term performance of the detectors and data quality are monitored continuously, and a set of high-quality devices installed in the Observatory monitors the atmospheric conditions during operation.

High-quality data have been collected continuously for about ten years, with a SD annual exposure of 5500 km^2 sr yr. The longitudinal profile reconstructed by the FD is providing a nearly calorimetric measurement of the primary energy, with total systematic uncertainty of 14% [2]. From the shower lateral distribution reconstructed using the WCD signals, a SD energy estimator is inferred. A high-quality subset of hybrid events recorded by both the SD and the FD is used to calibrate the SD energy estimator with the FD energy measurement, hence providing an almost model-independent energy calibration.

3. The spectrum

The Pierre Auger Observatory has collected data of excellent quality for more than 10 years, which has already led to a measurement of the flux of UHECRs above 3×10^{17} eV with unprecedented statistics.

The events used for the determination of the energy spectrum consist of 4 different sets of data: the SD-1500 vertical events with zenith angle up to 60°, the SD-1500 inclined events with zenith angle between 60° and 80°, the SD-750 vertical events and the hybrid events. The hybrid set of data contains events detected simultaneously by the fluorescence telescopes and by at least one WCD.

The first step in the procedure used for the determination of the spectrum is the evaluation of the energy of the events. The FD allows the measurement of the electromagnetic energy released by the shower in the atmosphere as a function of the atmospheric depth. The total primary energy is then derived by integrating this longitudinal profile over the depth range and adding an estimate of the so-called "invisible energy" carried into the ground by high-energy muons and neutrinos.

The SD samples the shower particles that reach the ground. The intensities of the signals registered in the WCD are used to quantify the shower size and the impact point of the shower axis on the ground.

The absolute calibration of the SD sets of data is inferred from a high-quality subset of hybrid events (full details in [3–5]).

The final step in the procedure used for the determination of the energy spectrum is a precise evaluation of the exposure. Above the energy for full detector efficiency, the calculation of the SD exposure is based on the determination of the geometrical aperture of the array for the corresponding zenith-angle interval and of the observation time. The choice of a fiducial trigger based on active hexagons allows one to exploit the regularity of the array, and to compute the aperture simply as the sum of the areas of all active hexagons. The calculation of the exposure for the hybrid set of data is more complex. It relies on a detailed time-dependent Monte Carlo simulation which exactly reproduces the data taking conditions and includes the response of the hybrid detector [6].

The energy spectrum reported in Fig. 2 has been obtained by combining the four independent sets of data. They are combined using a method that takes into account the systematic uncertainties of the individual measurements (see details in [7]).

The characteristic features of the combined energy spectrum, shown in Fig. 2, have been quantified by fitting a model that describes a spectrum by a power-law below the ankle $J(E) = J_0(E/E_{ankle})^{\gamma_1}$ and power-law with a smooth suppression at the highest energies:

$$J(E) = J_0 \left(\frac{E}{E_{ankle}}\right)^{-\gamma_2} \left[1 + \left(\frac{E_{ankle}}{E_s}\right)^{\Delta\gamma}\right] \times \left[1 + \left(\frac{E}{E_s}\right)^{\Delta\gamma}\right]^{-1}$$

Here, γ_1 and γ_2 are the spectral indexes below and above the ankle energy E_{ankle} respectively, E_s is the energy at which the differential flux falls to one-half of the value of

DOI: 10.1051/epjconf/201714505001

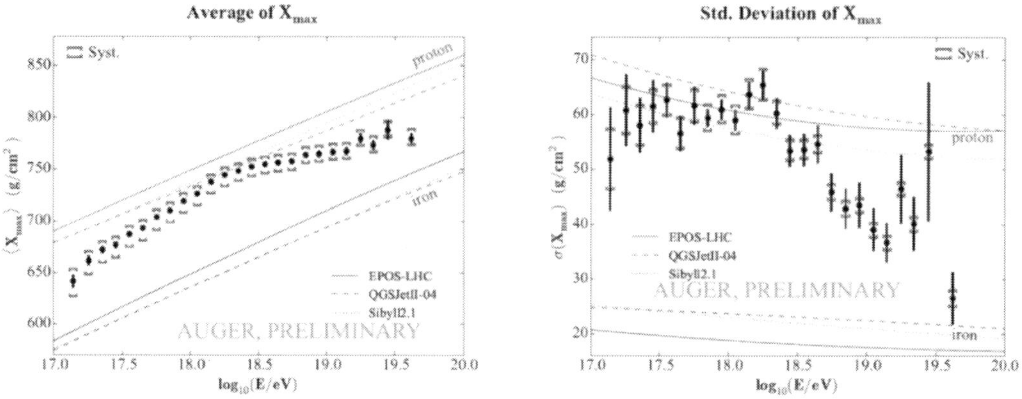

Figure 3. Average and RMS of the X_{max} compared to the model predictions for an all-proton and an all-iron composition.

Table 1. Best-fit parameters for the combined energy spectrum, statistical and systematic uncertainties being reported.

J_0 [eV^{-1} km^{-2} sr^{-1} yr^{-1}]	E_{ankle} [EeV]	E_s [EeV]	γ_1	γ_2	$\Delta\gamma$
$(3.30 \pm 0.15 \pm 0.20)x10^{-19}$	$4.82 \pm 0.07 \pm 0.8$	$42.1 \pm 1.7 \pm 7.6$	$3.29 \pm 0.02 \pm 0.05$	$2.69 \pm 0.02 \pm 0.1$	$3.14 \pm 0.2 \pm 0.4$

the power-law extrapolation from the intermediate region, $\Delta\gamma$ gives the increment of the spectral index beyond the suppression region, and J_0 is the normalization of the flux, taken as the value of the flux at $E = E_{ankle}$.

The result of the best fit is shown in Fig. 2 and the corresponding parameters are presented in Table 1, quoting both statistical and systematic uncertainties.

The energy spectrum can also be exploited to study the distribution of cosmic-ray sources by searching for a flux variation with declination (δ) of the incoming directions. This study is of particular interest to the discussion of the difference seen in the suppression region between the spectra measured by Auger and by the Telescope Array experiment [8], which, despite being still compatible within the quoted systematic uncertainties of both experiments, is not understood so far.

4. Mass composition

Different observables can be used to obtain information on the primary composition, the most direct of which is the depth of maximum development of the longitudinal shower profile (X_{max}), measured by the FD. X_{max} is related to the depth of the first interaction of the primary and to the subsequent development of the shower. For this reason, the interpretation in terms of composition is complicated by the large uncertainties in the hadronic interaction models used in the simulations (see [9] these proceedings). The average of the X_{max} for different energies of the primary and its RMS can be directly compared to the predictions of air shower simulations using recent post-LHC hadronic interaction models, as shown in Fig. 3.

Our measurements are clearly at variance with model predictions for pure composition; assuming no change in hadronic interactions at these energies, they point to a composition getting heavier above the ankle.

5. Search for Anisotropy

Complementary to the spectrum and mass measurements are the searches for anisotropy. Due to the energy losses during propagation through the Cosmic Microwave Background (CMB) the position of the astrophysical sources of UHECRs should be relatively nearby, which in case of light nuclei with $E > 40$ EeV would result in anisotropies in arrival directions, reflecting the distribution of the nearby extragalactic matter. For protons of this energy, deflections in the encountered magnetic fields are quite small, while for nuclei of atomic number Z they are Z times larger.

In the more recent analysis [10] 602 events have been selected with energy above 40 EeV recorded in the 10 years of operation of the Pierre Auger Observatory in the declination range of $-90°$ to $+45°$. Various kinds of tests were performed to search for anisotropy above a certain threshold in various angular windows as well as auto-correlations of arrival directions. Correlations were also sought with catalogs of plausible candidates of UHECR sources. The two largest departures from isotropy are found above 58 EeV and correspond to the direction within 15° of Centaurus A, the closest radio-loud AGN, and to the arrival direction within 18° of Swift-BAT AGNs closer than 130 Mpc and brighter than 10^{44} erg s^{-1}. None of the excesses is statistically significant (see Fig. 4). Therefore, either the primaries are light nuclei and the sources are numerous or the isotropy is caused by large deflections of nuclei with larger Z. A full sky study of arrival directions has been performed in collaboration with Telescope Array and IceCube [11]. The study included 231 events with energy above 52 EeV from the Pierre Auger Observatory, 87 events with energy above 57 EeV from the Telescope Array and the very high energy neutrinos from IceCube. No correlations at the discovery level were found.

Large scale anisotropy can reveal the effects of the global distribution of sources and the collective motion of cosmic rays. Using events recorded by the SD from 2004 January 1 to 2013 December 31 an analysis [12] was performed including also inclined events, reaching a coverage of about 85% of the sky. The set of data consists of about 70 000 events with energies above 4 EeV, where full efficiency for inclined events is attained. Two Rayleigh

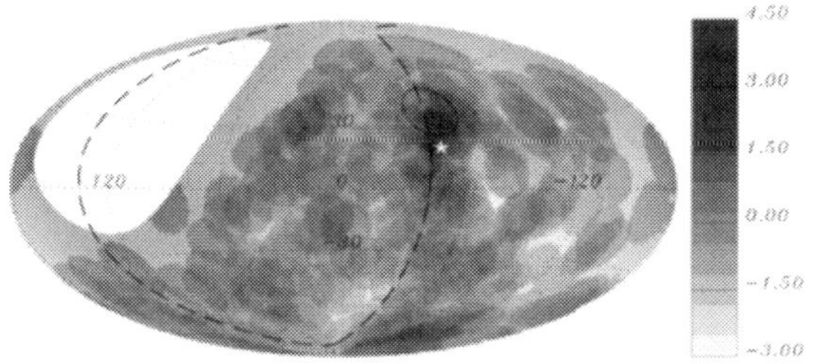

Figure 4. Map of the Li-Ma significances of overdensities in windows of 12° for events with $E \geq 54$ EeV. Dashed line: Super-Galactic Plane. White star: Centaurus A.

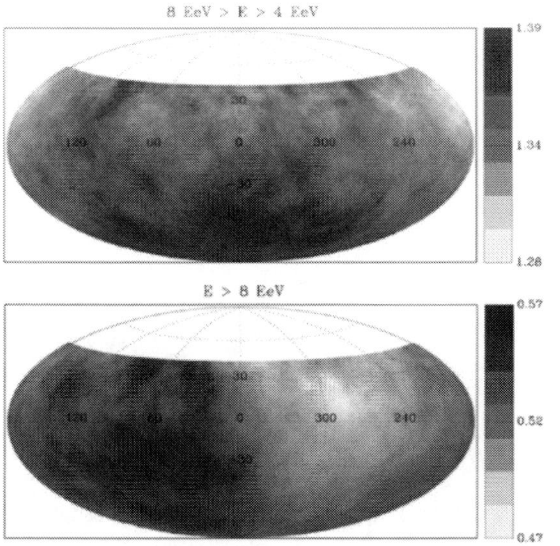

Figure 5. Sky map of flux for two bins of energy in $km^{-2}yr^{-1}sr^{-1}$ units. The data are smoothed in angular windows of 45°. (Galactic coordinates).

analyses were performed, in right ascension and azimuth angles in two energy bins, 4−8 EeV and above 8 EeV. While the lower energy bin does not show any significant departure from isotropy of arrival directions, a dipole was found at energies above 8 EeV of an amplitude 0.073 ± 0.015 pointing to $(\alpha, \delta) = (-95° \pm 13°, -39° \pm 13°)$. A sky map of the flux in equatorial coordinates is shown in Fig. 5. Observation of dipolar amplitudes in arrival directions of UHECRs is consistent with expectations for heavier nuclei suggested by the X_{max} distributions.

6. Astrophysical interpretation

The accurate measurement of the spectrum, the results from the study of the mass composition and the distribution of the arrival directions of the primaries, gives the possibility to infer some hypothesis on the origin and propagation of UHECRs [13].

Assuming that all UHECR sources emit hydrogen-1, helium-4, nitrogen-14 and iron-56 with a broken exponential rigidity cutoff, it is possible to infer some characteristics of the acceleration sites and put some constraint on the propagation process. Unfortunately, with the publicly available codes that simulate the propagation of the UHECRs from the sources to the Earth [14, 15], it is not possible to take into account the anisotropy study in the global fit procedure. For this reasons all the UHECR sources are assumed identical and uniformly distributed in the near universe. In this analysis both the propagation codes and different sets of parameters for the propagation processes are taken into account. The interaction of the primary with the atmosphere has been simulated with three different interaction codes [16–18]. The detector resolutions and acceptances are included. The data we attempt to fit consist of 15 measurements of the UHECR energy spectrum and 110 non-zero measurements of the X_{max} distribution. The free parameters of the fit are: the injection normalization factor J_0, the injection spectral index γ, the cutoff rigidity R_{cut} and the fractions of the different primaries at injection (three free parameters). In total there are 125 non-zero data points and 6 free parameters.

The minimization of the χ^2 distribution identifies a best fit solution for $\gamma = 0.94^{+0.09}_{-0.10}$ and $R_{cut} = 10^{18.67\pm0.03}$ with $\chi^2/d.o.f. = 178.5/119$. A second minimum in the χ^2 distribution is found at $\gamma = 2.03$ and $R_{cut} = 10^{19.84}$ for $\chi^2/d.o.f. = 235/119$ [13]. The corresponding simulated spectra and the mean and variance of the simulated X_{max} distributions are shown in Fig. 6 for the two solutions. The best fit solution of the Auger data predicts a very hard injection spectrum ($\gamma < 1$) and a rigidity cutoff that implies that the UHECR flux above $10^{19.5}$ eV is mostly limited by the maximum energy at the sources. The second minimum ($\gamma \sim 2$) with its larger rigidity cutoff is more in line with the standard models for UHECR acceleration.

In this analysis, the spectrum measured with high statistics and with different combinations and configurations of detectors has been combined with the mass composition sensitive parameters detected only with the fluorescence detector in a limited range of energy and with a number of events that is about 15% of the statistics collected for the spectrum. Moreover, a similar analysis limited to regions of the sky where there is an evidence for anisotropies in the cosmic ray flux is desirable. This is not possible with the mass sensitive data collected only by the fluorescence detector due to the limited statistics.

7. AugerPrime motivations

In order to solve the uncertainty still present in the interpretation of the cosmic ray cutoff ([19]) an upgrade of the Auger Observatory (named AugerPrime [20, 21]) has been planned. The main aim of AugerPrime is to provide additional measurements of observables sensitive to the mass composition of the most energetic cosmic rays, allowing the determination of the primary mass on a shower-by-shower basis. With AugerPrime it will be possible to study the mass composition not only with the fluorescence detector but also with the surface detector. The study of the origin of the flux suppression will provide fundamental constraints on the astrophysical sources, but also will allow us to determine more precise estimates of gamma-ray and neutrino fluxes at ultra-high energy. The high statistics of surface detector measurements will allow us to analyze the mass composition in different regions of the sky. Moreover, the AugerPrime upgrade will give us the possibility to measure the flux contribution of protons and elucidate the physics potential of existing and future cosmic ray, neutrino, and gamma-ray detectors.

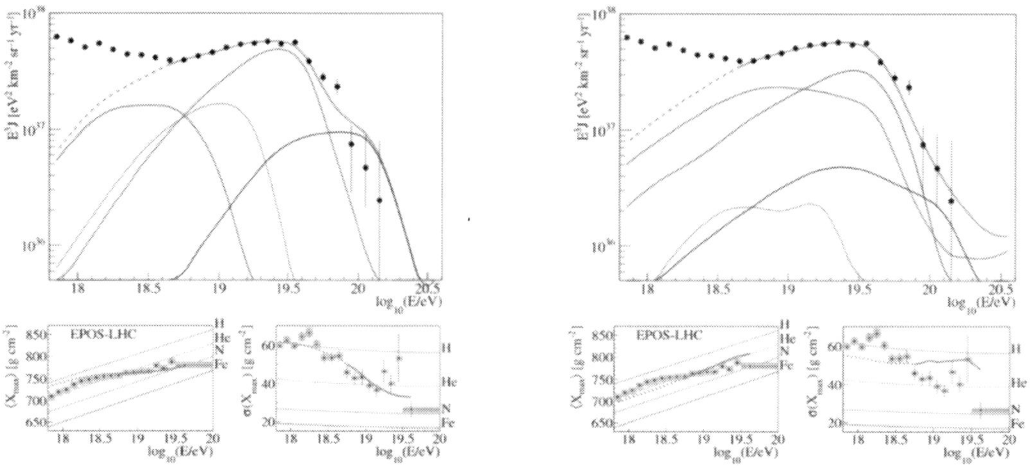

Figure 6. Top: simulated energy spectrum of UHECRs (multiplied by E^3) at the top of the Earths atmosphere with the best-fit parameters (left) and the second minimum of the likelihood for $\gamma \sim 2$ (right), compared with Auger data points. Partial spectra are grouped according to the mass number as follows: A=1 (red), 2<A<4 (grey), 5<A<26 (green), 27<A (blue), total (brown). Bottom: average and standard deviation of the X_{max} distribution as predicted in the two scenarios (brown). Pure 1 H (red), 4 He (grey), 14 N (green) and 56 Fe (blue) are shown as reference. Only the energy range where the brown lines are solid is included in the fit.

The determination of the primary mass composition of ultra-high energy cosmic rays is also deeply related to our understanding of extensive air showers and hadronic interactions (see [9] these proceedings). In the Auger data, there is a disagreement between the observed and expected muon numbers, therefore it is important to study the hadronic interaction in extensive air showers.

8. The AugerPrime upgrade

The AugerPrime upgrade consists of four main improvements explained in the following subsections.

8.1. The scintillator surface detector (SSD)

The scintillator surface detector consists of a plane of scintillation detector mounted above and triggered by the larger WCD detector below it. The combination of the two detectors allows us to obtain a good measurement of the density of muons in the air showers.

An SSD unit is a box of area 3.8 m × 1.3 m, containing two scintillator sub-modules, each composed of extruded polystyrene scintillator bars of about 1.6 m length, 5 cm width and 1 cm thickness [21]. The 4 m^2 scintillator planes are protected by light-tight, weatherproof enclosures, and mounted on top of the existing WCD with a strong support frame (see Fig. 7). The scintillator light will be read out with wavelength-shifting fibers inserted into straight extruded holes in the scintillator planes. The fibers are bundled and attached to a single light detector.

8.2. The underground muon detector (AMIGA)

The underground muon detector will provide a direct measurement of the muon content of a sample of showers observed by the upgraded Auger surface detector [22]. AMIGA will be used to verify and fine-tune the methods used to extract shower muon content using the SSD and WCD stations. AMIGA consists of 61 muon detectors deployed on a 750 m grid in the Infill area of the SD, instrumenting a total area of 23.5 km^2. Each detector consists of a plane of plastic scintillator of about 30 m^2 that will be buried about 2 meters underground.

Figure 7. One SSD module already installed in the Pierre Auger Observatory.

8.3. Change in fluorescence detector operation

The fluorescence detector [1] provides information about extensive air showers such as a model-independent energy reconstruction and longitudinal development profiles of the extensive air showers. The main limitation of the FD is its duty cycle, currently at the level of 15%. A significant increase of the duty cycle is possible by the extension of the FD operation to times at which a large fraction of the moon in the sky is illuminated. However, during such operations the PMT gain must be reduced by lowering the supplied high voltage (HV) to avoid high anode current and, therefore, an irreversible deterioration of the PMT sensitivity. The HV power used for the FD allows switching between two high voltage levels and the PMTs can be operated at the nominal gain (standard operation mode) and a lower gain (new operation mode with a large fraction of the moon in the sky).

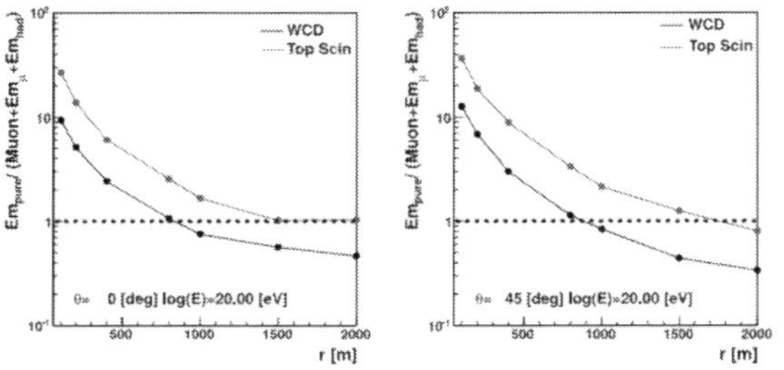

Figure 8. Ratios of different contributions to the integrated signal detected for air showers of 10^{20} eV at two zenith angles. Shown is the ratio between the electromagnetic component and the muonic component. The curve labeled "WCD" corresponds to the water-Cherenkov detectors of the Auger array, while the red one "Top Scin" corresponds to the scintillator detector.

9. AugerPrime performance

A thin scintillation detector, which is mounted above the larger water-Cherenkov detector, provides a robust and well-understood scheme for particle detection that is sufficiently complementary to the water-Cherenkov technique and permits a good measurement of the density of muons. This can be understood by comparing the signal contributions for different shower components as shown in Fig. 8. Over a wide range in lateral distance, the ratio between the integrated signal of electromagnetic particles (photons and electrons) and that of muons is more than a factor two higher in a scintillation detector without any shielding than in a water-Cherenkov detector.

The matrix formalism developed in [23] for a layered surface detector can be used for the reconstruction of the number of muons detected in a single upgraded detector station. The signal responses to particles of the muonic and electromagnetic shower components in the two detectors allow us to derive the muonic signal according to the expression:

$$S_{\mu,WCD} = aS_{WCD} + bS_{SSD}$$

where $S_{\mu,WCD}$ is the muonic signal in the WCD that can be converted in number of vertical equivalent muons (VEMs). S_{WCD} and S_{SSD} are respectively the total signal in the WCD and the total signal in the SSD. a and b are parameters inferred from detector simulation and are weakly dependent on the lateral distance from the shower core and on the mass of the primary.

To check the impact of this small dependence, the resolution and the mass-dependent bias in the reconstruction of the muon density for individual detector stations are given in Fig. 9. If stations with a lateral distance up to 1000 m are considered, the resolution is about 20–30% at the single detector level for showers in the flux suppression region. It is expected that the resolution can be improved for large lateral distances by using the constraint that the detector signal is dominated there by muons.

Another method that can be used to separate the muonic from the electromagnetic signals in the upgraded SD is based on the universality principle [24].

A universality-based analysis, or a sophisticated multivariate analysis, allows us to correlate the detector signals at different lateral distances and also takes

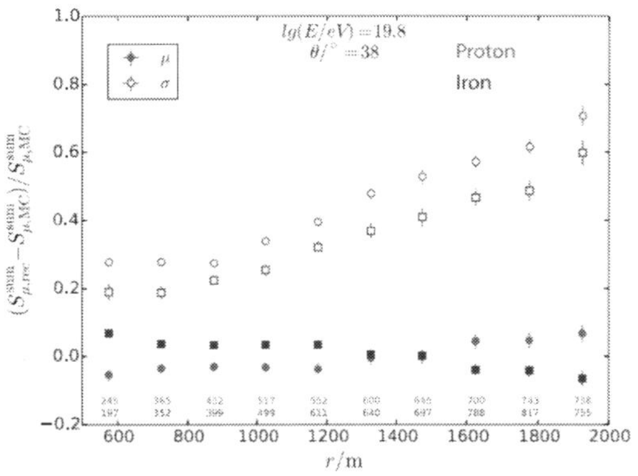

Figure 9. Reconstruction bias (solid symbols) and resolution (open symbols) of the muonic signal contribution for individual detector stations. The results for proton and iron showers are shown in red and blue, respectively. The number of detector stations analyzed for the different lateral distance intervals is also given.

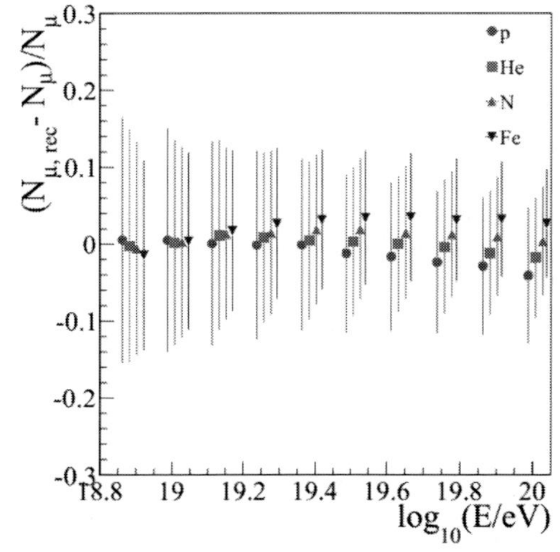

Figure 10. The reconstructed number of muons compared with the true number of muons as a function of energy for different cosmic ray primaries. Error bars represent the RMS of the distributions.

advantage of the arrival time (shower front curvature) and temporal structure of the signal measured in the detectors. The universality method predicts for the entire range of primary masses the air shower characteristics on the ground using only three parameters: the shower energy, the depth of shower maximum and the number of muons. Therefore, analyzing the structure of the signals in a ground based detector it is possible to infer the three parameters. The redundancy coming from the re-sampling of almost the same particles in two independent detectors (WCD and SSD) permits to eliminate possible degeneracies thanks to the different response to the muonic and electromagnetic components of the two detectors (see Fig. 8). The bias of the X_{max} reconstruction is less than 15 g/cm^2 with a resolution improving from 40 g/cm^2 at 10^{19} eV to 25 g/cm^2 at 10^{20} eV. The corresponding results for the muon number reconstruction are shown in Fig. 10.

DOI: 10.1051/epjconf/201714505001

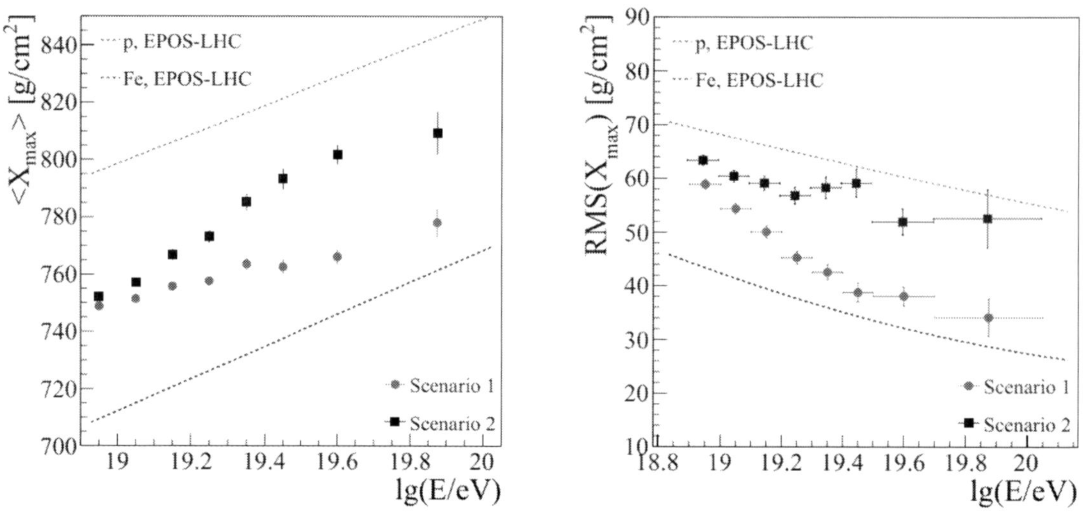

Figure 11. Predicted $X_{\max}$ and $\sigma(X_{\max})$ for the two scenarios in Sect. 6.

10. AugerPrime expected science impact

One of the key questions of the Pierre Auger upgrade is that of being able to discriminate between different compositions and, hence, physics scenarios in the energy range of the flux suppression.This is very difficult to demonstrate without knowing what composition to expect. For this reason two benchmark spectra have been chosen as representations of a maximum-rigidity scenario (scenario 1 corresponding to the best fit solution, Sect. 6) and of a photo-disintegration scenario (scenario 2 corresponding to the second minimum, Sect. 6).

Figure 11 shows the mean $X_{\max}$ and the corresponding $\sigma(X_{\max})$ for these spectra, using the SD data of the upgraded array. The $\sigma(X_{\max})$ contains the intrinsic air-shower fluctuations and the detector resolution. The same quantities as expected for pure proton and pure iron compositions are illustrated. The difference in the evolution of these two observables for the two models in the energy range of the cutoff can be distinguished with high significance. While the mean $X_{\max}$ and $\sigma(X_{\max})$ are very similar up to $10^{19.2}$ eV, which corresponds to the energy range covered by data of the fluorescence telescopes, the models predict significantly different extrapolations into the GZK suppression region and the two scenarios can be distinguished with high significance and statistics. In addition to these studies, the availability of muon information on an event-by-event basis allows for many ways of studying features of hadronic interactions. For example, the correlation between the number of muons and the depth of shower maximum can be used to study general features of muon production, including the search for exotic interaction scenarios at very high energy.

11. Conclusions

AugerPrime will collect high-quality data from 2018 until 2024. The statistic collected will be double compared with the existing Pierre Auger data sets, with the advantage that every event will have mass information and will allow us to better address some of the most pressing questions in UHECR physics. Obtaining additional composition-sensitive information will help to better reconstruct the properties of the primary particles at the highest energies. Moreover, it will improve the measurements in the important energy range just above the spectral ankle. Measurements with the new detectors will help to reduce systematic uncertainties related to the hadronic interaction models and to the reconstruction algorithms. This improved knowledge of air-shower physics will allow a reanalysis of existing data for improved energy evaluation and for improved mass composition studies.

References

[1] The Pierre Auger Collaboration, Nucl. Instrum. Meth. **A 798**, 172 (2015)
[2] V. Verzi for the Pierre Auger Collaboration, in *Proc. of ICRC 2013* (Proceedings, Rio de Janeiro Brazil, 2013) 0928
[3] The Pierre Auger Collaboration, JCAP **1408**, 08, 019 (2014)
[4] R. Pesce for the Pierre Auger Collaboration, in *Proc. of ICRC 2011* (Proceedings, Beijing China, 2011) 0214
[5] D. Ravignani for the Pierre Auger Collaboration, in *Proc. of ICRC 2013* (Proceedings, Rio de Janeiro Brazil, 2013) 0693
[6] The Pierre Auger Collaboration, Astropart. Phys. **34**, 368 (2011)
[7] I. Valiño for the Pierre Auger Collaboration, in *Proc. of ICRC 2015* (Proceedings of Science, The Hague Netherlands, 2015)
[8] Pierre Auger and Telescope Array Collaborations, I.C. Maris et al., in *Proc. of UHECR2014* (Utah USA, 2015)
[9] A. Yushkov for the Pierre Auger Collaboration, *in these proceedings*
[10] The Pierre Auger Collaboration, The Astrophysical Journal **804 (1)**, 15 (2015)
[11] IceCube, Telescope Array and Pierre Auger Collaborations, in *Proc. of ICRC 2015* (Proceedings of Science, The Hague Netherlands, 2015)

[12] The Pierre Auger Collaboration, The Astrophysical Journal **802**(2), 111 (2015)
[13] A. Di Matteo for the Pierre Auger Collaboration, in *Proc. of ICRC 2015* (Proceedings of Science, The Hague Netherlands, 2015) 0249
[14] R. Aloisio et al., JCAP **1210**, 007 (2012)
[15] E. Armengaud et al., Astropart. Phys. **42**, 463 (2007)
[16] S. Ostapchenko, Phys. Rev. **D74**, 014026 (2006)
[17] E.-J. Ahn et al., Phys. Rev. **D80**, 094003 (2009)
[18] K. Werner, F.-M. Liu, and T. Pierog, Phys. Rev. **C74**, 044902 (2006)
[19] D. Martello for the Pierre Auger Collaboration, Nucl. Instr. and Meth. **A742**, 16 (2014)
[20] R. Engel for the Pierre Auger Collaboration, in *Proc. of ICRC 2015* (Proceedings of Science, The Hague Netherlands, 2015) 0686
[21] Pierre Auger Collaboration, The Pierre Auger Observatory Upgrade AugerPrime Preliminary Design Report, in `arXiv:1604.03637` [astro-ph.IM] (2015)
[22] Pierre Auger Collaboration, Journal of Instrumentation, **11**, P02012 (2016)
[23] A. Letessier-Selvon et al., Nucl. Instrum. Meth. **A767**, 41 (2014)
[24] M. Ave et al., in *Proc. of ICRC 2015* (Proceedings of Science, The Hague Netherlands, 2015) 0378

Recent results from the Pierre Auger Observatory on the mass composition and hadronic interactions of ultra-high energy cosmic rays

Alexey Yushkov[1a] for the Pierre Auger Collaboration[2,3]

[1] Instituto de Tecnologías en Detección y Astropartículas (CNEA, CONICET, UNSAM), Centro Atómico Constituyentes, Comisión Nacional de Energía Atómica, Argentina
[2] Observatorio Pierre Auger, Av. San Martín Norte 304, 5613, Malargüe, Argentina
[3] Full author list: `http://www.auger.org/archive/authors_2016_08.html`

Abstract. Recent results on the mass composition and hadronic interactions of ultra-high energy cosmic rays (UHECR) obtained at the Pierre Auger Observatory are reviewed. Different studies based on shower observables recorded with the fluorescence and surface detectors of the Observatory indicate that the mass composition is getting lighter for the energies below ≈ 2 EeV and heavier afterwards. A fully consistent description of the data collected at the Observatory cannot be achieved with any of the up-to-date interaction models.

1. Introduction

A large variety of data, with unprecedented statistics, has been collected at the Pierre Auger Observatory [1] since the start of the operation in 2004. Analysis of the UHECR mass composition can be performed using different characteristics of the longitudinal and lateral profiles of electromagnetic (EM) and muon shower components with the methods having independent systematic uncertainties. The approaches relying on the studies of the EM component (analysis of the depth of the shower maximum $X_{\max}$) or using general properties of the extensive air showers (analysis of the correlation between $X_{\max}$ and the signal in surface stations or studies of the elongation rates) have smaller uncertainties due to hadronic interaction models compared to the approaches making use of absolute values related to the muon component. In this proceedings the results on the mass composition from the first and second types of the methods are compared and the conclusions on both the evolution of the primary mass with energy and on the drawbacks of the interaction models (EPOS-LHC [2], QGSJetII-04 [3], Sibyll 2.1 [4]) are presented.

The data reviewed in the paper are recorded using the fluorescence (FD) and surface detectors (SD) of the Pierre Auger Observatory, located in the Mendoza province, Argentina. The SD consists of 1660 water-Cherenkov detectors (WCD) placed in a triangular grid with a spacing of 1500 meters over the area of 3000 km^2. Twenty-seven telescopes of the FD are located at four sites on the perimeter of the SD. The FD operates only during clear moonless nights and its duty cycle is around 15%, while the duty cycle of the SD is $\approx$100%. Showers triggering both FD and SD are called hybrid events. For them, complementary information from both detectors allows one to reconstruct the shower parameters with higher precision and to perform cross-checks of the operation of the detectors. Three of the analyses presented below – measurements of the depth of shower maximum $X_{\max}$, of the correlation between $X_{\max}$ and signal in WCDs, and of the muon content in highly inclined events – exploit the advantages of the hybrid design of the Observatory. The measurements of risetime asymmetry and of muon production depth are performed using information from WCDs on particle arriving times sampled at the rate of 40 MHz.

Unless otherwise specified the data used are from the period from January 2004 till December 2012.

2. Measurement of the depth of shower maximum $X_{\max}$ [5, 6]

Direct determination of $X_{\max}$ and the calorimetric measurement of the energy of the primary particle are based on the detection of fluorescence light emitted by the molecules of atmospheric nitrogen due to their interactions with the shower particles. The parameters of the atmosphere, acting as a calorimeter, are thoroughly monitored at the Observatory and only events passing a pre-selection on good atmospheric conditions (cloud coverage, aerosol density) are accepted for the analysis. These events are subject to further selection ensuring unbiased measurement of the $X_{\max}$ distributions including specific requirements on the field of view of the FD telescopes and the quality cuts on the profile. For full details on the selection and reconstruction procedures the reader is referred to [5,6].

In June 2010 three High Elevation Auger Telescopes (HEAT) started to take data in addition to the standard 24 telescopes of the FD. The HEAT in combination with the standard telescopes at the Coihueco site (called further HeCo) have a field of view $2°-60°$ in elevation (compared to $2°-30°$ of the standard FD) and in this

[a] e-mail: `yushkov.alexey@gmail.com`

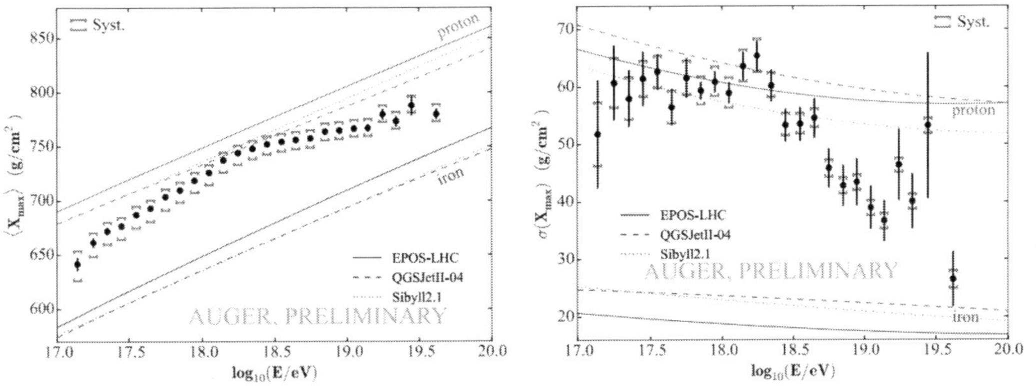

Figure 1. The mean and the standard deviation of the measured $X_{\max}$ distributions compared to the simulations with EPOS-LHC, QGSJetII-04and Sibyll 2.1for protons and iron.

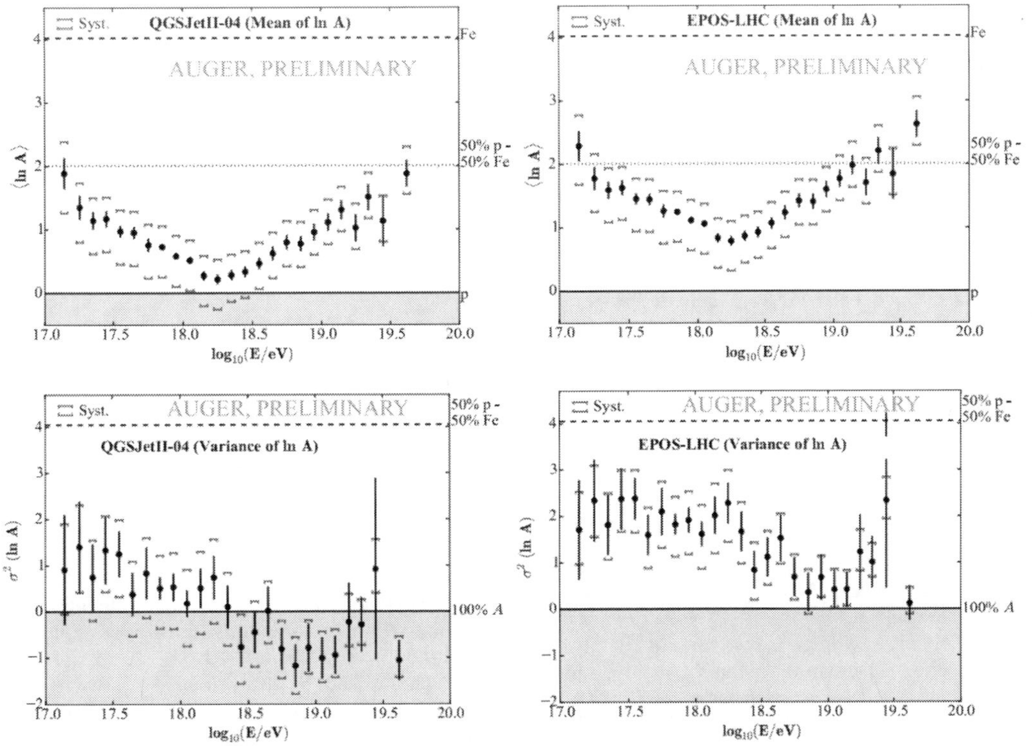

Figure 2. The evolution of the mean (top) and of the variance (bottom) of ln A determined from the measurements of $X_{\max}$.

way nearby showers with energies down to 10^{17} eV can be observed. To make the data set recorded by HeCo complementary and independent, events (1377 in total) with energies $< 10^{18.3}$ eV observed by HeCo during the period 01/06/2010 – 15/08/2012 are excluded from the data of the standard FD.

The final data set contains 5490 HeCo events with energies $10^{17.0}-10^{18.3}$ eV and 18382 events from the standard FD with energies $> 10^{17.8}$ eV. The $X_{\max}$ resolution is around 30 g/cm^2 at $10^{17.0}$ eV and improves with the energy to 15 g/cm^2 for $E > 10^{18.5}$ eV. The systematic uncertainty is below $(-15; +10)$ g/cm^{2}for most of the energies.

The comparison of the first two central moments of the measured $X_{\max}$ distributions with predictions of different hadronic interaction models for protons and iron nuclei is given in Fig. 1. The observed elongation rate (ER) is described well with a broken line, the energy of the break is $\lg(E_0/\text{eV}) \approx 18.3$ ($E_0 \approx 2$ EeV) and the slopes below and above the break are equal to ≈ 85 and ≈ 26 [g/cm^2/decade] correspondingly. The ER in simulations for all hadronic interaction models and all constant compositions lies in the range of 54–64 [g/cm^2/decade], thus the observed ERs indicate that the primary mass is decreasing below the break around 2 EeV and is increasing for the higher energies. Using the superposition model (see e.g. [7,8]) the mean logarithmic mass can be found from the measured $\langle X_{\max}\rangle$ using the relation $\langle X_{\max}\rangle \approx \langle X^p_{\max}\rangle - \langle \ln A\rangle D_{10}/\ln(10)$, where $\langle X^p_{\max}\rangle$ is the value for protons. The evolution of $\langle \ln A\rangle$ with energy is shown in Fig. 2 (top row). The observed transition from lighter to heavier composition implies that the primary beam is not pure and represents a mix of different nuclei. The degree of mixing (the variance of the logarithmic masses $\sigma^2_{\ln A}$) can be estimated using the measured width of the $X_{\max}$ distributions [9]: $\sigma^2(X_{\max}) \approx \langle\sigma^2_{\rm sh}\rangle + \sigma^2_{\ln A}\,(D_{10}/\ln(10))^2$, where $\sigma_{\rm sh}$ is the shower-to-shower $X_{\max}$ fluctuations for a fixed mass. The results on $\sigma^2_{\ln A}$, presented in Fig. 2 (bottom row), show different degrees of mixing in the primary beam depending on the hadronic interaction model used. Though for QGSJetII-04the variance of the masses takes negative values for some of the energies, the discrepancy is not larger than two standard statistical deviations and is non-significant if one takes into account the systematic errors on $X_{\max}$ measurements and on the FD energy scale.

3. Measurement of the correlation between $X_{\max}$ and $S(1000)$ [10]

The spread of the masses in the primary beam $\sigma(\ln A) = \sqrt{\langle \ln^2 A\rangle - \langle \ln A\rangle^2}$ can be estimated from the measurement of the correlation between $X_{\max}$and the signal in WCDs at 1000 m from the shower axis $S(1000)$ [10,11]. In simulations this correlation is close to zero or is slightly positive for pure compositions. Mixed compositions show a negative correlation: heavier primary nuclei have shallower $X_{\max}$ ($\Delta X_{\max} \sim -\Delta \ln A$) and larger $S(1000)$(due to a larger number of muons $N_\mu \sim A^{1-\beta}$, $\beta \simeq 0.9$ [12]), thus in a sample with a mixed composition the shallower showers on average will have larger signals. Such behavior is a general property of extensive air showers and thus the results of the correlation analysis are weakly sensitive to the choice of a particular interaction model.

In this analysis, to avoid a decorrelation due to the spreads of energies and zenith angles, $X_{\max}$and $S(1000)$ are scaled to a reference energy of 10 EeV and $S(1000)$ is additionally scaled to zenith angle of 38°. The scaled variables are denoted further as $X^*_{\max}$and S^*_{38}. The correlation between $X^*_{\max}$and $S^*_{38}$is evaluated using a ranking correlation coefficient $r_{\rm G}$proposed in [13]. The conclusions do not change when other correlation coefficients are used.

To have a data set with an unbiased mass composition the event selection is the same as in the $X_{\max}$analysis [6]. In addition, to have a reliable reconstruction of $S(1000)$, it is required that the SD station with the highest signal should have at least 5 active neighbour stations. For the energy range $\lg(E/{\rm eV}) = 18.5-19.0$ and the zenith angle range $0°-65°$, the final data set contains $n = 1376$ events. The statistical uncertainty can be approximated [10] by $\Delta r_{\rm G} \simeq 0.9/\sqrt{n}$, thus $\Delta r_{\rm G}({\rm data}) = 0.024$. The systematic error on $r_{\rm G}$is small $\lesssim 0.01$: invariance of the ranking correlation coefficients to any transformations leaving ranks of the events unchanged makes the correlation analysis almost insensitive to the systematic uncertainties in $X_{\max}$and $S(1000)$.

In Fig. 3 the distributions of $X^*_{\max}$and S^*_{38}in data and in simulations for protons and iron with EPOS-LHCare shown together with the corresponding values of the correlation coefficients. The correlation coefficients for protons for all models are non-negative: $r_{\rm G}$(EPOS-LHC) = 0.00, $r_{\rm G}$(QGSJetII-04) = 0.08 and $r_{\rm G}$(Sibyll 2.1) = 0.06. For heavier nuclei the correlation coefficients are larger

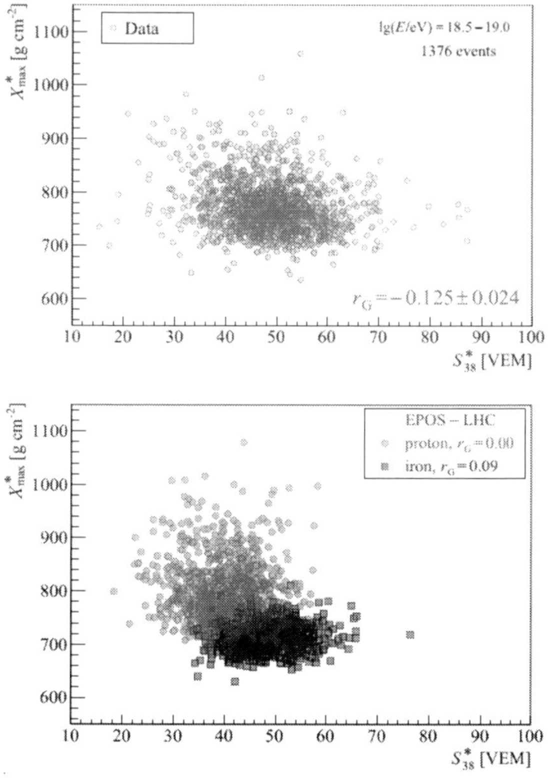

Figure 3. Distributions of $X^*_{\max}$ *vs.* S^*_{38} (see text) for $\lg(E/{\rm eV}) = 18.5-19.0$. Top: data; bottom: 1000 proton and 1000 iron showers simulated with EPOS-LHC.

than those for protons and for all p − He mixtures $r_{\rm G} \gtrsim 0$ as well. Thus, the negative correlation found in data $r_{\rm G}(X^*_{\max}, S^*_{38}) = -0.125 \pm 0.024$ (stat) cannot be reproduced using any pure composition and the data can be explained only by mixed compositions containing primary nuclei heavier than helium $A > 4$. The conclusions do not change when the key parameters of the interaction models (cross-section, multiplicity, elasticity, pion charge ratio) are modified by a factor ≈ 1.5.

The spread of the primary masses $\sigma(\ln A)$ can be estimated from Fig. 4 where the correlation value found in data is compared to the values in simulated mixtures with different fractions of protons, helium, oxygen and iron nuclei. The correlation $r_{\rm G}(X^*_{\max}, S^*_{38})$gets more negative for the larger spreads of the masses. For all three models (results for Sibyll 2.1, not shown in Fig. 4, are almost the same as for QGSJetII-04) the spread of masses corresponding to the correlation found in data, lies in the range $1.0 \lesssim \sigma(\ln A) \lesssim 1.7$.

4. Azimuthal asymmetry in the risetime of signals in WCDs [14]

Muons produce short and spiky time traces in WCDs, while for EM particles the distribution of the arrival times is larger and smoother. Since the relative contribution of muons to the SD signal increases with the growth of the mass of the primary nuclei, the risetime, $t_{1/2}$, defined as the

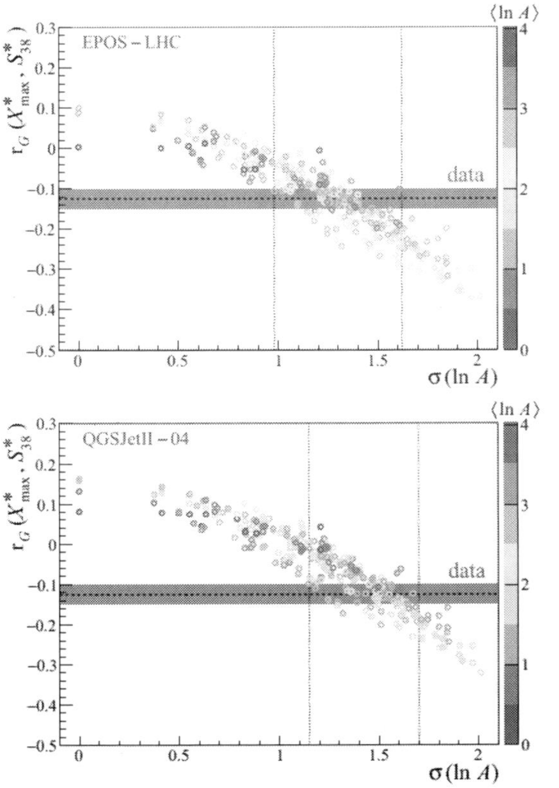

Figure 4. Dependence of the correlation coefficients r_G on $\sigma(\ln A)$ for EPOS-LHC (top) and QGSJetII-04(bottom) in mixtures of protons, helium, oxygen and iron nuclei. The shaded area shows the value for the data $r_G(X^*_{max}, S^*_{38}) = -0.125 \pm 0.024$ (stat). $\lg(E/\text{eV}) = 18.5-19.0$.

time of the increase of the total integrated signal in WCDs from 10% to 50%, is a mass-sensitive observable. Due to changes in the muon/EM signal ratio and in the thickness of the shower front with distance to the shower axis r and zenith angle θ, risetime is a function of r and θ. Risetime grows almost linearly $t_{1/2}(r) \propto r$ for a large range of r, thus $t_{1/2}/r$ is used to cancel out the distance dependence and to include WCDs from all distances in the analysis.

In inclined showers signal risetime in individual WCDs depends on their azimuth angle ζ in the shower plane: particles hitting 'early' stations ($|\zeta| < \pi/2$) cross smaller atmosphere thickness and have smaller angle to the shower axis compared to the particles hitting 'late' stations ($|\zeta| > \pi/2$). With the growth of zenith angle fewer EM particles reach late detectors due to absorption by the additional atmosphere (the signals thus become more muon-rich) and risetime in the late detectors becomes shorter compared to that in early detectors. As shown in Fig. 5 (top) the asymmetry in risetimes reaches its maximum at a certain zenith angle and decreases for larger angles where the muon component starts to give a dominant contribution to the signals.

The dependence of the risetime on azimuth angle can be parameterized as $\langle t_{1/2}/r\rangle(\zeta) = a + b\cos\zeta + c\cos^2\zeta$ (Fig. 5 (top)) and the risetime asymmetry can be characterized using the parameter $b/(a+c) \sim \left(\langle t_{1/2}/r\rangle(0) - \langle t_{1/2}/r\rangle(\pi)\right) / \left(\langle t_{1/2}/r\rangle(0) + \langle t_{1/2}/r\rangle(\pi)\right)$.

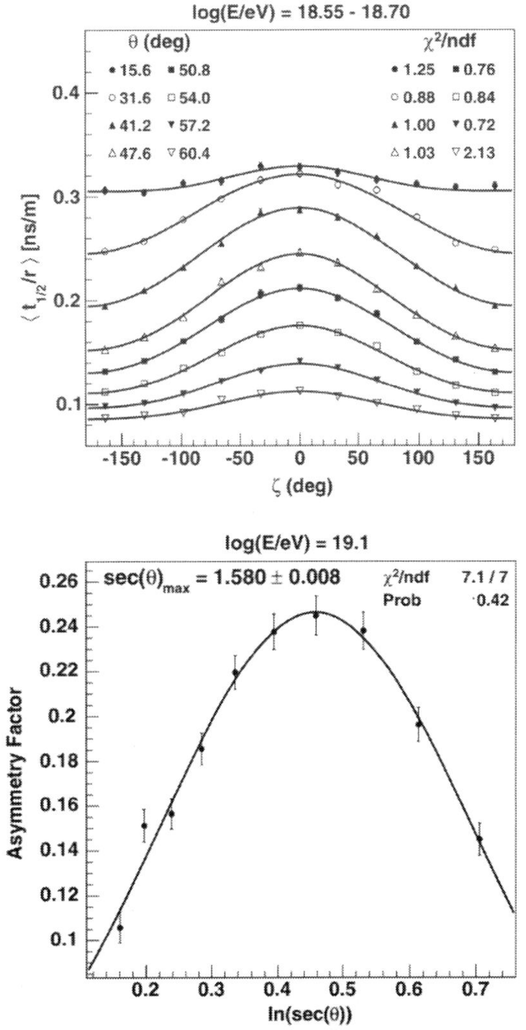

Figure 5. Top: dependence of $\langle t_{1/2}/r\rangle$ on the azimuth angle ζ fitted with a quadratic function (lines) for various zenith angles θ. $\lg(E/\text{eV}) = 18.55-18.70$. Bottom: dependence of the asymmetry factor $b/(a+c)$ on $\ln(\sec(\theta))$ fitted with a Gaussian (line). $\lg(E/\text{eV}) = 19.1$.

The behavior of the asymmetry parameter with $\ln(\sec(\theta))$ at $\lg(E/\text{eV}) = 19.1$ in stations within the range $r = 500 - 1000$ m is shown in Fig. 5 (bottom) together with a fit with a Gaussian function. The value of $\sec(\theta)_{max}$ at which the asymmetry parameter reaches its maximum provides information on the mass of the primary particle.

The behavior of $\sec(\theta)_{max}$ in the SD data is studied using events with the energies corresponding to 100% SD efficiency $E > 3 \times 10^{18}$ eV for the zenith angle range of $\theta = 30° - 62°$. For an accurate reconstruction of $t_{1/2}$only the stations with the 100% trigger probability (signal > 10 VEM) are used. The stations should not have saturated traces and the reliable determination of $t_{1/2}$ should be possible, these requirements limit the range of the WCDs distances to $r = 500-2000$ m. For the data recorded from January 2004 to October 2014 a total of 191534 signal traces from 54584 events are selected. The evolution of $\sec(\theta)_{max}$ with energy is analyzed in two ranges of the distances to the shower axis $r = 500-1000$ m and $r = 1000-2000$ m. Systematic error on $\sec(\theta)_{max}$ is ≈ 0.014.

The comparison of $\sec(\theta)_{max}$ in the data and simulations for protons and iron nuclei is shown in Fig. 6. The values for data are bracketed by the QGSJetII-04and EPOS-LHCpredictions for protons and iron. There are indications of an increase of the primary mass with energy, in agreement with other results reported in this proceedings, but uncertainties from the interaction models on the absolute value of $\sec(\theta)_{max}$ do not allow one to make robust conclusions of average mass of the primary nuclei. For QGSJetII-04such conclusions have an unphysical dependence on the range of the distances to the shower axis favouring lighter composition at larger distances $r = 1000-2000$ m.

From Fig. 7, where the values of $\sec(\theta)_{max}$ converted to $\langle \ln A \rangle$are presented, one can see that the average mass from the risetime asymmetry analysis lies between the values obtained from the measurements of X_{max} and the muon production depth (MPD, discussed in the next section). This might be due to the different role played by the muon and EM components in each of the analyses: for X_{max}determination the EM component is of major importance, MPD analysis relies on the reconstruction of properties of muons, and the $\sec(\theta)_{max}$ behavior results from a complicated interplay between both muon and EM components.

5. Measurements of the muon production depth X^{μ}_{max} [15]

Muons in the atmosphere travel along almost straight lines and the production height z of each muon along the shower axis can be reconstructed from a quite simple geometry shown in Fig. 8 [15]: $z \approx 1/2\left(r^2/c(t - \langle t_\varepsilon \rangle) - c(t - \langle t_\varepsilon \rangle)\right) + \Delta - \langle z_\pi \rangle$, here $\langle t_\varepsilon \rangle$ – mean kinematic delay; $\langle z_\pi \rangle$ – pion decay length. After conversion of heights z to the mass units $X^\mu = \int_z^\infty \rho(z')dz'$ one gets a distribution of the MPD for an event and the depth of the maximum muon production X^{μ}_{max} is determined from the fit with a Gaisser-Hillas function as shown in Fig. 8. To reduce the EM contamination of the traces in the WCDs only inclined showers with zenith angles $\theta = [55°; 65°]$ are used in the current analysis. Additionally, a cut on the primary energies $E > 20$ EeV is applied to have sufficient number of muons for the reconstruction of the profile, and a cut on the station distances to the core $1700\,\mathrm{m} < r < 4000\,\mathrm{m}$ ensuring at the same time similar proton and iron selection efficiencies and good X^{μ}_{max} resolution. The final data set contains 481 events, the systematic uncertainty on X^{μ}_{max} is around 17 g/cm^2, the resolution is 100 (80) g/cm^2 at $10^{19.3}$ eV for p (Fe) and 50 g/cm^2 at $10^{20.0}$ eV.

The evolution of the MPD in data and the predictions for protons and iron, obtained with EPOS-LHCand QGSJetII-04, are shown in Fig. 9. Comparison of the measured ER $D^{\mu}_{10} = d\langle X^{\mu}_{max}\rangle/d\lg(E/\mathrm{eV}) = -25 \pm 22\,(\mathrm{stat.}) \pm 21\,(\mathrm{sys.})$ [g/cm^2/decade] (Fig. 9) with the values for protons (35.9 ± 1.2 [g/cm^2/decade]) and iron (48.0 ± 1.2 [g/cm^2/decade]) indicates the transition from lighter to heavier composition with increasing energy. Due to large uncertainties on the absolute values of $\langle X^{\mu}_{max}\rangle$ the reconstructed $\langle \ln A \rangle$ strongly depends on the interaction model. For EPOS-LHCthe mass composition is heavier than iron (cf. Figs. 7, 9). The deeper X^{μ}_{max} for EPOS-LHCis due to a larger elasticity in proton-air and pion-air interactions, which results from a better, compared to QGSJetII-04, description of rapidity gap distributions measured at the LHC [16]. Thus the better agreement of $\langle \ln A \rangle$from $\langle X^{\mu}_{max}\rangle$ and $\langle X_{max}\rangle$ measurements for QGSJetII-04 (Fig. 7) might be rather accidental.

6. Muon content of highly inclined showers [17]

Using highly inclined events (zenith angles $> 62°$) an almost direct measurement of the muon shower content can be performed. The approach consists in production of the reference muon density lateral profiles $\rho_{\mu,19}(\vec{r};\theta,\phi)$ with a shape weakly depending on energy, on the type of the primary nuclei, on the hadronic interaction model and on the software used for shower simulations. For a given muon density ρ_μ, a muon scale factor N_{19} with respect to $\rho_{\mu,19}(\vec{r};\theta,\phi)$ is then introduced: $\rho_\mu = N_{19}\,\rho_{\mu,19}(\vec{r};\theta,\phi)$. By construction N_{19} does not depend on the zenith angle. Finally one compares scaling factors in data and in Monte-Carlo (MC) simulations $R_\mu = N^{\mathrm{data}}_{19}/N^{\mathrm{MC}}_{19}$.

The production of the reference profiles $\rho_{\mu,19}(\vec{r};\theta,\phi)$ for different zenith θ and azimuth ϕ angles is done using QGSJetII-03 protons at 10^{19} eV. Hybrid events with energies above $10^{18.6}$ eV (≈ 4 EeV), corresponding to 100% trigger efficiency of the SD, are accepted for the analysis, the zenith angles are limited to the [62°; 80°] range. In total 174 events remain after the selection.

The muon scales $\langle R_\mu \rangle$ in data and simulations for protons and iron are shown in Fig. 10. Similarly to the measurements of X_{max}, risetime asymmetry and MPD, the slope $d\langle R_\mu\rangle/d\lg(E/\mathrm{eV})$ in data differs from the MC predictions for constant compositions and is compatible with a transition to a heavier mass composition. The absolute values of $\langle R_\mu \rangle$ in data correspond to masses close to or heavier than iron, thus being in contradiction with the results on $\langle \ln A \rangle$from the analysis of X_{max}. Similar conclusions were obtained on the muon content of the vertical showers [18]. To quantify the discrepancy, the mean logarithmic mass from the analysis of X_{max} distributions can be converted into the mean logarithmic muon scale $\langle \ln R_\mu \rangle$ [17] thus leading to a one-to-one relation between $\langle \ln R_\mu \rangle$ and $\langle X_{max}\rangle$. The results presented in Fig. 11 show, that the muon content $\langle \ln R_\mu \rangle$ in data is (30–80)% larger compared to simulations for primary masses $\langle \ln A \rangle$ corresponding to the measured $\langle X_{max}\rangle$ Nevertheless, the minimum difference between data and EPOS-LHCis only $1.4\sigma_{sys}$.

7. Discussion

The results of studies of the UHECR mass composition at the Pierre Auger Observatory, presented in these proceedings, indicate that $\langle \ln A \rangle$is not constant for all energies above 10^{17} eV. Specifically, the ER of $X_{max}(d\langle X_{max}\rangle/d\lg(E/\mathrm{eV}))$ shows that the composition is getting lighter below 2 EeV and then heavier for higher energies. Conversion of the moments of X_{max}distributions to the moments of the logarithmic mass produces results that are consistent with a mixed composition within

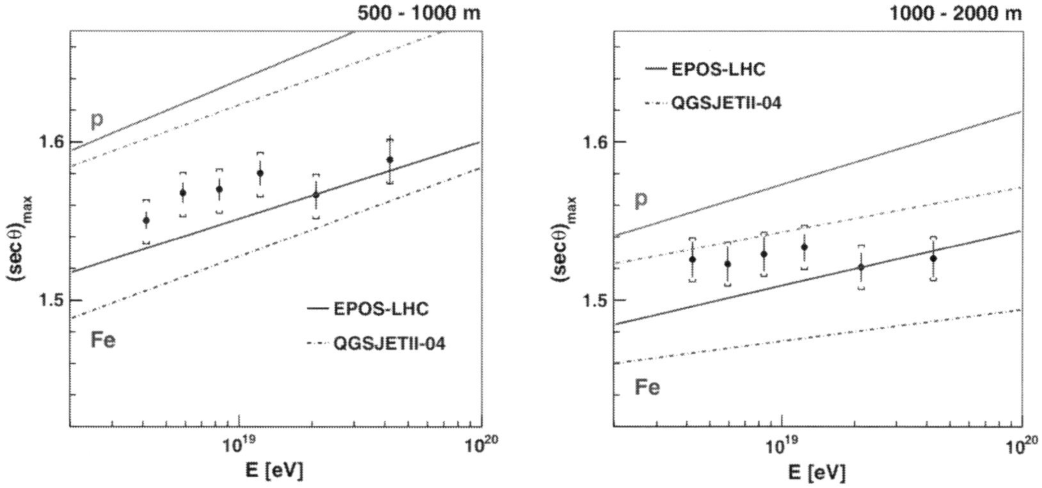

Figure 6. Dependence of $\sec(\theta)_{max}$ on energy in data (points) and simulations with EPOS-LHCand QGSJetII-04for protons and iron nuclei in the distance ranges $r = 500-1000$ m (left) and $r = 1000-2000$ m (right).

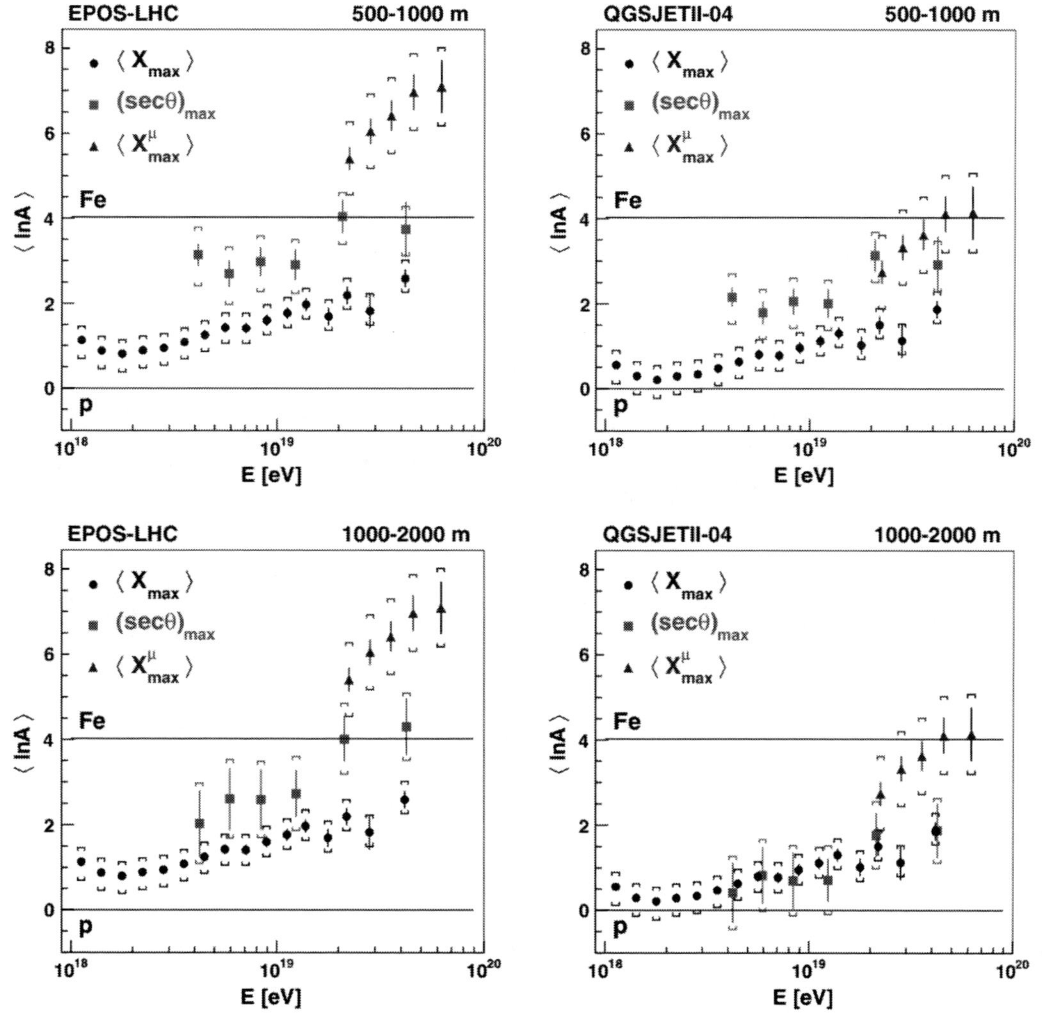

Figure 7. Comparison of the mean logarithmic masses from the measurements of X_{max}, $\sec(\theta)_{max}$ and X^{μ}_{max}. Distance ranges $r = 500-1000$ m and $r = 1000-2000$ m refer only to the analysis of $\sec(\theta)_{max}$.

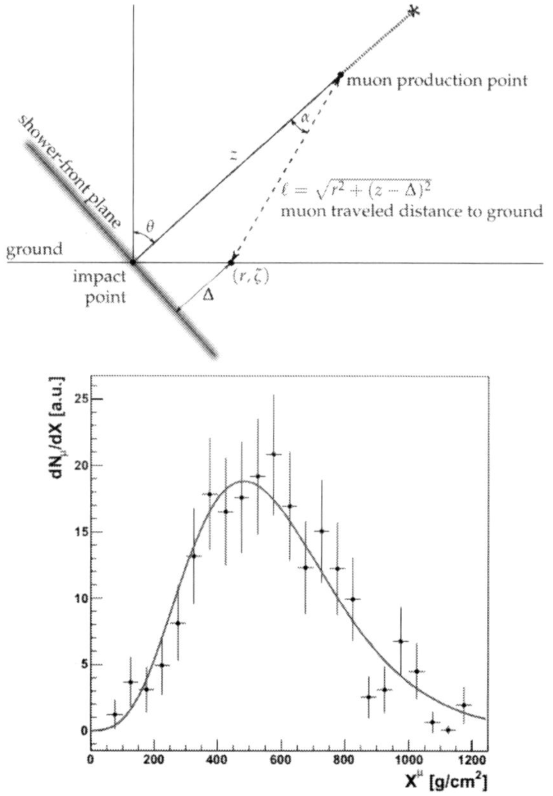

Figure 8. Top: the geometry used for the estimation of the muon production height z. Bottom: the muon production depth (MPD) profile of a real event with the energy of 33 ± 1 EeV fitted with the Gaisser-Hillas function (red line).

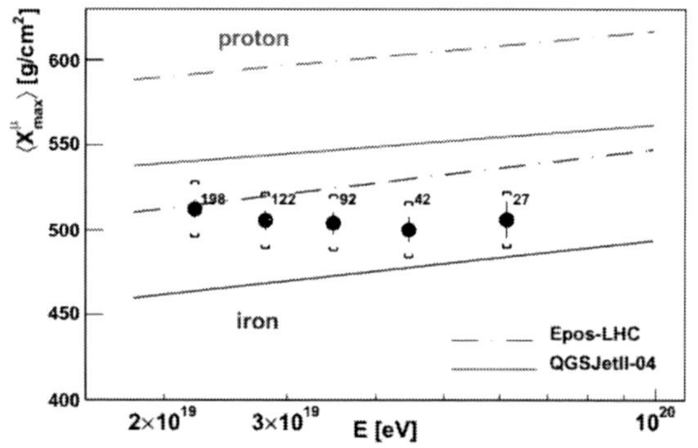

Figure 9. Mean MPD $\langle X^\mu_{max} \rangle$ in data (points) compared to simulations with EPOS-LHCand QGSJetII-04for protons and iron.

systematic uncertainties when EPOS-LHCis used. For QGSJetII-04the conclusions are less certain and even an unphysical variance of logarithmic masses ($\sigma^2_{\ln A} < 0$) is observed for $E \gtrsim 3$ EeV, though the values differ from zero by less than $2\sigma_{stat}$.

Analysis of the correlation between $X_{max}$and $S(1000)$ provides a strong evidence on the presence of several types of nuclei near the ankle (≈ 5 EeV) in the primary spectrum with the spread of the masses $\sigma(\ln A) > 1$ for all interaction models used (EPOS-LHC, QGSJetII-04, Sibyll 2.1). The data are described well by mixed

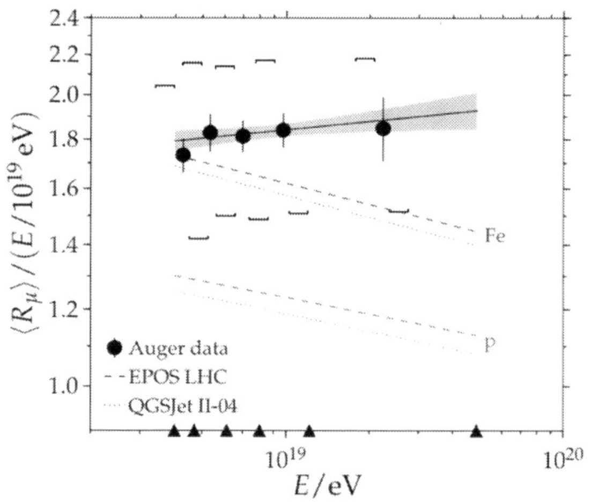

Figure 10. The muon scales $\langle R_\mu \rangle$ in data and in the MC simulations with EPOS-LHCand QGSJetII-04for protons and iron.

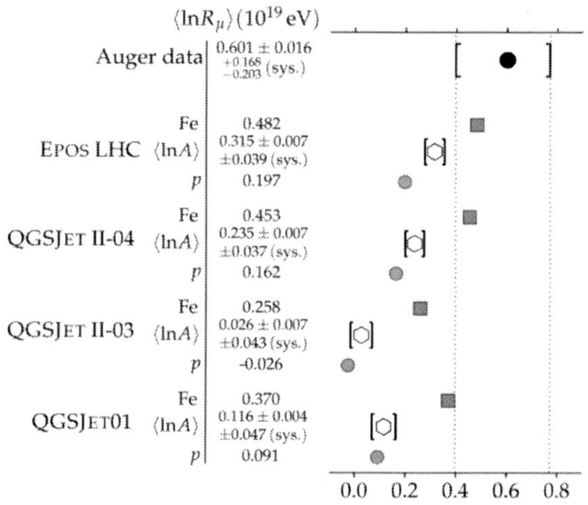

Figure 11. Comparison of $\langle \ln R_\mu \rangle$ at 10^{19} eV in data to the values in simulations for protons, iron and for the mixes with $\langle \ln A \rangle$from the measurements of X_{max}.

compositions containing nuclei heavier than helium $A > 4$. The results are robust with respect to modifications of the key parameters of the interaction models. In [19] the X_{max}distributions measured at the Pierre Auger Observatory were interpreted in terms of fractions of protons, helium, nitrogen and iron. For QGSJetII-04and Sibyll 2.1the composition was found to be close to $0.5\,\mathrm{p} - 0.5\,\mathrm{He}$ ($\sigma(\ln A) \approx 0.69$) for energies $\lg(E/\mathrm{eV}) = 18.5{-}19.0$. The correlation coefficient for $0.5\,\mathrm{p} - 0.5\,\mathrm{He}$ mix for these models is $r_G \approx 0.08$ and is in contradiction with the value found in data, indicating thus that QGSJetII-04and Sibyll 2.1do not describe X_{max}distributions and $(X_{max}, S(1000))$ correlation in a self-consistent way. For EPOS-LHCthe fits of X_{max}distributions produce a mixed composition including protons, helium and nitrogen with $\sigma(\ln A) \approx 1.2$ and $r_G \approx -0.094$ in a better agreement with the correlation analysis.

The ERs for the azimuthal asymmetry in the risetime and for the MPD differ from the ERs for constant compositions and are consistent with the increase of $\langle \ln A \rangle$ for the energies $E > 20\,\mathrm{EeV}$. The risetime asymmetry analysis reveals further possible deficiencies in QGSJetII-04leading to an unphysical dependence of the mass composition on the range of the WCDs distances to the shower axis. In its turn, the absolute value of $\langle X^{\mu}_{\mathrm{max}} \rangle$ is beyond the predictions for iron for EPOS-LHC, corresponding thus to masses $\langle \ln A \rangle > 4$.

Similarly to other analyses, the observed gain of the muon scale with energy in inclined showers $d\langle R_{\mu} \rangle / d\,\mathrm{lg}(E/\mathrm{eV})$ differs from predictions for constant compositions and indicates an increase of the primary mass for $E > 4\,\mathrm{EeV}$. Muon content in simulations, corresponding to the primary mass derived from analysis of X_{max} distributions, is found to be (30−80)% smaller than in data, but the difference with EPOS-LHCis only at the level of $1.4\sigma_{\mathrm{sys}}$. Similar conclusions for the vertical events were obtained in [18].

References

[1] A. Aab et al. (Pierre Auger), Nucl. Instrum. Meth. **A798**, 172 (2015), 1502.01323
[2] T. Pierog, I. Karpenko, J.M. Katzy, E. Yatsenko, K. Werner, Phys. Rev. **C92**, 034906 (2015), 1306.0121
[3] S. Ostapchenko, Phys. Rev. **D83**, 014018 (2011), 1010.1869
[4] E.J. Ahn, R. Engel, T.K. Gaisser, P. Lipari, T. Stanev, Phys. Rev. **D80**, 094003 (2009), 0906.4113
[5] A. Porcelli (Pierre Auger), PoS **ICRC2015**, 420 (2016), 1509.03732
[6] A. Aab et al. (Pierre Auger), Phys. Rev. **D90**, 122005 (2014), 1409.4809
[7] J. Matthews, Astropart. Phys. **22**, 387 (2005)
[8] K.H. Kampert, M. Unger, Astropart. Phys. **35**, 660 (2012), 1201.0018
[9] P. Abreu et al. (Pierre Auger), JCAP **1302**, 026 (2013), 1301.6637
[10] A. Aab et al. (Pierre Auger), Phys. Lett. **B762**, 288 (2016), 1609.08567
[11] P. Younk, M. Risse, Astropart. Phys. **35**, 807 (2012), 1203.3732
[12] J. Alvarez-Muñiz, R. Engel, T.K. Gaisser et al., Phys. Rev. **D66**, 033011 (2002), astro-ph/0205302
[13] R. Gideon, R. Hollister, JASA **82**, 656 (1987)
[14] A. Aab et al. (Pierre Auger), Phys. Rev. **D93**, 072006 (2016), 1604.00978
[15] A. Aab et al. (Pierre Auger), Phys. Rev. **D90**, 012012 (2014), [Erratum: Phys. Rev. D92, no.1, 019903(2015)], 1407.5919
[16] T. Pierog, B. Guiot, K. Werner, PoS **ICRC2015**, 337 (2016)
[17] A. Aab et al. (Pierre Auger), Phys. Rev. **D91**, 032003 (2015), [Erratum: Phys. Rev. D91, no. 5, 059901(2015)], 1408.1421
[18] A. Aab et al. (Pierre Auger), Phys. Rev. Lett. **117**, 192001 (2016), 1610.08509
[19] A. Aab et al. (Pierre Auger), Phys. Rev. **D90**, 122006 (2014), 1409.5083

EPJ Web of Conferences 145, 05003 (2017) DOI: 10.1051/epjconf/201714505003
ISVHECRI 2016

The ARGO-YBJ legacy to next generation wide field-of-view experiments

Giuseppe Di Sciascio [a]

INFN - Rome Tor Vergata, Italy

Abstract. The ARGO-YBJ experiment has been in stable data taking for more than 5 years at the YangBaJing Cosmic Ray Observatory (Tibet, P.R. China, 4300 m a.s.l., 606 g/cm^2). With a duty-cycle greater than 86% the detector collected about 5×10^{11} events in a wide energy range, from few hundreds GeV up to about 10 PeV. High altitude location and detector features make ARGO-YBJ capable of investigating a wide range of important issues in Cosmic Ray and Astroparticle Physics by imaging the front of atmospheric showers with unprecedented resolution and detail. In this contribution some of the latest physics results obtained by ARGO-YBJ in gamma-ray astronomy and in cosmic ray physics are summarized. The prospects of TeV gamma-ray observations with new ground-based wide field-of-view detectors are presented.

1. Introduction

The riddle of the origin of Cosmic Rays (hereafter CR) is unsolved since more than one century. The identification of the galactic sources able to accelerate particles beyond PeV ($=10^{15}$ eV) energies, the so-called *'PeVatrons'*, is certainly one of the main open problems of high energy astrophysics. In fact, even there is no doubt that galactic CR are accelerated in SuperNova Remnants (SNRs), the capability of SNRs to accelerate CRs up to the *'knee'* of the all-particle energy spectrum ($\sim3\times10^{15}$ eV) and above is still under debate. The determination of the maximum energy at which protons are accelerated inside their sources (the *'proton knee'*), as well as the measurement of the evolution of the heavier component across the knee, are the key for understanding acceleration mechanisms and the propagation processes in the Galaxy, and to investigate the transition from Galactic to extra-galactic CRs.

The study of CRs is based on two complementary approaches:

1. Measurement of energy spectrum, elemental composition and anisotropy in the CR arrival direction distribution, the three basic observables crucial for understanding origin, acceleration and propagation of the radiation.
2. Search of their sources through the observation of neutral radiation (photons and neutrinos), which points back to the emitting sources not being affected by the magnetic fields.

The integrated study of charged CRs and of gamma rays/neutrinos (the *'cosmic ray connection'*), which should trace high energy hadronic interactions mainly nearby the still unidentified acceleration sites, is one of the most important (and exciting) fields in the so-called *'multi-messenger astronomy'*.

In the PeV energy region an accurate measurement of the CR primary spectrum can be carried out only by ground-based Extensive Air Showers (hereafter EAS) arrays, detectors capable of simultaneously and continuously viewing all the overhead sky due to an effective field-of-view of about 2 sr and a duty-cycle of $\sim$100%. These two characteristics make them the key instruments for studying the highest energy gamma rays, extended sources and transient phenomena. The all-sky survey provides un *unbiased* map of the sky useful to enable the detection of unexpected sources, to provide testing ground for new theoretical ideas and to provide targets for in-depth observations by satellites or Cherenkov telescopes (IACTs).

The ARGO-YBJ experiment has been proposed to investigate the *'cosmic ray connection'* through a combined study of photon- and charged particle-induced EAS with the same detector. The combination of the high elevation of the site and the high granularity of the readout provides very important advantages in various aspects of CR physics. In fact, in addition to the decrease of the threshold energy down to few hundreds GeV, the location of the experiment above 4000 m a.s.l. ensures that shower fluctuations are small (we are working in the shower maximum region) and that all nuclei produce showers with nearly the same electromagnetic size. The low energy threshold is crucial to evaluate the absolute energy scale of the detector exploiting the Moon shadow technique and for the overposition with direct measurements carried out by balloon/satellite-born detectors allowing the cross-calibration between different energy scales. The independence of the size on the mass of the primary particle determines the size – energy relation to be better defined, allowing the study of the elemental composition around the knee in a very efficient and reliable way.

In this paper the latest results obtained by ARGO-YBJ in measuring the p+He energy spectrum, the CR anisotropy and the diffuse gamma-ray emission from the Galactic

[a] e-mail: disciascio@roma2.infn.it

Plane are summarized. We note that these topics can be studied only by EAS-arrays, and not by IACTs, due to their capability in studying the extended gamma emissions and in selecting showers induced by different nuclei. We also introduce two new projects for future wide field-of-view experiments in the Northern and in the Southern hemispheres.

2. The ARGO-YBJ experiment

ARGO-YBJ was a full coverage air shower detector constituted by a central carpet $\sim 74\times 78\,m^2$, made of a single layer of Resistive Plate Chambers (RPCs) with ~93% of active area, enclosed by a guard ring partially instrumented (~20%) up to $\sim 100\times 110\,m^2$. The apparatus has a modular structure, the basic data acquisition element being a cluster ($5.7\times 7.6\,m^2$), made of 12 RPCs ($2.85\times 1.23\,m^2$ each). Each chamber is read by 80 external strips of $6.75\times 61.80\,cm^2$ (the spatial pixels), logically organized in 10 independent pads of $55.6\times 61.8\,cm^2$ which represent the time pixels of the detector [1]. The readout of 18,360 pads and 146,880 strips is the experimental output of the detector. In addition, in order to extend the dynamical range up to PeV energies, each chamber is equipped with two large size pads ($139\times 123\,cm^2$) to collect the total charge developed by the particles hitting the detector [2]. The RPCs are operated in streamer mode by using a gas mixture (Ar 15%, Isobutane 10%, TetraFluoroEthane 75%) for high altitude operation [3]. The high voltage settled at 7.2 kV ensures an overall efficiency of about 96% [4]. The central carpet contains 130 clusters and the full detector is composed of 153 clusters for a total active surface of $\sim 6{,}700\,m^2$. The total instrumented area is $\sim 11{,}000\,m^2$. For each event the location and timing of every detected particle is recorded, allowing the reconstruction of the lateral distribution and the arrival direction. The trigger is based on a time correlation among the pad signals depending on their relative distance. In this way, all the shower events giving a number of fired pads $N_{pad} \geq N_{trig}$ in the central carpet in a time window of 420 ns generate the trigger. The whole system has been in stable data taking from November 2007 to January 2013, with the trigger condition $N_{trig} = 20$ and a duty cycle ≥86%. The trigger rate is ~3.5 kHz with a dead time of 4%.

Because of the small pixel size, the detector is able to record events with a particle density exceeding 0.003 particles m^{-2}, keeping good linearity up to a core density of about 15 particles m^{-2}. This high granularity allows a complete and detailed three-dimensional reconstruction of the front of air showers at an energy threshold of a few hundreds GeV. Showers induced by high energy primaries (> 100 TeV) are also imaged by the charge readout of the large size pads which allows to study the structure of the shower core region up to particle densities of $\sim 10^4/m^2$ [2].

Details on the analysis procedure (e.g., reconstruction algorithms, data selection, background evaluation, systematic errors) are discussed in [5–7]. The performance of the detector (angular resolution, pointing accuracy, energy scale calibration) and the operation stability are continuously monitored by observing the Moon shadow, i.e., the deficit of CRs detected in its direction [6,8]. The measured angular resolution is better than 0.5° for CR-induced showers with energy E > 5 TeV and the overall absolute pointing accuracy is ~0.1°. According to MonteCarlo simulations the angular resolution for γ-induced events results smaller by 30%–40%. The absolute rigidity scale uncertainty of ARGO-YBJ is estimated at 10% level in the range 1–30 TeV/Z [6,8].

2.1. Measurement of the Cosmic Ray light component (p+He) energy spectrum

A measurement of the CR primary p+He energy spectrum has been carried out with ARGO-YBJ from few TeV up to about 5 PeV exploiting different approaches:

- *'Digital-Bayes' analysis*, based on the strip multiplicity, i.e. the picture of the EAS provided by the strip/pad system, in the few TeV–300 TeV energy range. The selection of light elements (i.e. p+He) is based on the particle lateral distribution. The energy is reconstructed, on a statistical basis, by using a Bayesian approach [9, 10].
- *'Analog-Bayes' analysis*, based on the RPC charge readout [2], covers the 30 TeV–5 PeV energy range. The energy is reconstructed (as in the previous analysis), on a statistical basis, by using a Bayesian approach.
- *'Hybrid measurement'*, carried out by ARGO-YBJ and a wide field-of-view Cherenkov telescope (WFCTA), in the 100 TeV–3 PeV region. The selection of (p+He)-originated showers is based on the shape of the Cherenkov image and on the particle density in the core region measured by ARGO-YBJ [11,12].

We note that with ARGO-YBJ the selection of (p+He)-originated showers is performed not by means of an unfolding procedure after the measurement of electronic and muonic sizes, but on an event-by-event basis exploiting showers topology, i.e. the lateral distribution of charged secondary particles, in the shower core region. This approach is made possible by the full coverage of the central carpet, the high segmentation of the read-out and the high altitude location of the experiment that retains the characteristics of showers lateral distribution in the core region.

The energy spectrum of the p+He component measured by ARGO-YBJ with different analyses is shown in the Fig. 1. The p+He measured by CREAM [13] is also shown for comparison. The average spectrum can be described by a single power-law with a spectral index of -2.63 ± 0.06 up to about 600 TeV. The absolute flux at 400 TeV is $(1.79\pm 0.16)\times 10^{-11}\,GeV^{-1}\,m^{-2}\,sr^{-1}\,s^{-1}$. This result is consistent for what concerns the spectral index and absolute flux with the measurements carried out by CREAM. The flux difference is about 10% and can be explained with a difference in the experiments energy scale less than ±4% [11]. This result is very important to fix the energy scale of the experiment. Below 10 TeV the absolute energy scale of ARGO-YBJ is calibrated at 10% level exploting the displacement of the Moon shadow under the effect of the Geomagnetic Field. Above this energy the overposition with CREAM allows to check the energy scale at a few percent level up to the 100 TeV range.

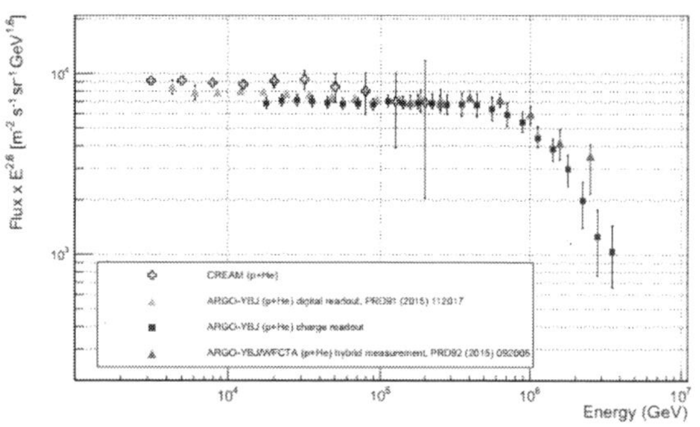

Figure 1. Light (p+He) component energy spectrum of primary CRs measured by ARGO-YBJ. The (p+He) energy spectrum measured by CREAM [13] is shown for comparison. The systematic uncertainty is shown by the error bars.

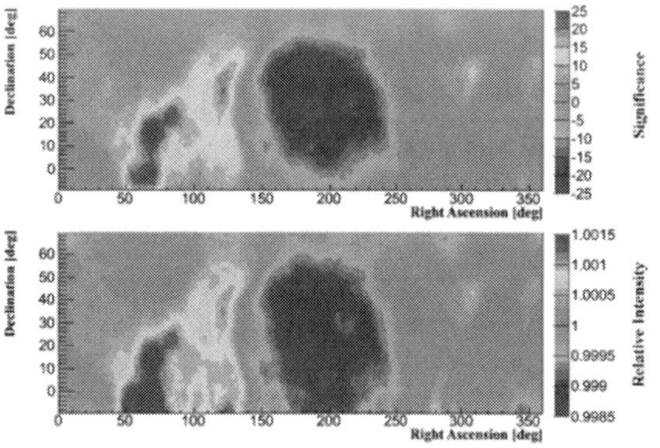

Figure 2. Large scale anisotropy observed by ARGO-YBJ [17]. The upper panel shows the significance map in standard deviations (s.d.), the lower panel gives the CR relative intensity. The sky maps are in the equatorial coordinate system. The corresponding proton median energy is about 1 TeV.

For higher energies the different analyses show clear evidence of a knee-like structure in the p+He spectrum starting from about 700 TeV. The energy spectra measured by ARGO-YBJ/WFCTA can be fitted with a broken power-law function with $E_k = 700 \pm 230$ TeV, $J(E_k) = (4.65 \pm 0.27) \times 10^{-12}$ GeV^{-1} m^{-2} s^{-1} sr^{-1}, $\beta_1 = -2.56 \pm 0.05$ and $\beta_2 = -3.24 \pm 0.36$ above the knee [12].

ARGO-YBJ confirms the observation of the knee in the all-particle energy spectrum at few PeV, but the clear observation of a knee in the p+He energy spectrum below the PeV suggests that the elemental composition in the few PeV range is heavier than Helium, at variance with the current interpretation of the electron and muon KASCADE data, which attributes the knee of the all-particle spectrum to the proton component [14–16].

These results demonstrate the possibility of exploring the CR properties in a wide energy range with a single ground-based detector without exploiting the measurement of the muon size, thus reducing the uncertainties due to hadronic interaction models at a few percent level.

2.2. Cosmic ray anisotropy

The *"Large Scale Anisotropy* (LSA) observed by ARGO-YBJ at about 1 TeV in 2008 and 2009, during the latest minimum of the solar activity, is shown in Fig. 2 [17].

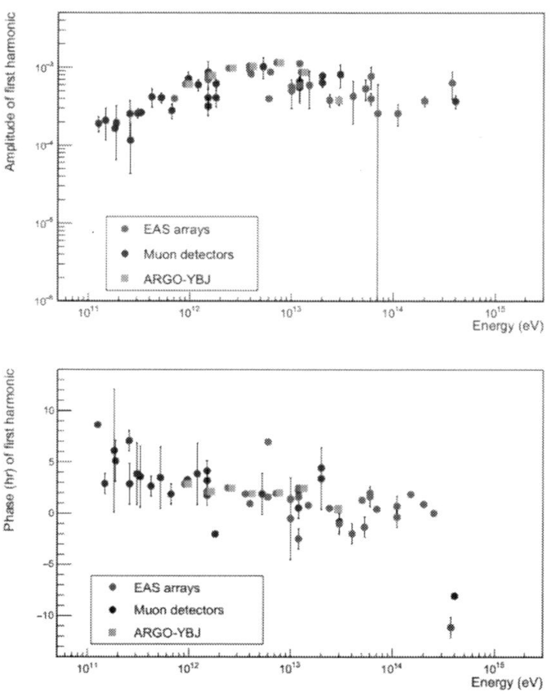

Figure 3. Amplitude and phase of the first harmonic (upper and lower plots, respectively) measured by ARGO-YBJ [17] compared with a compilation of data obtained by different experiments (muon telescopes or EAS-arrays) as a function of the CR primary energy (for details and references see [18]).

The existence of two distinct broad anisotropy regions in sidereal time, one showing an excess of CRs (called *"tail-in"*), distributed around 40° to 90° in Right Ascension (R.A.), the other a deficit (the *"loss cone"*), distributed around 150° to 240° in R.A., has been clearly observed by ARGO-YBJ. The center of the "tail-in" component is close to the direction of the heliospheric tail, which is opposite to the proper motion direction of the solar system. The center of the "loss cone" deficit component points to the direction of the north Galactic pole. These observations rule out the hypothesis that a Compton-Getting effect due to the motion of the heliosphere with respect to the local insterstellar medium (expected as a dipole with a maximum in the direction of the Galactic Center decl.$\simeq$49°, R.A.$\simeq$315° and a larger amplitude 3.5×10^{-3}) is a major source of the anisotropy [18, 19].

In Fig. 3 the amplitude and phase of the first harmonic (upper and lower plots, respectively) measured by different experiments are shown as a function of the primary CR energy. As can be seen from the plots

a) The amplitude of the CR anisotropy is extremely small (10^{-4}–10^{-3}).
b) A slow increase of the amplitude to a maximum at a few TeV is observed. After the maximum the anisotropy decreases to a minimum at $\sim$100 TeV. Evidence for a new increase for higher energies appears from data.
c) The phase of the first harmonic is nearly constant (slowly decreasing) around 0 hrs. A dramatic change of phase is observed around $\sim$100 TeV, suggesting a

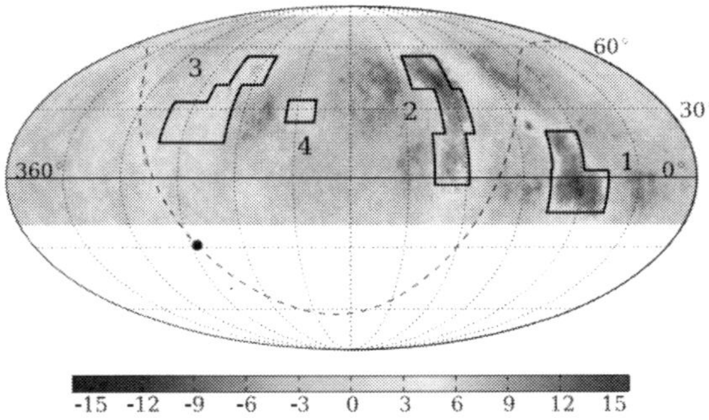

Figure 4. ARGO-YBJ sky-map in equatorial coordinates [20]. The color scale shows the statistical significance of the observation in s.d.. The dashed line represents the Galactic Plane and the black point the Galactic Center.

dipole opposite to the initial one. This observation clearly rules out that the Compton-Getting effect is a major source of the anisotropy.

d) The results obtained by ARGO-YBJ (green squares) are in fair agreements with other experiments.

With ARGO-YBJ we collected such large statistics to allow the investigation of anisotropic structures on smaller angular scale than the ones corresponding to the dipole and the quadrupole (*"Medium/Small Scale Anisotropy"*, MSA), showing that the CR intensity has quite a complicated structure unaccountable simply by kinetic models. Figure 4 shows the ARGO-YBJ sky map in equatorial coordinates as obtained with about $3.7 \cdot 10^{11}$ events reconstructed with a zenith angle $\leq 50°$ (selecting the declination region $\delta \sim -20° \div 80°$) [20]. According to the simulation, the median energy of the isotropic CR proton flux is $E_p^{50} \approx 1.8$ TeV (mode energy ≈ 0.7 TeV). The boxes represent the parametrization of the 4 regions of interest selecting the part of signal more than 3 s.d. [20].

The most evident features are observed by ARGO-YBJ around the positions $\alpha \sim 120°$, $\delta \sim 40°$ and $\alpha \sim 60°$, $\delta \sim -5°$. These regions are observed with a statistical significance of about 15 s.d. . On the left side of the sky map, several new extended features are visible, though less intense than the ones aforementioned. The area $195° \leq R.A. \leq 290°$ seems to be full of a few-degree excesses not compatible with random fluctuations (the statistical significance is up to 7 s.d.). We note that the region 4 is located in the "loss cone" of the LSA, near the North Galactic pole. The observation of regions 3 and 4 is reported by ARGO-YBJ for the first time. We note that the regions over which ARGO-YBJ observes significant MSA have total extension $\sim$0.8 sr, i.e one third of the ARGO-YBJ field of view in celestial coordinates.

2.3. Diffuse gamma-ray emission from the galactic plane

Diffuse gamma rays are produced by relativistic electrons by bremsstrahlung or inverse Compton scattering on background radiation fields, or by protons and nuclei via the decay of π^0 produced in hadronic interactions with interstellar gas. Thus, the space distribution of this emission can trace the location of the CR sources and the distribution of interstellar gas. On the other hand, the diffuse emission provides a background in the search for point sources.

In the GeV energy range, the EGRET data showed a significant excess compared to expectations (the 'EGRET GeV excess') [22] likely due to instrumental effects [23] because not confirmed by the more recent and accurate Fermi-LAT data [24,25]. A diffuse gamma ray flux at TeV energies from the $30° < l < 110°$ longitudinal range of the Galactic plane has been reported by Milagro [26–28].

The measured flux, once connected to the EGRET data by a power law with differential spectral index of –2.61, reveals a *"TeV excess"* in the diffuse gamma ray spectrum, being this flux 5–10 times higher than expectations [29]. However, as will be discussed in the following, the Milagro result does not take into account the contributions from the Cygnus Cocoon and from the overlapping point or extended sources TeV J2032+4130, VER J2019+407 and VER J2016+372 [30].

The events collected by ARGO-YBJ have been analysed to determine the diffuse gamma ray emission in the Galactic plane at Galactic longitudes $25° < l < 100°$ and Galactic latitudes $|b| < 5°$ [21]. This analysis was carried out in the energy range from $\sim$350 GeV to $\sim$2 TeV, connecting the region explored by Fermi/LAT with that investigated by Milagro. In particular, the analysis was focused on two selected regions of the Galactic plane, i.e., $40° < l < 100°$ and $65° < l < 85°$ (the Cygnus region), where Milagro observed an excess with respect to that predicted by current models.

In the Galaxy region $40° < l < 100°$, $|b| < 5°$, after masking the discrete sources and subtracting the residual contribution, an excess with a statistical significance of 6.1 s.d. above the background is found. The spectral analysis provides the three fluxes shown in Fig. 5, upper plot, at median energies 350 GeV, 680 GeV and 1.47 TeV (with uncertainties of about 30%). The fit to the ARGO-YBJ data with a power law gives a spectral index -2.90 ± 0.31, and the corresponding flux at 1 TeV is compatible with the extrapolation of the Fermi/LAT template [31].

On the other hand, as mentioned, the first measurement of the diffuse TeV (integral) flux from the Galactic plane made by Milagro [32] revealed a "TeV excess" in the diffuse gamma ray spectrum with respect to expectations [33]. This Milagro measurement, converted into differential flux, is only 34% greater than the value expected from the extrapolation of the Fermi/LAT template, and within the experimental uncertainties (see triangle with error bars in Fig. 5, upper plot). Moreover, considering that the Milagro result does not take into account the contributions from the Cygnus Cocoon (not yet discovered at the time of the measurement) [30] and from overlapping point and extended sources, the discrepancy with the Fermi/LAT predictions is almost cancelled out. Therefore, the full set of measurements with ground-based detectors is in agreement with direct observations by Fermi/LAT, and the evidence of any "TeV excess", requiring additional sources or particle production processes other than those producing Galactic cosmic rays, is ruled out.

In the Galactic region $65° < l < 85°$, $|b| < 5°$, after masking the discrete sources and the Cygnus Cocoon and subtracting the residual contribution, an excess of 4.1 s.d. is left. This direction points into our spiral arm at the remarkable Cygnus star-forming region, located at a

DOI: 10.1051/epjconf/201714505003

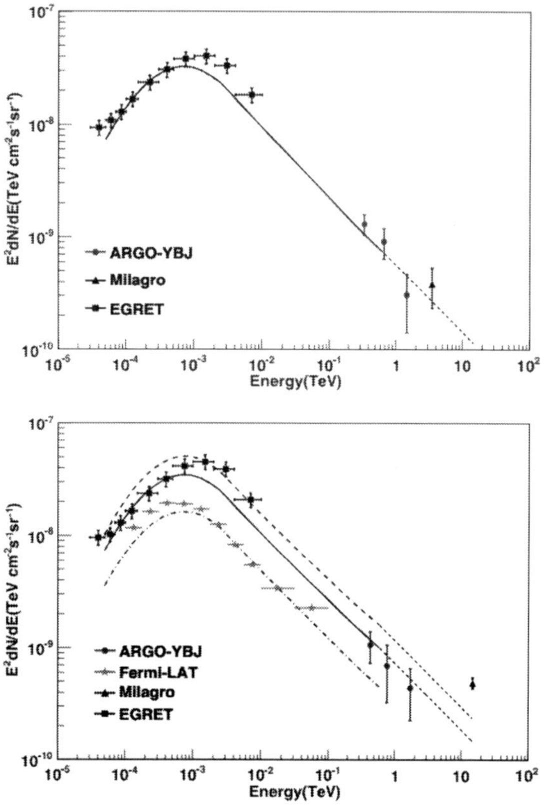

Figure 5. Upper plot: energy spectrum of the diffuse gamma ray emission measured in the Galactic region $40° < l < 100°$, $|b| < 5°$ [21]. Lower plot: energy spectra of the diffuse gamma ray emission measured by ARGO-YBJ and EGRET in the Galactic region $65° < l < 85°$, $|b| < 5°$, by Milagro in the region $65° < l < 85°$, $|b| < 2°$ and by Fermi/LAT in $72° < l < 88°$, $|b| < 15°$ [21]. The lines indicate the energy spectra expected from the Fermi/LAT template (with spectral index –2.6, which also rules its short-dashed extension) in the different sky regions investigated by the detectors [31].

distance of about 1.4 kpc. The spectral energy distribution of gamma ray emission is shown in Fig. 5, lower plot (filled stars) [21] together with the spectrum expected from the Fermi/LAT template (dot-dashed line) [31]. Milagro measured the diffuse gamma ray emission from the region $65° < l < 85°$, $|b| < 2°$ at a median energy of 15 TeV [28], obtaining the flux reported as a filled triangle in Fig. 5, lower plot. For comparison, the long-dashed line shows the expected energy spectrum for this region according to the Fermi/LAT template. The Milagro flux results about 75% higher than the Fermi template, suggesting the presence of an excess. The spectral analysis of ARGO-YBJ data provides the three fluxes shown in Fig. 5 lower plot at median energies 440 GeV, 780 GeV and 1.73 TeV (with uncertainties of about 40%). The fit to ARGO-YBJ data with a power law gives a spectral index -2.65 ± 0.44, and the corresponding flux at 1 TeV is about 10% lower than the extrapolation of the Fermi/LAT template. These data do not show any excess at energies around 1 TeV which corresponds to the excess found by Milagro at a median energy of 15 TeV. Again, this

Table 1. The spectral indices of gamma diffuse emission measured by ARGO-YBJ in different regions of the Galactic plane for $|b| < 5°$.

Locations	Statistical Significance	Spectral Index
$25° < l < 100°$	6.9 s.d.	-2.80 ± 0.26
$40° < l < 100°$	6.1 s.d.	-2.90 ± 0.31
$65° < l < 85°$	4.1 s.d.	-2.65 ± 0.44
$25° < l < 65°$ & $85° < l < 100°$	5.6 s.d.	-2.89 ± 0.33

discrepancy can be explained taking into account that the contribution of all the discrete gamma ray sources was not completely removed from the Milagro data. According to the ARGO-YBJ data, the 1 TeV flux associated to the Cygnus Cocoon is of the same order of the diffuse flux [30].

The spectral indices measured by ARGO-YBJ in different regions of the Galactic plane are reported in Table 1 [21]. The statistical significance of the observations in standard deviations (s.d.) is shown in the second column. The TeV flux averaged over the Cygnus region $65° < l < 85°$ cover the energy range 400 GeV –2 TeV and follow a power law with spectral index –2.65±0.44, a value very close to that found for TeV emission from the Cygnus Cocoon and for the p+He component measured by ARGO-YBJ, indicating the presence of young CRs accelerated by a nearby source [30].

3. What's Next?

3.1. The LHAASO experiment

A new project, developed starting from the experience of the high altitude experiment ARGO-YBJ, strategically built to study with unprecedented sensitivity the energy spectrum, the elemental composition and the anisotropy of CRs in the energy range between 10^{12} and 10^{17} eV, as well as to act simultaneously as a wide aperture (~2 sr), continuosly-operated gamma-ray telescope in the energy range between 10^{11} and 10^{15} eV is the LHAASO experiment [34,35]. The remarkable sensitivity of LHAASO in CR physics and gamma astronomy would play a key-role in the comprehensive general program to explore the *"High Energy Universe"*.

The first phase of LHAASO will consist of the following major components (see Fig. 6):

- 1 km^2 array (LHAASO-KM2A) for electromagnetic particle detectors (ED) divided into two parts: a central part including 4931 scintillator detectors 1 m^2 each in size (15 m spacing) to cover a circular area with a radius of 575 m and an outer guard-ring instrumented with 311 EDs (30 m spacing) up to a radius of 635 m.
- An overlapping 1 km^2 array of 1146 underground water Cherenkov tanks 36 m^2 each in size, with 30 m spacing, for muon detection (MD, total sensitive area ~42,000 m^2).
- A close-packed, surface water Cherenkov detector facility with a total area of about 78,000 m^2 (LHAASO-WCDA).
- 18 wide field-of-view air Cherenkov telescopes (LHAASO-WFCTA).

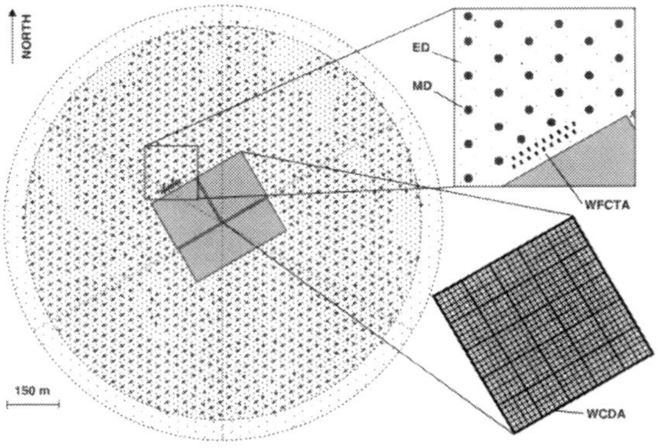

Figure 6. Layout of the LHAASO experiment. The insets show the details of one pond of the WCDA and of the KM2A array constituted by two overimposed arrays of electromagnetic particle detectors (ED) and of muon detectors (MD). The telescopes of the WFCTA, located at the edge of a pond, are also shown.

LHAASO is under installation at high altitude (4410 m asl, 600 g/cm^2, 29°21′31″ N, 100°08′15″ E) in the Daochen site, Sichuan province, P.R. China. The commissioning of one fourth of the detector will be implemented in 2018. The completion of the installation is expected by the end of 2021.

In Table 2 the characteristics of the LHAASO-KM2A array are compared with other experiments. As can be seen, LHAASO will operate with a coverage of $\sim$0.5% over a 1 km^2 area. The sensitive area of muon detectors is unprecedented and about 17 times larger than CASA-MIA, with a coverage of about 5% over 1 km^2.

LHAASO will enable studies in CR physics and gamma-ray astronomy that are unattainable with the current suite of instruments:

1) LHAASO will perform an *unbiased sky survey of the Northern sky* with a detection threshold of a few percent Crab units at sub-TeV/TeV energies and around 100 TeV in one year (Fig. 7). This sensitivity grants a high discovery potential of flat spectrum Geminga-like sources not observed at GeV energies. This unique detector will be capable of continuously surveying the γ-ray sky for steady and transient sources from about 100 GeV to 1 PeV.
From its location LHAASO will observe at TeV energies and with high sensitivity about 30 of the sources catalogued by Fermi-LAT at lower energy, monitoring the variability of 15 AGNs (mainly blazars) at least.
2) The sub-TeV/TeV LHAASO sensitivity will allow to observe AGN flares that are unobservable by other instruments, including the so-called TeV orphan flares.
3) LHAASO will study in detail the high energy tail of the spectra of most of the γ-ray sources observed at TeV energies, opening for the first time the 100–1000 TeV range to the direct observations of the high energy cosmic ray sources. *LHAASO's wide field-of-view provides a unique discovery potential.*
4) LHAASO will map the Galactic *diffuse gamma-ray emission* above few hundreds GeV and thereby measure the CR flux and spectrum throughout the Galaxy with high sensitivity. The measurement of the space distribution of diffuse γ-rays will allow to trace the location of the CR sources and the distribution of interstellar gas.
5) The high background rejection capability in the 10–100 TeV range will allow LHAASO to measure the *isotropic diffuse flux of ultrahigh energy γ radiation* expected from a variety of sources including Dark Matter and the interaction of 10^{20} eV CRs with the 2.7 K microwave background radiation. In addition, LHAASO will be able to achieve a limit below the level of the IceCube diffuse neutrino flux at 10–100 TeV, thus constraining the origin of the IceCube astrophysical neutrinos.
6) LHAASO will allow the reconstruction of the energy spectra of different CR mass groups in the 10^{12}–10^{17} eV with unprecedented statistics and resolution, thus tracing the light and heavy components through the knee of the all-particle spectrum.
7) LHAASO will allow the measurement, for the first time, of the CR anisotropy across the knee separately for light and heavy primary masses.
8) The different observables (electronic, muonic and Cherenkov components) that will be measured in LHAASO will allow a detailed investigation of the role of the hadronic interaction models, therefore investigating if the EAS development is correctly described by the current simulation codes.
9) LHAASO will look for signatures of WIMPs as candidate particles for DM with high sensitivity for particles masses above 10 TeV. Moreover, axion-like particle searches are planned, where conversion of gamma-rays to/from axion-like particles can create distinctive features in the spectra of gamma-ray sources and/or increase transparency of the universe by reducing the Extragalactic Background Light (EBL) absorption. Testing of Lorentz invariance violation as well as the search for Primordial Black Holes and Q–balls will also be part of the scientific programme of the experiment.

In the next decade CTA-North and LHAASO are expected to be the most sensitive instruments to study Gamma-Ray Astronomy in the Northern hemisphere from about 20 GeV up to PeV.

3.2. The LATTES project

To maximize the scientific return for Galactic sources, a future wide field-of-view γ-ray experiment should be located at sufficiently Southern latitude to continuously monitor the Galactic Center and the Inner Galaxy. To lower the energy threshold down to hundred GeV range such an instrument would require extreme altitude location (>5000 m asl), high secondary photons conversion capability and a full coverage approach (with a coverage factor >90%) with high segmentation of the readout.

The LATTES project will consist of a carpet of RPCs on top of Water Cherenkov Detectors (WCD) of small dimensions [36].

The basic element, shown in Fig. 8, is composed by one WCD, with a rectangular horizontal surface of 3 m × 1.5 m

 DOI: 10.1051/epjconf/201714505003

Table 2. Characteristics of different EAS-arrays.

Experiment	Altitude (m)	e.m. Sensitive Area (m^2)	Instrumented Area (m^2)	Coverage
LHAASO	4410	5.2×10^3	1.3×10^6	4×10^{-3}
TIBET ASγ	4300	380	3.7×10^4	10^{-2}
IceTop	2835	4.2×10^2	10^6	4×10^{-4}
ARGO-YBJ	4300	6700	11,000	0.93 (central carpet)
KASCADE	110	5×10^2	4×10^4	1.2×10^{-2}
KASCADE-Grande	110	370	5×10^5	7×10^{-4}
CASA-MIA	1450	1.6×10^3	2.3×10^5	7×10^{-3}
		μ Sensitive Area (m^2)	Instrumented Area (m^2)	Coverage
LHAASO	4410	4.2×10^4	10^6	4.4×10^{-2}
TIBET ASγ	4300	4.5×10^3	3.7×10^4	1.2×10^{-1}
KASCADE	110	6×10^2	4×10^4	1.5×10^{-2}
CASA-MIA	1450	2.5×10^3	2.3×10^5	1.1×10^{-2}

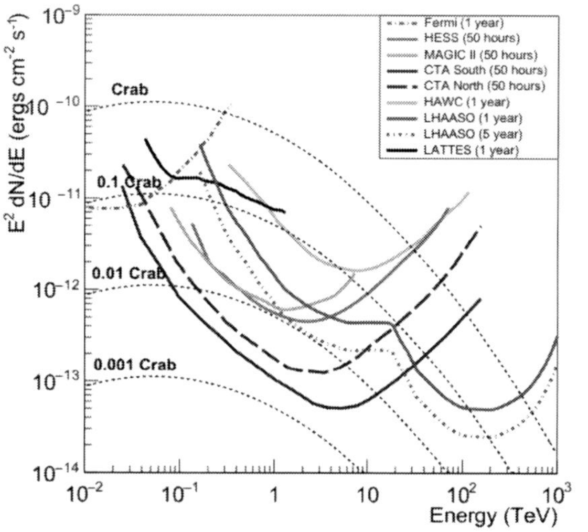

Figure 7. Differential sensitivity of LHAASO to a Crab-like point gamma-ray source compared to other experiments (multiplied by E^2). The Crab Nebula spectrum, extrapolated to 1 PeV, is reported as a reference together with the spectra corresponding to 10%, 1% and 0.1% of the Crab flux.

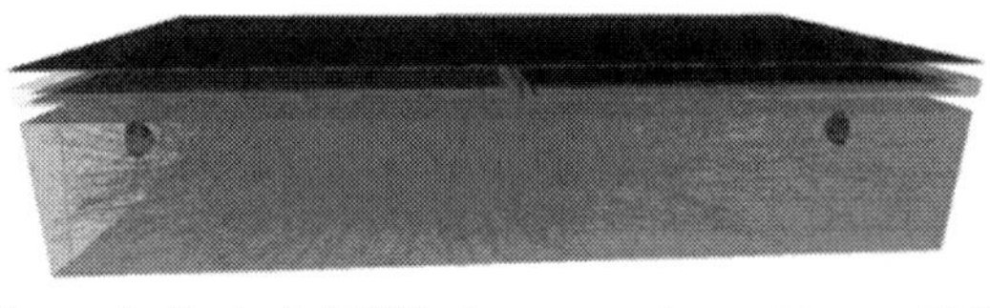

Figure 8. Basic LATTES detector station, with one WCD covered with RPCs and a thin slab of lead [36]. The green lines show the tracks of the Cherenkov photons inside the water.

and a depth of 0.5 m, covered by two RPCs, each with a surface of $(1.5 \times 1.5)\,m^2$, with a lead layer on top (5.6 mm) to exploit the secondary photons conversion. The WCD read-out will be provided by two 15 cm PMTs at both ends of the smallest vertical face of the tank. The read-out of the RPCs will be provided by 16 charge collecting pads. This hybrid detector is expected to improve the trigger selection at low energies and the rejection of the background of charged nuclei. The shower energy will be reconstructed from the total signal, defined as the sum of the number of photoelectrons in all WCD stations. The proposed experiment will consist of an array of 60×30 stations, covering an effective area of about $10{,}000\,m^2$ located at 5200 m asl. The different detectors will be separated by a small distance (roughly 0.5 m) to allow access to PMTs and RPCs. An angular resolution better than 2° is expected in the 100 GeV range.

The 1-year expected sensitivity, at level of 15% of the Crab Nebula flux in the 100–400 GeV energy range, is shown in Fig. 7.

4. Conclusions

The ARGO-YBJ detector exploiting the full coverage approach and the high segmentation of the readout imaged the front of atmospheric showers with unprecedented resolution and detail. The physics of Galactic CRs has been studied with a combined measurement of photon- and charged particle-induced showers.

With ARGO-YBJ we demonstrated that RPCs can be safely operated at extreme altitudes for many years. The benefits in the use of RPCs in ARGO-YBJ were:

1. high efficiency detection of low energy showers (energy threshold ∼300 GeV) by means of the dense sampling of the central carpet;
2. unprecedented wide energy range investigated by means of the digital/charge read-outs (∼300 GeV → 10 PeV);
3. good angular resolution and unprecedented details in the core region by means of the high granularity of the read-outs.

Open problems in cosmic ray physics push the construction of a new generation of EAS arrays to study, in the $10^{11}-10^{18}$ eV energy range, photon- and charged-induced events at the same time.

LHAASO is the most ambitious project for a new generation multi-component wide field-of-view experiment in the Northern hemisphere.

A new EAS array to study the 100 GeV γ-sky with high sensitivity and to monitor the Galactic Center should be a high priority. The LATTES project is a hybrid detector to exploit both the ARGO-YBJ full coverage approach with RPCs and the HAWC/LHAASO technique to reject the background of charged cosmic rays with a water Cherenkov detector. Such an experiment, if located in the Southern hemisphere at 5200 m asl with an instrumented area of 10,000 m^2, can reach a sensitivity of order of 10% of the Crab Nebula flux in one year in the 100 GeV energy range thus monitoring the Galactic Center and complementing the coming Cherenkov Telescope Array.

References

[1] G. Aielli et al., NIM **A562**, 92 (2006)
[2] B. Bartoli et al., Astropart. Phys. **67**, 47 (2015)
[3] C. Bacci et al., NIM **A443**, 342 (2000)
[4] G. Aielli et al., NIM **A608**, 246 (2009)
[5] G. Aielli et al., Astrophys. J. Lett. **714**, L208 (2010)
[6] B. Bartoli et al., Phys. Rev. D **84**, 022003 (2011)
[7] B. Bartoli et al., Astrophys. J. **734**, 110 (2011)
[8] B. Bartoli et al., Phys. Rev. D **85**, 022002 (2012)
[9] B. Bartoli et al., Phys. Rev. D **85**, 092005 (2012)
[10] B. Bartoli et al., Phys. Rev. D **91**, 112017 (2015)
[11] B. Bartoli et al., Chinese Phys. C **38**, 045001 (2014)
[12] B. Bartoli et al., Phys. Rev. D **92**, 092005 (2015)
[13] Y.S. Yoon et al., Astrophys. J. **728**, 122 (2011)
[14] T. Antoni et al., Astropart. Phys. **24**, 1 (2005)
[15] W.D. Apel et al., Astropart. Phys. **31**, 86 (2009)
[16] W.D. Apel et al., Astropart. Phys. **47**, 54 (2013)
[17] B. Bartoli et al., Astrophys. J. **809**, 90 (2015)
[18] G. Di Sciascio and R. Iuppa, *On the observation of the Cosmic Ray Anisotropy below 10^{15} eV*, in "Homage to the Discovery of Cosmic Rays, the Meson-Muons and Solar Cosmic Rays", Chapter 9, pagg. 221-257 (Nova Science Publishers, Inc., New York, 2013). Preprint: `arXiv:1407.2144`
[19] G. Di Sciascio, *ASTRA* Proc. **2**, 27 (2015)
[20] B. Bartoli et al., Phys. Rev. D **88**, 082001 (2013)
[21] B. Bartoli et al., Astrophys. J. **806**, 20 (2015)
[22] S.D. Hunter et al., Astrophys. J. **481**, 205 (1997)
[23] F.W. Stecker et al., Astropart. Phys. **29**, 25 (2008)
[24] A.A. Abdo et al., Phys. Rev. Lett. **103**, 251101 (2009)
[25] A.A. Abdo et al., Astrophys. J. **710**, 133 (2010)
[26] R. Atkins et al., Phys. Rev. Lett. **95**, 251103 (2005)
[27] A. Abdo et al., Astrophys. J. **658**, L33 (2007)
[28] A.A. Abdo et al., Astrophys. J. **688**, 1078 (2008)
[29] F. Aharonian et al., Rep. Prog. Phys. **71**, 096901 (2008)
[30] B. Bartoli et al., Astrophys. J. **790**, 152 (2014)
[31] M. Ackermann et al., Astrophys. J. Suppl. **209**, 34 (2013)
[32] R. Atkins et al. Phys. Rev. Lett. **95**, 251103 (2005)
[33] T. Prodanović, B. D. Fields & J. F. Beacom, Astropart. Phys. **27**, 10 (2007)
[34] G. Di Sciascio, in *CRIS 2015 proceedings*, Preprint `arXiv:1602.07600`
[35] Z. Cao et al., Chin. Phys. C **34**, 249 (2010)
[36] P. Assis et al., preprint `arXiv:1607.03051`

Recent highlights from the Telescope Array

J.N. Matthews [a]

University of Utah, High Energy Astrophysics Institute, Department of Physics and Astronomy, Salt Lake City, UT 84112, USA

Abstract. The Telescope Array studies the properties of ultra high energy cosmic rays by measuring the extensive air showers which they induce. We do this using a variety of techniques including an array of scintillator detectors to sample the footprint of the air shower when it reaches the Earth's surface and telescopes to measure the fluorescence and Čerenkov light of the air shower. From this we determine the energy spectrum and chemical composition of the primary particles. We also search for sources of cosmic rays and anisotropy. We have found evidence of a possible source of ultra high energy cosmic rays in the northern sky. The experiment and its recent measurements are discussed.

1. Introduction

The Telescope Array collaboration was forged by members of the High Resolution Fly's Eye (HiRes) and Akeno Giant Air Shower Array (AGASA) collaborations. Its purpose is to study ultra high energy cosmic rays, their spectrum, composition, and anisotropy/sources. Additional goals included understanding the differences in the observations of HiRes and AGASA as well as the transition from galactic to extra-galactic sources. The collaboration has, over time, grown to include members from the US, Japan, Russia, South Korea, and Belgium.

The Telescope Array is located in central Utah in the US at about 39.30° north latitude and 112.91° west longitude. The high energy component of the experiment consists of 38 telescopes, (9728 PMTs) at three telescope stations overlooking an array of 507 scintillator surface detectors (SD). The SD array has an area greater than $700\,\mathrm{km}^2$. The main body of the Telescope Array was complete and operational as of 2008. The layout of the TA experiment is shown in Fig. 1.

We use the atmosphere as a part of our calorimeter and observe the energy deposited by the primary cosmic ray and the extensive air shower it initiates. Telescopes view the longitudinal development of the air shower as it passes through the atmosphere. The three telescope stations are located on the perimeter of the SD array (on a $\sim$30 km triangle) and view the sky over it. The northern telescope station has 14 telescopes, each with a $5.2\,\mathrm{m}^2$ mirror. It re-utilizes telescopes from the High Resolution Fly's Eye, HiRes-I site. The southern two stations each have 12 telescopes, with $6.8\,\mathrm{m}^2$ mirrors. The new telescopes at these sites use FADC electronics, while the HiRes-I telescopes are using older sample-and-hold electronics.

The cameras in the telescopes utilize 256 PMTs for pixels, each viewing about 1° of sky. The PMTs are hexagonal and are arranged in a 16×16 hexagonal close pack (honey-comb) array. Each site views from $3-31°$ in elevation and has full azimuthal view over the SD array.

The telescopes are looking for an amount of light similar to a 50–100 W UV light bulb, but 20–40 km away and moving at the speed of light. Due to the extreme sensitivity of the telescopes and cameras, the telescopes are only operated on clear moonless nights. This results in a duty cycle of about 10%.

The scintillator surface detector (SD) array, samples the density of air shower particles when they reach the Earth's surface. The 507 detectors each have two layers of half-inch (1.25 cm) thick scintillator extruded with grooves in it. Wavelength shifting optical fibers are placed into the grooves. They gather the light, shift the wavelength, and guide it to two PMTs, one for each optically separated layer. Each layer samples $3\,\mathrm{m}^2$ in area.

The scintillator surface detectors are deployed on a 1.2 km square grid. They are powered by solar panels and batteries for 24/7 continuous operations. They are self-calibrated by using the muon background and are read out by 2.4 GHz radio. Three radio towers divide the detectors among them. Each second the towers poll all of the detectors in their part of the array for signals, then a computer at the tower scans this for signs of a shower footprint. If a shower is found, then the tower requests that all data from the time period be sent back and recorded.

Each telescope station operates separately and independently from the SD array. The data from each can be analyzed separately or in conjunction with the other parts. Single telescope only data is known as "mono", from multiple telescopes "stereo" and telescope with SD is "hybrid". Having more information is always better since it constrains the data and reduces the uncertainties. However, one does not always have that luxury – for example the SD data set is the largest since it operates 24/7 (nearly 100% duty cycle) while the telescopes have a duty cycle of about 10% due to sun light, moon light, and weather conditions. Hence the SD only data is used where one can tolerate the slightly larger uncertainties such as in the spectrum and anisotropy measurements. Hybrid and stereo have their own places, these are required to reduce uncertainties enough to make composition measurements.

[a] e-mail: `jnm@physics.utah.edu`

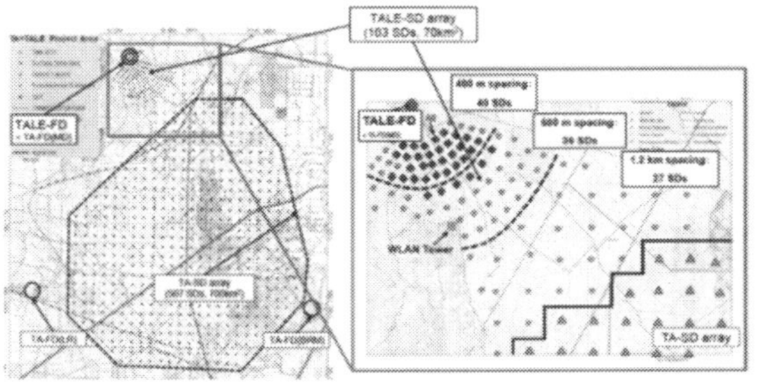

Figure 1. Layout of the Telescope Array detector site. The main array is shown on the left with a surface detector (SD) array of 507 scintillation counters surrounded by three fluorescence detector (FD) stations (MD: Middle Drum, BRM: Black Rock mesa, LR: Long Ridge) looking inward over the SD. The expanded inset on the right shows the layout of the TA Low-energy Extension (TALE). The TALE addition to Telescope Array consists of 10 new high-elevation (31°−59°) fluorescence telescopes located at the MD site, and an in-fill SD array of 103 counters arranged with spacings that grow with distance from the TALE FD, in order to make optimal coverage at energies below 10^{18} eV.

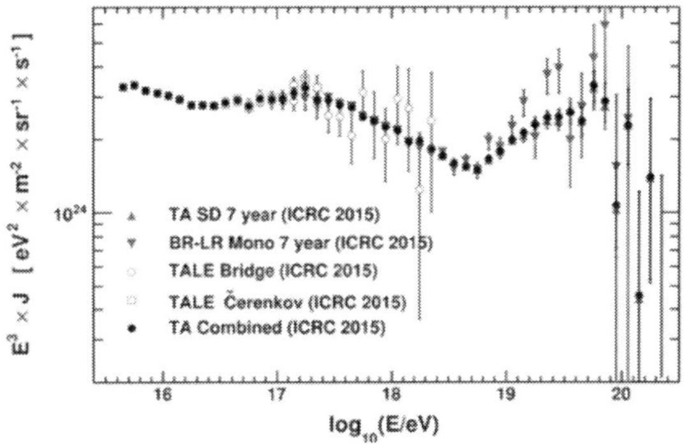

Figure 2. The combined spectrum from the Telescope Array Collaboration, including the first 7 years data from the surface detector spectrum (upright triangles), the 7 year monocular spectrum from the Black Rock and Long Ridge fluorescence detectors (inverted triangles), the TALE fluorescence spectrum (open circles), and the TALE Čerenkov spectrum (open squares). The averaged spectrum combining these four measurements is shown by the black circles.

2. Energy spectrum

The scintillator array has the greatest number of events. Telescope Array previously published an energy spectrum from the first four years of SD operation [1]. The energy spectrum of ultra high energy cosmic rays is now updated to include data from the first seven years of data and is shown in Fig. 2. The Telescope Array measurements are in excellent agreement with the previous measurements by the HiRes experiment. Using the first 7 years data set, the break point for the GZK cut-off [2,3] is found to be $10^{19.78\pm0.06}$ eV. Integrating an unbroken line from $10^{19.78}$ to 10^{21} eV, one would expect to find ∼99 events in this range, while we actually observe 44 events. Hence, the significance of the cut-off is $\sim 6\sigma$.

A "dip" or "ankle" structure is also seen at $10^{18.70\pm0.02}$ eV. The simplest explanation for the combination of spectral features is that of interaction between extragalactic cosmic ray protons and the cosmic microwave background radiation (CMBR) attenuating the flux via photo-pion production above the cut-off (via a Δ resonance), and by e^+e^- pair production above the ankle [4]. The Telescope Array spectrum shows an apparent break at just above 10^{17} eV consistent with the "Second Knee" feature [5]. Another dip feature is seen just above the 10^{16} eV, which has been previously reported by other experiments [6,7].

When the Auger spectrum is scaled up by 10% in energy (well within the energy uncertainties of both experiments, the Auger spectrum is also in good agreement with the Telescope Array and the HiRes spectra at lower energies. However, for $E > 10^{19.3}$ eV, the two spectra appear to diverge. The Auger spectrum has a significantly lower cut-off energy. Understanding this discrepancy is one area of joint study between the two groups.

3. Composition

We measure the composition with the Xmax technique. That is, we measure the depth of shower maximum in the atmosphere and we use Monte Carlo models to correlate this with chemical composition of the primary cosmic ray. Protons have a smaller cross-section and interact deeper than heavier nuclei. The distribution also has more width to it. However, there is considerable overlap to the distributions, so one mostly looks at mean Xmax behavior and width behavior. These shift as a function of energy and this is called elongation.

To measure composition, one requires hybrid or stereo data in order to constrain the geometry and thereby to make a well understood measurement of Xmax, the depth of shower maximum. One advantage of stereo data is that it also provides a redundant measurement of Xmax which allows a resolution verification measurement. With hybrid, one extends the timing information in the angle vs. time plots which one uses to measure the pointing direction of the incoming cosmic ray. In addition, one gets the core location of the shower footprint on the Earth's surface. These additional constraints tie down the geometry.

As of 2010, the two main measurements of composition of ultra high energy cosmic rays came from HiRes and Auger. The HiRes experiment found a composition which was very light, consistent with protons between $10^{18.2}$ eV and $10^{19.7}$ eV. The HiRes data is, in fact, in quite good agreement with the QGSjet-01 proton model. The Auger experiment found that the composition was light in the lower energy range, but shifted to a heavier and heavier composition for $E > \sim 10^{18}$ eV. The Auger surface detector is composed of thick water tanks. These have different sensitivities for electrons and muons. The Auger experiment has decided that the EPOS hadronic model does the best job of modeling its data. Unfortunately, there is a fairly wide dispersion between the various hadronic models and where they place the mean Xmax of protons, helium, nitrogen, and iron.

The HiRes measurement had been made using a stereo measurement from two telescope sites about 12km apart. The Auger measurement had been done using hybrid measurements. The Telescope Array was now in a position

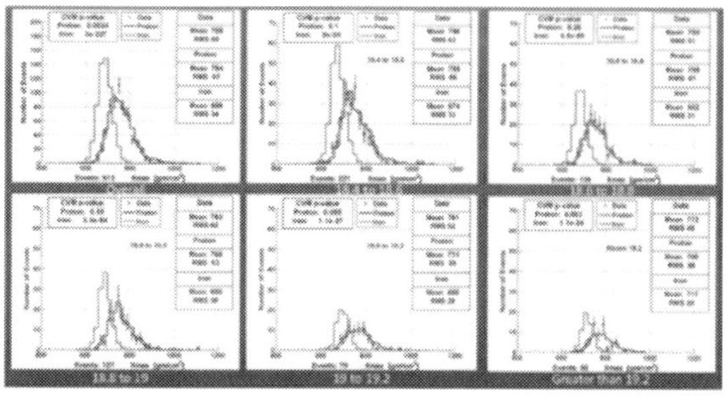

Figure 3. The comparison of the Telescope Array hybrid data with the QGSjet II.03 model. The top left histogram show the comparison of the Telescope Array data (black points) with the QGSjet II.03 model (blue for the proton model and red for the iron model) over all energies. The data is clearly more proton-like than iron-like. This is quantified by performing a Cramer von-Mises test (CvM) which yields a p-value of 0.0024 for protons, but a p-value of 10^{-227} for iron. The other five panels of the plot show the same plot, but broken up into energy bins. For all bins, including $E > 10^{19.2}$ eV, the data clearly agrees much more closely with the protons than the iron. Even in that sub-set of data the CvM p-value for comparison to the proton model is 0.05, while it is 10^{-24} for iron.

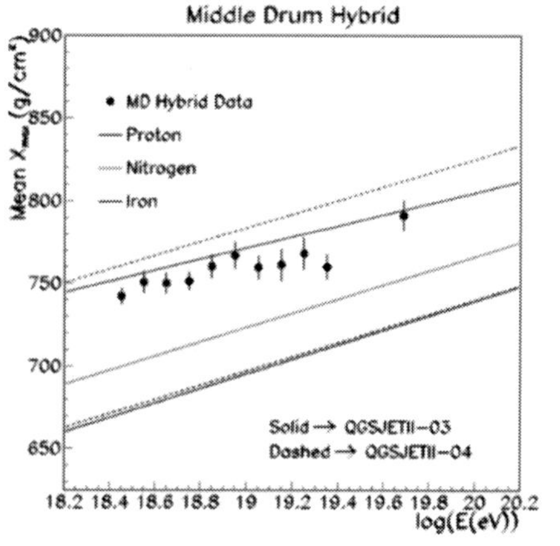

Figure 4. The Elongation Plot for the hybrid data from the Middle Drum site. The mean Xmax is plotted as a function of energy. The data is shown as the black points. The Monte Carlo results for two different hadronic models QGS-jet II.03 and QGSjet II.04 are shown for a) protons (red), b) nitrogen (green), and c) iron (blue). The data most closely resembles the QGSjet II.03 proton Monte Carlo.

to make the hybrid measurements to see if this contributed to any of the difference.

At the Telescope Array, we instituted a pattern recognition cut to verify that we were seeing both the rise and the fall of the shower development profile. With these cuts, the Xmax resolution is ~25 gm/cm^2 over all energies. Figure 3 shows the data-Monte Carlo comparisons for the Telescope Array hybrid Xmax data with the QGSjet-II.03 model for protons and iron. In all cases, especially when one divides the set up into energy bins, the data looks much more like protons than iron.

Plotting the mean as a function of energy results in what is called an elongation plot. The elongation plot for the hybrid data from the Middle Drum site is shown in Fig. 4. As in the data-Monte Carlo comparisons for the

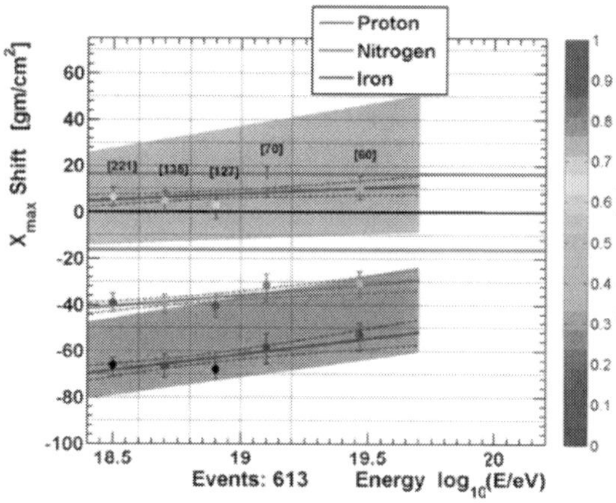

Figure 5. Shift plot showing how much one would have to shift Xmax in a given energy bin to make the mean Xmax of data align with the QGSjet II.03 Monte Carlo. Three examples are given for protons (red), iron (blue), and nitrogen (magenta). After each given shift, the goodness of fit is calulated and assigned an s-value. One can see that the proton Monte Carlo requires a small shift (well within uncertainties) and that the goodness of fit is also very good, though dropping a bit at the highest energies. The iron MC requires a shift of about 60 gm and even after that, the result has a very poor fit. The nitrogen MC does not fare much better than the iron MC. It requires a shift of about 40 gm and also has poor fit comparisons. The pink and blue bands show the spread due to using other hadronic models. Unfortunately, this spread is quite large.

energy slices, the data looks most like the QGSjet-II.03 proton Monte Carlo.

Another way to display this information is in what we call a shift-plot. We determine how much we would have to shift the Xmax distribution to get the best agreement between data and Monte Carlo. In Fig. 5 one can see that to get agreement with the proton Monte Carlo requires a shift of only a few grams, well within the measurement's uncertainty. However, a shift of more than 60 grams is required for the iron Monte Carlo. In addition, the points on the plot are colored by their s-value, or probability of agreement in the two shapes after the shifts in mean. One can see that over all energies the proton Monte Carlo has high s-values, though dropping a bit in the highest energy bin. However, over all energies, the iron Monte Carlo shows very low s-values. The data rejects iron. The same shift plot is shown for nitrogen which requires a shift of about 40 grams and also yields very low s-values over the energy range.

The two hybrid and the stereo measurements of Xmax are in good agreement with one another as well as indicating a predominately light composition. The interpretation of a light composition is fairly robust, as illustrated by the scales shown in Fig. 6. To leading order, the departure of the data points from the proton line is roughly proportional to ln(A) where A is the nuclear mass number. Even if the scale were set to a hypothetical model that has a proton rail that is 33% deeper, the data points shown in the plot would only reach that expected for helium in this scenario. However, protons and helium are very different astrophysically. The two have much different interaction lengths. Protons with energies approaching

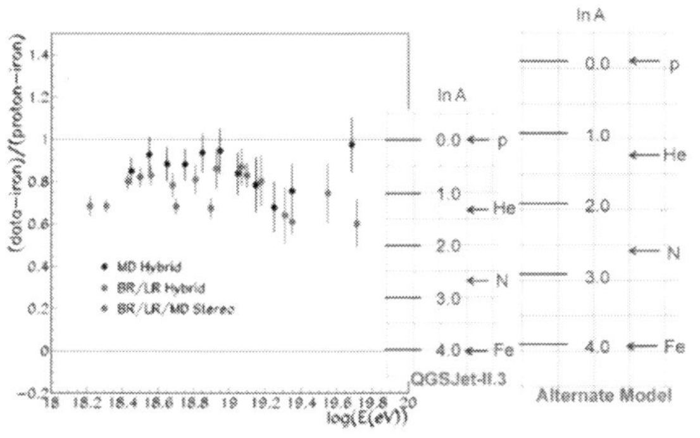

Figure 6. Summary of average X_{MAX} from Telescope Array fractional as the fractional position between the expectations of pure iron (lower lines at zero) and pure proton (upper line at 1.0) using the QGSjet-II.03 hadronic model. At right is the illustrative interpretation, in terms of lnA of the summary plot against QGSjet-II.03 and an alternative model where the proton rail is about 33% deeper in X_{MAX}.

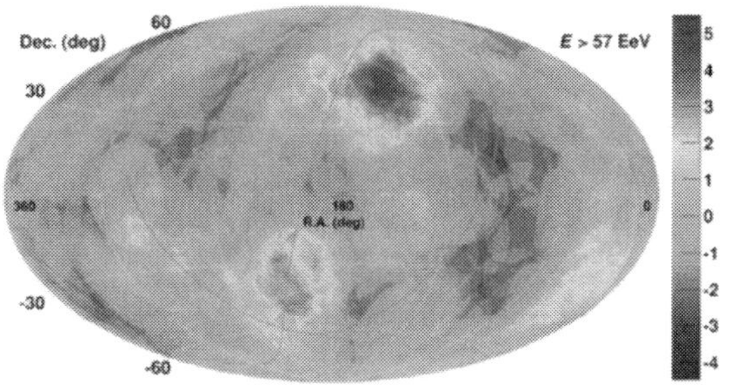

Figure 7. Full sky map combining the Telescope Array and Pierre Auger data events with $E > 5.7 \times 10^{19}$ eV. The events have oversampling with a 20° radius circle. The Telescope Array data set includes 109 events, representing the first 7 years of data collection. The Auger data set includes 157 events, representing 10 years of data. No correction was made for the energy scale difference between the Telescope Array and Pierre Auger data sets.

10^{20} eV have an interaction length of hundreds of Mpc, while helium with $E > 10^{19}$ eV has an interaction length which rapidly falls to the 10's of Mpc and lower. Simply, the helium can not reach us except from nearby sources.

4. Anisotropy

The Telescope Array collaboration has previously published that we have observed indications of a potential hotspot in the arrival direction of ultra high energy cosmic rays. [9] This publication used the data from the first five years of SD operations. Using the Auger criteria for their published AGN correlations, we observe 72 events with $E > 5.7 \times 10^{19}$ eV. These events are then over-sampled using 20° circles. We find that 19 of the 72 events overlap a region we call the hotspot which is centered at 146.7° in right ascension and 43.2° declination. Thus, 26% of events are in about 6% of the observable area. We expect a background of 4.5 isotropic events in this area. The Li-Ma significance of this is about 5.2σ. Taking into account that this spot could appear anywhere in the sky gives us a chance probability estimate of 3.4σ.

The hotspot is about 19° off of the super-galactic plane, near Ursa Major. The Ursa Major super-cluster is extended more than ±10° from the super-galactic plane. Therefore, we can not rule out some sort of relationship between the hotspot and this super-cluster.

When we update the measurement by adding two additional years of data, there are 37 new events in the data set for a total of 109 events with $E > 5.7 \times 10^{19}$ eV. There are 4 new events in the hotspot (3 in the 6th year and 1 in the 7th year). The expected background for this is 2.3 events. The global excess chance probability is 3.7×10^{-4}, which corresponds to 3.4σ, the same as for the first 5 years. A K-S test shows that the rate of arrival of events to the hotspot is consistent with the fluctuations expected from a Poisson distribution with a mean of 3.4 events/year. The rate of event arrivals is also inconsistent with chance excess from an isotropic distribution, a Poisson average of 0.9 events/year, at about the 2.6σ. One interesting thing to do is to make a full-sky map combining the Telescope Array and Pierre Auger events with $E > 5.7 \times 10^{19}$ eV. See Fig. 7. The plot is shown in Hammer-Aitoff projection, in Equatorial coordinates. No correction was made for the energy scale difference between the Telescope Array and Pierre Auger data sets. The thin gray line just above and left of the Telescope Array (upper) hotspot is the supergalactic plane. The pre-trial significance of this is 5.2σ (post-trial chance probability of $\sim$3.4σ). The Auger "warm spot" (lower) is located on the supergalactic plane at Centaurus-A and has a pretrial significance of $\sim$3.6σ. Neither the Telescope Array, nor Pierre Auger data shows any sign of excess in the direction of Virgo.

The Auger group reported correlation of their high energy, $E > 5.7 \times 10^{19}$ eV, events with the 472 AGNs from the Veron catalog with $Z < 0.018$. The Telescope Array group has looked periodically for this correlation, but have yet to find any significant correlation. The number of Telescope Array events correlating with AGNs is about 1σ greater than the background.

We have also looked for correlation with the Large Scale Structure (LSS). This is done by comparing to the matter distribution observed by the 2MASS Galaxy redshift catalog (XSCz). The sky was divided into five intensity regions, these were then compared by a Kolmagorov-Smirnov test comparing to the expected flux distribution. For events with energies, $E > 10^{18}$ eV and $E > 4 \times 10^{19}$ eV, the test can not distinguish between the Large Scale Structure and isotropy. However, for events $E > 5.7 \times 10^{19}$ eV, the KS test indicated that the data is marginally incompatible with an isotropic source distribution and it is compatible with the LSS simulation.

5. Future plans

The plan is to quadruple the size of the Telescope Array. We are adding 500 scintillator surface detectors to the North East and South East of the present Telescope Array site. See Fig. 8. The detectors will be on a grid a 2.08 km spacing and so will focus on the highest energy region. This part of the expansion has been funded in 2015 by the Japanese funding agencies. There will also be two new telescope stations which will view the sky over this expansion. This was recently funded (summer 2016) by the US NSF. The first 100 scintillator detectors already arrived

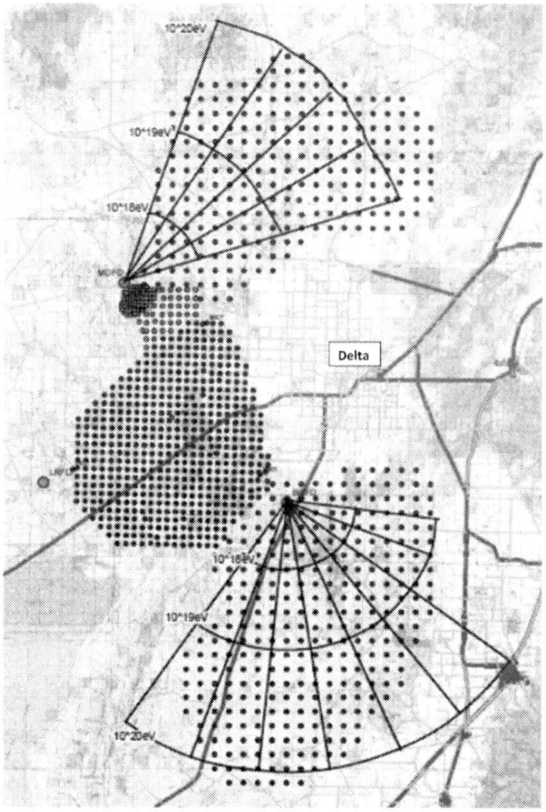

Figure 8. Map showing the Telescope Array with the high energy expansion, TA×4. The main Telescope Array with 1.2 km spacing is shown with the red points. The TALE, Telescope Array Low Energy Extension is shown as yellow points. This is a graded array with 1.2, 0.6 and 0.4 km spacing. The TA×4 expansion is shown to the NE and SE of the main Telescope Array with the dark points at 2.08 km spacing. The large blue points indicate the telescope stations. New telescopes will be added to the northern (Middle Drum) and south-eastern (Black Rock Mesa) telescope sites to view the sky over the TA×4 scintillator array expansion. The field of view of the 12 new telescopes is shown as the black fan with radii drawn to indicate the rough viewing distance for a vertical shower with energy 10^{18}, 10^{19}, and 10^{20} eV. The white space (with many red lines – roads) to the right of the main Telescope Array is the town of Delta.

in Delta, Utah and 70 more are currently being assembled in the Akeno Observatory of the University of Tokyo. The first set of scintillator detectors will be deployed early in 2017. With this enlarged Telescope Array detector, we will be able collect the equivalent of 19 current Telescope Array SD years of data by 2020. With the new data from the TA×4, we will be able to quickly verify the hotspot and to begin searching for sub-structure within the hotspot. If there are really two separated sources there, we would begin to be able to resolve this.

6. Summary

The Telescope Array has measured the energy spectrum, composition, and arrival direction of ultra high energy cosmic rays in the northern hemisphere. The low energy extension, TALE is coming on line and we are pushing the energy threshold of the experiment down below 10^{16} eV. The Telescope Array and TALE have already measured the energy spectrum between 6×10^{15} eV to more than 10^{20} eV with a single set of cross calibrated detectors. By looking at the composition and the energy spectrum in this range, we hope to identify the galactic to extragalactic transition. The surface array portion of TALE was recently funded and this will give us the capability to do hybrid composition in this energy range. The composition of cosmic rays is consistent with a light composition for $E > 10^{18.2}$ eV from both hybrid and stereo measurements. We have also reported on a hotspot near the direction of Ursa Major with a 3.4σ significance. We need more data to verify the source and to better to resolve it. More data is needed, but more data is coming with the expansion of TA×4.

The Telescope Array experiment is supported by the Japan Society for the Promotion of Science through Grants-in-Aids for Scientific Research on Specially Promoted Research (21000002) "Extreme Phenomena in the Universe Explored by Highest Energy Cosmic Rays" and for Scientific Research (19104006), and the Inter-University Research Program of the Institute for Cosmic Ray Research; by the U.S. National Science Foundation awards PHY-0307098, PHY-0601915, PHY-0649681, PHY-0703893, PHY-0758342, PHY-0848320, PHY-1069280, PHY-1069286, PHY-1404495 and PHY-1404502; by the National Research Foundation of Korea (2007-0093860, R32-10130, 2012R1A1A2008381, 2013004883); by the Russian Academy of Sciences, RFBR grants 11-02-01528a and 13-02-01311a (INR), IISN project No. 4.4502.13 and Belgian Science Policy under IUAP VII/37 (ULB). The foundations of Dr. Ezekiel R. and Edna Wattis Dumke, Willard L. Eccles and the George S. and Dolores Dore Eccles all helped with generous donations. The State of Utah supported the project through its Economic Development Board, and the University of Utah through the Office of the Vice President for Research. The experimental site became available through the cooperation of the Utah School and Institutional Trust Lands Administration (SITLA), U.S. Bureau of Land Management, and the U.S. Air Force. We also wish to thank the people and the officials of Millard County, Utah for their steadfast and warm support. We gratefully acknowledge the contributions from the technical staffs of our home institutions. An allocation of computer time from the Center for High Performance Computing at the University of Utah is gratefully acknowledged.

References

[1] T. Abu-Zayyad et al. (TA Collaboration), Astrophys. Journal Letters **768**, L1 (2013)
[2] K. Greisen, Phys. Rev. Lett. **16**, 222 (1966)
[3] G.T. Zatsepin and V.A. Kuz'min, JETPL **4**, 78 (1966)
[4] V. Berezinsky, A.Z. Gazizov, and S.I. Grigorieva, Physics Letters B **612**, 147 (2005)
[5] D.R. Bergman and J.W. Belz, J.Phys. **G34**, R359 (2007)
[6] W.D. Apel et al. (KASCADE-Grande Collaboration) Astropart. Phys. **36**, 183 (2012)
[7] M.G. Aartsen et al. (IceCube Collaboration) Phys. Rev. D **88**, 042004 (2013)
[8] R.U. Abbasi et al. (TA Collaboration), Astropart. Phys. **64**, 49 (2015)
[9] R.U. Abbasi et al. (TA Collaboration), Astrophysical Journal Lett. **790**, L21 (2014)
[10] T. Abu-Zayyad et al. (TA Collaboration), Astrophysical Journal **777**, 88 (2013)

High energy physics and very high energy astrophysics

Lawrence W. Jones [a]

University of Michigan, Ann Arbor, Michigan, USA

Abstract. The current research programs at the CERN LHC and at the U.S. Fermi National Accelerator Laboratory (FNAL) and other activities are briefly reviewed. The programs which are related to cosmic ray issues and which form a link between the cosmic ray community and the particle accelerator community are noted. The AMS-02 satellite program is also noted.

1. Introduction

What do we know and what are our problems? Recently, the Higgs Boson was discovered at CERN, with global recognition and enthusiasm. The "Standard Model" of elementary particle physics is now well established, and (at least by theorists) well understood. Gravity Waves have now been observed by LIGO, the Laser Interferometer Gravity Wave Observatory (actually, two detectors; one in the State of Washington, and the other in the State of Louisiana). The neutrino is now understood to have 3 "flavors" and 3 different masses and the neutrino oscillates between these states. Dark matter remains a puzzle. A dark matter substance is required to understand the orbits of stars in galaxies, and also even the bending of light rays; e.g. light from distant astronomical objects passing close to a galaxy. There is currently no clue concerning the nature of dark matter. A popular guess is that it is composed of a flux of massive and very weakly interacting elementary particles, but to date there has not been any observed evidence for such particles. The question of Dark Energy is even more puzzling. The galaxies in the universe appear to be flying apart – accelerating – as if some force were driving them, however we have no idea what this force could be. A more relevant question in this Symposium is the source of cosmic rays; where do they come from, and what accelerates them? To be sure, explosions, collisions, or other unusual phenomena at galactic centers are good candidates, but much has yet to be learned.

It turns out that there have been several relevant conferences and meetings recently in the U.S. and in Europe. At the beginning of August there was a meeting in Chicago; "ICHEP 2016", the 38th International Conference on High Energy Physics. There was also a meeting in Michigan recently; "Cosmo-16", with much discussion of topics like dark matter and dark energy. In June 2016, there was the annual meeting of the Fermilab (FNAL) Users Organization, and the lectures there were a good presentation of the current activities at Fermilab. A recent issue of the CERN periodical, the "CERN Courier", contains useful reviews of the research programs of the four major experimental programs at the LHC (Large Hadron Collider); CMS, ATLAS, LHCb, and ALICE.

[a] e-mail: `lwjones@umich.edu`

2. FERMILAB and CERN, current activities and projects

Since the Tevatron (the 2 TeV c.m. anti-proton, proton collider) was terminated (following some decades of excellent research), Fermilab has focused its activities on muon and neutrino physics. Indeed, physicists are still producing good physics from the extensive data collected at the Tevatron. Also, many Fermilab physicists are active in CERN LHC experiments, in particular CMS, which has a major analysis program and analysis facility at Fermilab. Members of the Fermilab staff are also active in cosmic ray programs, such as the Pierre Auger experiment in Argentina, as well as in searches for Dark Matter and Dark Energy.

Although managed and financed by the U.S. Department of Energy, Fermilab also has many physicists from other countries involved in its activities. One program which includes a significant Russian involvement is related to the SeaQuest experiment. This employs a proton beam of 120 GeV and uses the Drell-Yan process to probe the anti-quark structure of target nucleons. The asymmetry between anti-down quarks and anti-up quarks in the nucleon sea is measured in targets of liquid hydrogen and liquid deuterium, as well as other target materials.

One project with this facility is to produce a polarized proton beam to study spin effects. However, the generation of a polarized beam at high energies is a problem. A proton beam at low energy can be polarized easily, but in a multi-GeV circular accelerator, maintaining the polarization is a problem. When the protons circulate around an alternating gradient ring thousands of times, the polarization is lost.

In 1978, Ya.S. Derbenev and A.M. Kondratenko (then at Novosibirsk) conceived of a way to preserve the polarization. The solution is to rotate the spin of the protons 180 degrees twice per revolution. Note that, in a horizontal accelerator, the spins are aligned vertically, parallel to the dominant magnetic field. Derbenev argued that the rotation of the spins could be accomplished by a toroidal magnet, and with these magnets and the spin rotations, the spins were retained during acceleration. Ernest Courant labeled these magnets "Siberian Snakes".

Besides the work at Michigan relevant to polarized beams for SeaQuest, a Russian group is also working on having a polarized proton beam in the NICA accelerator at Dubna. This group includes A.S. Belov, Kovalenko, A.M. Kondratenko, Yu Filatov, V.S. Morozov, M.N. Ukhanov, and Troshin. The Michigan group is headed by Alan Krisch, and the Michigan and Russian groups remain in close contact (including monthly international discussions) on topics and problems related to polarized beams. Although these programs are not related to cosmic ray physics, because of the close Michigan – Russian liaison and physics objectives, they merit recognition at this international conference.

2.1. CERN LHC program

The CERN Large Hadron Collider is colliding proton beams of 7.5 TeV, producing collisions of 13 TeV (center-of-mass), equivalent to cosmic ray proton energies of about 100 PeV colliding with stationary protons. Four major experiments are in operation there.

CMS is the largest of the LHC experiments. This collaboration is seeking evidence for more massive versions of the W and Z bosons, as well as heavier, integral-spin versions of the six known quarks.

LHCb has been an experiment studying b-quarks and other heavy quarks in greater detail. Although a further upgrade of this detector is planned, they will continue their studies in greater detail and complexity with the up-graded LHC.

ATLAS is also a very large detector, like CMS, which observes close to 4π steradians final-state solid angle (as does CMS). One of its current objectives is the search for evidence of Dark Matter. Astronomical observations confirm the existence of Dark Matter, but we have no idea of its nature. Assuming that Dark Matter is composed of weakly-interacting particles with a rest mass accessible at the LHC, the ATLAS data enables searches for evidence of their production.

ALICE is a detector with a primary objective of studying proton – lead (p – Pb) collisions, and comparing them with proton – proton (p – p) collisions, in the distribution of reaction products, particle jets, etc. As with LHCb, the ALICE detector array subtends a smaller solid angle than the ATLAS and CMS detectors, but its results are relevant and interesting.

At the meeting of this Symposium at CERN in 2014, I noted that a useful LHC experiment would be the study of collision of protons with oxygen, nitrogen, or carbon nuclei, to replicate the interactions of proton cosmic rays in the air. Such data would be very useful for the cosmic ray community in better understanding proton-initiated air showers, to aid in the interpretation of air shower data from ground-level detectors (e.g. the distribution and density of gamma ray, electron, and muon distributions). Unfortunately, none of the LHC groups has undertaken this study, but it would still merit the efforts of the cosmic ray community to encourage such a study by one of the LHC groups. Such data, at a modest requirement of running time and effort by an LHC program (e.g. ATLAS or CMS) would be very valuable for the cosmic ray community in their use of air shower data in programs such as AUGER.

3. Muon physics

An area of physics which Fermilab is now pursuing is the study of muons. Protons accelerated to about 120 GeV produce pions in their interaction in a static target; these pions, with a mean lifetime of about 28 nano-seconds, decay in flight into muons, with a lifetime of about 2.2 microseconds, and the resulting muon flux can be contained, controlled, and used. For example, Fermilab currently has a program to make a more accurate measurement of "G–2" (the Gyromagnetic ratio of the muon minus 2.000). While this has been measured at Brookhaven as well as at CERN, starting in the 1960s, a more accurate measurement will be possible at Fermilab. The Brookhaven storage-ring permanent magnet has been shipped to Fermilab, and will be used – as it was at Brookhaven – for these measurements.

Among other possible muon physics, there is also discussion about the possibility of building a "muon collider"; positive and negative muons would be collected and injected into a storage ring, circulating in opposite directions, and the properties of their interactions would be studied. As with electron – positron colliders, interesting heavy mesons would be produced and studied.

4. Neutrino physics

From recent studies, it is known that there are 3 "flavors" of neutrinos; Electron neutrinos, Muon neutrinos, and Tau neutrinos. And there are also 3 neutrino masses (each less than about 2 electron volts), the mass states and the flavor states are not the same. Neutrinos oscillate between their different modes, e.g. the neutrinos from nuclear interactions in the Sun are electron neutrinos, but when studied at the Earth, only about a third of the Solar neutrino flux are electron neutrinos; comparable numbers of muon neutrinos and tau neutrinos are also observed. Understanding this complex situation of the relationship between the three mass states and the three flavor states, is unusual and complex. It is the goal of current neutrino experiments to understand and explain this unusual situation.

There are two neutrino experiments which have Fermilab as their neutrino source. NOvA is a Fermilab neutrino experiment in which muon neutrinos are generated by pion decays, which are produced by the 120 GeV proton beam, pass through a 300 ton detector on the Fermilab site and are directed toward a 14,000 ton detector in a mine located in northern Minnesota, 810 kilometers distant. The detectors utilize 4×6 cm liquid scintillator–filled PVC cells as their detectors. While the detector on the Fermilab site detects almost 100% muon neutrinos, those detected in Minnesota are a mixture of the three different flavors. The data from NOvA are enabling the knowledge of the matrices relating the massive and flavor neutrinos.

Another Fermilab neutrino experiment is DUNE (Deep Underground Neutrino Experiment) at the Sanford site in South Dakota. This experiment, still under construction at the Sanford site, uses four liquid Argon detectors, each containing 10,000 tons of liquid argon. The liquid Argon serves as a detector for the Cherenkov radiation produced by muons and electrons from neutrino interactions. As with NOvA, DUNE is located deep underground to minimize the flux of cosmic ray muons (and other particles). The Sanford site is a former gold mine, about 1.5 km deep, and is located 1300 km from Fermilab. At Fermilab, as with NOvA, the neutrinos are generated from pion decays, with a high flux of pions produced by the interactions of 120 GeV protons. There will be 4 similar – but smaller – liquid argon detectors on the Fermilab site. The DUNE program is an international collaboration with many participants from Europe and elsewhere.

Although unrelated to Fermilab, it is worth noting here the underwater neutrino experiments, which have a rich history. All neutrino experiments require an enormous volume of material, because of the very small interaction cross section of the neutrinos. So for many years, neutrino experiments have comprised of detector arrays deployed deep under water, in lakes and/or seas. The detectors (usually phototube arrays) observe the Cherenkov radiation produced by electrons or muons, which in turn are generated by neutrino interactions. In the last century, DUMAND was a concept for a neutrino detection experiment to be located in the Pacific Ocean, off the coast of Hawaii. It never materialized, regrettably. ANTARES was located in the Mediterranian Sea, as is KM3NET (still under construction). Lake Baikal is the home to Baikal-GVD (where GVD stands for "Gigaton Volume Detector").

A successor to the deep under-water neutrino detectors is IceCube, a cubic kilometer array of photodetectors in the Antarctic ice (near the South Pole). The detector is a matrix of 5160 photo-detectors, each of which is a spherical module containing several photomultipliers. These are deployed within the cubic kilometer of ice, which is 1.5–2.5 km below the surface. While neutrinos from cosmic rays interacting in the Earth's atmosphere are observed, the primary objective is to observe cosmic neutrinos which have come directly from their sources; the only component of cosmic ray particles which come directly from their sources (e.g. not deflected by galactic magnetic fields or obstructed by nuclear interactions). Since 2012, IceCube has observed 54 neutrino events (via muons and electron showers) from 30 to 2000 TeV.

One neutrino phenomenon reported some years ago, from a neutrino experiment at Los Alamos, was the possible existence of a "Sterile Neutrino", e.g. a neutrino which did not have a flavor like the three known neutrinos, and did not transform. Other experiments have sought evidence for these Sterile Neutrinos but none have seen evidence for it. The IceCube experiment has recently reported strong negative evidence for such a neutrino.

5. Dark matter

It is certainly true that a major mystery in particle physics and cosmology currently is the nature of Dark Matter. The existence of Dark Matter comes from the observations of the stellar dynamics of galaxies; the fact that the orbital velocities of stars in the outer reaches of a galaxy require a larger galactic mass than is observed. Also, the bending of light (from more distant objects) passing a galaxy requires more than the visible mass. This missing mass can be most easily explained as due to a population of massive, but weakly-interacting particles. As weakly-interacting particles, without strong or electromagnetic interactions (like neutrinos), they would impact galactic dynamics but would otherwise have no observable properties. However, if they had weak interactions (like neutrinos), they should rarely induce an observable particle interaction, hence this is the basis for current searches by particle physicists.

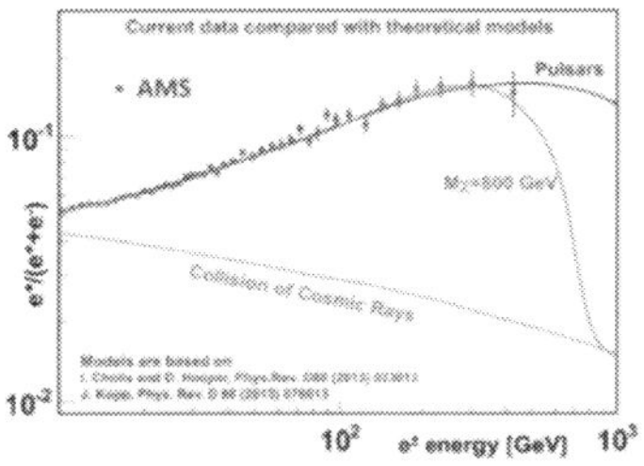

Figure 1. Positron fraction, $e^+/(e^- + e^+)$, of the electron plus positron flux between 10 and 1000 GeV.

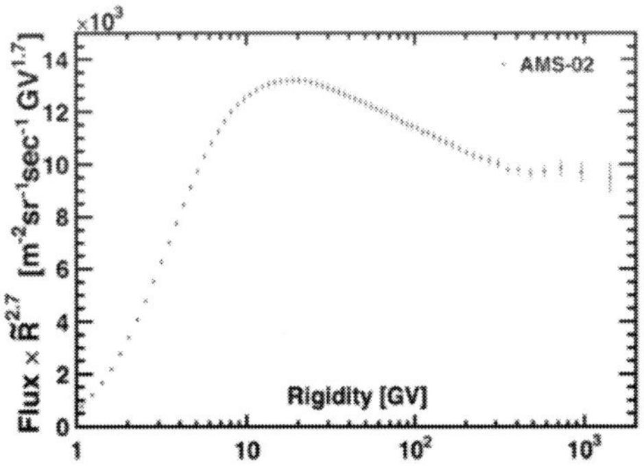

Figure 2. Precision measurement of the proton flux in primary cosmic rays from rigidity 1 GeV to 1.8 TeV, based on 300 million events.

There have been and are currently many searches for evidence of weakly-interacting paricles which could be the substance of Dark Matter. Fermilab physicists have a major Dark Matter search program in progress, using a very large detector in the South Dakota Sanford mine. Named LUX (for Liquid Underground Xenon), it is a 370 kg volume of liquid Xenon, and located at 1480 m underground in the Sanford mine. Operating since 2012, the search for WIMPs (Weakly Interacting Massive Particles) has come up empty.

Of course, other similar searches are in progress at other sites, and by other groups. Dark Matter remains one of the current three major mysteries in the physics of Cosmology-Astrophysics; the other two are the nature of Dark Energy (the source of the continued acceleration-expansion of the Universe) and the details of the evolution of the early Universe (its minute size, dramatic explosion, etc.).

6. AMS-02

One interesting program not otherwise included among the presentations at this Symposium is a report on the AMS-02 satellite experiment aboard the International Space Station. This is a major program directed by Professor Samuel Ting of M.I.T. Ting's original objective was to explore whether heavy nuclei (high atomic number, A) had equal matter and antimatter components. Observations of electromagnetic radiation (light, X-rays, infra-red, radio, etc.) are insensitive to whether their source is matter or antimatter. Ting argued that, since galaxies are separate and spaced apart, it would be possible that perhaps half of the galaxies are composed of anti-matter, and therefore, observing energetic nuclear cosmic rays, identifying

DOI: 10.1051/epjconf/201714505005

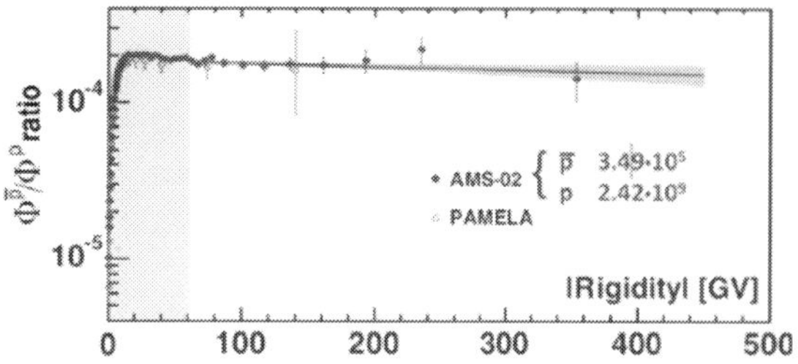

Figure 3. Flux ratio of anti-protons to protons up to 400 GeV.

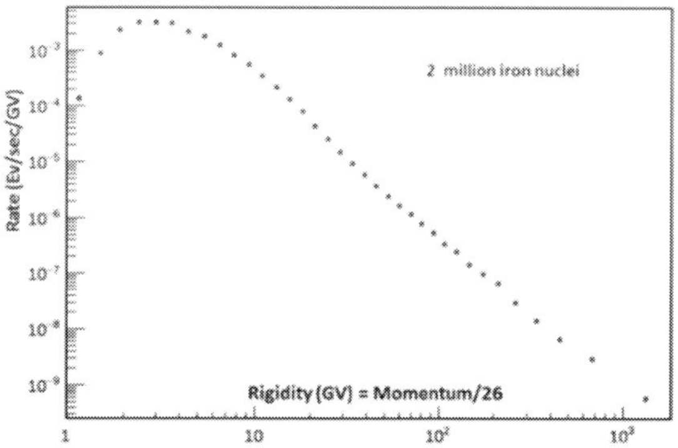

Figure 4. Flux of 2 million iron nuclei measured by AMS vs. energy (momentum) divided by the nuclear charge, Z = 26 between 1 GeV and 1 TeV.

their electric charge and mass, would enable the search for primordial antimatter. The community of traditional cosmic ray physicists were skeptical of the idea that half of the universe would be antimatter, arguing that the galaxies were not so far separated that there wouldn't be evidence for matter-antimatter annihilation between galaxies. Such annihilation would be observable, e.g. as sources of energetic gamma radiation, etc. To be sure, the matter – anti-matter asymmetry of the Universe remains a problem.

But a cosmic ray magnetic spectrometer, designed to study the mass, energy, and electric charge of cosmic ray nuclei, would be a useful tool for better understanding the nature of primary cosmic rays, therefore NASA (National Aeronautics and Space Administration, in America) approved Ting's concept.

He then set out to build this spectrometer, which would be located on the International Space Station, a large satellite program that was already in orbit. Ting labelled his detector the "Anti-Matter Spectrometer", or AMS. However NASA did not like a U.S. space program to employ "Anti" in its name, so Ting's program was re-named the "Alpha Magnetic Spectrometer" (hence still AMS). The detector is a magnetic spectrometer, using a permanent magnet. Originally, Ting had considered a super-conducting magnet, however the desire to collect data for several years made an electro-magnet impractical. Ting's group built a prototype magnetic spectrometer, AMS-01, which was flown earlier, and demonstrated the feasibility of the program. The AMS-02 was then built and flown, and is still on the International Space Station collecting data. AMS-02 has dimensions of $3 \times 4 \times 5$ meters, and weighs 7.5 tons. The detectors include a Transition Radiation Detector, an Electromagnetic Calorimeter, a Silicon Tracker, a Time of Flight counter, and a Ring Imaging Cherenkov counter system; these add up to about 300,000 channels of detectors. The data from the detectors are processed by 650 on-board computers. A good description of the AMS-02 device is in the Physical Review Letters paper [1]. It should be noted that the AMS-02 international collaboration includes over a hundred physicists, who are all co-authors on these publications, so – in references to these papers – only Ting's name (as the program spokesman) is cited.

It may be noted that, although Ting's home institution is M.I.T., Ting has long, close attachments to CERN (in Geneva, Switzerland), as former spokesman for the L3 experiment at the CERN electron-positron collider, LEP. Hence CERN has a building on their Prevessin cite which is occupied by AMS, and in which the data collection and analysis are located.

Attached here, as Figs. 1, 2, 3, and 4, are graphs of data from AMS-02: the positron/electron+positron fraction, the proton spectrum, the proton – anti-proton ratio and the iron nuclei spectrum. The published papers presenting these and other data are cited in references [2–5].

References

[1] S. C. C. Ting, Phys. Rev. Letters **110**, 141102 (2013)
[2] S. C. C. Ting, Phys. Rev. Letters **113**, 121102 (2014)
[3] S. C. C. Ting, Phys. Rev. Letters **113**, 221102 (2014)
[4] S. C. C. Ting, Phys. Rev. Letters **114**, 171103 (2015)
[5] S. C. C. Ting, Phys. Rev. Letters **117**, 091103 (2016)

New stage in high-energy gamma-ray studies with GAMMA-400 after Fermi-LAT

N.P. Topchiev[1,a], A.M. Galper[1,2], V. Bonvicini[3], O. Adriani[4], I.V. Arkhangelskaja[2], A.I. Arkhangelskiy[2], A.V. Bakaldin[5], S.G. Bobkov[5], M. Boezio[3], O.D. Dalkarov[1], A.E. Egorov[1], M.S. Gorbunov[5], Yu.V. Gusakov[1], B.I. Hnatyk[6], V.V. Kadilin[2], V.A. Kaplin[2], M.D. Kheymits[2], V.E. Korepanov[7], A.A. Leonov[1,2], F. Longo[3], V.V. Mikhailov[2], E. Mocchiutti[3], A.A. Moiseev[8], I.V. Moskalenko[9], P.Yu. Naumov[2], P. Picozza[10], M.F. Runtso[2], O.V. Serdin[5], R. Sparvoli[10], P. Spillantini[4], Yu.I. Stozhkov[1], S.I. Suchkov[1], A.A. Taraskin[2], M. Tavani[11], Yu.T. Yurkin[2], and V.G. Zverev[1]

[1] Lebedev Physical Institute, 119991 Moscow, Russia
[2] National Research Nuclear University MEPhI (Moscow Engineering Physics Institute), Kashirskoe highway 31, Moscow 115409, Russia
[3] Istituto Nazionale di Fisica Nucleare, Sezione di Trieste, 34149 Trieste, Italy
[4] Istituto Nazionale di Fisica Nucleare, Sezione di Florence, 50019 Sesto Fiorentino, Florence, Italy
[5] Scientific Research Institute for System Analysis, 117218 Moscow, Russia
[6] Taras Shevchenko National University, Kyiv 01601, Ukraine
[7] Lviv Center of Institute of Space Research, Lviv 79060, Ukraine
[8] NASA Goddard Space Flight Center and CRESST/University of Maryland, Greenbelt, Maryland 20771, USA
[9] Hansen Experimental Physics Laboratory and Kavli Institute for Particle Astrophysics and Cosmology, Stanford University, Stanford, CA 94305, USA
[10] Istituto Nazionale di Fisica Nucleare, Sezione di Rome "Tor Vergata", 00133 Rome, Italy
[11] Istituto Nazionale di Astrofisica IASF and Physics Department of University of Rome Tor Vergata, 00133 Roma, Italy

Abstract. Fermi-LAT has made a significant contribution to the study of high-energy gamma-ray diffuse emission and the observations of 3000 discrete sources. However, one third of all gamma-ray sources (both galactic and extragalactic) are unidentified, the data on the diffuse gamma-ray emission should be clarified, and signatures of dark matter particles in the high-energy gamma-ray range are not observed up to now. GAMMA-400, the currently developing gamma-ray telescope, will have angular ($\sim 0.01°$ at 100 GeV) and energy ($\sim 1\%$ at 100 GeV) resolutions in the energy range of 10–1000 GeV which are better than Fermi-LAT (as well as ground gamma-ray telescopes) by a factor of 5–10. It will observe some regions of the Universe (such as the Galactic Center, Fermi Bubbles, Crab, Cygnus, etc.) in a highly elliptic orbit (without shading the telescope by the Earth) continuously for a long time. It will allow us to identify many discrete sources, to clarify the structure of extended sources, to specify the data on the diffuse emission, and to resolve gamma rays from dark matter particles.

1. Current gamma-ray study challenges

1.1. Analysis of gamma-ray results according to the Fermi-LAT and ground-based telescope data

Since 2008 Fermi-LAT is operating in a near-Earth orbit in the scanning mode and surveying full sky every three hours. Up to now, three catalogs of gamma-ray sources have been published based on the Fermi-LAT observational results: 1FGL [1] and 2FGL [2] for the energy range from 100 MeV to 100 GeV, 3FGL [3] for the energy range from 100 MeV to 300 GeV. Moreover two catalogs of high-energy gamma-ray sources were published: 1FHL [4] for energies above 10 GeV and 2FHL [5] for the energy range of 50 GeV–2 TeV. Figure 1 [6] shows the percentage of the different types of gamma-ray sources according to 3FGL. It is seen that 33% of gamma-ray sources are unidentified. In [3] the exposure of the source observations according to 3FGL is presented. It is seen that during four years of operation Fermi-LAT observed the Galactic center during only 12% or 1/8 of the total operation time.

Based on results of gamma-ray observations at energies above 100 GeV by ground-based facilities VERITAS [7] MAGIC [8] H.E.S.S. [9] and others, the TeVCat catalog (`http://tevcat.uchicago.edu/`) of discrete gamma-ray sources was created, which contains about 180 sources and mainly matches with 3FGL.

It is important to note that the observational data from Fermi-LAT and ground-based facilities were obtained for the energy ranges which overlap poorly for many gamma-ray sources. Sometimes they don't overlap at all. **Hence, the frontier range around 100 GeV is still very interesting for investigations.** In addition, the angular resolution of Fermi-LAT, existing ground-based telescopes, and even the planned CTA [10] in the region of around 10–300 GeV is only $\sim 0.1°$. **Therefore a**

[a] e-mail: `tnp51@yandex.ru`

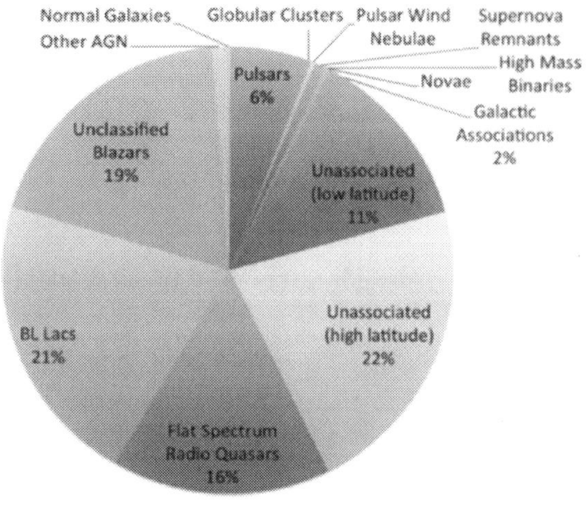

Figure 1. The percentage of the different types of gamma-ray sources [6] according to the 3FGL.

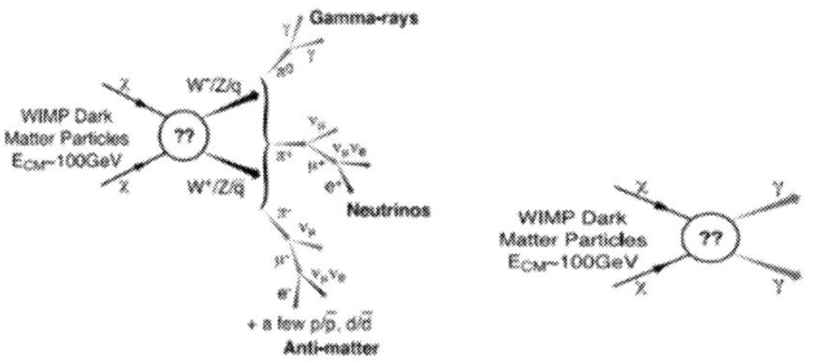

Figure 2. Possible channels of WIMP annihilation $\chi\chi \rightarrow b\bar{b}, \tau^+\tau^-, W^+W^-, \mu^+\mu^-, q\bar{q}, ZZ$ with gamma-ray production [12].

much better angular resolution is required in order to identify many gamma-ray sources.

1.2. Indirect searches of dark matter

Another very interesting and important goal in the studies of the gamma-ray sky is indirect searches of dark matter (DM). In general, an exact physical nature of DM is a top puzzle in modern astrophysics. Many candidates for the DM role are being proposed. However, WIMPs with mass between several GeV and several TeV are still considered as the most probable candidate [11]. WIMPs can annihilate or decay with the production of gamma rays. This emission can have both a continuum energy spectrum or monoenergetic lines. This depends on which annihilation channel realizes in nature. The continuum spectrum would come in the case of annihilation into particle pairs like $\chi\chi \rightarrow b\bar{b}, \tau^+\tau^-, W^+W^-, \mu^+\mu^-, q\bar{q}, ZZ$ or others and gamma-ray lines would be produced in the case of direct annihilation into photons $\chi\chi \rightarrow \gamma\gamma, \gamma Z, \gamma H$ (Fig. 2 [12]).

To resolve gamma-ray lines from the background it is necessary to have a high-energy resolution. Figure 3 shows the expected energy spectrum for the annihilation of 300-GeV WIMPs producing gamma rays ($\gamma\gamma, \gamma Z, \gamma H$ lines), which can be resolved from the background by various telescopes with energy resolutions of 10%, 5%, and 0,5% [13]. Note that the energy resolution of Fermi-LAT and ground-based facilities is only 10–15% at energies of 10–300 GeV. **Thus, as seen from Fig. 3, future telescopes need to have 1–2% energy resolution.**

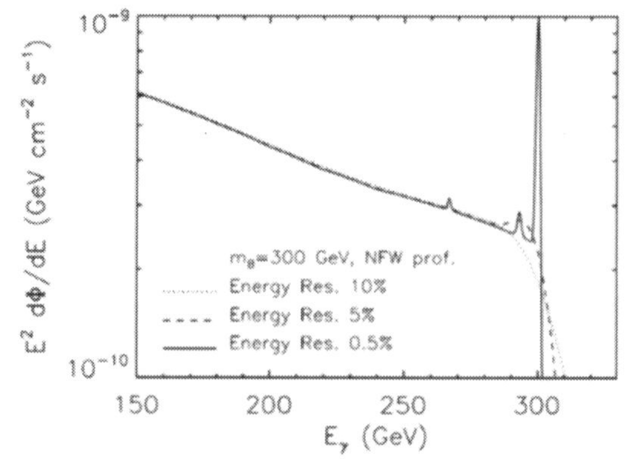

Figure 3. Expected energy spectrum for the annihilation of 300-GeV WIMP producing gamma rays ($\gamma\gamma$, γZ, γH lines), which can be resolved from the background by various telescopes with energy resolutions of 10%, 5%, and 0,5% [13].

2. THE GAMMA-400 GAMMA-RAY TELESCOPE

Thus, to resolve unidentified gamma-ray sources and search for potential gamma-ray lines from DM we need a gamma-ray telescope with an angular resolution of several hundredth degrees and an energy resolution of a few percent for an energy of ~100 GeV. This is going to be GAMMA-400, which will be installed onboard the Russian space observatory [14–18].

The GAMMA-400 main scientific goals are: a search for dark matter by means of gamma-ray astronomy; precise measurements of the Galactic Center, Fermi Bubbles, Crab, Vela, Cygnus, Geminga, Sun, and other regions, extended and point gamma-ray sources, diffuse gamma rays with unprecedented angular (**$\sim$0.01° at E_γ >100 GeV) and energy ($\sim$1% at E_γ >100 GeV)** resolutions.

2.1. The GAMMA-400 physical scheme

The physical scheme of the GAMMA-400 gamma-ray telescope is shown in Fig. 4. The GAMMA-400 energy range for gamma-ray studies is from $\sim$20 MeV to $\sim$1000 GeV. The GAMMA-400 field of view (FoV) is $\pm 45°$.

GAMMA-400 consists of plastic scintillation anticoincidence top and lateral detectors (ACtop and AClat), a converter-tracker (C), plastic scintillation detectors (S1 and S2) for the time-of-flight system (ToF), calorimeter (CC), plastic scintillation detector (S3).

The anticoincidence detectors surrounding the converter-tracker are used to distinguish gamma rays from a significantly larger number of charged particles (e.g., in the region of 10–100 GeV, the flux ratios for gamma rays to electrons and protons are $\sim 1 : 10^2 : 10^5$.

All scintillation detectors consist of two independent 1-cm layers. The time-of-flight system, where detectors

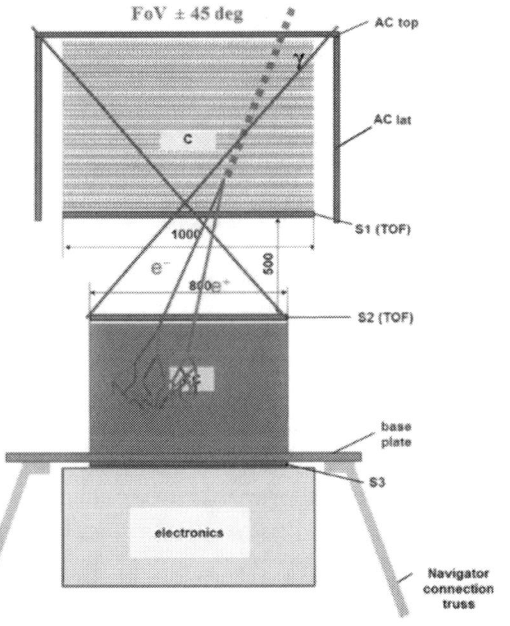

Figure 4. The GAMMA-400 physical scheme.

S1 and S2 are separated by approximately 500 mm, determines the top-down direction of arriving particles. The additional scintillation detector S3 improves hadron and electromagnetic shower separation.

The converter-tracker consists of 22 layers of double (x, y) silicon strip coordinate detectors (pitch of 0.08 mm). Twenty layers are interleaved with tungsten conversion foils and the final two layers have no tungsten. This configuration allows us to measure gamma rays down to approximately 20 MeV. The total converter-tracker thickness is about 1 radiation length, X_0. The converter-tracker information is used to precisely determine the conversion point and the direction of each incident particle.

The calorimeter CC measures the particle energy and consists of CsI(Tl) crystals. The total calorimeter thickness is $\sim 21X_0$ or $\sim 1.0\lambda_0$ (where λ_0 is the nuclear interaction length) when detecting vertical incident particles and $\sim 43X_0$ or $\sim 2.0\lambda_0$ when detecting laterally incident particles. Using the deep calorimeter allows us to extend the energy range up to several TeV for gamma rays, and to reach an energy resolution of approximately 1% above 100 GeV.

2.2. The GAMMA-400 gamma-ray observatory

The GAMMA-400 gamma-ray observatory will be installed onboard the Navigator space platform, which is designed and manufactured by the Lavochkin Association.

Using the Navigator space platform gives the GAMMA-400 experiment a highly unique opportunity for the near future gamma- and cosmic-ray science, since it allows us to install a scientific payload (mass of 2500 kg, power consumption of 2000 W, and telemetry downlink of 100 GB/day, with a lifetime more than 7 years), which will provide GAMMA-400 with the means to significantly contribute as the next generation instrument for gamma-ray astronomy and cosmic-ray physics.

The GAMMA-400 experiment will be initially launched into a highly elliptical orbit (with an apogee of 300,000 km and a perigee of 500 km, with an inclination of 51.4°), with 7 days orbital period. Under the action of gravitational disturbances of the Sun, Moon, and the Earth after ~6 months the orbit will transform to about an approximately circular one with a radius of ~200000 km and will not suffer from the Earth's occultation and shielding by the radiation belts. A great advantage of such an orbit is the fact that the full sky coverage will always be available for gamma-ray astronomy, since the Earth will not cover a significant fraction of the sky, as is usually the case for a low-Earth orbit. Therefore, the GAMMA-400 source pointing strategy will hence be properly defined to maximize the physics outcome of the experiment. The launch of the GAMMA-400 space observatory is planned for the middle of the 2020s.

2.3. Comparison of GAMMA-400 with Fermi-LAT and ground-based facilities

GAMMA-400 has numerous advantages in comparison with Fermi-LAT:

- a highly elliptical orbit (without the Earth's occultation and away from the radiation belts) allows us to observe with the full aperture of ±45° different gamma-ray sources continuously over a long period of time with an exposure greater by a factor of 8 than for Fermi-LAT operating in the sky-survey mode;
- thanks to a smaller pitch (by a factor of 3) and analog readout in the coordinate silicon strip detectors, GAMMA-400 has an excellent angular resolution above ~20 MeV;
- due to the deep ($\sim 21X_0$) calorimeter, GAMMA-400 has an excellent energy resolution and can detect gamma rays more reliably up to several TeV for vertically incident events;
- owing to the better gamma-ray separation from cosmic rays (in contrast to Fermi-LAT, the presence of a special trigger with event timing, time-of-flight system, two-layer scintillation detectors), GAMMA-400 is significantly well equipped to separate gamma rays from the background of cosmic rays and backscattering events.

GAMMA-400 will also have better angular and energy resolutions in the energy region 10–1000 GeV in comparison with current and future space- and ground-based instruments (VERITAS [7], MAGIC [8], H.E.S.S. [9], CTA [10], and HAWC [19] (Figs. 5, 6) and will allow us to fill the gap at energies of ~100 GeV between the space- and ground-based instruments.

3. The anticipated GAMMA-400 scientific output

3.1. Galactic plane

GAMMA-400 will study continuously over a long period of time different regions of the Galatic plane (Fig. 7), for example, the Galactic center, Fermi Bubbles, Crab, Vela, Cygnus, Geminga with FoV of ±45°. In particular, using the gamma-ray fluxes obtained by Fermi-LAT, we can expect that GAMMA-400, when observing the Galactic

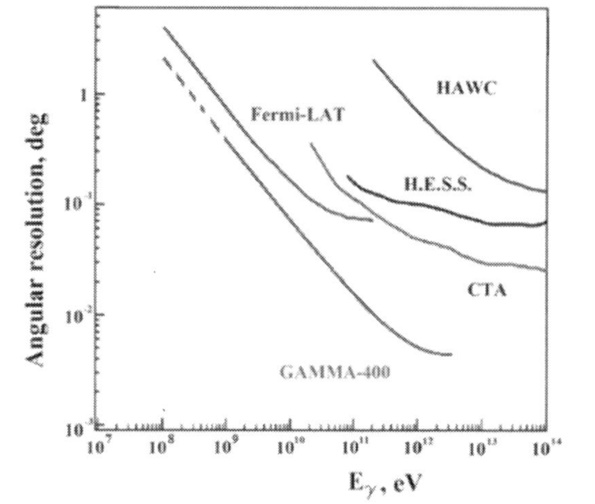

Figure 5. Comparison of angular resolutions for GAMMA-400, Fermi-LAT, H.E.S.S., HAWC, and CTA.

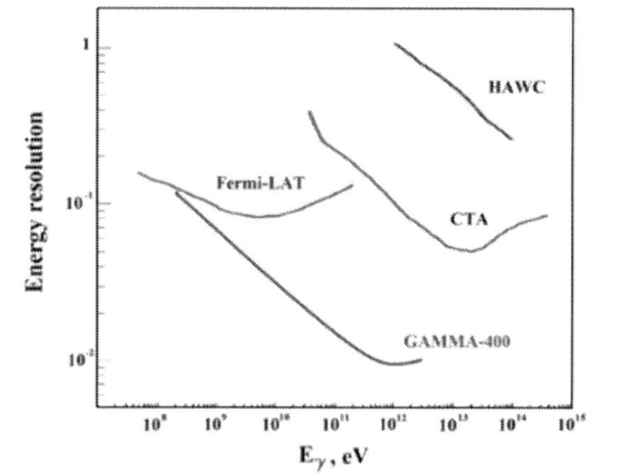

Figure 6. Comparison of energy resolutions for GAMMA-400, Fermi-LAT, H.E.S.S., HAWC, and CTA.

center with an aperture of $\pm 45°$ during 1 year will detect: 57400 photons for $E_\gamma > 10\,\mathrm{GeV}$; 5240 photons for $E_\gamma > 50\,\mathrm{GeV}$; 1280 photons for $E_\gamma > 100\,\mathrm{GeV}$; 535 photons for $E_\gamma > 200\,\mathrm{GeV}$.

3.2. Dark matter searching

When detecting gamma rays from dark matter the intensity from gamma rays is calculated by

$$\frac{d\phi}{d\Omega dE} = \frac{1}{2}\frac{\langle\sigma v\rangle}{m_\chi^2}\frac{dN_\gamma}{dE} \times \frac{J}{4\pi}$$

$$[\mathrm{erg\,s^{-1}\,cm^{-2}\,GeV^{-1}\,sr^{-1}}]$$

$$J = \int_{los} \rho^2(\vec{r})dl$$

Here, the first term contains the information only on dark matter particle: $\langle\sigma v\rangle$ is the particle annihilation cross section, m_χ is the particle mass, dN_γ/dE is the spectrum of gamma rays from annihilation (`http://www.marcocirelli.net/PPPC4DMID.html`). The second term is determined using the distribution of dark matter density $\rho(r)$ in the object under study.

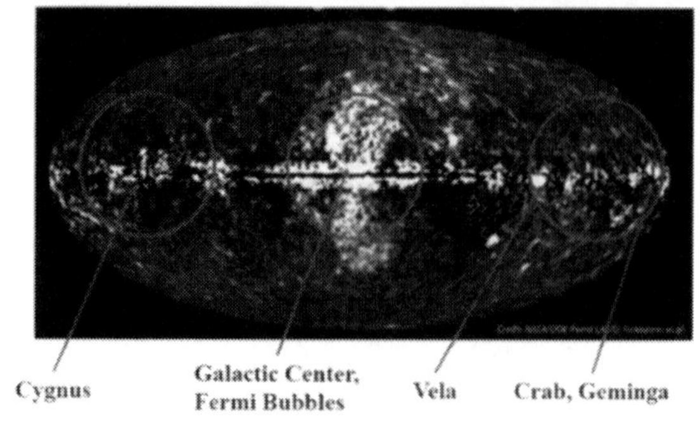

Figure 7. Galactic center, Fermi Bubbles, Crab, Cygnus, Vela, Geminga, and other regions will be observed with the GAMMA-400 FoV of $\pm 45°$.

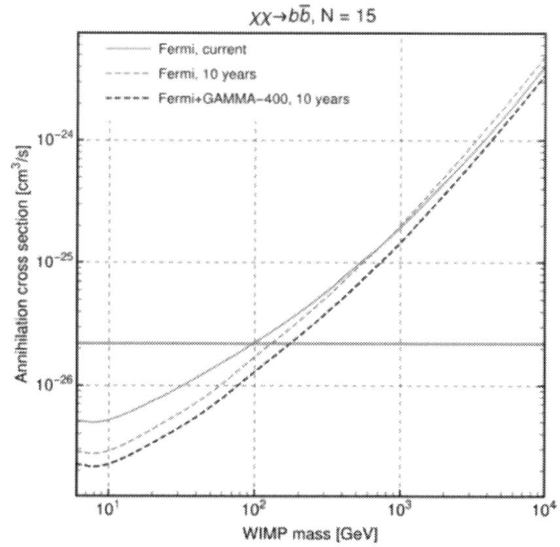

Figure 8. Current and predicted constraints for dark matter parameters according to observations of 15 known dwarf galaxies.

The main targets to search for gamma rays from dark matter are:

The Milky Way. The center of Milky Way is, apparently, the best potential source of dark matter emission possessing the largest J-factor [11]. Moreover, recently, an anomalous excess of gamma-ray emission in the GeV energy range was revealed near the Galactic center (the region of about one degree) [20], which can be well described by dark matter with a mass of several tens of GeV and annihilation cross section of about standard thermal $10^{26}\,\mathrm{cm^3/s}$. However, this observed excess can have another interpretation - the presence of a population of millisecond pulsars [21]. Therefore, the new GAMMA-400 observational data can help solve this problem.

Milky Way satellites have been considered for a long time as the strongest sources of constraints for dark matter, because they have sufficiently large J-factors and at the same time have considerably less gamma-ray background in comparison with the Galactic center. Current constraints for the dark matter parameters obtained by Fermi-LAT are shown by solid lines in Fig. 8 (for the typical quark-antiquark channel [22]). The prediction for constraints for a 10-year operation of Fermi-LAT is also shown in Fig. 8.

At present, ~15 dwarf galaxies are reliably investigated for constructing constraints. GAMMA-400 will able to specify the constrain area.

Other objects. Other potentially interesting objects are other galaxies and their clusters, where dark matter may be present and can emit gamma rays. GAMMA-400 with the highest energy resolution of 1% will have a unique sensitivity for detecting dark matter.

This work was supported by the the Space Council of Russian Academy of Sciences and the Russian Federal Space Agency.

References

[1] A. Abdo, et al., ApJ **188**, 405 (2010)
[2] P. Nolan, et al., ApJS **199**, 31 (2012)
[3] F. Acero, et al., ApJS **218**, 23 (2015)
[4] M. Ackermann, et al., ApJS **209**, 1 (2013)
[5] M. Ackermann, et al., ApJS **222**, 1 (2016)
[6] R. Bühler. `arXiv:1509.00012` (2015)
[7] R.E. Ong, et al., ASR **53**, 1483 (2014)
[8] D. Mazin, et al., `arXiv:1410.5073` (2014)
[9] A. Balzer, et al., `arXiv:1311.3486` (2013)
[10] CTA Consortium. Exp. Astron. **32**, 193 (2011)
[11] G. Bertone. Particle dark matter – observations, models and searches (Cambridge Univ. Press, 2010)
[12] G. Bertone, et al., `arXiv:1009.5107` (2010)
[13] D. Nekrassov, et al., `arXiv:1106.2752` (2011)
[14] V. Dogiel, et al., Space Sci. Rev. **49**, 215 (1988)
[15] A. Galper, et al., ASR **51**, 297 (2013)
[16] A. Galper, et al., AIP Conf. Proc. **1516**, 288 (2013)
[17] N. Topchiev, et al., Bull. RAS. Phys. **79**, 417 (2015)
[18] N. Topchiev, et al., J. Phys.: Conf. Ser. **675** (2016) 032009
[19] S. Westerhoff, et al., ASR **53**, 1492 (2014)
[20] K. Abazajian and M. Kaplinghat. Phys. Rev. **D86**, 083511 (2012) [`arXiv:1207.6047`]
[21] R. Bartels, S. Krishnamurthy, and C. Weniger. `arXiv:1506.05104` (2015)
[22] E. Charles, et al., `arXiv:1605.02016` (2016)

GRAINE balloon experiment in 2015

Precise observations of cosmic gamma rays by a high-resolution emulsion telescope

Hiroki Rokujo[1a] and GRAINE collaboration[2]

[1] Nagoya University, Furo-cho, Chikusa-ku, Nagoya 464-8602, Japan
[2] Aich University of Education, ISAS/JAXA, Utsunomiya University, Okayama University of Science, Kobe University, and Nagoya University, Japan

Abstract. Observations of cosmic gamma rays are important for studying high energy phenomena in the universe. Since 2008, the Large Area Telescope on the Fermi satellite has surveyed the whole gamma-ray sky in the sub-GeV/GeV energy region, and accumurated a large amount of data. However, observations at the low galactic latitude remains difficult because of a lack of angular resolution, increase of background flux originating from galactic diffuse gamma rays, etc. The Gamma-Ray Astro-Imager with Nuclear Emulsion (GRAINE) is a gamma-ray observation project with a new balloon-borne emulsion gamma-ray telescope. Nuclear emulsion is a high-resolution 3D tracking device. It determines the incident angle with 0.1° resolution for 1 GeV gamma rays (1.0° for 100 MeV), and has linear polarization sensitivity. GRAINE aims at precise observation of gamma-ray sources, especially in the galactic plane, by repeating long-duration balloon flights with large-aperture-area (10 m^2) high-resolution emulsion telescopes. In May 2015, we performed a balloon-borne experiment in Alice Springs, Australia, in order to demonstrate the imaging performance of our telescope. The emulsion telescope that has an aperture area of 0.4 m^2 was employed in this experiment. It observed the Vela pulsar (the brightest gamma-ray source in the GeV sky) at an altitude of 37 km for 6 hours out of the flight duration of 14 hours. In this presentation, we will report the latest results and the status of the GRAINE project.

1. Introduction

1.1. Sub-GeV/GeV gamma-ray observation

Observations of cosmic gamma rays emitted from black Holes, pulsars, super nova remnants, etc. are important to understand such high-energy objects or phenomena in the universe. AGILE [1] launched in 2007 and the Large Area Telescope on the Fermi Gamma-ray Space Telescope (Fermi-LAT) [2] launched in 2008 have surveyed the sub-GeV/GeV gamma-ray sky. They have achieved good results and contributed to the development of gamma-ray astronomy: detection of more than 3000 gamma-ray sources [3]; discovery of cosmic-ray proton acceleration in super nova remnants [4, 5].

On the other hand, new issues have come to light. Observations at the low galactic latitude remain difficult because of a lack of angular resolution, an increase of background flux originating from galactic diffuse gamma rays, etc. Unassociated sources are still included in the high rate in gamma-ray sources detected in the galactic plane [3]. Furthermore, an unexpected gamma-ray excess in the galactic center region was reported [6].

Polarimetry data of high-energy photons from pulsars etc. is important to study the mechanism of gamma-ray emission. Several observations in the hard X-ray region have been initiated, but substantially sensitive observations in the gamma ray region have not been carried out so far. To make more progress in gamma-ray astronomy, not only a quantitative increase in the observation data but also qualitative improvement is required.

1.2. GRAINE project

The Gamma-Ray Astro-Imager with Nuclear Emulsion (GRAINE) project aims at precise observations of gamma-ray sources by a balloon-borne gamma-ray telescope. The high-angular resolution gamma-ray telescope, called emulsion telescope, consists of nuclear emulsion films.

Nuclear emulsion is a high precision tracking detector. It can measure the incident angle and the position of charged particles with milli-radian and sub-micron resolution in spite of the thin material ($\sim 10^{-3}$ in the radiation-length unit). Read-out process by a microscope is needed after exposure and development because of an integrating detector, but automated and fast scanning systems have been developed and used practically in accelerator neutrino experiments.

Because nuclear emulsion can determine angles of electrons and positrons at the beginning point of pair creation interaction ($\gamma \rightarrow e^+ + e^-$), the angular resolution for gamma rays (10 MeV–10 GeV) is about one order of magnitude higher than that of Fermi-LAT (Fig. 1).

In addition, nuclear emulsions can measure an azimuthal angle of a plane consists of electron and positron tracks, so it has sensitivity of a linear polarization of gamma rays [7]. Nuclear emulsion has no time resolution, but gamma-ray incident timing is given by a new technique

[a] e-mail: `rokujo@flab.phys.nagoya-u.ac.jp`

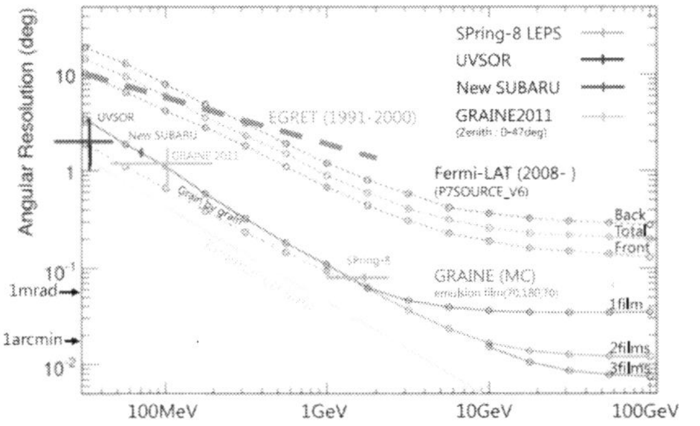

Figure 1. Angular resolution of the emulsion gamma-ray telescope (lines show simulation results and dots with error bars show experimental results), Fermi-LAT (dotted lines), and EGRET (a broken line).

Table 1. Performance of emulsion gamma-ray telescope.

Energy range	10 MeV–> 10 GeV
Angular resolution[1]	0.93°@100 MeV 0.1°@1 GeV
Polarization sensitivity	Yes
Energy resolution	10%–20%
Effective area[1 2]	2.1 m^2@100 MeV 2.8 m^2@1 GeV
FOV	>2.2 sr
Time resolution	< 1 s
Dead time	No

developed for GRAINE, named emulsion multi-stage shifter (described in detail in [8]).

The emulsion telescope doesn't consist of any electronic counter, so it is easy to enlarge the aperture area without deterioration of the resolution caused by limitation of the number of read-out channels. We started developing a large-area telescope (aperture area ~10 m^2) and promoting long-duration balloon flights (~200 hours), similar to the JACEE or RUNJOB balloon-borne experiments, for scientific observation. The target objects are sources on the galactic plane/center, bright and extended sources (for example, SNR W44), galactic pulsars, gamma-ray bursts, etc. Table 1 shows the performance of the emulsion gamma-ray telescope.

2. Balloon-borne experiment in 2015

2.1. GRAINE 2015

In the first balloon experiment, GRAINE 2011, the technical feasibility was demonstrated by a small-scale emulsion telescope and a star camera system [9]. The second experiment, GRAINE 2015, aims at detecting gamma-ray sources, and demonstrating the imaging performance of the emulsion telescope in the 100 MeV energy region. By a medium duration balloon flight (12–24 hours) in Alice Springs, Australia, observation of the Vela pulsar, the brightest gamma-ray source in the sub-GeV/GeV gamma-ray sky, for 6.5 hours (within the field of view of the emulsion telescope) is expected.

[1] At nomal incidence.

[2] Effective area was considered with 10 m^2 aperture area, transmittance at 5 g/cm^2 atmospheric depth, conversion efficiency at 0.54 radiation lengths and detection efficiency for each energy.

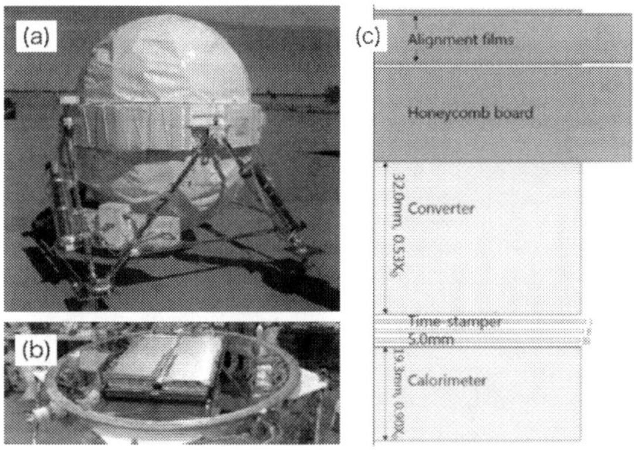

Figure 2. (a) Picture of gondola. A spherical container (the inside diameter is 1.6 m) is a balloon-style pressure vessel. Three star cameras as the attitude monitors were mounted on the trusses (one of them was on the other side). (b) Inside view of the pressure vessel. At the center of a ring, the emulsion gamma-ray telescope was put. The total aperture area of four units was 3780 cm^2. (c) Cross sectional view of the emulsion gamma-ray telescope. It consists of alignment films, the converter, and time stamper, and the calorimeter.

2.2. Detector

Figure 2 shows pictures and a cross sectional view of the emulsion gamma-ray telescope employed in GRAINE 2015. We enlarged the aperture area of the second telescope to 3780 cm^2, which is 29 times larger than that of the first telescope. The converter consists of a hundred high-sensitive emulsion films. The thicknesses of the plastic layer and the both-side emulsion layers in a film are 180 μm and 70 μm, respectively. The total thickness and the radiation length are 32 mm and 0.53 X_0, respectively. 34% of vertical-incident gamma rays convert to electron-positron pairs, and these tracks are recorded in emulsion films. The area of a converter film is 37.8 cm × 25 cm. Four units with the same structure were employed.

An alignment unit (two or three emulsion films kept vacuum-packed with an aluminum honeycomb panel) was put on the top of each converter unit. The alignment unit is the standard surface of the detector system, and each angle of track recorded in the converter is calibrated by high momentum tracks penetrating both the alignment unit and the converter.

The multi-stage shifter system as time stamper was put at the bottom of the converter. 2–4 emulsion films were mounted on each movable stage. Three stages are driven by stepping motors. In the whole of the observation, they slide cyclically like an analog clock, and create independent combinations of the stage position. During the Vela-observation period, we changed the operation mode and made the speed of stage faster to obtain ~10 msec resolution (in the normal operation mode, the time resolution is expected to be below one second).

The energy measurement for multi-GeV gamma-ray events was performed by the analysis of the calorimeter. It has the sandwich structure of sixteen emulsion films and

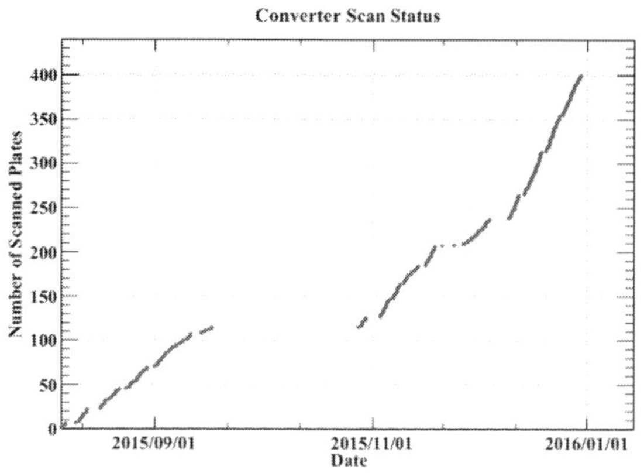

Figure 3. Performance of data taking of the converter by HTS. The vertical axis indicates the number of scanned films. The area of one film is 37.8 cm^2 × 25 cm^2.

fifteen 1 mm-thick stainless steel plates. The thickness and radiation length were 19.3 mm and 0.90 X_0, respectively.

The balloon also carried three star trackers as attitude monitors, a balloon-style pressure vessel to maintain the vacuum-packed emulsion chamber, and several sensors (for GPS, temperature, pressure, etc.) as its payload.

2.3. Balloon flight

On May 12th, 2015, the balloon was launched from the Alice Springs balloon-launching station, Australia. The balloon left the ground at 6:33 Australian Central Standard Time (ACST), and reached 37.2 km altitude at 8:50. At 14:15 the Vela pulsar entered the field of view of the emulsion telescope. It observed the target for about six hours, then the balloon released the gondola at 20:22. The total flight duration in this experiment was 14.4 hours, with 11.5 hours of level flight at 36.0–37.4 km altitude and 4.7–3.8 hPa residual atmospheric pressure. The gondola landed about 130 km north of Longreach at 20:55. On the next morning, all payloads were recovered. More detail information is described in [10].

3. Flight data analysis

3.1. Data Taking by emulsion scanning system

Data taking of emulsion films was done by the latest scanning system, Hyper Track Selector (HTS), which was developed in Nagoya University [11]. The first practical data taking of HTS went smoothly and it took about three months to finish 41 m^2 of the converter films and the time stamper films. Figure 3 shows the performance of scanning by HTS.

After alignment process of segmented tracks (each track has an angle, a position, and a darkness information on a coordinate of each film), tracks running on a coordinate of the emulsion chamber were reconstructed. The track-finding efficiency of HTS was higher than 95% at $\tan\theta < 2.0$ angular region as a result of evaluating the filling rate of tracks penetrating multiple films [12].

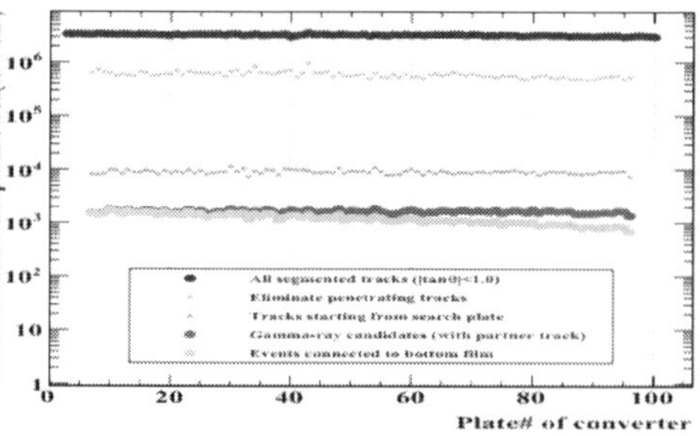

Figure 4. Result of gamma-ray event ($\gamma \to e^+ + e^-$) selection by using a part of scan data from the converter (13 cm × 9 cm × 100 films).

3.2. Gamma-ray event analysis

3.2.1. Event detection

Figure 4 shows a result of gamma-ray event ($\gamma \to e^+ + e^-$) selection by using a part of scan data from the converter (13 cm × 9 cm × 100 films). 3 $\times 10^6$ segmented tracks at $|\tan\theta| < 1.0$ angular region in each projection were scanned by HTS from a film. First, data of eight adjoining films was extracted, then tracks penetrating the volume were eliminated. Second, the following tracks were selected: tracks starting from the fourth film; tracks running to the eighth film (the bottom film in the extracted volume). This process corresponds to veto by three films of the upstream. Third, the paired topology, another independent track runs abreast nearby the track, was requested. Here, the number of tracks was reduced by $\sim 2 \times 10^{-3}$. Fourth, these electron and positron candidate tracks were connected downward from the detected film nearby the vertex, that called track follow down process. The gamma-ray event candidates which reached the bottom film (100th film of the whole of the converter unit) become the effective events for timestamp analysis. Figure 5 shows 3D views of the above processes.

3.2.2. Energy reconstruction

The energies of detected gamma-ray events below GeV were reconstructed by the converter data. The amount of multiple coulomb scattering of an electron (a positron) running in the converter was measured as RMS of angular difference and converted into momentum. Figure 6 shows a distribution of reconstructed gamma-ray energy from measured momentum of electrons and positrons. For high energy events without significant scattered angle, momentum can be determine by additional analysis of the calorimeter data.

3.2.3. Comparison with simulation data

We checked the response of the gamma-ray selection by Monte Carlo simulation (MC). MC data was generated by geant4.10.01. Primary gamma rays were exposed to the detector, then the response of HTS (track-finding efficiency and position and angular accuracy) were applied to electron and positron tracks. A simple power low spectrum was adopted as the energy distribution of incident gamma rays. The topological selection was performed as same as the flight data analysis mentioned above.

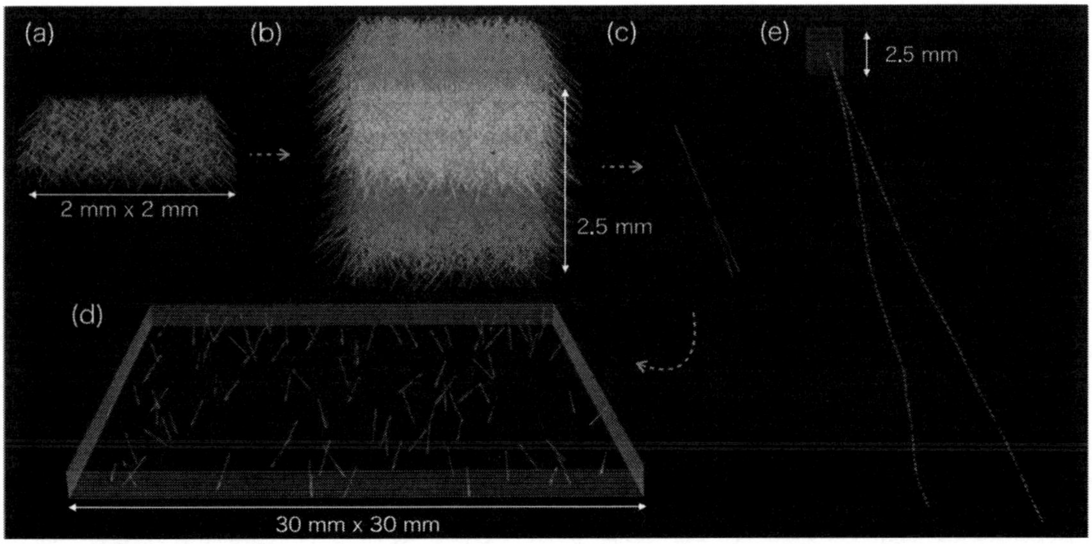

Figure 5. 3D views of scanned track data. (a) indicates 2 mm × 2 mm area in a GRAINE 2015 film. the density is ~400 tracks/mm². (b) indicates the volume of eight films. (c) is a typical gamma-ray event detected by the selection process. (d) is a wide view (30 mm × 30 mm) as a result of gamma-ray detection. (e) shows a typical event as a result of the track follow down.

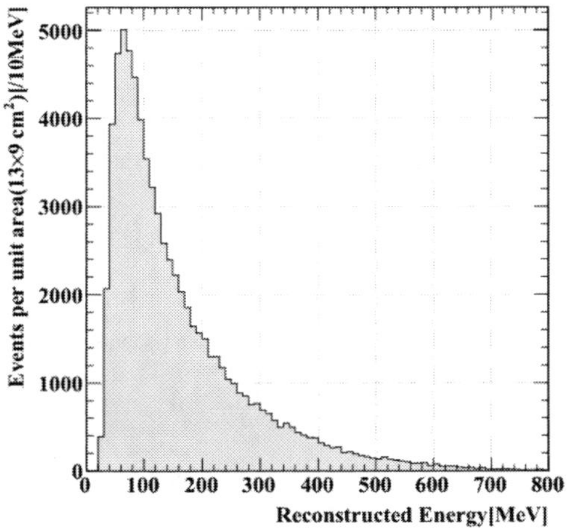

Figure 6. Distribution of reconstructed gamma-ray energy.

Figure 7 shows distributions of the opening angle between electron and positron tracks, reconstructed momentum, the ratio of the division of energy, and the invariant mass. MC data reproduced the tendencies of the flight data. In the current conventional selection, the event-pickup efficiency in the converter was estimated at 65% and 83% for 100 MeV and 200 MeV gamma rays, respectively.

3.3. Background measurement at balloon altitude

The detected gamma-ray events at the converter were connected to the time stamper, and the incident time were given to them. The main backgrounds for the observation

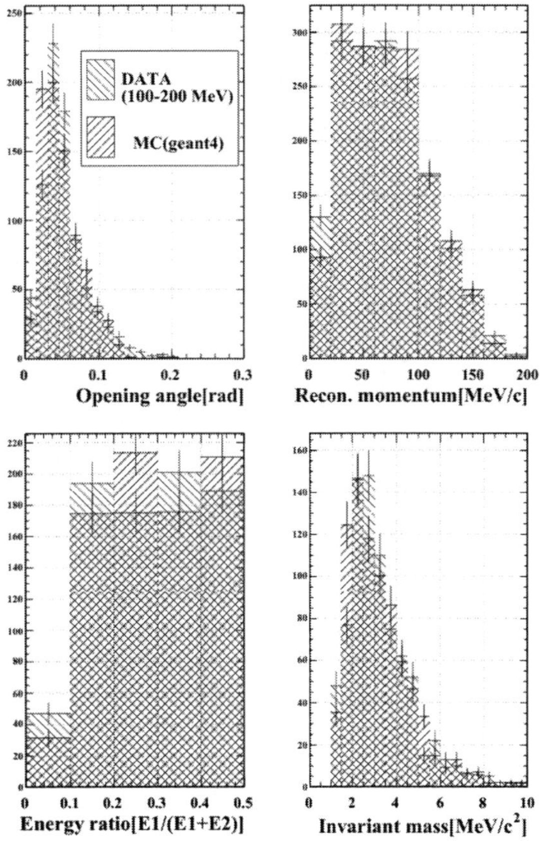

Figure 7. Comparisons of kinematical distributions between flight data and MC.

DOI: 10.1051/epjconf/201714506002

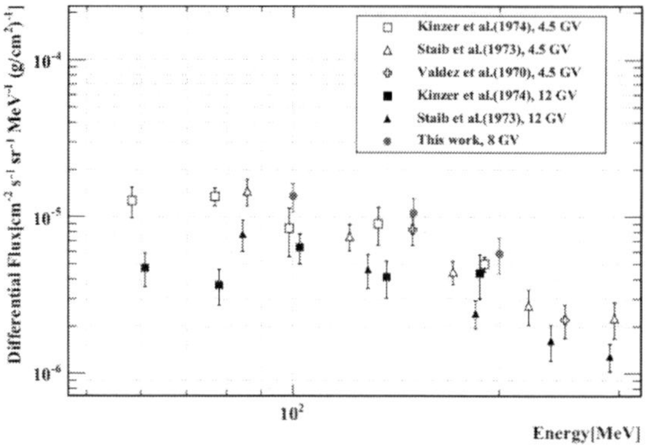

Figure 8. Preliminary result of measurement of atmospheric gamma-ray flux.

of cosmic gamma-ray sources consist of the external component (atmospheric gamma ray), and the internal component (secondary gamma ray induced in the detector by proton or electron). The latter can be rejected by identifying the hadron interaction vertex or the electron running abreast with the same incident time as the gamma ray. Here, the atmospheric gamma-ray flux was measured at the Vela-observation period (the altitude 36.0–37.4 km). Gamma-ray events were selected from upstream events, which started from the 95th or upper films in the converter, to avoid the contamination of the internal component. Figure 8 shows a preliminary result of measurement of the atmospheric gamma-ray flux in 100–200 MeV energy region. We are obtaining the tendency similar to the past measurements and studying the detector response to reduce the systematic error to 5% or less.

4. Summary and prospects

The second balloon experiment of GRAINE project was performed on May12th, 2015, to demonstrate the imaging performance of the emulsion gamma-ray telescope in the 100 MeV energy region. In the experiment, a 3780 cm^2 aperture telescope was employed. We succeeded in the balloon flight and recovery, and got the flight data (emulsion films). So far, 75% of data process (data taking by the scanning system, and gamma-ray event detection in the offline process) were finished. Currently, tuning of the detector response and the measurement of background is ongoing.

We will map gamma-ray events on the celestial coordinate by matching the attitude information and discuss detection of the Vela pulsar. After that, we will enlarge the aperture area and start the scientific observation with long-duration balloon flights.

The scientific balloon(DAIKIKYU) flight opportunity was provided by ISAS, JAXA. This work was supported by JSPS KAKENHI Grant Number 26247039, 26105510, and 16K17691.

References

[1] M. Tavani et al., Astron. Astrophys. **502**, 995 (2009)
[2] W.B. Atwood et al., Astrophys. J. **697**, 1071 (2009)
[3] F. Acero, et al., The Astrophysical Journal Supplement Series **218**(2), 23 (2015)
[4] A. Giuliani, et al., The Astrophysical Journal Letters **742**(2), L30 (2011)
[5] M. Ackermann, et al., Science **339**(6121), 807–811 (2013)
[6] T. Daylan, et al. Physics of the Dark Universe **12**, 1–23 (2016)
[7] K. Ozaki, et al., Nucl. Instrum. Meth. A **833**, 165–168 (2016)
[8] S. Takahashi et al., Nucl. Instrum. Meth. A **620**, 192 (2010)
[9] S. Takahashi et al., Prog. Theor. Exp. Phys. **2015**, 043H01 (2015)
[10] S. Takahashi et al., Prog. Theor. Exp. Phys. **2016**, 073F01 (2016)
[11] M. Yoshiomto et al., to be published in Prog. Theor. Exp. Phys.
[12] K. Ozaki et al., Journal of Instrumentation **10**, P12018 (2015)

Cosmic ray nuclei detection in the balloon borne nuclear emulsion gamma ray telescope flight in Australia (GRAINE 2015)

Atsushi Iyono[1,a], Saya Yamamoto[1,b], Shigeki Aoki[2], Toshio Hara[2], Kenji Kuretsubo[2], Toshitsugu Marushima[2], Haruka Matsumoto[2], Fukashi Mizutani[2], Keita Ozaki[2], Emi Shibayama[2], Atsumu Suzuki[2], Satoru Takahashi[2], Yurie Tateishi[2], Misato Yabu[2], Kyohei Yamada[2], Koichi Kodama[3], Kaname Hamada[4], Hiroaki Kawahara[5], Ryosuke Komatani[5], Masahiro Komatsu[5], Motoaki Miyanishi[5], Misaki Morishita[5], Kunihiro Morishima[5], Mitsuhiro Nakamura[5], Naotaka Naganawa[5], Toshiyuki Nanano[5], Akira Nishio[5], Kimio Niwa[5], Naoto Otsuka[5], Hiroki Rokujo[5], Osamu Sato[5], and Masahiro Yoshimoto[5]

[1] Dept. of Fundamental Science, Faculty of Science, Okayama University of Science, Okayama 700-0005, Japan
[2] Faculty of Human Development, Kobe University, Kobe-city 657-8501, Japan
[3] Aichi University of Education, Kariya-city 448-8542, Japan
[4] JAXA/ISAS, Sagamihara-city 252-5210, Japan
[5] Dept of Physics, Faculty of Science, Nagoya 464-8602, Japan

Abstract. Nuclear emulsion plates for studying elementary particle physics as well as cosmic ray physics are very powerful tracking tools with sub-micron spatial resolutions of charged particle trajectories. Even if gamma rays have to be detected, electron-positron pair tracks can provide precise information to reconstruct their direction and energy with high accuracy. Recent developments of emulsion analysis technology can digitally handle almost all tracks recorded in emulsion plates by using the Hyper Track Selector of the OPERA group at NAGOYA University. On the other hand, the potential of time resolutions have been equipped by emulsion multilayer shifter technology in the GRAINE (Gamma Ray Astro-Imager with Nuclear Emulsion) experiments, the aims of which are to detect cosmic gamma rays such as the Vela pulsar stellar object by precise emulsion tracking analysis and to study cosmic ray particle interactions and chemical compositions. In this paper, we focus on the subject of cosmic ray nuclei detection in the GRAINE balloon flight experiments launched at Alice Springs, Australia in May 2015.

1. Introduction

The observation of high-energy cosmic gamma rays provides the origins and acceleration mechanism of high-energy cosmic rays in the universe. Recently Fermi-LAT [1,2] reported observations of more than three thousand gamma ray point sources including 30 percent unidentified objects. To explore these sources in detail, a more precise angular resolution has been needed in the gamma ray energy range of 10 MeV–100 GeV.

Nuclear emulsion films are precise tracking detectors which record the charged particle trajectories with less than one micrometer accuracy, three-dimensionally. To detect gamma rays in nuclear emulsion detectors, the leading point of the electron-positron pair created by a gamma-ray has been measured with precise submilliradian angular resolution in emulsion films; the gamma ray arrival direction and momentum can be precisely determined.

On May 12, 2015, a Japan-Australia JAXA collaborative balloon-borne experiment (GRAINE 2015) [3] was successfully launched from the Alice Springs balloon-launching station to Longreach with a 3780 cm^2-aperture telescope for a duration of 14.4 hours, including 11.5 hours of level flight at an altitude of 36.0 km to 37.4 km (4.8 g/cm^2 to 3.9 g/cm^2). The primary purpose of this flight was to confirm the overall system performance and to observe the morphology of well-known and brightest gamma-ray objects such as the Vela pulsar with the gamma-ray emulsion telescope.

The nuclear emulsion films had been exposed to cosmic radiation at this height directly, and we expected comic ray nuclei components including He nuclei and heavier ones. Cosmic ray nuclei provide important information on the origin, propagation and acceleration mechanism of cosmic rays in terms of primary-secondary nuclei ratio. In the GRAINE balloon flight [3], the detection of these cosmic ray nuclei components was examined with image analysis technologies as well as time-stamper systems.

2. Apparatus

The structure of the emulsion gamma-ray telescope [3,4] which consists of a converter, time-stamper, calorimeter, and an attitude monitor is shown in Fig. 1. The materials of the converter part make the incident gamma rays convert into electron-positron pairs and these electrons have been registered as three dimensional tracks in the stacks of nuclear emulsion films. The time-stamper, which consists of triple emulsion film shifters as hour, minute

[a] e-mail: iyono@das.ous.ac.jp
[b] e-mail: S15SM08SY@ous.jp

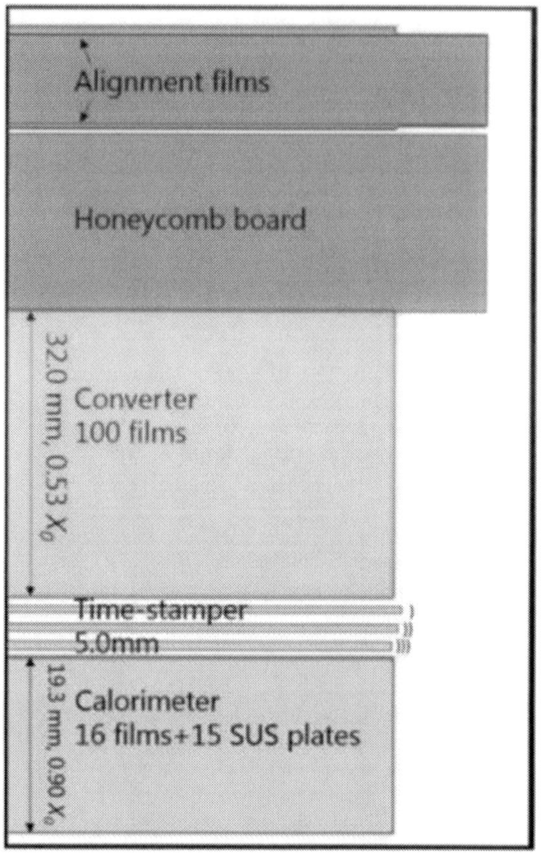

Figure 1. Emulsion telescope structure which consists of the converter, time-stamper and calorimeter.

Figure 2. Emulsion chambers in the pressure vessel. Three motors shift emulsion films in the time-stamper part in the three different time period such as hour, minute and second hands of an analog clock.

and second hands of an analog clock shown in Fig. 2, provides an arrival time (or time-stamp) information for all charged tracks. After the recognition and reconstruction of electron pairs, the arrival time and direction as well as the momentum were determined. The arrival time obtained by time-stamping of electron pairs is combined with three star camera images deployed on the balloon gondola shown in Fig. 3, and their arrival directions into the celestial coordinate were determined. These cameras monitored the attitude of gondola to observe the stars positions. The

Figure 3. Gondola with thermal shielding. The three star cameras are deployed at the truss structure and view three different directions.

trajectory of charged particles measured in the emulsion films enables an estimation of their momentum by using the multiple Coulomb scattering method up to several hundred MeV/c.

3. Emulsion film analysis

GRAINE2015 emulsion films were recovered after completing the balloon flight at Longreach, and sent to Sydney University for chemical developments within less than one month. The design values of emulsion film thickness are summarized in Table 1. Further chemical treatment for these developed films was handled at Kobe university. The typical thickness of the emulsion part of films after development was very thin (such as $\sim 40\,\mu$m). For further image recognition procedures with high efficiency, we need the original emulsion thickness when exposing them to cosmic radiation. The glycerin swelling process has been carried out for all emulsion films to swell the gelatin part of nuclear emulsion by glycerin infiltration. The recovered thickness was achieved up to 90% of the nominal thickness.

Table 1. The thickness of emulsion films in each part of chamber.

Part of detector	Thickness of emulsion layer [μm]	Thickness PS base [μm]	Total thickness [μm]
Converter	70	180	320
Time-stamper	42	180	264
	40	40	120
Calorimeter	45	180	270

PS: polystyrene base film.

In the emulsion scanning process, GRAINE has used the high performance and high speed track selector HTS (Hyper Track Selector shown in Fig. 4) developed

DOI: 10.1051/epjconf/201714506003

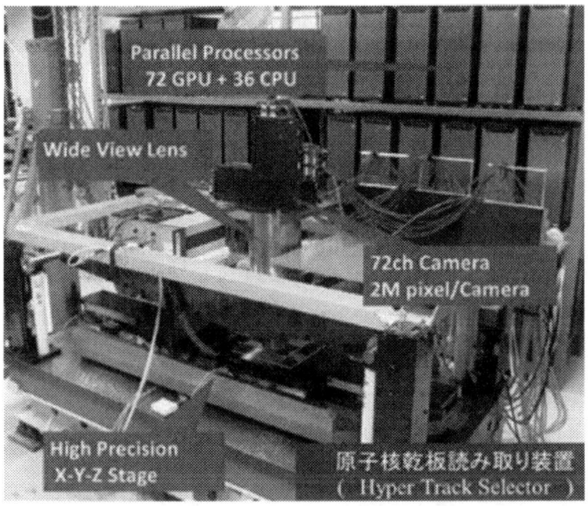

Figure 4. Hyper Track Selector (HTS) implemented at the Fundamental particle laboratory in Nagoya University.

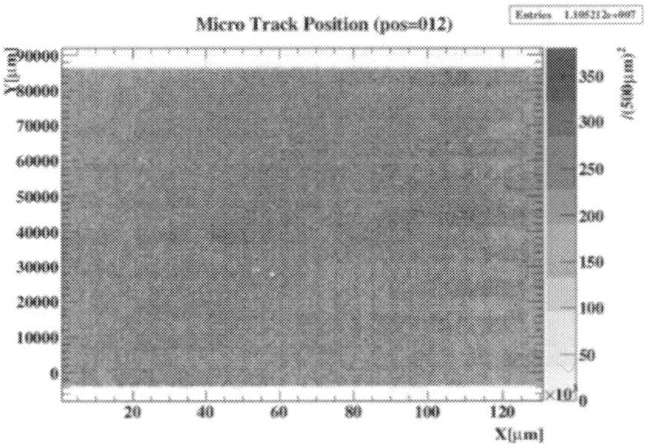

Figure 5. HTS enables general scanning of emulsion films extended to 8 cm × 12 cm area within $\tan\theta \leq 2$. The map shows the track density per $500 \times 500\,\mu m^2$.

and implemented at the Fundamental Particle Laboratory (F-lab.) [5,6] in Nagoya University. HTS achieves over $5000\,cm^2$ per hour scanning speed. Figure 5 is the identified track density map in one HTS scanning command. This capability provides the general scan method for the emulsion analysis with wide zenith angle allowance for cosmic ray nuclei studies. To recognize charged tracks in the emulsion films, HTS sliced both upper and bottom emulsion layers into 32 digitized images in GPU memory space. The GPU pipeline logically shifted them in two dimensional directions and summed the image overlapping along the specified zenith angle and orientation angle. The Pulse Height (PH) and Pulse Height Volume (PHV) were defined as the number of overlapped pixels and the total number of PH values above threshold values (typically 7 or 8), respectively. The HTS scanning algorithm is shown in Fig. 6. The nominal allowance of zenith angle was extended up to 63 degrees ($\tan\theta \leq 2.0$) in the GRAINE2015 analysis and the track recognition efficiency became more than 95%. The extension of angular allowances and quick scan capability in HTS made the cosmic ray nuclei analysis in GRAINE experiments possible.

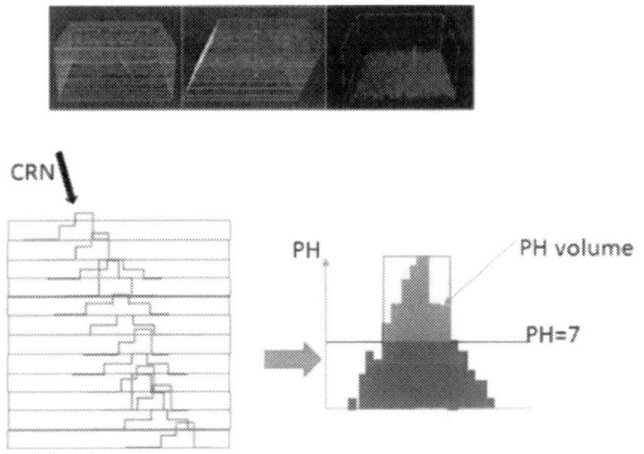

Figure 6. Hyper Track Selector (HTS) logical algorithm to search for tracks and definitions. The symbol:CRN represents incident charged particle to emulsion layer and PH and PH volume are the number of overlapped pixels and summed pixel number above PH greater than 7, respectively.

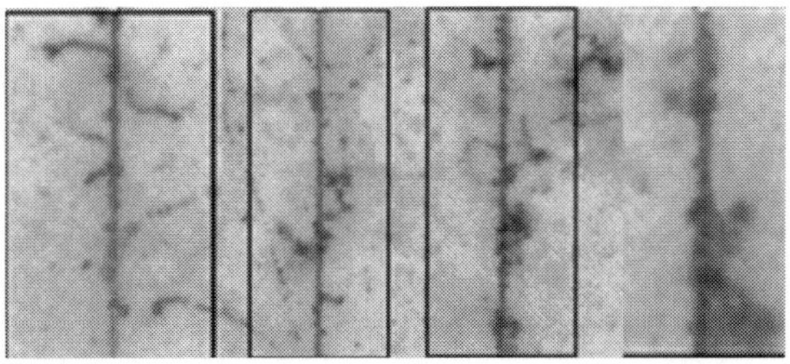

Figure 7. 4 example raw images for high PHV tracks. Images are processed by applying dark peak fold filter and accumulated along the track directions.

4. Pulse height volume (PHV) analysis

The pulse height volume defined in Sect. 3 could provide the ionization information of each track's (Z/β). We selected 50 tracks for both $\tan\theta < 0.6$ and $\tan\theta > 1.0$ with the highest PHV signals and obtained their raw images by using the HTS manual mode operations shown in Fig. 7. Most of them were associated with knock-on electrons (δ rays), which indicated the incidence of cosmic ray heavier nuclei. Because the number of δ rays were very limited, we did not go into detail for its relation to PHV data at present.

We analyzed about 10^6 tracks in each emulsion film and followed them among 100 emulsion film stacks. The tracking efficiencies between three films were typically 95%. We examined the PHV distribution in one emulsion film. The PHV data for single charged tracks depend on their incident zenith angles because all grains overlapped for the case of $\tan\theta = 0$ and sparsely overlapped for horizontal tracks. The PHV distributions for each zenith angle group are shown in Fig. 8. The mode values of PHV distributions decreased as their zenith angles increased, due to the grain separation of single charged tracks. On the other hand, the PHV highest values became larger as the zenith angle increased, because the horizontal tracks blurred some multiple sliced images. We demonstrated the normalized PHV distribution by fixing the mode value equal to 50, the second peak PHV value above 100 and

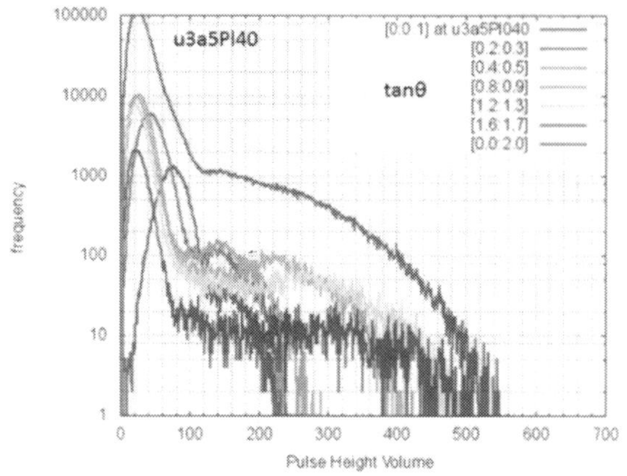

Figure 8. PHV distribution at one emulsion layer. The horizontal axis represents PHV. Each color represents a zenith angle group.

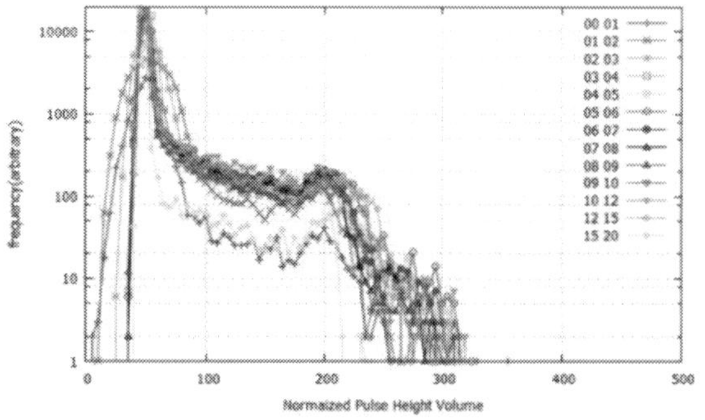

Figure 9. Normalized PHV distribution at emulsion layer. Horizontal axis represents nomalized PHV. Each color represents a zenith angle ($\times 10$) group.

minimum PHV values. The obtained nomalized PHV distributions for several zenith angle groups are shown in Fig. 9. Around the PHV mode, these distributions scattered with zenith angle. This could come from the grain overlapping effects. But nomalized PHV distributions were well described with one shape above PHV$\geq$100. This could provide some universality of the relation of PHV to zenith angle. In these distributions, we clearly obtained a second peak for each zenith angle group. The zenith angle dependency of the nomalization parameters (minimum, mode and maximum PHV values) were plotted as a function of zenith angles in Fig. 10. We have still not concluded these feature of PHV zenith angle dependency to identify this second PHV peak as cosmic ray nuclei components so far.

5. Pulse height volume distributions at sea level and flight level

The GRAINE project has developed and utilized emulsion film shifters in order to determine charged track arrival timing described above. This system enables us to distinguish between the accumulation of background muon tracks at sea level and cosmic rays in flight by using track arrival timing. We set up emulsion shifters at special coordinates such as an origin when waiting for

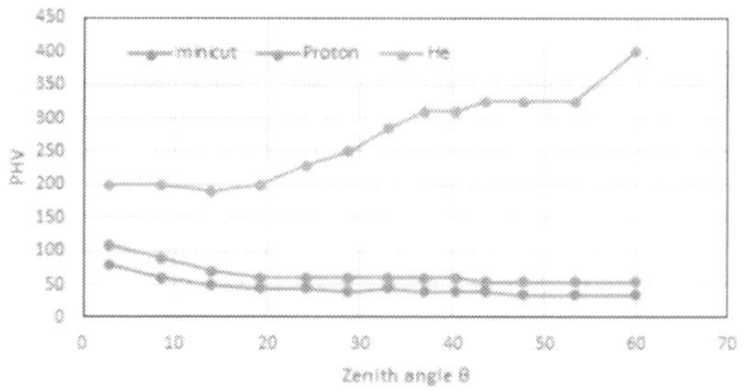

Figure 10. The parameters of the normalized PHV distribution dependence on zenith angle. The minimum, mode and maximum PHV values are indicated by minicut, Proton and He respectively.

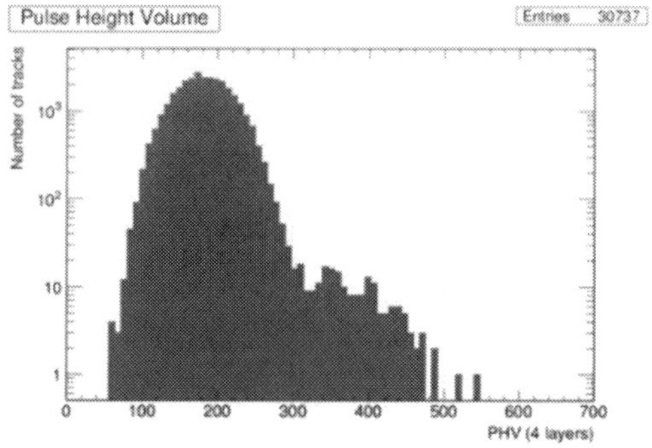

Figure 11. PHV distributions at ground level.

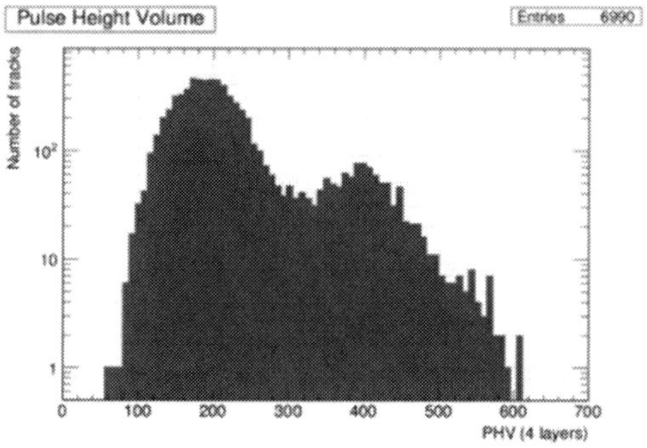

Figure 12. PHV distributions at ground level.

the balloon launch. In the flight, three shifter stages moved in each proper period like hour, minute and second hands of an analog clock. The PHV distribution of the tracks which penetrated the chambers at sea level is shown in Fig. 11. The horizontal axis represents the PHV summation between 4 emulsion layers, and then PHV values were twice as large as the previous ones. In this distribution, the mode PHV (~ 180) came mainly from single charged particles such as EAS muons. Above PHV equal to 300, the second peak of PHV distribution was unseen and the maximum PHV does not reach 500. On the other hand, the PHV distribution obtained in the balloon flight shown in Fig. 12 differed from the previous one especially above PHV $=$ 300. We have clearly seen the second peak at around PHV $=$ 400 and the maximum values extend to PHV $=$ 600. This difference came from cosmic ray Helium and heavier nuclei contributions in the high PHV range. This result proved the potential of cosmic ray nuclei measurement in GRAINE projects.

6. Conclusion

The GRAINE projects successfully completed a balloon flight at Alice Springs, Australia of 14 hours duration. The overall system worked well within the design accuracies. The emulsion film handling before and after the balloon flight was controlled very well in both Australia and Japan. The swelling process before HTS image recognition process has been established in order to control the emulsion layer thickness suitable for HTS scanning efficiency.

The pulse height volume obtained by the HTS track recognition process was able to be used for cosmic ray nuclei identification. The PHV distributions provided cosmic ray nuclei composition information. Finally, the time-stamper system can distinguish the ground level EAS muons and flight-components of direct cosmic rays, by using the arrival timing of cosmic rays.

The qualitative characteristics of pulse height volume (PHV) were successfully derived from these analyses so far. But we have to understand the quantitative estimation of PHV dependencies on the incident zenith angles. We still continue data analyses of track data in the GRAINE2015 balloon flight.

References

[1] F. Acero et al., Astrophys. J. **218**, 23, (2015)
[2] C. Gene et al., Astrophys. J.L. **L13**, 797 (2014)
[3] S. Takahashi et al., Prog. Theor. Exp. Phys. 073F01, 12 (2016)
[4] H. Rokujo et al., Proc. 34th Int. Cosmic Ray Conference The Hague, PoS(ICRC2015)1021, (2015)
[5] K. Niwa et al., Proc. Int. Cosmic Ray Symp. High Energy Phenomena Tokyo, 149 (1974)
[6] M. Yoshimoto, Master's thesis, Nagoya University, Japan (2015) (in Japanese)

Particle spectra and mass composition in the ultra-high energy region in the framework of the Galactic origin of cosmic rays

A.A. Lagutin[a], N.V. Volkov, A.G. Tyumentsev, and R.I. Raikin

Altai State University, Radiophysics and Theoretical Physics Department, 61 Lenin ave, 656049 Barnaul, Russia

Abstract. The possibility for a self-consistent description of all the basic features of the observed cosmic ray spectra and primary composition variations in the energy range of $10^{15} \div 10^{20}$ eV within the Galactic origin scenario is examined.

We assume the existence of Galactic sources that accelerate particles up to $\sim 3 \cdot 10^{18} Z$ eV and take into account a highly inhomogeneous (fractal-like) distribution of matter and magnetic fields in the Galaxy that leads to extremely large free paths of particles ("Lévy flights"), along with an overwhelming contribution to the cosmic ray fluxes observed above $\sim 10^{18}$ eV from particles reaching the Solar System without scattering. Our scenario was refined on the basis of recent experimental results on primary mass composition. Model predictions, which could be verified with the improved high-precision measurements in the nearest future are discussed.

1. Introduction

Physical mechanisms governing the features of energy spectra of cosmic rays (CRs) and their composition variations in the energy range that spans from PeV to tens of EeV are still under debate. The so-called standard scenario assumes that the Galactic component of CRs accelerated by supernova remnants extends up to energies of a few 10^{17} eV with gradually heavier composition due to the rigidity-dependent maximum acceleration energy of nuclei. Then Galactic CRs merge into a light and flatter extragalactic component, which terminates at $\sim 5 \cdot 10^{19}$ eV due to the GZK effect.

Recent measurements by several new generation experiments and also the results of some new methods of experimental data analysis [1–5] have severely challenged the standard scenario. In particular, an ankle-like feature in the light CR component measured by KASCADE-Grande at $E > 10^{17}$ eV [1] together with evidence for a light-mass fraction (protons and helium) of about 80 percent at $10^{17} \div 10^{17.5}$ eV [2] cause difficulties for an explanation in the framework of the standard model without strong additional assumptions. At ultra-high energies the chemical composition is still ambiguous. Mixed composition including nuclei heavier than helium is reported by the Pierre Auger Observatory (PAO) on the basis of various improved methods and techniques [3,4], while the Telescope Array (TA) interprets their data as being compatible with a largely protonic composition [5]. Thus, the non-GZK origin of the energy spectrum suppression at the highest energies (e.g., the acceleration limits in Galactic cosmic ray sources above 10^{18} eV) can not be excluded.

In our papers [6–8] it was shown that basic structures in the all particle CR spectrum (knee, ankle and cut-off) can be reproduced under the assumption that essentially all CRs, including those at the highest energies, originate from Galactic sources, which can accelerate particles up to the cut-off energy. The highly inhomogeneous (fractal-like) distribution of matter and magnetic fields in the Galaxy that leads to extremely large free paths of particles (so called "Lévy flights"), along with large contribution of non-scattered particles to cosmic ray fluxes above $\sim 10^{18}$ eV is the main element of our model–the anomalous diffusion model. At the present stage the new more precise data, mainly the spectrum features measured separately for light and heavy components along with the improved composition estimations, provide an opportunity for refinements and updates of the model.

In the present paper we revise the possibility of the self-consistent description of the CRs energy spectra and composition results within the Galactic cosmic ray origin scenario based on the anomalous diffusion model and discuss the crucial model predictions, which could be verified with improved high-precision measurements in the near future.

2. Key elements of the proposed scenario

The following basic principles and assumptions were initially included in the proposed scenario.

1. The existence of Galactic sources that accelerate particles with atomic charge Z up to $E_0 \sim 4 \cdot 10^{18}$Z eV. Consequently, the observed suppression of the primary CR spectrum at $\sim 5 \cdot 10^{19}$ eV is not an extragalactic GZK feature; it reflects the acceleration limit of Galactic CR sources.
2. The highly inhomogeneous (fractal-like) distribution of matter and magnetic fields in the Galaxy that

[a] e-mail: lagutin@theory.asu.ru

leads to the anomalous diffusion of CRs manifested, in particular, by abnormally large free paths of particles (so-called "Lévy flights") with a power-law distribution.

3. An overwhelming contribution to cosmic ray fluxes observed above $\sim 10^{18}$ eV from particles reaching the Solar system without scattering.

Note, that the main argument contrary to any model assuming the Galactic origin of ultra-high energy cosmic rays is that particles originating from Galactic sources could hardly be as isotropic as presently observed. However, in recent works [9,10] it was shown that the isotropy can be understood in the context of Galactic production if a sufficiently careful treatment of CR propagation in the Galaxy is undertaken.

The spatial distribution of sources also suggests the separation of the observed CR flux into three components as follows:

$$J(\mathbf{r},t,E) = J_G(\mathbf{r},E) + J_L(\mathbf{r},t,E) + J_{NS}(\mathbf{r},E).$$

Here

- J_G is the global spectrum component determined by the multiple old ($t \geq 10^6$ yr) distant ($r \geq 1$ kpc) sources.
- J_L is the local component, i.e., the contribution of nearby ($r < 1$ kpc) young ($t < 10^6$ yr) sources. The spatial and temporal coordinates of the local sources are presented in [11].
- J_{NS} is the flux of non-scattered particles.

We assume that the non-scattered component is also formed by nearby ($r < 1$ kpc) sources, defining the spectrum in the ultra-high energy region, and provides the observed flattening of the spectrum at $E \geq 10^{18}$ eV.

The flux of non-scattered particles, J_{NS}, is determined by the injected flux ($S_0 E^{-\gamma}\exp(-E/E_0)$, where $E_0 = 4\cdot 10^{18} Z$ eV), multiplied by a factor describing the probability of reaching the observer at a distance, r, from the source for a particle with a certain energy and atomic number without scattering. This factor has a power-law asymptotic with respect to r (the Lévy flight probability; see the next section for the explicit form) and also suppressed at energies $E < 3\cdot 10^{17} Z$ eV reflecting the fact that for the nucleus with gyroradius less than the typical Galactic inhomogeneity size this inhomogeneity would be opaque.

3. Parameters of the cosmic ray anomalous diffusion model

The highly inhomogeneous character of matter distribution and associated magnetic fields in the Galaxy should be adequately incorporated into the cosmic ray diffusion model. A physically reasonable way for the generalization of the normal diffusion model is to abandon the assumption about statistical homogeneity of the distribution of matter in favour of its fractal distribution. A principle consequence of this generalization is the power-law distribution of free paths, r, in such a medium $p(\mathbf{r},E) \propto A(E,\alpha) r^{-\alpha-1}$, $r \to \infty$, $0<\alpha<2$—so-called Lévy flights. Besides, the intermittent magnetic field of the fractal-like interstellar medium (ISM) leads to a higher probability of a long stay of particles in inhomogeneities, leading to a presence of the so-called Lévy traps. In the general case, the probability density function $q(t,E)$ of time t, during which a particle is trapped in the inhomogeneity (Lévy trap), also has a power-law behaviour: $q(t,E) \propto B(E,\beta) t^{-\beta-1}$, $t \to \infty$, $\beta < 1$.

Generalization of the homogeneous normal diffusion model to the case of inhomogeneous (fractal-like) ISM, has been made for the first time in our papers [12,13]. Later, it was shown [11,14–16] that an anomalous cosmic ray diffusion model allows to describe the main features of nuclei, electron and positron spectra observed in the Solar system. Particularly, in the anomalous diffusion model the key feature of the all particle energy spectrum—the knee at $3\cdot 10^{15}$ eV—appears naturally without additional assumptions.

The equation for the density of particles with energy E at the location $\mathbf{r}$ and time t, generated in a fractal-like medium by Galactic sources with a distribution density $S(\mathbf{r},t,E)$ can be written as [11,14]

$$\frac{\partial N(\mathbf{r},t,E)}{\partial t} = -D(E,\alpha,\beta)\mathrm{D}_{0+}^{1-\beta}(-\Delta)^{\alpha/2}N(\mathbf{r},t,E) + S(\mathbf{r},t,E). \tag{1}$$

Here $\mathrm{D}_{0+}^{1-\beta}$ denotes the Riemann-Liouville fractional derivative [17] and $(-\Delta)^{\alpha/2}$ is the fractional Laplacian ("Riesz operator") [17]. The anomalous diffusion coefficient $D(E,\alpha,\beta) \sim A(E,\alpha)/B(E,\beta) = D_0(\alpha,\beta)E^{\delta}$.

The solution of Eq. (1) for a point impulse source with a power-law injection spectrum and emission time T $S(\mathbf{r},t,E) = S_0 E^{-\gamma}\delta(\mathbf{r})\Theta(T-t)\Theta(t)$ ($\Theta(\tau)$ is the step function) has the form [11,14]

$$N(\mathbf{r},t,E) = \frac{S_0 E^{-\gamma}}{D(E,\alpha,\beta)^{3/\alpha}} \times \int\limits_{\max[0,t-T]}^{t} d\tau\, \tau^{-3\beta/\alpha}\Psi_3^{(\alpha,\beta)}\left(|\mathbf{r}|(D(E,\alpha,\beta)\tau^{\beta})^{-1/\alpha}\right), \tag{2}$$

where $\Psi_3^{(\alpha,\beta)}(\rho)$ is the density of the fractional stable distribution [18,19]

$$\Psi_3^{(\alpha,\beta)}(\rho) = \int\limits_0^{\infty} g_3^{(\alpha)}(r\tau^{\beta}) q_1^{(\beta,1)}(\tau)\tau^{3\beta/\alpha}d\tau.$$

Using the representation $N = N_0 E^{-\eta}$ and the property $d\Psi_m^{(\alpha,\beta)}(\rho)/d\rho = -2\pi\rho\Psi_{m+2}^{(\alpha,\beta)}(\rho)$ of the scaling function [18], one can easy find the spectral exponent η for observed particles:

$$\eta = -\frac{d\log N}{d\log E} = \gamma + \frac{\delta}{\alpha}\Xi, \tag{3}$$

where

$$\Xi = 3 - \frac{2\pi r^2}{D(E,\alpha,\beta)^{2/\alpha}} \times \frac{\int\limits_{\max[0,t-T]}^{t} d\tau\, \tau^{-5\beta/\alpha}\Psi_5^{(\alpha,\beta)}\left(|\mathbf{r}|(D(E,\alpha,\beta)\tau^{\beta})^{-1/\alpha}\right)}{\int\limits_{\max[0,t-T]}^{t} d\tau\, \tau^{-3\beta/\alpha}\Psi_3^{(\alpha,\beta)}\left(|\mathbf{r}|(D(E,\alpha,\beta)\tau^{\beta})^{-1/\alpha}\right)}. \tag{4}$$

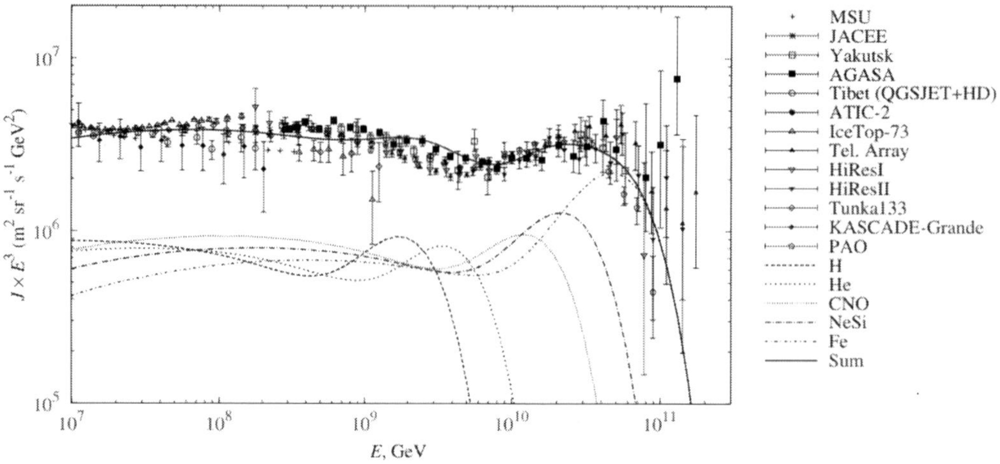

Figure 1. All-particle spectrum obtained in the anomalous diffusion approach. Our results are compared with results from experiments: MSU [26], JACEE [27], Yakutsk [28], AGASA [29], Tibet [30], ATIC-2 [31,32], IceTop-73 [33], Telescope Array [34], HiRes I [35], HiRes II [36], Tunka-133 [37], KASCADE-Grande [38], PAO [39].

Let E_k be a solution of the equation $\Xi(E) = 0$. One can see from (3) and (4) that at $E = E_k$ the spectral exponent for observed particles η is equal to the spectral exponent for particles generated by the source: $\eta(E_k) = \gamma$. Since the exponent $\eta|_{E \ll E_k} = \gamma - \delta$ is less than γ at $E \ll E_k$, but the exponent $\eta|_{E \gg E_k} = \gamma + \delta/\beta$ is greater than γ at $E \gg E_k$, E_k can be called the "knee" energy.

From experimental values of $\eta|_{E \ll E_k}$ and $\eta|_{E \gg E_k}$ one can derive the main parameters of the model (γ, δ) versus the spectral exponent β of the "Lévy waiting time":

$$\delta = \left(\eta|_{E \gg E_k} - \eta|_{E \ll E_k}\right) \frac{\beta}{1+\beta}, \quad \gamma = \eta|_{E \ll E_k} + \delta.$$

Since $\eta|_{E \gg E_k} - \eta|_{E \ll E_k} \sim 0.6$, $\eta|_{E \ll E_k} \sim 2.55 \div 2.65$ [20], $\delta \sim 0.27$, the last equations permit to retrieve self-consistently both spectral exponents γ and β:

$$\gamma \sim 2.85 \div 2.95, \quad \beta \sim 0.8.$$

We note that a similar steep spectrum of accelerated particles has been observed from the supernova remnants W44 [21] and IC 443 [22] with the Fermi Large Area Telescope, and W49B with H.E.S.S. and Fermi-LAT [23]. The value $\gamma = 3$ has been found in [24] for RX J1713.7-3946. At energies higher than 400 GeV, VERITAS and Fermi-LAT observe gamma-ray emission from Tychos SNR with power-law index ~ 2.92 [25].

To evaluate the parameter α, general results for the particles' spectral exponent γ, obtained in the framework of the diffusive shock acceleration theory extended to the case of anomalous transport with "Lévy flights" and "Lévy traps", have been used (see [Lagutin A.A., 2017, to be published]). It was shown that for $\beta = 0.8$ the spectral exponent $\gamma = 2.8 \div 2.9$ corresponds to nondiffusive transport with $\alpha \sim 1.7$.

The technique for determining the anomalous diffusivity was described in [11]. In this work, we used the following value $D_0 \approx 1.5 \cdot 10^{-3}$ pc$^{1.7}$y$^{-0.8}$.

Table 1. Anomalous diffusion model parameters.

Parameter	Value
α	1.7
β	0.8
$D_0(\alpha, \beta)$	$1.5 \cdot 10^{-3}$ pc$^{1.7}$ y$^{-0.8}$
δ	0.27
δ_L	$\delta/2$
γ	2.85
E_0	$4 \cdot 10^{18}$ Z eV
T	10^4 y

As a result we have the cosmic ray spectrum

$$J(\mathbf{r}, t, E) = \frac{v}{4\pi}\Bigg[S_G E^{-\gamma-\delta/\beta} + \frac{S_0 E^{-\gamma}}{D(E,\alpha,\beta)^{3/\alpha}} \times \sum_{\substack{r_j < 1\ \mathrm{kpc} \\ t_j < 10^6\ \mathrm{yr}}} \int_{\max[0, t_j - T]}^{t_j} d\tau\, \tau^{-3\beta/\alpha} \Psi_3^{(\alpha,\beta)} \times \left(|\mathbf{r}_j|(D(E,\alpha,\beta)\tau^{\beta})^{-1/\alpha}\right) + S_{NS} \sum_{r_j < 1\ \mathrm{kpc}} E^{-\gamma+\delta_L} |\mathbf{r}_j|^{-\alpha}\Bigg] \exp\left(-\frac{E}{E_0}\right). \tag{5}$$

A set of anomalous diffusion model parameters adopted in this paper is given in Table 1.

4. Results

In the framework of the model described above we have analysed the energy spectra and mass composition behaviour in the energy range up to the cut-off observed at $\sim 5 \cdot 10^{19}$ eV. The energy spectra for all particles and elemental groups of nuclei are shown in Fig. 1 compared with the data of different experiments. Along with good overall agreement with the representation of all the basic spectrum features, several key model predictions should be noted.

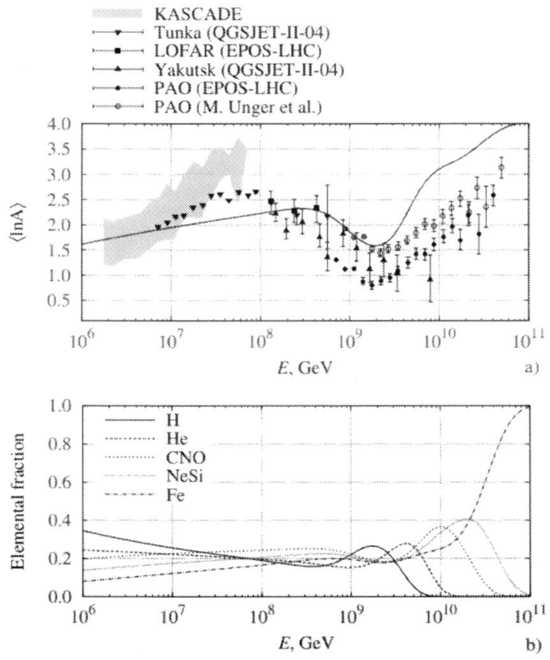

Figure 2. Mean logarithmic mass (a) and elemental fractions (b) of CRs predicted by the anomalous diffusion model (lines). A shaded area and symbols show experimental results from [40] (KASCADE, Tunka, LOFAR, Yakutsk), [41] (PAO) and [42] (PAO corrected by M. Unger et al. for systematic uncertainties in energy and depth of maximum scales).

As already mentioned, the anomalous diffusion provides the knee in the energy spectrum as a pure propagation effect without any specific assumptions about maximum energy that a particle can achieve in a source. Another important feature of the model is that at the energy of the knee the spectral exponent, η, is equal to the injection spectra exponent γ. Thus, if anomalous diffusion is, indeed, the main mechanism of the knee feature, then the injection spectra exponent could be estimated from the analysis of the experimentally measured energy spectra around the knee.

The energies of knee retrieved for five elemental groups of nuclei (H, He, CNO, NeMgSi, Fe) are shown in Table 2. One can see that the knee energies are almost proportional to Z. We note that our estimations for the H spectrum are consistent with the results [20] where the knee-like feature in the (p+He) spectrum is observed at energies around 700 ± 230 TeV.

Mean logarithmic mass and elemental fractions of CRs from 10^{15} to 10^{20} eV are shown in Fig. 2. According to our model, the mass composition became heavier with energy up to $\sim 4 \cdot 10^{17}$ eV. It is seen that at $10^{17.5}$ eV our data is consistent with LOFAR estimations [2].

The non-scattered particles of atomic charge Z contribute to the total flux starting from the energy $3 \cdot 10^{17} Z$ eV, at which their gyroradius became large enough to escape from the typical Galactic inhomogeneity and reach the observer. This results in a decreasing mean mass of CRs in the energy region $(4 \cdot 10^{17} \div 2 \cdot 10^{18}$ eV) matching the PAO data [4] quite well. The lightest composition ($\langle \ln A \rangle \sim 1.6$) is observed at $\sim 2 \cdot 10^{18}$ eV.

Table 2. The knee energies for different elemental groups of nuclei.

Nuclei	Knee energy, eV
H	$(6.1 \div 6.5) \cdot 10^{14}$
He	$(1.8 \div 2.2) \cdot 10^{15}$
CNO	$(5.8 \div 6.3) \cdot 10^{15}$
NeMgSi	$(0.8 \div 1.2) \cdot 10^{16}$
Fe	$(2.6 \div 3.0) \cdot 10^{16}$

At ultra-high energies we have a gradually heavier composition with progressive cut-offs in each group of nuclei at $E_0 = 4 \cdot 10^{18} Z$ eV. At 10^{19} eV our predictions give a heavier composition in comparison with the Auger data obtained from $X_{\max}$ distributions [4], but agree qualitatively with recent estimations [3] made on the basis of the advanced $X_{\max}$ and ground signal correlation analysis.

Finally it should be noted that our model contains an ankle in the light component similar to that observed by KASCADE-Grande [1] and more commonly the multiple ankle-like features in spectra of elemental groups of nuclei at $E > 3 \cdot 10^{17} Z$ eV. This should be considered as a crucial model prediction, which can be verified with the improved high-precision future measurements.

5. Conclusions

We have made a revision of the Galactic cosmic ray origin scenario in the framework of the anomalous diffusion model. It was shown that recent data on the cosmic ray energy spectrum and mass composition can be well reproduced by our model over the entire primary energy range with a reasonable set of key assumptions.

The knee energies were retrieved for different elemental groups of nuclei showing the rigidity-dependent behaviour similar to the standard scenario prediction. It is shown that, if the knee feature is caused by the anomalous diffusion of particles through the fractal-like ISM, the source injection spectra exponent could be estimated from the experimentally observed energy spectra being equal to the spectral exponent at the knee energy.

The following basic model predictions are expressed.

- The suppression of the all-particle spectrum at $E = 5 \cdot 10^{19}$ eV is due to the nuclei fluxes cut-offs caused by the rigidity-dependent energy limitation of the Galactic sources.
- The composition becomes heavier with energy from the knee to $\sim 10^{17.5}$ eV and reaches a maximum of mean logarithmic mass $\langle \ln A \rangle \sim 2.4$.
- In the energy region $(4 \cdot 10^{17} \div 2 \cdot 10^{18}$ eV) the mean logarithmic mass decreases reaching the minimum value of $\langle \ln A \rangle \sim 1.6$.
- The rapid weighting of the mass composition is observed at $E > 2.5 \cdot 10^{19}$ eV up to the pure iron composition at the cut-off.
- Multiple ankle-like spectrum features at energies above $3 \cdot 10^{17} Z$ eV reflecting the contribution of the non-scattered CR component are expected.

This work was supported in part by the Ministry of Education and Science of The Russian Federation (state assignment for the fundamental and applied research performed at Altai State University).

References

[1] W.D. Apel et al. (KASCADE-Grande collaboration), Phys. Rev. D **87**, 081101(R) (2013)
[2] S. Buitink et al. (LOFAR collaboration), Nature **531**, 70–73 (2016)
[3] A. Aab et al. (Pierre Auger collaboration), Phys. Lett. B **762**, 288–295 (2016)
[4] M. Settimo, Journal of Physics: Conference Series **718**, 052037 (2016)
[5] R.U. Abbasi et al. (Telescope Array), Astropart. Phys. **64**, 49–62 (2015)
[6] A.A. Lagutin, A.V. Yushkov, A.G. Tyumentsev, IJMP A **20**, 6834–6836 (2005)
[7] A.A. Lagutin, A.G. Tyumentsev, N.V. Volkov, and R.I. Raikin, Bulletin of the Russian Academy of Sciences: Physics **79**, 322–325 (2015)
[8] N. Volkov, A. Lagutin, A. Tyumentsev, R.Raikin, J. Phys.: Conf. Ser. **632**, 012028 (2015)
[9] R. Kumar, D. Eichler, ApJ **781**, 47 (2014)
[10] D. Eichler, N. Globus, R. Kumar, E. Gavish, ApJL **821**, L24 (2016)
[11] A.A. Lagutin, A.G. Tyumentsev, Bulletin of Altai State University **5**, 4 (2004)
[12] A.A. Lagutin, Y.A. Nikulin, V.V. Uchaikin, Preprint ASU (2000/4)
[13] A.A. Lagutin, Y.A. Nikulin, V.V. Uchaikin, Nucl. Phys. B (Proc. Suppl.) **97**, 267 (2001)
[14] A.A. Lagutin, V.V. Uchaikin, NIM B **201**, 212 (2003)
[15] A.D. Erlykin, A.A. Lagutin, A.W. Wolfendale, Astropart. Phys. **19**, 351 (2003)
[16] A.A. Lagutin, N.V. Volkov, A.S. Kuzmin, A.G. Tyumentsev, Bull. of RAS: Physics **73**(5), 581 (2009)
[17] S.G. Samco, A.A. Kilbas, O.I. Marichev, *Fractional integrals and derivatives—Theory and Applications* (Gordon and Breach, New York, 1993)
[18] V.V. Uchaikin, V.M. Zolotarev, *Chance and Stability* (VSP, Netherlands, Utrecht, 1999)
[19] V.M. Zolotarev, V.V. Uchaikin, V.V. Saenko, JETP **88**(4), 780 (1999)
[20] B. Bartoli et al. (ARGO-YBJ and LHAASO collaborations), Phys. Rev. D **92**, 092005 (2015)
[21] A.A. Abdo, M. Ackermann, M. Ajello et al., Science **327**, 1103 (2010)
[22] A.A. Abdo, M. Ackermann, M. Ajello et al., ApJ **712**, 459 (2010)
[23] H.E.S.S. Collaboration, H. Abdalla, A. Abramowski, F. Aharonian et al. arXiv:1609.00600
[24] T. Tanaka, Y. Uchiyama, F. Aharonian et al., ApJ **685**, 988 (2008)
[25] S. Archambault, A. Archer, W. Benbow et al. ApJ, **836**, 23 (2017)
[26] Yu.A. Fomin, G.B. Khristiansen, G.V. Kulikov et al., Proc. 22 ICRC **2**, 85–88 (1991)
[27] K. Asakimory, T.H. Burnett, M.L. Cherry et al., ApJ. **502**, 278–283 (1998)
[28] B.N. Afanasiev et al., Proc. 24 ICRC, Rome **2**, 756 (1995)
[29] S. Yoshida, N. Hayashida, K. Honda et al., Astropart. Phys. **3**, 105–123 (1995)
[30] M. Amenomori et al. (Tibet ASγ collaboration), ApJ **678**, 1165–1179 (2008)
[31] A.D. Panov et al. (ATIC-2 collaboration), Bull. of the Russian Academy of Sci., Physics **71**, 494–497 (2007)
[32] A.D. Panov et al. (ATIC-2 collaboration), Bull. of the Russian Academy of Sci., Physics **73**, 564–567 (2009)
[33] M.G. Aartsen et al. (IceTop collaboration), Phys. Rev. D **88**, 042004 (2013)
[34] T. Abu-Zayyad et al. (Telescope Array) Astropart. Phys. **61**, 93–101 (2015)
[35] R.U. Abbasi et al. (HiRes I collaboration), Phys. Rev. Lett. **92**, 151101 (2004)
[36] R.U. Abbasi et al. (HiRes II collaboration), Phys. Rev. Lett. **100**, 101101 (2008)
[37] V.V. Prosin et al. (Tunka collaboration), Nucl. Instr. and Meth. A **756**, 94 (2014)
[38] W.D. Apel et al. (KASCADE-Grande collaboration), Astropart. Phys. **47**, 54–66 (2013)
[39] A. Aab et al. (Pierre Auger collaboration), Contrib. to the 33 ICRC (2013)
[40] S. Thoudam et al., A&A **595**, A33 (2016)
[41] A. Aab et al. (Pierre Auger collaboration), Phys. Rev. D **90**, 122005 (2014)
[42] M. Unger, G.R. Farrar, L.A. Anchordoqui, Phys. Rev. D **92**, 123001 (2015)

The PRISMA-LHAASO project: Status and overview

Victor Alekseenko[1], Anastasia Bagrova[1,2], Shuwang Cui[3,a], Yayun He[3], Bingbing Li[3], Xinhua Ma[4], Egor Pozdnyakov[5], Oleg Shchegolev[1], Yuri Stenkin[1,2], and Vladimir Stepanov[1]

[1] Institute for Nuclear Research, Russian Academy of Sciences, Russia
[2] Nuclear Research National University "Moscow Engineering Physics Institute", Russia
[3] The College of Physics and Information Engineering, Hebei Normal University, China
[4] Key Laboratory of Particle Astrophysics, Institute of High Energy Physics, Chinese Academy of Sciences, China
[5] CJSC NPF "Luminofor", Russia

Abstract. The status of and main features of the PRISMA-LHAASO project are presented. We discuss here the science project and some preliminary results obtained with the PRISMA prototype.

1. Introduction: The PRISMA-LHAASO project science

There exist (or existed) very few experiments specially designed to solve the 60 years old "knee problem" in cosmic ray. The best of them, namely, KASCADE and Tibet AS gave very precise and interesting but contradictory results and did not solve the problem [1]. It became even less clear. Neither the knee origin nor the cosmic ray energy spectrum and its mass composition are known today. These problems could not be solved in our opinion by a traditional Extensive Air Shower (EAS) method studying electromagnetic and muon components. New approaches [2] based on the new ideas and multi-component EAS analysis will help us to solve this complicated problem.

The idea of a novel type of array for EAS study in 2001 [3] has been developed in 2008 to the PRISMA (PRImary Spectrum Measurement Array) project [4,5]. It is based on a simple idea: hadrons are the main EAS component forming its skeleton and resulting in all its properties at an observational level [6], so the hadron component should also be the main component to be measured in experiments. During its longitudinal developing in the atmosphere the EAS passes through several stages from the first cosmic ray interaction to almost full degradation deep in the atmosphere. Only the hadron content can be used to recognize the stage of EAS development. The development of a novel type of EAS array detector (en-detector) [7], able to record the hadronic component through thermal neutron detection and electronic component as well, can help to solve the problem.

The experiment has to be performed at high altitude where the number of secondary hadrons reaches maximal. The detector looks like an ordinary EAS detector but with a specific thin inorganic scintillator sensitive to thermal neutrons and having low sensitivity to charged particles. A thin layer of scintillator consists of an alloy of natural boron salt with inorganic scintillator ZnS(Ag). The recording efficiency of the compound to thermal neutrons is close to 20%. By spreading these detectors over a large area on the Earth's surface one can obtain a kind of hadronic calorimeter of practically unlimited area. To make the energy threshold of the array lower it has to be installed at high altitude. We proposed to locate the PRISMA array at 4400 m above sea level in the central part of the LHAASO array.

The PRISMA-LHAASO project is mostly aimed to solve the "knee problem" in the cosmic ray spectrum and to measure any changes of cosmic ray mass composition at energies in the PeV region. The best way to do so could be direct cosmic ray spectrum measurements. Unfortunately, it cannot be performed due to the very low intensity of cosmic rays with energies above 1 PeV. That is why physicists are pressed to use an indirect EAS method. However, as a payment for that, one has to make very complicated and model dependent recalculations from the measurements to primary parameters. Solving the inverse task one should be sure that: i) a solution exists and ii) the measured parameters are connected with real ones by known dependencies. Both points are not known *a priori*. Solving the direct task one also has to introduce many parameters by hand, concerning the details of the model used, cosmic ray mass composition, existence or presence of the "knee" in the primary spectrum etc. [8].

Traditionally EAS arrays measure the electron component first of all. This is not the best but the simplest and most convenient choice because the electron component is the most numerous. However, the electron component is the secondary EAS component sensitive to EAS longitudinal development, which is formed by the cascading of high-energy hadrons. These two components are in a dynamic equilibrium. However, the equilibrium exists only while hadrons exist. When the cascading hadrons are fully exhausted (note that the number of such hadrons is high only at the maximum of shower development and then exponentially decreases with depth

[a] e-mail: cui_shuwang@163.com

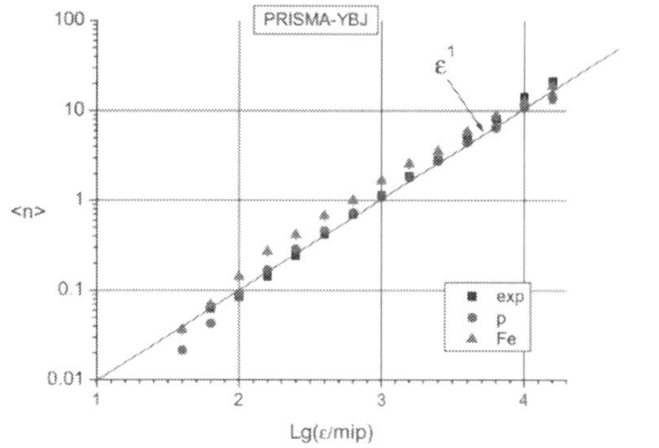

Figure 1. Correlation plot between electromagnetic EAS component (total array energy deposit ϵ) and the mean number of recorded thermal neutrons $\langle n \rangle$.

in the atmosphere), the equilibrium violation occurs. This occurs at energy 100 TeV/nucleon. It changes the EAS properties dramatically and results in a visible break in the electromagnetic components (including Cherenkov light) [2, 8]. Interpretation of the data obtained with traditional EAS arrays is very complicated and ambiguous. Therefore, the best way is to record the primary EAS component, namely the hadronic one, along with other components. The PRISMA experiment will realize this approach measuring hadrons (thermal neutrons) over the whole array area $100 \times 100\,m^2$. In fact, it can be regarded as a kind of huge area single layer hadronic calorimeter, which can be extended without any problem.

2. The PRISMA-LHAASO array details

The array will consists of 25 independent clusters, 16 en-detectors each. The cluster geometry is very simple: a homogeneous rectangular grid with 5 m spacing covering $15 \times 15\,m^2$. Each cluster will work independently triggering by a 2-fold coincidence of any 2 hit detectors with a threshold of 5 particles (m.i.p.). All clusters will be synchronized with each other and so the full array data combination and processing will be made off-line. We will use two signals from each detector: from the last and the 5th dynodes to extend the pulse height measurement range up to $\sim$10000 particles. The dynamic range for neutron measurements is close to 1000 in each detector (or 16000 in total) inside a time gate of 20 ms after EAS passage. All pulses will be digitized using 32-channels and 16-bits FADC. For all powerful showers the full 20 ms waveforms are stored on disc with $1\,\mu$s time resolution. This allows us to check and analyze the selected events visually.

The daily energy deposit spectra of each detector are stored to hard disc to control its gain and stability. Moreover, the time series of the counting rate for events selected by pulse shape analysis as "neutrons" and those as "charged", are stored every 5 min. These data are used for the environmental neutron flux variations study. The latter also allows us to carry out geophysical research, such as neutron and radon diurnal and seasonal waves, thunderstorms, tidal waves, seismic activity, free Earth's oscillations, etc. (see below).

3. Some results of EAS study with a prototype

The PRISMA prototype array made with 4 en-detectors (PRISMA-YBJ) has been running in Tibet (Yangbajing, 4300 m a.s.l.) for more than 3 years and some interesting results have been obtained with it up to date. One can see the correlation plot between EAS electromagnetic and neutron components in Fig. 1 in comparison with the experiment Monte-Carlo simulations for primary protons and iron nuclei. No normalization is applied and the agreement is good. Unfortunately, this plot depends on the absolute values and is weakly sensitive to a mass composition and therefore, it can not be successfully used for such measurements. We used CORSIKA code (ver. 6900, QGSJET, Gheisha) for EAS simulation and GEANT4.10 for the en-detector response simulations [9, 10]. It is seen that at least at high altitude close to EAS maximum, the dependence can be well fitted with a linear function.

Many other results obtained with our prototypes can be found elsewhere [11, 12].

4. Geophysical researches with environmental neutrons flux

Environmental neutrons are produced by two natural sources: by cosmic rays in air and in the upper layers of soil and by natural radioactivity (mostly due to (α,n)-reactions on light nuclei) throughout the Earth's crust. Fast produced neutrons are then moderated by media up to thermal energy and live there up to nuclear capture. Neutron lifetime depends on the media's chemical composition, temperature and water (or any hydrogenous material) content. The natural radioactive chain daughter product inert gas radon-222, having 3.8d half-life, can migrate long distances in air and in soil (rock, concrete, etc.) and even accumulate in some places thus changing the neutron flux in underground locations. It is also sensitive to local seismic activity. Therefore, the flux (or concentration) of thermal neutrons in the media is sensitive to the media parameters such as its temperature, humidity, porosity (seismic activity), etc. Therefore, measuring neutron flux variations for a long time could be used to control the above media parameters.

We plan to use the en-detectors of PRISMA-LHAASO array for continuous environmental thermal neutron flux monitoring and its variations, a study needed not only for EAS experiment background estimation but also for some geophysical applications. The following geophysical items will be investigated through thermal neutron studies with the en-detectors installed underground or on the surface:

- neutrons during thunderstorms (surface)
- Earth's crust Moon tidal effects (surface and underground)
- Seasonal radon-neutron waves at high altitude (surface)
- Free Earth's oscillations (underground)
- Forbush effect and Sun-Earth interconnections (surface and underground)
- Ground Level Enhancement (GLE) effect (surface).

The additional geophysical studies on the Earth's surface could be performed cost free using PRISMA-LHAASO

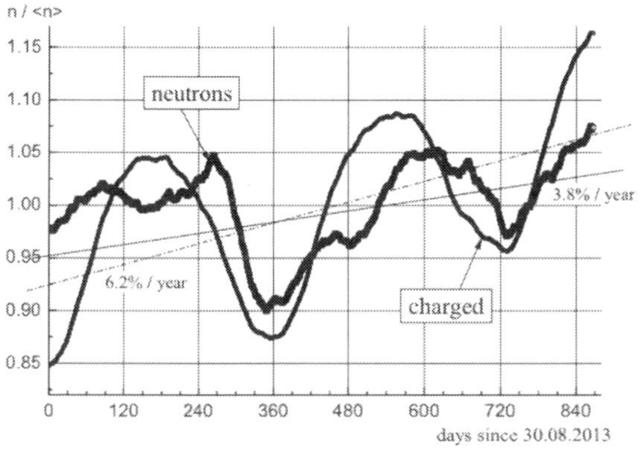

Figure 2. Long-term smoothed variations of thermal neutron background at Yangbajing.

detectors. The items need an underground detector location and could use existing underground or basement rooms to decrease the cosmic ray background and to emphasize the radon-neutron source. Otherwise, it will need additional investments.

An example of thermal neutron variations study can be seen in Fig. 2 where a normalized and 60-day smoothed counting rate is shown for both neutrons and for "charged" for about a 3y-period of observation at Yangbajing. Seasonal variations of about 8% amplitude wave anti-correlated with the rainy season is clearly visible in both curves [13]. These data allows us to estimate the neutron background at high altitude. As we have found the neutron background has also a long-term trend, which we connected with the decrease of solar activity in the 24th solar cycle, resulting in an increase of low energy cosmic rays producing neutrons [14].

More results concerning geophysical studies can be found elsewhere [13,15,16].

5. Summary

A novel type of EAS full-scale array (PRISMA-LHAASO) will be made at high altitude in association with the LHAASO project in the next 4 years. Preliminary results obtained with the prototypes and M-C simulations make us sure that en-detectors are very informative and very stable and the future array could give us absolutely new and fascinating results on cosmic rays near and above so-called "knee" in cosmic ray spectrum. We hope that the problem of the "knee" will be solved soon with the PRISMA-LHAASO.

This work was supported in Russia by RFBR (grants 14-02-00996, 16-32-00054 and 16-29-13067_ofi_m), RAS Presidium Program "Fundamental properties of matter and astrophysics", and in China by NSFC (No. 10975046, No. 11375052).

References

[1] M. Amenomori et al., Astroph. J. **678**, 1165 (2008)
[2] Yu.V.Stenkin, Mod. Phys. Lett. A **8**, 18, 1225–1234 (2003)
[3] Yu.V. Stenkin and J.F. Valdes-Galicia, Proc. of 27th ICRC, 1453 (2001)
[4] Yu.V. Stenkin, Nucl. Phys. B (Proc. Suppl.) **175-176**, 326 (2008)
[5] Yu.V. Stenkin, Nucl. Phys. B (Proc. Suppl.) **196**, 293–296 (2009)
[6] G.T. Zatsepin, Docl. AN SSSR **67**, 993 (1949)
[7] Yuri V. Sten'kin. LARGE SCINTILLATOR DETECTOR FOR THERMAL NEUTRON RECORDING. In: Nuclear Track Detectors: Design, Methods and Applications ISBN: 978-1-60876-826-4, Editor: Maksim Sidorov and Oleg Ivanov © 2010 Nova Science Publishers, Inc. **10**, 253–256.
[8] Yu. V. Stenkin, Physics Atomic Nucl. **71**, 1, 98 (2008)
[9] Y.V. Stenkin et al., Chinese Physics C **37**, 1, 015001 (2013)
[10] O.B. Shchegolev and Yu.V. Stenkin, Bulletin of the Lebedev Physics Institute **43**, 7, 223 (2016)
[11] B. Bartoli et al., Astroparticle Physics **81**, 49–60 (2016)
[12] O. B. Shchegolev et al., J. Phys.: Conf. Ser. **718**, 052038 (2016)
[13] Yu.V. Stenkin et al., Pure and Appl. Geophys., in press; `arXiv:1605.01283 [physics.geo-ph]`
[14] Yu.V. Stenkin et al., Bulletin of Russian Academy of Sciences, in press (2017)
[15] V. Alekseenko et al., Phys. Rev. Lett. **114**, 125003 (2015)
[16] V. Alekseenko et al., J. Phys.: Conf. Ser. **409**, 012190 (2013)

New installation for inclined EAS investigations

E.A. Zadeba[1,a], N.V. Ampilogov[1], N.S. Barbashina[1], A.G. Bogdanov[1], A.A. Borisov[1,2], D.V. Chernov[1], L.I. Dushkin[1], R.M. Fakhrutdinov[1,2], R.P. Kokoulin[1], K.G. Kompaniets[1], A.S. Kozhin[1,2], V.V. Ovchinnikov[1], A.S. Ovechkin[1], A.A. Petrukhin[1], V.V. Shutenko[1], N.S. Volkov[1], V.S. Vorobjev[1], and I.I. Yashin[1]

[1] National Research Nuclear University MEPhI (Moscow Engineering Physics Institute), 115409 Moscow, Russia
[2] RF SSC Institute of High Energy Physics, Protvino 142281, Russia

Abstract. The large-scale coordinate-tracking detector TREK for registration of inclined EAS is being developed in MEPhI. The detector is based on multiwire drift chambers from the neutrino experiment at the IHEP U-70 accelerator. Their key advantages are a large effective area (1.85 m^2), a good coordinate and angular resolution with a small number of measuring channels. The detector will be operated as part of the experimental complex NEVOD, in particular, jointly with a Cherenkov water detector (CWD) with a volume of 2000 cubic meters and the coordinate detector DECOR. The first part of the detector named Coordinate-Tracking Unit based on the Drift Chambers (CTUDC), representing two coordinate planes of 8 drift chambers in each, has been developed and mounted on opposite sides of the CWD. It has the same principle of joint operation with the NEVOD-DECOR triggering system and the same drift chambers alignment, so the main features of the TREK detector will be examined. Results of the CTUDC development and a joint operation with NEVOD-DECOR complex are presented.

1. Introduction

The aim of the large-scale coordinate-tracking detector is solving the problem of the excess of muon bundles that increases with the energy of the primary cosmic rays [1,2], that can be caused by both cosmo- or nuclear-physical reasons [3]. The only characteristic that responds differently to changes in the composition of cosmic rays and the inclusion of new physical processes is the energy of the muon component of extensive air showers [3] which, up to now, has not been investigated enough. Such studies are performed at the experimental complex NEVOD-DECOR [4]; however, the coordinate detector DECOR does not cover the entire aperture of the Cherenkov water detector (CWD) and does not exclude the possibility of several muons passing between the individual supermodules of the detector. Besides, the size of its cells limits the possibility of separating two or more particles at small distances (less than 3 cm). The new coordinate-tracking detector [5] based on drift chambers will increase the coverage of the side aperture of the Cherenkov water detector NEVOD and significantly improve the resolution of close tracks.

2. Drift chambers

Drift chambers (DC) came to accelerator particle experiments and cosmic ray studies in the late 70-ies of the last century, after proportional chambers. Their main advantage is the ability to ensure the measurement accuracy of charged particle tracks significantly better than the characteristic distance between sensitive elements (signal wires). This property provides the possibility to create large-scale installations. One such setup created in the 80-ies was the IHEP-JINR neutrino detector at the U-70 accelerator, for which a large area multiwire drift chamber was developed [6].

Overall sizes of the chambers are $4000 \times 508 \times 112\,mm^3$ with a sensitive area of 1.85 m^2 that is 91% of the side chamber area. There are four sense wires in the middle of the chamber alternately shifted by ± 0.75 mm parallel to the drift direction to solve right-left ambiguity. The distance between the sense wires is 10 mm. There are two guard wires to remove the edge effect. The sense and guard wires are surrounded by the cathode wires. A uniform electric field is created in the drift gap by field-forming wires with a uniformly distributed potential. The chamber gas volume is limited by the aluminium alloy case, 1.5 mm thick, which serves simultaneously as an electric screen and the chamber frame. The chamber ends are limited by plexiglass plugs, where wires, gas inlet and outlet, amplifiers, high voltage supply circuit and cable connectors are assembled. Blind holes in the plugs are provided for the installation of the drift chambers so they are firmly isolated from the setup framework. Due to a nice uniformity of the electric field inside the chamber, electron drift velocity can be assumed to be constant and we can use a linear relation between drift time and coordinate. Configuration of signal wires enables reconstruction of a projection of particle track to the plane orthogonal to the wires. Thus, for reconstruction in space at least two non-parallel drift chambers are required. The chamber is filled with a gas mixture of Ar (94%) and CO_2 (6%) at a small overpressure of about 10–20 mbar. The spatial accuracy of the IHEP drift chamber is 0.6 mm and its angular resolution is about 0.03 rad. Right-left ambiguity is resolved in 98% of events.

[a] e-mail: EAZadeba@mephi.ru

Figure 1. Photographs of the CTUDC planes.

The signals from the wires are processed by an on-board amplifier-shaper that forms 75–100 ns pulses (depending on the input signal) in the LVDS levels.

3. CTUDC

3.1. Placement

The coordinate-tracking unit based on drift chambers (CTUDC) consists of two vertical coordinate planes installed at different sides of the CWD in the short galleries of the third floor of the NEVOD building, one floor above the DECOR supermodules (Fig. 1). Such a location allows the registration of near horizontal tracks by CTUDC (triggered by CWD) or also by joint operation with DECOR, that will significantly increase the range of muon track zenith angles from 85°–95° to 80°–100°.

Each plane consists of 8 drift chambers installed in two rows, overlapping by 30 cm to exclude dead zones in the chamber ends; it causes a 4° angle between the planes and CWD wall. The chambers are mounted on a special frame (Fig. 1) that allows precise adjustment of the DCs in all directions.

The distance between the registration system of the setup and the chambers is different for planes. It causes the difference in the length of cables and requires additional commutation blocks. The effective area of the plane is 14.8 m^2, the area of two DECOR supermodules located a floor below is 17.5 m^2, so the total area of the coordinate detectors for registration of near-horizontal particles coming along the CWD almost doubles.

3.2. Power supply

The drift chambers are supplied with two high voltages: 12 kV (field-forming wires) and 2.2 kV (signal wires), and with ±6 V low-voltage. The high voltage is provided by the multi-channel HV power supply controlled by a PC via a USB bus. The software of the CTUDC controls the current on the wires providing a smooth rise of the voltage during turning on the supply. A special microcontroller gets commands from software and turns on a relay to enable a light alarm outside the gallery of CTUDC. The most dangerous process is turning on/off 12 kV, since it causes electrostatic potential on the chambers. The light alarm flashes at 1 Hz during such process until the current stops rising (it always glows at a steady voltage). Such capabilities of the supply allow getting reliable noise characteristics of the chambers, because every point in such a measurement should be obtained with a steady current [7].

Two linear low-voltage supplies provide ±6 V to amplifiers situated on the DCs, their total consumption is about 6 A.

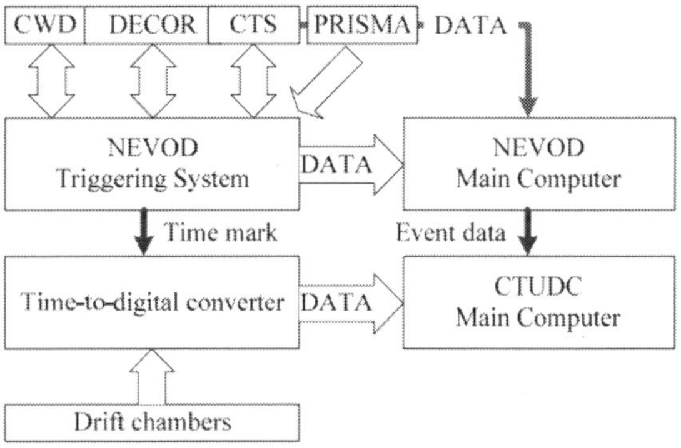

Figure 2. Principal scheme of the joint operation of CTUDC and NEVOD TS.

3.3. Gas supply

The gas preparation system is situated on the first floor of the building. Iit consists of two gas tanks (Ar and CO_2), ramp, two controlled flow meters, gas mixer and the PC, which manages the modes of gas supply. The mixture is supplied to the chambers via PVC tubes. Software of the system allows tracking of volumes of gas passing through chambers and to turn off the flow if one component of the mixture is exhausted. The CTUDC has a sequential gas tube connection, but every chamber or a group of chambers can be connected separately to the gas system.

3.4. Registration system

CTUDC is designed for joint operation with the triggering system of the experimental complex NEVOD (NEVOD TS) that binds registration systems of CWD, DECOR, calibration telescope system (CTS) and array of neutron detectors PRISMA. The triggering system has a rather fast data handling: the period between the passage of a particle through the working volume of CWD and the trigger formation is about 500 ns. On the other hand, the maximum drift time of electrons in the drift chamber is 6 μs, so the registration system and DAQ of CTUDC cannot be directly integrated into the NEVOD TS and should be implemented separately with the possibility of off-line inter-connection between the NEVOD and CTUDC data.

The registration system of CTUDC (Fig. 2) consists of a CTUDC main PC (MPC), VME crate with an optical bridge and a 128-channel time-to-digital converter (TDC) CAEN V1190. The TDC resolution is 100 ps while the time resolution of the drift chambers is about 5 ns. The

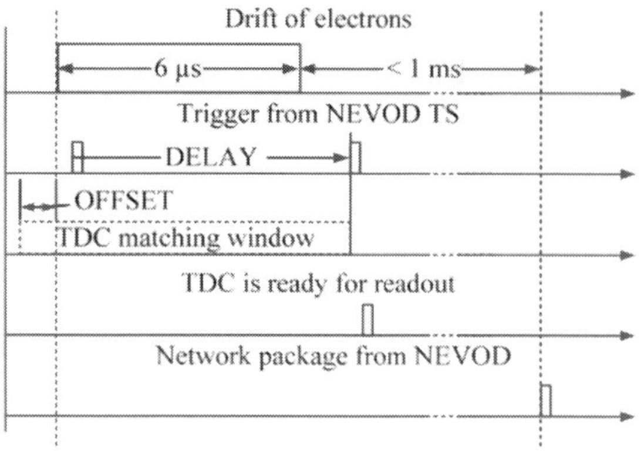

Figure 3. Timing diagram of the joint operation of CTUDC and NEVOD TS.

internal memory of the device allows storing large amount of events to the moment of TDC readout. The ability to work in a continuous mode allows measuring signal and noise rate per-channel with more than 90% live time.

The timing diagram of the CTUDC registration system is shown in Fig. 3. The NEVOD TS fixes an event if one of the experimental complex setups is triggered: CWD, DECOR, CTS and PRISMA. Each setup has its own trigger conditions. For CWD it is lighting up more than 60 measuring modules [8], for DECOR it is triggering at least two supermodules. For CTS there are three conditions: triggering of several detectors of the top plane, triggering of several detectors of the bottom plane, triggering of a single detector in each plane. For PRISMA it is triggering of two neutron detectors by charged particles (the threshold corresponds to the simultaneous passage of five particles within the scintillator detector). The NEVOD TS gathers information from all setups for every event, no matter if they gave no trigger.

In the main mode of operation, the trigger for CAEN TDC should be received after all hits of the event, so the signal from NEVOD TS is delayed for 8 μs. The matching window of TDC is chosen to be certainly longer than the delay value, so it is 12 μs. This leads to the fact that the starting point of counting the electron drift time is different from zero in the TDC data. The position of that point in the data is called the offset. In fact, it consists of two parts: basic offset for all channels ($\sim$1.5 μs) and individual offset caused by the difference in the length of the signal wires of the right and left planes of CTUDC and different twisting step in them.

For several types of events, MPC of NEVOD sends a network packet via Ethernet with information about the last event for further matching of CTUDC and NEVOD TS data, the most important information in the packet is the event number.

3.5. Software

CTUDC software is divided into server and client programs. The first one is constantly running on a MPC of CTUDC, it provides exposition, controls VME crate with TDC, HV supply and light alarm. The client can be launched on multiple computers in a shared server network; it can control modes of the exposition and the high voltage supply. It is also used as a program for remote monitoring of the setup. Communication between applications is held via UDP TCP/IP Protocol in a synchronous mode. Conflicts between client programs for the server time were not observed.

Because of the relatively high frequency of events in the NEVOD TS, network packets cannot be sent for each event that could result in an overload of the internal network, but only for two types of events: triggering of one supermodule DECOR in each short gallery, and triggering of CWD and DECOR at the same time. The total frequency of such events is around 1 Hz. After the reception of the network packet a readout of TDC events occurs. Then the server software compares the number of read events with the difference in the event numbers in the last two network packets. If these numbers match, then all read events are assigned with a corresponding intermediate number. Otherwise, this set of events is logged in a special file for individual processing. It happens on average to 4% of events.

Currently, a new mode of exposition is being tested: when the difference in numbers is found the system waits for the third network packet and then the comparison is performed for the events between the first and the third package. This reduces the number of dropped events to 0.5%.

Every 20 minutes, NEVOD is monitored for 30 seconds, the MPC of CTUDC receives a network packet at the beginning of this process and software starts the 30 seconds measurement of noise from drift chambers. The results are stored on the server that allows to monitor the performance of the DCs and to make corrections for specific pieces of data.

4. Processing of CTUDC data

Primary processing of CTUDC data includes calibration of drift chambers (forming accurate configuration files for further processing) and passportization of experimental runs.

4.1. Calibration

The calibration of the unit is performed after every replacement of drift chambers and amplifiers in them. It includes the correction of data, the position of the drift chambers and the selection of individual offsets for the measuring channels.

A spatial calibration is performed by cross-calibration of CTUDC with DECOR supermodules [9]. The position of the drift chambers is varied by software until the position corresponding to the best matching of tracks in both systems will be found.

The determination of individual offsets for each wire is performed according to calibration series. Statistics for one channel in such a run is about 100000 hits. Figure 4 shows the distribution of hits in drift time per channel. The background of the distribution is due to noise; the duration of the main part of the distribution is 5.5–6 μs, which corresponds to the maximum drift time of the electrons and depends on the current state of the gas mixture in the chamber. The position of the left edge of the distribution corresponds to the individual offset of the channel. However, this definition cannot be considered absolutely reliable because of the rather large value of bins

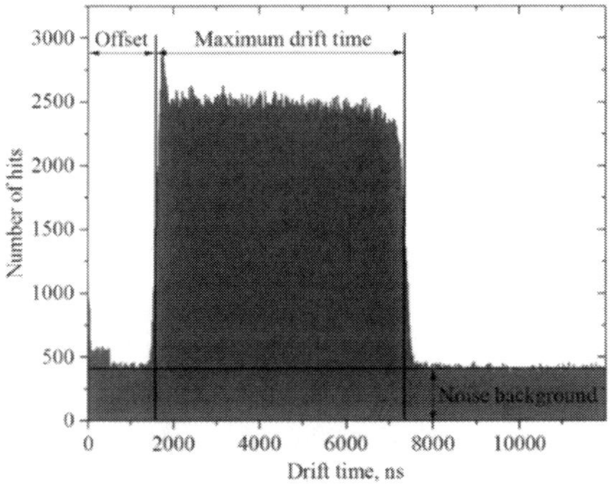

Figure 4. Distribution of hits in drift time.

(about 10 ns), and due to the large influence of the jitter of the receipt of the trigger signal. Therefore, precision accuracy in the value of the individual offset is determined by a sequential search method.

For each chamber about 10,000 single tracks are selected to a separate file. Individual offsets of the channels are varied in increments of 2.5 ns relative to each other and for each combination software makes the reconstruction of all tracks using the least squares method. The algorithm chooses the combination in which the number of tracks with the sum of squares of deviations of the track from the experimental points is less than a pre-selected threshold (10 mm^2) is maximized. The basic offset is obtained by the same method, but it uses events nearly perpendicular to the chamber plane, the criterion is smoothness of the distribution of the coordinate of intersection of reconstructed track with the drift chamber plane. If the offset is chosen incorrectly, the peak or a gap will occur in the area corresponding to the center of the chamber in such distribution.

4.2. Determination of current chamber characteristics

The value of the drift velocity is determined individually for each chamber in each experimental run and is texted in the passport of the given run. Not only is the width of the distribution of the drift time important, but also its form. During degradation of the gas mixture the distribution of the hit drift times (Fig. 4) changes its form: it falls in the region of large drfit times and looks trapezoidal, i.e., the efficiency of the chamber drops for the registration of particles that have passed far from the center of the chamber. Such defective chambers should be flushed with 600 liters of the gas mixture. The average time of operation of the chamber without additional venting is about 3 months.

One of the most important characteristics of the drift chambers is the efficiency of the signal wires. Efficiency is defined as the ratio of the number of events without the participation of the signal wire to the sum of this value with the number of full events (with all triggered wires) in DC. It was 99.95% at the test bench.

In the experiment, this value is 99.3% on average. The decrease in the efficiency is due to the increase of the threshold shaper-amplifiers, which was required because of a higher level of electromagnetic interference in the CTUDC in comparison with a test bench. The threshold level for the amplifiers is set so that the noise rate on the chamber channels does not exceed 2 kHz.

4.3. Passportization

To handle joint events of CTUDC and NEVOD TS the offline linking of events is carried out. Joining is made in several stages. The first step is to separate the CTUDC data on the parts corresponding to the experimental runs of the NEVOD TS (approximately 40 hours of exposition). Then DECOR tracks are reconstructed in NEVOD data. The third stage is forming of a single file with the events that have a structure given in Table 1.

Table 1. Format of joint events.

Statistics of the NEVOD RUN	
Header of the event	
CTUDC	RUN and event number, time
	Number of hits
	Triple vector of hits (Chamber : Wire : Time)
	Event statistics
NEVOD	RUN and event number, time
	Number of trigger signals from CWD and CTS
	Number of modules with amplitude information
	Status registers of TS
	Pressure, temperature, etc.
DECOR	Number of tracks
	Vector of track information (Vertex : Vector)
...	
RUN trailer	

The passportization of the experimental runs is carried out after joining. The passport includes information about the correspondence of events, efficiency of the measuring channels and statistics on the frequency of different classes of events in CTUDC. The drift velocities are also specified and are later used in event handling.

5. First results

For the moment, more than 2500 hours of debugging and measuring series have been held at the CTUDC. According to their data, a cross-calibration was made with the coordinate-tracking detector DECOR and the individual offsets for the measuring channels were determined.

Figure 5 shows the distribution of events according to the setups that triggered NEVOD TS. The graph shows that the least likely response in CTUDC is accompanied by events from CTS and DECOR (usually these are single muon events). Most of the events from PRISMA are usually accompanied by CTUDC response. Figure 6 shows the same distribution for the average number of triggered events in DC at different trigger types. Events of the PRISMA setup typically have a greater density of charged

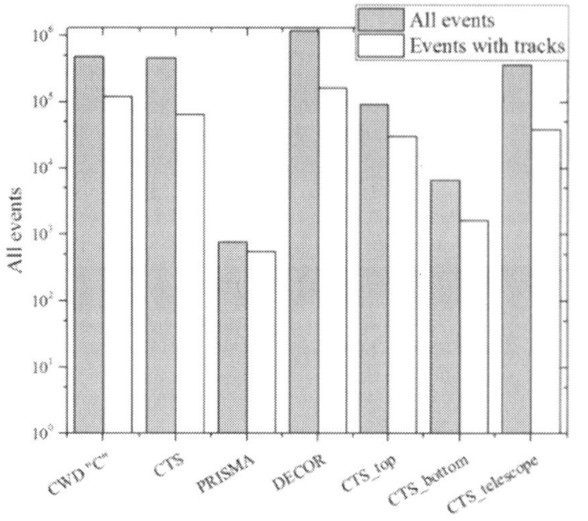

Figure 5. Number of events per each trigger for one experimental run.

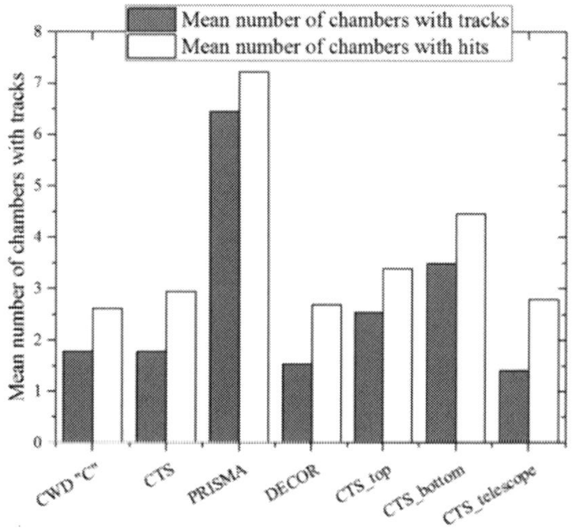

Figure 6. Mean number of hit chambers in events with different triggers.

particles (due to a high threshold on the detectors). In such events, on average, up to 7 drift chambers have tracks. On the other hand, events with CTS trigger are characterized by a large difference between the number of chambers with full tracks and chambers having at least one hit. This suggests that most of the hits on this trigger are accidental (noise).

The setup is able to register single muons and muon bundes with density up to 10 particles per m^2. Figure 7 shows the distribution of registered events in track multiplicity.

Processing of the data uses three methods of reconstruction of multiple events: the sequential search method, the method of finding the straight line and the histogram method. The last method determines the number of tracks under the same zenith angle in the event. It is most

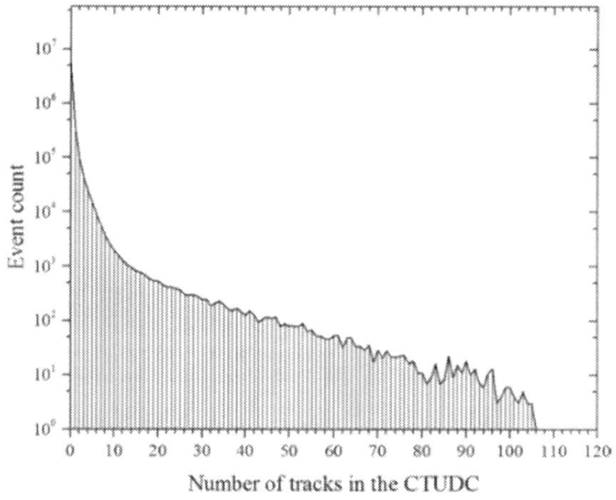

Figure 7. Distribution of multiplicity of events in CTUDC.

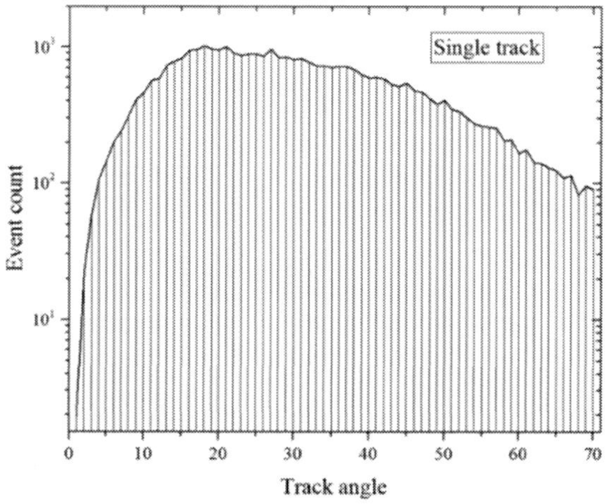

Figure 8. Distribution of single track events in the angle of their projections inside the drift chamber.

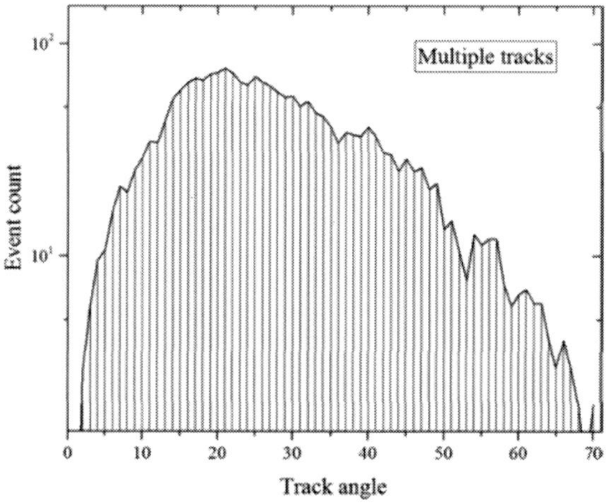

Figure 9. Distribution of multiple track events in the angle of their projections inside the drift chamber.

suitable for the determination of multiplicity: because of the remoteness of the points of generation of muons, their trajectories are quasiparallel. The influence of the dead time of measuring channels ($\sim$100 ns) and the number of secondary particles produced in the concrete walls of the NEVOD building are also taken into account in determining the number of particles in the bundle.

Figure 8 shows the distribution of events according to the value of the angle between the projection of the track on the plane othogonal to signal wires and a plane passing through them. The 90° angle corresponds to vertical tracks, the zero angle corresponds to tracks with zenith angle of 90° and perpendicular to the DC plane.

Figure 9 shows the same distribution for multiple events. Both distributions have maxima in the region of 20°, that is due to the fact that the events are selected according to the trigger from CWD and DECOR so it means that particles moved horizontally through the chambers to the water volume or in the opposite direction.

5.1. Conclusion

The new coordinate-tracking unit on the drift chambers (CTUDC) is developed in MEPhI. The installation greatly improves the capabilities of the experimental complex NEVOD in the registration of near-horizontal extensive air showers.

The setup successfully passed adjustments and calibration stages and is currently gathering experimental data.

References

[1] N.S. Barbashina et al., Nucl. Phys. B (Proc. Suppl.) **165**, 317 (2007)
[2] A.G. Bogdanov et al., Physics of Atomic Nuclei **73**, 1852 (2010)
[3] A.A. Petrukhin, Nucl., Phys. B (Proc. Suppl.) **212–213**, 228 (2011)
[4] E.A. Kovylyaeva et al., Bull. Lebedev Phys. Inst. **41**, 218 (2014)
[5] E.A. Zadeba et al., JINST **9**, C08018 (2014)
[6] N.I. Bozhko et al., Nucl. Instrum. Meth. A **243**, 388 (1986)
[7] E.A. Zadeba et al. J. Phys.: Conf. Ser. **632**, 012031 (2015)
[8] O. Saavedra et al., J. Phys.: Conf. Ser. **409**, 012009 (2013)
[9] E.A. Zadeba et al. J. Phys.: Conf. Ser. **675**, 032039 (2016)

Scintillation array of the Experimental Complex NEVOD for EAS investigations

I.A. Shulzhenko[1,a], M.B. Amelchakov[1], N.S. Barbashina[1], A.G. Bogdanov[1], D.S. Brovtsev[1], A. Chiavassa[1,2,3], N.E. Fomin[1], D.M. Gromushkin[1], N.N. Kamlev[1], S.S. Khokhlov[1], R.P. Kokoulin[1], K.G. Kompaniets[1], O.I. Likiy[1], V.V. Ovchinnikov[1], A.A. Petrukhin[1], O. Saavedra[2], V.V. Shutenko[1], and I.I. Yashin[1]

[1] National Research Nuclear University MEPhI (Moscow Engineering Physics Institute), 115409 Moscow, Russia
[2] Dipartimento di Fisica dell' Università degli Studi di Torino, 10125 Torino, Italy
[3] Sezione di Torino dell' Istituto Nazionale di Fisica Nucleare – INFN, 10125 Torino, Italy

Abstract. In 2014, the creation of a new cluster type shower array NEVOD-EAS was started around the Cherenkov Water Detector NEVOD and coordinate detector DECOR. The shower array will allow to determine the size, axis position and arrival direction of EAS registered by these detectors. In 2015–2016, the NEVOD-EAS central part, including 4 clusters of scintillation detector stations located on an area of about 10^4 m^2, was deployed and started operation. This article presents the description of the NEVOD-EAS arrangement, detecting elements and data acquisition system, as well as the first results on EAS detection.

1. Introduction

For studying muon bundles in a wide range of zenith angles and energies with the complex of Cherenkov Water Detector (CWD) NEVOD [1,2] and coordinate detector DECOR [3], the method of local muon density spectra (LMDS) [4] has been developed. However, the estimations of primary particle energy using the LMDS technique has rather low accuracy ($\sigma(\lg E_0) \sim 0.4$). It is due to the contribution of the extensive air showers (EAS) registered at different distances from the setup to the events with a fixed local muon density.

For an independent estimation of EAS parameters (arrival direction, size, axis position) in the energy range from 10^{15} to 10^{17} eV using classical technique and, therefore, for calibration of the LMDS method and increasing its accuracy, a new cluster type shower array NEVOD-EAS is being constructed around the NEVOD-DECOR experimental complex.

2. NEVOD-DECOR experimental complex

The NEVOD-DECOR experimental complex (Figure 1) is a multi-purpose facility for effective registration of almost all components of cosmic rays at ground level. The setup is located in the National Research Nuclear University MEPhI (Moscow, Russia) in a special four-story building with a water reservoir (inner dimensions $9 \times 9 \times 26$ m^3) and is equipped with technical and technological systems necessary for the setup operation.

The basic installation of the complex is the CWD NEVOD. Its detecting system represents a spatial lattice of quasi-spherical modules (QSM). Each QSM includes six PMTs with flat photocathodes directed along the coordinate axes. The lattice of QSMs allows detection of Cherenkov radiation from all directions with practically equal efficiency and is formed by a set of vertical strings each containing 3 or 4 QSMs. The distances between the modules are 2.5 m along the detector, and 2.0 m across it and over the depth. Currently, the detecting system includes 91 QSMs (in total 546 PMTs). The signal read-out is produced from the 12th and 9th dynodes that provides a wide dynamic range from 1 to 10^5 photoelectrons and ensures operation in the calorimetric mode.

To calibrate the QSM photomultipliers during continuous experimental runs, the calibration telescope system (CTS) is used. This system contains 40 upper scintillation counters (on the water tank top) and 40 lower counters (on the tank bottom). Counter dimensions are $40 \times 20 \times 2$ cm^3. Vertical pairs of counters ensure calibration of PMTs of modules in four surrounding strings. Other combinations allow to obtain calibration curves for various distances from muon tracks.

The coordinate detector DECOR (see Fig. 1) is a system of plastic streamer tube chambers deployed around the CWD NEVOD. It includes eight vertically suspended eight-layer assemblies (supermodules) of such chambers with a total sensitive area of 70 m^2. The chambers contain 16 tubes with 9×9 mm^2 inner cross section and 3.5 m length integrated in one PVC box. A special three-component gas mixture (Ar + CO_2 + n-pentane) provides operation of chambers in the limited streamer mode. Each layer of chambers is equipped with two-coordinate external strip readout system (256X + 256Y channels) that allows to localize tracks with about 1 cm accuracy. The angular accuracy of track reconstruction is better than 0.7° and 0.8° for zenith and azimuth angles, respectively.

[a] e-mail: IAShulzhenko@mephi.ru

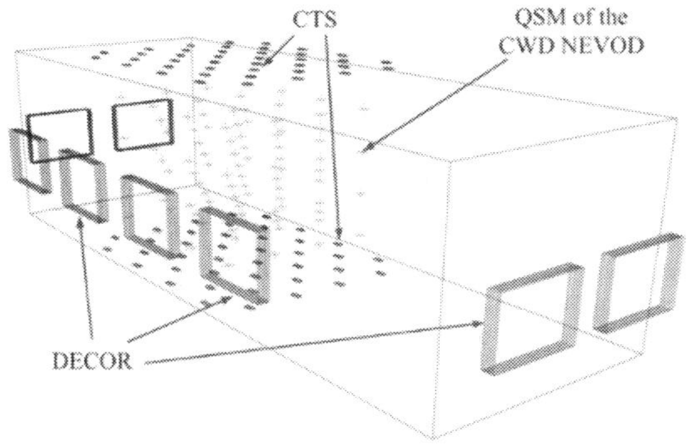

Figure 1. NEVOD-DECOR experimental complex.

3. NEVOD-EAS shower array

3.1. Experimental arrangement

In 2014, the creation of the NEVOD-EAS scintillation shower array was started. It is organized by the cluster principle and includes 12 independent clusters located around the experimental complex NEVOD-DECOR on the roofs of MEPhI laboratory buildings and on the ground. The layout of the NEVOD-EAS array is shown in Figure 2. Each cluster of the array consists of 16 scintillation counters of the EAS electron-photon component particles combined in 4 detector stations placed in the vertices of a rectangle with typical dimensions of about $15 \times 15\,m^2$. The characteristic distance between the centres of the clusters is about 50 m. The total area of the array is about $2 \times 10^4\,m^2$.

Since the clusters are located not in the same plane, but at different altitudes, new approaches to the EAS detection and organization of data acquisition system, as well as to experimental data analysis have been developed.

3.2. Scintillation counters and detector stations

Scintillation counters of the NEVOD-EAS array (see Fig. 2) were previously used in the KASCADE-Grande [5] (Germany) and EAS-Top [6] (Italy) experiments. A counter consists of plastic scintillator NE102A with the dimensions of $800 \times 800 \times 40\,mm^3$ and photomultiplier Philips XP3462 which are installed inside the pyramidal housing. The distance between the PMT and scintillator is 30 cm. The counter housing provides a possibility to install one additional PMT. The results of studies of the PMT and scintillator characteristics are presented in [7] and [8].

The detector station (DS) (see Fig. 2) of the array includes 4 scintillation counters. Three of the four DS counters are equipped with one ("standard") PMT for time and EAS particle density measurements. The standard PMT gain is adjusted for every counter in order to obtain the counter response to 1 particle (vertical equivalent muon, VEM) of about 13 pC. That provides a counter dynamic range of up to 100 particles. The fourth DS counter is equipped with two photomultipliers (a "standard" and an "additional" ones). The additional PMT ensures a wide linearity range at high particle densities (~ 10000 particles/m^2) and has a gain 90 times lower than the standard one. The response of the additional PMT to 1 particle is about 0.15 pC. In the array, the DS counters are mounted inside special external housings. The total DS area is about $2.5\,m^2$.

Detector stations also include a PMT power supply based on DC/DC-converter which transforms constant voltage $+12$ V into high voltage in the range from -100 to -2000 V. For measuring ambient temperature and accounting for temperature effects on the scintillator light yield, the temperature sensor is installed inside the DS housing.

Analog signals from the DS counters are transferred to the cluster Local Post (LP) of primary data processing via coaxial cables.

3.3. DAQ-system

The data acquisition system of the NEVOD-EAS scintillation array is organized by a cluster principle and includes 2 main structural elements: the cluster and the Central DAQ Post (CP). The functional scheme of the NEVOD-EAS DAQ system is presented in Fig. 3.

3.3.1. Cluster of the shower array

Each cluster is an autonomous part of the setup which provides EAS registration, selection of events according to triggering conditions (registration thresholds, coincidence multiplicity and time gate) specified at the beginning of the experimental run, digitizing of amplitude information, timestamping of events and data transfer to the CP of the array. Basic elements of the NEVOD-EAS cluster are: the detecting elements (4 DS) and the Local Post of primary data processing.

Analog signals from the cluster detector stations (5 signals–4 from standard PMTs and 1 from additional one–per DS) are fed to the cluster LP. It includes: summator-multiplexor (SM), industrial I/O-module for the SM control and processing of temperature sensors' data, block of primary data processing (BPDP), some network equipment for communication with CP and several power supplies ensuring operation of all cluster elements. All LP electronics are enclosed inside the water-proof thermally stabilized box.

The BPDP digitizes DS signals, selects events according to various intra-cluster triggering conditions and transmits data to the Central DAQ Post. It includes two 2-channel boards of amplitude analysis (BAA) and a controller based on the FPGA Xilinx Spartan-6 which ensures all processor functions. Each BAA channel has 2 inputs: active (AI) and passive (PI). Signal digitizing is performed using a 12-bit fADC with a sampling frequency of 200 MHz. The registration threshold is set on the AI using 8-bit DAC. When the signal on the AI exceeds the registration threshold, the REQUEST signal is sent to the FPGA. If the number of REQUEST signals from BPDP measuring channels inside the specified time gate fulfils the specified multiplicity, FPGA sends a CONFIRMATION signal to the FIFO buffers of all channels. Then data are timestamped with an accuracy of 10 ns, packed and sent to the CP.

Four-channel SM performs summation of analog signals from the standard PMTs. Each SM channel has 4 inputs and serves 1 DS. The summed signal from 4 standard PMTs of the DS is fed to the AI of the

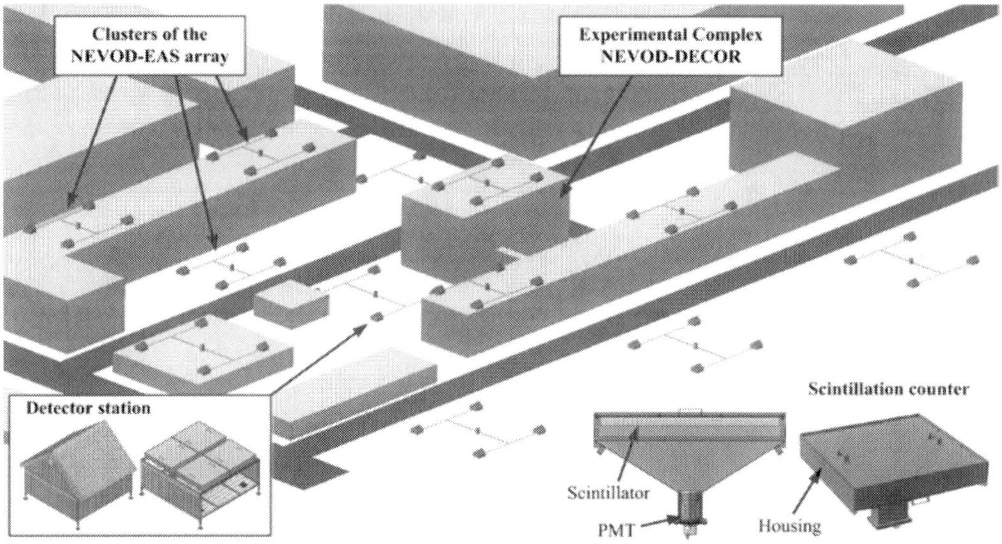

Figure 2. The layout of NEVOD-EAS shower array around the NEVOD-DECOR experimental complex. Left bottom corner: detector station housing and layout of counters inside the detector station. Right bottom corner: scheme of the counter.

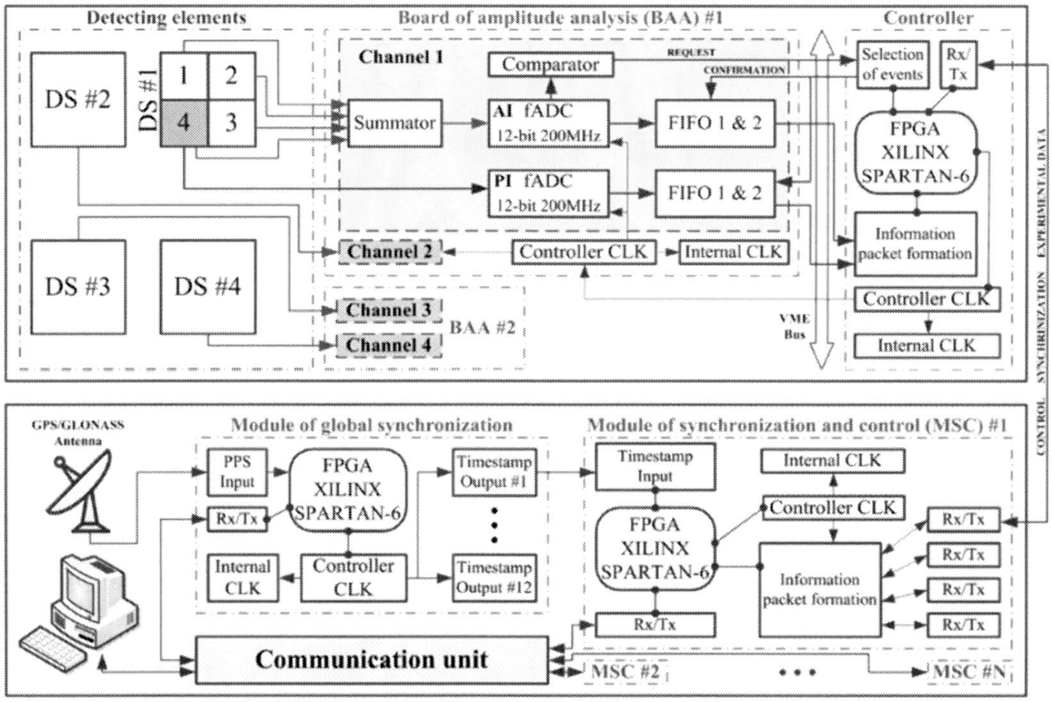

Figure 3. Functional scheme of the NEVOD-EAS shower array data acquisition system.

corresponding BAA channel. The additional PMT signal is fed to the corresponding PI of the BAA. The SM provides an opportunity to activate or deactivate defined inputs for measuring response spectra from the selected counter of the DS during the cluster operation monitoring. The SM is controlled using I/O-module which sets the mask of active SM inputs. This module has 8 analog inputs for the temperature sensors data.

All communications with the CP are performed using TCP/IP protocol via the fiber-optical line.

3.3.2. Central DAQ Post

The Central DAQ Post ensures synchronization and control of all Local Posts of the NEVOD-EAS array, as well as the experimental and monitoring data reception

Figure 4. Photograph of the 3rd cluster of the NEVOD-EAS array on the roof of the building of the NEVOD-DECOR experimental complex.

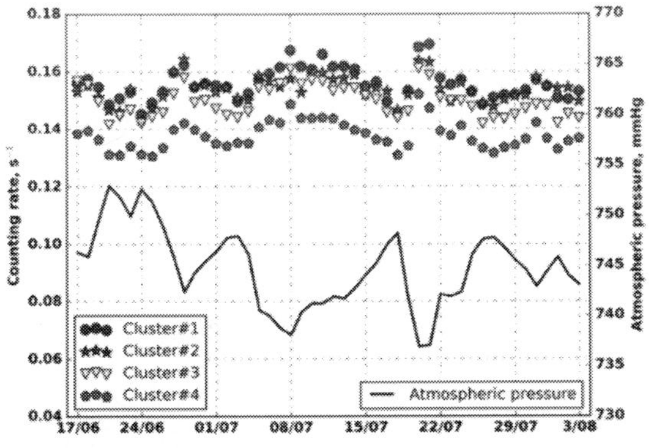

Figure 5. Counting rates of the NEVOD-EAS central part clusters and atmospheric pressure during the experimental run 17/06/16-03/08/16.

and storing. The CP (see Fig. 3) includes: personal computer (PC), GPS/GLONASS antenna, several modules of synchronization and control (MSC), module of global synchronization (MGS) and communication unit providing network connection between all CP elements.

Every MSC has 4 optical interfaces for the network connection to the BPDP of 4 NEVOD-EAS clusters and 1 Ethernet interface for the connection to central PC via the communication unit. At the beginning of an experimental run the specialized software on the central PC sends information packets to the MSC containing values of registration thresholds, coincidence multiplicities and time gates for 4 clusters. Then MSC broadcasts these options to certain clusters which configure their measuring channels and start registration. If one of the clusters registers an event, it sends information on the event time and contents of all ADC FIFO buffers to the MSC. The MSC, in its turn, adds information about the number of triggered clusters to the information packet and sends it to the central PC for processing and storing.

All electronics of the NEVOD-EAS array DAQ-system operate in the synchronous mode at the same clock frequency of 100 MHz. So, the synchronization accuracy of NEVOD-EAS clusters is 10 ns.

Time synchronization of clusters is performed using the global positioning system (GPS/GLONASS) antenna and the MSC and MGS modules. The antenna which is directly connected to the MGS generates the PPS signal with a frequency of 1 Hz. Upon receiving the PPS signal, the MGS send the synchronisation packet

Table 1. Dimensions and coordinates of the clusters of the NEVOD-EAS central part.

Cluster No.	Cluster dimensions, m^2	Coordinates of the cluster centres, m		
		X	Y	Z
1	13.3×12.4	−28.4	−7.8	−7.0
2	13.3×12.4	−28.4	23.8	−7.0
3	25.1×13.5	0.0	0.0	0.0
4	15.0×15.0	33.3	7.8	−14.5

containing the number of cycles of its internal clock to the special interfaces of the MSC modules. The MSC modules compare the obtained value to the number of cycles of their own internal clocks. Then MSC modules send the number of cycles of the MGS internal clock to the BPDP controllers of the clusters which also compare the obtained value to the number of cycles of their own and ADCs' internal clocks.

If some deviations are revealed during this comparison, all internal clocks of the NEVOD-EAS array DAQ-system are stopped. Upon the next PPS signal from the antenna all clocks are started again in the way described above.

The MSC also allows to measure time of the synchronization packet transmission from the CP to the LP of all clusters via optical cables. This value determines the time shift in the LP local time which had to be accounted for the correction of the cluster event time. All processor functions of the MSC and MGS modules are performed by the FPGA Xilinx Spartan-6.

3.4. Central part of the NEVOD-EAS array

In 2015–2016, the central part of the NEVOD-EAS scintillation array was created and started. The central part includes 4 clusters. Three clusters (1, 2 and 3) are located on the roofs of MEPhI laboratory buildings and the 4th one is on the ground surface. The total area of the NEVOD-EAS central part is about 10^4 m^2. The dimensions of clusters and their coordinates relative to the center of the roof of the NEVOD-DECOR building are presented in Table 1. The photograph of one of the NEVOD-EAS clusters is shown in Fig. 4.

The first continuous experimental run at the central part of the NEVOD-EAS array lasted 47 days (17/06/16–03/08/16). The triggering condition for all clusters was a 4-fold coincidence of DS during the time gate with a duration of 120 ns. Registration threshold for all detector stations was about 0.5–1.0 particles (VEM). The counting rates of 4 central part clusters and the atmospheric pressure during this run are shown in Fig. 5. It is seen that the first 3 clusters have almost the same counting rates while the counting rate of the 4th cluster is a little bit lower. This is due to its location close to one of the laboratory buildings which acts as a shield for EAS electron-photon component. The clear anti-correlation of the cluster counting rates and the atmospheric pressure can be observed. The estimated barometric coefficient is about–0.8%/mmHg.

3.5. EAS detection

As the NEVOD-EAS array clusters are located not in the same plane, but at different altitudes, a special cluster approach to the reconstruction of registered extensive air

DOI: 10.1051/epjconf/201714507003

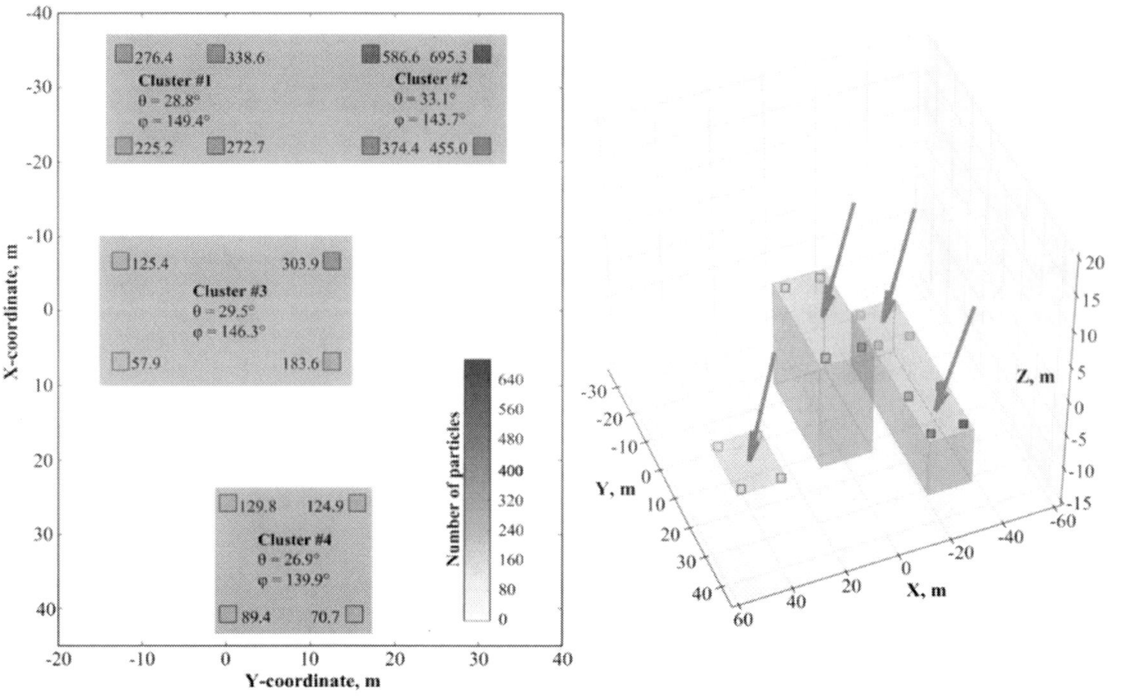

Figure 6. Reconstruction of the NEVOD-EAS array central part response to the extensive air shower.

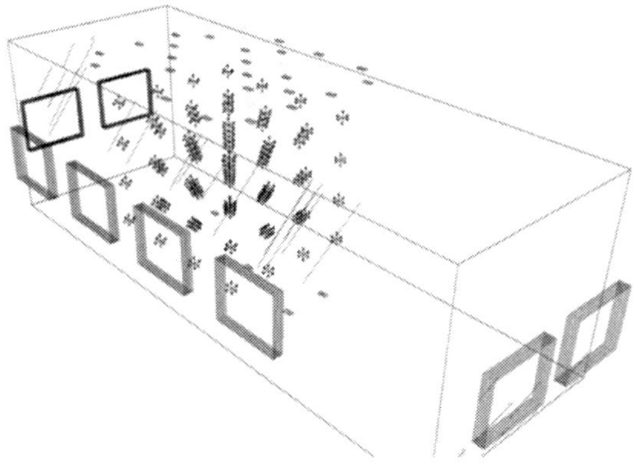

Figure 7. Reconstruction of the NEVOD-DECOR experimental complex response to the extensive air shower.

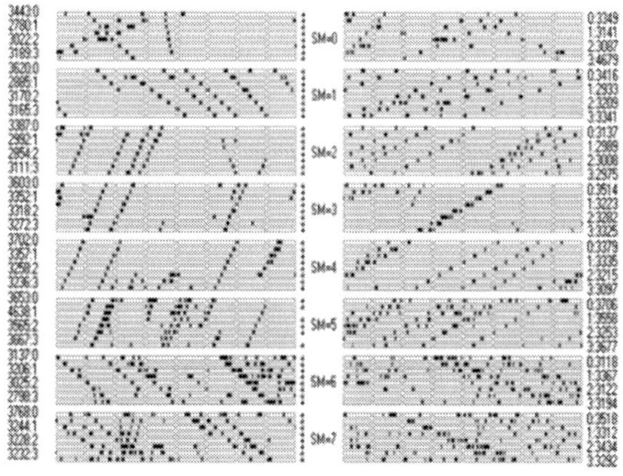

Figure 8. Response of the supermodules of the coordinate detector DECOR to the extensive air shower.

showers has been developed. In particular, the EAS arrival direction is determined in the following way. The events in which not less than 3 clusters have been triggered during the time interval of about 0.5 μs determined by the characteristic size of the setup ($\sim$100 m) are selected. The intra-cluster triggering condition is a 4-fold coincidence of detector stations within a 120 ns time gate. For each cluster the arrival direction is calculated on the assumption of a flat EAS front. The resulting direction in the event is calculated by averaging directions reconstructed according to the data of triggered clusters.

The number of particles registered by each detector station is determined as the ratio between the DS response charge and its response charge to 1 particle (VEM, $\sim$13 pC). At high particle densities (> 100 particles/m^2) the data of the DS additional PMTs are taken into account.

Figure 6 shows an example of the reconstruction of the NEVOD-EAS array central part response to the extensive air shower. Grey rectangles illustrate the cluster locations. Rectangles represent the detector stations. the number of particles registered by each DS is shown by the tone and printed near the rectangles. For all clusters the reconstructed azimuthal and zenith angles of the EAS arrival direction are indicated. The left part of Fig. 6 is the 3-dimensional view of this event. The arrows correspond to the EAS front arrival direction reconstructed according to the data of each cluster. The highest particle density is observed in the DS #1 of the 2nd cluster. The particle density decreases in the direction from the 2nd cluster to the 4th one.

This extensive air shower has also been registered by the NEVOD-DECOR. Figure 7 shows geometric reconstruction of the experimental complex response. Crosses and circles represent triggered QSMs and PMTs of the CWD NEVOD, correspondingly. Rectangles are for the triggered counters of the CTS system. Straight lines show the tracks of muons reconstructed according to the data of the coordinate detector DECOR.

Figure 8 shows the response of the DECOR supermodules to this extensive air shower. Dark points represent triggered strips. In the left part of the figure are Y-strips (azimuth angle measurements) of supermodules. In the right part – X-strips (projected zenith angle measurements).

According to the data of the DECOR detector the azimuthal and zenith angles of the considered EAS arrival direction are 144.4° and 29.5°, respectively. The results of the arrival direction reconstruction by the 3rd cluster of the NEVOD-EAS array ($\varphi = 146.3°$, $\theta = 29.5°$) are in a good agreement with these values.

4. Conclusion

In 2015–2016, the central part of the NEVOD-EAS cluster type shower array which includes 4 independent clusters (in total, 64 scintillation counters) located around the experimental complex of the CWD NEVOD and coordinate detector DECOR at the area of about 10^4 m^2 has been created and started.

The distributed structure of the NEVOD-EAS array DAQ-system allows to deploy registering elements of the setup even at different altitudes and provides implementation of a newly developed cluster approach to the reconstruction of EAS parameters. The first experimental run has proved the possibility of using of such a cluster approach.

This work was performed at the Unique Scientific Facility "Experimental complex NEVOD" with the support of the Ministry of Education and Science of the Russian Federation (grant RFMEFI59114X0002, MEPhI Academic Excellence Project 02.a03.21.0005 of 27.08.2013) and the Russian Foundation for Basic Research (grant 16-29-13028-ofi-m).

References

[1] V.M. Aynutdinov et al., Astrophysics and Space Science **258**, 105 (1998)
[2] V.V. Kindin et al., Physics Procedia **74**, 435 (2015)
[3] N.S. Barbashina et al., Instrum. Exp. Tech. **43**, 743 (2000)
[4] R.P. Kokoulin et al., Nucl. Phys. B. (Proc. Suppl.) **196**, 106 (2006)
[5] M. Aglietta et al., Nucl. Phys. B (Proc. Suppl.) **16**, 493 (1990)
[6] A. Chiavassa et al., 28th ICRC, Tsukuba **2** (2003), p. 992
[7] I.A. Shulzhenko et al., Bull. Russ. Acad. Sci. Phys. **79**, 389 (2015)
[8] I.I. Yashin et al., J. Phys.: Conf. Ser. **632**, 012029 (2015)

Search for ultra high energy cosmic ray sources

A.A. Mikhailov[1,a] and V.A. Kolosov[1]

Yu.G. Shafer Institute of Cosmophysical Research and Aeronomy, 31 Lenin Ave., 677980 Yakutsk, Russia

Abstract. Earlier we suggested a new method to search for sources and anisotropy of the arrival directions of ultra high energy particles. Here using this method we analyzed the data of the Yakutsk EAS array and found that particle fluxes with energies $E < 4.10^{19}$ eV come to the Earth from our Galaxy and the arrival directions of particles correlate with known pulsars. The chance probability of correlating particles with pulsars and the anisotropy of the distribution of particles is 10^{-4}–10^{-5}.

We also analyzed the Telescope Array (TA) data with energies $E < 4.10^{19}$ eV by our method and found that the maximum of the distribution of particles points to extragalactic sources but the direction of the maximum does not coincide with the maximum direction of the distribution of particles found by the authors of the Telescope Array data (the maximums of the arrival directions received by two different methods of analyses differ by about 10%).

1. Introduction

From 1974 near Yakutsk, Russia, an extensive air shower array (EAS) of area 18 km^2 began to register extensive air showers. The Yakutsk array is situated in North Asia near the North Ice ocean at a distance of 50 km south-west from the city of Yakutsk in the Lena river valley near the village of Oktyomsy. The geographical coordinates are: latitude 61.7° N and longitude 129.4° E at a height of 100 m above sea level, i.e., 1020 g.cm^2.

The Yakutsk array consists of 49 surface scintillation plastic detectors each of area $2 \times 2\,m^2$, 9 surface scintillation plastic detectors of area 2 m^2 for electrons and photons, 5 underground scintillation plastic detectors of 20 m^2 for muons with an energy threshold of $E_\mu > 1 \times \sec\theta$ GeV (where θ is the zenith angle) and 45 Cherenkov light detectors for registering electrons and photons. The primary particle energy is estimated by a calorimetric method using the Cherenkov light of an EAS. Since 1990 the surface detectors formed a uniform net of 18 km^2 area (until 1990 the array area was 10 km^2) with a distance of 500 m between neighboring units. In the center of the array is a muon detector of 192 m^2 to register muons with energy $E_\mu > 0.5 \times \sec\theta$ GeV. Timing signals for each detector are distributed over a microwave link: a resolution of 10^{-7} s is achieved. A full description of the array has been given in references [1,2].

We analyzed the distribution of particles in the celestial sphere of the Yakutsk array data [3] by classic methods (harmonics analyses) and by our suggested method. We considered the arrival directions of extensive air shower for particles with zenith angles $\theta < 60°$ and cores located inside the array perimeter. The threshold energy of the primary particles was $E \geq 5.10^{18}$ eV and the energy and arrival direction were determined with an accuracy of ~30% and ~5–7° respectively.

[a] e-mail: mikhailov@ikfia.ysn.ru

2. The distribution of particles in right ascension

We considered the Yakutsk array data from 1974 until 2011 for energies $E \geq 5.10^{18}$ eV. At first we analysed the arrival distribution of particles in the celestial sphere by harmonic analysis.

Figures 1 and 2 show the distribution of particles in right ascension detected by the Yakutsk array for the period 1974–2011 for different energy intervals, where the shower axea are situated inside the perimeter of the array:

- for the energy interval, $E = (5-8)10^{18}$ eV (Fig. 1), the amplitudes and phases in right ascension of the first harmonic of Fourier analysis are: first harmonic, r_1, and phase, ϕ_1, of the maximum: $r_1 = 0.037 \pm 0.04$, $\phi_1 = 76° \pm 58°$, second harmonics are $r_2 = 0.04 \pm 0.04$, $\phi_2 = 42° \pm 48°$, number of particle n=1469. The chance probability, P, to have amplitudes $r_1 \geq 0.037$ according to Linsley [4] is P ~0.6;
- for energies $E \geq 8.10^{18}$ eV (Fig. 2), the amplitude and phase in right ascension of the 1st harmonic of Fourier analysis are $r_1 = 0.14 \pm 0.04$, $\phi_1 = 0.6° \pm 19°$, the number of particles n = 938; the chance probability, P, to have amplitudes $r_1 \geq 0.14$ is P ~0.01.

In our opinion, the method of analysing by harmonic functions is not effective because of the results of summing the number of particles in declination (this summing is demanded by the method), and some local maximums and minimums in declination can smooth them out. Because of the smooth data, a received anisotropy of particle arrival directions perhaps differs from reality. In addition, this method only analyzes data along one coordinate – Right Ascension. Therefore, this analysis can only be used for a preliminary data analysis. Harmonic analyses point to galactic origin (Table 1).

We therefore suggested a new method [5,6] which analyzes data using two coordinates: declination and right ascension. Using this method we analyzed the extensive

Table 1. Amplitudes and phases of the analysis of particle distributions in Right Ascension.

Energy ev	N Number of particles	A_1% 1st harm.	A_2% 2nd harm.	ϕ_1 deg.	ϕ_2 deg.	P chance probability
$(5–8)10^{18}$	1469	3.6	4.4	76°	42°	0.6
$(8–40)10^{18}$	938	0.14	0.02	7°	7.6°	0.012

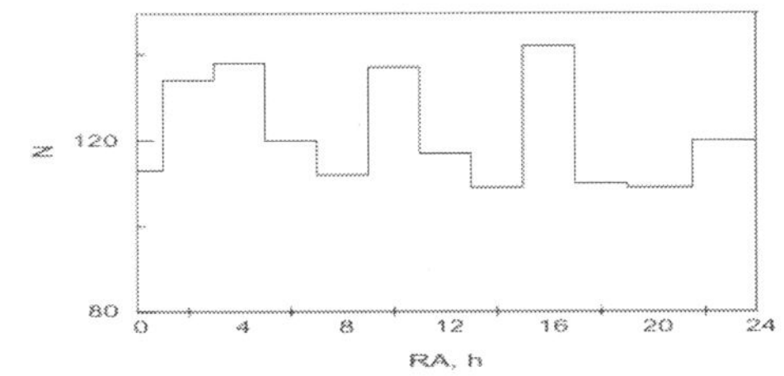

Figure 1. The distribution of particles in right ascension with energies $E = (5 \div 8)10^{18}$ eV; Number of particles N = 1469.

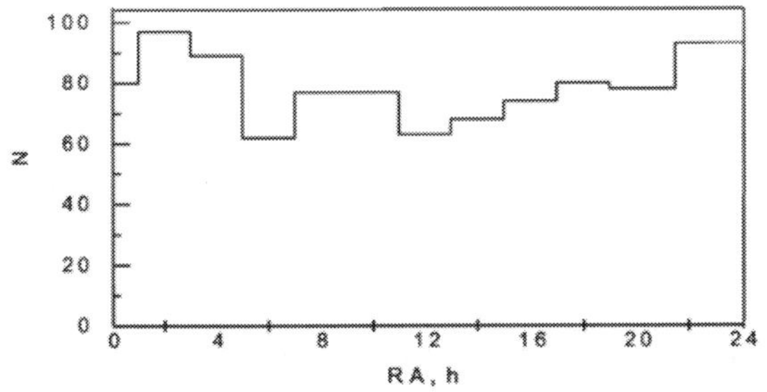

Figure 2. The distribution of particles in right ascension with energies $E \geq 8.10^{18}$ eV; Number of particles N = 938.

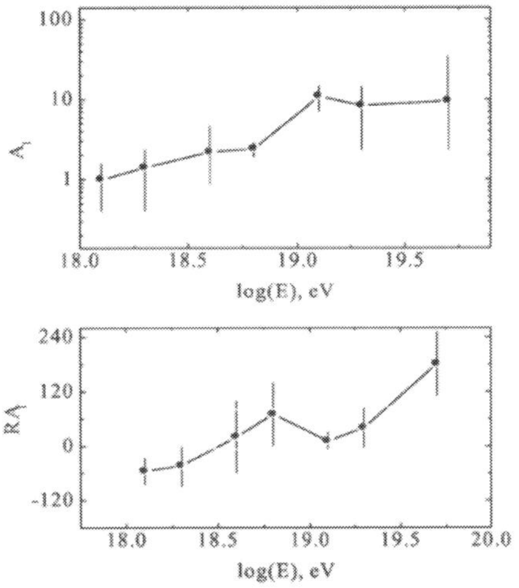

Figure 3. Amplitudes and phases of the 1st harmonic in the shower distribution depending on their energy.

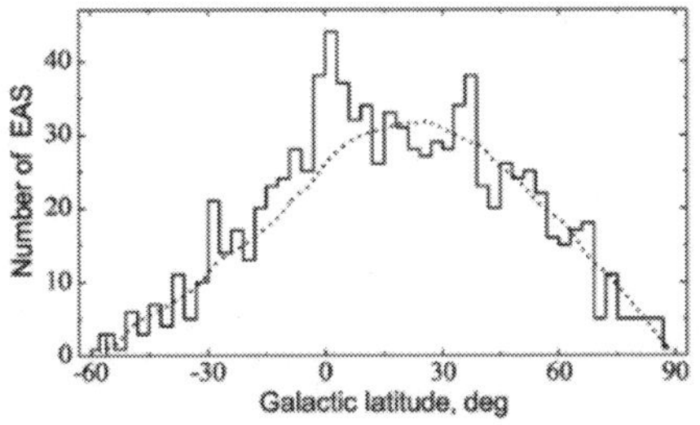

Figure 4. The distribution of particles with energy $E \geq 8.10^{18}$ eV in galactic latitude

air shower data of Yakutsk and the Telescope Array. We considered the Yakutsk data registered for 1974–2011 with shower axes inside the perimeter of the array. As a result of our analyses we concluded [1,7–15] that cosmic rays up to energy $E \sim 4.10^{19}$ eV are galactic and their sources are pulsars [16].

3. The particle flux from the galactic plane

We consider the distribution of particles with energies $E \geq 8.10^{18}$ eV in galactic latitude, b, in 3° bins. The distribution of particles in galactic latitude is shown in Fig. 4; the expected numbers of events at the observation level is the dashed line. Note that the expected numbers of events have been determined by simulating 10^6 random events using the Monte-Carlo method. Events are simulated in the horizontal coordinate system taking into account the zenith-azimuth distribution of particles $dn(\theta) \sim \cos(\theta)\sin(\theta)d\theta$ and in azimuth angle ϕ uniformly. Then, the distribution of the events was transferred into the galactic system coordinates.

As seen in Fig. 4, at latitudes $b \sim 0°$ an excess is seen of the observed number of showers compared with those expected. In the latitude interval $0 \leq b < 3°$ the excess of the number of showers is

$$(n_{obs} - n_{exp})/\delta n_{exp} \text{ is } \sim 3.5\sigma$$

where the observed number of showers $n_{\text{obs}} = 45$. This distribution of phases, first harmonics maximum of particles indicates that, most likely, particles up to $\sim 10^{19}$ eV are galactic in origin. There are also two excesses of particles at latitudes $-30° \leq b \leq -27°$ and $33° \leq b \leq 39°$.

Results of harmonic analyses and the distribution of particles in galactic latitude also indicate that particles up to 4.10^{19} eV are galactic in origin.

4. Method to search for sources and anisotropy

We suggest a new method to search for sources and anisotropy of the arrival directions of ultrahigh energy cosmic rays [6]: around possible sources (pulsars) we drew circles with an angular distance, R, equal to or greater than the deviation angles of particles from the sources to Earth. Then we find the number of particles inside these circles and choose those circles as sources where the number of particles inside the circles, <R, is above a selected number of particles, > n, (usually n is the mean number of particles).

If the circles around proposed sources overlap, then we unite them into one region. Furthermore we determine the chance probability, P, to observe a number of particles, n_1, in the considered region of the celestial sphere. In our initial study we found a correlation of particle

 DOI: 10.1051/epjconf/201714507004

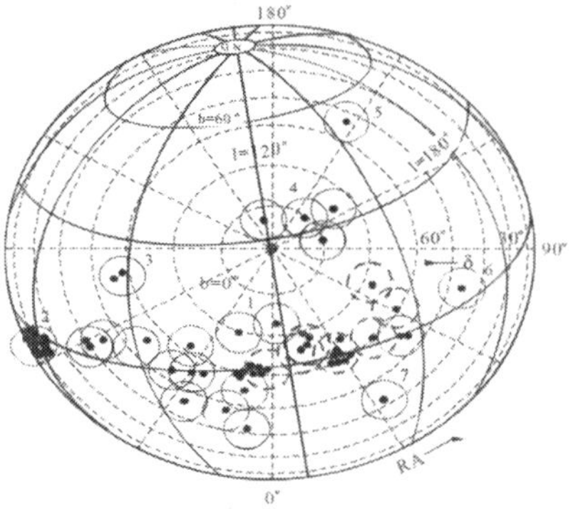

Figure 5. Yakutsk, $E \geq 8.10^{18}$ eV, correlation of particles with pulsars, circles around pulsars at R = 6° and number of particles inside circles n ≥ 10.

arrival directions with pulsars, suggested as most probable sources of particles, by our analysis data at angular radius ≤ 6° for energies $E \geq 10^{19}$ eV for the Yakutsk EAS data [7]. We therefore take this angular distribution as a pattern for the correlation between particles and galactic sources, pulsars.

5. Correlation of particles with pulsars

From 1974 to 2011 the Yakutsk EAS array registered 938 particles for E > 8.10^{18} eV inside the perimeter of the array. We select those pulsars [16] as sources of cosmic rays if the number of particles inside the radius R< 6° from the considered pulsar was n≥10. If circles around pulsars overlap then we unite them into one region. The distribution of particles is shown on the map of equal exposure of the celestial sphere where an equal number of particles is expected from equal regions [1]. We have found 7 regions: inside the 1st region (Fig. 5) – 39 pulsars, inside the 2nd region – 16 pulsars, the 3rd region–2 pulsars, the 4th region–4 pulsars and separately pulsars 5–7 (Fig. 6) and inside these regions 183, 17, 11, 33, 14, 11, 10 particles respectively. The chance probabilities to find such number of events from 938 particles are: $P_1(938 \geq 183) \sim 6.10^{-5}$, $P_2(938 \geq 17) \sim 1.8 \times 10^{-3}$, $P_3(938 \geq 11)$, ~0.6 , $P_4(938 \geq 33) \sim 0.5$, $P_5(938 \geq 11) \sim 0.6$, $P_6(938 \geq 33) \sim 0.5$, $P_7(938 \geq 14) \sim 0.04$, $P_6(938 \geq 11) \sim 0.1$ and $P_7(938 \geq 10) \sim 0.1$.

Note that chance probabilities are only significant around the 1st and 2nd groups of pulsars (Fig. 6) at radius R¡6° and are 3σ more than expected for the case of isotropy. The maximum number of events $20 \leq n \leq 23$ at R< 6° are observed around the 1st groups of pulsars: PSR 0141 + 6009, 0146 + 6145, 0147 + 5922, 0157 + 6212, 0215 + 6218. These pulsars are in the galactic plane and have been considered earlier as possible sources of particles [1].

A flux of particles was observed from the side center of the Galaxy (Fig. 5, number 2). All 14 particles arrive at large zenith angles $60° \geq \theta \geq 38°$.

We considered the Telescope array EAS data for energies $E \geq 5.7 \times 10^{19}$ eV [3] and found 31 particles at R< 6° from pulsars. The chance probability to find such a number of particles from 72 is P ~0.1. Thus we did not find any correlation with pulsars in the arrival directions of particles from the TA array data.

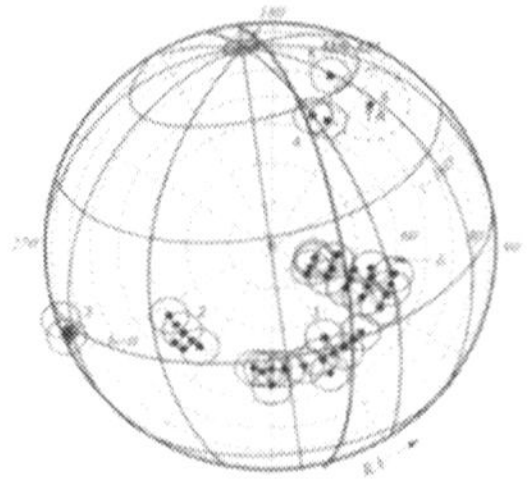

Figure 6. Yakutsk, $E \geq 8.10^{18}$ eV, anisotropy: distribution of points in the celestial sphere • within which at radius R < 6°–≥16 particles were observed. 1, 2, 3, 4–data of Yakutsk array. 5–circles around the galaxy Mkr 421, 6-TA–maximum particles found by the Telescope Array data, R< 20°.

Table 2. The energy and co-ordinates of particles which came from the Galactic Center during 1974-2014.

Energy ev	N Number of particles	δ	RA
$> 4 \cdot 10^{19}$	14	~ 20°	290°

6. Anisotropy of ultrahigh energy particles

We have estimated the anisotropy of the arrival directions of ultrahigh energy particles. According to our method [5,6] it draws lines of a coordinate grid at declination, δ, and right ascension, RA, on the celestial sphere through 5° x 5°, beginning with declination $\delta = 0°$ and right ascension RA = 0°. We have counted up the number of observed particles at R< 6° around points where the above-stated coordinates are crossed.

We selected those points around which at R < 6° the observed number of particles was n ≥ 16 events for the Yakutsk array data (Fig. 7). We found in the 1st region 115 particles after the connection of regions R < 6° which form as a result of cross declinations and right ascensions 5° x 5°; in the 2nd region–24 particles, in 3rd region–14 particles, in 4th region–32 particles. The probabilities to find such a number of events from 938 particles in these regions are: $1 - P(938 \geq 115) \sim 2.10^{-5}$, $2 - P(938 \geq 24) \sim 10^{-2}$, $3 - P(938 \geq 14) \sim 3.10^{-4}$, $4 - P(938 \geq 32) \sim 8.10^{-2}$.

It is seen in Fig. 6 that the maximum of the distribution of particles is around the coordinates: $\delta = 60°$, RA = 25° – 40° (number of particles in this region for R < 6°, $n \geq 18$). This maximum of the particle distribution coincides with the position of the following pulsars: PSR 0141 + 6009, 0146 + 6145, 0147 + 5922, 0157 + 6212, 0215 + 6218). 23 particles were observed at R< 6° around the peak of this maximum at $\delta = 60°$ and RA = 25° and the chance probability is $P(938, \geq 23) \sim 2.10^{-4}$.

The number of particles in the Yakutsk data is not enough to make any conclusion (at energy $E > 4.10^{19}$ eV there were < 50 particles).

In addition we also analyzed the TA data. We did not find any excess of particles in the TA data around the galaxy Mkr 421 (Fig. 7) (our colleagues [3] found

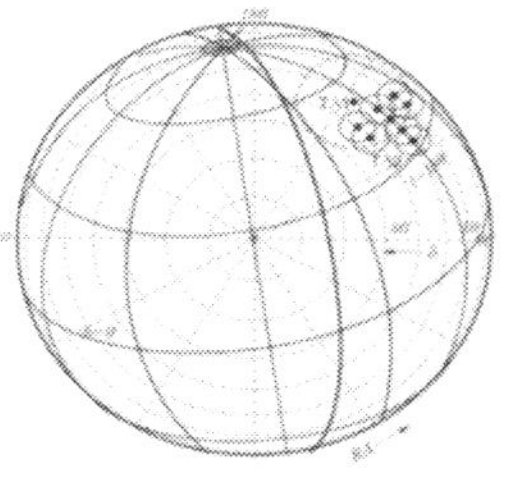

Figure 7. Telescope array data, $E \geq 4.10^{19}$ eV, TA1–maximum distribution of particles found by the authors of the Telescope Array data, R< 20°. Around 8 points at R < 6° maximum distribution of particles by our method, n ≥ 3 particles.

an excess of the particle flux from this galaxy). Only 5 particles at R< 6° around this galaxy from 72 particles were observed and the chance probability by our method is only P ∼0.6.

According to the TA data 19 particles were observed at a radius R< 20° (dashed curve around the TA hotspot, Fig. 4) centered around the coordinate $\delta = 43.2°$, RA 146.7° [3]. We estimate the chance probability using the TA data [3] and our method is only P∼ 10^{-2}. Our method differs from the method used by the authors of the TA collaboration.

We therefore decided to search for any anisotropy in the TA data using our method. Curved lines were drawn on a coordinate grid 5° × 5° around the whole celestial sphere and we selected those points where coordinates crossed and demanded that the number of observed particles was $n \geq 3$ from these points at distances R< 6°. Such a number of particles was observed around only 8 points at declinations $\delta = 40°$, 50°, 55° and right ascensions 125°, 130°, 135°, 140° crossing (Fig. 3, 8 points, TA2). Around these 8 points at radius R< 6° there are 8 particles and the chance probability, P, to observe 8 particles from 72 by our method [5] is $P(72, \geq 8) \sim 3.10^{-5}$.

7. Conclusion

We have found correlations in the arrival directions of ultrahigh energy particles with $E < 4.10^{19}$ eV with the positions of pulsars and an anisotropy of the arrival directions of these particles from the side of the galactic plane in the Yakutsk array data and the Telescope Array data at energy $E > 5.7 \cdot 10^{19}$ eV at high galactic latitudes. Our analysis of the data shows that the cosmic rays observed up to $4 \cdot 10^{19}$ eV are galactic and above this energy most likely extragalactic.

References

[1] Dyakonov M.N., Efimov N.N., Egorov T.A., Mikhailov A.A., et al. Kozmicheskoe izluchenie predelno visokoi energii. Novosibisk, Monograph, Nauka, 1991, 251
[2] Afanasiev B.N., Glushkov A.V., Egorova V.P, Mikhailov A.A. et al., Some Characteristics of EAS and Primary Cosmic Rays on Yakutsk Array data. In: Proceedings of the Tokyo Workshop on Techniques for the Study of the Extremely High Energy Cosmic Rays, edited by M. Nagano, Tokyo, 1996, 32
[3] Abbasi R.U., Abe M., T. Abu-Zayad et al., 2014: `Arxiv: 1404.5890v3`
[4] Linsley J., Phis. Rev. Lett., 1975, **34**, 1530–1533
[5] Efimov N.N., Mikhailov A.A. Astropart. Phys. 1994, **2**, 329
[6] Mikhailov A.A., Analysis of the Arrival Directions of Ultrahigh Energy Particles. Discovery of Cosmic Rays, book, Denver, June 26–28, 2012, ed. J.Ormes
[7] Mikhailov A.A., Izvestia RAN, 1999, ser.fiz. **63**, 556
[8] Mikhailov A.A., Astronom. J. 1979, **56**, 505
[9] Mikhailov A.A., Syrovatsky S.I. Pisma v AJ, 1979, **56**, 505
[10] Berezinsky V.S., Mikhailov A.A. Pisma v JETF, 1984, **10**(4), p. 269
[11] Berezinsky V.S., Mikhailov A.A. Pisma v AJ, 1984, **10**(4), p. 269
[12] Erlykin A.D., Mikhailov A.A., Wolfendale A.W., J. Phys. G., Nucl, Part. Phys., 2002, 8
[13] Becker P.A., G.S., Bisnovatyi-Kogan, D., Casadei, D., Mikhailov A.A. et al., Frontiers in Cosmic Rays Research, Monograph, NY, Nova Science Publishers, Inc. NY, 2010, 161
[14] Malov I.E., Mikhailov A.A.,Avramenko A., Lorenzo Iorio et al., Pulsars: Theory, Categories and Application., Monograph, 2011, Nova Science Publishers, Inc. NY, 31
[15] Dorman L.I., Dorman I.V., Ahluwalia H.S., Mikhailov A.A. et al., Homage the Discovery of Cosmic Rays, the Meson-Muon and Solar Cosmic Rays. Monograph, New York, Nova Science Publishers, Inc. NY, 301
[16] Manchester R.N., Hobbs G.B., Teoh A., & Hobbs M., AJ., 2005, **129**, 1993
[17] V.A.Kuzmin, V.A. Rubakov, Yadernaya Fizika, 1998, **61**, 1122
[18] Anda R., Aguirre C., Tachi K., Proc.17-th ICRC, Paris, 1981, **2**, 164
[19] Bray A.D., Horton L., McCusker C.B.A. et al., Proc. 17-th ICRC, 1981, **2**, 168

Search for neutrino generated air shower candidates with energy $\geq 10^{19}$ eV and Zenith angle θ

Stanislav Knurenko[a], Igor Petrov[b], and Artem Sabourov[c]

Yu. G. Shafer Institute of Cosmophysical Research and Aeronomy. 31 Lenin ave., Yakutsk, Russia

Abstract. The description of the methodology and results of searching for air showers generated by neutral particles such as high energy gamma quanta and astroneutrinos are presented. For this purpose, we conducted a comprehensive analysis of the data: the electron, the muon and the EAS Cerenkov light, and their response time in scintillation and Cherenkov detectors. Air showers with energy more than $5 \cdot 10^{18}$ eV and zenith angle $\theta \geq 55°$ are selected and analyzed. Search results indicate a lack of air shower events formed by gamma-rays or high-energy neutrinos, but it does not mean that such air showers do not exist in nature; for example, experiments that recorded showers having a marked low muon content, i.e., "Muonless", are likely to be candidates for showers produced by neutral primary particles.

1. Introduction

Air showers from neutrinos can be formed in any part of the atmosphere because of its physical properties, passing quite a long way through matter. As a rule–it's strongly inclined showers formed near the detector, i.e., young showers. The basis of such showers is the electron-photon component that is scattered at large angles and, therefore, has high latency particles relatively to particles formed in the shower core. Therefore, in such events a large number of peaks of the electrons in the signal scanned from the scintillation detector should be expected [1–3].

The aim of our work was to search for extensive air shower (EAS) candidates produced by atmospheric neutrinos. In order to do this a large number of air showers with time sweep responses from scintillation detectors of different sizes and different energy thresholds have been analyzed [1,2].

Preliminary results of the data analysis of the Yakutsk array indicates the absence of EAS events produced by neutrinos, but it does not mean that such EAS do not exist. It requires a further collection of strongly inclined showers and careful analysis. This requires improving the methodology for recording and processing showers.

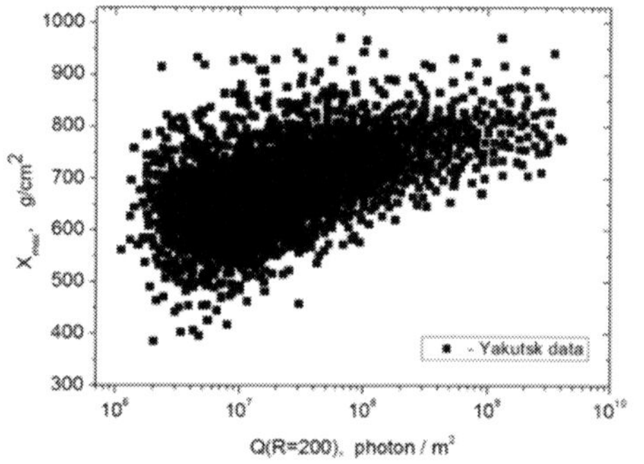

Figure 1. Dependence of X_{max} on the classification parameter Q(200)–EAS Cherenkov light flux density at a distance of 200 m from the shower axis. Data obtained in 1973–2014.

2. Search methodology of high energy gamma rays and astro-neutrinos

2.1. Longitudinal development of air showers: depth of maximum development X_{max}

Longitudinal shower development at the Yakutsk array were reconstructed through the registration of EAS Cherenkov light [4,5], using the mathematical apparatus used when solving inverse problems [6,7]. This allowed us to find connection parameters describing the cascade curve X_{max}, N_{max}, the width of the cascade, q, with directly measured air shower characteristics at sea level, the shower energy E, the total number of charged particles and the shape of the lateral distribution of the Cherenkov lightR $= lg$ (Qi / Qj), where Qi and Qj are the flux densities of Cherenkov light at selected distances from the shower axis [8]. Using this method, an EAS database with reconstructed cascade curves was created. In addition, we selected air showers by X_{max} to study the depth of the maximum offset $dX_{max}/dlgE$ and its fluctuation $\sigma_{X_{max}}$ for various energy intervals. Figure 1 shows the dependence of the distribution of X_{max} of individual showers from the classification parameter of the shower Q(200)–the flux density of Cherenkov light at 200 m distance from the shower axis. Air shower energies can be determined by the formula (1) obtained by the energy balance method [9].

$$E_0 = (1.78 \pm 0.44) \cdot 10^{17} \cdot \left(\frac{Q(200)}{10^7}\right)^{1.01 \pm 0.04} \quad (1)$$

[a] e-mail: knurenko@ikfia.ysn.ru
[b] e-mail: igor.petrov@ikfia.sbras.ru
[c] e-mail: tema@ikfia.ysn.ru

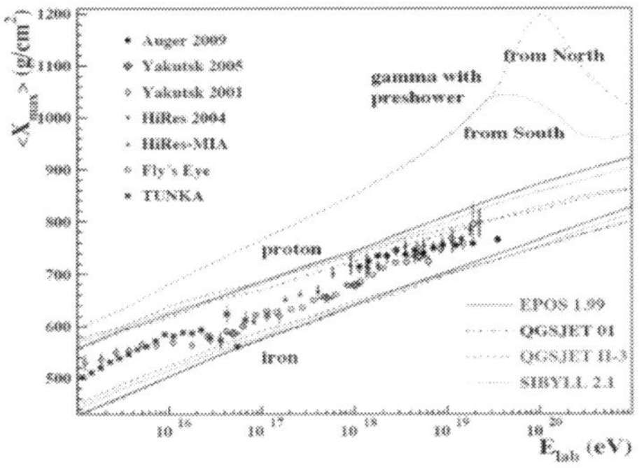

Figure 2. Dependence of X_{max} from energy. Experimental data comparison with calculations for different hadron interaction models (proton, iron nuclei and high-energy gamma ray).

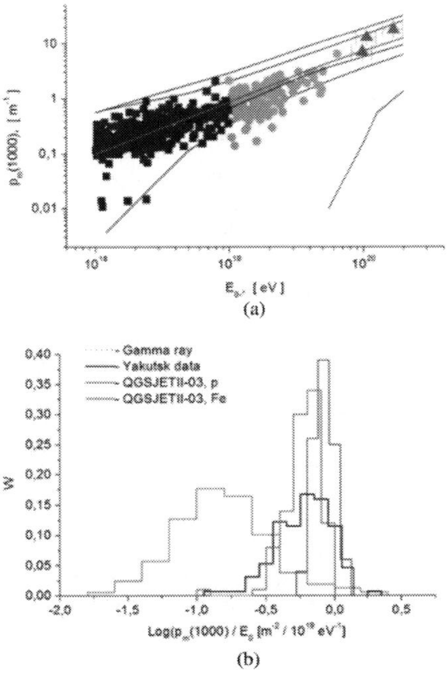

Figure 3. (a) Energy dependency of $\rho_\mu(1000)$ for observed events with energy 10^{18}–10^{19} eV (squares) and 10^{19}–10^{20} eV (triangles). Expected $\pm1\sigma$ bounds of the distributions are indicated for proton, iron and gamma ray by different curves. (b) Fluctuations of $\rho_\mu(1000)/10^{19}$ value (eV) in showers with $E_0 > 10^{19}$ eV compared to simulation results (QGSjetII-03 + UrQMD) for protons, iron and photons.

Averaged Yakutsk data together with data from other arrays are shown in Fig. 2. In addition, there are calculations for some hadron interaction models for primary nuclei (proton and iron) and high-energy gamma rays. Figure 2 shows that the height of showers produced by gamma rays is lower by 150–180 g/cm^2 than proton produced showers with energy 10^{19} eV. In fact, the cascade curve X_{max} from gamma rays is near sea level at depth ~950 g/cm^2. In this case, there is a narrow cascade consisting mainly of electrons and photons with a very low muon content. This distinguishes showers produced by gamma rays from those produced by protons or nuclei of

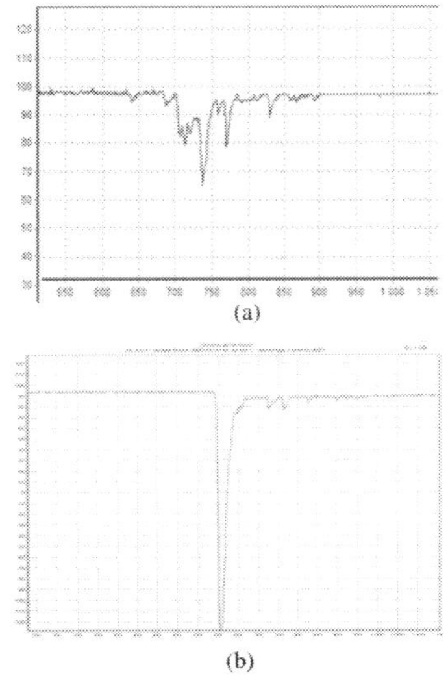

Figure 4. (a) Vertical air shower pulse. $E_0 = 1.7 \cdot 10^{19}$ eV, $\theta = 18°$, R = 1298 m. Detector with area s = 2 m^2 and threshold $\varepsilon_{thr} \geq 10$ MeV. (b) Inclined air shower event. $E_0 = 2.4 \cdot 10^{19}$ eV, $\theta = 56°$, $\psi = 200°$, R = 1000 m.

any other element. We can assume that X_{max} can be used as the first criterion to search for EAS produced by gamma rays.

2.2. Number of muons in air showers

As mentioned in Sect. 2.1 gamma ray produced showers consist of a small number of muons. If the array detects muons then from the number of muons in the shower one can judge the nature of the primary particle i.e., its atomic weight (Fig. 3a and 3b).

In the Yakutsk array the proportion of muons in the shower is determined by the relation of the muon flux density at distances of 600 m and 1000 m to the total charged component $\rho_\mu/\rho_{\mu+e}$, since these parameters are measured with better accuracy than the total number of muons, N_μ, and charged particles, $N_{\mu+e}$, in showers with total energy $E_0 \geq 10^{18}$ eV. This will be the second criterion to search for showers produced by neutral particles that include gamma rays and neutrinos.

For these purposes, there are muon detectors at the Yakutsk array with areas of 1 m^2, 20 m^2 and 190 m^2, which record muons with a threshold energy $\varepsilon_{thr} \geq 1$ GeV for vertical showers and $\varepsilon_{thr} \geq 2$ GeV for inclined showers $\theta \geq 60°$ [10]. The muon detectors are located on the array in such a way that in each shower they are measuring distances in the range from 100 to 1500 m, which allows us to estimate the fraction of muons at a particular distance from the shower axis. In particular, this analysis uses the estimated proportion of muons in showers with energies $\geq 10^{18}$ eV by the ratio of the normalized energy of the muon flux density to the density of charged particles measured by the surface scintillation detectors. Experimental results are shown in Fig. 3a, Fig. 3b [11–13]. Lines are calculations performed by the QGSJETII-03 model for protons, iron nuclei and primary gamma rays. The calculations take into account the instrumental errors and for this reason, in

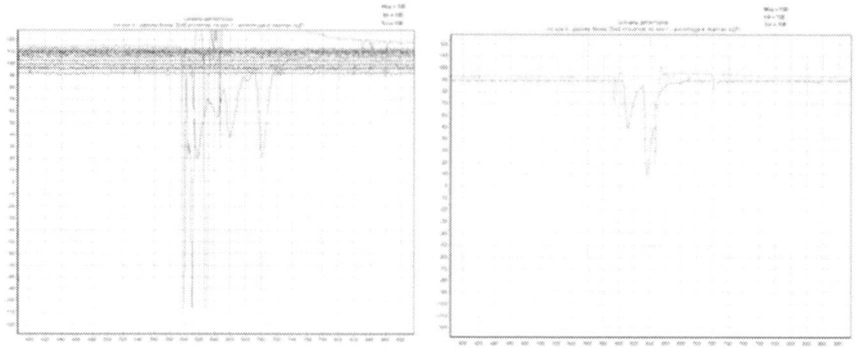

Figure 5. Distribution of pulses from electron and muon asymmetry in the air shower disc. Air shower event from 22.03.2009, $\theta = 32.4°$, $\lg E_0 = 19.45$. a) R = 549 m b) R = 886 m.

Fig. 3a, double lines indicate the boundaries at 1σ expected areas measurements of muons in the case of different mass composition of primary particles.

Figure 3a shows that there is an overlapping region for gamma ray calculations (solid line) and the experimental data. In Fig. 3b there is also a similar overlap, which shows muon flux fluctuations at 1000 m from the shower axis, normalized to the shower energy (dots). This suggests the existence of high-energy gamma rays in the flow of cosmic particles, which produce EAS in the Earth's atmosphere.

2.3. Signal time sweep of surface and underground scintillation detectors

Air showers produced by different primary particles have a maximum development at different depths in the atmosphere. Because of this, some of the secondary particles (mostly electrons) lose energy to ionization in the air and are eliminated from the cascade process. So at sea level only certain types of particles will arrive: electrons, photons and muons in the case of inclined showers and only muons in the case of strongly inclined showers, which can be seen in the signal time sweep of the scintillation detectors. For primary gamma rays and neutrinos, the depth of the maximum is going to be near the level of observation and we can expect a scintillation detector response inherent in the electron-photon component of the shower. This is another criterion by which we can select showers produced by neutral particles.

Examples of scintillation detector responses in the case of vertical and inclined showers are given in Fig. 4a and Fig. 4b. These figures show that the shape of the pulse is different in each case. Vertical showers have many peaks, where each peak corresponds to a group or a single particle. As can be seen, all of the particles are distributed in time, i.e., they arrive with different delays with respect to the first particle. Most likely, it is due to the electrons scattered in the shower subcascades.

From Fig. 4b it is seen that for strongly inclined showers the pulse structure is different from vertical showers. It is a clear single, narrow pulse. The compactness of the arrival of these particles indicates that these particles are produced in the first interactions of the primary particle with air nuclei and in the course of decay processes of $\pi\pm$–mesons, i.e., they are muons. The proof that single narrow pulses are muons passing through the scintillation detectors may be seen in Fig. 5, which shows the pulses from particles passing through different amounts of material in the path to the detector, for example in over and under axial stations.

From Fig. 5, it follows that the number of peaks in the case of over axial stations are significantly reduced and is connected with the absorption of the electron-photon component of the shower because of the difference in particles path in the atmosphere. Thus strongly inclined air shower consist mainly of muons.

It must be noted that the signal shape asymmetry is significantly expressed in showers with zenith angles $\theta = 40°–55°$. For large zenith angles, only a narrow single pulse is observed in the signal scan from air showers.

3. Experimental data analysis

In this work, we selected EAS events with $\theta \geq 60°$ and energy above $5\cdot10^{18}$ eV for the pulse shape analysis. It was necessary for the shower axis to be within a circle with a radius of 1000 meters from the center of the Yakutsk array, where many of the monitoring stations and all underground stations for muon detection are located. In this case, the accuracy of the shower axis determination ($\sigma_x = 25$ m, $\sigma_y = 35$ m) was about the same and the best for all selected events. This also applies to the determination of the basic parameters of showers, such as the energy flux density of charged particles and muons at distances of 600 m and 1000 m, the zenith ($\sigma_\theta = 1.5°$) and azimuth ($\sigma_\varphi = 3.5°$) angles. Selection of showers not only included the requirement of the pulse shape, but also its amplitude. Therefore, in the analysis we included showers with a pulse amplitude 5σ more than the average noise signal on the scan. The share of muons was determined by the ratio of the density of muons registered at distances of 600 m and 1000 m from the shower axis with respect to the flow of charged particles, registered at the same distances. The accuracy of determining each component is in the range of (15–25)% [14, 15].

Showers selected according to the criteria formed the basis of the data for analysis. A separate list was made up of showers with energies above $1\cdot10^{19}$ eV. The sample made was from data taken between 2000 and 2014 and included showers registered with both surface and underground detectors (charged particles and muons). An example of the lateral distribution of charged particles and muons of strongly inclined showers is shown in Fig. 6a.

Table 1. Strongly inclined showers with energy E $\geq 1 \cdot 10^{19}$ eV and $\theta \geq 60°$.

Date	cos θ	θ	lg E	Station number	number of peaks	$\rho_\mu/\rho_{\mu+e}$	QGSjetII-03
11.05.2000	0.266	74.6	19.63	33	1	0.91±0.09	0.84±0.03/0.89±0.03/0.20±0.03
11.01.2002	0.322	71.2	19.08	27	1	0.92±0.10	0.84±0.03/0.89±0.03/0.20±0.03
11.02.2002	0.330	70.7	19.04	27	1	0.94±0.11	0.84±0.03/0.89±0.03/0.20±0.03
14.04.2002	0.319	71.4	19.10	32	1	1.11±0.16	0.84±0.03/0.89±0.03/0.20±0.03
11.05.2006	0.437	64.1	19.05	42	1	1.06±0.09	0.84±0.03/0.89±0.03/0.20±0.03
12.10.2006	0.445	63.6	19.17	49	1	1.17±0.12	0.84±0.03/0.89±0.03/0.20±0.03
03.01.2007	0.495	60.3	19.43	35	1		0.84±0.03/0.89±0.03/0.20±0.03
15.05.2007	0.498	60.1	19.29	26	1		0.84±0.03/0.89±0.03/0.20±0.03
10.03.2008	0.425	64.8	19.18	49	1		0.84±0.03/0.89±0.03/0.20±0.03
21.02.2009	0.257	75.1	19.31	18	1	1.04±0.07	0.84±0.03/0.89±0.03/0.20±0.03
21.03.2009	0.280	73.7	19.11	28	1		0.84±0.03/0.89±0.03/0.20±0.03
10.04.2009	0.401	66.4	19.17	41	1		0.84±0.03/0.89±0.03/0.20±0.03
25.05.2009	0.298	72.7	19.24	20	1		0.84±0.03/0.89±0.03/0.20±0.03
24.12.2009	0.402	66.3	19.21	44	1		0.84±0.03/0.89±0.03/0.20±0.03

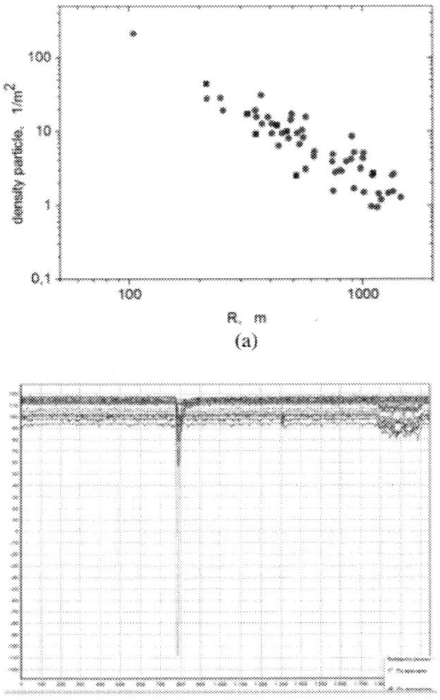

Figure 6. (a) Lateral distribution of the flux density of charged particles (red dots) and muons with $\varepsilon_{thr} \geq 2$ GeV (blue dots) in the shower with $E_0 = 1.5 \cdot 10^{19}$ eV and zenith angle $\theta = 64.8°$. (b) Air shower signal scan, registered on 10.03.2008 by scintillation detectors (colored in darker colors) and Cherenkov light detectors (brighter colors).

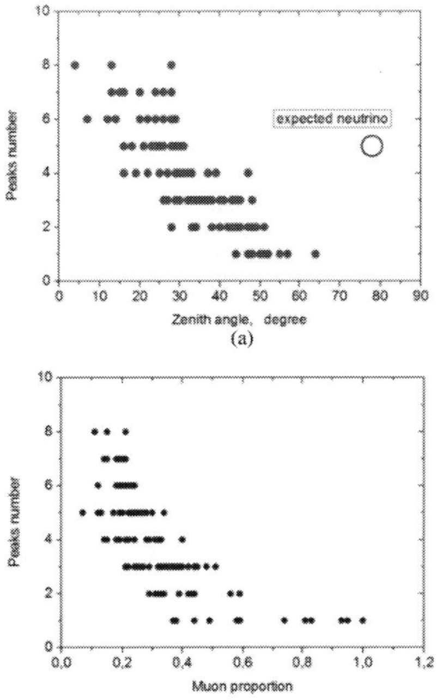

Figure 7. Number of peaks in the time sweep of scintillation detectors. a) Surface scintillation detector with threshold 10 MeV plotted against zenith angle b) underground scintillation detector with $\varepsilon_{thr.} \geq 1$ GeV.

It is not difficult (Fig. 7) to notice that the readings from the surface and underground detectors are equal, i.e., the same component of air showers is registered high-energy muons. On the signal scan, there are pulses from multiple scintillation detectors (colored in dark color) and by Cherenkov light detectors (brighter colors) (Fig. 7b). The shower front arrives at the detectors at the same time– this can be seen on the pulse rise time. The amplitudes of the Cherenkov light detector is greater than the amplitude registered by the scintillation detectors. Practically there are not many particles in the shower disk after passing through the thickness of two atmospheres, and these particles are most likely high-energy muons. They arrive at sea level in a compact group, which forms a single and sufficiently narrow pulse. As can be seen, there is a strong variation in the detector readings for strongly inclined showers. This is due to the influence of the magnetic field on charged particles and the accuracy of locating the shower axis because of the significant asymmetry of the lateral distribution of the muons (Fig. 5a, Fig. 5b).

Characteristics of the largest showers are given in Table 1. These and other showers were included in this analysis to search for EAS candidates produced by neutral particles.

In Table 1, $\rho_\mu/\rho_{\mu+e}$ is for the distance 600 m from shower axis. In the case of QGSjetII-03, calculations are for p/Fe/γ for energy E $= 10^{19}$ eV and $\theta \geq 60°$.

Figure 7 shows the dependence of the number of peaks in the time sweep response of the scintillation detector from the zenith angle of the shower arrival (Fig. 7a) and from the proportion of muons at a distance of 600 meters from the EAS axis (Fig. 7b).

It can be seen that the number of peaks in the signal scan (Fig. 7a) strongly depends on the zenith angle. In the case of vertical showers with $\theta \leq 30°$ and moderate distances from the shower axis, the number of peaks ranges from eight to four, and starting from zenith angles $\theta \geq 59°$ there is only a single peak in the signal scan. The proportion of muons increases significantly with zenith angle and reaches almost 100% for $\theta \geq 60°$. The observed pattern can be used as a sort of search criteria for air shower produced by neutral particles. For example if an EAS has a depth of maximum near sea level, $X_{max} = (800–1000)$ g/cm^2, the number of muons is small and there are a large number of peaks in the signal scan at zenith angle $\theta \geq 60°$ we can assume that this shower was produced by a neutral particle–gamma ray or neutrino - with a high probability. In Fig. 7a, this region is marked by a large circle. Thus, this work suggests that the search criteria for primary particles with characteristics different from protons and other nuclei may be suitable for such an analysis.

4. Conclusion

1944 EAS events with energy above $5 \cdot 10^{18}$ eV and $\theta \geq 60$ ° were registered over 15 years of continuous observations at the Yakutsk array. With the methodology suggested above, a comprehensive analysis of charged particles, muons and Cherenkov light data was carried out including time sweep in underground and surface scintillation detectors. The dependence of the shape of the scintillation detectors response in the time sweep on the zenith angle was studied. A characteristic feature was that there were many peaks in vertical showers and only a single peak in strongly inclined showers [2]. Use of the Cherenkov component of the shower in the analysis and reconstruction of the longitudinal shower development [5, 8] indicated a direct relationship of the peaks in the time sweep of the scintillation detector with the longitudinal development of the shower and number of muons in the shower. All this is well illustrated by the data shown in Figs. 7a and 7b.

A multi component analysis of air showers using the above described criteria, found no showers produced by gamma rays or neutrinos. At the same time, QGSJETII-03 model calculations for primary protons, iron nuclei and gamma rays (Figs. 2, 3a and 3b) tells us that if fluctuations of the muon measurements within 1σ are taken into account, the probability of detecting air showers produced by neutral particles exist. "Muonless" showers detected at the Yakutsk array can be considered as candidates for such showers [16].

References

[1] S. Knurenko, Z. Petrov and Yu. Yegorov, J. Phys.: Conf. Ser. **409**, 012090 (2013) doi: 10.1088/1742-6596/409/1/012090
[2] S. Knurenko and A. Sabuorov, Proc. of the 33th International Cosmic Ray Conference, Rio-de-Janeiro, 0055 (2013)
[3] G.I. Rubtsov et al. EPJ Web of Conferences **53**, 05001 (2013) doi: 10.1051/epjconf/20135305001
[4] M.N. Dyakonov, S.P. Knurenko, V.A. Kolosov et al. Nuclear Instruments and Meth-s in Phys. Rec. Amst. **A248**(4), p. 224–226 (1986)
[5] S.P. Knurenko, V.A. Kolosov, Z.E. Petrov. Proc. 27-th ICRC, Hamburg (Germany) **1**, p. 157–160 (2001)
[6] A.N. Tihonov, V.Ya. Arsenin. *Methods for solving ill-posed problems* (Science, Moscow, 1979), p. 286
[7] V.A. Kochnev. Proc. of conference "The use of computers in control problems", 62–71 (1985)
[8] S.P. Knurenko. Ph. D. work, Yakutsk (2003)
[9] S.P. Knurenko, A.A. Ivanov, I.Ye. Sleptsov, A.V. Sabourov. JETP Lett. **83**, 563–567 (2006)
[10] V.P. Artamonov, B.N. Afanasyev, A.V. Glushkov. Bulletion of RAS. Physics Ser. **58**, 92–97 (1994)
[11] S.P. Knurenko et al. Nuclear Physics B (Proc. Suppl.) **151**, 92–95 (2006)
[12] S.P. Knurenko, A.A. Ivanov, M.I. Pravdin, A.V. Sabourov and I.Ye. Sleptsov. Nucl. Phys. B (Pros. Suppl.) **175-176**, 201–206 (2008)
[13] S.P. Knurenko, A.V. Sabourov. Nuclear Physics B (Proc. Suppl.) **196**, 319–322 (2009)
[14] M.N. Dyakonov, A.A. Ivanov, S.P. Knurenko et al. Proc. 17th ICRC **6**, 78–81 (1981)
[15] Stanislav Knurenko et al. EPJ Web of Conferences **99**, 04001 (2015)
[16] V.A. Kolosov et al. Proc. 31th ICRC, 287 (2009)

Recent pioneers

Lawrence W. Jones[a]

University of Michigan, Ann Arbor, Michigan, USA

Abstract. Recognition of outstanding Russian cosmic ray physicists and earlier meetings of this Symposium.

1. Introduction

I have had the pleasure of attending almost all of these International Symposia on Very High Energy Cosmic Ray Interactions since their origin in 1980. We had the privilege of hosting the 1992 meeting at my home institution, the University of Michigan in Ann Arbor, Michigan, U.S.A. The meetings were always informative, pleasant, and lively, with a diverse international attendance of physicists from around the world. To be sure, there was certainly some overlap in topics discussed and those in attendance at the International Cosmic Ray Conferences held during the summer on alternate years. However, our ISVHECRI meetings have always focused more narrowly on particle interactions and air showers, and the participation at the meetings is smaller and more intimate.

2. Pioneers

The community of physicists at the meetings was always pleasant and stimulating. Essentially everyone had been involved in a significant program, usually a cosmic ray experiment. In the last century (before the year 2000), almost all of the significant experiments were on the ground, including experimental sites at mountain elevations. To be sure, there were also balloon experiments, which were of limited weight and area.

At and near sea level, the most common experiments were (and still are) air shower arrays, where the data collect particles (electrons, muons, and gamma rays, mostly) from interactions in the atmosphere above. However at high mountain elevations, primary interactions are less uncommon, and the interactions of cosmic ray protons can be (and, in the last century, were) studied.

Two experimental programs of interest in those days were the studies in Bolivia on Mt. Chacaltaya (at an elevation of 5200 meters) and in the Russian Pamir mountains (at 4370 meters). This Russian experiment was named "Pamir".

The Russian physicist who directed the Pamir program was Serguei Slavatinsky, whose home base was the Lebedev Physical Institute in Moscow. Without going into details on the physics results from Pamir, it is appropriate to note that it was a very active and productive program in the 1980s. I must also note that Slavatinsky was a most pleasant and delightful friend; he was an active participant in these ISVHECRI Symposia, as well as at the ICRC meetings. Sadly, he died several years ago, so is no longer with us. However, in view of his significant leadership of the Pamir program with his reporting and communication, he certainly merits recognition as one of the Russian "Pioneers".

I will add, as an appropriate citation to his work and communication to the cosmic ray community, reference to the paper he presented at our 1992 Michigan meeting of the ISVHECRI. His was the first paper presented there; entitled "Ordinary and Unordinary Results of Experiment "Pamir" on Interactions of Hadrons with Energies at 10–100 PeV" [1].

Of course there are many other significant international physicists, who merit recognition. For example, Sir Arnold Wolfendale from England was an earlier regular attendee at these meetings, and always had relevant and interesting contributions. Gaurang Yodh, from the U.S. but of Indian origin, has also been a lively contributer; he is currently directing HAWC, a very productive mountain Cherenkov air-shower array in Mexico. Oscar Saavedra, from Bolivia (and now in Italy) has championed the Mt. Chacaltaya program. Nor should we forget the valuable contributions made over the years by A.E. Chudakov, G. Zatsepin, S.I. Nikolsky, J. Linsley all sadly deceased.

References

[1] S.A. Slavatinsky, pp. 3–27 AIP Conference Proceedings 276; Very High Energy Cosmic Ray Interactions VIIth International Symposium (1992)

[a] e-mail: lwjones@umich.edu

The hadronic interaction model Sibyll – past, present and future

Ralph Engel[1,a], Felix Riehn[1,2,4,b], Anatoli Fedynitch[3], Thomas K. Gaisser[4], and Todor Stanev[4]

[1] Institute for Nuclear Physics (IKP), Karlsruhe Institute of Technology (KIT), P.O. Box 3640, 76021 Karlsruhe, Germany
[2] LIP, Av. Elias Garcia 14-1, 1000-149 Lisboa, Portugal
[3] DESY, Platanenallee 6, 15738 Zeuthen, Germany
[4] Bartol Research Institute & Department of Physics and Astronomy, University of Delaware, Newark, DE 19716, USA

Abstract. Sibyll is one of the first microscopic interaction models that was specifically developed for interpreting cosmic ray data. It combines non-perturbative concepts of simulating hadronic particle production with predictions derived from perturbative QCD calculations, focusing on forward particle production of relevance in studying cosmic ray interactions. In this contribution we briefly recall the history of Sibyll and then, in this context, describe improvements made in the different versions of the Sibyll model. The discussion focuses on the basic concepts and ideas of these improvements rather than going into detail or giving a comprehensive description of the models. We also discuss shortcomings, conceptual problems, and uncertainties in modeling hadronic interactions and make some suggestions how to address these open questions in the future.

1. Introduction

In the early days of cosmic ray physics the measurements of inclusive fluxes and air showers were compared to analytic and semi-analytic calculations to derive information on the energy of the primary particles, their flux, composition, and also to learn about general features of hadronic multiparticle production. A system of coupled cascade equations had to be solved for different initial conditions (see, for example, [1–4]). A number of approximations and simplifications were needed to keep the problem of calculating secondary particle fluxes and particle cascades treatable at the time. With the availability of increasingly powerful computers, solving cascade equations numerically became more and more common [5–7]. But even then it was not possible to calculate the secondary fluxes in full detail, and shower-by-shower fluctuations could not be predicted. Ultimately, the full complexity of the problem can only be treated with Monte Carlo methods, which are suited to deal with the large number of degrees of freedom involved in these calculations.

Today a variety of dedicated program packages is publicly available for calculating the results of cosmic ray interactions in the atmosphere. These include programs in which Monte Carlo generated tables of hadronic interactions are used to numerically solve cascade equations (for example, MCEq [8]), programs in which the Monte Carlo method is combined with numerical solutions of cascade equations (for example, SENECA [9] and CONEX [10]), and fully Monte Carlo based programs (for example, AIRES [11,12], CORSIKA [13], and COSMOS [14]).

Very important elements of these code packages are hadronic interaction models, often referred to as event generators. Similar to the evolution found in cascade calculations, the first interaction models were just analytic parametrizations of the secondary particle yields and phase space distributions. One of the notable very early Monte Carlo models of hadronic interactions is the Hillas splitting algorithm (see [15] and also appendix of [4]), which was developed for one of the first Monte Carlo packages for air showers, MOCCA [15,16]. This splitting algorithm is still used in a slightly modified form in AIRES. With time the complexity of hadronic interaction models increased quickly and we now have very sophisticated and powerful Monte Carlo event generators available. These event generators are based on microscopic models and certain aspects of QCD, the theory of strong interactions, and represent the state-of-the art of our understanding of high-energy hadronic multiparticle production. Sibyll [17–21] is one of these event generators whose development started already in the late 1980ies.

After recalling the motivation for developing Sibyll we will discuss the major steps of arriving at the currently available version 2.3 of this interaction model (Sect. 2). Recent improvements triggered by the progress in understanding hadronic interactions in general, and the new LHC and fixed-target measurements in particular, will be summarized in Sect. 3. After presenting examples of comparisons to accelerator data, predictions for air showers and inclusive fluxes will be shown and discussed in the context of those obtained with other interaction models. Finally, a critical outlook is given in Sect. 4.

2. The past – a short history of Sibyll

In the following we will recall some fundamental hypothesis and experimental observations to introduce the

[a] e-mail: `ralph.engel@kit.edu`
[b] e-mail: `friehn@lip.pt`

scientific contest and the motivation for developing the Sibyll model.

2.1. Regge theory

In the early days of high energy physics it was common wisdom that the total and elastic interaction cross sections of hadrons decrease with the center-of-mass energy $\sqrt{s}$. This behavior was well understood in terms of Regge theory [22] by assuming maximally analytic scattering amplitudes and accounting for exchanges of many hadronic particles in a scattering process. Qualitatively, the predicted relation is $\sigma \sim s^{\alpha_i(0)-1}$, with $\alpha_i(0)$ being a constant, called the Regge intercept of the Regge trajectory $\alpha_i(t)$, which depends on the quantum numbers of the exchanged hadrons. These Regge intercepts are related to the spin-mass relation of resonances (Chew-Frautschi plot [23]) and the largest measured values of $\alpha_i(0)$ are in the range of $0.45 - 0.55$.

With the first data from the Serpukhov accelerator it became clear that this seemingly self-consistent picture, based entirely on hadronic degrees of freedom, had to be extended. The Serpukhov data of the late 1960ies indicated that hadronic cross sections start to be energy-independent above ~ 50 GeV/c lab. momentum ($\sqrt{s} \sim 10\,\mathrm{GeV}$). This motivated Pomeranchuk to postulate the existence of a new Regge trajectory with $\alpha(0) \approx 1$, now commonly referred to as the pomeron. In 1971, with the data of the CERN Intersecting Storage Ring (ISR), it became obvious that the total and elastic cross sections of hadrons slowly increase with energy. In the Reggeon model, this can only be explained by assuming a super-critical pomeron, i.e. $\alpha(0) > 1$. Since then there have been many attempts to find the resonances belonging to the pomeron trajectory, which are possibly short-lived glue balls. Whether glue balls indeed exist is still an open question [24].

At this time also a number of phenomenological scaling hypotheses were commonly tested and found to be approximately satisfied. The most important ones are Feynman scaling [25] of particle production cross sections and Koba-Nielsen-Olesen (KNO) scaling [26] of particle multiplicities.

If Feynman scaling applies, the inclusive production cross section $ab \to cX$ is expected to approach, in the high-energy limit, a universal function

$$E\frac{\mathrm{d}\sigma_{ab\to c}}{\mathrm{d}^3 p} \to f_{ab\to c}(x_F, p_\perp), \tag{1}$$

with $x_F = p_\parallel / p_{\parallel,\max} \approx p_\parallel/\sqrt{s}$ and $p_\perp$ being the transverse momentum of c. An important consequence of Feynman scaling is the prediction of an energy-independent height of the rapidity distribution $\mathrm{d}N_{ab\to c}/\mathrm{d}y$ of particle c for $x_F \to 0$

$$E\frac{\mathrm{d}\sigma_{ab\to c}}{\mathrm{d}^3 p} = \frac{\mathrm{d}\sigma_{ab\to c}}{\mathrm{d}y\mathrm{d}^2 p_\perp} \to f_{ab\to c}(x_F \to 0, p_\perp). \tag{2}$$

Another important phenomenological scaling hypothesis, which is similar to that of Feynman scaling, is the idea limiting fragmentation [27]. Here the assumption is that the distributions of leading particles, i.e. particles with momenta similar to those of the projectile or target, depend only on the projectile and target particles and approach a universal function at very high energy.

KNO scaling manifests itself by energy-independent moments, γ_q, of the multiplicity distribution

$$\gamma_q = \frac{\langle (n-\bar{n})^q \rangle}{\bar{n}^q}, \tag{3}$$

where $\bar{n}$ denotes the average particle multiplicity and the average is taken over all events of the data sample.

All these phenomena could be reasonably well described, on a phenomenological level, by Regge-inspired models. The energy dependence of the total and elastic cross sections could be parametrized and, with the help of the triple-Regge graph [28] and corresponding unitarity cuts, it was even possible to understand inclusive particle distributions. Still, most of the predictions were of a qualitative nature and measurements were needed to make them more quantitative.

2.2. Parton model and perturbative QCD

After the introduction of quarks for the classification of hadrons in 1964 [29,30] the concept of constituents of hadrons was extended to high-energy scattering phenomena within the parton model. Originally, the parton model was developed by Feynman [31–33]. to describe scaling laws found in inelastic lepton-nucleon scattering by introducing point-like constituents in hadrons, but soon many other applications were found. Finally, with the formulation of Quantum Chromodynamics (QCD) [34] in 1973 a selfconsistent theory of strong interactions was available.

QCD was enthusiastically adopted with the realization that asymptotic freedom of partons is one of its fundamental properties. It was realized that Feynman's partons were the quarks and gluons of QCD and soon first predictions of parton-parton scattering, calculated in perturbation theory [35,36], and estimates for high-$p_\perp$ particle and jet production [37] were made and found to be in reasonable agreement with measurements. Also scaling laws for the fragmentation of partons to (leading) hadrons were derived [38] and evolution equations for the densities of partons in hadrons were written down [39].

In the early 1980ies rare processes involving large momentum transfer (hard processes), such as jet production, could be calculated reliably and many experiments at colliders were built to study these processes. At the same time, very little progress was made in understanding the bulk of hadronic particle production that does not proceed through processes of high momentum transfer (soft processes). Even though the partonic view of high-energy interactions was very successful for hard processes, Regge-type models were still needed to describe soft hadron production processes. In particular, it became clear that it is very difficult to combine the hadronic view of soft processes with the partonic view of hard processes in a consistent and quantitative way [40].

2.3. Minijet Model

In the second half of the 1980ies the data of the CERN SPS collider confirmed the growth of the total and elastic $p\bar{p}$ cross sections [41,42] as already seen at the ISR. New observations were the strong violation of Feynman scaling at central pseudorapidities [43] and also the violation of KNO scaling [44]. For example, the pseudorapidity

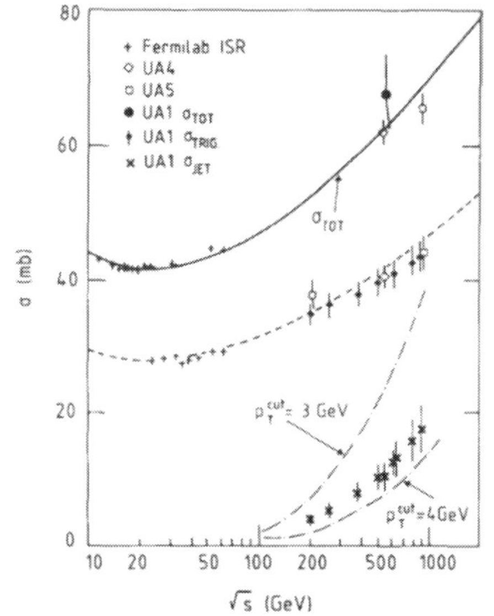

Figure 1. Inclusive cross section for minijet production measured by the UA1 Collaboration [46]. The data are compared to measurements and model curves for the elastic and total $\bar{p}p$ cross sections. Also shown are predictions based on perturbative QCD for parton-level jets for different transverse momentum thresholds (dash-dotted lines).

density of charged particles $dN_{ch}/d\eta$ changes from 2.2 at $\sqrt{s} = 200\,\mathrm{GeV}$ to about 3 at $\sqrt{s} = 900\,\mathrm{GeV}$ for $\eta \approx 0$. In contrast, the leading particle distributions, as far as accessible at the SPS experiments, were found to scale as expected for limiting fragmentation [43,45].

Another surprising result was the rapid increase with energy of the inclusive cross section for jet production [46]. This is shown in Fig. 1. This observation led to a controversial discussion about down to what scales of momentum transfer can perturbative calculations still be expected to be reliable. Calculations of cross sections for so-called minijet production (i.e. jets with a transverse momentum of 3 − 5 GeV) were met with scepticism and the debate is still ongoing.

Extrapolating the measured jet rates or relying on calculated inclusive cross sections led to the conclusion that hard parton-parton scattering should be an abundant process in hadronic scattering at high energy [47]. Therefore it was very natural to assume that, if the until-then theoretically unexplained rise of the total cross section does originate from minijet production, a new process is setting in just in this energy range. After all, parton-parton scattering is a fundamental and undisputed prediction of QCD.

One of the key problems with this assumption was, however, the fact that the cross section for jet production increases with energy much faster than the total cross section, see Fig. 1. This problem is addressed in the Minijet Model [48–52], which was developed in the late 1980ies. It was the first model offering a consistent microscopic picture of the relation between the inclusive jet production rate and the growth of the elastic and total cross sections.

In the minijet model, the apparent contradiction between the rapid growth of the inclusive jet cross section and the only moderate growth of the total cross section

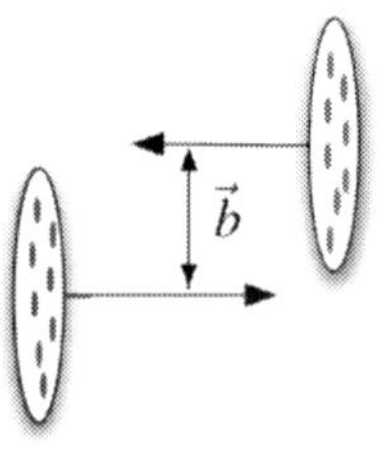

Figure 2. Geometric picture of two hadrons colliding at high energy. The hadrons are Lorentz-contracted along the beam axis. The distribution of partons in a hadron is described by their density in transverse space. The number of parton pairs expected to interact depends on the parton densities in transverse space and the impact parameter $|\mathbf{b}|$ of the collision.

is resolved by allowing for multiple, independent parton pairs interacting in a single hadronic collision. Keeping in mind that jet cross sections σ_{jet} calculated within perturbative QCD are inclusive cross sections one has the relation

$$\sigma_{jet} = \langle n_{int} \rangle \, \sigma_{ine}, \qquad (4)$$

with $\langle n_{int} \rangle$ being the average number of interacting parton pairs and σ_{ine} denoting the inelastic cross section. If, for simplicity of the argument, the energy dependence is written as $\sigma_{ine} \sim s^{0.08}$ and $\sigma_{jet} \sim s^{0.4}$ a rapid growth of the mean number of interactions $\langle n_{int} \rangle \sim s^{0.32}$ is predicted. Already a very simple and intuitive geometric model of hadrons is sufficient to calculate the probability distribution of the number, n_{int}, of interacting parton pairs and, hence, the relation between the inelastic and jet cross sections. This is illustrated in Fig. 2. Neglecting possible correlations between the partons in the colliding hadrons (the partonic state is frozen-in for the time of the hadronic collision) one obtains the well-known eikonal model for multiple interactions (for a pedagogical presentation, see [4]). The eikonal formulas can also be derived within field theory. This more rigorous derivation also predicts the correct complex phases for the resulting scattering amplitudes, see [53,54] and references therein.

The minijet model, combined with the eikonal approach for estimating the amount of multiple partonic interactions, is a model that has only a very small number of parameters. These are the transverse momentum threshold that is needed for regarding partonic interactions as hard processes to which asymptotic freedom applies (and also for calculating the inclusive minijet cross section), the parton density functions $f(x, Q^2)$, and the parton distributions in transverse space. To make the model complete soft interactions have to be added, which cannot be calculated in perturbation theory.

There are a number of very remarkable successes of the minijet model [48,55]. First of all, the rise of the observed cross sections can be understood and predictions for higher energies can be calculated, see [50,56–60] for examples of early works. Secondly, the multiplicity distribution of secondary particles, including the increasing violation of KNO scaling is correctly predicted [48,61,62]. Thirdly, the model predicts a rise of the central pseudorapidity plateau and, hence, violation of Feynman scaling at $|x_F| \approx 0$. Additional features of the model are strong forward-backward multiplicity correlations in rapidity space [63] and the expectation of only very minor scaling violations of the distributions of leading secondary particles, which are related to effects of energy-momentum conservation.

These successes of the minijet model are, to a large extent, related to the concept of multiple interactions [64]. Other multiple interaction models, such as the Dual Parton Model [65] and the Quark-Gluon Strings Model [66], which were originally developed for soft interactions only, also describe many of the observed features of multiparticle production up to $\sqrt{s} \sim 200\,\text{GeV}$. In comparison to these models, the minijet model is an attempt to relate as many features of multiparticle production as possible to hard parton-parton interactions and to minimize additional assumptions needed for soft processes. With hard interactions becoming increasingly important at high energy, this approach promises a framework for extrapolations to very high energies without many parameters.

2.4. Early versions of Sibyll up to Sibyll 1.7

Already three years after the development of the minijet model, a first implementation of multiple hard interactions in the hadronic event generator PYTHIA became available in 1987 [67]. In PYTHIA, only the cross section of inelastic events with (possibly multiple) hard parton-parton interactions is calculated following the minijet model. This cross section is considered as a part of the larger total inelastic cross section, which is taken from an external parametrization. No attempt is made to fully implement the minijet model, i.e. to also derive the total and elastic cross sections from the implemented partonic cross sections.

Sibyll was the first Monte Carlo event generator with a full implementation of the minijet model. Elastic and total cross sections as well as partial cross sections for a given number of multiple interactions are calculated in a self-consistent way. Key assumptions of the model were a constant soft cross section, which had to be added in the eikonal to the inclusive cross section for hard interactions, and the assumption of an energy-independent transverse momentum threshold $p_\perp^{\rm cut}$ for calculating the minijet cross section. Many aspects of Sibyll are very strongly inspired by the Lund approach for modeling hadronic interactions. The string fragmentation used in Sibyll, for example, is an implementation of the Lund model [68,69] of these days.

Early versions of the code date back to 1987 and were used in various research projects, for example, see [70]. In a collaboration with J. Engel, a library for nuclear targets and projectiles [17] was developed in the early 1990ies, adding a sophisticated abrasion-ablation model for nuclear fragmentation to Sibyll. Proton/pion/kaon interactions with the nuclei of air are simulated based on the Glauber model for nuclear interactions [71]. Keeping in mind the application for air shower simulations, the interaction of primary nuclei is implemented using the semi-superposition model. This means that the depth distribution of interaction points, the number and mass distribution of the nuclear fragments, and number of interacting nuclei is correctly calculated using the Glauber approach combined with a nuclear fragmentation model. But at each given interaction point N interacting nucleons of the projectile are replaced by N protons and neutrons whose interaction with air nuclei is simulated separately [17]. Finally, version 1.5 of Sibyll was made publicly available and described in detail in Ref. [18]. Minor additions such as including neutral pions as possible projectiles led to an increase of the version number to 1.7. In 1993 Sibyll was included as one of the first microscopic interaction models in the shower simulation code MOCCA [15,16], and later CORSIKA [13], and AIRES [11,12]. A multitude of shower simulations were done with Sibyll, including all the simulations for designing the Pierre Auger Observatory [72,73].

2.5. Sibyll 2.1

Although Sibyll 1.7 was a very successful model in the 1990ies, it was far from being perfect.

Already the comparison to data of fixed-target and collider experiments made in the initial publication of Sibyll showed some shortcomings of the model [18]. For example, the predicted distribution of the secondary particle multiplicity was not as wide as found in data. Other problems were related to the overall growth of the particle multiplicity with energy and the distribution of secondary particles around $x_F \sim 0.8 - 0.9$, the transition region between non-diffractively and diffractively produced secondaries.

Independent of the discrepancies found in comparisons with accelerator data, the first data on deep-inelastic scattering of the HERA collider indicated a behavior of parton densities at low x similar to $f(x, Q^2) \sim x^{-1.3}$, which is very different from what had been commonly used before. The minijet cross section in Sibyll 1.x had been calculated with the Eichten-Hinchliffe-Lane-Quigg [74] (proton) and Duke-Owens [75] (pion) parton densities, which satisfy $f(x, Q^2) \sim x^{-1}$.

Last but not least, there were also indications from the KASCADE air shower array [76] that the showers simulated with Sibyll either had too many electrons or too few muons at ground level for obtaining a consistent description of shower data through the knee energy region.

These observations triggered the development of an improved version of Sibyll that started in 1997. The model for diffraction dissociation was replaced by a more consistent treatment of diffraction using a two-channel eikonal model based on the Good-Walker approach [77]. Multiple soft interactions were implemented in the generation of the string configurations and hadronic final states to consistently interpret the eikonal amplitude according to the Abramovski-Gribov-Kancheli cutting rules [78]. The parametrizations of parton densities were updated to those by Glück, Reya, and Vogt [79,80], which were in good agreement with HERA data at this time.

The modified parton densities led to a much steeper increase of the predicted minijet cross section and it became clear that, even by adjusting model parameters very generously, the slow rise of the total cross section could not be described without modifying the original ideas of the minijet model. Either the total cross section would grow too fast or the secondary particle multiplicity would exceed by far that found at the Tevatron. Qualitatively, the correlation of these quantities follows directly from Eq. (4). An eikonal model satisfies unitarity, i.e. the inclusive cross section for minijet production always stays the same.

Unitarity is such a fundamental property of scattering amplitudes that the tension between the energy dependence of the jet cross section, total cross section, and the particle multiplicity has to be interpreted as a failure to calculate inclusive minijet cross section (the other quantities are measured, and (4) holds in any model that satisfies unitarity). There are many reasons why a calculation at transverse momenta as low as $p_\perp \sim 2-3\,\mathrm{GeV/c}$ could fail. For example, collinear factorization as used in most perturbative calculations is expected to break down and, instead of leading $\log(Q^2/\Lambda^2_{\rm QCD})$ terms, one should re-sum $\log(1/x)$ terms, or the extrapolation of parton densities to low x is failing for such low parton virtualities. Also calculations at leading-log accuracy are know to be subject to large uncertainties stemming from the arbitrary choice of the factorization and renormalization scales.

Such problems in calculating the minijet cross section had already been anticipated many years earlier, see [47] for a review. The large density of partons at low x leads to collective effects, i.e. partons cannot be considered to be independent objects anymore. Whether these collective effects lead to a saturation or just a reduction of the parton densities is the subject of an ongoing debate.

In the high-energy limit, i.e. $\ln(1/x)$, $\ln(Q^2/\Lambda^2_{\rm QCD}) \to \infty$ (double-leading-log approximation), the gluon density $g(x, Q^2)$ is given by

$$xg(x,Q^2) \sim \exp\left[\frac{48}{11-\frac{2}{3}n_f}\ln\frac{\ln\frac{Q^2}{\Lambda^2_{\rm QCD}}}{\ln\frac{Q_0^2}{\Lambda^2_{\rm QCD}}}\ln\frac{1}{x}\right]^{\frac{1}{2}} \sim \frac{1}{x^{0.4}}, \tag{5}$$

with n_f being the number of active quark flavors. Then, assuming a transverse size of a gluon being about the size of its cross section $\sigma \sim \alpha_s(p_T^2)/p_\perp^2$, one has as condition for all gluons fitting next to each other in a proton of radius R_p [81]

$$\frac{\alpha_s(p_T^2)}{p_T^2}\cdot xg(x,p_T^2) \le \pi R_p^2, \tag{6}$$

where α_s is the strong coupling constant. If the gluons are not distributed democratically in transverse space by forming, for example, hot spots around the valence quarks, the non-linear effects suppressing the number of independent partons will set in much earlier than indicated by Eq. (6).

In Sibyll, Eq. (6) is used as guidance for introducing an energy-dependent transverse momentum threshold applied in calculating the minijet cross section

$$p_\perp^{\rm min}(s) = p_\perp^0 + \Lambda \exp\left[c\sqrt{\ln(s/\mathrm{GeV}^2)}\right], \tag{7}$$

with $p_\perp^0$ and Λ being model parameters. All interactions producing only partons with $p_\perp < p_\perp^{\rm min}(s)$ are considered as soft interactions. The previously constant cross section for soft interactions had to be made energy-dependent to account for the fraction of partonic interactions that has been shifted in the classification from hard to soft. Inspired by Regge phenomenology for soft processes, this energy dependence is assumed to be a power law $\sigma_{\rm soft} \sim s^\Delta$ with Δ being a parameter that depends on $p_\perp^0$ and Λ. Δ is determined by fitting total and elastic cross sections (typical values are $\Delta \approx 0.1$).

The fit to the the pp and $p\bar{p}$ cross sections is shown in Fig. 3. Thanks to the improved model for diffraction

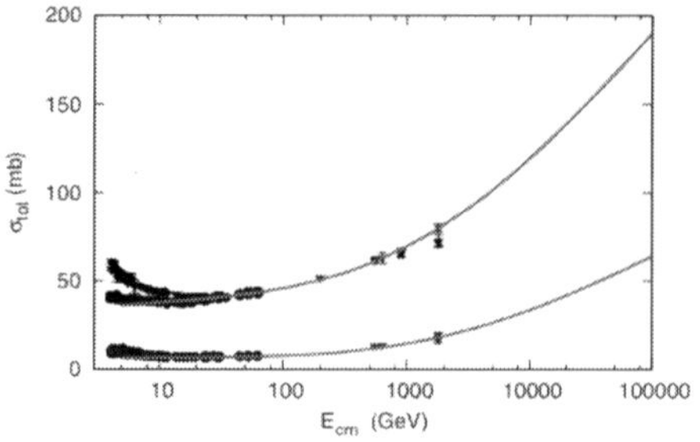

Figure 3. Elastic and total cross sections of pp and $p\bar{p}$ interactions. Shown are data together with the cross section fit of Sibyll 2.1 [19] as a function of the center-of-mass energy $E_{\rm cm} = \sqrt{s}$.

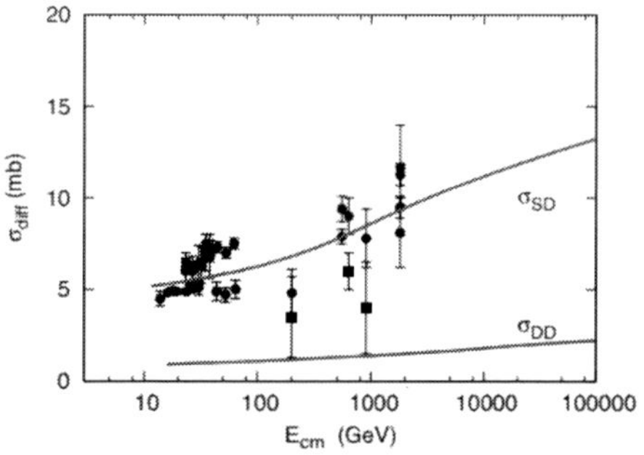

Figure 4. Single- and double-diffractive cross sections of pp and $p\bar{p}$ interactions. Shown are measurements for diffractively produced masses of $M_D^2 < 0.05\,s$ together with the predictions of Sibyll 2.1 [19] as a function of the center-of-mass energy $E_{\rm cm} = \sqrt{s}$.

dissociation, a good description of single- and double-diffractive cross sections is also obtained, see Fig. 4.

This construction of matching soft and hard interactions illustrates a generic problem of hadronic interaction models. Strictly speaking, partons are only well defined objects for hard interactions. Nevertheless, there has to be a smooth transition to soft interactions, for which the parton language does not apply. The currently applied makeshift solution is to generate string configurations for soft interactions that would correspond to hard interactions with very low $p_\perp$, i.e. two strings per interaction. Another problem is the significantly different energy dependence of the soft and hard processes. A smooth transition between the two categories of processes can only be maintained if the transverse momentum threshold for hard processes is increasing with energy.

Finally it should be mentioned that the inclusion of multiple soft interactions in Sibyll is nothing other than a particular implementation of Gribov's Reggeon calculus [82]. In this sense the model implemented in the new version of Sibyll goes beyond the classic minijet model of the late 1980ies and becomes in some aspects similar to the Dual Parton Model realizations in PHOJET [83,84] and DPMJET [85,86].

In 1999 first predictions of the new version of Sibyll were shown [90] and the code was released to the public in 2000 as Sibyll 2.1. Because of the conceptually non-satisfying treatment of the matching of soft and hard interactions this model was never considered as

DOI: 10.1051/epjconf/201714508001

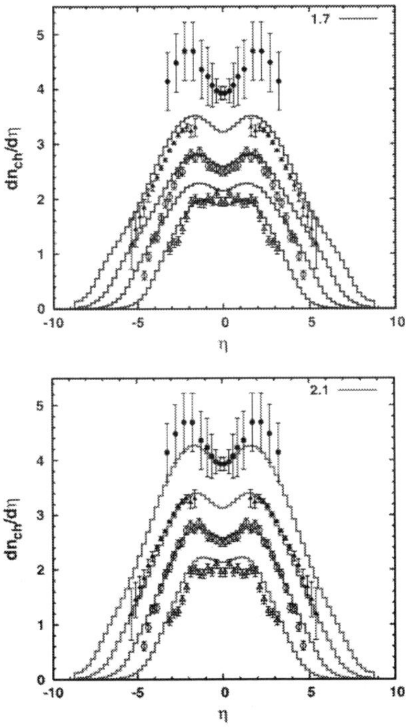

Figure 5. Pseudorapidity distribution of charged particles in $p\bar{p}$ collisions. ISR and SPS data are compared to predictions of Sibyll 1.7 and Sibyll 2.1 [19]. Data points from top are $\sqrt{s} =$ 1800 GeV [87], 630 GeV [88], 200 GeV [43], and 53 GeV [89].

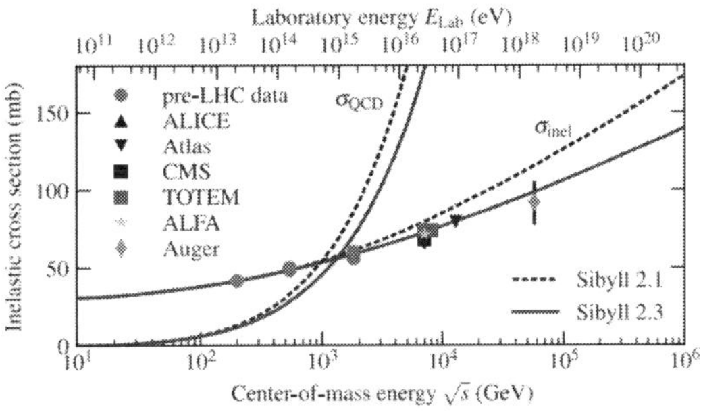

Figure 6. Inelastic cross section for $p\bar{p}$ interactions. Data are compared to calculations made with Sibyll versions 2.1 and 2.3. The new LHC measurements [92–97] are shown individually. At high energy the cross section derived from p–air interactions by the Auger Collaboration is also given [98]. In addition the inclusive cross section for minijet production of the two model versions is shown as dashed lines.

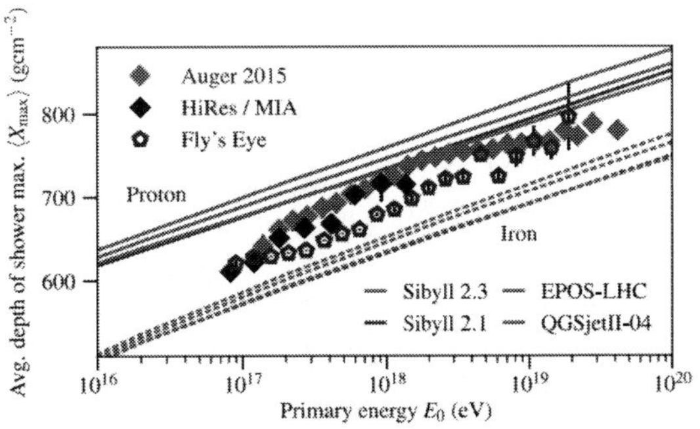

Figure 7. Mean depth of shower maximum. Data [110–112] are compared with different model predictions.

a long-term version of Sibyll. The successes of the model in describing many collider measurements and also providing reasonable interpretations of air shower data showed, however, that the model was well-suited for the purposes it had been developed for. Motivated by the continuous use of the model in many calculations, a comprehensive description of the model was finally published in 2009 [19].

One of the very striking improvements in describing data when switching from Sibyll 1.7 to Sibyll 2.1 is shown in Fig. 5. While this figure shows data to which the model was tuned, predictions for higher energies are compared with first LHC measurements in Ref. [91]. The Sibyll extrapolations were very close to these measurements.

3. The present – Sibyll version 2.3

The very encouraging agreement of Sibyll predictions with the first minimum bias data of LHC experiments give strong support to the underlying ideas of this model, even though it is much less sophisticated than other hadronic interaction models such as EPOS [99–101], QGSJET [102–105], and DPMJET [85,86]. On the other hand, the LHC data led to the discovery of new shortcomings and helped reducing the uncertainties of the model extrapolation to high energy.

Motivated by the LHC data and also new fixed-target measurements a number of improvements have been implemented in Sibyll. A new version of the code, Sibyll 2.3, was released in 2016. In the following an overview of some of the model changes are given. A complete description of the model will be given in a forthcoming publication [106].

The discrepancy between the different measurements of the total $p\bar{p}$ cross section [107–109] at Tevatron led to a considerable uncertainty in the cross section extrapolation. The fit of Sibyll 2.1 made in 2000 was favoring the higher value of the CDF Collaboration, see Fig. 3. LHC data have greatly reduced this uncertainty and the Sibyll predictions were found to be too high. By changing the shape of the distribution of the partons in transverse space, see Fig. 2, the Sibyll cross section has been adjusted to fit the LHC data. A comparison of the previous and current cross section predictions of the Sibyll model is shown in Fig. 6. Although a reasonable description of the LHC data on elastic and total cross sections could be obtained, a deviation was found for the slope of the differential elastic cross section $d\sigma/dt$ at $|t| \to 0$. For a better description of elastic scattering a full treatment of the complex phase of the scattering amplitude is needed, which is beyond the scope of the Sibyll model.

Another important improvement of the model was the implementation of the Good-Walker approach [77] to inelastic diffractive scattering in form of a two-channel model also for interactions with nuclei. This lead to an increase of the cross section for projectile diffraction dissociation on nuclei. Together with the reduced inelastic cross section this results in a shift of the predicted mean depth of shower maximum deeper into the atmosphere, see Fig. 7.

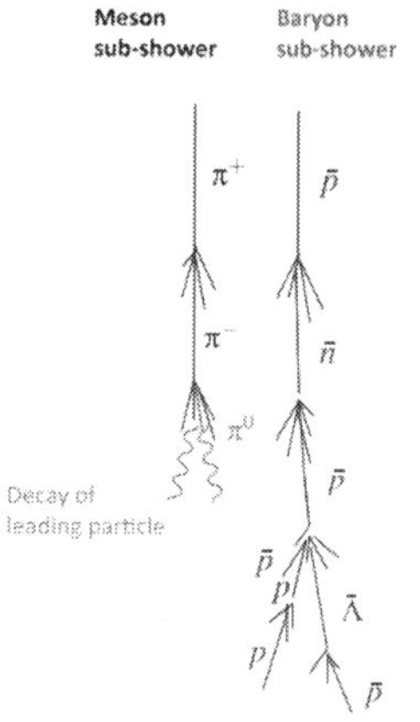

Figure 8. Illustration of processes of direct relevance to muon production in air showers. Right: Baryon number conservation makes baryons a very efficient source of secondary hadrons. Left: There is about a 30% chance probability that the leading particle of a pion-air interaction is a neutral pion. Hence sub-showers initiated by charged pions have a high probability to be terminated by production of a leading π^0.

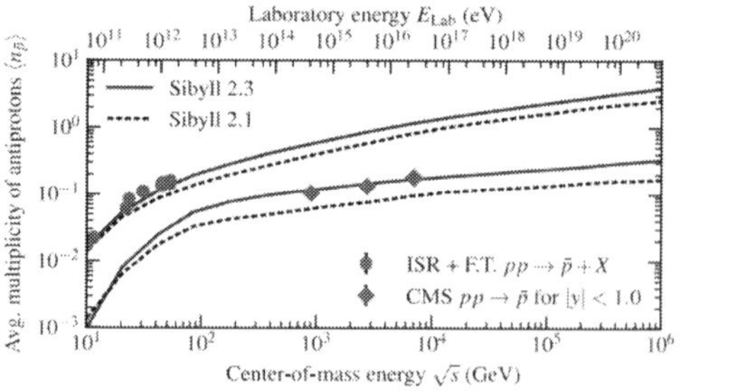

Figure 9. Multiplicity of secondary antiprotons in pp collisions. Shown are results extrapolated to the full phase space of secondary particles (upper curves and points) and data for a central rapidity window (lower curves and points).

Further improvements of Sibyll 2.3 with respect to Sibyll 2.1 are the sampling of the transverse momentum of partons at the string ends from an exponential distribution in transverse mass $m_\perp = \sqrt{p_\perp^2 + m^2}$ instead of $p_\perp$ and a better sampling of the momentum fractions of the partons entering hard scattering processes. These changes allow a better description of a multitude of LHC data sets, but are not expected to be of great importance for air shower predictions.

Air shower calculations with Sibyll always predicted a smaller number of muons at ground than one would obtain with EPOS or QGSJET, see [113]. A number of air shower measurements indicate, however, that even simulations with EPOS or QGSJET do predict fewer muons than observed in the data [114–117]. Therefore special emphasis has been put in Sibyll 2.3 on model features that are related to the production of muons in the GeV energy range.

The energy transfer from the hadronic core of an air shower to the electromagnetic component by π^0 production is a very important parameter that influences the number of muons arriving at ground. The more energy is kept in each hadronic interaction in the hadronic core of the shower, i.e. given to hadrons that interact again or decay into secondaries that include muons, the larger is the number of muons at ground. The importance of baryon-antibaryon pair-production in this respect had been noted already in 1973 [118]. Thanks to baryon number conservation, all baryons produced in a shower propagate and re-interact until their energy is too low for hadronic particle production. The multiplicity of secondary baryons is small and, hence, secondary baryons are unimportant in the first generations of an air shower. They serve, however, as a very efficient source of secondary pions at low energy when the charged pions forming the bulk of the hadronic core begin to decay, see Fig. 8. Enhanced baryon pair-production leads to an enhancement of the number of low-energy muons in air showers. This has been demonstrated in a detailed study made with EPOS [119]. Simulations show that baryon-induced sub-showers produce about 30% more muons than pion showers. Indeed, the average number of produced baryon pairs is too low in Sibyll 2.1 and had to be re-tuned, see Fig. 9.

Another aspect of hadronic interactions of direct relevance to muon production is the production of hadronic resonances of high energy in the fragmentation of beam remnants [120]. The surprising experimental observation is that the leading particle in $\pi^\pm p$ interactions is often a ρ^0 meson instead of a π^0. In contrast to the π^0, the ρ^0 meson decays directly to two charged pions and the energy of the leading particle is kept in the hadronic shower component [121]. The enhanced production of leading ρ^0 mesons is not reproduced in models that rely on string fragmentation.

In string fragmentation models, the production ratio of states of different spin and the same valence quarks is given by a model parameter and independent of the production kinematics. This can be seen in Fig. 10 (top). To a good approximation there is a constant ratio between the π^0 and ρ^0 curves in Sibyll 2.1. The data show a very different behavior. Adding the possibility to have excited states formed by the remnant of a hadron in an interaction allows the implementation of an enhanced production of ρ^0 mesons in forward direction. This is shown for Sibyll 2.3 in Fig. 10 (bottom).

The impact of the tuned baryon-antibaryon and leading ρ^0 production on the number of muons at ground is shown in Fig. 11. The Sibyll predictions change by a factor of 1.4 at low energy and 1.6 high energy. Now Sibyll is the model that predicts the highest muon multiplicity in air showers. From these results it is clear that measurements of baryon pair and leading ρ^0 production will be needed for pion-nucleus interactions to make the prediction of muon numbers in air showers more reliable, see [122].

Finally, for completeness, it should be mentioned that a phenomenological treatment of charm production was implemented in Sibyll 2.3 – based on the approach developed in Ref. [123] – and tuned to fixed-target and LHC data. A realistic model for charm production is very important for calculating prompt lepton production in the atmosphere [8, 124]. The Sibyll predictions for atmospheric lepton fluxes will be discussed in detail in Ref. [125].

4. The future

No attempt will be made to predict the future development of Sibyll. Only a number of general observations will

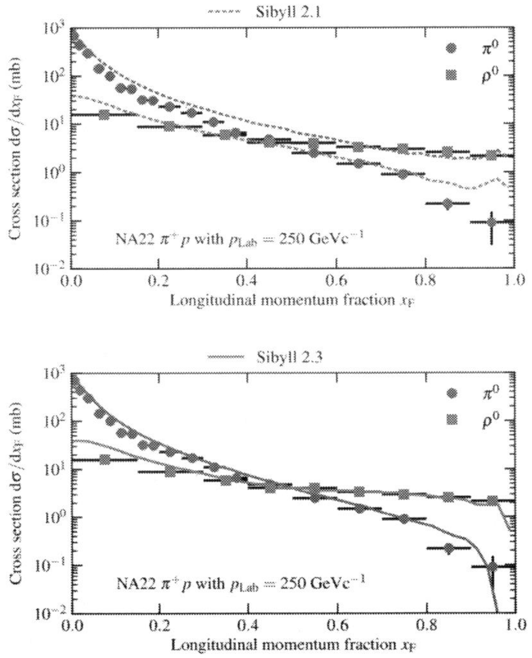

Figure 10. Inclusive π^0 and ρ^0 production in $\pi^+ p$ interactions at 250 GeV lab momentum. The top panel shows the predictions of Sibyll 2.1 which was not tuned to describe these data sets. The results of Sibyll 2.3 are shown in the bottom panel after extending the treatment of hadron remnants in the model.

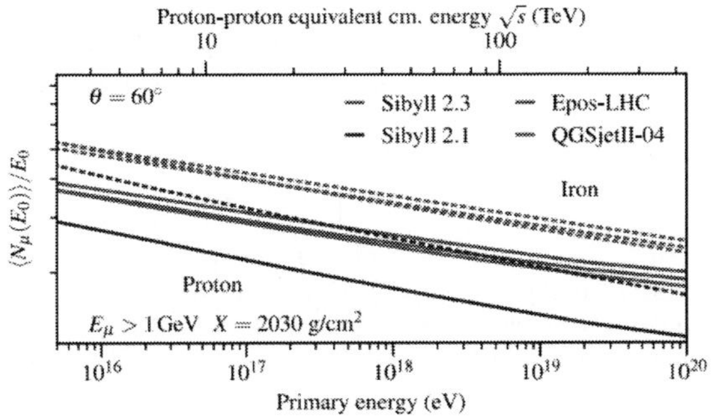

Figure 11. Mean number of muons in air showers of 60° zenith angle as observed at sea level. The muon energy threshold is 1 GeV.

be pointed out focusing on open questions and possible strategies for finding answers.

4.1. High-precision data and model predictions

In the last decade an exceptionally large number of minimum bias measurements with very small systematic uncertainties have been published by LHC and fixed-target experiments. The LHC as an energy frontier has pushed particle production measurements into the energy region in which it is dominated by minijets and collective effects have been found not only in heavy ion but also proton-proton collisions. The awareness of the need for minimum bias measurements with light nuclei has increased and many different fixed-target data sets were published, for example, see [126–129]. Moreover, model builders have also profited from the need of long-baseline neutrino experiments for high-statistics measurements of particle production using different targets including light nuclei [130–133].

This large number of new data has triggered the development of a new generation of hadronic interaction models, often referred to as post-LHC models, which are of much higher quality than the previous generation. At the same time the existence of these rich and very diverse data sets has caused a dilemma for builders of hadronic interaction models. With the lack of not being able to calculate particle production from first principles it cannot be expected that any model, as sophisticated as it might be, will describe all the data reasonably well and provide a consistent view of hadronic interactions.

In light of the large amount of accelerator data available the careful selection of data sets used for optimizing hadronic interaction models for cosmic ray interactions becomes more and more important. Selecting the data according to the covered phase space and their relevance to air shower physics is an important first step. But even the best selection of data used for tuning has its limitations because many data sets provide only indirect information on forward particle production, if at all. There are further limitations stemming from the complexity of the physics of hadronic multiparticle production. For example, multiplicities and transverse momentum spectra of particles in heavy ion collisions should not be used for the tuning of models that do not include a treatment of collective effects such as transverse flow. While such collective effects will probably not be important in air shower physics, they will have to be considered for maximum use of the information provided by LHC data.

The advantage of restricting Sibyll to the modeling of phenomena needed for understanding extensive air showers and, hence, being a very simple and flexible model, turns here to a disadvantage, limiting the number of data sets it can be compared to. With the amount of highly specialized measurements continuously increasing it will be more and more difficult to tune the model to these data sets without introducing new phenomena in the model even if these are not really needed for air shower simulation. One solution to this problem would be, of course, the measurement of forward particle distributions, integrated over transverse momentum and covering large phase space ranges. Those distributions could be used directly for model tuning and air shower simulation.

4.2. Conceptual problems and open questions

Theoretical approaches and approximations for describing particle production are available for different energy regions and kinematic regimes. These include resonance and effective field theories close to the particle production threshold, Regge theory [22] for low-energy and low-pt processes, perturbative QCD combined with collinear [134] or $k_\perp$ factorization [135] for hard processes in different limits of $\log(p_\perp^2/\Lambda_{\rm QCD}^2)$ vs. $\log(1/x)$, the Color Glass Condensate approach [136] for high parton densities, and hydrodynamics of the Quark-Gluon plasma [137, 138] at very high energy densities. Typically, hadronic interaction models include only a small number of these approaches and the transition is between these kinematic regimes and covered phase space regions is implemented in a pragmatic way, driven by the goal of reaching a good

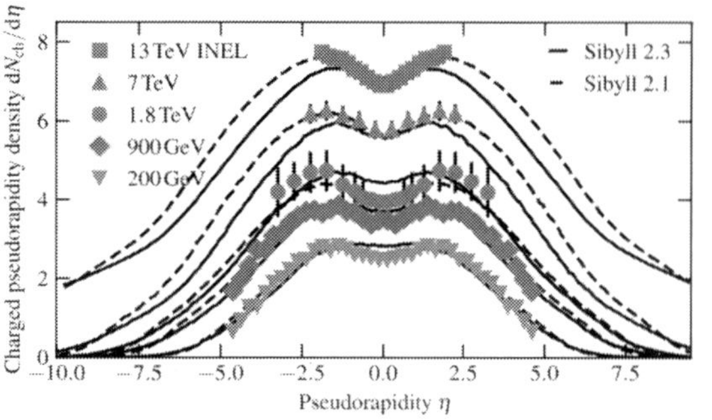

Figure 12. Pseudorapidity distribution of charged particles. Collider data from CMS, CDF and UA5 [43, 87, 139, 140]. are compared to simulations made with Sibyll versions 2.1 and 2.3. At high energy Sibyll 2.3 systematically predicts a lower particle density in forward and backward directions than measured.

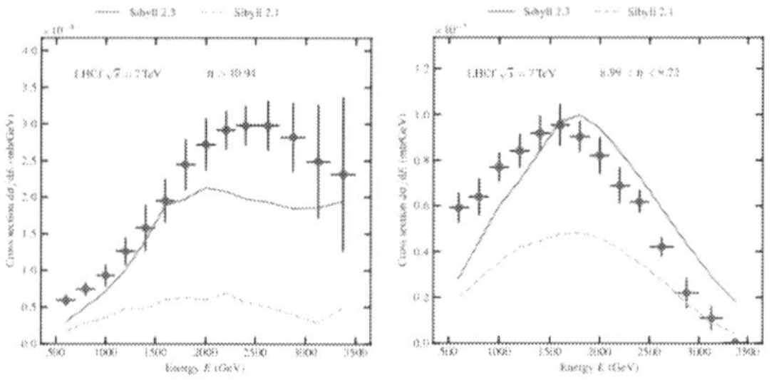

Figure 13. Leading neutron production as measured with the LHCf experiment [142]. The data are compared with the predictions of two versions of Sibyll.

description of data. More theoretical work is needed to better understand the transition between the theoretical treatments.

One prominent example of the need to combine two different approaches is the transition from soft to hard processes. At low energy, in the energy range of fixed-target experiments, multiparticle production is dominated by soft processes and hadronic degrees of freedom are most appropriate for model building. In the energy range from $\sqrt{s} \sim 100 - 1000\,\mathrm{GeV}$, a transition to partonic degrees of freedom takes place. And at high energies, most clearly seen at LHC for the first time, parton-parton interactions are dominating large parts of phase space in multiparticle production. Tuning a model and obtaining a very good description of low-energy data does not automatically imply a good description of high-energy data.

The assumptions on the description of the transition from hard to semi-hard and then to soft processes play an important role at high energy and are an essential element of model building. This is illustrated in Fig. 12 showing the pseudorapidity distribution of charged particles over a wide range of collision energies. At low energy soft interactions are still producing the bulk of the particles. This is no longer the case at LHC energies. The discrepancies between data and Sibyll 2.3 at $|\eta| \sim 2$ indicate some shortcoming in the modeling of the transition between soft and hard processes, in particular of the parton densities at very low x, or possibly of the underlying soft processes in the presence of hard parton-parton interactions. That these shortcomings do not necessarily affect the very forward direction is demonstrated in Fig. 13, in which the forward neutron production at LHC is compared to Sibyll predictions. There the improvement of data description with Sibyll 2.3 wrt. version 2.1 is clearly visible.

In Sibyll, the geometric criterion (6) is used for defining a threshold for the applicability of the simple multiple-scattering model. A straightforward generalization to nuclei is the use of the sum of the local parton densities of all nucleons in Eq. (6), taken at the relevant impact parameter of the interaction. This means that, in general, the transverse momentum threshold for hard interactions should vary from collision to collision, and the parameters for soft interactions should change too. For comparison, in EPOS, the parton densities are modified at low x by hand, i.e. the number of partons is reduced, in dependence on the local energy density, irrespective of how many nucleons are participating in a collision [100]. In QGSJET II (soft) pomeron-pomeron interactions (so-called enhanced pomeron graphs) are re-summed to all order, leading to a rather involved theoretical description of hadronic interactions [104, 105, 141]. All these approaches for implementing a transition from hard to soft processes imply the violation of QCD factorization at semi-hard, minijet scales while preserving factorization for high-$p_\perp$ processes. The diversity of the approaches is a sign of our lack of understanding semi-hard processes.

It has been speculated already for a long time that central particle production will ultimately also change leading particle distributions once the scattering process the black disk limit [143]. This effect is analogous to the suppression of leading particle production in dependence on the number of participating nucleons (i.e. centrality) in hadron-nucleus interactions. It will be important to demonstrate this effect at LHC by, for example, combining the central data taken by ATLAS with that of the forward experiment LHC on an event-by-event basis. Furthermore, because we have little knowledge on the physics relation between central and forward particle production, we still cannot reliably predict the leading particle distributions in hadronic interactions involving nuclei even if these distributions were known (measured) in pp and πp interactions. Only the direct measurement of proton interactions with light nuclei can help to reduce these uncertainties. With oxygen being used as carrier ion for heavy ion injection, pO interactions should be a technically feasible option at LHC without too much accelerator tuning needed.

The authors thank their colleagues of the Pierre Auger, IceCube and KASCADE-Grande Collaborations for many inspiring discussions. In particular, the very fruitful interaction with Tanguy Pierog is gratefully acknowledged. We also thank Bryan Pattison for comments on the manuscript. This work is supported in part by the German Ministry of Education and Research (BMBF), grant No. 05A14VK1, and the Helmholtz Alliance for Astroparticle Physics (HAP), which is funded by the Initiative and Networking Fund of the Helmholtz Association.

References

[1] B. Rossi, K. Greisen, Rev. Mod. Phys. **13**, 240 (1941)
[2] J. Nishimura, Handbuch der Physik **46/2**, 1 (1965)
[3] P. Lipari, Astropart. Phys. **1**, 195 (1993)

[4] T.K. Gaisser, R. Engel, E. Resconi, *Cosmic Rays and Particle Physics*, 2nd edn. (Cambridge University Press, 2016), ISBN 9780521016469
[5] L.G. Dedenko, Proc. of 9th Int. Cosmic Ray Conf., London **1**, 662 (1965)
[6] T.K. Gaisser, R.J. Protheroe, K.E. Turver, T.J.L. McComb, Rev. Mod. Phys. **50**, 859 (1978)
[7] G. Bossard et al., Phys. Rev. **D63**, 054030 (2001), `hep-ph/0009119`
[8] A. Fedynitch, R. Engel, T.K. Gaisser, F. Riehn, T. Stanev, EPJ Web Conf. **99**, 08001 (2015), Proc. of 18th Int. Symposium on Very High Energy Cosmic Ray Interactions (ISVHECRI 2014) Geneva, Switzerland, August 18-22, 2014, `1503.00544`
[9] H.J. Drescher, G.R. Farrar, Phys. Rev. **D67**, 116001 (2003), `astro-ph/0212018`
[10] T. Bergmann et al., Astropart. Phys. **26**, 420 (2007), `astro-ph/0606564`
[11] S.J. Sciutto (1999), `astro-ph/9911331`
[12] S.J. Sciutto (2001), `astro-ph/0106044`
[13] D. Heck, J. Knapp, J.N. Capdevielle, G. Schatz, T. Thouw, Wissenschaftliche Berichte, Forschungszentrum Karlsruhe **FZKA 6019** (1998)
[14] K. Kasahara et al., `http://cosmos.n.kanagawa-u.ac.jp/cosmosHome`
[15] A.M. Hillas, Proc. of 17th Int. Cosmic Ray Conf., Paris **8**, 193 (1981)
[16] A.M. Hillas, Nucl. Phys. Proc. Suppl. **52B**, 29 (1997)
[17] J. Engel, T.K. Gaisser, T. Stanev, P. Lipari, Phys. Rev. **D46**, 5013 (1992)
[18] R.S. Fletcher, T.K. Gaisser, P. Lipari, T. Stanev, Phys. Rev. **D50**, 5710 (1994)
[19] E.J. Ahn, R. Engel, T.K. Gaisser, P. Lipari, T. Stanev, Phys. Rev. D **80**, 094003 (2009), `0906.4113`
[20] F. Riehn, R. Engel, A. Fedynitch, T.K. Gaisser, T. Stanev, PoS **(ICRC2015)**, 558 (2015), `1510.00568`
[21] F. Riehn, R. Engel, A. Fedynitch, T.K. Gaisser, T. Stanev, EPJ Web Conf. **99**, 12001 (2015), `1502.06353`
[22] P.D.B. Collins, *An Introduction to Regge Theory and High-Energy Physics* (Cambridge Univ. Pr., 1977)
[23] G.F. Chew, S.C. Frautschi, Phys. Rev. Lett. **8**, 41 (1962)
[24] E. Klempt, A. Zaitsev, Phys. Rept. **454**, 1 (2007), `0708.4016`
[25] R.P. Feynman, Phys. Rev. Lett. **23**, 1415 (1969)
[26] Z. Koba, H.B. Nielsen, P. Olesen, Nucl. Phys. **B40**, 317 (1972)
[27] J. Benecke, T.T. Chou, C.N. Yang, E. Yen, Phys. Rev. **188**, 2159 (1969)
[28] A.H. Mueller, Phys. Rev. **D2**, 2963 (1970)
[29] M. Gell-Mann, Phys. Lett. **8**, 214 (1964)
[30] G. Zweig, preprint CERN-TH-401 (1964)
[31] R.P. Feynman, Proc. of High Energy Collisions: Third International Conference at Stony Brook pp. 237–249 (1969), N.Y. Gordon & Breach
[32] J.D. Bjorken, E.A. Paschos, Phys. Rev. **185**, 1975 (1969)
[33] S.D. Drell, T.M. Yan, Annals Phys. **66**, 578 (1971)
[34] H. Fritzsch, M. Gell-Mann, H. Leutwyler, Phys. Lett. **B47**, 365 (1973)
[35] B.L. Combridge, J. Kripfganz, J. Ranft, Phys. Lett. **B70**, 234 (1977)
[36] B.L. Combridge, Nucl. Phys. **B151**, 429 (1979)
[37] J.F. Owens, E. Reya, M. Gluck, Phys. Rev. **D18**, 1501 (1978)
[38] W. Ochs, Nucl. Phys. **B118**, 397 (1977)
[39] G. Altarelli, G. Parisi, Nucl. Phys. **B126**, 298 (1977)
[40] P.D.B. Collins, A.D. Martin, Rept. Prog. Phys. **45**, 335 (1982)
[41] G.J. Alner et al. (UA5), Z. Phys. **C32**, 153 (1986)
[42] C. Augier et al. (UA4/2), Phys. Lett. **B344**, 451 (1995)
[43] G.J. Alner et al. (UA5), Z. Phys. **C33**, 1 (1986)
[44] G.J. Alner et al. (UA5), Phys. Lett. **B138**, 304 (1984)
[45] G.J. Alner et al. (UA5), Phys. Rept. **154**, 247 (1987)
[46] C. Albajar et al. (UA1), Nucl. Phys. **B309**, 405 (1988)
[47] L.V. Gribov, E.M. Levin, M.G. Ryskin, Phys. Rept. **100**, 1 (1983)
[48] T.K. Gaisser, F. Halzen, Phys. Rev. Lett. **54**, 1754 (1985)
[49] G. Pancheri, Y.N. Srivastava, Phys. Lett. **B182**, 199 (1986)
[50] L. Durand, H. Pi, Phys. Rev. Lett. **58**, 303 (1987)
[51] L. Ametller, D. Treleani, Int. J. Mod. Phys. **A3**, 521 (1988)
[52] A. Capella, J. Tran Thanh Van, J. Kwiecinski, Phys. Rev. Lett. **58**, 2015 (1987)
[53] J.H. Weis, Acta Phys. Polon. **B7**, 851 (1976)
[54] K. Werner, Phys. Rep. **232**, 87 (1993)
[55] T.K. Gaisser, T. Stanev, Phys. Lett. **B219**, 375 (1989)
[56] T.K. Gaisser, U. Sukhatme, G.B. Yodh, Phys. Rev. **D36**, 1350 (1987)
[57] L. Durand, H. Pi, Phys. Rev. **D40**, 1436 (1989)
[58] L. Durand, H. Pi, Phys. Rev. **D43**, 2125 (1991)
[59] J. Dias de Deus, J. Kwiecinski, Phys. Lett. **B196**, 537 (1987)
[60] B. Margolis, P. Valin, M.M. Block, F. Halzen, R.S. Fletcher, Phys. Lett. **B213**, 221 (1988)
[61] T.K. Gaisser, F. Halzen, A.D. Martin, Phys. Lett. **B166**, 219 (1986)
[62] J. Dias de Deus, J. Kwiecinski, M. Pimenta, Phys. Lett. **B202**, 397 (1988)
[63] K. Alpgard et al. (UA5), Phys. Lett. **B123**, 361 (1983)
[64] K.A. Ter-Martirosyan, Phys. Lett. **B44**, 377 (1973)
[65] A. Capella, U. Sukhatme, C.I. Tan, J. Tran Thanh Van, Phys. Rept. **236**, 225 (1994)
[66] A.B. Kaidalov, Phys. Atom. Nucl. **66**, 1994 (2003)
[67] T. Sjöstrand, M. van Zijl, Phys. Rev. **D36**, 2019 (1987)
[68] B. Andersson, G. Gustafson, G. Ingelman, T. Sjostrand, Phys. Rept. **97**, 31 (1983)
[69] T. Sjöstrand, Int. J. Mod. Phys. **A3**, 751 (1988)
[70] T.K. Gaisser et al., Phys. Rev. **D47**, 1919 (1993)

[71] R.J. Glauber, G. Matthiae, Nucl. Phys. **B21**, 135 (1970)
[72] A. Etchegoyen et al. (1996), FERMILAB-PUB-96-024, `http://www.auger.org/admin/DesignReport/`
[73] A. Aab et al. (Pierre Auger), Nucl. Instrum. Meth. **A798**, 172 (2015), `1502.01323`
[74] E. Eichten, I. Hinchliffe, K.D. Lane, C. Quigg, Rev. Mod. Phys. **56**, 579 (1984), [Addendum: Rev. Mod. Phys.58,1065(1986)]
[75] D.W. Duke, J.F. Owens, Phys. Rev. **D30**, 49 (1984)
[76] T. Antoni et al. (KASCADE), Nucl. Instrum. Meth. **A513**, 490 (2003)
[77] M.L. Good, W.D. Walker, Phys. Rev. **120**, 1857 (1960)
[78] V.A. Abramovski, V.N. Gribov, O.V. Kancheli, Sov. J. Nucl. Phys. **18**, 308 (1974)
[79] M. Glück, E. Reya, A. Vogt, Z. Phys. **C53**, 651 (1992)
[80] M. Glück, E. Reya, A. Vogt, Z. Phys. **C67**, 433 (1995)
[81] E.M. Levin, M.G. Ryskin, Phys. Rep. **189**, 267 (1990)
[82] V.N. Gribov, Sov. Phys. JETP **26**, 414 (1968), [Zh. Eksp. Teor. Fiz.53,654(1967)]
[83] R. Engel, Z. Phys. **C66**, 203 (1995)
[84] R. Engel, J. Ranft, Phys. Rev. **D54**, 4244 (1996), `hep-ph/9509373`
[85] J. Ranft, Phys. Rev. **D51**, 64 (1995)
[86] S. Roesler, R. Engel, J. Ranft, Proc of 27th Int. Cosmic Ray Conf., Hamburg **2**, 439 (2001)
[87] F. Abe et al. (CDF), Phys. Rev. **D41**, 2330 (1990)
[88] R. Harr et al., Phys. Lett. **B401**, 176 (1997), `hep-ex/9703002`
[89] K. Alpgard et al. (UA5), Phys. Lett. **B112**, 183 (1982)
[90] R. Engel, T.K. Gaisser, T. Stanev, P. Lipari, Proc. of 26th Int. Cosmic Ray Conf., Salt Lake City **1**, 415 (1999)
[91] D. d'Enterria, R. Engel, T. Pierog, S. Ostapchenko, K. Werner, Astropart. Phys. **35**, 98 (2011), `1101.5596`
[92] G. Aad et al. (ATLAS), Nucl. Phys. **B889**, 486 (2014), `1408.5778`
[93] M. Aaboud et al. (ATLAS), Phys. Rev. Lett. **117**, 182002 (2016), `1606.02625`
[94] B. Abelev et al. (ALICE), Eur. Phys. J. **C73**, 2456 (2013), `1208.4968`
[95] S. Chatrchyan et al. (CMS), Phys. Lett. **B722**, 5 (2013), `1210.6718`
[96] G. Antchev et al. (TOTEM), Europhys. Lett. **95**, 41001 (2011), `1110.1385`
[97] G. Antchev et al. (TOTEM), Phys. Rev. Lett. **111**, 012001 (2013)
[98] P. Abreu et al. (Pierre Auger), Phys. Rev. Lett. **109**, 062002 (2012), `1208.1520`
[99] F.M. Liu, J. Aichelin, K. Werner, M. Bleicher, Phys. Rev. **C69**, 054002 (2004), `hep-ph/0307008`
[100] K. Werner, F.M. Liu, T. Pierog, Phys. Rev. **C74**, 044902 (2006), `hep-ph/0506232`
[101] K. Werner, T. Pierog, AIP Conf. Proc. **928**, 111 (2007), `0707.3330`
[102] N.N. Kalmykov, S.S. Ostapchenko, Sov. J. Nucl. Phys. **50**, 315 (1989)
[103] N.N. Kalmykov, S.S. Ostapchenko, A.I. Pavlov, Nucl. Phys. Proc. Suppl. **52B**, 17 (1997)
[104] S. Ostapchenko, Phys. Rev. **D74**, 014026 (2006), `hep-ph/0505259`
[105] S. Ostapchenko, Phys. Lett. **B636**, 40 (2006), `hep-ph/0602139`
[106] F. Riehn et al. (2017), in preparation
[107] N.A. Amos et al. (E710), Phys. Rev. Lett. **63**, 2784 (1989)
[108] F. Abe et al. (CDF), Phys. Rev. **D50**, 5550 (1994)
[109] C. Avila et al. (E811), Phys. Lett. **B445**, 419 (1999)
[110] T. Abu-Zayyad et al. (HiRes-MIA), Astrophys. J. **557**, 686 (2001), `astro-ph/0010652`
[111] R.U. Abbasi et al. (HiRes), Astrophys. J. **622**, 910 (2005), `astro-ph/0407622`
[112] A. Aab et al. (Pierre Auger), Phys. Rev. **D90**, 122005 (2014), `1409.4809`
[113] R. Engel, D. Heck, T. Pierog, Ann. Rev. Nucl. Part. Sci. **61**, 467 (2011)
[114] T. Antoni et al. (KASCADE), Astropart. Phys. **24**, 1 (2005), `astro-ph/0505413`
[115] R. Engel (Pierre Auger), Proc. of 30th Int. Cosmic Ray Conf., Merida **4**, 385 (2007), `arXiv:0706.1921 [astro-ph]`
[116] A. Aab et al. (Pierre Auger), Phys. Rev. **D91**, 032003 (2015), `1408.1421`
[117] A. Aab et al. (Pierre Auger), Phys. Rev. Lett. **117**, 192001 (2016), `1610.08509`
[118] P.K.F. Grieder, Proc. of 13th Int. Cosmic Ray Conf., Denver **4**, 2467 (1973)
[119] T. Pierog, K. Werner, Phys. Rev. Lett. **101**, 171101 (2008), `astro-ph/0611311`
[120] H.J. Drescher, Phys. Rev. **D77**, 056003 (2007), `arXiv:0712.1517 [hep-ph]`
[121] S. Ostapchenko, EPJ Web Conf. **52**, 02001 (2013)
[122] A. Hervé et al. (NA61) (2017), these proceedings
[123] E.J. Ahn, R. Engel, T.K. Gaisser, P. Lipari, T. Stanev (2011), Proc. of 16th Int. Symposium on Very High Energy Cosmic Ray Interactions (ISVHECRI 2010): Batavia, Illinois, USA, June 28-July 2, 2010, `1102.5705`
[124] T.K. Gaisser, M. Honda, Ann. Rev. Nucl. Part. Sci. **52**, 153 (2002), `hep-ph/0203272`
[125] A. Fedynitch et al. (2017), in preparation
[126] C. Alt et al. (NA49), Eur. Phys. J. **C45**, 343 (2006), `hep-ex/0510009`
[127] C. Alt et al. (NA49), Eur. Phys. J. **C49**, 897 (2007), `hep-ex/0606028`
[128] T. Anticic et al. (NA49), Eur. Phys. J. **C68**, 1 (2010), `1004.1889`
[129] B. Baatar et al. (NA49), Eur. Phys. J. **C73**, 2364 (2013), `1207.6520`
[130] N. Abgrall et al. (NA61/SHINE), Phys. Rev. C **84**, 034604 (2011), `1102.0983`
[131] N. Abgrall et al. (NA61/SHINE), Phys. Rev. **C85**, 035210 (2012), `1112.0150`
[132] N. Abgrall et al. (NA61/SHINE), Phys. Rev. **C89**, 025205 (2014), `1309.1997`
[133] N. Abgrall et al. (NA61/SHINE), Eur. Phys. J. **C74**, 2794 (2014), `1310.2417`
[134] J.C. Collins, D.E. Soper, G.F. Sterman, Nucl. Phys. **B261**, 104 (1985)
[135] S. Catani, M. Ciafaloni, F. Hautmann, Nucl. Phys. **B366**, 135 (1991)

[136] F. Gelis, E. Iancu, J. Jalilian-Marian, R. Venugopalan, Ann. Rev. Nucl. Part. Sci. **60**, 463 (2010), 1002.0333
[137] D.H. Rischke, Prog. Part. Nucl. Phys. **52**, 197 (2004), nucl-th/0305030
[138] E. Shuryak, Prog. Part. Nucl. Phys. **62**, 48 (2009), 0807.3033
[139] V. Khachatryan et al. (CMS), Phys. Lett. **B751**, 143 (2015), 1507.05915
[140] V. Khachatryan et al. (CMS), Phys. Rev. Lett. **105**, 022002 (2010), 1005.3299
[141] S. Ostapchenko, Phys. Rev. **D81**, 114028 (2010), 1003.0196
[142] O. Adriani et al. (LHCf), Phys. Lett. **B750**, 360 (2015), 1503.03505
[143] H.J. Drescher, A. Dumitru, M. Strikman, Phys. Rev. Lett. **94**, 231801 (2005), hep-ph/0408073

Detection and analysis of atmospheric muons using the ALICE detector at the LHC

Bruno Alessandro[a] for the ALICE Collaboration

I.N.F.N. Sezione di Torino, Italy

Abstract. ALICE is a general purpose experiment designed to investigate nucleus-nucleus collisions at the CERN Large Hadron Collider (LHC). Located 52 meters underground, with 28 meters of overburden rock, it has also been used to detect the muonic component of the extensive air showers produced by cosmic-ray interactions in the upper atmosphere. A program of cosmic-ray data taking, with specific triggers for atmospheric muons, was started in 2010 in periods when there is no beam circulating in the LHC. Several million events have been recorded to date. The large size and excellent tracking capability of the ALICE Time Projection Chamber are exploited to detect and reconstruct these muons. In this paper the analysis of the multiplicity distribution of the atmospheric muons detected by ALICE between 2010 and 2013 is presented, along with the comparison with Monte Carlo simulations. Special emphasis is given to the study of high multiplicity events containing more than 100 reconstructed muons. The comprehension of the frequency of these events was an unsolved problem since the pioneering studies performed by ALEPH and DELPHI experiments at LEP. In our work the ALICE measurements show that such high multiplicity events demand primary cosmic rays with energy above 10^{16} eV. Their frequency can be successfully described by assuming a heavy mass composition of primary cosmic rays above this energy and using the most recent interaction models to describe the development of the air shower resulting from the primary interaction.

1. Introduction

ALICE (A Large Ion Collider Experiment) [1] is a general-purpose, Heavy-Ion detector at the CERN LHC. It focuses on the study of the properties of the Quark-Gluon Plasma (QGP) created in strongly interacting matter at extreme energy densities in high energy nucleus-nucleus collisions. The underground location of the experiment allowed ALICE to also develop a cosmic-ray physics program, exploiting the excellent tracking capabilities of the Time Projection Chamber (TPC) to detect and reconstruct the muons produced in extensive air showers.

The use of collider detectors to study the atmospheric muons was pioneered by the LEP experiments ALEPH, DELPHI and L3. All results reported by them were consistent with standard hadronic interaction models, except the observation of high muon multiplicity events [2,3]. These experiments concluded that the high multiplicity events occur with a frequency which is almost an order of magnitude above expectation, even when assuming that the primary cosmic rays are purely composed of iron nuclei.

A development of this program is possible at the LHC, where the experiments are expected to operate for many years with the possibility of collecting a very large sample of cosmic-ray data. In this context ALICE began a cosmic-ray physics program taking data between 2010 and 2013.

Around 22.6 million events with at least 1 cosmic muon were recorded in 30.8 days of live time.

The analysis of these data and in particular the study of the muon multiplicity distribution, with the purpose of addressing the problem of the rate of high muon multiplicity events, is discussed in this paper.

2. The ALICE detectors for cosmic-ray physics

The ALICE experiment is located at Point 2 of the LHC tunnel, 52 m underground and with 28 m of overburden rock. This depth of rock completely absorbs the electromagnetic and hadronic components of the cosmic-ray induced air shower, and poses an energy threshold of 16 GeV for vertical muons [4].

ALICE is a typical collider experiment with a cylindrical symmetry around the beam pipe. A solenoid magnet houses the central barrel detectors, where different techniques are exploited to detect and reconstruct all particles coming out from the primary interaction vertex. A forward muon arm, consisting of an absorber, a large dipole magnet and planes of gaseous detectors, is located on one side outside the solenoid magnet. A detailed description of the ALICE detector is given in [1]. The detectors used for cosmic-ray data taking are described below.

The Alice COsmic Ray DEtector (ACORDE) is an array of 60 scintillator modules located on the three top octants of the ALICE magnet, covering 10% of their surface. Each module consists of two superimposed plastic scintillators. The trigger is given by the coincidence of the signal from two different modules in a 100 ns window. The rate is about 1 Hz.

[a] e-mail: alessandro@to.infn.it

The Silicon Pixel Detector (SPD) forms the two innermost coaxial cylinders of the Inner Tracking System (ITS) around the beam pipe, centered at the nominal position of the interaction point with a radius of 39 mm and 76 mm, respectively. The trigger configuration used for cosmic detection requires a coincidence between signals coming from the top and bottom halves of the outermost layer, which generated a trigger rate of around 0.16 Hz.

The Time Of Flight (TOF) detector is a cylindrical array of 1638 Multi-gap Resistive Plate Chambers pads arranged in 18 sectors, completely surrounding the TPC. The trigger configuration used for the present analysis requires a coincidence between a signal in a sector of the upper part and a signal in either a sector of the opposite lower part forming a back-to-back alignment with respect to the central axis, or in one of the two sectors adjacent to the opposite lower sector. The rate of this trigger configuration is about 80 Hz.

Cosmic ray data were acquired with a combination (logical OR) of at least two out of the three trigger conditions (ACORDE, SPD and TOF) depending on the run period.

The ALICE Time Projection Chamber (TPC) is the largest detector of its type, and is the main ALICE tracking device with excellent capabilities for high-track densities. The TPC has an inner radius of 80 cm, an outer radius of 280 cm and a total length of 500 cm along the beam axis. It is filled with a mixture of Ne-CO_2-N_2 and is read out by multi-wire proportional chambers at both end caps. The total area for cosmic muon detection is about 26 m^2, however after applying a minimum length cut to the reconstructed muon tracks the effective area reduces to about 17 m^2.

3. Data selection and reconstruction

The data used for the cosmic-ray analysis presented here [5] were collected between 2010 and 2013 during periods without circulating beams in LHC. Most of the data were taken with magnetic field of 0.5 T with both polarity, while a smaller sample was collected without magnetic field. The integrated live time amounts to 30.8 days, during which ~22.6 M events with at least 1 reconstructed muon in the TPC were accumulated. The majority of them are single muon events. A total of 7487 multi-muon events were collected, having at least 5 muons detected in the TPC.

The TPC tracking algorithm was designed to reconstruct tracks coming out from the interaction region, working inwards from the outer radius. As a consequence a cosmic muon crossing the TPC gets reconstructed as two separate tracks, called *up* and *down*, belonging to the two halves of the TPC cylinder. A specific algorithm was developed to match the two track segments as a single one. Real and Monte Carlo events of different multiplicities were used to optimize the parameters of this matching algorithm. Moreover, to avoid possible reconstruction inaccuracies associated with the most inclined tracks, the zenith angle of all events was restricted to $0 < \theta < 50°$.

TPC tracks were required to consist of at least 50 clusters (out of a maximum of 159) and, in events where the magnetic field was on, to have a minimum momentum of 0.5 GeV/*c*, to eliminate all possible background from

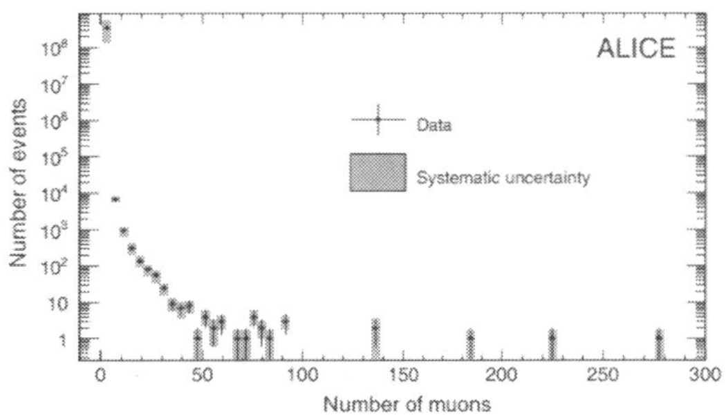

Figure 1. Atmospheric muon multiplicity distribution of the whole sample of data (2010–2013) corresponding to 30.8 days of data taking.

$e^{\pm}$. For multi-muon events a parallelism cut was also applied, which requires the angular difference between two tracks to be $\cos\Delta\psi > 0.99$ to accept the tracks. Finally, to match up and down tracks a maximum distance of closest approach of 3 cm in the TPC middle plane was imposed. A muon reconstructed with two TPC tracks (up and down) is called a *matched muon*. A track satisfying all cuts but without the counterpart in the other sector (up or down) is still accepted as a muon but flagged as a *single-track muon*. Most single-track muons are particles crossing the TPC near the edges where part of their trajectory may fall outside the sensitive volume.

4. The muon multiplicity distribution

The first measurement was the muon multiplicity distribution (MMD) to analyse its behavior and to confirm or not the existence of high muon multiplicity (HMM) events as observed by the Aleph and Delphi experiments. The MMD obtained from the whole data sample and corrected for trigger efficiency is shown in Fig. 1 [5]. The systematic uncertainties were estimated by varying the parameters of the track reconstruction and matching algorithms.

The data show a smooth distribution up to a muon multiplicity of ~ 70 and then 5 events with a multiplicity greater than 100. Events with $N_\mu > 100$ are defined HMM events.

In order to understand the MMD, simulated events equivalent to 30.8 days of live time were generated using CORSIKA [6] as event generator and QGSJET [7] for the hadronic interaction model. CORSIKA 6990 with QGSJET II-03 was used to study the MMD and HMM events, CORSIKA 7350 with QGSJET II-04 was used to further check and confirm the frequency of HMM events. Two samples, pure proton (p), representing a light composition and pure iron (Fe), representing an extremely heavy compositions, were generated. The primary cosmic-ray energy was restricted to the interval $10^{14} < E < 10^{18}$ eV following the usual power law energy spectrum $E^{-\gamma}$, with a spectral index $\gamma = 2.7$ below the knee ($E_k = 3 \times 10^{15}$ eV) and $\gamma = 3.0$ above. The total all-particle absolute flux was extracted from [8]. For each shower the core was randomly scattered at the surface level over an area 205×205 m^2 centered above the nominal LHC interaction point in ALICE.

The comparison between the MMD in the intermediate range $7 < N_\mu < 70$ and the simulated distribution fitted

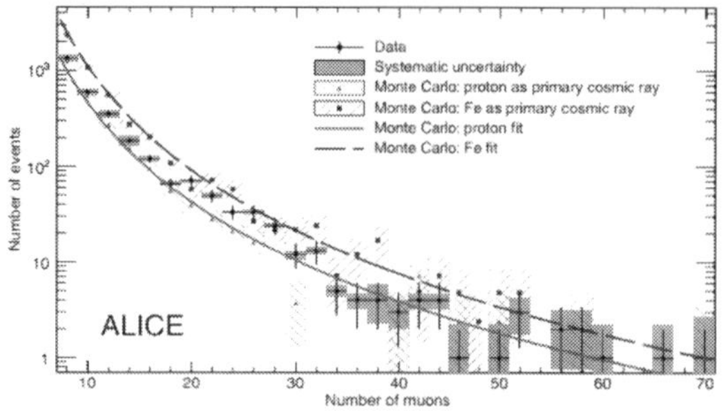

Figure 2. Measured MMD compared with values and fits obtained from simulations with p and Fe primaries.

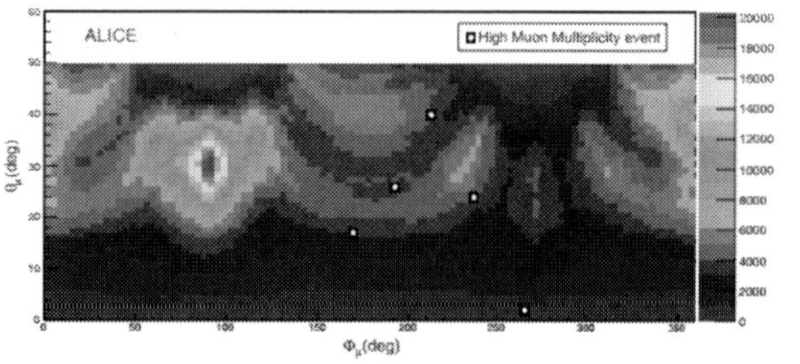

Figure 3. Zenith vs. Azimuth angle distribution of the muons in ALICE with superimposed the location of the five high muon multiplicity events (white circle on a black square).

with a power-law function is shown in Fig. 2 [5], where errors are shown separately (statistical and systematic) for data, and summed in quadrature for Monte Carlo. Below $N_\mu \sim 30$ the data, as expected, are between the pure p composition (approaching it at low multiplicity) and the pure Fe composition (at higher multiplicity). Above ~ 30 the low statistics does not allow to draw any firm conclusion, though the experimental points, considering their errors, are inside the region limited by the p and Fe curves.

5. The high muon multiplicity events

In 30.8 live days 5 HMM events were recorded, corresponding to a rate of 1.9×10^{-6} Hz.

To obtain a sort of radiography of the ALICE environment we have studied the angular distribution of the muons reconstructed in the TPC. Figure 3 depicts the θ (zenith angle) vs. Φ (azimuth angle) scatter plot of the arrival direction of the muons detected with the data and the location of the 5 HMM events (white circle on a black square). It can be noted that these events do not belong to any of the areas of the access shafts (red and green zones), where the energy threshold of the muons is lower and more muons can reach the TPC.

The highest multiplicity event reconstructed in the TPC was found to contain 276 muons, corresponding to a muon areal density of 18.1 m^{-2}. The display of this event is shown in Fig. 4.

To estimate the rate of these events, while limiting the fluctuations in the number of HMM simulated events, a live time equivalent of 1 year was simulated. A simplified Monte Carlo (which does not simulate the rock overburden and the detector response, but simply extrapolates all muon

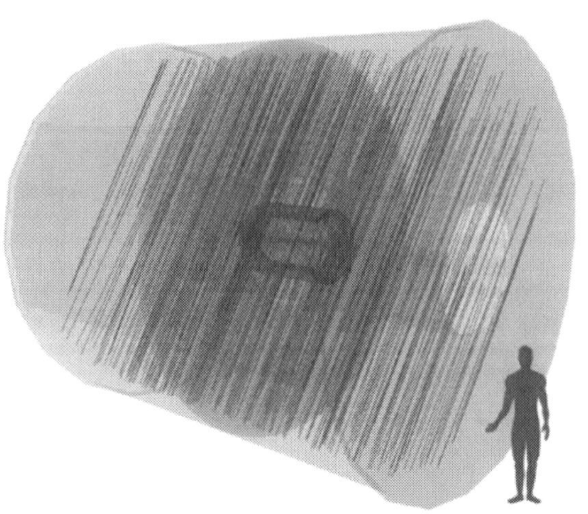

Figure 4. Event display of the highest multiplicity event found with 276 reconstructed muons crossing the TPC.

Table 1. Comparison of the HMM event rate between data and Monte Carlo.

HMM events	CORSIKA 6990 QGSJET II-03		CORSIKA 7350 QGSJET II-04		Data
	p	Fe	p	Fe	
Period [days per event]	15.5	8.6	11.6	6.0	6.2
Rate [$\times 10^{-6}$ Hz]	0.8	1.3	1.0	1.9	1.9
Uncertainty (sys+stat) (%)	25	25	22	28	49

tracks down to the ALICE level) demonstrated that only primaries with $E > 10^{16}$ eV contribute to these events. Therefore the full simulation was restricted to primaries with $10^{16} < E < 10^{18}$ eV, still with two extreme primary compositions, pure p and pure Fe. To further reduce the statistical fluctuations, 5 simulations of 1 year live time each were performed, using the same EAS sample and randomly varying the shower core in the 205×205 m^2 area. Given that the TPC acceptance is some 3000 times smaller, this ensures that the samples are statistically independent. By averaging the 5 samples the number of HMM events in 1 year and its statistical error are estimated. The uncertainties are dominated by statistical errors on real data, and by systematic errors on Monte Carlo. There are two sources of systematic errors in the simulation, the uncertainties in the generation parameters and in the muon reconstruction algorithm. Both were carefully estimated and found to amount to $\sim 20\%$.

In Table 1 [5] the results from Monte Carlo simulations are compared with data. The rate of HMM events can be well reproduced by the latest interaction models and a primary flux extrapolated from the direct measurements at 1 TeV. Pure Fe primary composition seems in closer agreement with measured rate, though the large uncertainty of the latter prevents a definite conclusion about the elements originating these events. This is consistent with the fact that HMM events stem from primaries with energy $> 10^{16}$ eV, where the composition is expected to be dominated by heavier elements.

DOI: 10.1051/epjconf/201714509001

6. Conclusions

In 2010–2013 the ALICE experiment collected 30.8 live days of cosmic-ray data. The MMD distribution at low and intermediate multiplicity is well reproduced by Monte Carlo simulations using CORSIKA 6990 with QGSJET II-03 model. The measurements by ALICE presented here suggest a mixed ion primary cosmic-ray composition with an average mass increasing with energy. In the same period 5 HMM events ($N_\mu > 100$) were recorded. The observed rate is consistent with the predictions of CORSIKA 7350 with QGSJET II-04 models using a pure Fe primary composition and energies $E > 10^{16}$ eV. For the first time the rate of HMM events has been well reproduced using conventional hadronic interaction models and reasonable primary flux model.

References

[1] K. Aamodt et al. (ALICE Coll.), JINST **3**, S08002 (2008)
[2] V. Avati et al. (ALEPH Coll.), Astrop. Phys. **19**, 513 (2003)
[3] J. Abdallah et al. (DELPHI Coll.), Astrop. Phys. **28**, 273 (2007)
[4] B. Alessandro et al. (ALICE Coll.), J. Phys. G: Nucl. Part. Phys. **32**, 1295 (2006)
[5] J. Adam et al. (ALICE Coll.), JCAP **01**, 032 (2016)
[6] D. Heck et al. FZKA-6019 (1998)
[7] S. Ostapchenko, Nucl. Phys. Proc. Suppl. **151**, 143 (2006)
[8] J. R. Hörandel, Astrop. Phys. **19**, 193 (2003)

Results of the LHCf experiment and the forward measurements at the LHC

T. Sako[1,2,a], O. Adriani[3,4], E. Berti[3,4], L. Bonechi[3], M. Bongi[3,4], G. Castellini[3,5], R. D'Alessandro[3,4], M. Haguenauer[6], Y. Itow[1,2], T. Iwata[7], K. Kasahara[7], K. Masuda[1], E. Matsubayashi[1], H. Menjo[8], Y. Muraki[1], P. Papini[3], S. Ricciarini[3,5], N. Sakurai[9], K. Sato[1], M. Shinoda[1], T. Suzuki[7], T. Tamura[10], A. Tiberio[3,4], T. Torii[7], A. Tricomi[11,12], W.C. Turner[13], M. Ueno[1], and Q.D. Zhou[1]

[1] Institute for Space-Earth Environmental Research, Nagoya University, Nagoya, Japan
[2] Kobayashi-Maskawa Institute for the Origin of Particles and the Universe, Nagoya University, Nagoya, Japan
[3] INFN Section of Florence, Florence, Italy
[4] University of Florence, Florence, Italy
[5] IFAC-CNR, Florence, Italy
[6] Ecole Polytechnique, Palaiseau, France
[7] RISE, Waseda University, Shinjuku, Tokyo, Japan
[8] Graduate School of Science, Nagoya University, Nagoya, Japan
[9] Tokushima University, Tokushima, Japan
[10] Kanagawa University, Kanagawa, Japan
[11] INFN Section of Catania, Italy
[12] University of Catania, Catania, Italy
[13] LBNL, Berkeley, California, USA

Abstract. The Large Hadron Collider is intensively used to test hadronic interaction models used in cosmic-ray physics because its maximum (designed) collision energy $\sqrt{s}$ = 14 TeV corresponds to the interaction of a 10^{17} eV cosmic-ray proton hitting the atmosphere. In this paper various types of forward experiments at the LHC, where the particles relevant to the air shower development are observed, are reviewed. Recent results of a dedicated experiment for precise forward measurement, the LHC forward, are discussed in detail. A future possibility at the LHC, light ion collisions, is also discussed.

1. Introduction

Interpretations of atmospheric air shower observations rely on Monte Carlo simulations, where we must assume a hadronic interaction model (see [1] for a review). However, because of the lack of knowledge both in theory and experiment of the high-energy hadronic interaction, there is an uncertainty in the modeling affects in the interpretations. The scope of this paper is to review the relation between the properties of hadronic interactions and air shower development, then focus on the measurements of forward particle production. Recent collider results, especially forward measurements at the LHC, are reviewed. The results of LHCf are discussed in detail.

2. Air shower and accelerator experiments

2.1. Air shower development and hadronic interactions

The properties of hadronic interactions can be categorized in a few steps from the cosmic-ray point of view.

1. *The total inelastic cross section*, σ_{ine}, determines the mean free path of the interaction, and is directly related to the maximum depth of air shower, $< X_{max} >$.
2. After an inelastic interaction multi-particle production takes place, where the *inelasticity* determines the fractional energy carried by mesons, or *elasticity*, the fractional energy carried by the leading baryon.
3. Energy sharing between the mesons, *energy spectra, or differential cross sections*, are also important.
4. Nuclear effects to incorporate the atmosphere as the target and nuclear projectile.
5. Meson-air interactions are also important, but are not considered in this paper.

The effects of some properties, which can be parameterized, to the atmospheric air showers were studied in [2].

2.2. Forward particle measurements at colliders

Figure 1 shows the number and energy densities of the secondary particles in $\sqrt{s}$=14 TeV p-p inelastic interactions as functions of the pseudorapidity η. It is found that most of the particles are produced in the central rapidity, $|\eta| \sim 0$, but most of the energy flows in the forward, large $|\eta|$, directions. To understand the air shower development measurements of the energy flow,

[a] e-mail: sako@isee.nagoya-u.ac.jp

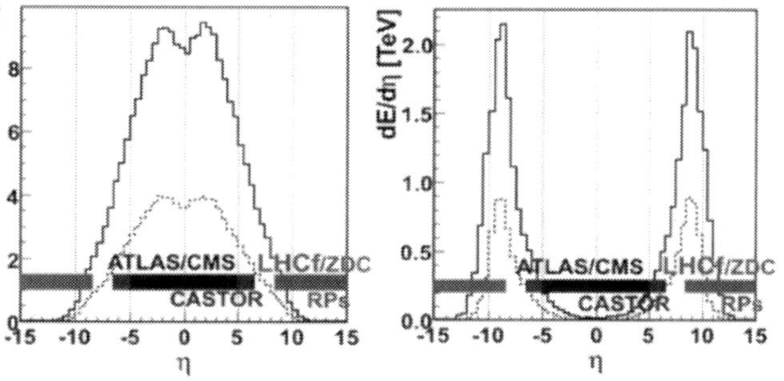

Figure 1. Number (left) and energy (right) densities of the secondary particles as functions of pseudorapidity η in $\sqrt{s}$=14 TeV p-p collisions. Pseudorapidity coverages of some LHC experiments are shown by horizontal bands.

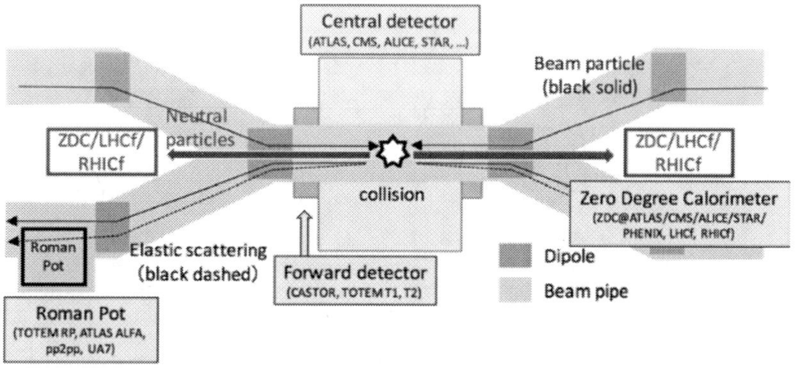

Figure 2. Schematic configuration of the forward detectors in the collider environment.

meaning the particle production in the forward direction, is essential. However, because forward directions are not covered by the main part of the general purpose detectors like ATLAS and CMS, dedicated experiments and detectors are required.

Forward experiments are classified in three cases according to their configurations as shown in Fig. 2.

1. As a part of the central detector but installed far from the interaction point and as close to the beam pipe as possible to detect inelastically scattered forward particles. CMS CASTOR [3] and TOTEM T1/T2 [4] at the LHC, for example, are categorized in this class.
2. Tracking detectors inserted into the vacuum beam pipes to measure the particles elastically scattered in very small angles. The insertion technique is called the Roman Pot, and TOTEM RP, ATLAS ALFA [5] at the LHC and pp2pp [6] at the RHIC are in this category. UA7 [7] at Sp$\bar{p}$S also used roman pots, but this experiment inserted calorimeters to measure inelastically produced photons.
3. Calorimeters installed between the beam pipes connected to the storage rings can measure neutral particles inelastically produced around zero degree. The detectors, called Zero Degree Calorimeters, were designed to measure the number of spectator neutrons in heavy ion collisions that are sensitive to the impact parameter of each collision. Similar types of ZDCs are installed in the ALICE [8], ATLAS [5] and CMS [3] experiments at the LHC, and PHENIX and STAR at the RHIC [9]. LHCf [10] and its extension RHICf [11] are a kind of ZDCs, but very different structure. LHCf was designed to precisely measure individual photons to identify π^0, but with a smaller aperture. Installation slots of ZDC are available only in particle-particle colliders, but not in the particle-antiparticle collider like the Tevatron.

3. Forward measurements at LHC

3.1. Roman pot experiments

One of the most important measurements at LHC for cosmic-ray physics is the determination of σ_{ine}. The central detectors measure σ_{ine} by counting the number of interactions with as wide a coverage as possible, namely minimum bias events, but they still need a model-based extrapolation to estimate the number of interactions in which the produced particles escape from the detector acceptance [12–15]. The missing events are produced in the diffractive process and its theoretical prediction is difficult.

Instead of counting *most of* the inelastic interactions at the central rapidity, the roman pot experiments measure the elastic interactions at a very small scattering angle. By measuring the differential cross section $d\sigma/dt$ at small t, where t is the momentum transfer of protons, the roman pot experiments can determine the total elastic cross section $\sigma_{ela} = \int(d\sigma/dt)dt$. Extrapolation of the differential cross section gives the value $\frac{d\sigma}{dt}|_{t=0}$, and this is connected with the total cross section σ_{tot} through the optical theorem. Finally the total inelastic cross section is determined as $\sigma_{ine} = \sigma_{tot} - \sigma_{ela}$.

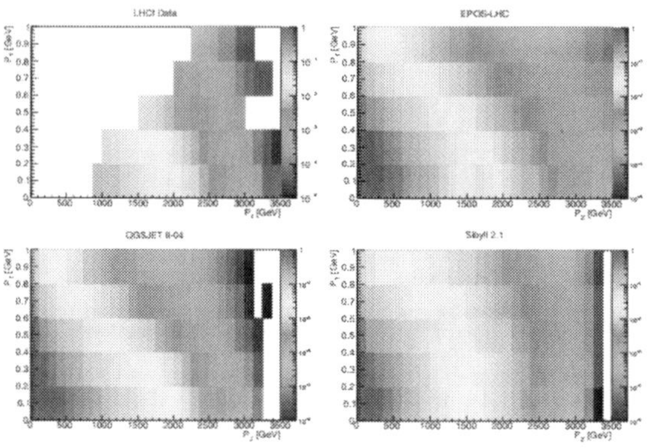

Figure 3. Invariant cross sections of π^0 production in the p_z-p_T phase space in 7 TeV p-p collisions. Invariant cross section $(1/\sigma_{ine})Ed^3\sigma/dp^3$ GeV^{-2} is shown in the color scale. LHCf result, predictions of EPOS-LHC, QGSJET II-04 and Sibyll 2.1 models are shown in the left-top, right-top, left-bottom and right-bottom panels, respectively.

Because σ_{ine} had two different values in the Tevatron measurements [16–18], model predictions in the very high-energy region diverged significantly depending on the reference value used from the Tevatron results [19]. The improvement at the LHC measurements is not only that they are at the highest energy points, but also the results are consistent between experiments with small experimental uncertainties. Thanks to the convergence at the highest energy data, the post-LHC models like EPOS-LHC and QGSJET II-04 predict very similar σ_{ine} beyond the LHC energy, and the difference in the $< X_{max} >$ prediction is significantly reduced [19].

3.2. The LHCf experiment

The Large Hadron Collider forward (LHCf) experiment has two detectors called Arm1 and Arm2 at either side of IP1, 140 m away from the interaction point. Each detector is composed of two position sensitive electromagnetic calorimeters. Because dipole magnets are located between the LHCf detectors and the interaction point (Fig. 2), charged particles are swept away and only neutral particles, predominantly photons and neutrons, arrive at the detectors. The angular acceptance of the detectors is η >8.4. According to Fig. 1, it is found that the LHCf covers a large fraction of energy flow. The energy and position resolutions of the detectors are <5% and <200 μm, respectively, for photons with energy above 100 GeV and ~40% and <1 mm, respectively, for hadrons with energy above 500 GeV. By identifying the energies and positions of two photons, and assuming they decay at the interaction point, LHCf can identify π^0 with energies greater than 600 GeV.

LHCf so far obtained collision data for $\sqrt{s}$=0.9, 2.76, 7 and 13 TeV p-p collisions, $\sqrt{s_{NN}}$=5.02 and 8.16 TeV p-Pb collisions[1]. The forward photon spectra at 0.9 and 7 TeV [20,21], forward neutron spectra at 7 TeV [22], forward π^0 spectra at 7 TeV [23] were already published. Forward π^0 spectra in the 5.02 TeV p-Pb collisions were compared with the π^0 spectra interpolated from the 2.76 and 7 TeV p-p collision data. Nuclear effects in terms of the nuclear modification factor were consistent with model predictions [24].

[1] 8.16 TeV data was obtained in Nov. 2016 after the ISVHECRI conference.

π^0 cross section at 7 TeV

Cross sections of π^0 production in the LHC Run1 data are summarized in [25]. A data-model comparison of π^0 production cross section in p_z-p_T phase space in the 7 TeV p-p collisions is shown in Fig. 3. The LHCf result in the left-top panel is compared with the predictions by EPOS-LHC (right-top), QGSJET II-04 (left-bottom) and Sibyll 2.1 (right-bottom) models. The ratio of the model predictions to the experimental result are shown in Fig. 4. A very flat ratio found in the QGSJET II-04 (left-bottom) means the spectral shape has very good agreement with the data, but the light blue to blue color means the cross section is underestimated over the whole kinematic region. EPOS-LHC (right-top) shows good agreement below 2.5 TeV, above which energy the cross section is overestimated. Because the cross section at the high energy end is very small, the effect of this discrepancy to the air shower development will be small. Sibyll 2.1 shows a trend similar to EPOS-LHC. However, the overestimate is found in high p_T rather than high energy. As shown in Fig. 3, Sibyll 2.1 holds larger cross section at high p_T than the other models. Though the LHCf data constrains only at high energy, the flat p_T dependence of Sibyll is likely excluded by the LHCf data. The LHCf 13 TeV data can extend the p_T coverage and can enforce this conclusion.

Feynman scaling

Cross sections of π^0 at 2.76 TeV and 7 TeV were compared in the same x_F and p_T region as shown in Fig. 5 [25]. Here the Feynman x, x_F, is defined as $2p_z/\sqrt{s}$. The comparison shows a good agreement of the differential cross sections at the 20% level, indicating the existence of Feynman scaling [26]. On the other hand, when we compare the forward neutron cross sections between ISR at 30–60 GeV [27,28], PHENIX at 200 GeV [29] and LHCf at 7 TeV [22], the ISR and PHENIX results show a good scaling, but the LHCf result indicates a break of the scaling [30]. To test the Feynman scaling

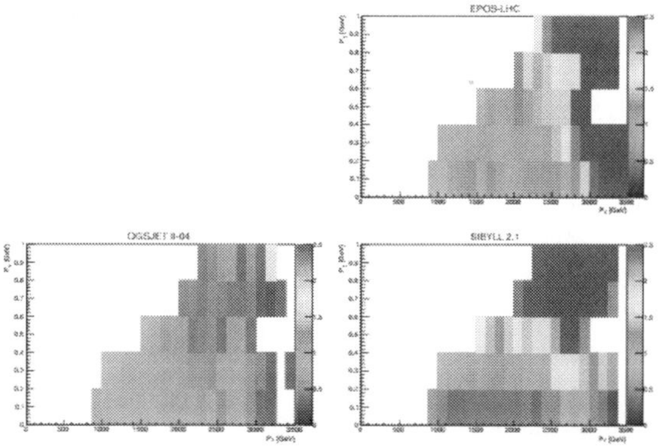

Figure 4. π^0 production cross sections by three models relative to the LHCf result. Ratio is shown in the color scale.

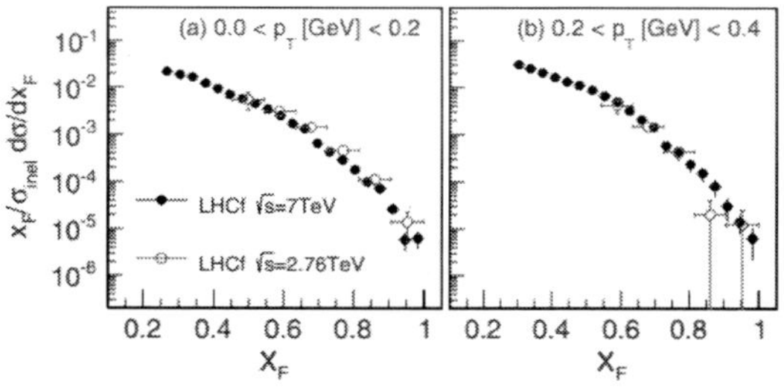

Figure 5. Cross sections of π^0 production measured at 2.76 TeV and 7 TeV. Horizontal axis, x_F, is normalized energy to the beam energy (see text). Data are compared in the same p_T ranges, a) $0.0<p_T<0.2$ GeV, b) $0.2<p_T<0.4$ GeV.

or its breaking, experiments at wider $\sqrt{s}$ coverage are important. Analysis of the LHCf 13 TeV data and the RHICf experiment at $\sqrt{s}$=510 GeV [11] will extend such comparisons. Studies of $\sqrt{s}$ dependence are important to interpolate and extrapolate the accelerator data to the energy ranges of the knee and the GZK, respectively.

Prospect for ATLAS-LHCf joint analysis

The origin of forward particle production is roughly classified in two cases; diffraction and non-diffraction. Contribution from two different origins in the LHCf photon observations are studied by using different interaction models as shown in Fig. 6 [31]. It is interesting that the two processes contribute to the LHCf result almost equally and the detail is model dependent. Because the fundamental physics is very different between the two processes, experimental classification of the LHCf data provides a new clue to improve the models. In [31], a simple method to classify the LHCf data using the information of the ATLAS tracker is proposed. ATLAS and LHCf have successfully taken common data in the 13 TeV operation, and new analysis of forward-central correlation is in discussion.

3.3. Other forward measurements

The CMS and TOTEM experiments report the particle number density and energy density in a wide pseudorapidity range. The number density in the interval $0<\eta<7$ at $\sqrt{s}$=8 TeV shows generally good agreement with the model predictions [32]. Energy density distributions at $\sqrt{s}$=7 TeV and 13 TeV are reported in [33] and [34], respectively. Although the predictions by the cosmic-ray models show good agreements with the 7 TeV data, a systematic difference is found in the 13 TeV results.

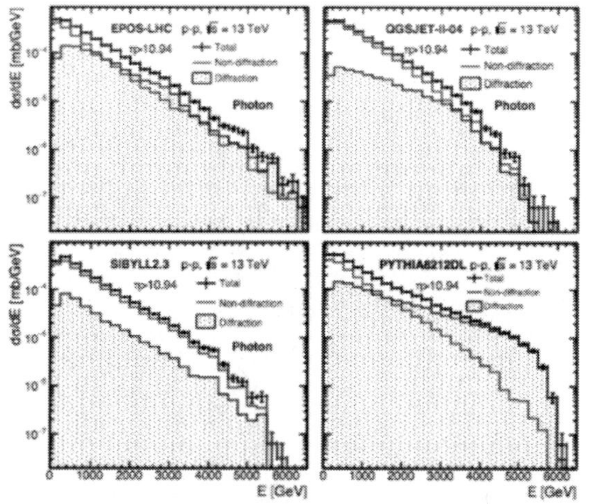

Figure 6. Cross sections of forward photon production in 13 TeV collisions predicted by EPOS-LHC (left-top), QGSJET II-04 (right-top), Sibyll 2.3 (left-bottom) and PYTHIA 8212DL (right-bottom). Black histograms are inclusive cross sections while blue and red show the contributions from diffractive and non-diffractive processes, respectively.

The ATLAS experiment reports a comparison between the 13 TeV data and model predictions for the *number of hits in the minimum bias trigger scintillators (MBTSs)* [15]. The MBTSs cover the beam pipe in the forward direction with an azimuthal segmentation. Differently from the other minimum bias data, the predictions by EPOS-LHC and QGSJET II-04 have a significant discrepancy from the data. Because the number of hits is related to the correlation in the multi-particle production, this data will give a new view point to test the hadronic interaction models.

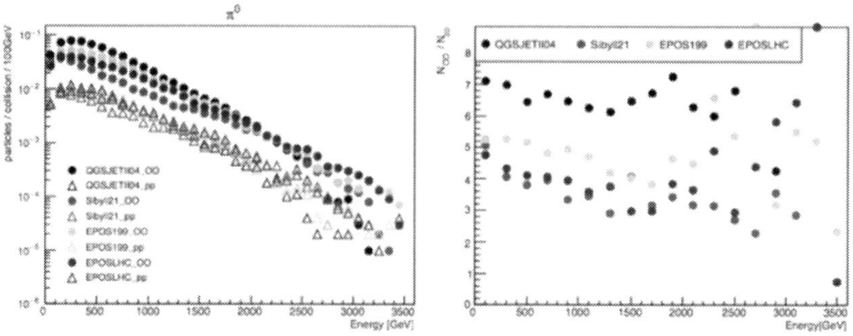

Figure 7. Left: forward π^0 production rates in the $\sqrt{s_{NN}} = 7$ TeV p-p collisions (open triangles) and O-O collisions (filled circles) at $\eta > 8.78$. Right: ratio of the production rates in the O-O collisions divided by the rates in the p-p collisions. Black, red, green and blue markers show the results of QGSJET II-04, Sibyll 2.1, EPOS 1.99 and EPOS-LHC interaction models, respectively.

4. Possibility of Oxygen collisions at LHC

The ultimate goal to use the LHC for cosmic-ray physics is to collide light ions with various kinds of ions, ranging from protons to Iron. A possibility of light ion operation at the CERN accelerator complex was studied and presented in [35]. They concluded the acceleration and collisions of the Oxygen beam is technically feasible when low luminosity collisions are accepted. A study of the existing model uncertainty in the p-O collisions was given in [19]. Although the post-LHC models, EPOS-LHC and QGSJET II-04, have good agreement for the p-p collisions in $dN/d\eta$, they have clear differences in predictions of p-O collisions.

Light ions represent not only the atmospheric target, but also cosmic-ray primary particles. A study of the O-O collisions was also made by the LHCf group. Figure 7 (left) shows the model predictions of the forward π^0 production rates in $\sqrt{s_{NN}}$=7 TeV p-p collisions and O-O collisions at $\eta > 8.78$ The p-p collisions are calculated to extract the model dependence of nuclear effects by dividing the O-O results with the p-p results. The ratios are shown in the right panel of Fig. 7. It is found that the EPOS-LHC and QGSJET II-04 show a factor 1.5–2 difference in the ratio, or in the nuclear effect.

5. Summary

The LHC experiments are extensively studying minimum bias events with various types of forward detectors. Some interaction models have been improved using the LHC Run1 data and they improved the interpretation of the cosmic-ray data. New data from 13 TeV collisions become available from the LHC. Not only the multiplicity distribution, but also various new comparisons with data to enable further improve the models. Comparisons of data in a wide $\sqrt{s}$ range are also important to access energies beyond the LHC. Discussions for future light ion collisions must be continued.

References

[1] K.-H. Kampert and M. Unger, Astropart. Phys. **35** (2012) 660–678
[2] R. Ulrich, R. Engel, M. Unger, Phys. Rev. D **83** (2011) 054026
[3] The CMS Collaboration, JINST **3** (2008) S08004
[4] The TOTEM Collaboration, JINST **3** (2008) S08007
[5] The ATLAS Collaboration, JINST **3** (2008) S08003
[6] S. Bültmann et al., Nucl. Instrum. Methods A **535** (2004) 415–420
[7] The UA7 Collaboration, Phys. Lett. B **242** (1990) 531–535
[8] The ALICE Collaboration, JINST **3** (2008) S08002
[9] C. Adler et al., Nucl. Instrum. Methods A **499** (2002) 433–436
[10] The LHCf Collaboration, JINST **3** (2008) S08006
[11] The RHICf Collaboration, `arXiv:1409.4860`
[12] The ATLAS Collaboration, Nature Commun. **2** (2011) 463
[13] The CMS Collaboration, Phys. Lett. B **722** (2013) 5
[14] The ALICE Collaboration, Eur. Phys. J. C **73**(6) (2013) 2456
[15] The ATLAS Collaboration, ATLAS-CONF-2015-038
[16] The E710 Collaboration, Phys. Rev. Lett. **68** (1992) 2433–2436
[17] The CDF Collaboration, Phys. Rev. D **50** (1994) 5550–5561
[18] The E811 Collaboration, Phys. Lett. B **445** (1999) 419–422
[19] K. Akiba et al., J. Phys. G: Nucl. Part. Phys. **43** (2016) 110201
[20] The LHCf Collaboration, Phys. Lett. B **715** (2012) 298–303
[21] The LHCf Collaboration, Phys. Lett. B **703** (2011) 128–134
[22] The LHCf Collaboration, Phys. Lett. B **750** (2015) 360–366
[23] The LHCf Collaboration, Phys. Rev. D **86**, 092001 (2012)
[24] The LHCf Collaboration, Phys. Rev. C **89**, 065209 (2014)
[25] The LHCf Collaboration, Phys. Rev. D **94**, 032007 (2016)
[26] R. P. Feynman, Phys. Rev. Lett. **23**, 1415 (1969)
[27] J. Engler, et al., Nucl. Phys. B **84** (1975) 70
[28] W. Flauger, F. Mönnig, Nucl. Phys. B **109** (1976) 347–356
[29] The PHENIX Collaboration, Phys. Rev. D **88** (2013) 032006

[30] K. Kawade, PhD thesis, Nagoya University (2014); CERN-THESIS-2014-315
[31] Q.D.Zhou, Y. Itow, H. Menjo and T. Sako, `arXiv:1611.07483v2 [hep-ex]`
[32] The TOTEM Collboration, EPJC (2015) **75**, 126
[33] The CMS Collboration, JHEP **11** (2011) 148
[34] The CMS Collboration, CMS-PAS-FSQ-15-006
[35] D. Manglunki, presented in the "Results and prospects of forward physics at the LHC: Implications for the study of diffraction cosmic ray interactions, and more," `https://indico.cern.ch/event/223562/contributions/470231/`

Recent hadron production measurements performed by the NA61/SHINE collaboration

Alexander E. Hervé for the NA61/SHINE Collaboration[a]

Institut für Kernphysik, Karlsruher Institut für Technologie, Postfach 3640, 76021 Karlsruhe, Germany

Abstract. NA61/SHINE is a fixed target experiment which uses secondary beams produced at the CERN SPS. Hadron-hadron interactions have been recorded at beam momenta between 13 and 350 GeV/c with a wide-acceptance spectrometer.
Recent measurements released by the NA61/SHINE experiment are of significant importance for a precise prediction of the J-PARC neutrino beam used for the T2K experiment and for interpretation of EAS data. These measurements also provide a large amount of input for tuning of hadron production models in Monte-Carlo generators.
The latest measurements of pion-carbon and proton-carbon interactions will be reviewed.

1. The NA61/SHINE experiment

NA61/SHINE[1] [1] is a multi-purpose fixed target experiment to study hadron production in hadron-nucleus and nucleus-nucleus collisions at the CERN Super Proton Synchrotron (SPS). Among its physics goals are precise hadron production measurements for improving calculations of the neutrino beam flux in the T2K and Fermilab neutrino oscillation experiment [2] as well as for more reliable simulations of hadronic interactions in air showers. Moreover, p+p, p+Pb and nucleus+nucleus collisions are measured to study the properties of the onset of de-confinement and search for the critical point of strongly interacting matter (see e.g. Ref. [3]).

The layout of the NA61/SHINE detector is sketched in Fig. 1. The NA61/SHINE Collaboration uses large time-projection-chambers (TPCs) inherited from the NA49 experiment [4] to measure the charge and momentum of particles. The momentum resolution, $\sigma(1/p) = \sigma(p)/p^2$, is about $10^{-4}\,(\text{GeV/c})^{-1}$ at full magnetic field and the tracking efficiency is better than 95%. A set of scintillation and Cherenkov counters as well as beam position detectors upstream of the spectrometer provide timing reference, identification and position measurements of the incoming beam particles. Particle identification is achieved by measuring the energy loss along the tracks in the TPCs and by determining their velocity from the time of flight provided by large scintillator walls placed downstream of the TPCs. The centrality of nucleus-nucleus collisions can be estimated using the measurement of the energy of projectile spectators with a calorimeter [5] located behind the time of flight detectors. For nucleon-nucleus collisions, the centrality is determined by counting low momentum particles from the target (so called 'gray protons') with a small TPC around the target [6].

Data taking with the NA61/SHINE experiment started in 2007. After a first run with proton on carbon at 31 GeV/c, the data acquisition system was upgraded during 2008 to increase the event recording rate by a factor of $\approx$10 [7]. In recent years, a wealth of data has been recorded by the experiment at beam momenta ranging from 13 to 350 GeV/c with various beam particles and targets. In this paper we present results obtained from the data relevant for neutrino oscillations and the interpretation of air showers at ultra-high energies.

2. Measurements of p+C interactions for the improvement of neutrino flux calculations

Data for the T2K experiment were collected in 2007, 2009 and 2010 by using a 31 GeV/c proton beam and 2 targets: a thin carbon target and a T2K replica target. Measurements with the thin target are used for constraining primary proton-carbon interactions which account for around 60% of the T2K neutrino flux. In contrast to the thin target, measurements with the T2K replica target include primary interactions and re-interactions in the target which account for more than 90% of the T2K neutrino flux. Results of the measurements performed in 2007 were published [8–11] and used for the T2K beam simulation. Results of the thin and T2K replica target measurements performed in 2009 have been recently published [12, 13] and will be discussed in the next two subsections. Analysis of the T2K replica target data taken in 2010 is ongoing.

2.1. Hadron spectra in proton-carbon at 31 GeV/c

In the analysis of the data, different particle identification techniques were used in order to maximize coverage of the phase space and cross-check the results. For charged hadrons, we used 3 different analysis techniques: dE/dx identification, combined dE/dx and time-of-flight

[a] `http://shine.web.cern.ch/content/author-list`
[1] SHINE = SPS Heavy Ion and Neutrino Experiment.

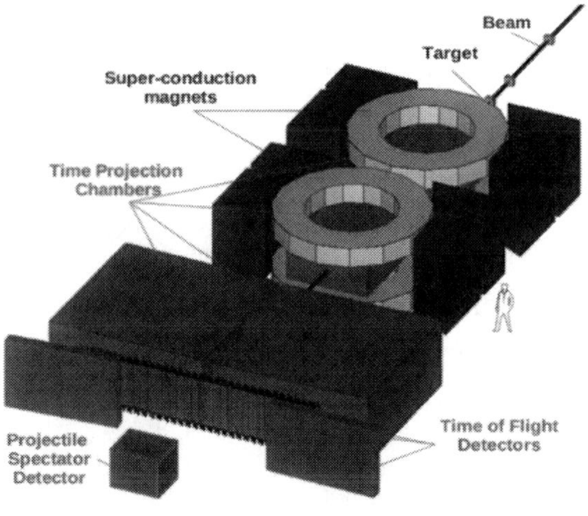

Figure 1. Schematic layout of the NA61/SHINE experiment.

identification and no identification (just for π^-). Examples of the first two methods can be seen in Fig. 2. This allows us to measure spectra in the regions of the phase space where one technique fails (for example in the region where Bethe-Bloch curves overlap). Spectra of the neutral hadrons have been extracted by fitting the invariant mass of the decay candidates in the detector. All results have been fully corrected with the Monte Carlo. Systematic uncertainty for the $\pi^\pm$ spectra is around 5%, while for other hadrons it is up to 15%. A detailed description of the different analysis techniques and overview of the systematic uncertainties can be found in [12]. The results are presented in the form of differential multiplicities as required for the T2K collaboration. Comparisons of the results with various MC model predictions (VENUS 4.12 [14,15], EPOS 1.99 [16], GiBUU 1.6 [17,18] and several GEANT4 physics lists) were also made. In Fig. 3 $K^\pm$ spectra are presented and compared with the model predictions. In this case, EPOS 1.99 gives the most accurate description of the spectra. Spectra of $\pi^\pm$, $K^\pm$, p, K^0_S and Λ with the MC predictions can be found in [12]. It is important to note that none of the models give accurate predictions for all particle spectra. Presented results are being used in the T2K beam simulation. The total uncertainty of the T2K neutrino flux after the re-weighting of the flux with the thin target measurements is reduced down to 10%. To reach the T2K goal of 5%, accurate re-weighting which includes re-interactions in the T2K target is necessary. For this reason, hadron spectra exiting the surface of the T2K replica target are needed.

2.2. Spectra of charged pions in the proton + T2K replica target interactions at 31 GeV/c

Measurements of the hadron spectra coming from the surface of the T2K replica target have one major difference with respect to the thin target measurements. Since neutrino flux depends on the position along the target surface from which neutrino parent particles are coming, the target is divided into 5 longitudinal bins and the downstream target face. TPC tracks were extrapolated towards the target surface and stored in the longitudinal bins. For each longitudinal bin, analysis similar to the analysis of the thin target dataset was performed. The results are shown in the form of differential multiplicities in the bins of momentum, polar angle and longitudinal position along the target surface. Spectra of the π^+ exiting the target surface together with FLUKA 2011 prediction are shown in Fig. 4. Each column represents a different longitudinal bin. The largest systematic contribution is due to the track extrapolation and can be up to 10%. Full results for the $\pi^\pm$ and detailed description of the analysis can be found in [13].

The ratio of ν_μ flux re-weighted with the thin target procedure over the T2K replica target flux at SK is shown in Fig. 5. At low neutrino energy the thin target procedure predicts a 10% larger value of the flux which is however consistent with the replica target prediction within one standard deviation. At high neutrino energy, the ratio tends to one. This is due to the fact that within this energy range the dominant part of the ν_μ spectra comes from kaons and thus is not affected by the re-weighting procedure on pion spectra. Extraction of the $K^\pm$ spectra from this dataset is not possible because of the low statistics, but analysis of the high-statistics T2K replica target dataset recorded in 2010 is ongoing.

3. Measurements of π+C interactions for the interpretation of cosmic ray air showers

During the 2009 run, NA61/SHINE also recorded data with negatively charged pion beams at 158 and 350 GeV/c on a thin carbon target. These measurements will provide particle-production data for the tuning of hadronic interaction models used in air-shower simulations. The first preliminary results have already been released [19–22] and are summarized below.

3.1. Spectra of ρ^0 mesons

The measurement of resonances in π + C is useful to constrain the production of ρ^0 meson, which is important to predict the number of muons observed in air showers due to the leading particle effect (see e.g. Ref. [23]).

In the inclusive $\pi^+\pi^-$ invariant mass spectra there is a large combinatorial background, which dominates over the effective mass distributions of individual resonances. The method used to estimate the background is the so called charge mixing, which uses the $(\pi^+\pi^+ + \pi^-\pi^-)$ mass spectra as an estimate of the background.

The fitting procedure uses templates of the $\pi^+\pi^-$ mass distribution for each resonance. These templates are constructed by passing simulated π+C interactions, generated with the Epos1.99 [16] hadronic interaction model using CRMC [24] (v1.5.3), through the full NA61 detector Monte Carlo chain. All the cuts that are applied to the data are also applied to the templates. This method of using templates allows fitting, in addition to resonances with 2-body decays, resonances with dominant three body decays, such as the ω, and resonances with non-$\pi^+\pi^-$ decays, such as the K^{*0}.

An example of one of these fits can be seen in Fig. 6. The templates in the fit are the background found from

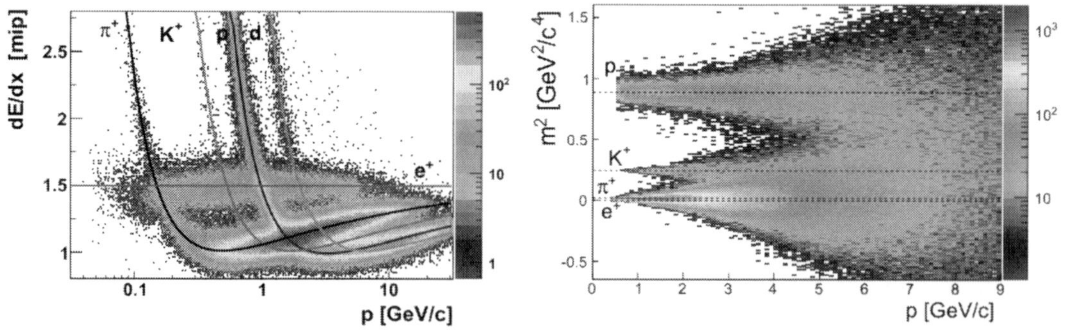

Figure 2. Examples of PID capabilities of the NA61/SHINE spectrometer for positively charged particles. Left: specific energy loss in the TPCs as a function of momentum. Curves show parameterizations of the mean dE/dx calculated for different particle species. Right: mass squared, derived from the ToF-F measurement and the fitted path length and momentum, versus momentum. The lines show the expected mass squared values for different particles.

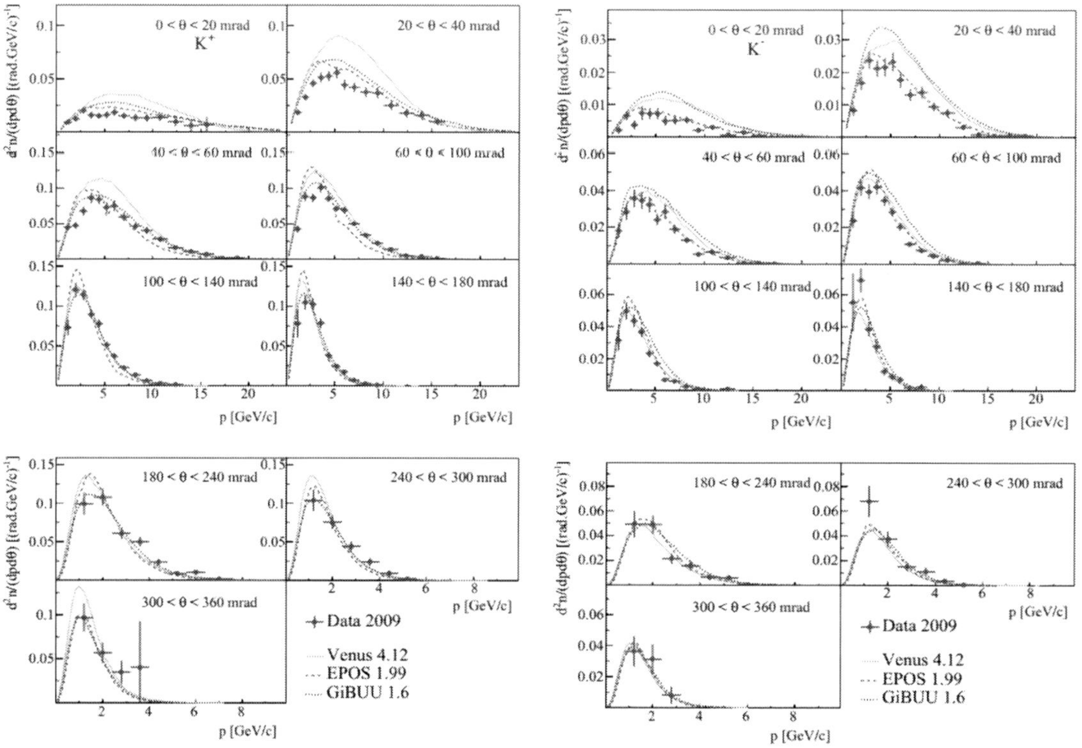

Figure 3. Comparison of measured K$^+$ (left) and K$^-$ (right) spectra with model (VENUS, EPOS, GIBUU) predictions. Distributions are normalized to the mean K$^+$ (K$^-$) multiplicity in all production p+C interactions. The vertical error bars on the data points show the total (stat. and syst.) uncertainty. The horizontal bars indicate the bin size in momentum.

charge mixing and the following resonances: ρ^0, K^{*0}, ω, f$_2$, f$_0$ (980), a$_2$, η and K^0_S.

The average multiplicity of ρ^0 mesons is presented in Fig. 7. Also shown are predictions by Epos1.99 [16], DPMJet3.06 [25], Sibyll2.1 [26], QGSJetII-04 [27] and EposLHC [28]. It can be seen that there is an underestimation of the ρ^0 yield for almost all hadronic interaction models, with the exception of QGSJetII-04 for $x_F > 0.8$. It is interesting to note that while QGSJetII-04 and EposLHC were tuned to NA22 π^++p data [29], there is an underestimation in π^-+C.

The systematic error is dominated by the background estimates, up to 14%, whereas the other errors are less than 4%. Other sources of uncertainty, such as using templates from a different model, are found to be much smaller.

3.2. Spectra of charged pions

The measured average multiplicities of charged pions are shown in Fig. 8 and the measurements are compared to predictions of pion production in π^-+C at 158 GeV/c from hadronic interaction models in Fig. 9. As can be seen, none of the generators describe the data well.

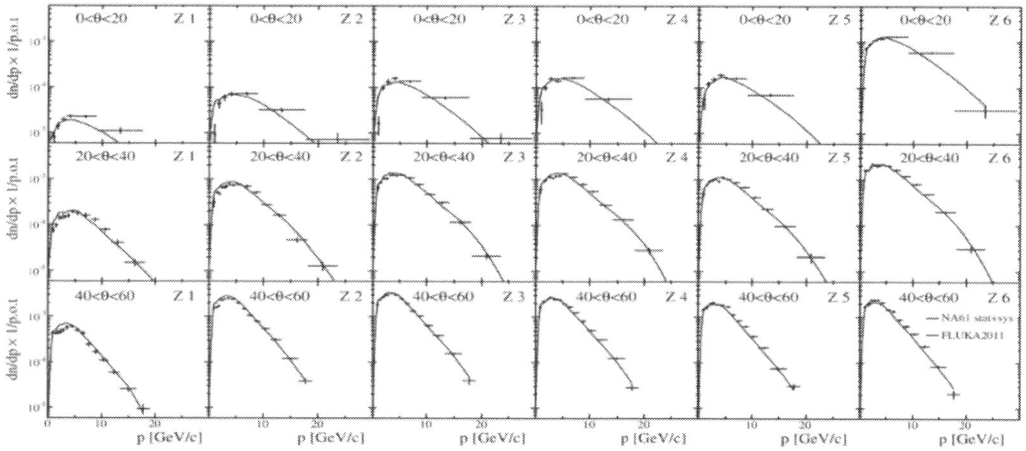

Figure 4. Spectra of positively charged pions at the surface of the T2K replica target, in the polar angle range from 0 to 60 mrad, and for the six longitudinal bins as a function of momentum. The normalization is per proton on target. The prediction from FLUKA2011 is overlaid.

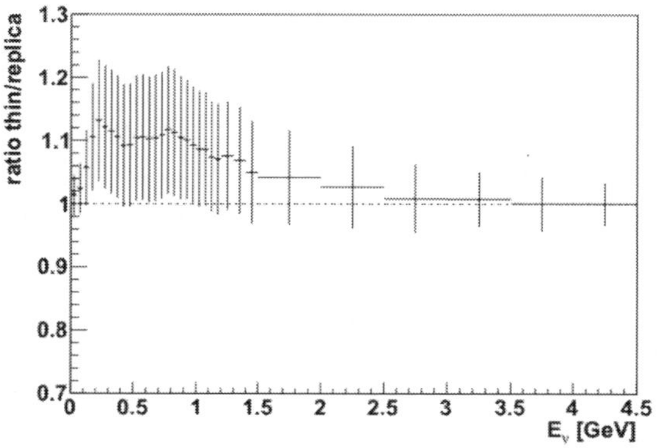

Figure 5. Ratio of the ν_μ flux at SK re-weighted with the thin target procedure to the p+(T2K RT) at 31 GeV/c flux. For the latter, pion spectra presented in [13] have been used in the re-weighting. The dominant part of the ν_μ spectra at high E_ν comes from kaons and thus is not affected by the re-weighting of the pion spectra.

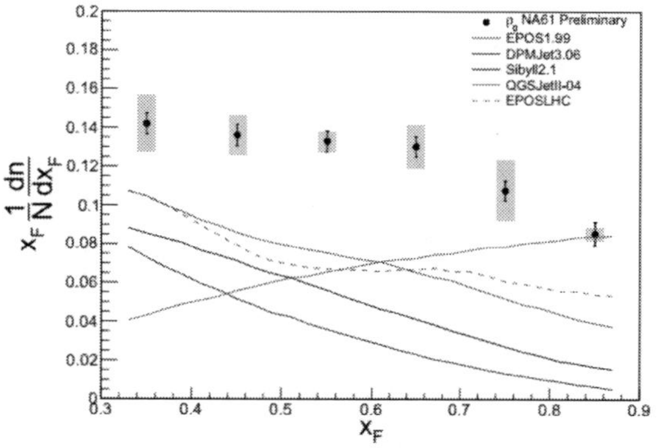

Figure 7. Average multiplicity of the ρ^0 meson in π+C at $p_{beam} = 158$ GeV/c as a function of Feynman-x. The bars show the statistical errors; the bands indicate systematic errors. The lines depict predictions of hadronic interaction models: red – Epos1.99, blue – DPMJet3.06, black – Sibyll2.1, green – QGSJetII-04, dashed red – EposLHC.

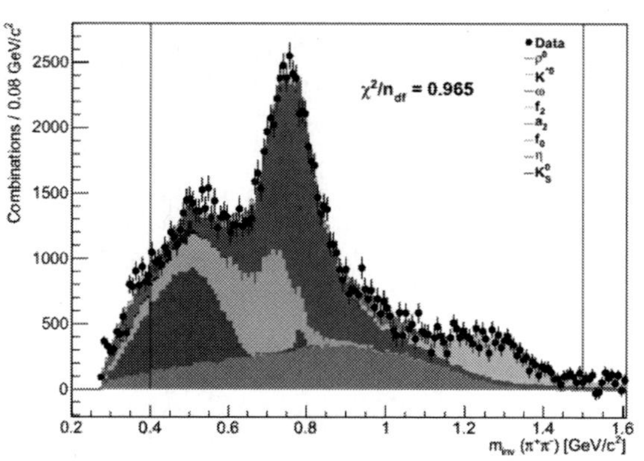

Figure 6. $\pi^+\pi^-$ invariant mass distribution in π^-+C interactions at 158 GeV/c in the range $0.4 < x_F < 0.5$. Dots with error bars denote the data and the fitted resonance templates are shown as filled histograms. The vertical lines indicate the range of the fit.

These spectra were obtained within a fiducial phase space of the NA61/SHINE detector, for which the detection and selection efficiency for charged tracks is close to unity, and then corrected for the detector acceptance, selection efficiency, feed-down from weak decays and re-interactions in the target. The latter two corrections are currently estimated using model predictions (Epos1.99, QGSJetII-04, DPMJet3.06) and they are typically well below 5%, but can reach up to 20% at low particle momenta. Overall, the systematic uncertainty of the corrected number of tracks, $\Delta n'$, is estimated to be $\lesssim 7\%$.

The average multiplicity of particles produced within a p/pT-bin is then obtained by dividing $\Delta n'$ by the total number of events in which an interaction occurred, N_{prod}. N_{prod} is estimated by extrapolating the number of recorded interaction triggers to full phase space. The correction amounts to $(92.5 \pm 3.5)\%$ at 158 GeV/c and $(92.5 \pm 4.0)\%$ at 350 GeV/c, where the uncertainty was

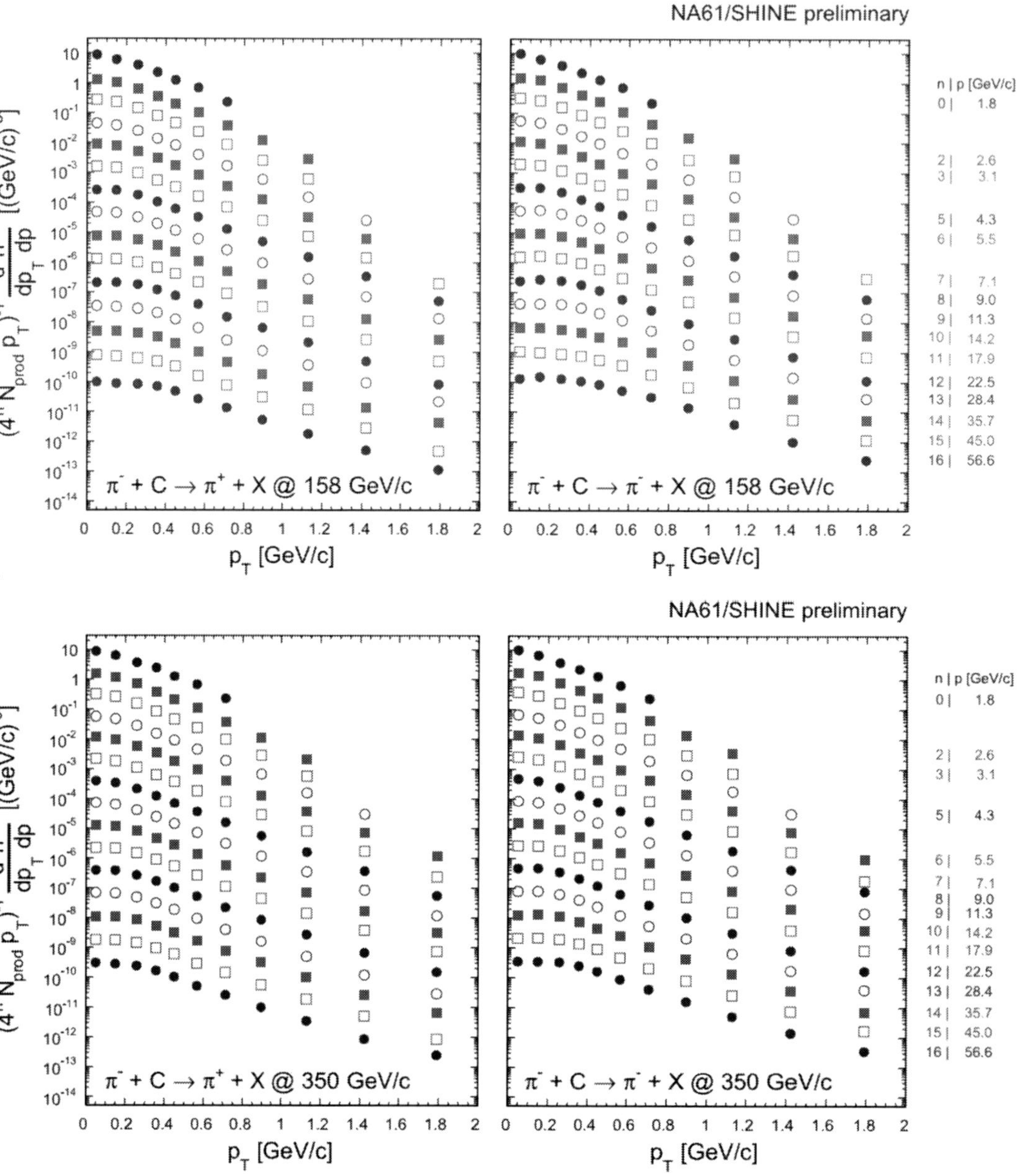

Figure 8. Inclusive spectra of charged pions in π^-+C interactions at beam energies of 158 and 350 GeV/c. For better visibility, the spectra from the nth momentum bin are multiplied by a factor of $^1/_{4^n}$. The momentum increases from top to bottom as indicated in the legend on the right. Systematic errors are less than 7%

derived by running different generators to evaluate the correction.

4. Conclusions and outlook

In this article, results from pion-carbon and proton-carbon interactions, measured with the multi-purpose experiment NA61/SHINE at the CERN SPS, have been summarised. These results are of importance for the modelling of cosmic ray air showers and neutrino flux predictions.

Measurements of charged pion, kaon, proton, K^0_S and Λ production in p+C interactions at 31 GeV/c released by NA61/SHINE are of significant importance and were already employed for precise predictions of the J-PARC neutrino beam used for the first stage of the T2K experiment. The recently published results of the 2009 measurements presented here will further improve the precision of these predictions.

Preliminary results from the π-carbon interactions suggest the high-energy interaction models need further tuning to reproduce the charged pion and ρ^0 spectra.

Future measurements for neutrino physics will include precise measurements of hadron emission from a replica target of the Long-Baseline Neutrino Facility (LBNF) at Fermilab required for the Deep Underground Neutrino Experiment (DUNE). The NA61/SHINE measurements

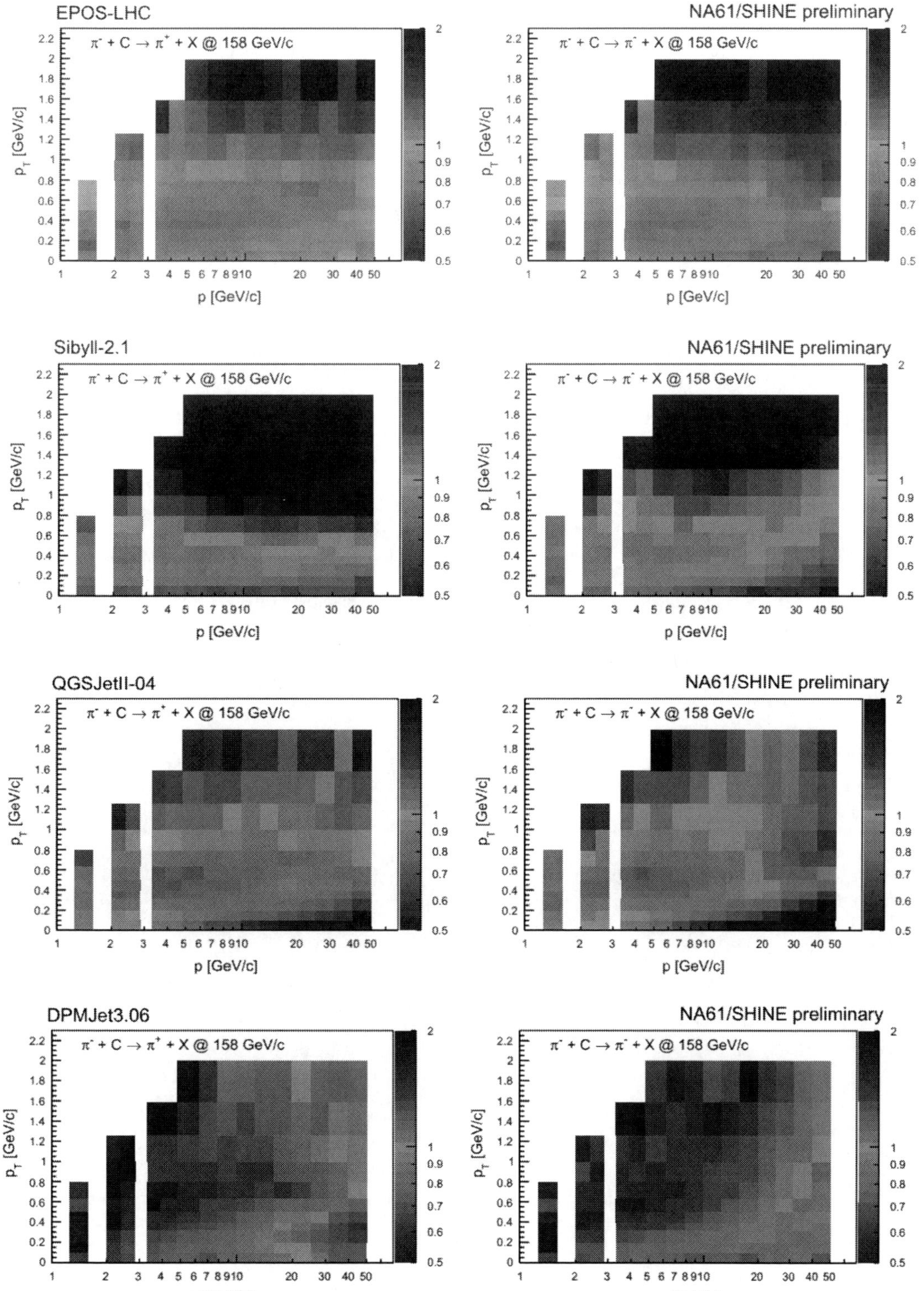

Figure 9. Comparison of the measured production spectra of charged pions to predictions from hadronic interaction models used for the interpretation of cosmic ray data [25–28]. The colors denote the ratio of data to simulation and the color scale is truncated at 0.5 and 2. The two empty p-bins at $p \lesssim 2$ GeV/c and $p \gtrsim 3$ GeV/c are momenta excluded from the analysis due to the ambiguity in identification of pions.

with thin and replica targets for T2K demonstrated a need for precise data on hadron productions from the surface of targets of future long baseline neutrino oscillation experiments, e.g. DUNE and Hyper-K.

The rich data set of NA61/SHINE will also be very useful to reduce uncertainties in air shower calculations and increase the knowledge on interactions in the late stages of the shower development below a TeV.

This work was supported by the Hungarian Scientific Research Fund (grants OTKA 68506 and 71989), the János Bolyai Research Scholarship of the Hungarian Academy of Sciences, the Polish Ministry of Science and Higher Education (grants 667/N-CERN/2010/0, NN 202 48 4339 and NN 202 23 1837), the Polish National Center for Science (grants 2011/03/N/ST2/03691, 2012/04/M/ST2/00816, 2013/11/N/ST2/03879, 2014/13/N/ST2/02565), the Foundation for Polish Science — MPD program, co-financed by the European Union within the European Regional Development Fund, the Federal Agency of Education of the Ministry of Education and Science of the Russian Federation (SPbSU research grant 11.38.242.2015), the Russian Academy of Science and the Russian Foundation for Basic Research (grants 08-02-00018, 09-02-00664 and 12-02-91503-CERN), the Ministry of Education, Culture, Sports, Science and Technology, Japan, Grant-in-Aid for Scientific Research (grants 18071005, 19034011, 19740162, 20740160 and 20039012), the German Research Foundation (grant GA 1480/2-2), the EU-funded Marie Curie Outgoing Fellowship, Grant PIOF-GA-2013-624803, the Bulgarian Nuclear Regulatory Agency and the Joint Institute for Nuclear Research, Dubna (bilateral contract No. 4418-1-15/17), Ministry of Education and Science of the Republic of Serbia (grant OI171002), Swiss Nationalfonds Foundation (grant 200020117913/1) and ETH Research Grant TH-01 07-3.

References

[1] N. Abgrall et al. [NA61/SHINE Collaboration], JINST **9** (2014) P06005
[2] K. Abe et al. [T2K Collaboration], Nucl. Instrum. Meth. A **659** (2011) 106
[3] M. Gazdzicki, M. Gorenstein, P. Seyboth, Acta Phys. Polon. **B42** (2011) 307
[4] S. Afanasev et al. [NA49 Collaboration], Nucl. Instrum. Meth. A **430** (1999) 210
[5] M.B. Golubeva, F.F. Guber, A.P. Ivashkin, A.B. Kurepin, V.N. Marin, A.S. Sadovsky and O.A. Petukhov, Phys. Atom. Nucl. **75** (2012) 673
[6] K. Marton, G. Kiss, A. László and D. Varga, Nucl. Instrum. Meth. A **763** (2014) 372
[7] A.László et al., Nucl. Instrum. Meth. A **798** (2015) 1
[8] N. Abgrall et al. [NA61/SHINE Collaboration], Phys. Rev. C **84** (2011) 034604
[9] N. Abgrall et al. [NA61/SHINE Collaboration], Phys. Rev. C **85** (2012) 035210
[10] N. Abgrall et al. [NA61/SHINE Collaboration], Phys. Rev. C **89** (2014) 025205
[11] N. Abgrall et al. [NA61/SHINE Collaboration], Nucl. Instrum. Meth. A **701** (2013) 99
[12] N. Abgrall et al. [NA61/SHINE Collaboration], Phys. Rev. C **76** (2016) 84
[13] N. Abgrall et al. [[NA61/SHINE Collaboration], Eur. Phys. J. C **76** (2016) 617
[14] K. Werner, Nucl. Phys. A **525** (1991) 501
[15] K. Werner, Phys. Rept. **232** (1993) 87
[16] T. Pierog and K. Werner, Phys. Rev. Lett. **101** (2008) 171101
[17] O. Buss et al., Phys. Rept. **512** (2012) 1
[18] K. Gallmeister and U. Mosel, Nucl. Phys. A **826** (2009) 151
[19] M. Unger [NA61/SHINE Collaboration], EPJ Web Conf. **52** (2013) 01009
[20] L. Zambelli (2013) Ph.D. Thesis, University of Paris VII
[21] B. A. Popov, EPJ Web Conf. **99** (2015) 02002
[22] A. Herve [NA61/SHINE Collaboration], PoS ICRC **2015** (2016) 330 [`arXiv:1509.06586` [`nucl-ex`]]
[23] H.-J. Drescher, Phys. Rev. D **77** (2008) 056003
[24] `https://web.ikp.kit.edu/rulrich/crmc.html`
[25] S. Roesler, R. Engel and J. Ranft, [`arXiv:hep-ph/0012252`]
[26] E.J. Ahn et al., Phys. Rev. D **80** (2009) 094003
[27] S. Ostapchenko, Phys. Rev. D **83** (2011) 014018
[28] T. Pierog, I. Karpenko, J. M. Katzy, E. Yatsenko and K. Werner, Phys. Rev. C **92**, no. 3 (2015) 034906
[29] N.M. Agababyan et al. [EHS/NA22 Collaboration], Z. Phys. C **46** (1990) 387

ATLAS and ultra high energy cosmic ray physics

James Pinfold[a]

For the ATLAS Collaboration, Physics Department, University of Alberta, Edmonton, Alberta, Canada

Abstract. After a brief introduction to extended air shower cosmic ray physics the current and future deployment of forward detectors at ATLAS is discussed along with the various aspects of the current and future ATLAS programs to explore hadronic physics. The emphasis is placed on those results and future plans that have particular relevance for high-energy, and ultra high-energy, cosmic ray physics. The possible use of ATLAS as an "underground" cosmic muon observatory is briefly considered.

1. Introduction

Over the past ten years or so ATLAS has been involved in developing the detector coverage in the forward direction in order to expand the physics horizons of the ATLAS detector in the arena of hadronic physics. A discussion of recent developments in the forward detector system of the ATLAS detector will be presented in section one. In section two we will present the latest results in the ATLAS study of the p-p total, inelastic and elastic cross-section and in section three the latest p-p charged particle distributions. In section four we will briefly consider some implications of ATLAS hadronic physics results for of high energy cosmic ray physics. Penultimately, in section five, we will touch on the possible use of the ATLAS detector as an underground cosmic muon observatory.

1.1. High energy cosmic ray physics and hadronic physics at the LHC

Understanding the sources and propagation of high energy cosmic rays are core issues of astroparticle physics. There is strong evidence that supernova remnants can accelerate cosmic rays up to energies of $\sim$Z $\times 10^{14}$ eV (where Z is the charge of the cosmic ray nucleus). However, the sources and accelerating mechanisms for particles with higher energies, up to $\sim 10^{20}$ eV are not known [1]. Accelerating particles to such energies requires exceptional astrophysical objects and/or novel accelerating mechanisms [2]. Ultra-high energy cosmic rays are messengers from the extreme Universe that provide a window to particle physics at energies up to $\sqrt{(s)} \sim$ 400 TeV.

As can be seen from the energy spectrum of cosmic rays shown in Fig. 1 the study of cosmic rays of energies higher than about 10^{14} eV is hampered by the low flux of the particles incident on the Earth. Thus, one has to use very large detector areas that use the upper atmosphere as a target and/or a calorimeter and resort to the large aperture observation of extensive air showers produced by cosmic rays entering the atmosphere [3]. The challenge to this approach is the need of detailed air shower simulations for deriving the primary particle energy and composition from measured air shower observables.

A compilation of more recent measurements of the flux of cosmic rays is given in Fig. 2. Measurements at the LHC allow us to access, for the first time, energies beyond the 'knee' in the cosmic ray spectrum where there is a sudden steepening of the power law of the flux at about 3×10^{15} eV, which is not yet completely understood. Additionally, the very high energy cosmic ray flux exhibits an ankle at about 3×10^{18} eV (thought to be related to the transition between galactic and extragalactic cosmic rays [4,5]) and a very strong suppression at energies above 8×10^{19} eV.

One possible cause of this cut off at the highest cosmic ray energies is GZK flux suppression [6], arising from scattering of ultra high energy cosmic primaries off CMB photons. However, it could be that the mass composition indicates that the upper end of the energy spectrum is more likely related to the maximum injection energy of the particles at the sources [7].

To study the sources of cosmic particles at energies above the knee and the astrophysical origin of the main features of the cosmic ray spectrum we need to derive the mass composition of cosmic rays. Thus, being able to reliably deduce the mass composition of cosmic rays from air shower measurements is of fundamental importance and is currently hampered mainly by the large uncertainties in predicting hadronic multi-particle production at high energy [4,8]. Thus, performing further measurements at accelerators is the only way to make progress.

Cosmic ray physics interaction models have been tuned to describe particular data sets taken at certain collider energies and also the energy dependence of multiparticle production. While cosmic ray models were developed before LHC was turned on, their predictions bracket many distributions measured in minimum-bias mode at LHC. This success in the development of the phenomenology developed for soft hadronic interactions can be improved considerably by tuning the models to match the LHC data. In the following document we will discuss LHC measurements that can further improve the understanding of hadronic multiparticle production with direct relevance to cosmic ray physics and air shower simulations.

[a] e-mail: jpinfold@ualberta.ca

DOI: 10.1051/epjconf/201714510001

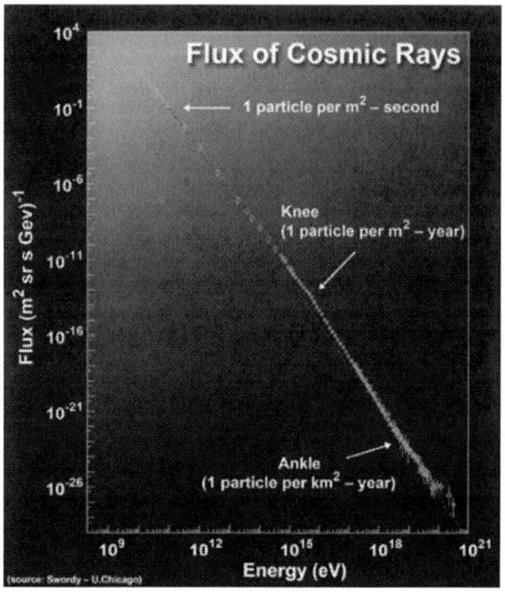

Figure 1. The cosmic ray energy spectrum.

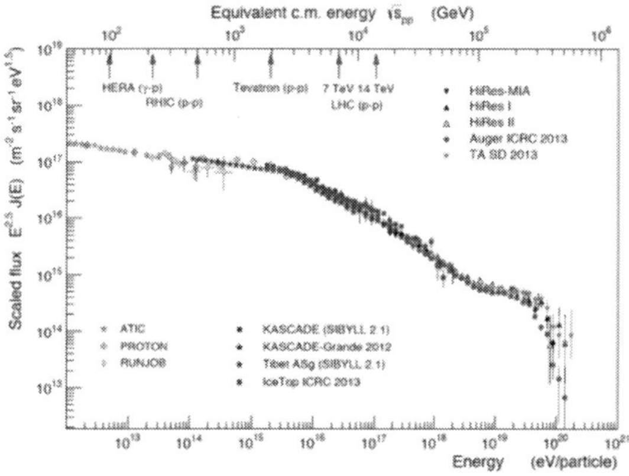

Figure 2. Compilation of representative measurements of the flux of cosmic rays at Earth (from [4], updated). The equivalent energies of LHC proton-proton collisions at different c.m.s. energies are shown at the upper axis.

2. The ATLAS forward detector system

The ATLAS Forward Detector system is shown in Fig. 3 indicating their distance from the ATLAS Interaction Point (IP) and the pseudorapidity (η) range that they cover. In this report we will consider the following forward detectors: the New Minimum Bias Trigger Scintillator (MBTS) [9]; the Absolute Luminosity for ATLAS (ALFA) [10] and ATLAS Forward Protons (AFP) [11]. These detectors, in combination with the main ATLAS detector, provide the results discussed in this paper.

2.1. The new MBTS detector

The first Minimum Bias Trigger Scintillators (MBTS) delivered the minimum bias triggers for for the low luminosity LHC's Run I, from 2009–2013 for proton-proton, lead-lead and lead-proton collisions. An upgraded version of MBTS was installed for Run II of the LHC (2015–2018). This new MBTS will provide the key ingredients for the physics measurements at larger LHC collision energies of the charge multiplicity, proton-proton cross section, rapidity gap measurements, etc. The upgrade strategy included a replacement of radiation damaged scintillator and a modification of the read out scheme for Run II conditions.

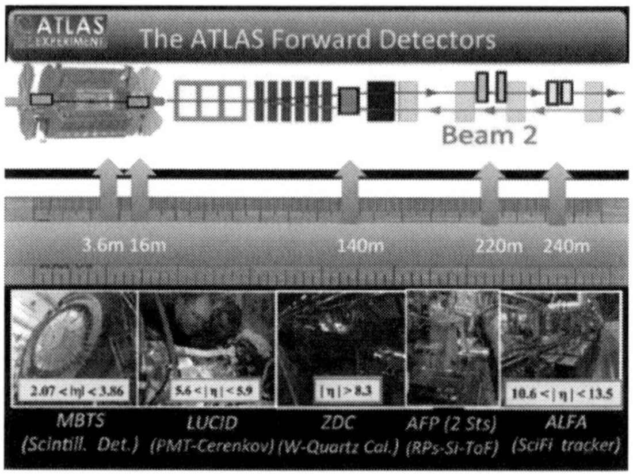

Figure 3. The Forward Detector system of the ATLAS experiment.

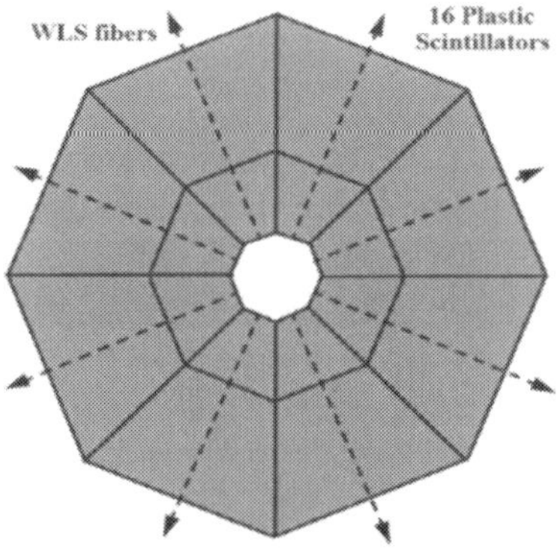

Figure 4. The basic structure of the Run I and Run II MBTS detector.

It is estimated that in Run I, the MBTS received a dose in the range of 0.1 to 0.4 Gy. Since in Run II, the same Soft QCD measurements have to be performed for the new LHC energy $\sqrt{s} = 13$ TeV, and at potentially higher luminosity, the decision was made to replace the MBTS with a new detector based on the existing design and trying to keep the same general structure and materials used in the Run I MBTS detector.

The basic structure of the MBTS is shown in Fig. 4. The readout scheme for Run II was modified in the following ways: the outer ring detectors light connectors have been merged two by two, while the inner rings have been kept as they were in Run I. The WLS fibres path has been modified and run closer to the detector edges, reducing the dips in the efficiency observed in Run I. The η coverage has been slightly modified, with the outer and inner ring covering $2.08 \leq |\eta| \leq 2.76$ and $2.76 \leq |\eta| \leq 3.86$, respectively. Side A of the Run II MBTS detector was

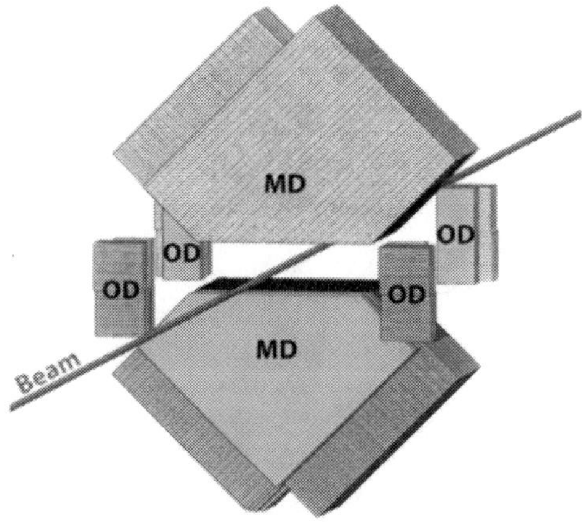

Figure 5. A sketch of the detector principle: MD labels the main detectors above and below the beam while OD stands for the overlap detectors. The lines across MDs and ODs indicate the orientation of the scintillating fibres. The areas without a fibre structure mark the trigger tiles.

installed in February 2014, while the C side was installed in July 2014.

2.2. The ALFA detector

The ALFA detector (Absolute Luminosity For ATLAS) is designed to measure elastic proton (small-angle) scattering. Two tracking stations are placed on each side of the central ATLAS detector at distances of 238 m and 241 m from the interaction point. The tracking detectors are housed in Roman Pots (RPs) which can be moved to within millimetres of the circulating proton beams. Combined with special beam optics, for example high β allows the detection of protons at scattering angles down to $10\,\mu$rad. Each station has an upper and lower RP connected by bellows to the primary LHC vacuum. Elastically scattered protons are detected in the main detectors (MDs) while dedicated overlap detectors (ODs) measure the distance between upper and lower MDs. The arrangement of the upper and lower MDs and ODs, with respect to the beam, is illustrated in Fig. 5.

The ALFA tracking detectors are based on staggered layers of square cross-section scintillating fibres, read out by Multi-Anode Photo-Multiplier Tubes (MAPMTs). The fibres of the Main Detectors (MDs) are inclined at 45° with respect to the vertical. The sensitive region has the shape of a truncated square. In the ODs the fibres are placed horizontally measuring the vertical coordinate only. Both, MDs and ODs, are equipped with scintillator tiles for triggering. They are made of 3 mm thick standard plastic scintillator and read out via bundles of clear plastic fibres of 0.5 mm diameter.

2.3. The AFP detector

The ATLAS Forward Proton (AFP) detector will identify events in which one or both protons emerge intact from the proton-proton collisions at the LHC. The installation of AFP is being performed in two stages: an AFP tracking system at one side of the IP, was already installed during the end-of-year 2015–2016 shutdown. The full two-arm

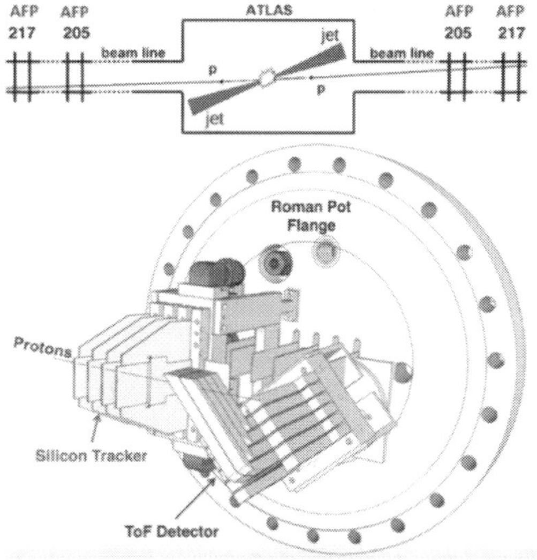

Figure 6. Layout of the AFP stations at both sides of the ATLAS IP at 205 and 217 m. Bottom: Design of the 217 m AFP detector including tracking and time-of-flight systems (the final version will comprise four ToF LQbars per train).

system, with both tracking and timing detectors at both sides of the IP, is currently being completed during the extended end-of-year 2016–2017 shutdown. The final AFP detector will consist of two Roman-pot stations at 205 m and 217 m, each side of the ATLAS IP, as depicted in Fig. 6 top.

Using a Roman Pot system, tracking and timing detectors can be placed 2–3 mm from the beam, 205 m and 217 m away from the ATLAS interaction point. The 3-D silicon tracker will provide momentum measurement, while the time of flight system is used to reduce the background from multiple proton-proton collisions. The study of soft and hard diffractive events at low luminosities ($\mu \approx 1$) is the core of the AFP physics program.

3. Total p-p elastic, inelastic and total cross-sections

A measurement of the total p-p cross section; elastic, inelastic and total cross-sections was performed with the ALFA sub-detector of ATLAS at the LHC at $\sqrt{s}$ of 7 TeV and 8 TeV. In the case of the 8 TeV measurement [12] an integrated luminosity of $500\,\mathrm{fb}^{-1}$ was accumulated in a special run with high-β^* beam optics to measure the differential elastic cross section as a function of the Mandelstam t. A fit to the differential elastic cross section in the $-t$ range from $0.014\,\mathrm{GeV}^2$ to $0.1\,\mathrm{GeV}^2$ was used to extrapolate to $t \to 0$. The measured differential cross-section and the fit of the theoretical prediction for that distribution is shown in Fig. 7. The total cross-section can be calculated using the Optical Theorem:

$$\sigma_{tot} = 4\pi\, Im[f_{el}(t \to 0)] \quad (1)$$

which relates the total p-p cross-section σ_{tot} to the elastic-scattering amplitude extrapolated to the forward direction $f_{el}(t \to 0)$, with t being the four-momentum transfer

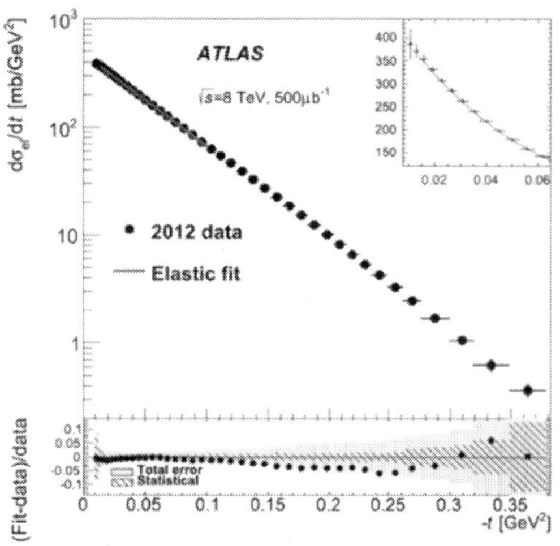

Figure 7. The fit of the theoretical prediction to the differential elastic cross section with σ_{tot} and the slope of the exponential function describing the elastic cross section, B, as free parameters. In the lower plot the points represent the relative difference between fit and data, the yellow (light grey) area represents the total experimental uncertainty and the hatched area the statistical component. The red line (grey) indicates the fit range; the fit result is extrapolated in the lower plot outside the fit range. The upper right insert shows a zoom of the data and fit at small t.

squared. The result of the fit is:

$$\sigma_{tot}(pp \to X) = 96.07 \pm 0.18(stat.)$$
$$\pm 0.85(exp.) \pm 0.31(extr.)\text{mb}, \quad (2)$$

where the first error is statistical, the second accounts for systematic uncertainties and the last arises from uncertainties in the extrapolation $t \to 0$. The measurements at 8 TeV are significantly more precise than the previous measurements at 7 TeV because of the smaller luminosity uncertainty and a larger data sample.

Using the total cross section, the total integrated elastic cross section, σ_{el}, can be calculated, provided that the Coulomb amplitude is neglected, using the formula:

$$\sigma_{el} = \frac{\sigma_{tot}^2}{B}\frac{1+\rho^2}{16\pi\hbar^2c^2} \quad (3)$$

The result is:

$$\sigma_{el} = 24.33 \pm 0.04(stat.) \pm 0.39(syst.)\text{mb} \quad (4)$$

The total inelastic cross-section is determined by subtraction of the total elastic cross section from the total cross section. The resulting value is:

$$\sigma_{inel} = 71.73 \pm 0.15(stat.) \pm 0.69(syst.)\text{mb} \quad (5)$$

The measurements of ATLAS and TOTEM are compared to measurements at lower energy and to a global fit [13] in Fig. 8 for σ_{tot}.TOTEM also reported evidence of non-exponential behaviour of the differential cross-section [14]. The TOTEM Collaboration measured σ_{tot} using the luminosity-independent method, using the same LHC fill used by ATLAS-ALFA. The result they obtained was $\sigma_{tot} = 101.7 \pm 2.9$ mb [21], 1.9σ higher than the ATLAS measurement presented here.

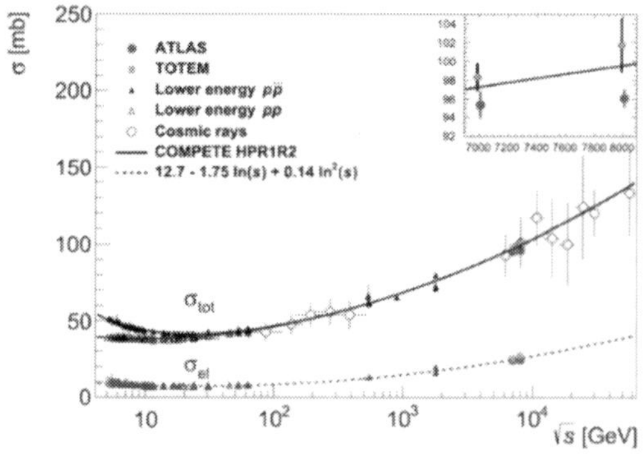

Figure 8. Comparison of total and elastic cross-section measurements presented here with other published measurements [13, 15–19] and model predictions as a function of the centre-of-mass energy.

3.1. Measurement of the inelastic cross-section with the MBTS detector

The inelastic proton-proton cross section was measured utilizing $60\,\mu b^{-1}$ of p-p collisions at a center-of-mass energy of 13 TeV with the ATLAS detector at the LHC [20]. Inelastic interactions were selected using the ATLAS MBTS detector ($2.07 < |\eta| < 3.86$) with the trigger requirement that there is at least one MBTS hit above threshold. The counters are not sensitive to elastic pp scattering and diffractive dissociation processes in which neither proton dissociates into a system, X, of mass $M_X >$ 13 GeV, or equivalently, $\xi = M_X^2/s > 10^{-6}$. The cross-section measurement is reported in this fiducial region, $\xi > 10^{-6}$, and after extrapolation to the total inelastic cross section using models of inelastic interactions.

Two sets of events passing the MBTS trigger requirements were selected: an inclusive set and a single-sided set. The inclusive selection demands at least two MBTS counters. Requiring two hits rather than one substantially reduces background due to collision-induced radiation and activation. To constrain the diffractive component of the cross-section and reduce the uncertainty in extrapolation to σ_{inel}, an additional single-sided selection is defined, requiring hits in at least two counters on one side of the detector and no hits on the other. In the data, 4,159,074 events pass the inclusive selection and 442,192 events pass the single-sided selection. A plot of the ratio of single sided events to inclusive events as a function of $f_D = (\sigma_{SD} + \Sigma_{DD})/\sigma_{inel}$ is given in Fig. 9. To facilitate background studies, data were also collected with the same selection when no proton bunch ("empty") or a single proton bunch from only one of the two beams ("single beam") was passing through ATLAS.

A cross-section of 68.1 ± 1.4 mb is measured in the fiducial region $\xi = M_X^2 > 10^{-6}$, where m_X is the larger invariant mass of any two final state hadronic systems separated by the largest rapidity gap in the event. In this range the MBTS detector is highly efficient. For diffractive events this corresponds to cases where at least one proton dissociates to a system with $M_X > 13$ GeV. The fiducial cross-section is determined from:

$$\sigma_{ineal}^{fid}(\xi > 10^{-6}) = \frac{N - N_{BG}}{\epsilon_{trig} \times \mathcal{L}}\frac{1 - f_{\xi<10^{-6}}}{\epsilon_{sel}} \quad (6)$$

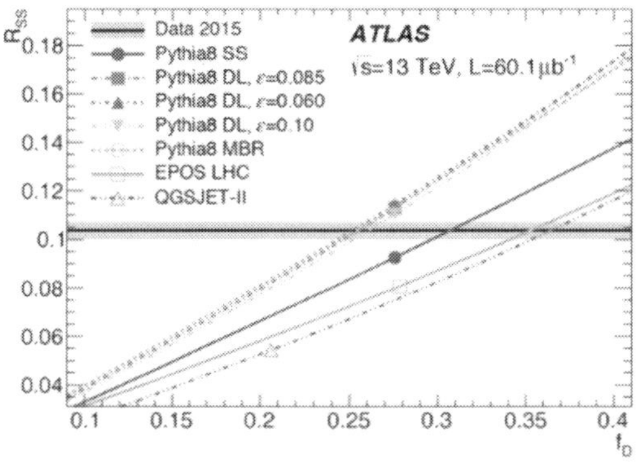

Figure 9. The ratio of the number of single-sided to inclusive events (RSS) as a function of the fraction of the cross section that is diffractive according to each model (fD). The default value of f_D in each model is shown with a marker

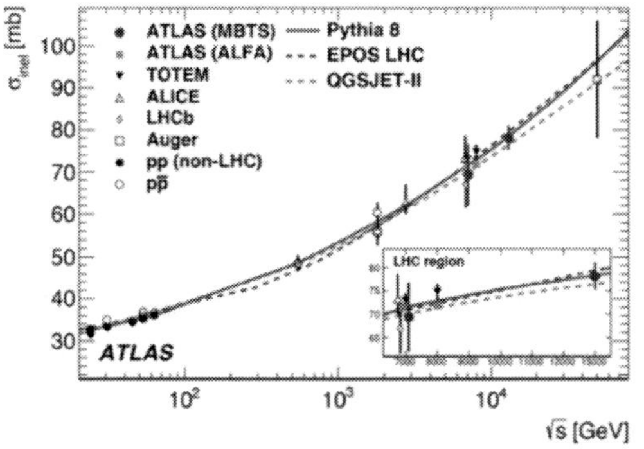

Figure 10. The inelastic proton-proton cross section versus $\sqrt{s}$. Measurements from other hadron collider experiments [13,15, 21–23] and the Pierre Auger experiment [16] are also shown. Some LHC data points have been slightly shifted in the horizontal position for display purposes. The data are compared to the Pythia8, Epos LHC and QGSJet-II MC generator predictions. The uncertainty in the ATLAS ALFA measurement is smaller than the marker size.

where N is the number of events observed passing the inclusive selection, N_{BG} is the number of background events, ϵ_{trig} and ϵ_{sel} are the trigger and event selection efficiencies, respectively, $1 - f_{\xi<10^{-6}}$ accounts for the migration of events with $\xi < 10^{-6}$ into the fiducial region and $\mathcal{L}$ is the integrated luminosity corresponding to the data sample. An inelastic cross-section of $\sigma_{inel} = 78.1 \pm 2.9$ mb is obtained by extrapolating the fiducial cross-section to the full phase space

The LHC experiments have measured σ_{inel} with $\sqrt{s}$ [13] from $\sqrt{s} = 6$ to 13 TeV [15,21,24–28]. ALICE has measured σ_{inel} at $\sqrt{s} = 2.76$ TeV [22]. The Pierre Auger Collaboration measured the inelastic p-air cross section at $\sqrt{s} = 57$ TeV and extracted σ_{inel} using the Glauber model [16]. The measured cross section agrees well with a variety of theoretical predictions and is clearly consistent with the continued rise of the inelastic cross section with centre-of-mass energy, Fig. 10.

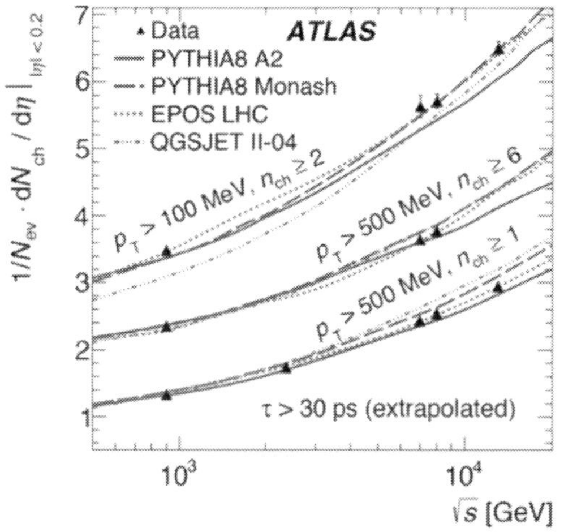

Figure 11. The average primary charged-particle multiplicity in p-p interactions per unit pseudorapidity η for $|\eta| < 0.2$ as a function of $\sqrt{s}$. The data are shown as black triangles with vertical error bars representing total uncertainty. They are compared to various MC predictions shown as coloured lines.

4. Charged particle distributions in p-p interactions

ATLAS has performed measurements of distributions of charged particles produced in proton-proton collisions with a centre-of-mass energy $\sqrt{s}$ of of 8 TeV [29] and 13 TeV [30]. The multiplicity, pseudorapidity, and transverse momentum distributions of charged particles are plotted in different kinematic regions and charged particle multiplicity, including measurements of final states at high multiplicity. The results are corrected for detector effects and are compared to the predictions of various Monte Carlo event generator models which simulate the full hadronic final state.

As an example, a plot of the charged particle multiplicity for different transverse momentum and minimum number of charged tracks is given in Fig 11. The values for the other p-p centre-of-mass energies shown in this plot are taken from previous ATLAS analyses [29,31]. The value for particles with $p_T > 500$ MeV at $\sqrt{s} = 13$ TeV is taken from Ref. [30]. The results have been extrapolated to include charged strange baryons (charged particles with a mean lifetime of 30 ps $< \tau <$ 300 ps).

For the 13 TeV analysis the data recorded by the ATLAS detector corresponded to an integrated luminosity of 151 μb^{-1}. The particles are required to have a transverse momentum greater than 100 MeV and an absolute pseudorapidity $\eta < 2.5$. The measurements performed using proton-proton collisions at $\sqrt{s} = 8$ TeV were obtained in 2012 in a special run with a small number of interactions per beam crossing (below 0.004) corresponding to an integrated luminosity of 160 μb^{-1}. In both cases a minimum-bias trigger obtained from the MBTS detector was utilised to select a data sample.

For both studies, the charged-particle multiplicity, its dependence on transverse momentum and pseudorapidity as well as the dependence of the mean transverse momentum on multiplicity are measured in events containing at least two charged particles satisfying the above kinematic criteria. The results are corrected for detector effects and compared to the predictions from several Monte Carlo event generators.

5. Some implications of ATLAS hadronic physics results for cosmic ray physics

The hadronic physics results from ATLAS shows a very good, in the case of EPOS-LHC, to satisfactory agreement with the various standard Monte Carlo models used in the particle and astroparticle physics arenas. However, all of the models are at least broadly descriptive of the observed ATLAS hadronic physics program data. The ATLAS experiment has recorded hadronic data over the last few years at a centre-of-mass energy that is equivalent to a cosmic ray proton energy of $\sim 10^{17}$ eV - beyond the knee region ($\sim 4 \times 10^{15}$ eV). Consequently, we can rule out the hypothesis that the knee in the cosmic ray spectrum is due to the presence of some strong indication of new hadronic physics threshold, where some 20% of the cosmic ray primary energy would need to be transferred to invisible channels. Such new physics is not indicated in the hadronic data obtained from ATLAS and CMS.

The hadronic results from ATLAS and other LHC detectors such as CMS and LHCf are extremely useful in refining the models used in high-energy air shower physics reducing the required extrapolation factors to a factor of $\sim$40. Possible future Collider results with p-O (oxygen) or p-N (Nitrogen) would be desirable to extend this program. An example of the use of LHC data in the improvement of our understanding of high energy cosmic ray physics is provided by measurements of the inelastic cross-section by ATLAS, CMS, TOTEM and Auger. These measurements have indicated a slower energy rise of the inelastic cross-section than was previously predicted by some popular models of the hadronic physics of air showers utilized by cosmic ray physicists. This new input lead to a reduction in the p-Air cross-section used in the models resulting in an average deepening of the extended air shower maximum in the atmosphere. A consequence of this finding was that the estimation of the composition of primary cosmic rays moved towards heavier elements as the energy of the cosmic ray primary energy increases.

More recently we have seen that information on hadronic physics doesn't only flow from Collider to Cosmic ray Physics. For example, the measurement of the inelastic p-N - and subsequent estimate of the p-p cross-section at $\sqrt{s} = 57$ GeV - performed by Auger, is also a useful tool in assessing various hadronic models at LHC energies. Indeed, it seems that some models, for example EPOS 1.99 (and version above), normally utilized largely for studies of cosmic ray air showers, are proving to be a valuable complement to the models traditionally used for studies at Colliders.

6. The possible use of ATLAS as an underground cosmic muon observatory

Underground measurements, made by Collider detectors, of multiplicities, or charge ratios of muons produced by cosmic rays in the atmosphere, were first published by

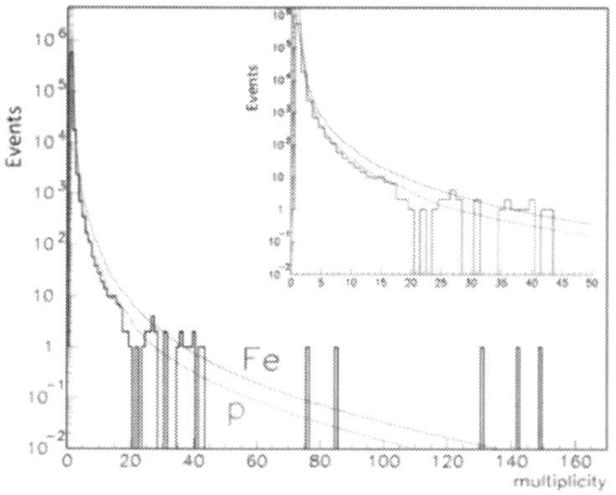

Figure 12. Multiplicity distribution of muons in the TPC compared to CORSIKA simulations for p and Fe as primary particle.

the LEP experiments ALPHA, DELPHI and L3 [32] that together formed the Cosmo-LEP Collaboration.

In particular, multi-muon events where the muon tracks were parallel and grouped together, called muon bundles, were first highlighted by the Cosmo-LEP activity. They found that muon bundles with very large muon multiplicities measured underground were not described by the then current MC generators. In fact, this was the only LEP result that did not agree with Standard Model expectations! Even the combination of extreme assumptions of highest measured flux value and pure iron spectrum, failed to describe the abundance of high multiplicity events. An example of the discrepancy between measurement and Monte Carlo prediction in this arena is shown in Fig. 12. At the LHC ALICE has already reported similar measurements using their ACORDE sub-detector, as shown in Fig. 13 [33]. The horizontal detection area for the ALICE ACORDE cosmic ray detector of 12.5 m^2 is defined by the ALICE TPC. Again, it can be seen that the MC simulations are struggling to explain the highly multiplicity tail of the muon-bundle multiplicity plot. The ATLAS muon system covers a cylinder 20 m in diameter and 25 m long with a horizontal cross section of approx. 500 m^2. It consists of roughly 12000 m^2 of total detector area and roughly 1 million channels of data readout Several types of muon detectors but 30 mm diameter straw tubes are the main element. Thus, ATLAS provides an unprecedented capability for precision cosmic muon measurement within a magnetic field.

The ACME group of the ATLAS collaboration [34] is preparing a proposal to use the ATLAS-LHC detector in conjunction with a surface array of scintillator detectors, as a facility to study comic ray muon physics. Operating ATLAS in this mode will provide unique data on multi-muon production in cosmic ray air showers in a particularly interesting energy regime. Together with the simultaneous and complementary measurement of the electromagnetic structure of the air showers, ACME will be able to determine the composition of cosmic ray primaries over the energy range 10^{14} eV to a few times 10^{16} eV. In addition, it will provide a sensitive instrument for studying the structure of cosmic ray air showers and to search for new physics in the form of exotic phenomena such as Centauro and anti-Centauro events,

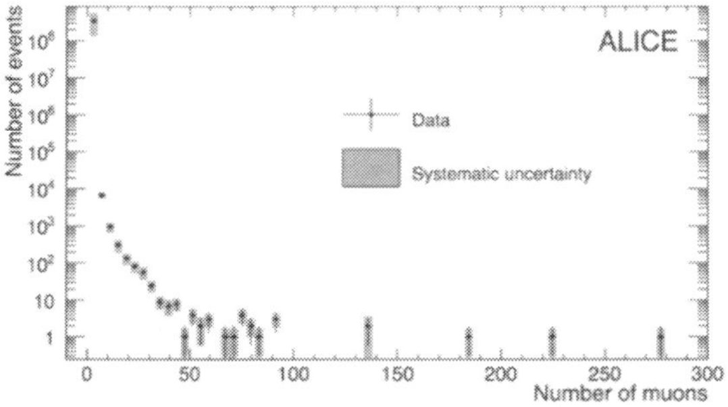

Figure 13. Muon multiplicity distribution of the whole sample of data (2010–2013) corresponding to 30.8 days of data taking.

and, significantly extend the study of the high multiplicity muon excess previously observed, for example, with the Cosmo-LEP detectors. Lastly, the surface array can be used to flag large cosmic ray showers that may contribute backgrounds to exotic physics processes produced in ATLAS collisions.

7. Conclusion

Current models of hadronic interactions are broadly descriptive of the observed ATLAS hadronic physics program data. The ATLAS data recorded over the last year or so were taken at a centre-of-mass energy that is equivalent to a cosmic ray proton energy 10^{17} eV just beyond the knee region (10^{15} eV). Consequently, we can rule out the hypothesis that the knee in the cosmic ray spectrum is due to the effect of some new hadronic physics threshold, where some 20% of the cosmic ray primary energy would need to be transferred to invisible channels. Future ATLAS running at 13 TeV will allow us to further refine our hadronic interaction models. Importantly the turn on of the AFP detector will enable important studies of single and double diffraction, a key input to our overall understanding of hadronic physics at the LHC. Last, but not least, the use of the ATLAS detector as a high precision cosmic-muon observatory would enable the LHC to contribute directly to our understanding of high energy cosmic ray physics across the knee region of the cosmic ray energy spectrum.

References

[1] K. Kotera and A.V. Olinto, Ann. Rev. Astron. Astrophys. **49**, 119 (2011)
[2] A.M. Hillas, Ann. Rev. Astron. Astrophys. **22**, 425 (1984)
[3] J. Blümer, R. Engel, and J. R. Hörandel, Prog. Part. Nucl. Phys. **63**, 293 (2009)
[4] V. Berezinsky, A.Z. Gazizov, and S.I. Grigorieva, Phys. Rev. D **74**, 043005 (2006)
[5] D. Allard, E. Parizot, E. Khan, S. Goriely, and A.V. Olinto, Astron. Astrophys. **443**, L29 (2005)
[6] K. Greisen, Phys. Rev. Lett. **16**, 748 (1966); G. T. Zatsepin and V. A. Kuzmin J. Exp. Theor. Phys. Lett. **4**, 78 (1966)
[7] J. Abraham, Phys. Rev. Lett. **104**, 091101 (2010); D. Allard, Astropart. Phys. **39–40**, 33 (2012); P. Abreu et al., JCAP **1302**, 026 (2013)
[8] K.-H. Kampert and M. Unger, Astropart.Phys. **35**, 660 (2012)
[9] A. Sidoti et al., ATLAS Collaboration, JINST **9**, C10020 (2014)
[10] S. Abdel Khalek et al., ALFA/ ATLAS Collaboration, JINST **11**, P11013 (2016)
[11] S. Grinstein, AFP/ATLAS Collaboration, Nucl. Part. Phys. Proc. **273–275**, 1180 (2016)
[12] M. Aabound et al., The ATLAS Collaboration, Phys. Lett. B **761**, 158 (2016)
[13] K.A. Olive, et al., Particle Data Group, in: Review of Particle Physics, Chin. Underground measurements, made by Collider detectors, of multiplicities or charge ratios of muons produced by cosmic. C **38**, 090001 (2014)
[14] G. Antchev, et al., TOTEM Collaboration, Nucl. Phys. B **899**, 527 (2015)
[15] G. Antchev, et al., TOTEM Collaboration, Europhys. Lett. **101**, 21002 (2013)
[16] P. Abreu, et al., Pierre Auger Collaboration, Phys. Rev. Lett. **109**, 062002 (2012)
[17] G. Aielli, et al., ARGO-YBJ Collaboration, Phys. Rev. D **80** (2009) 092004
[18] M. Honda, et al., Phys. Rev. Lett. **70**, 525 (1993)
[19] R.M. Baltrusaitis, et al., Phys. Rev. Lett. **52**, 1380 (1984)
[20] M. Aaboud et al., ATLAS Collaboration, Phys. Rev. Lett. **117**, 182002 (2016)
[21] G. Antchev, et al., TOTEM Collaboration, Phys. Rev. Lett. **111**, 012001 (2013)
[22] B. Abelev et al., ALICE Collaboration, Eur. Phys. J. C **73**, 2456 (2013)
[23] LHCb Collaboration, R. Aaij et al., JHEP **02**, 129 (2015)
[24] M. Aaboud et al., ATLAS Collaboration, Nucl. Phys. B **889**, 486 (2014)
[25] G. Antchev et al., TOTEM Collaboration, Eur. Phys. J. **C76**, 661 (2016)
[26] M. Aaboud et al., ATLAS Collaboration. Phys. Lett. B **761**, 158 (2016)
[27] G. Aad, ATLAS Collaboration, Nature Commun. **2**, 463 (2011)
[28] S. Chatrchyan, CMS Collaboration, Phys. Lett. B **722**, 5 (2013)
[29] M. Aaboud et al., ATLAS Collaboration, Eur. J. Phys. **C76**, 403 (2016)
[30] M. Aaboud et al., ATLAS Collaboration, Eur, J. Phys. **C76**, 502 (2016)
[31] G. Aad et al., ATLAS Collaboration, New J. Phys. **13**, 053033 (2011)
[32] C. Avati et al., CORAL Collaboration, CERN/2001-003, SPSC/P231 (2001) and references therein
[33] J. Adam et al., ALICE Collaboration, JCAP **1601**, 032 (2016)
[34] J.L. Pinfold, Proceedings of the 13th ICATPP Conference, Villa Olmo, Como, Italy, 3-7 October 2011, DOI: `http://dx.doi.org/10.1142/9789814405072_0158`

Forward energy measurement with CMS

Lev Kheyn[a] for the CMS Collaboration

Skobeltsyn Institute of Nuclear Physics, Lomonosov Moscow State University, Russia

Abstract. Energy flow is measured in the forward region of CMS at pseudorapidities up to 6.6 in pp interactions at 13 TeV with forward (HF) and very forward (CASTOR) calorimeters. The results are compared to model predictions. The CMS results at different centre-of-mass energies are compared using the pseudorapidity variable shifted by the beam rapidity, thus studying applicability of the hypothesis of limiting fragmentation.

1. Introduction

An almost hermetic calorimetric instrumentation of the CMS experiment extends from $\eta = -6.6$ to 5.2 (for neutral particles also at $|\eta| > 8.1$). This makes it possible to perform energy measurements in a wide range of pseudorapidities. Such measurements allow to explore basic characteristics of the underlying event and multiple parton interactions (MPI) as a function of pseudorapidity and center-of-mass energy. For models used in cosmic ray physics, data at collider energies are an important reference necessary for extrapolating to highest energies.

At the LHC, energy flow measurements have been performed by ATLAS, LHCb, and CMS collaborations. ATLAS and LHCb performed measurements in pp collisions at $\sqrt{s} = 7$ TeV in the pseudorapidity ranges $|\eta| < 4.8$ [1] and $1.9 < |\eta| < 4.9$ [2], respectively. CMS performed studies in pp collisions at 0.9 and 7 TeV [3], as well as in collisions of lead ions at 2.76 TeV [4] and proton-lead-ion collisions at 5.02 TeV [5].

In this work, we present results of measurements of the dependence of the energy flow on pseudorapidity at $3.15 < |\eta| < 6.6$ performed with use of the HF and CASTOR calorimeters [6], and the energy distribution in the CASTOR calorimeter [7].

2. CMS detector

A central part of the Compact Muon Solenoid (CMS) apparatus is a superconducting solenoid of 6 m internal diameter, providing a strong magnetic field of 3.8 T. Within the field volume are an inner silicon pixel and strip tracker that measures charged particles in the range $|\eta| < 2.5$, a homogeneous lead tungstate crystal electromagnetic calorimeter, and a brass and scintillator hadron calorimeter. Muons are measured in gas-ionization detectors embedded in the steel return yoke. Extensive forward calorimetry complements the coverage provided by the barrel and endcap detectors. For more details on CMS see [8].

Cherenkov calorimetry was chosen in CMS for the forward detectors. The HF calorimeters cover $2.9 < |\eta| < 5.2$ and use quartz fibers embedded within a steel absorber. The HF is segmented in 13 towers in η, each with a size given by $\Delta\eta \simeq 0.175$, except for the lowest and highest-$|\eta|$ towers with $\Delta\eta \simeq 0.1$ and $\Delta\eta \simeq 0.3$, respectively. The azimuthal segmentation of all towers is $10°$, except for the one at highest-$|\eta|$, which has $\Delta\varphi = 20°$. The very forward angles are covered by the CASTOR ($-6.6 < \eta < -5.2$) calorimeter which is made of quartz plates embedded in tungsten absorbers. The CASTOR has 16 azimuthal sectors (22.5° each) and 14 longitudinal modules (2 modules in the electromagnetic section, 10 radiation lengths each, and 12 modules in the hadronic section, twice the depth of electromagnetic modules each). The overall longitudinal depth of both CASTOR and HF corresponds to 10 nuclear interaction lengths.

3. Event selection

Two different strategies are used for selection of inelastic collision events: an inclusive selection, allowing significant fraction of diffractive events, and a non-single-diffractive selection where single-diffractive contribution is effectively suppressed. This is achieved by requiring activity in HF either on at least one side (for the former) or on both sides (for the latter) with respect to the nominal interaction point of CMS.

In the analysis, activity in HF is identified by presence of a reconstructed calorimeter tower with energy well above noise level. For selection of the inclusive inelastic events, the cut 5 GeV is used. For the non-single-diffractive selection, due to reduced sensitivity to noise, a lower cut of 4 GeV is chosen.

The data are corrected to the hadron level using Pythia8 Monash, Pythia8 MBR, EPOS-LHC and QGSJET-II-04 MC generators together with a simulation of the CMS detector based on GEANT4. The hadron level energy is a sum of the energies of stable particles, with exclusion of muons and neutrinos, within the pseudorapidity range of the measurement ($3.15 < \eta < 6.6$) without any threshold. Corrections are applied on a bin-by-bin basis. The correction factor is calculated as a ratio of the average MC predictions at the hadron level to the average predictions at the detector level. Final correction factor is the average of the correction factors for four different MC samples.

[a] e-mail: kheyn@mail.cern.ch

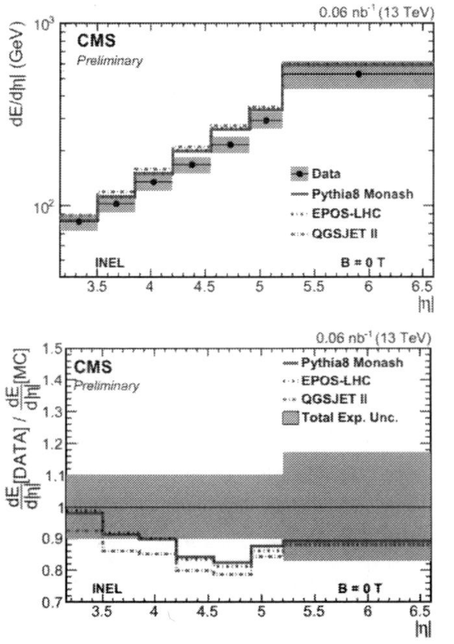

Figure 1. A comparison of the energy flow measurements as a function of pseudorapidity at 13 TeV to MC models [6]. The gray band shows the total systematic uncertainty correlated across pseudorapidity bins. The lower panel shows the ratio of measured data to MC predictions.

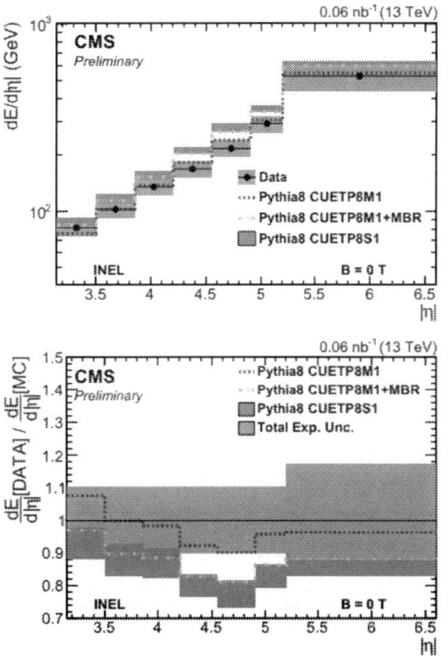

Figure 2. A comparison of the energy flow measurements as a function of pseudorapidity at 13 TeV for soft-inclusive-inelastic events to MC models at particle level [6]. The gray band shows the total systematic uncertainty correlated across pseudorapidity bins. The red band corresponds to the envelope of the uncertainties of the parameters of Pythia8 CUETP8S1 tune. The lower panel shows the ratio of measured data to MC predictions.

Selection of events on the hadron level is done differently for soft-inclusive-inelastic sample and for non-single-diffractive sample. Soft-inclusive-inelastic events are selected with use of the cut $\xi = \max(\xi_X, \xi_Y) > 10^{-6}$. Here $\xi_X = M_X^2/s$ and $\xi_Y = M_Y^2/s$ where M_X and M_Y are invariant masses of two systems of stable particles that are separated by the largest rapidity gap in the event. Non-single-diffractive events are selected with a requirement of at least one particle except muon or neutrino within η acceptance of the HF calorimeters at both sides. Also, for comparison to previously published results [3], non-single-diffractive events are selected with a requirement of at least one charged particle at both sides of the CMS in the pseudorapidity range $3.9 < |\eta| < 4.4$.

A dominant systematic effect which limits accuracy of the measurement of the forward energy flow is the global energy scale uncertainty of the forward calorimeters, which is estimated to be $\pm 10\%$ and $\pm 17\%$ for HF and CASTOR calorimeters, respectively.

4. Results and discussion

The energy and transverse energy densities, $dE/d\eta$ and $dE_T/d\eta$, measured using the CMS HF and CASTOR forward calorimeters and corrected to the particle level, are presented in the pseudorapidity range $3.15 < |\eta| < 6.6$.

A comparison of the measured energy flow to MC model predictions for the soft-inclusive inelastic selection is shown in Figs. 1–2. The gray band represents the total systematic uncertainty correlated across pseudorapidity bins. The statistical uncertainties are negligible and not shown. The red band indicates the envelope of the uncertainties of the parameters of Pythia8 CUETP8S1 tune. In the upper panels the comparison is shown in absolute values, while in the lower panels the ratios quantify the agreement of the measurements and MC predictions.

There is a large spread of the model predictions in the region of the HF coverage, $3.15 < |\eta| < 5.20$. At low $|\eta|$ predictions are consistent with the data, most of the models overestimate the measurements towards higher $|\eta|$ values, and become again closer to the data in the CASTOR bin, $5.2 < |\eta| < 6.6$. The best description of the data is provided by the Pythia8 tune CUETP8M1.

In Figs. 3–4, energy flow measurements are confronted to the MC model predictions for the non-single-diffractive-enhanced event selection. In the upper panels the comparison is shown in absolute values, while in the lower panels the ratios quantify the agreement of the measurements and MC predictions. For non-single-diffractive-enhanced events, the spread of the model predictions is smaller than that for soft-inclusive-inelastic events. The EPOS-LHC and QGSJETII-04 MC generators provide good overall description of the data in the whole $|\eta|$ range of the measurements. The Pythia8 tune CUETP8M1 provides satisfactory description of the data, significantly deviating from the measurements only at low $|\eta|$.

In the lower panel of Figs. 3–4, the ratio is also shown of the energy flow for non-single-diffractive- enhanced to that for soft-inclusive-inelastic events. No significant difference in the slope of the pseudorapidity dependence is observed.

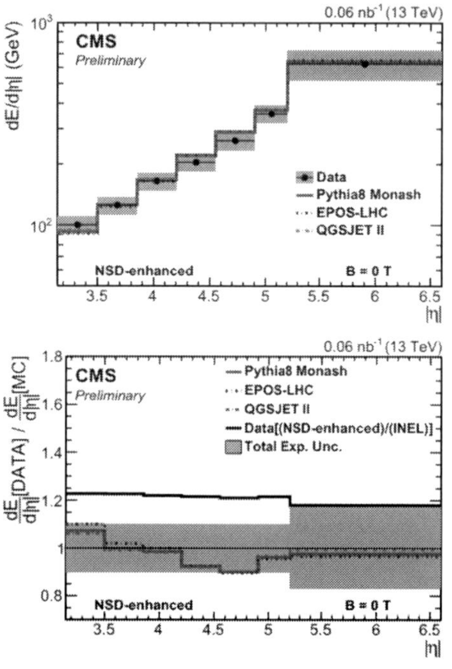

Figure 3. A comparison of the energy flow measurements as a function of pseudorapidity at 13 TeV for non-single-diffractive-enhanced events to MC models at particle level [6]. The gray band shows the total systematic uncertainty correlated across pseudorapidity bins. The lower panel shows the ratio of measured data to MC predictions and, as a solid black line, ratio of the results for non-single-diffractive-enhanced (NSD-enhanced) and soft-inclusive-inelastic (INEL) events.

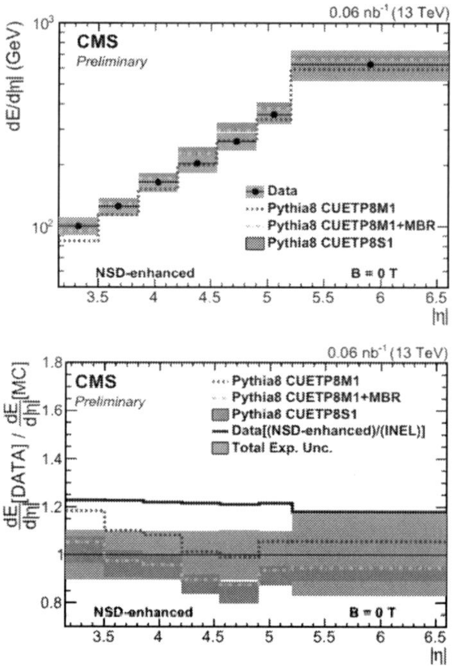

Figure 4. A comparison of the energy flow measurements as a function of pseudorapidity at 13 TeV for non-single-diffractive-enhanced events to MC models at particle level [6]. The gray band shows the total systematic uncertainty correlated across pseudorapidity bins. The red band corresponds to the envelope of the uncertainties of the parameters of Pythia8 CUETP8S1 tune. The lower panel shows the ratio of measured data to MC predictions and, as a solid black line, ratio of the results for non-single-diffractive-enhanced (NSD-enhanced) and soft-inclusive-inelastic (INEL) events.

Figure 5 presents a comparison of the transverse energy density, $dE_T d\eta'$, with published CMS data at lower $\sqrt{s}$ and results at 13 TeV with models as a function of shifted pseudorapidity variable $\eta' = \eta - y_{beam}$. Non-single-diffractive events are selected with requirement of at least one charged particles in the range $3.9 < |\eta| < 4.4$ on each side of CMS. Here y_{beam} stands for the beam rapidity.

The models provide overall reasonable description of the data. The results at different energies are shown to be consistent in terms of shifted pseudorapidity variable, $\eta - y_{beam}$, and consistent with the hypothesis of limiting fragmentation.

In Figure 6, the distribution of the total energy in the CASTOR calorimeter is shown for soft-inclusive-inelastic events. Despite significant systematic uncertainties, mainly due to the energy scale, model predictions show clear differences with the measured data. The bump at a few hundred GeV is also visible in all model predictions. It is best described by QGSJETII-04. The Pythia8 tunes tend to overestimate the contribution of this soft part to the spectrum. The high energy tail on the other hand is described by all models within the experimental uncertainties. Only QGSJETII-04 has a too flat total and hadronic spectrum in the range $0.5\,\text{TeV} < E < 1.8\,\text{TeV}$ and a too steeply falling spectrum above 1.8 TeV. The data is also very sensitive to MPI and the underlying event.

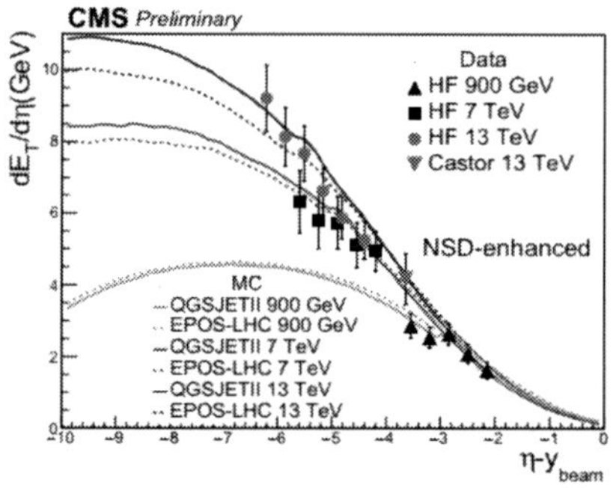

Figure 5. A comparison of the measured hadron level transverse energy density $dE_T d\eta'$ to the published CMS measurements and several models at different $\sqrt{s}$ as function of the shifted pseudorapidity variable, $\eta - y_{beam}$ [6]. The hadron level definition is according to the CMS published results [3].

5. Summary

A measurement of the energy flow is presented in the pseudorapidity range $3.15 < |\eta| < 6.6$ in proton-proton collisions for the centre-of-mass energy of $\sqrt{(s)} = 13\,\text{TeV}$. The energy flow is studied as a function of pseudorapidity for soft-inclusive-inelastic and non-single-diffractive-enhanced events. The results are compared to several Pythia8 tunes, EPOS-LHC and QGSJETII-04 hadronic interaction models, commonly used in cosmic ray physics, and earlier pp data at $\sqrt{s} = 900\,\text{GeV}$ and 7 TeV. The considered models provide in general reasonable

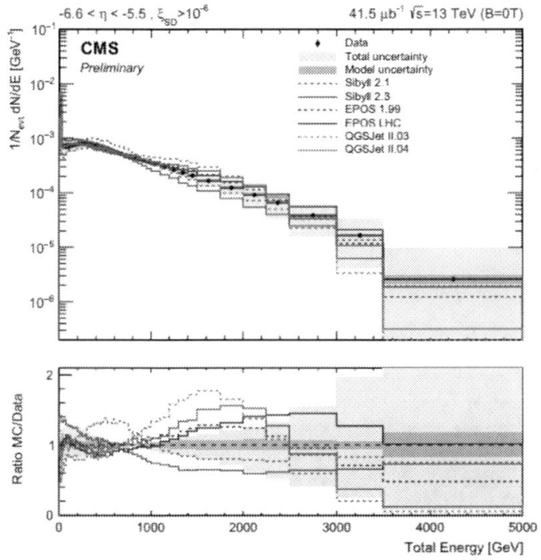

Figure 6. Normalized total energy spectrum in the CASTOR for events with $\xi > 10^{-6}$ [7].

description of the measured energy flow at the highest energy achieved so far at the LHC. The spread of the model predictions for soft-inclusive-inelastic events is large and becomes smaller for non-single-diffractive-enhanced events. The results at different energies are shown to be consistent in terms of shifted pseudorapidity variable $\eta' = \eta - y_{beam}$, and reveal overall consistency with the hypothesis of limiting fragmentation in a wide range of pp collision energies from 900 GeV to 13 TeV.

Model predictions show clear differences from the measurements in the distribution of the total CASTOR calorimeter energy.

References

[1] ATLAS Collaboration, JHEP **11** (2012) 033, arXiv:1208.6256
[2] LHCb Collaboration, Eur. Phys. J. C **73** (2013) 2421, arXiv:1212.4755
[3] CMS Collaboration, JHEP **11** (2011) 148, JHEP02 (2012) 055, arXiv:1110.0211. [Erratum: JHEP02, 055 (2012)]
[4] CMS Collaboration, CMS Physics Analysis Summary CMS-PAS-HIN-12-006, 2012
[5] CMS Collaboration, CMS Physics Analysis Summary CMS-PAS-HIN-14-014, 2015
[6] CMS Collaboration, CMS Physics Analysis Summary CMS-PAS-FSQ-15-006, 2015
[7] CMS Collaboration, CMS Physics Analysis Summary CMS-PAS-FSQ-16-002, 2016
[8] CMS Collaboration, JINST **3** (2008) S08004

Unexpected behaviour of cross sections of high energy protons

I.M. Dremin[1,2,a]

[1] Lebedev Physics Institute, Moscow 119991, Russia
[2] National Research Nuclear University "MEPhI", Moscow 115409, Russia

Abstract. Nobody expected that protons will try to keep their entity in their collisions at higher and higher energies. However, the comparison of experimental results on proton-proton interactions from ISR to LHC energies clearly shows this tendency with a definitely increasing share of the elasic scattering cross section. This unexpected behaviour immediately leads to the unexpected corollary about the shape of the spatial interaction region of the two protons. The unquestionable principle of unitarity combined with available experimental data on elastic scattering is used to get new conclusions about this shape. Its evolution at present energies from the ISR to the LHC and predictions about its behaviour at ever higher energies are considered. The shape can transform rather drastically if the share of elastic processes keeps rising. The possible origin of the effect and its interrelation to the strong interaction dynamics are speculated. Cosmic ray studies can help in clarifying this behaviour.

Experimental data on proton-proton interactions in high energy collisions show quite a special and unexpected behaviour of the share of elastic scattering compared to inelastic processes with increasing energy. It decreases at the beginning (at comparatively low energies) but then starts to increase. From Intersecting Storage Rings (ISR) energies of 20–60 GeV up to higher energies 7–13 TeV at the Large Hadron Collider (LHC) it increases by more than a factor 1.5! According to intuitive classical ideas we would expect a stable tendency with increasing share of the break-down of protons compared to their survival probability.

If a cup falls to the floor, it breaks into pieces but sometimes stays intact. The harder it hits the floor, the less chance to be unbroken.

If two high energy protons collide, many new particles (mostly pions) are produced, but sometimes they scatter elastically and retain their entity. It is surprising enough that at very high collision energies the proportion of elastic processes increases with increasing energy from the ISR to the LHC.

One can assume that either the asymptotic freedom or the extremely short time of flight of high energy protons through each other are in charge of such a surprising effect. The unquestionable principle of unitarity combined with the available experimental data on elastic scattering is used to get new conclusions about the shape of the interaction region of colliding protons. Its evolution at present energies is considered. Some predictions about its behaviour at even higher energies are described with different assumptions on the relative roles of elastic scattering and inelastic processes. The shape can transform rather drastically if the proportion of elastic processes keeps rising. This unexpected property leads to an unexpected corollary. The possible origin of the effect and its interrelation to the strong interaction dynamics are speculated.

Our main goal here is to get some knowledge about the spatial region of interactions of high energy protons at current energies, to draw a pictorial view of its evolution with increasing energy and to discuss possible theoretical and experimental implications of these findings.

The most stringent and reliable information (albeit rather limited!) about the interrelation of elastic and inelastic processes comes from the unitarity of the S-matrix

$$SS^{+} = 1 \qquad (1)$$

or for the scattering matrix T ($S = 1 + iT$)

$$2\mathrm{Im}T_{ab} = \Sigma_n \int T_{an}T^{*}_{nb}d\Phi_n, \qquad (2)$$

where a, b, n denote the number of particles. The whole n-particle phase space Φ_n is integrated over. For the elastic scattering amplitude $a = b = 2$, the unitarity condition relates the amplitude of elastic scattering $f \propto T_{22}$ to the amplitudes of n-particle inelastic processes T_{2n} declaring that the total probability of all outcomes of the interaction (elastic and inelastic) must be equal to 1[1]. Information about the elastic scattering of protons comes from the measurement of the differential cross section $d\sigma/dt$ at some energy, s, as a function of the transferred momentum, t, at its experimentally accessible values. It is related to the scattering amplitude, $f(s, t)$, in the following way

$$\frac{d\sigma}{dt} = |f(s, t)|^2. \qquad (3)$$

The variables s and $-t$ are respectively the squared total energy, $2E$, and squared transferred momentum of

[a] e-mail: dremin@td.lpi.ru

[1] The non-linear contribution from the elastic amplitude appears in the right-hand side for $n = 2$.

DOI: 10.1051/epjconf/201714510003

Table 1. The energy behaviour of ζ, $G(s,0)$ and $\sigma_{inel}/\sigma_{el}$.

$\sqrt{s}$, GeV	4.11	4.74	7.62	13.8	62.5	546	1800	7000
ζ	0.98	0.92	0.75	0.69	0.67	0.83	0.93	1.00-1.04
$G(s,0)$	1.00	0.993	0.94	0.904	0.89	0.97	0.995	1.00
					ISR	$Sp\bar{p}S$	FNAL	LHC
$\sigma_{inel}/\sigma_{el}$					5			3

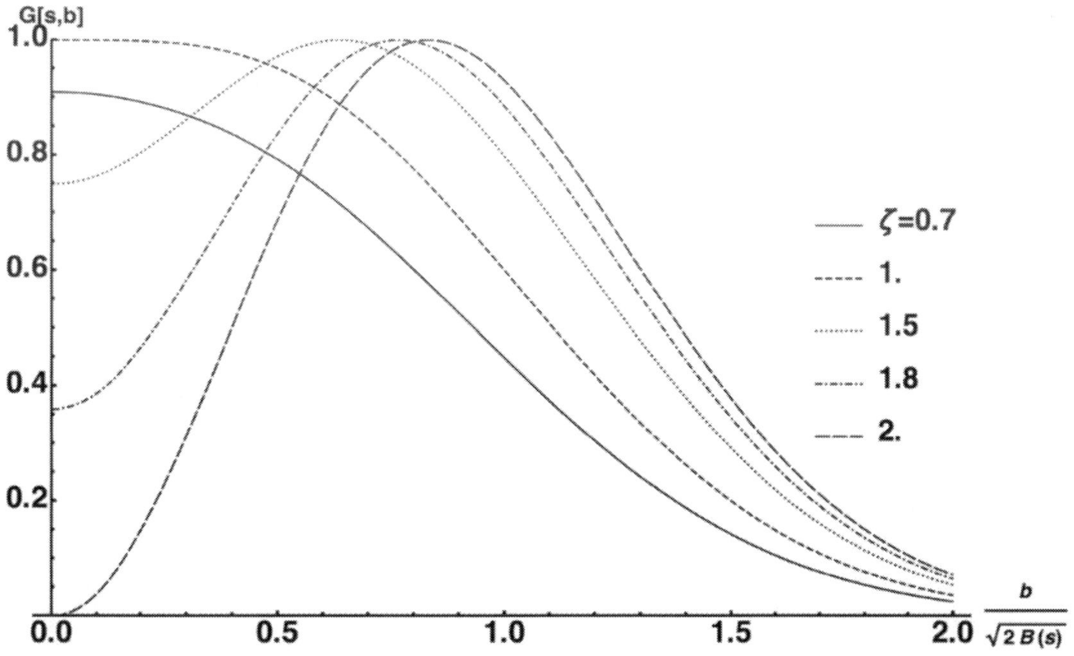

Figure 1. The energy evolution of the shape of the inelastic interaction region for different values of the survival probability $\zeta/4$. The values $\zeta = 0.7$ and 1.0 correspond to ISR and LHC energies and agree well with the result of detailed fitting to the elastic scattering data [4–6]. A further increase of ζ leads to the change of the BEL regime (Blacker, Edgier, Larger - [7]) by the toroid-like shape with a dip at $b = 0$ (TEH regime–Toroidal Elastic Hollow- [8–11]) The values $\zeta = 1.5$ are proposed in [12,13] and $\zeta = 1.8$ in [14] as describing the asymptotical regimes. The value $\zeta = 2$ corresponds to the so-called "black disk" regime with equal elastic and inelastic cross sections ($\sigma_{el} = \sigma_{inel} = 0.5\sigma_{tot}$).

the two colliding protons in the center-of-mass system $s = 4E^2 = 4(p^2 + m^2)$ (p is the proton momentum) and $-t = 2p^2(1-\cos\theta)$ at the scattering angle, θ. In the s-channel this unquestionable condition is usually expressed in the form of the well known integral relation (for more details see [1,2]). This relation is quite complicated for arbitrary values of the transferred momentum, t. However, for forward scattering at $t = 0$ it leads to the widely used optical theorem showing the normalization of the imaginary part of the amplitude $\mathrm{Im}f(s)$ by its direct connection with the total cross section σ_{tot}:

$$\mathrm{Im}f(s,0) = \sigma_{tot}(s)/4\sqrt{\pi} \tag{4}$$

and to the general statement that the total cross section is the sum of cross sections of elastic and inelastic processes

$$\sigma_{tot} = \sigma_{el} + \sigma_{inel}, \tag{5}$$

i.e., that the total probability of all processes equals 1.

One can use the Fourier – Bessel transform of the amplitude, f, to reduce the integral relation to the more simple algebraic one. This transformation retranslates the momentum data to the shortest transverse distance between the trajectories of the centers of colliding protons called the impact parameter, b, and is written as

$$i\Gamma(s,b) = \frac{1}{2\sqrt{\pi}}\int_0^\infty d|t| f(s,t) J_0(b\sqrt{|t|}). \tag{6}$$

Then the unitarity condition in the b-representation reads

$$G(s,b) = 2\mathrm{Re}\Gamma(s,b) - |\Gamma(s,b)|^2 \tag{7}$$

(for reviews see, e.g., Refs. [2,3]). This relation establishes the connection between the distributions of the intensity of all processes in the transverse configuration space:

$$\frac{d^2\sigma_{inel}}{db^2} = \frac{d^2\sigma_{tot}}{db^2} - \frac{d^2\sigma_{el}}{db^2}. \tag{8}$$

The left-hand side in Eqs. (7), (8) describes the transverse impact-parameter profile of inelastic collisions of protons. It satisfies the inequalities $0 \leq G(s,b) \leq 1$ and determines how absorptive the interaction region is at the given impact parameter (with $G = 1$ for the full absorption and $G = 0$ for the complete dominance of elastic scattering). The profile of elastic processes is determined by the subtrahend in Eqs. (7), (8). Thus we get a spatial view of the whole process if the elastic scattering amplitude f is integrated in Eq. (6).

In the following, we use the exponential parametrization of the imaginary part of the amplitude, f, to proceed with analytical calculations and argue that it is very precise:

$$\text{Im} f(s,t) = \frac{\sigma_{tot}(s)}{4\sqrt{\pi}} \exp[B(s)t/2]. \quad (9)$$

Formally, this approximation is not valid for differential cross sections at large transferred momenta. However, for our purposes only the integral contribution of f at large $|t|$ to Eq. (6) is important. It is negligibly small there compared to the peak of the diffraction cone.

For central collisions $b = 0$ the unitarity condition (7) is written as

$$G(s, b = 0) = \zeta(2 - \zeta) \quad (10)$$

where

$$\zeta(s) = \frac{4\sigma_{el}(s)}{\sigma_{tot}(s)}. \quad (11)$$

Thus, according to the unitarity condition (Eq. 10) the darkness of the inelastic interaction region for central collisions (absorption) is defined by the only experimentally measured parameter $\zeta(s)$ depending on energy. It has the maximum $G(s, 0) = 1$ for $\zeta = 1$ (see Table 1 for the LHC and Figure 1). Any decline of ζ from 1 ($\zeta = 1 \pm \epsilon$) results in the parabolic decrease of the absorption ($G(s, 0) = 1 - \epsilon^2$), i.e., to an even much smaller decline from 1 for small ϵ. The elastic profile, equal to ζ^2 in central collisions, also reaches the value 1 for $\zeta = 1$.

Let us rewrite Eq. (10) as

$$\zeta(s) = 1 \pm \sqrt{1 - G(s, 0)}. \quad (12)$$

The critical value $\zeta = 1$ reveals itself in the usage of different signs in front of the square root term (different branches of the unitarity condition) for $\zeta < 1$ and $\zeta > 1$.

The shape of the inelastic profile at any b is given by

$$G(s, b) = \zeta \exp\left(-\frac{b^2}{2B}\right)\left[2 - \zeta \exp\left(-\frac{b^2}{2B}\right)\right]. \quad (13)$$

It is shown in Fig. 1 for different ζ. The drastic change of the shape of the inelastic interaction region is seen. It becomes completely dark for central collisions up to LHC energies and can develop a hollow in there at higher energies. The elastic scattering of protons can become prevailing in the central region. It can have some implications for extensive air showers at energies above 10^{15} eV in the laboratory system that correspond to ISR energies where the effect of the increasing share of elastic scattering becomes observable. The detailed analyses of such extensive air showers can, probably, say something about "escaping" high energy protons.

To conclude, it has been shown that an unexpected property leads to an unexpected corollary. This is the uncommon shape of the interaction region and the transition to another branch of the unitarity condition.

No reliable explanation of these findings exists yet.

References

[1] K.A. Olive (PDG Collaboration), China Phys. C **38**, 090513 (2014)
[2] I.M. Dremin, Physics-Uspekhi **56**, 3 (2013)
[3] I.M.Dremin, Physics-Uspekhi **58**, 61 (2015)
[4] U. Amaldi et al., Phys. Lett. B **44**, 112 (1973)
[5] I.M. Dremin, V.A. Nechitailo, Nucl. Phys. A **916**, 241 (2013)
[6] A. Alkin, E. Martynov, O. Kovalenko, S.M. Troshin, Phys. Rev. D **89**, 091501 (2014) (R)
[7] R.Henzi, P. Valin, Phys. Lett B **132**, 443 (1983)
[8] I.M. Dremin, JETP Lett. **99**, 243 (2014)
[9] I.M. Dremin, S.N. White, The interaction region of high energy protons; `arXiv 1604.03469`
[10] S.N. White, Talk at the conference QCD at Cosmic Energies, Chalkida, Greece, May 2016; http://www.lpthe.jussieu.fr/cosmic2016/TALKS/White.pdf
[11] I.M. Dremin, Int. J. Mod. Phys. A **31**, 1650107 (2016)
[12] A.K. Kohara, E. Ferreira, T. Kodama, Eur. Phys. J. C **74**, 3175 (2014)
[13] D.A. Fagundes, M.J. Menon, P.V.R.G. Silva, Nucl. Phys. A **946**, 194 (2016)
[14] S.M. Roy, A two component picture for high energy scattering: unitarity, analitycity and LHC data; `arXiv:hep-ph, 1602.03627`

Applying universality in the development of cascade processes for the research of high energy cosmic particles in space experiments

Anastasiya Fedosimova, Pavel Kharchevnikov, Igor Lebedev[a], and Abzal Temiraliev

Institute of Physics and Technology, Almaty, Kazakhstan

Abstract. An approach for measuring the energy of high-energy particles by a thin calorimeter, is presented. The method is based on the universality in the development of cascade processes. For measurements of the primary energy of cosmic ray particles, the correlation analysis of the dependence of the number of secondary particles, N_e, at the observation level and the relation of the number of particles, dN, at two levels, divided by an absorber layer, is used. It is shown, that the use of correlation curves ($logN_e$ versus dN) allows to essentially reduce errors in determining the energy of the primary particle, which are connected with the uncertainty in the nature of the primary nucleus and with fluctuations in the development of the cascade process. Uncertainties of energy reconstruction on the basis of the correlation curves methodology, is less than 10 percent.

1. Introduction

Research of the characteristics of cosmic rays (CR) on the basis of "direct" measurements outside the atmosphere on spacecraft or high-altitude balloons can allow us to solve many problems of particle physics, cosmic ray physics and astrophysics. Consequently, this region has a significant interest and dynamism to the development of theoretical and experimental studies worldwide [1].

Direct experiments on high-altitude balloons, JACEE and RUNJOB, which were aimed to study the elemental composition of cosmic rays, gave very similar spectra of protons, but very different spectra of helium nuclei [2]. Data on heavier nuclei have weak statistics due to the relatively high energy threshold of the applied methodology. Great hopes were pinned on the ATIC [3] and CREAM [4] experiments, however, contradictory data were obtained from the spectral indices of the basic elements of cosmic rays, which do not provide a consistent picture of the processes occurring in the sources of cosmic rays and their propagation to the Earth.

The main advantage of direct experiments is the ability to measure the charge of the incident particle. It is much harder to use an energy detector for particles with energies of $E > 10^{12}$ eV outside the atmosphere.

For particles at relatively low energies, $E < 10^{12}$ eV, one can use different techniques such as threshold Cherenkov counters, the method of the RICh detector or the technique of a magnetic mass spectrometer. At higher energies, principal threshold effects, limiting their application, arise. The only best option remaining for energy measurements of different nuclei in a wide energy range (at $E > 10^{12}$ eV) is, at present, the ionization calorimeter method [5].

[a] e-mail: `lebedev692007@yandex.ru`, `lebedev@sci.kz`

The technical realization of modern ionization calorimeters can be various, but the idea remains invariable: the primary particle enters into a dense substance (for example, iron or lead), in which numerous nuclear and electromagnetic interactions take place giving rise to a cascade of secondary particles.

For measuring the characteristics of the cascade the dense substance is sandwiched between special detectors.

By measuring the signals from these detectors the cascade curve (dependence on the number of particles in the cascade, N_e, versus penetration depth, d, of the cascade in the calorimeter) is formed. If the maximum of a cascade curve is measured, then the energy of a primary particle can be determined.

The main problem with this method of energy measurement is the need for a massive installation (dozens of tons) because the calorimeter must have a sufficiently large depth to build the cascade curve. The huge weight of the installation greatly complicates the possibility of using such a device in space experiments.

In this regard, a more promising approach, for determining the energy on the basis of direct measurements of CR, is the use of a thin calorimeter. In a thin calorimeter the whole cascade of secondary particles is not measured, and only the beginning of the cascade is recorded. The energy is determined on the basis of the analysis of the size of the cascade, because the number of particles in the cascade is almost proportional to the energy of the primary particle [6].

The most interesting example of a thin calorimeter is the detector of the space experiment NUCLEON [7]. To improve the energy reconstruction a combined approach is used: the number of secondary particles in the cascade and additionally their spatial density, are measured.

The accuracy of the energy determination is about 50 percent. The weight of the installation is about 200 kg.

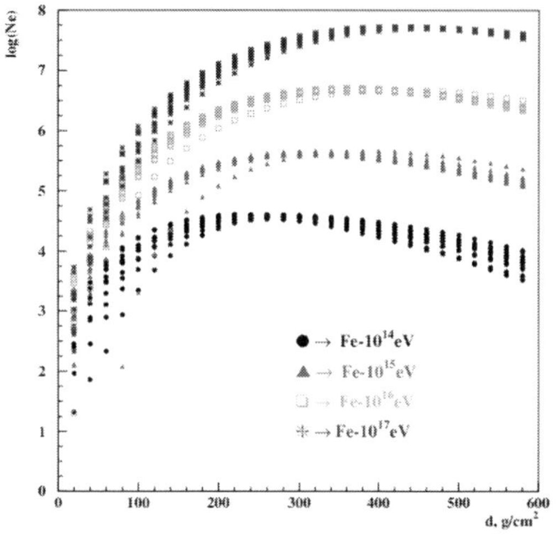

Figure 1. Cascade curves, initiated by iron nuclei at energies 10^{14} eV, 10^{15} eV, 10^{16} eV, 10^{17} eV in air.

In the approach, proposed in our work, the assumption about the universality of the development of the cascade, is used. What is meant?

2. The method of correlation curves

The main problem for energy measurements by a thin calorimeter is due to fluctuations of N_e.

Let's look at Fig. 1, which shows cascade curves, initiated by iron nuclei at energies 10^{14} eV, 10^{15} eV, 10^{16} eV, 10^{17} eV in air. The showers were simulated using CORSIKA QGSJET [9].

As it is seems from Fig. 1, the cascade curves essentially fluctuate and practically merge (not separable) at small values of penetration depth, d. This fact does not allow us to use them for determining the energy of a primary particle on the basis of a thin calorimeter, i.e., on the basis of the limited quantity of measurements on an ascending branch of a cascade curve.

Most experimental groups define the primary energy, E, on the basis of measuring the number of secondary particles, N_e, at the observation level, using the dependence:

$$N_e = aE^b \tag{1}$$

where the parameters a, b, depend on the penetration depth and mass of the primary particle. Statistically the equation works correctly.

However, on the ascending branch of the cascade curve the energy is defined above the real value for quickly developing cascades, and a lower real energy for slowly developing cascades.

What is the reason for such huge fluctuations?

First of all, the development speed of the shower depends on fluctuations of the penetration depth before the first interaction.

In addition, fluctuations in shower development depend on peculiarities of nucleus-nucleus interaction and the initial geometry of the collision. For the research of such peculiarities in multiparticle production, different approaches and methods are used [10–13].

In this paper specific features of nucleus-nucleus interaction are not analyzed. Only the number of particles produced in the first interaction is important.

For example, showers with the multiplicity in the first interaction of M = 100 particles, develop faster than showers with M = 10 particles. After the first interaction, each secondary particle can interact once more. But, when the number of participating particles is larger, the fluctuations in the shower development is less, because fluctuations of individual interactions more or less compensate each other.

The main idea of the approach, proposed in our work, is based on two main positions:

1. The assumption about the universality of the development of the cascade is the following.

   Among the various causes of fluctuations, the greatest effect is achieved from fluctuations of the first (first few) interaction.

   When the number of participating particles is large, the fluctuations of individual interactions more or less compensate each other. Therefore, all the cascades, formed by primary particles of the same mass and energy, starting from a certain moment, develop almost identically.

   This assumption was confirmed by simulation results.
2. The value of the penetration depth is not suitable as a parameter to determine the energy of the primary particles. And so, a new set of correlation parameters must be found which do not depend on fluctuations in the cascade process development.

   The influence of fluctuations in cascade development on the results of measurement of the energy can be essentially reduced by using correlation methods of the analysis of the development of the cascade. It allows to considerably increase the energy measurement accuracy.

   The method of correlation curves has recommended itself successfully in the analysis of extensive air showers [14].

   A modified technique is applied in the given work. The method is developed and tested for the decision of a problem of the measurement of the energy of high-energy particles by thin calorimeters based on computer calculations.

In the technique of correlation curves some internal correlations, which are independent on fluctuations in cascade development, are analyzed.

As a result of the research of the various parameters, characterizing the development of the cascade process, it is revealed that the determination of the energy accuracy can be essentially increased if we use the correlation curve of the dependences of particles number, $logN_e(d1)$, at the observation level, $d1$, versus a relation of the number of particles $dN = logN_e(d1) - logN_e(d2)$ at two levels, ($d1$ and $d2$), divided by a layer of an absorber.

In Fig. 2 the correlation curves for the same showers, as in Fig. 1, are presented.

Correlation curves, as seen from the figure, represent a more ordered picture. Fluctuations of an ascending branch

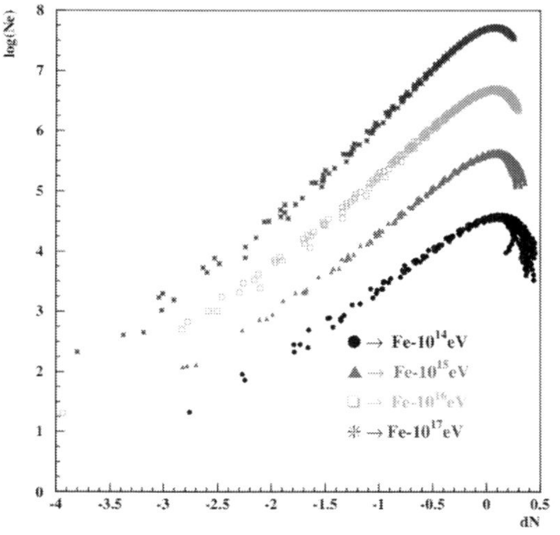

Figure 2. Correlation curves, initiated by iron nuclei at energies 10^{14} eV, 10^{15} eV, 10^{16} eV, 10^{17} eV in air. The thickness of the absorber is $100\,g/cm^2$.

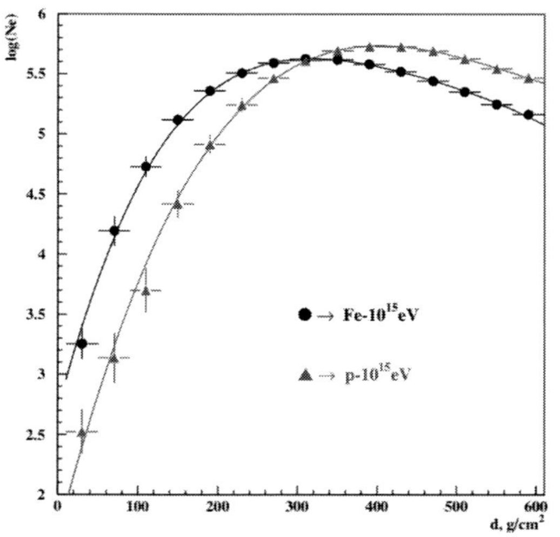

Figure 3. Average cascade curves, initiated iron nuclei and proton at energy 10^{16} eV in the air.

of a correlation curve are not as considerable, as in the case of cascade curves.

Figures 1 looks like a mess at the beginning of the cascade curves but Fig. 2 shows a rather ordered structure in the correlation curves. The negative region of dN corresponds to small values of the penetration depth.

The second most important parameter, influencing the accuracy of the energy measurement, is the uncertainty of the nature of the primary nucleus.

In Fig. 3 average cascade curves of interactions of iron nuclei and protons at a fixed energy 10^{16}eV with the nuclei of air atoms, are presented.

The cascade curves, formed by a proton, are shifted in area of the bigger depths of penetration in comparison with Fe cascade curves. This fact leads to underestimating the energy of proton cascades and to an overestimation for Fe cascades.

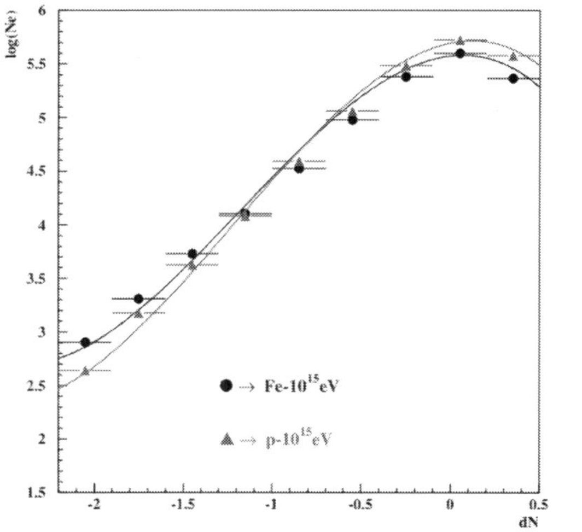

Figure 4. Average cascade curves, initiated by iron nuclei and protons at energy 10^{16} eV in air. The thickness of the air absorber is $100\,g/cm^2$.

Using correlation curves allows to essentially reduce energy definition errors, connected with the uncertainty of the primary nucleus.

Average correlation curves are presented in Fig. 4. As seen from the figure, correlation curves practically coincide for different nuclei. Maximum points of proton and iron correlation curves are shifted in one point $dN = 0$, which is independent from the depth of penetration.

Summing up the given section we once again underline the following: the accuracy of the energy definition on the basis of a thin calorimeter can be essentially increased if we use the correlation curve dependences of the number of particles at the observation level versus a difference of the number of particles at the two levels, divided by an absorber layer.

3. Procedure of the energy reconstruction

The analysis procedure consisted of several basic stages:

1. The simulation of cascades, formed by primary protons and iron nuclei with fixed energies.
2. Fitting the $logN$-dN distributions by polynomial functions of the third order for each fixed energy as a function of $logN$ versus dN in the form:

$$logN(dN) = a_0 + a_1dN + a_2dN^2 + a_3dN^3 \quad (2)$$

3. Fitting the coefficients of a_0, a_1, a_2, a_3 in the dependence on the energy and the construction of the dependence $logN$ on dN and E_0 in the form:

$$logN(dN, E_0) = a_0(E_0) + a_1(E_0)dN + \ldots + a_3(E_0)dN^3 \quad (3)$$

4. The simulation of test cascades, formed by primary protons and iron nuclei with arbitrary energies, and reconstruction of the primary energy on the basis of the dependence of Eq. (3).

To determine the energy of i-th test cascade, using a set of curves (Eq. (3)), it is necessary to vary dN and E_0 so that at the same time minimize the following differences

$$|logN_m - logN(dN, E_0)| < \epsilon \tag{4}$$

$$|dN_m - dN| < \epsilon \tag{5}$$

where $logN_m$ and dN_m are "measured" characteristics of the test cascade at the observation level.

By the definition of the energy of the test cascades it was found that uncertainties in the energy reconstruction, $< |(E_i - E_0)/E_0| >$, in the energy estimation by the formulae 2–5, is less than 10 percent.

The accuracy of the energy determination by the standard technique of cascade curves, in the case of registration of the cascade to the depth of its maximum development, is about 50 percent [6].

This clearly argues in favor of the proposed methodology for determining the primary energy, as it significantly reduces the required thickness of the calorimeter and improves the accuracy of the measurements.

4. Conclusion

An approach for measuring the energy of high-energy particles by a thin calorimeter, is presented.

It is shown, that the use of correlation curves ($logN_e$ versus dN) allows to essentially reduce the errors in the determination of the primary particle energy, which are connected with the uncertainty in the nature of the primary nucleus and with fluctuations in the development of the cascade process.

Uncertainties of energy reconstruction on the basis of the correlation curves methodology, is less than 10 percent.

Further, for a possible technical realization of the project of a thin calorimeter it will be necessary to solve problems connected with the calculation of the installation response, the estimation of the influence of the interaction model on the reconstruction results, etc.

However, now the main result is received: on the basis of computer simulation the correlation parameters, which are independent on fluctuations in shower development and which are almost independent on the uncertainty of the primary nucleus, allow us to determine the energy of the primary nuclei with very high accuracy for thin calorimeters.

References

[1] R. Sparvoli, Nucl. Phys. B - Proc. Suppl. **239–240**, 115–121 (2013)
[2] E.-S. Seo, Astropart. Phys. **76**, 29–40 (2012)
[3] H.S. Ahn et al., Ap.J. **714**, 89–93 (2010)
[4] D. Panov, N.V. Sokolskaya, V.I. Zatsepin, JP: Conf.Ser. **409**, 012036 (2013)
[5] M.I. Panasyuk et al., *The mystery of the origin of galactic cosmic rays* Bull. Moscow Univ., ser.3 **6**, 25–30 (2012)
[6] V.S. Murzin, *Astrophysics of cosmic rays* (MGU, Moscow, 2006) 668p.
[7] V.L. Bulatov at al., *Testing of technological prototype of «NUCLEON» installation on the pion beam* PTE **1**, 34–41 (2010)
[8] D. Heck, J. Knapp, J.N. Capdeville, G. Schatz, T. Thouw, *CORSIKA: A Monte Carlo Code to Simulate Extensive Air Showers* (FZKA 6019, Karlsruhe, 1998) 89p.
[9] M.I. Adamovich et al., *Azimutal correlations of secondary particles in* ^{32}S *induced interactions with Ag(Br) nuclei at 4.5 GeV/c/nucleon* Part. and Nucl. Lett. **101**, 76 (2000)
[10] M.I. Adamovich et al., *Factorial Moments of* ^{28}Si *Induced Interactions with Ag(Br) Nuclei* APH N.S. Heavy Ion Phys. **113** 213 (2001)
[11] I.A. Lebedev et al., J. Phys. G **23**, 637 (1997)
[12] T.N. Kvochkina et al., J. Phys. G **26**, 35 (2000)
[13] A.I. Fedosimova et al., J. Phys.: Conf. Ser. **668**, 012067 (2016)
[14] W.D. Apel et al., *Applying Shower Development Universality to KASCADE data* Astropart.Phys. **29**, N6, 412-419 (2008)

RHIC forward experiment to study $\sqrt{s}$ dependence of forward particle production

T. Sako[1,2,a], O. Adriani[3,4], L. Bonechi[3], R. D'Alessandro[3,4], Y. Goto[5,6], Y. Itow[1,2], K. Kasahara[7], M. Kim[5,8], Y. Makino[1], E. Matsubayashi[1], H. Menjo[9], I. Nakagawa[5,6], J. Park[5,10], N. Sakurai[11], K. Sato[1], R. Seidl[5,6], M. Shinoda[1], T. Suzuki[7], K. Tanida[12], S. Torii[7], A. Tricomi[13,14], M. Ueno[1], and Q.D. Zhou[1]

[1] Institute for Space-Earth Environmental Research, Nagoya University, Nagoya, Japan
[2] Kobayashi-Maskawa Institute for the Origin of Particles and the Universe, Nagoya University, Nagoya, Japan
[3] INFN Section of Florence, Florence, Italy
[4] University of Florence, Florence, Italy
[5] RIKEN Nishina Center for Accelerator-Based Science, Saitama, Japan
[6] RIKEN BNL Research Center, Brookhaven National Laboratory, New York, USA
[7] RISE, Waseda University, Shinjuku, Tokyo, Japan
[8] Korea University, Seoul, South Korea
[9] Graduate School of Science, Nagoya University, Nagoya, Japan
[10] Seoul National University, Seoul, South Korea
[11] Tokushima University, Tokushima, Japan
[12] Japan Atomic Energy Agency, Ibaraki, Japan
[13] INFN Section of Catania, Italy
[14] University of Catania, Catania, Italy

Abstract. The Relativistic Heavy Ion Collider forward experiment is ready to take data in the RHIC Run2017 $\sqrt{s} = 510$ GeV p-p collisions using the LHCf Arm1 detector. New accelerator data are valuable to verify the Feynman scaling of π^0 production and to study the evolution of the break of scaling in neutron production. Using the transversely polarized proton beam, asymmetric production of forward neutrons is precisely measured. That is useful to understand the fundamental meson exchange in the proton-proton collisions.

1. Introduction

Hadronic interaction models used in the analyses of air shower observations are extensively tested using the forward and minimum bias events at the Large Hadron Collider (LHC) [1].The LHC provides the highest collision energy corresponding to the cosmic-ray interaction energy of 10^{17} eV. However, measurements at lower energies using the pre-LHC colliders are not available such as the LHC case. Even though the interaction is better understood at the LHC energy range, experimental data are not enough to extend the knowledge beyond the LHC energy, or even below.

To extrapolate the accelerator results to the arbitrary collision energy, which is required for cosmic-ray analysis, verification of scaling or $\sqrt{s}$ dependence is important. The LHCf experiment [2] reported that the differential cross sections of forward π^0 production in x_F-p_T phase space hold the Feynman scaling [3] between $\sqrt{s} = 2.76$ TeV and 7 TeV p-p collisions [4]. A similar scaling was also reported [5] for forward neutron production between the ISR [6] and PHENIX [5] experiments at $\sqrt{s} = 30$–60 GeV and 200 GeV, respectively. On the other hand, the LHCf result at $\sqrt{s} = 7$ TeV indicates a break of the scaling [7,8]. To extend the verification of the scaling in π^0 production or to study the evolution of the break of scaling in neutron production, 510 GeV p-p collisions at the Relativistic Heavy Ion Collider (RHIC) at the Brookhaven National Laboratory is a unique opportunity.

In this paper, a new experiment *RHIC forward* [9] ready to take data in 2017 is reviewed. The overview, operation plan and expected results of the experiment are introduced in Sect. 2.1, Sect. 2.2 and Sect. 2.3, respectively. Another scientific target, the measurement of the single-spin asymmetry of the forward particles, is also introduced in Sect. 2.4.

2. The RHIC forward experiment

2.1. Experimental overview

The RHICf experiment uses one of the LHCf detectors, LHCf Arm1 [2]. The RHICf detector is installed in the West side of the STAR experiment, 18 m away from the interaction point. The detector is located between two beam pipes connected to the arc of the storage ring as shown in Fig. 1 (left) and the Zero Degree Calorimeter [10] is located behind the detector. Because there is a dipole magnet between the interaction point and the detector, only neutral particles emitted around zero degree arrive at the detector. The detector is composed of two compact sampling calorimeters with position sensitive layers as

[a] e-mail: sako@isee.nagoya-u.ac.jp

DOI: 10.1051/epjconf/201714510005

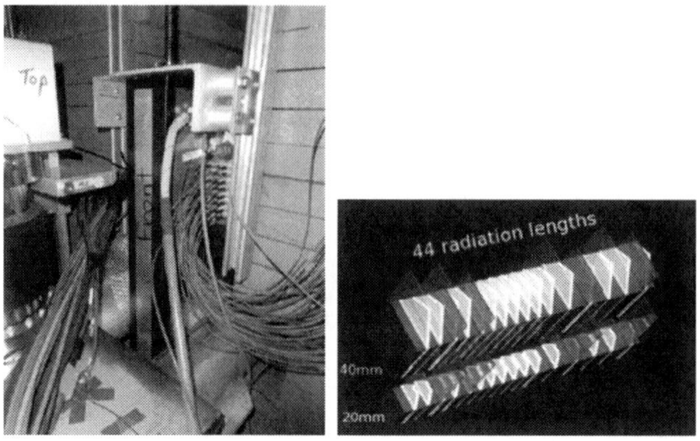

Figure 1. Left: a photo of the RHICf detector installed at 18 m from the STAR interaction point in the RHIC. Right: schematic view of the RHICf calorimeters.

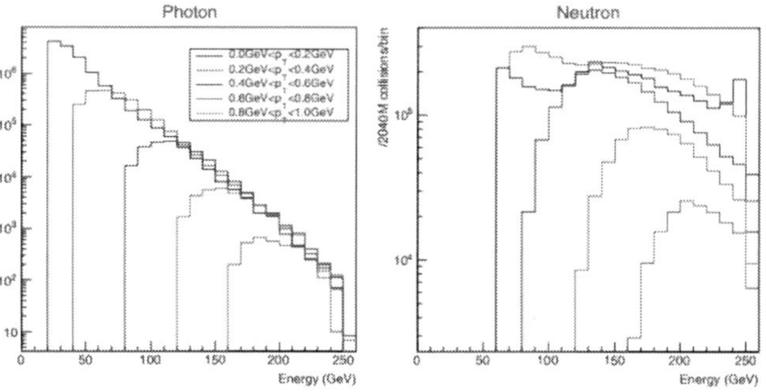

Figure 2. Expected observed spectra of single photon (left) and neutron (right) events after 12 hours of data taking. Data acquisition efficiency is taken into account, but the resolution is not considered.

shown in Fig. 1 (right) and covers pseudorapidity $\eta > 6$ by changing the position vertically. The performance of the RHICf detector (the LHCf Arm1 detector) was well studied using the test beams at the CERN SPS North area site [11, 12]. The energy range of the beam test, 100–250 GeV electron beams and 150–350 GeV hadron beams, is perfectly matching with the RHIC energy.

2.2. Operation plan

To measure the production angle of each particle precisely, RHICf requests more parallel beams than the usual RHIC operation, namely $\beta^* = 10$ m. This reduces the luminosity of RHIC down to 2.0×10^{31} cm^{-2} s^{-1}. This is ideal when considering the data acquisition speed of the RHICf and the event statistics. As discussed in Sect. 2.4, the proton beams will be radially polarized.

The RHICf detector will measure photons and neutrons of energy above about 50 GeV. Exact threshold will be determined at the actual operation condition and is different between photons and neutrons. By detecting photon pairs hitting two calorimeters simultaneously, π^0's are identified. Because the data acquisition speed of the RHICf is limited to 1 kHz, only 5% of single particle events are recorded but all π^0 events are recorded at 100 Hz.

A week of special physics operation time is approved for the RHICf in early June 2017, at the end of the 510 GeV p-p run. Two days for the special beam setup and two days for physics data taking are scheduled. During the RHICf operation period, STAR will record the collision data according to the trigger of the RHICf. Test of common data record and its offline confirmation is ongoing between the RHICf and STAR collaborations.

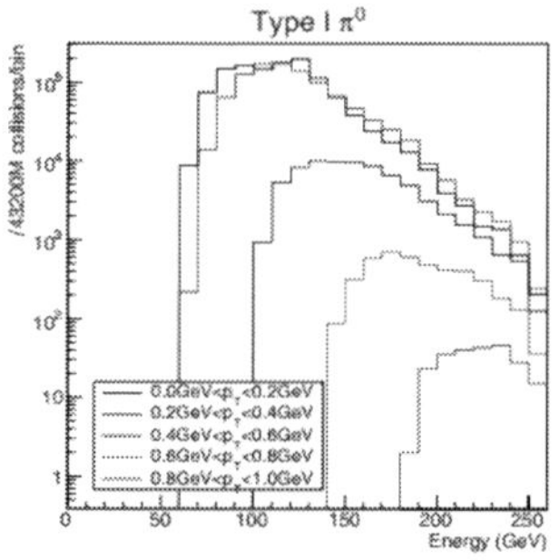

Figure 3. Expected observed spectra of π^0 events after 12 hours of data taking. All photon pair events are assumed to be recorded.

2.3. Expected results

Figure 2 shows the expected spectra of photons and neutrons in 12 hours of operation calculated using the EPOS-LHC model. The limit of the data acquisition speed is taken into account while the energy resolution is not considered. Operation at three vertical positions of the detector are assumed. Peculiar structure in the neutron spectra is due to the geometrical effect of the calorimeter

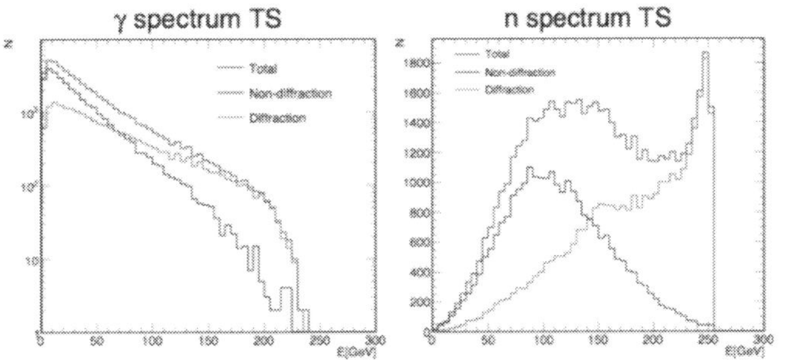

Figure 4. Expected spectra of photons (left) and neutrons (right) around 0° using PYTHIA 8212. Red and black histograms show the spectra originated from the diffractive and non-diffractive processes while purple show the total inclusive spectra.

shape and the positional scan. It is found, except for high-energy and high-p_T photons, <1% statistical accuracy is obtained with a 10 GeV binning in a wide energy and p_T region. A special trigger is in preparation to enhance the statistics of the high-energy photons.

Spectra of π^0s expected in the same operation time are shown in Fig. 3. Another special trigger allows all events with photons in two calorimeters to be recorded. It is found that in most of the region 1 - few % statistical accuracy is achieved. This will be slightly smaller than the expected uncertainty in the luminosity determination.

Because RHICf will take data with STAR, joint analyses will make new possibilities in the physics analyses. A joint analysis with the ZDC will improve the energy resolution of the neutron measurement. The advantages of the RHICf detector and the ZDC, high position resolution and high energy resolution, respectively, complement each other. Correlation between the forward and the central regions is useful to separate diffractive and non-diffractive interactions. Fig. 4 shows spectra of very forward photons and neutrons to be measured by RHIC calculated by PYTHIA 8212. The purple histograms indicate the inclusive spectra measured by the RHICf alone while the red and black histograms show the contributions from diffractive and non-diffractive interactions, respectively. It is interesting that the contributions from the two processes are similar but the shape is very different. It is shown that the information in the central tracker is powerful to distinguish two processes in the case of the ATLAS-LHCf common analysis [13], and the same method is applicable to the STAR-RHICf case.

2.4. Single-spin asymmetry measurement

Experiments at RHIC discovered an unexpectedly high azimuthal asymmetry of forward neutron production when transversely polarized proton beams collide [5,14]. Combining the data taken from three different $\sqrt{s}$, it is suggested that the amplitude of the asymmetry is proportional to p_T [15]. This p_T scaling of the forward neutron asymmetry is theoretically explained considering the interference in the scattering amplitudes of pion and a_1 mesons [16]. However, because of the limited position resolution in the measurement, p_T coverage was limited to <0.4 GeV and the coverages of the different datasets do not overlap each other. This prevents a definitive conclusion on the production mechanism of forward neutrons. To cover a wide p_T in a single $\sqrt{s}$ dataset, excellent position resolution of the RHICf detector is suitable. Thanks to the 1 mm position resolution of the RHICf detector [12] instead of the 10 mm resolution in the former measurement and the vertical position scan, RHICf can cover from 0.05 GeV to 1.2 GeV only in the single $\sqrt{s} = 510$ GeV.

Because the RHICf detector has a wide acceptance in the vertical direction, it is sensitive to the asymmetry in the vertical. To maximize the vertical asymmetry, collisions of radially polarized beam is required. Because this is not a usual setup of the RHIC polarized beams, RHICf requires 1 day for beam setup of radial polarization. As seen in Fig. 2, in parallel to the cross section measurements for cosmic-ray physics, a sufficient number of neutrons are observed. Using the same dataset shown in Fig. 2 and assuming a 50% polarization, the asymmetry amplitude can be determined with the statistical accuracy <0.2% at p_T<0.6 GeV and <1% at p_T<1.2 GeV.

3. Summary

A new experiment, RHICf, is ready to take data in early June 2017 in the RHIC Run2017 $\sqrt{s} = 510$ GeV p-p collisions. Using the former LHCf Arm1 detector, test of the Feynman scaling over a very wide $\sqrt{s}$ will be available. $\sqrt{s}$ dependence of the high-energy hadronic interaction is important to extrapolate the accelerator data to cosmic-ray physics. Using the unique polarized beam at the RHIC, RHICf can also measure the single-spin asymmetry of forward neutron production more precisely than previous measurements. This helps an understanding of the fundamental meson exchange in the p-p collisions.

References

[1] T. Sako et al., in this proceedings (2017)
[2] The LHCf Collaboration, JINST **3** (2008) S08006
[3] R.P. Feynman, Phys. Rev. Lett. **23**, 1415 (1969)
[4] The LHCf Collaboration, Phys. Rev. D **94**, 032007 (2016)
[5] The PHENIX Collaboration, Phys. Rev. D **88**, (2013) 032006
[6] J. Engler, et al., Nucl. Phys. B **84** (1975) 70; W. Flauger, F. Mönnig, Nucl. Phys. B **109** (1976) 347–356
[7] The LHCf Collaboration, Phys. Lett. B **750** (2015) 360–366
[8] K. Kawade, PhD thesis, Nagoya University (2014); CERN-THESIS-2014–315

[9] The RHICf Collaboration, `arXiv:1409.4860` [physics.ins-det]
[10] C. Adler et al., Nucl. Instrum. Methods A **499** (2002) 433–436
[11] T. Mase et al., Nucl. Instrum. Meth., A671 129-136 (2012); Y. Makino et al., JINST, submitted (2017)
[12] K.Kawade et al., JINST **9**, P03016 (2014)
[13] Q.D.Zhou, Y. Itow, H. Menjo and T. Sako, `arXiv:1611.07483v2` [hep-ex]
[14] Y. Fukao et al., Phys. Lett. B **650** (2007) 325–330
[15] The PHENIX Collaboration, J. Phys. Conf. Ser. **295**, 012097 (2011)
[16] B. Z. Kopeliovich, I. K. Potashnikova, I. Schmidt and J. Soffer, Phys. Rev. D **84**, 114012 (2011)

Latest results of the Tunka Radio Extension

D. Kostunin[1,a], P.A. Bezyazeekov[2], N.M. Budnev[2], O. Fedorov[2], O.A. Gress[2], A. Haungs[1], R. Hiller[1], T. Huege[1], Y. Kazarina[2], M. Kleifges[3], E.E. Korosteleva[4], O. Krömer[3], V. Kungel[1], L.A. Kuzmichev[4], N. Lubsandorzhiev[4], T. Marshalkina[2], R.R. Mirgazov[2], R. Monkhoev[2], E.A. Osipova[4], A. Pakhorukov[2], L. Pankov[2], V.V. Prosin[4], G.I. Rubtsov[5], F.G. Schröder[1], R. Wischnewski[6], and A. Zagorodnikov[2] (Tunka-Rex Collaboration)

[1] Institut für Kernphysik, Karlsruhe Institute of Technology (KIT), Karlsruhe, Germany
[2] Institute of Applied Physics, Irkutsk State University (ISU), Irkutsk, Russia
[3] Institut für Prozessdatenverarbeitung und Elektronik, Karlsruhe Institute of Technology (KIT), Germany
[4] Skobeltsyn Institute of Nuclear Physics, Lomonosov University (MSU), Moscow, Russia
[5] Institute for Nuclear Research of the Russian Academy of Sciences, Moscow, Russia
[6] Deutsches Elektronen-Synchrotron (DESY), Zeuthen, Germany

Abstract. The Tunka Radio Extension (Tunka-Rex) is an antenna array consisting of 63 antennas at the location of the TAIGA facility (Tunka Advanced Instrument for cosmic ray physics and Gamma Astronomy) in Eastern Siberia, nearby Lake Baikal. Tunka-Rex is triggered by the air-Cherenkov array Tunka-133 during clear and moonless winter nights and by the scintillator array Tunka-Grande during the remaining time. Tunka-Rex measures the radio emission from the same air-showers as Tunka-133 and Tunka-Grande, but with a higher threshold of about 100 PeV. During the first stages of its operation, Tunka-Rex has proven, that sparse radio arrays can measure air-showers with an energy resolution of better than 15% and the depth of the shower maximum with a resolution of better than 40 g/cm^2. To improve and interpret our measurements as well as to study systematic uncertainties due to interaction models, we perform radio simulations with CORSIKA and CoREAS. In this overview we present the setup of Tunka-Rex, discuss the achieved results and the prospects of mass-composition studies with radio arrays.

1. Introduction

The study of cosmic rays of ultra-high energies sheds light on the most powerful processes in the Universe. The fine structures of the primary cosmic ray spectrum and the mass composition yield information on the type of cosmic accelerators and their location. For example, at energies of about EeV, a transition from galactic to extragalactic cosmic ray sources is expected [1]. To distinguish between galactic and extragalactic sources, the precise determination of fluxes of different primary nuclei is required. Modern optical detectors, namely, non-imaging air-Cherenkov arrays and fluorescence telescopes reach energy resolutions of about 10% and a resolution for the depth of the shower maximum ($X_{\max}$) of about 20 g/cm^2. However, the duty cycle of such detectors is less than 15% [2,3].

Digital radio arrays, as a novel technique, which allows for measurements of air-showers produced by primary cosmic rays with energies above 100 PeV. A broad description of radio emission from air-showers, the technique of its detection and of historical and modern experiments is given in Ref. [4].

Modern detectors, such as LOFAR [5], AERA [6] and Tunka-Rex [7] have already proven the feasibility of this technique, and shown that radio detection has a resolution competitive to optical techniques. Triggered by an external particle array, a radio detector becomes a scalable, cost-effective extension, which provides precise measurements of ultra-high energy cosmic rays around-the-clock. In the present paper we focus on the setup of Tunka-Rex, discuss the achieved results and prospects of mass-composition studies with radio arrays.

2. The Tunka Radio Extension

The Tunka Radio Extension (Tunka-Rex) array has been commissioned in 2012 and originally consisted of 18 antennas distributed over an area of 1 km^2 [8]. The detector layout is mostly determined by the Tunka-133 [2] clusters, the original air-Cherenkov array of the TAIGA facility [9] located nearby southern tip of Lake Baikal in Siberia. At the moment the cosmic-ray detectors of TAIGA consists of Tunka-133, Tunka-Grande [10] and Tunka-Rex. Tunka-Rex now contains 63 antenna stations including six satellite stations extending the area to 3 km^2. The common layout of the three experiments is shown in Fig. 1.

Each Tunka-Rex antenna station consists of two perpendicular short aperiodic loaded loop antennas (SALLA) [11,12] rotated by an angle of $\pm 45°$ with respect to magnetic North. Each antenna contains a low-noise amplifier (LNA) and a load suppressing the downward direction, which makes the antenna upward-looking and decreases the uncertainty due to ground conditions to the level of only 3%. A Tunka-Rex antenna station and the SALLA gain pattern are shown in Fig. 2. Before

[a] e-mail: dmitriy.kostunin@kit.edu

DOI: 10.1051/epjconf/201714511001

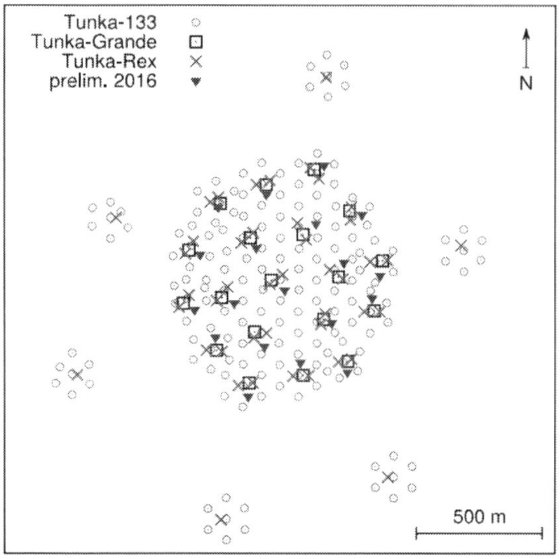

Figure 1. Layout of cosmic-ray detectors of the TAIGA [9] facility. The core consists of 19 clusters, each of them is equipped with 3 Tunka-Rex antenna stations, and 6 satellite clusters with one Tunka-Rex antenna station per cluster. Triangles depict preliminary positions of the antennas deployed in 2016 (precise position measurements will be performed soon).

digitalization, the signals are analogically processed with a filter-amplifier with an effective band of 30-76 MHz. Each Tunka-Rex antenna station is connected either to the Tunka-133 or the Tunka-Grande local data acquisition and shares the same ADC boards.

All clusters are synchronized with the central DAQ via optical fibers. We have checked the stability of the synchronization with a beacon-based method [13, 14]. The relative timing is stable to better than a nanosecond during a single run, however, after reset we obtain jumps of about 5 ns. Taking background into account, the resulting timing uncertainty is about 7 ns. Depending on the trigger mode, the entire cluster is triggered by air-Cherenkov (clear winter moonless nights) or particle detectors (the rest of the time), and traces from the Tunka-133 PMTs (when operating), scintillators and antennas are recorded simultaneously in traces of 1024 samples with 5 ns sampling rate and a bitdepth of 12 bits. As a result, TAIGA features duplex (particles + radio) and triplex (particles + radio + air-Cherenkov) measurements of cosmic rays with energies of 10^{16}–10^{18} eV.

2.1. Antenna calibration

The calibration of the Tunka-Rex antenna was performed in the following way: the directivity of the SALLA antenna was simulated with the NEC2 code [15], and then normalized to an amplitude calibration made with the commercial reference source VSQ1000+DPA4000 by Schaffner Electrotest GmbH (now Teseq). The same method was also used to calibrate the LOPES antennas [16], as well as the LOFAR ones [17], what makes these three experiments having consistent absolute calibration scale. The hardware response and temperature dependence of the LNA and filter-amplifier were measured under laboratory conditions, and the calibration of the

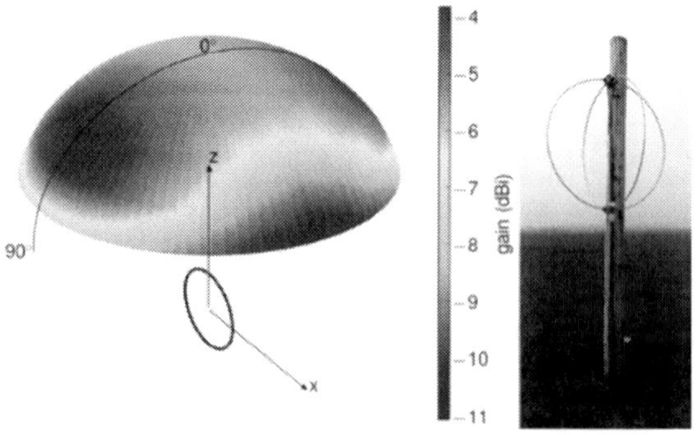

Figure 2. *Left:* Gain pattern of the SALLA antennas of Tunka-Rex. *Right:* A Tunka-Rex antenna station. Two perpendicular SALLAs are mounted on a wooden pole at a height of about 3 m (upper end of SALLA).

ADC was done on the board already deployed at the local DAQ of the clusters. As result, the overall uncertainty on the absolute amplitude reconstruction is 22%, with a dominating contribution of 16% from the calibration scale uncertainty of the reference source, and a number of smaller contributions given by environmental conditions, antenna production and deployment.

3. Event reconstruction

Since all clusters operate independently, single-cluster events are merged into shower events during offline analysis. At the first step, only events containing at least three antenna stations of signal-to-noise ratio (SNR) in power $S^2/N^2 > 10$ are selected. The amplitude of the signal S is defined as the maximum of amplitude of the Hilbert envelope of the vectorial sum of the two measured polarizations inside of the signal window. The position of the signal window is constant and defined by the hardware delay between the radio signal arrival time and the particle or air-Cherenkov trigger, while the width of the signal window is defined by timing uncertainties and different shower geometries. The noise level N is defined as RMS of the amplitude in a noise window. An example of radio and air-Cherenkov traces recorded at the same cluster are given in Fig. 3. Since Tunka-Rex is operating close to the threshold, the contribution of the background cannot be neglected. This contribution is taken into account for the estimation of timing and amplitude uncertainties, and the measured amplitudes is corrected for a systematic background bias using a function depending on SNR.

At the next step, the arrival direction is reconstructed with a plane wave front model, and is compared to the one measured by the master detector (Tunka-133/Tunka-Grande). Since the angular resolution of Tunka-Rex is about 1–2°, all events with direction deviating from the master reconstruction by more than 5° are rejected and excluded from analysis.

After the first quality cuts, the position of the shower core is reconstructed. In the triplex mode, the shower core and axis is taken from the Tunka-133 reconstruction, since it has much larger density than the other cosmic-ray detectors of TAIGA, and the resolution is better than 5 m. In the duplex mode we plan to combine the reconstructions from Tunka-Rex and Tunka-Grande, since both of them

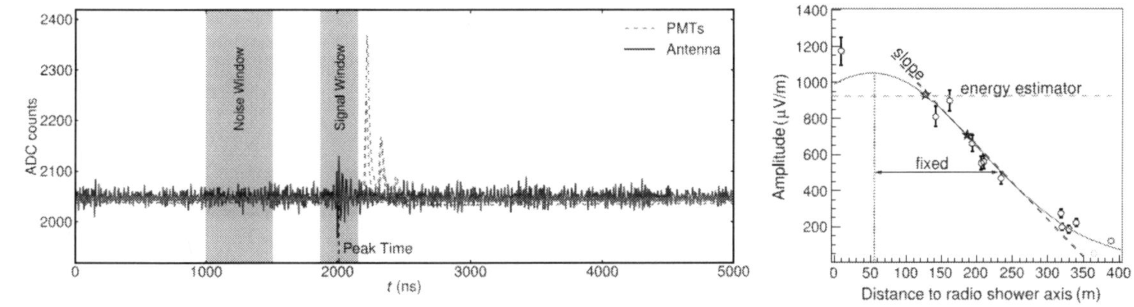

Figure 3. Left: example traces recorded by the Tunka-133 local DAQ. The radio signal is recorded earlier than the PMT pulses, mostly due to longer cables of Tunka-133. Right: sketch describing the reconstruction of the primary energy and distance to the shower maximum. The lateral distribution is fitted by a Gaussian-shape function with fixed width. The primary energy is proportional to the amplitude at 120 m distance, and the distance to the shower maximum is calculated from the slope at 180 m from the shower axis.

are expected to feature a resolution of about 20–30 m due to their spacing, and we expect that the combined reconstruction will improve this value.

After the reconstruction of the shower core, the amplitudes from the detector surface are projected to the shower axis forming the lateral distribution. Any antenna station passed SNR cut appears further on the lateral distribution than two antenna stations without the signal is considered as outlier and rejected as false positive.

To reconstruct the primary energy E_{pr} and the depth of the shower maximum X_{max}, the lateral distribution is corrected to remove the dependence on a geomagnetic and azimuth angles, which is introduced by the interference of the geomagnetic and charge excess effects [18]. The resulting distribution is fitted with a lateral distribution function (LDF), containing two free parameters. These two parameters, namely normalization and slope, are used for the reconstruction of E_{pr} and X_{max}, respectively. A sketch describing the idea of the LDF approach is given in Fig. 3. Since X_{max} is very sensitive to the shape of the LDF, we apply additional quality cuts to select high-quality events: the event must contain at least one antenna further than 200 m from the shower axis (to increase the sensitivity to the LDF slope), and the resulting fit uncertainty of X_{max} must be less than 50 g/cm^2.

It is worth noting, that the radio technique is only sensitive to the electromagnetic component of air-showers, which means that the method has additional uncertainties due to unknown primary particles. These uncertainties are discussed in Sect. 5.4.

Up to now, we have finished the reconstruction of measurements during 2012–2014, when Tunka-Rex was operating jointly with Tunka-133. To compare the reconstructions of the detectors, we selected events with core positions inside the dense part of detector, i.e. within 500 m radius around the center. To avoid implicit tuning in the cross-check of Tunka-Rex and Tunka-133, the half of Tunka-133 reconstruction of E_{pr} and X_{max} was blinded, and opened only after the final reconstruction of Tunka-Rex. The comparing set includes 148 events with reconstructed energy and 42 events with reconstructed X_{max}. The obtained resolution of Tunka-Rex is 15% for energy and 40 g/cm^2 for X_{max}, while no significant absolute shift between the reconstructions of Tunka-Rex and Tunka-133 was observed. Since all of the high-energy events have passed the quality cuts, we can use them for

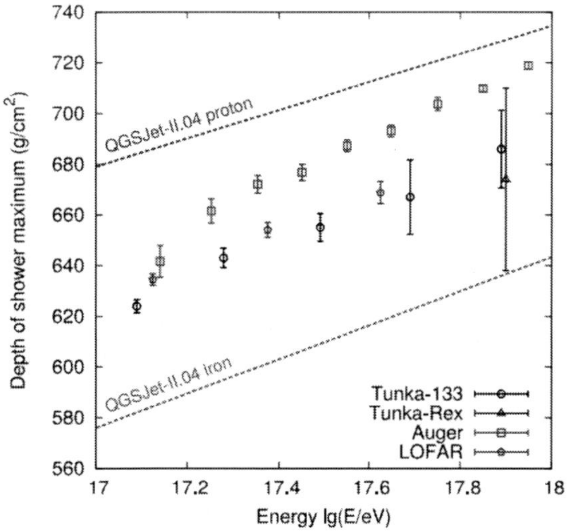

Figure 4. Mean depth of the shower maximum versus the primary energy reconstructed by modern cosmic-ray experiments: Tunka-133 [2], Auger [22], LOFAR [5] and Tunka-Rex. Errorbars indicate statistic uncertainties only, and all measurements agree within additional systematic uncertainties.

a mass-composition study. The mean X_{max} value obtained for $\lg(E_{pr}/\mathrm{eV}) = 17.9 \pm 0.1$ is given in Fig. 4 based on 8 events.

It is worth noting, that these results were obtained with the Tunka-Rex configuration consisting of one antenna per cluster. Meanwhile, starting from 2016 the array features three antennas per cluster.

To better understand systematic uncertainties, atmospheric effects will have to be taken into account [19]. For the effective frequencies of Tunka-Rex these uncertainties can be in the order of 2% and 5 g/cm^2 for the energy [20] and shower maximum [5] reconstructions, respectively. In future, we plan to include Global Data Assimilation System (GDAS) data to our analysis to decrease these uncertainties [21].

3.1. Energy reconstruction with a single antenna station

Besides the main goals of energy and shower maximum reconstruction we have shown that a single antenna station

DOI: 10.1051/epjconf/201714511001

can provide useful information when the shower core and axis are known [23]. Assuming a mean value for the steepness of the radio LDF of about $\eta_0^{-1} \approx 120\,\mathrm{m}$, and a threshold of $S_{\mathrm{th}} \approx 90\,\mu\,\mathrm{V/m}$, using reconstructed shower geometry from Tunka-133, we obtained a reliable energy reconstruction using single-antenna events. Particularly, the number of events was increased by more than three times, while the energy resolution has slightly decreased to 20%.

This result indicates the feasibility of equipping surface particle detectors with simple radio extensions, which allow for the determination of the electromagnetic energy deposit of high-energy air-showers above 10^{17} eV.

4. The radio amplitude as measure for the absolute energy scale

The independent energy reconstruction by Tunka-Rex is based on an absolute amplitude calibration of the antennas and on normalization parameters obtained with CoREAS. As shown above, the absolute energy scales of Tunka-Rex and Tunka-133 experiments are in very good agreement. Since the radio emission from air-showers is well understood and its generation and propagation depend weakly on the atmospheric condition, it can be used as a universal tool to compare or cross-check the energy scale of experiments located in different places and exploiting different techniques. To test this statement, we have selected KASCADE-Grande [24] with its radio extension LOPES [25] and compared it with Tunka-133 and Tunka-Rex measurement, respectively [26]. Since Tunka-Rex and LOPES were calibrated with the same reference source, most systematic uncertainties of the amplitude calibration cancel out in the comparison.

We realize this comparison with two different approaches. The first one is to compare the ratio, κ, of measured radio amplitudes and the energy reconstructed by the master experiment. Then the relative shift between the masters is defined as $f_{\mathrm{amp}} = \kappa_{\mathrm{Tunka-Rex}}/\kappa_{\mathrm{LOPES}}$. This method relies only on direct radio measurements, and the reconstruction procedure are chosen as similar as possible (i.e. the same bandwidth and the same LDF treatment), moreover, the reconstruction is corrected for the different observation depths and magnetic fields of the two locations.

The second method is implemented via CoREAS simulations: two simulations – one with proton and one with iron primary were produced for each event, and then the measured and simulated radio amplitudes were compared to each other. Then, the mean ratios F^{p} (F^{Fe}) between simulated and measured amplitudes are determined, and the scale shift between the master experiments is defined as $f_{\mathrm{sim}} = F_{\mathrm{Tunka-Rex}}/F_{\mathrm{LOPES}}$. The main uncertainty of these method is given by the hadronic model used in the simulation and uncertainties due to angular dependence of antenna gain.

The result of both methods is that the energy scales of Tunka-133 and KASCADE-Grande are consistent to about 10% limited by systematic uncertainties of the LOPES experiment, and the mean KASCADE-Grande energy scale is lower than Tunka-133 by about 5%. A similar shift is obtained by a straight-forward fit of the energy spectra of Tunka-133 and KASCADE-Grande. The spectra and results of the scale comparison are given in Fig. 5.

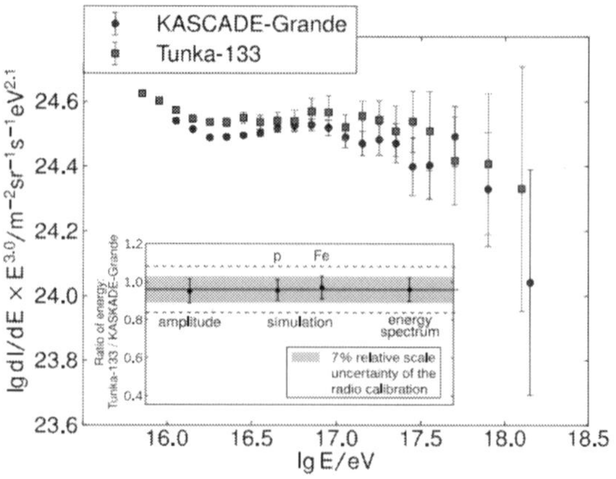

Figure 5. Energy spectra of cosmic rays from KASCADE-Grande [24] and Tunka-133 [2]: normalized flux per energy; and the results from the comparison of the energy scales between the experiments Tunka-Rex and LOPES (small box) in the energy range of 10^{17} to 10^{18} eV. With a systematic increase of KASCADE-Grande energies by 4% (or a corresponding decrease of Tunka-133 energies) the average flux per energy of both experiments can be brought to agreement in this energy range [26].

This result can be applied to study finer features of the energy spectrum with higher accuracy, e.g. it allows to define the positions of knee-like structures with lower uncertainty.

5. Ongoing activity

In this section we discuss theoretical work performing in the frame of Tunka-Rex experiment with a purpose of improving the reconstruction of air-shower events.

5.1. Lateral distribution function

Hereafter we perform calculations in the frame of the geomagnetic coordinate system based on the shower axis vector $\hat{\mathbf{V}}$ and the vector of the Earth's magnetic field $\hat{\mathbf{B}}$ (a hat over a vector denotes a unit vector: $\hat{\mathbf{B}} = \mathbf{B}/|\mathbf{B}|$):

$$\hat{\mathbf{e}}_x = \hat{\mathbf{V}} \times \hat{\mathbf{B}}, \tag{1}$$

$$\hat{\mathbf{e}}_y = \hat{\mathbf{V}} \times (\hat{\mathbf{V}} \times \hat{\mathbf{B}}), \tag{2}$$

$$\hat{\mathbf{e}}_z = \hat{\mathbf{V}}\,. \tag{3}$$

This way, the shower front is lying in the plane $(\hat{\mathbf{e}}_x, \hat{\mathbf{e}}_y)$. Let us also define useful angles: the geomagnetic angle between shower axis and magnetic field $\alpha_g = \angle(\mathbf{V}, \mathbf{B})$ and the geomagnetic azimuth $\phi_g = \angle(\hat{\mathbf{e}}_x, \mathbf{r})$, where $\mathbf{r}$ is the coordinate of an antenna station.

The main parameterization used by Tunka-Rex is the parameterization describing the distribution of radio amplitudes with respect to the shower axis, i.e. the LDF:

$$\mathcal{E}(r, \phi_g) = \hat{\mathrm{K}}^{-1}(\phi_g)\mathcal{E}_2(r), \tag{4}$$

$$\hat{\mathrm{K}}(\phi_g) = \left(\varepsilon^2 + 2\varepsilon\cos\phi_g\sin\alpha_g + \sin^2\alpha_g\right)^{-\frac{1}{2}}, \tag{5}$$

$$\mathcal{E}_2(r) = \mathcal{E}_0\exp\left(a_1(r - r_0) + a_2(r - r_0)^2\right), \tag{6}$$

where $\mathcal{E}(r, \phi_g)$ is the amplitude at the antenna station with coordinates (r, ϕ_g). This amplitude is described with two significant terms: first, the azimuthal asymmetry ε is corrected by the term $\hat{K}(\phi_g)$; second, the lateral distribution $\mathcal{E}_2(r)$, which is a Gaussian-like function with the normalization $\mathcal{E}_0$, with a width proportional to a_2 and the slope proportional to a_1. The maximum of $\mathcal{E}_2(r)$ reflects Cherenkov-like features of the radio emission. The parameter r_0 is arbitrarily chosen in a way to obtain maximum correlation of $\mathcal{E}_0$ and a_1 with the primary energy and the distance to the shower maximum, respectively.

5.2. Estimation of core position

After the upgrade of Tunka-Rex, we are now implementing an independent reconstruction of the position of shower core with radio standalone. The position of the core (x_0, y_0) is added to LDF parameters $r = r(x_0, y_0)$ and $\phi_g = \phi_g(x_0, y_0)$. Since Tunka-Rex operates near the threshold, currently the core position is reconstructed in three steps:

1. The initial core position is estimated as the center of mass of the radio amplitudes during the arrival-direction reconstruction (requires at least 3 antennas);
2. Parameters a_1 and a_2 are fixed to default values, and the LDF is fitted with three free parameters: $\mathcal{E}_0$, x_0 and y_0 (requires at least 4 antennas);
3. The LDF is fitted again including a_1 and a_2, and the limits for $\mathcal{E}_0$, x_0 and y_0 are defined from the previous stage (requires at least 6 antennas).

With this procedure we expect to obtain resolutions of about 20–30 m for the dense part of the detector (inner circle of 500 m around the center of the array). To improve these numbers we plan to apply stricter quality cuts on the signal reconstruction for this particular procedure.

5.3. Limitation of the one-dimensional approach

For the time being, all methods for $X_{\max}$ reconstruction are based on the simple relation between single parameters of the LDF and $X_{\max}$ or the distance to $X_{\max}$. For example, Tunka-Rex, LOFAR [27] and AERA [6] use Gaussian-like parameterizations and exploit the correlation of $X_{\max}$ with the slope and width of LDF, respectively.

In this section we describe the relation between the position of the depth of the shower maximum and the slope of the LDF, give a more strict consideration of it, and discuss its possible hidden features.

We use the following assumptions: the distribution of the electrons behaves as the Gaisser-Hillas function and the density of the Earth's atmosphere falls exponentially with increasing altitude, namely we use the CORSIKA parameterization of the standard atmosphere [28]. The simple form of the amplitude of the radio signal, $\mathcal{E}_\nu(r)$, with frequency, ν, at distance, r, from the shower axis is [29]:

$$\mathcal{E}_\nu(r) = \kappa \int\limits_{h_1^\nu(r,n_r)}^{h_2^\nu(r,n_r)} \frac{N(h)}{h} \mathrm{d}h\,, \tag{7}$$

where κ is the normalization coefficient (the dependence on geomagnetic angle, α_g, has already been taken into account), $N(h)$ is the number of electrons at the altitude, h, and $h_{1,2}^\nu(r, n_r)$ are the integration limits depending on the distance to shower axis, r, and refractive index, n_r.

Assuming that the refractive index, $n_r(h)$, is proportional to the density of the atmosphere, one can recalculate it from the atmospheric parameterization and use it as an input to estimate the integration limits $h_{1,2}^\nu(r, n_r)$ for *vertical* air-showers:

$$h_{1,2}^\nu(r, n_r) = \left(\frac{r}{r_{1,2}(\nu, n_r)}\right)^{\alpha_{1,2}(\nu, n_r)}\,, \tag{8}$$

This parameterization is obtained by numerical solution of the equation

$$\Delta t(r, h^\nu) = \Delta t(r, h_c^\nu) + \frac{1}{2\pi\nu}\,, \tag{9}$$

where

$$\Delta t(r, h^\nu) = \cos\left(\arctan\left(\frac{r}{h^\nu}\right)\right) \int\limits_0^{h^\nu} \frac{n_r(h')}{c} \mathrm{d}h' - \frac{h^\nu}{c}\,, \tag{10}$$

where c is the velocity of light, and h_c is defined as solution of equation

$$\frac{\partial}{\partial h_c} \Delta t(r, h_c) = 0\,. \tag{11}$$

For inclined air-showers everything is scaled by $\cos\theta$ (where θ is the zenith angle) at first approximation, but here we do not consider these cases.

The curves denoting the behavior of $h_{1,2}^\nu(r, n_r)$ are in Fig. 6. One can see that for lateral distances far from the Cherenkov bump ($r > 100$ m), the upper limit goes to infinity (becomes higher than the top of the atmosphere) and the lower limit goes to the position of the shower maximum and above.

In our case, the integration limits in Eq. (7) are as follows: $h_2^\nu(r, n_r) \to \infty$, $h_1^\nu(r, n_r) > h_{\max}$, i.e. we integrate over the upper tail of the electron distribution. This way, the value of the integral Eq. (7) has the following form:

$$\mathcal{E}_\nu(r > r_c) \propto \exp\left(-\frac{(r/r_1)^{\alpha_1}}{h_{\max}} f_{\mathrm{int}}(h_{\max}, \ldots)\right)\,. \tag{12}$$

As was expected, we obtained an exponential-like behavior of the LDF, where the exponent is defined by the altitude of the shower maximum $h_{\max}$. It is worth noting that the power of r (defined as $\alpha_1 \approx 3/2$, which is obtained from the fit of Eq. (8) to curves shown in Fig. 6) is between 1 and 2, i.e. both exponential and Gaussian describe the tail of the LDF only approximately.

The term f_{int}, which includes constants from the interaction models, has a weaker dependence on the shower geometry and scales with $h_{\max}$ due to the non-isotropical atmosphere. Simplifying $f_{\mathrm{int}} = 1$ one still conserves the correlation between the slope of $\mathcal{E}$ and $h_{\max}$ (see Fig. 6), which means that the radio signal has a high sensitivity to the position of the shower maximum. However, this one-dimensional slope method does not give additional information on the type of primary particle, which would go beyond $X_{\max}$.

One can see from Eq. (7), that the height of the shower maximum $h_{\max}$ is encoded in the slope of the LDF which means, that the radio signal has a high sensitivity to the position of the shower maximum. On the other side, this one-dimensional slope method does not give additional information on the type of primary particle.

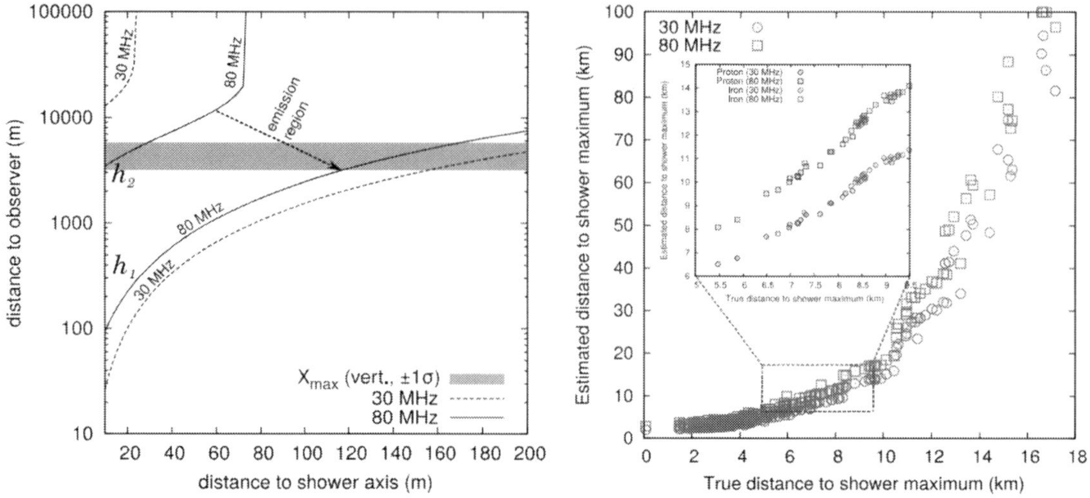

Figure 6. Left: sketch describing the radio emission from air-showers. The X axis indicates the position of the observing antenna, the Y axis indicates the altitude (the observer is placed at zero altitude), the colored band indicates possible positions of the shower maximum for typical air-showers (one sigma spread around the mean due to shower-to-shower fluctuations for vertical air-showers of energies $10^{17} - 10^{18}$ eV), and the lines indicate the bounds of the emission region for different frequencies. These lines are calculated using the atmosphere refractivity, n_r, as input. One can see, that the emission region reduces with increasing frequency and the point of intersection with the shower maximum will define the Cherenkov ring observed for high frequencies. For lower frequencies one can see the exponential fall-off of amplitudes after the intersection of h_1 with the shower maximum line. Right: correlation between the true geometrical distance to the shower maximum and the estimation made with the simplest slope method neglecting particle interactions and propagation constants. We obtained no difference between slopes of proton and iron induced air-showers. The dependence is not linear mostly due to the different density of the atmosphere at the different altitudes (i.e. the geometrical mean path length of electron is different); for real data analysis we use a more sophisticaed method where all reconstruction constants are tuned against CoREAS simulations to obtain a linear dependence [7, 18].

5.4. Systematic uncertainties

To study uncertainties given by hadronic interactions and shower-to-shower fluctuations, we performed simulations with recently released CORSIKA v7.5. Both, QGSJET-II.04 and EPOS-LHC yield almost the same radio amplitude with a difference less than one given by shower-to-shower fluctuations.

For the detailed study we use QGSJET-II.04, with which we simulated events with parameters similar to the events reconstructed in 2012–2014 with four different primary particles: proton, helium, nitrogen and iron. The energy resolution (taking noise into account) for each particle is about 10%, while the shift in the absolute values of the energy is about 12% between proton and iron. This is due to the fraction of the primary energy going into the electromagnetic cascade, the same feature is also observed with the optical methods, such as air-Cherenkov and fluorescence. This shift is much larger, than shower-to shower fluctuations for these particles: 5% and 1.3% for proton and iron, respectively. The reconstruction of the absolute value of the shower maximum is not much affected by the primary composition, since it is reconstructed with the simple slope method, and the one-dimensional slopes are the same for showers of different primary particles with the same shower maxima.

5.5. Hints from the charge-excess asymmetry

In the work [18] it is shown that the charge-excess asymmetry has a non-trivial dependence on the distance to the shower axis, particularly, the asymmetry features a local maximum depending on the distance to the shower maximum (see Fig. 7). This structure was obtained by analyzing the polarization of CoREAS simulations at individual antenna positions.

As mentioned above, in the Tunka-Rex reconstruction the LDF is corrected for charge-excess asymmetry ε. To show the consistency of the polarization and LDF approaches, we express ε in terms of the LDF. Since the Tunka-Rex LDF is azimuthal symmetric, we have used LDF developed for AERA [6] experiment. The expression for asymmetry has the following form:

$$\varepsilon(r) = \sin\alpha_g \frac{\mathcal{E}_{2G}(r) - \tilde{\mathcal{E}}_{2G}(r)}{\mathcal{E}_{2G}(r) + \tilde{\mathcal{E}}_{2G}(r)}, \quad (13)$$

with

$$\begin{cases} \mathcal{E}_{2G}(r) = \mathcal{E}_{2G}(\boldsymbol{r}_x) \\ \tilde{\mathcal{E}}_{2G}(r) = \mathcal{E}_{2G}(-\boldsymbol{r}_x), \end{cases} \quad (14)$$

where $\mathcal{E}_{2G}(r)$ is the AERA parameterization and vector $\boldsymbol{r}_x$ is along the Lorentz force (perpendicular to the shower axis and geomagnetic field). Taking mean values for the parameters of $\mathcal{E}_{2G}(r)$, one can obtain the corresponding curve $\varepsilon(r)$ and compare it to one obtained with the polarization approach. The comparison is presented in Fig. 7. One can see, that both definitions are in good agreement, which leads to an interesting conclusion: the asymmetry (or charge-excess) information can be extracted from the more simple measurement of the total radio amplitude, instead of precise measurements of the components of the electrical field. Measuring the

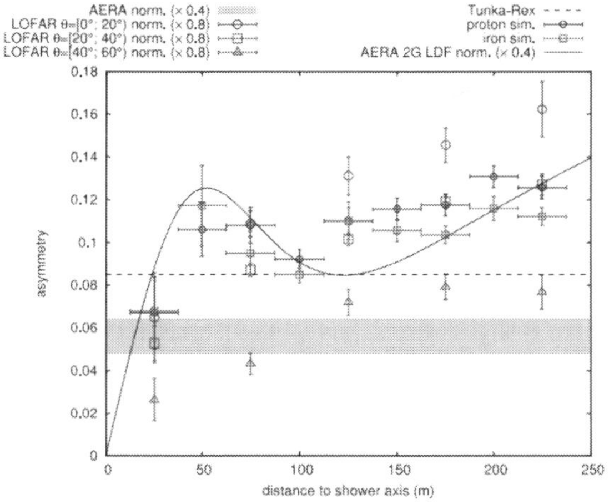

Figure 7. Askaryan asymmetry normalized to the geomagnetic field in the Tunka Valley as a function of the distance to the shower axis. Points indicate the polarization measurements by LOFAR [30] (green) and CoREAS simulations [18] (blue and red). The black solid line indicates the LDF asymmetry $\varepsilon(r)$ from Eq. (13). The blue band is the polarization measurements by AERA [31] (with uncertainties), the dashed line is the theoretical prediction for Tunka-Rex [18].

amplitude asymmetry requires higher number of stations per events, but lower signal-to-noise ratios.

As was shown in Ref. [18], the behavior of the asymmetry is connected to the distance to the shower maximum, i.e. an accurate measurement of the asymmetry by either means should be sensitive to the mass composition. The idea of a one-antenna analysis can also be applied to a polarization study of the asymmetry: knowing the geometry of the air-shower and the behavior of $\varepsilon(r)$ one can study the mean shower maximum via the mean asymmetry.

Finally, the asymmetry contains information not only on the total number of the charge particles, but also on the dynamics of their creation.

5.6. Signal recognition with neural networks

The recognition of the signal and the determination of its amplitude are the most fundamental problems of the basic event reconstruction. Since the measured signal is a sum of a true radio signal from the air-shower and the background of environment and hardware, the quality of the signal reconstruction is a function of the SNR. When the SNR is relatively small (conditionally, SNR < 100), the influence of the background cannot be neglected. On average, the amplitude of the measured signal is higher than the amplitude of the true one, and the average fraction between them is expressed as $\mathcal{E}_{\mathrm{t}} = \mathcal{E}_{\mathrm{m}}\sqrt{1 - k/\mathrm{SNR}}$, where k is a constant, which depends on the definition of SNR. However, this amplitude can be lower, since the background is uncorrelated with the signal [32,33].

In this year we started the investigation of the applicability of neural networks for the signal reconstruction. We designed a neural network, which gets input traces of 200 counts and predicts the amplitude of the true signal. We prepared a dataset of about 10000 events, randomly divided in two parts, and used the first one for the training and the second one for the control check.

The control check has shown that the resolution of amplitude reconstruction is about 22%, which corresponds to our standard reconstruction for signals near threshold. Thus, further investigations are required before this neutral-network approach can be implemented in our standard reconstruction.

6. Conclusion

The Tunka Radio Extension is a modern experiment which measures radio emission from air-showers induced by primary cosmic rays with energies above 100 PeV. Tunka-Rex has proven the feasibility and competitiveness of the radio detection technique. Operated as a sparse array with spacing of about 200 m between antennas, it has reached a precision of 15% and 40 g/cm^2 for the primary energy and the depth of the shower maximum, respectively.

After being upgraded during the last two years, Tunka-Rex has now reached triple of its original density and a new trigger from the recently deployed scintillator array Tunka-Grande has been implemented. This increases the operation time and quality of events, e.g. we now expect more than 1000 events per year instead of about 100 during the first stages of operation.

The construction of Tunka-Rex was funded by the German Helmholtz Association and the Russian Foundation for Basic Research (grant HRJRG-303). Moreover, this work has been supported by the Helmholtz Alliance for Astroparticle Physics (HAP), by Deutsche Forschungsgemeinschaft (DFG) grant SCHR 1480/1-1, and by the Russian grant RSF 15-12-20022.

References

[1] R. Aloisio, V. Berezinsky, A. Gazizov, Astroparticle Physics **39–40**, 129 (2012), cosmic Rays Topical Issue
[2] V.V. Prosin et al., J. Phys. Conf. Ser. **718**, 052031 (2016)
[3] A. Porcelli, PoS **ICRC2015**, 420 (2016)
[4] F.G. Schröder, Prog. Part. Nucl. Phys. **submitted**, arXiv:1607.08781 (2016)
[5] S. Buitink et al., Nature **531**, 70 (2016), 1603.01594
[6] Pierre Auger Coll., Phys. Rev. D **93**, 122005 (2016)
[7] P.A. Bezyazeekov, et al. - Tunka-Rex Coll., JCAP **01**, 052 (2016)
[8] P.A. Bezyazeekov, et al. - Tunka-Rex Coll., Nucl. Inst. Meth. A **802**, 89 (2015)
[9] N. Budnev et al., J. Phys. Conf. Ser. **718**, 052006 (2016)
[10] N.M. Budnev, A.L. Ivanova, Kalmykov et al., Bull. Russ. Acad. Sci. Phys. **79**, 395 (2015), [Izv. Ross. Akad. Nauk Ser. Fiz. **79**, no.3, 430 (2015)]
[11] O. Krömer et al. (LOPES Collaboration), Proc. of the 31st ICRC, Łódź, Poland (2009), `http://icrc2009.uni.lodz.pl/proc/html/`
[12] P. Abreu et al. (Pierre Auger), JINST **7**, P10011 (2012), 1209.3840
[13] F. Schröder et al., NIM A **615**, 277 (2010)
[14] A. Aab et al. (Pierre Auger), JINST **11**, P01018 (2016), 1512.02216

[15] G. Burke, A. Poggio, Tech. rep., Lawrence Livermore National Laboratory, Livermore U.S.A. (1977)
[16] W.D. Apel et al. (LOPES), Astropart. Phys. **75**, 72 (2016), 1507.07389
[17] A. Nelles et al., JINST **10**, P11005 (2015), 1507.08932
[18] D. Kostunin et al., Astropart. Phys. **74**, 79 (2016)
[19] J. Abraham et al. (Pierre Auger), Astropart. Phys. **32**, 89 (2009), [Erratum: Astropart. Phys. **33**, 65 (2010)], 0906.5497
[20] C. Glaser, M. Erdmann, J.R. Hrandel, T. Huege, J. Schulz, JCAP **1609**, 024 (2016), 1606.01641
[21] P. Abreu et al. (Pierre Auger), Astropart. Phys. **35**, 591 (2012), 1201.2276
[22] A. Porcelli for the Pierre Auger Coll., Proceedings of Science, p. 420 (ICRC2015)
[23] R. Hiller, et al. – Tunka-Rex Coll., EPJ Web of Conferences proc. of ARENA 2016, in press (2016)
[24] W.D. Apel, et al. (KASCADE-Grande Collaboration), Astroparticle Physics **36**, 183 (2012)
[25] W.D. Apel et al. (LOPES Collaboration), Phys.Rev. **D90**, 062001 (2014), 1408.2346
[26] W.D. Apel et al. (Tunka-Rex, LOPES), Phys. Lett. **B763**, 179 (2016)
[27] A. Nelles, S. Buitink, H. Falcke et al., Astropart. Phys. **60**, 13 (2014)
[28] D. Heck et al., FZKA Report 6019, Forschungszentrum Karlsruhe (1998)
[29] H.R. Allan, Progress in Elementary Particle and Cosmic Ray Physics **10**, 171 (1971)
[30] P. Schellart et al., JCAP **1410**, 014 (2014)
[31] Pierre Auger Coll., Phys. Rev. D **89**, 052002 (2014)
[32] H.R. Allan, R.W. Clay, J.K. Jones, Nature **225**, 253 (1970)
[33] F. Schröder et al., NIM A **662**, Supplement 1, S238 (2012), proc. of ARENA 2012

Development of the radio astronomical method of cosmic particle detection for extremely high-energy cosmic ray physics and neutrino astronomy

Igor Zheleznykh[1,2,a], Rustam Dagkesamanskii[3], Leonid Dedenko[4], and Grigorii Dedenko[5]

[1] Institute for Nuclear Research, Russian Academy of Sciences, Russia
[2] Moscow Institute of Physics and Technology, Russia
[3] Lebedev Physics Institute, Russian Academy of Sciences, Pushchino Radio Astronomy Observatory, Russia
[4] Moscow State University, Department of Physics, Russia
[5] Nuclear Research National University "Moscow Engineering Physics Institute", Russia

Abstract. The proposal to use ground based radio telescopes for detection of Askaryan radio pulses from particle cascades arising when extremely high-energy (EHE $> 10^{20}$ eV) cosmic rays (including neutrinos) interact with the lunar regolith of multi gigaton mass was made at the end of 1980s in the framework of the Russian (Soviet) DUMAND Program. During more than a quarter of century a number of lunar experiments were carried out mainly in the 1–3 GHz frequency range using the large radio telescopes of Australia, USA, Russia and other countries but these experiments only put upper limits to the EHE cosmic rays fluxes. For this reason, it would be of great interest to search for nanosecond radio pulses from the Moon in a wider interval of frequencies (including lower ones of 100–350 MHz) with larger radio detectors – for example the giant radio telescope SKA (Square Kilometer Array) which is constructed in Australia, New Zealand and South Africa. In this paper possibilities are discussed to use one of the most sensitive meter-wavelength ($\sim$ 110 MHz) Large Phased Array (LPA) of $187 \times 384\,\mathrm{m}^2$ and the wide field of view meter-wavelength array of the Pushchino Radio Astronomy Observatory as prototypes of low frequency radio detectors for lunar experiments. The new scheme for fast simulation of ultrahigh and extremely high-energy cascades in dense media is also suggested. This scheme will be used later for calculations of radio emission of cascades in the lunar regolith with energies up to 10^{20} eV and higher in the wide frequency band of 0.1– a few GHz.

1. Introduction: DUMAND and a new generation of neutrino telescopes

In the late fifties – early sixties it was proposed that a number of fundamental problems of Particle Physics and Astrophysics (Astronomy) could be studied using underground (underwater) installations for detecting high energy (HE) neutrinos produced in the Earth's atmosphere and in space [1–3]. The first HE atmospheric neutrinos were detected in 1965 with the help of scintillation detectors of tens m^2 located in deep gold mines in India (A. Wolfendale, M. Menon, S. Miyake, et al.) and in South Africa (F. Reines, W. Kropp, H. Sobel, et al.).

In 1966, F. Reines and J. Sellschop suggested to build COLOSSUS – a large deep underground detector of 10^5 tons to register neutrinos from the atmosphere and also beyond [4]. In 1973–1976 the idea to construct the large underwater optical neutrino telescope **DUMAND** of cubic kilometer size had been developed by F. Reines, J. Learned, H. Blood, H. Bradner, A. Roberts, G. Wilkins, et al. (see [5]).

The hydro-acoustical detection of electron-hadron cascades of ultrahigh-energies (UHE $> 10^{15}$ eV) and extremely high-energies was also considered in 1976–1979 (G. Askaryan and B. Dologoshein, T. Bowen, J. Learned).

The great value of the DUMAND program 1980–1995 (in particular the DUMAND Workshop in Hawaii in 1976) for developing deep underwater neutrino telescopes in the Ocean near Hawaii, in Baikal, in the Mediterranean Sea but also for the development and construction of the large neutrino detector IceCube in Antarctica and smaller underground water-filled detectors such as IMB and Kamiokande was emphasized by J. Learned and Ch. Spiering in their recent article [6].

It is significant that the creation of the deep underground water Cherenkov detectors of 5×10^3 m^3 and more had become a real compromise between COLOSSUS and DUMAND and that a first step to developing a huge Antarctic under-ice optical neutrino telescope IceCube had been made by F. Halzen and J. Learned in 1988 [7].

In **Sect. 2** we should like to pick out specific features of the **Russian (Soviet) DUMAND** program led by M.A. Markov in 1980–1994, in particular the development of new branches of DUMAND:

- the Radio Antarctic Muon and Neutrino Detection **(RAMAND)** to search for astrophysical sources of ultrahigh-energy (UHE $> 10^{14}$ eV)neutrinos using the Antarctic ice as a neutrino target;

[a] e-mail: `zhelezny@minus.inr.ac.ru`

DOI: 10.1051/epjconf/201714511002

- the Radio Astronomical Method of the Hadron and Neutrino Detection project **(RAMHAND)** to study EHE cosmic rays bombarding the Moon.

In **Sect. 3** the results of the first RAMHAND-type experiments are considered and great prospects of possible experiments with larger radio arrays SKA, LOFAR, LPA (Pushchino) are discussed. The new scheme for fast simulation of ultrahigh-energy and extremely high-energy cascades in dense media is suggested (see **Sect. 4**).

2. The Russian (Soviet) DUMAND and development of large-scale Radio Neutrino Telescopes

After the Workshops DUMAND-78 at La-Jolla and DUMAND-79 (August 1979) at Khabarovsk and Lake Baikal, Fred Reines and Moisey Markov agreed (in October 1979) that the USSR would join the International DUMAND Hawaii Project. But such a collaboration was not approved by the US authorities (see more about that in [5]).

In 1980 Markov headed the Soviet DUMAND program which included more than 10 Institutes from Kaliningrad (former Konigsberg) up to Vladivostok. The goal of the Soviet DUMAND project was the development of the deep underwater optical and acoustical methods of UHE and EHE neutrino detection in the World Ocean (in particular in the Mediterranean Sea) and in Lake Baikal. And from the beginning there were discussed alternative methods of UHE and EHE neutrino detection.

One of arguments to support the construction of larger-scale (multi gigatons) detectors was of basic interest to study the upper boundary of the neutrino energy spectrum and spectra of CR particles. The existence of neutrinos with energies higher than 10^{20} eV might be related to the fundamental problem of the Physics of Elementary Particles – the possible existence of super-heavy particles. It was emphasized that UHE and EHE neutrinos may arise not only in secondary processes of decays of pions, kaons, charmed particles produced in the interactions of UHE and EHE protons and nuclei with the pulsar shells, interstellar gas, relic photons, etc. They may result from decays and interactions of super-heavy particles. Some of the first "top-bottom" models of the EHE particle (super GZK neutrino) production, considered in 1979–1981, were as following: the decays of Mini-Black Holes [8], interactions (decays) of such hypothetical super heavy particles as magnetic monopoles, GUT particles etc. [9].

The first proposal to detect radio emission of electron-photon cascades, induced in the crust of the Earth by EHE neutrinos from the lower hemisphere and continuing to propagate in the atmosphere in a direction close to horizontal, was made in 1981 [9] (see also [10]).

Askaryan's idea of the 1960s [11] to detect coherent Cherenkov radio emission of cascades produced by cosmic rays in dense dielectric media was revived in 1983 by Gusev and Zheleznykh to be used for developing the radio-wave method of UHE ($> 10^{14}$ eV) cosmic neutrino detection [12]. It was suggested to build a large-scale neutrino detector (telescope) **RAMAND** using the ice in Antarctica as a neutrino target, taking advantage of very weak radio-wave absorption of ice at low temperatures at frequencies below a few GHz [12]. The proposed design involved deploying a surface grid of antennae listening to radio pulses in the frequency band from a few hundred MHz up to 1 GHz, produced by cascades from UHE neutrinos in the ice volume of 10^9–10^{11} m^3. After experiments performed at the Soviet Antarctic Vostok station, the R&D stage of RAMAND, an ice radio-wave detector with an effective volume up to 10 km^3 was approved in 1988 by the Soviet authorities. A hexagonal form of the array with 1280 local stations placed on the 10 km^2 detector area was chosen [13]. The main goal of the RAMAND project was to search for neutrinos with energies above a hundred TeV from astrophysical sources [14]. Unfortunately, RAMAND's work at Vostok was abruptly discontinued in 1991.

Some years later the Radio Ice Cherenkov Experiment (RICE [15]) with antennae placed at depths of 200–400 m in ice and the balloon-born Antarctic Impulsive Transient Antenna (ANITA [16]) experiments had been successfully conducted in Antarctica by D. Besson et al. and P. Gorham et al. Both experiments published limits on the UHE and EHE neutrino flux in the energy range 10^{16}–10^{21} eV. The constraints on the sensitivity of these experiments to the neutrino flux come from the relatively small effective volume of RICE, and relatively short exposure time, limited by the time of the balloon flight, in the case of ANITA.

An alternative design for an Antarctic UHE detection experiment that combines the strengths of RICE and ANITA, while having higher sensitivity and relatively low cost was proposed in [17]. In this paper, a possibility to place antennae on tethered balloons at a few kms above the ice surface was considered. The tethered scheme affords a large field of antenna view (compared to RICE), a high duty cycle (compared to ANITA), and can run in coincidence with a surface array (such as RAMAND).

Two huge radio neutrino detectors are currently being constructed as detectors for UHE and EHE astrophysical neutrinos based on the Cherenkov-Askaryan radio emission from neutrino-induced cascades of 10^{16}–10^{20} eV. The Askaryan Radio Array (ARA) [18] will be deployed in ice at the South Pole, covering 200 km^2, with neutrino sensitivity of 10^{16}–10^{19} eV to detect in particular GZK neutrinos. The Antarctic Ross Ice-Shelf Antenna Neutrino Array (ARIANNA) [19] (similar in concept to RAMAND) is a large surface antenna array.

In 1989 the Radio Astronomical Method of Hadron and Neutrino Detection (RAMHAND) project was proposed to use the Moon and even other celestial objects as giant targets for radio detectors of EHE hadrons and neutrinos [20]. The Moon had been considered as a perspective large-scale target and it was suggested to use large ground-based radio telescopes to search for Cherenkov-Askaryan radio pulses from cascades produced in the lunar regolith by EHE ($> 10^{20}$ eV) neutrinos and other cosmic particles. The possibility to detect radio emission of cascades of lower energies ($> 10^{16}$ eV) using a lunar satellite was also discussed.

3. RAMHAND: The past and the prospects

The first attempt to use a ground-based radio telescope as a detector and the lunar regolith as a target was undertaken by T.H. Hankins, R.D. Ekers and J.D. ÓSullivan in 1996 [21]. The observations were made at 1.5 GHz frequency range using a 64-meter radio telescope of the Parks radio observatory (Australia). A 10-hour search for neutrino interacting with the lunar regolith had only put an upper limit to the isotropic EHE-neutrinos flux. The following attempts of much longer exposures on the Moon's limb undertaken by several teams around the world (see, for example, [22,23]) had put similar but much stronger restrictions to the flux of EHE neutrinos. For example, our 120-hour observations using the Kalyazin 64-m radio telescope (Russia) in the 1.4 GHz and 2.3 GHz frequency ranges [24] showed that the rate of the radio pulses resulting in interactions of neutrinos with energy above 10^{20} eV with atoms of the lunar regolith does not exceed $(200\,\text{hrs})^{-1}$ or $1.4\ 10^{-6}\,\text{sec}^{-1}$.

In subsequent years many more attempts were undertaken to register the bursts of Cherenkov radio emission from the Moon using the Westerbork synthesis radio telescopes (The Netherlands), some part of the VLA antennae (USA), again the Parks 64-m radio telescope, and other radio astronomy tools [25–29]. The new essentially stronger upper limit to the EHE neutrino flux was obtained from these experiments. These results were used to exclude some of the most exotic cosmological models.

At the same time a tendency could be mentioned in using lower frequencies for observations in these relatively recent RAMHAND-type experiments. New estimates of the optimal frequency band for such observations are based not only on the spectrum of the cascade's emission in the lunar regolith. They also take into account an increasing of the target volumes as well as less scattering of the cascade's emission at lower frequencies. Of course, an additional reason is the large dimensions of the existing low frequency radio telescopes. For example, the giant Low Frequency Array (LOFAR), constructed in Europe during the last decade [29] is now considered as a very high sensitive detector in RAMHAND-type experiments [30].

During the two last decades radio astronomers from Europe, Australia, South Africa and some other countries actively work on the SKA-project. First phases of this Square Kilometer Array SKA-LOW and SKA-MID are under construction now in Australia and in South Africa, respectively. Both multi-element radio telescopes should start working in the beginning of the 2020 s decade. Radio astronomers and HE-physicists hope to use both arrays as detectors of the Cherenkov-Askaryan radio pulses from the Moon [31,32]. Modelling of the cascades, their radio emission and corresponding detecting processes show that both arrays could be very effective tools not only for many kinds of classic radio astronomical tasks, but also for RAMHAND-type experiments.

Nevertheless, despite such promising prospects, astronomers are continuing their efforts to perfect the methods of detecting Cherenkov bursts from the Moon, and to apply them on the existing already radio telescopes. We are also working in this field, trying to improve our registration system and computer programs, which we use for on-line and off-line data reduction and analysis. First of all, we are going to analyze the data obtained during the monitoring of the large part of the sky with the Large Phased Array of the Pushchino Radio Astronomy Observatory (Fig. 1). This meridian radio telescope is one of the most sensitive meter wave length

Figure 1. Large Phased Array of the Pushchino RAO ASC LPhI.

Figure 2. This 16-element modulus is 1/8 part of the future pathfinder in the Pushchino RAO.

radio telescopes. Its effective area is around $45000\,\text{m}^2$ in t,he zenith (corresponding to a declination around +55°) and decreases in the other directions as Cos Z. Beginning from mid-2013 this radio telescope has been used for day by day monitoring, at 111 MHz, of the large part of sky between declinations δ =-8° and +42°. We plan to carefully analyze all the tracks through the meridian plane by the Moon and hope to get around a 10-hour exposure using the data of the past three years. Multi-beam patterns of the phased array allow distinguishing various kinds of interference from the useful signal. We will try also to use another beam-forming system of the array and improve the time resolution of the data, but it will only be done for future observations.

Another possibility to get a new estimate of the extremely high energy cosmic rays and neutrinos, which we are planning to realize in the coming two or three years, is the implementation of monitoring the short radio bursts from the Moon at the small model of the meter-wave radio telescope with a wide field of view.

The model is constructed in Pushchino RAO (Fig. 2). It will be a wide-band phased array for the 40–65 MHz frequency range and will have around $1000\,\text{m}^2$ effective area. The field of view of this antenna will be around 2 steradians.

We plan to use it for monitoring different kinds of sporadic radio bursts as well as different transients and rather fast variations of the cosmic radio emission. The Moon will track through the radio telescope's field of view for 3 or 4 hours per day when its declination will be above +15°. The model will consist of eight 16-element modulus and will have 128-beam pattern. Certainly, we could not realize a good sensitivity with this model. It means that there will be a chance to only register particles above 10^{22} eV. At the same time we could hope to get with it a rather large exposure (around 300–350 hours per year). The registration system that will be worked out for this model could be later used for monitoring the Moon with other, much more sensitive, radio telescopes (for example, the Indian Ooty radio telescope).

4. Simulations of cascades

The new scheme for fast simulation of UHE cascades with energies E > 10^{15} eV and EHE cascades with energies up to 10^{20} eV in dense media is suggested. At a later time this scheme will be used for calculations of radio emission of cascades with energies up to 10^{20} eV and higher in the wide frequency band of ~0.1–a few GHz.

Electron-photon cascades up to energies 10^{13}–10^{15} eV may be simulated using GEANT-4 [33] with the standard cross sections of electromagnetic processes. Longitudinal and lateral charge excess profiles, track lengths of charged particles and other characteristics should be simulated. Results of these simulations should be approximated and stored. To reduce the CPU time for simulation of cascades at UHE energies a hybrid scheme was developed. This hybrid scheme had been suggested to estimate both the radio and acoustical signals caused by neutrino-induced cascades with energies up to 10^{20} eV in dense media (e.g. in ice, water and the lunar regolith). As a part of this hybrid scheme some original one-dimension Monte Carlo code was suggested to be used at higher energies. This original one-dimension Monte Carlo code will be exploited to simulate cascades up to 10^{20} eV using approximations of results estimated with help of the GEANT4 package. This Monte Carlo code exploits the Migdal cross sections [34] to take into account the Landau-Pomeranchuck effect [35] now called the Landau-Pomeranchuck-Migdal (LPM) effect. This Monte Carlo code follows both electrons and gamma quanta up to some threshold energy E_{th}. The Monte Carlo approach should be initially used to take into account fluctuations in the development of the individual cascades. It should be mentioned that these fluctuations are very large at high energies because of the LPM effect. Due to this effect the mean interaction length may increase many times and become much larger than the cascade radiation length. For this reason, the cascade curve would have a stochastic form. A program which realizes this Monte Carlo approach for the Migdal cross sections for processes of pair creation and bremsstrahlung has been developed and tested. It should be mentioned that to calculate these cross sections accurately, very small energy bins (order of 10^{-4}) have been used. It was found that the cascade length increases as $E^{0.395}$ and at an energy of 10^{20} eV it can exceed the value of 1 km (to be compared with the standard Bethe-Gaitler length of several dozen meters).

Below this threshold energy, E_{th}, approximated results of GEANT4 simulations are used to estimate the longitudinal and lateral charge excess profiles of charged particles in cascades induced by cosmic neutrinos. Then the radio emission of such cascades may be calculated. Some profiles for the total number of particles (electrons + positrons) as well as the charge excess (electrons − positrons) in showers initiated in the lunar regolith (d=1.41 g/cm^3) by gamma quanta with various energies have been calculated using a kinetic energy threshold of 0.1 MeV. Some results are shown in Figs. 3–5. Figure 3 shows a longitudinal distribution for both the charge excess $\delta(z)$ and its relative fraction $\delta n(z)$ for a gamma shower with energy E= 100 GeV. It is seen that the charge excess at the maximum of the shower z=100 cm is slightly below 30%.

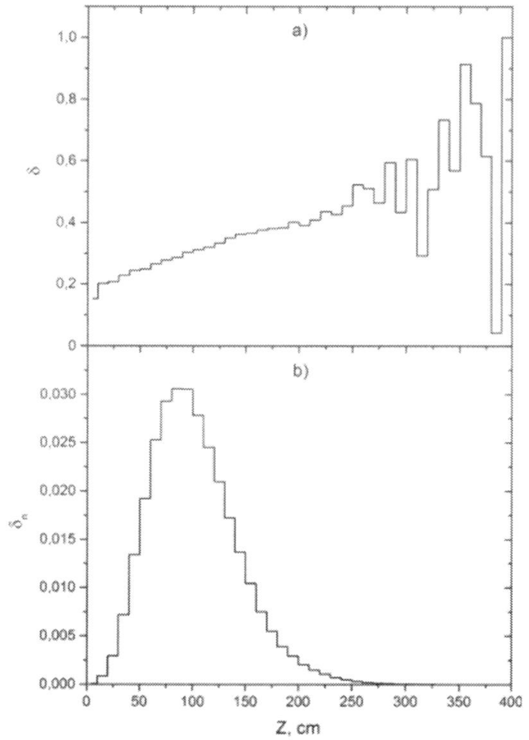

Figure 3. Longitudinal distribution of charge excess and its fraction.

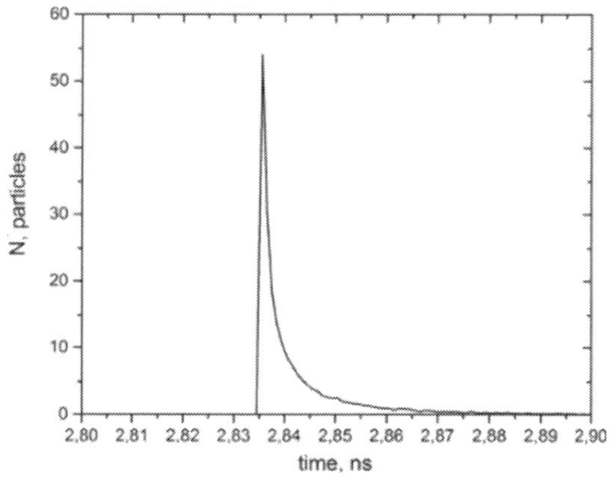

Figure 4. Time distribution of the shower front width.

The time distribution of particles which are crossing the position of shower maximum is shown in Fig. 4 for

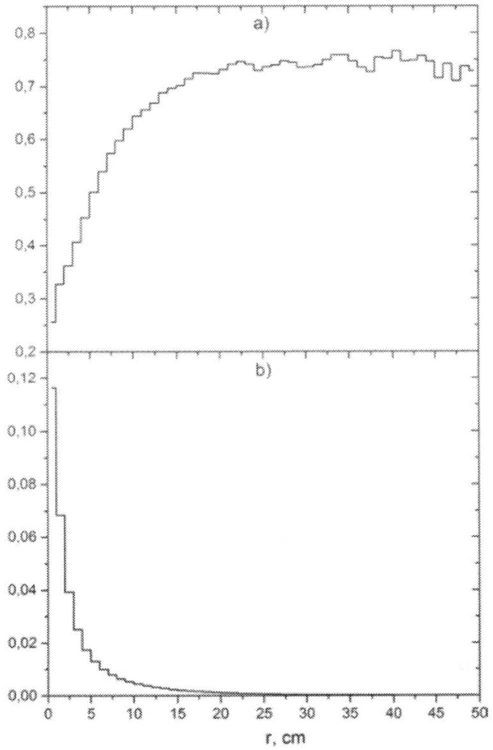

Figure 5. Lateral distribution of charge excess and its fraction.

a gamma shower with energy E=100 GeV. It is seen that the total width of this distribution is slightly below 0.04 ns which corresponds to nearly ~1 cm width of the shower front.

The radial distribution of the charge excess, $\delta(r)$, and its relative fraction, $\delta n(r)$, are shown in Fig. 5 for a gamma shower with energy E= 100 GeV. It is seen that the charge excess at r~1 cm is nearly equal to 30%. So, it is possible to conclude that both the time and radial distributions show widths of the shower charge excess of the order of 1-5 cm. So, detection of radio emission with frequencies ($\nu = c/\lambda$) below 3 GHz is possible. The charge excess distribution displays 25–30% negative excess near the maximum position.

5. Conclusion

At the beginning of the 2000 s V.L. Ginsburg wrote in his book "About Science, Myself and Others": "At last, it is literally on an exit there is Neutrino Astronomy of High Energies > 10^{12} eV"[36]. And the first astrophysical HE neutrinos had been registered by a team of the **gigaton optical HE neutrino telescope** IceCube in Antarctica in 2013. It happened as a result of the very successful development of the optical method of deep under-ice neutrino detection during 25 years.

The development of multi gigaton radio detectors for UHE (> 10^{15}) and EHE (> 10^{18}) Neutrino Astronomy, using the Antarctic ice (RAMAND, 1983–1991) and the lunar regolith (RAMHAND, since 1989), had started in the framework of the Russian (Soviet) DUMAND program more than 30 years ago.

It is possible to now say that it is literally on an exit there is UHE and EHE Neutrino Astronomy: the new generation of huge radio neutrino telescopes under construction in Antarctica **ARA and ARIANNA are designed as discovery instruments for astrophysical neutrinos of E > 10^{16}–10^{17} eV**.

Using the new larger-scale radio astronomical telescopes (SKA and others) for lunar experiments would give great possibilities for studying fundamental problems of Cosmic Ray (Astroparticle) Physics and Neutrino Astronomy in the interval of energies **more than 10^{20} eV**.

This work was supported by RFBR (grant 16-29-130740). The authors are deeply indebted to Bryan Pattison for valuable discussions.

References

[1] M.A. Markov, Proc. 10th Int. Conf. on High-Energy Physics at Rochester, 579 (1960)
[2] I.M. Zheleznykh, Thesis, Moscow State University, Physical Faculty (1958)
[3] M.A. Markov, I.M. Zheleznykh, Nucl. Phys. **27**, 385 (1961)
[4] F. Reines, J. Sellschop, Scientific American, February (1966)
[5] A. Roberts, Rev. Mod. Phys. **64**, 259 (1992)
[6] J. Learned, Ch. Spiering, CERN Courier, July/August (2016)
[7] F. Halzen, J. Learned, "High energy neutrino detection in deep polar ice", preprint U. Wisconsin, June (1988)
[8] M.A. Markov, I.M. Zheleznykh, Proc. 1979 DUMAND Workshop at Khabarovsk and lake Baikal, ed. J. Learned, Honolulu, 177 (1980)
[9] L.G. Dedenko, M.A. Markov, I.M. Zheleznykh, Proc. Neutrino-1981, vol. 1, eds. R Cence, E. Ma and A. Roberts, Maui, Hawaii, 292 (1981)
[10] M.A. Markov, I.M. Zheleznykh, Nucl. Instr. Meth. A **248**, 242 (1986)
[11] G.A. Askaryan, Sov. Phys. JETP **14**, 441 (1962)
[12] G.A. Gusev, I.M. Zheleznykh, JETP Letters **38**, 505 (1983)
[13] A.L. Provorov et al., Proc. 3rd Int. Workshop on Neutrino Telescopes, Venice, ed. Milla Baldo-Ceolin, 337 (1991)
[14] A.L. Provorov, I.M. Zheleznykh, Astroparticle Physics **4**, 55 (1995)
[15] I. Kravchenko, et al., Physical Review D **73**, 082002 (2006)
[16] P. Gorham, et al., Physical Review Letters **103**, 051103 (2009)
[17] D. Besson, R. Dagkesamanskii, I Kravchenko, et al., Nuclear Instr. & Methods in Phys. Res. A **662**, 50 (2012)
[18] P. Alison, et al., Astroparticle Physics **70**, 62 (2015)
[19] S.W. Barwick, et al., Astroparticle Physics **12**, 62 (2015)
[20] R.D. Dagkesamanskii, I.M. Zheleznykh, Sov. Phys. JETP **50**, 259 (1989)
[21] T.H. Henkins, R.D. Ekers, J.D. OŚullivan, Monthly Notices of the Roy. Astron. Soc. **283**, 1027 (1996)
[22] P.W. Gorham, C.L Hebert, K.M. Liewer, et al., Phys. Rev. Lett. **93**, 041101 (2004)

[23] A.R. Beresnyak, R.D. Dagkesamanskii, I.M. Zheleznykh, et al., Astron. Rep. **49**, 127 (2005)
[24] C.W. James, R.D. Ekers, J. Alvarez-Muniz, et al., Phys. Rev. D **81**, 042003 (2010)
[25] R.E. Spencer, A. Macfarlane, O. Mills, L. Piccirillo, Proc. 10th EVN Symposium, 097 (2010)
[26] T.R. Jaeger, R.L. Mutel, K.G. Gayley, Astropart. Phys. **43**, 293 (2010)
[27] S. Buitink, O. Scholten, J. Bacelar, et al., Astron. & Astrophys. **521**, A47 (2010)
[28] J.D. Bray, R.D. Ekers, C.W. James, et al., Proc. 32nd ICRC (2011)
[29] N.P. van Haarlem , M.W. Wise, A.W. Gunst, et al., LOFAR: The Low Frequency Array, `arXiv: 1305.3550`
[30] K. Singh, M. Mevius, O. Scholten, et al. (LOFAR Coll.), Nuclear Instr. & Methods in Phys. Res. A **664**, 171 (2012)
[31] J.D. Bray, J. Alvarez-Muniz, S. Buitink, R.D. Dagkesamanskii, R.D. Ekers, et al., "Lunar detection of ultra-high-energy cosmic rays and neutrinos with the Square Kilometre Array", Proceedings of Science, AASKA14, 144, in Advancing of Astrophysics with Square Kilometer Array, 1, 961 (2015). See also `arXiv: 1408.6069v2 [astro-ph.IM]` 19, Dec. 2014
[32] C. James, on behalf of the SKA High Energy Cosmic Particles (lunar) Focus Group, Proceedings of the ARENA 2016, Groningen, The Netherlands
[33] Geant4 User's Guide for Toolkit Developers `http://geant4.web.cern.ch/geant4/support/userdocuments`
[34] A.B. Migdal, Phys. Rev. **103**(6), 1811 (1956)
[35] L.D. Landau, I.Ya. Pomeranchuk, Dokladi AN USSR **92**, 535 (1953)
[36] V.L. Ginzburg, "About science, myself and others", Moscow, Fizmatlit (2003)

Search for EAS radio-emission at the Tien-Shan shower installation at a height of 3340 m above sea level

A. Beisenova[1], E. Boos[1], A. Haungs[2], T. Sadykov[1], N. Salihov[3], A. Shepetov[4], Y. Tautayev[1,5,a], L. Vildanova[4], and V. Zhukov[4]

[1] Institute of Physics and Technology, Almaty, Kazakhstan
[2] Karlsruhe Institute of Technology, Karlsruhe, Germany
[3] Institute of Ionosphere, Almaty, Kazakhstan
[4] P.N. Lebedev Physical Institute of the Russian Academy of Sciences, Moscow, Russia
[5] Kazakh National University by al-Farabi, Almaty, Kazakhstan

Abstract. The complex EAS installation of the Tien Shan mountain cosmic ray station which is situated at a height of 3340 m above sea level includes the scintillation and Cherenkov detectors of charged shower particles, an ionization calorimeter and a set of neutron detectors for registering the hadronic component of the shower, and a number of underground detectors of the penetrative EAS component. Now it is intended to expand this installation with a promising method for detecting the radio-emission generated by the particles of the developing shower. The facility for radio-emission detection consists of a three crossed dipole antennae, one being set vertically, and another two – mutually perpendicularly in a horizontal plane, all of them being connected to a three-channel radio-frequency amplifier of German production. By the passage of an extensive air shower, which is defined by a scintillation shower detector system, the output signal of antenna amplifier is digitized by a fast multichannel DT5720 ADC of Italian production, and kept within computer memory. The further analysis of the detected signal anticipates its operation according to a special algorithm and a search for the pulse of radio-emission from the shower. A functional test of the radio-installation is made with artificial signals which imitate those of the shower, and with the use of a N1996A type wave analyzer of Agilent Technologies production. We present preliminary results on the registration of extensive air shower emission at the Tien Shan installation which were collected during test measurements held in Summer 2016.

1. Introduction

The registration of radio-emission generated by particles of high-energy extensive air showers (EAS) has a number of advantages over other methods of primary cosmic ray investigation, both of which are based on direct registration of shower particles (a relative simple and cheap radio-detector system compared with wide-spread electronic detector systems, the large spatial volume of the space controlled by a single radio antenna, a sensitivity to the characteristics of longitudinal EAS development), and the methods connected with the registration of Cherenkov and fluorescent EAS emissions in ultraviolet and optic diapasons (an unrestricted duty cycle of radio installation which is independent of daytime and weather conditions). Because of these reasons, it is quite natural that the first registration attempts of radio-frequency (RF) signal from EAS were only tried in the 1960–1970 decades. These works were based on the results of theoretical studies of the possible generation mechanism of radio-emission by shower particles: the coherent electromagnetic radiation of negative charge excess which is present in developing EAS [1], and the model of a moving electric dipole which should arise because of the separation of the differently charged shower particles in a geomagnetic field [2]. As a result of these early investigations, the EAS signals were indeed found in the frequency range of 30–100 MHz, but the researchers met a number of difficulties in the interpretation of the registered signals which were connected with both insufficient qualities of those days' electronics and a bad reproducibility of the results because of electromagnetic interferences and weather conditions. A comprehensive review of previous studies on the detection of EAS radio signal can be found in [3].

The interest in the radio-method of EAS investigation reappeared in the early 2000 s when the experiments CODALEMA (France) [4] and LOPES (Germany) [5] started their work. The availability of this method was proved by the simplicity and low cost of radio devices built using modern electronic modules and perspective application of contemporary information technologies for analyzing the registered data, as well as its effectiveness for the study of EAS events caused by ultra-high energy cosmic ray particles (10^{19}–10^{20} eV) whose statistics remains rather scarce up to the present time.

A convenient place for the development of the radio-method of EAS particle registration is the Tien Shan mountain cosmic ray station (43°15′ N, 76°57′ E, 3340 m above sea level) whose installations for the complex EAS investigation in the range of primary energies of 10^{14}–10^{18} eV permit both a direct detection of the charged shower particles [6] at the station's level,

[a] e-mail: tautaev@mail.ru

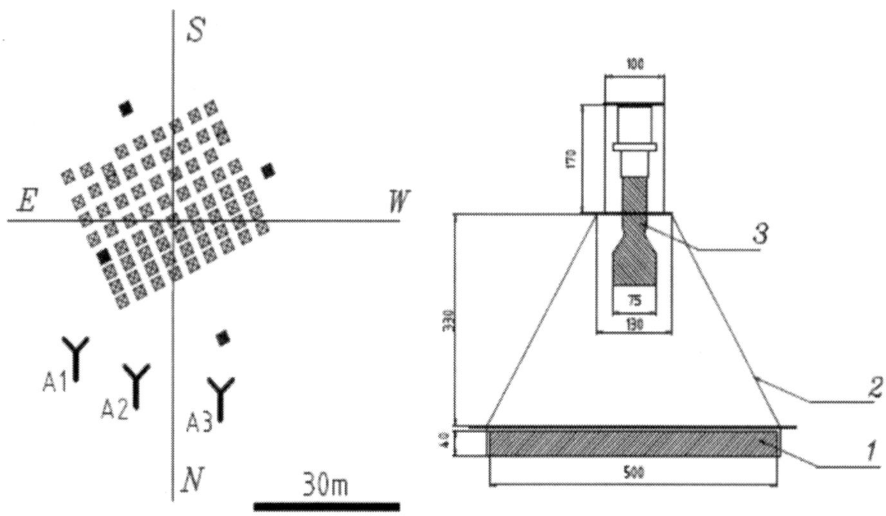

Figure 1. Left: the particle detector disposition scheme in the center part of the Tien Shan shower installation. Squares – the scintillation detectors of the central carpet, A1 – A3-the positions of the radio-antennae. Right: the internal arrangement of a scintillation detector (1–a $500 \times 500 \times 40\,mm^3$ plastic scintillator, 2–the pyramid-shaped reflector of scintillation light, 3–PMT; dimensions are shown in mm).

and registration of Cherenkov light emitted above in atmosphere [7]. Inclusion of a system for registering EAS electromagnetic radiation in the RF range into this complex gives a possibility of mutual calibration of all three EAS investigation methods, and permits to set accuracy limits on EAS parameters definition procedures based on their simultaneous employment. Later on, with the development of experimental techniques and enlargement of the detector system the study of EAS up to 10^{19} eV primary energy should be possible. This opens up the possibility of studying many yet unclear problems, such as the primary composition, anisotropy, sources of ultrahigh energy cosmic ray particles, specific features of their interaction with matter etc, which stay before the contemporary cosmic ray physics.

With the help of the Karlsruhe Institute of Technology, Germany which donated the necessary apparatus, activity started in Summer 2016 at the Tien Shan station to register EAS RF signals in the 25–85 MHz frequency range. The description of the experimental installation and preliminary results of the test measurements are the subject of this paper.

2. Experimental set-up

The wide-spread shower installation used for the registration of EAS charged particles at Tien Shan is built on the basis of plain polystyrene scintillators with $0.5 \times 0.5 \times 0.05\,m^3$ size. For collection homogeneity of scintillation light, the scintillators are placed within a pyramid-shaped light reflector casing and viewed by photomultiplier tubes (PMT) of types FEU49, FEU84, FEU115. A pair of electric pulse signals taken from the PMT anode with different electronic amplification ($a \sim 30\times$ amplification high-sensitive signal, and a $1\times$ low-sensitive one) are transmitted in parallel and in analogue form to the data collection and acquisition center; simultaneous registration of the signals of two amplitude diapasons ensures the whole dynamic range about $\sim(5\text{–}7) \times 10^4$ of scintillation amplitudes measured linearly and without noticeable saturation effects.

In the data acquisition center, the scintillation amplitudes which were present at the PMT outputs at the time of an EAS passage are digitized by a special multichannel amplitude-to-digital conversion (ADC) system which gives the momentary distribution profile of the density of charged shower particles over the installation plane. The ADC system of the Tien Shan station is built on the basis of 12-bit ADC chips of AD7888 type [8] and permits simultaneous digitizing of up to some thousands of analogue signals for the time that does not exceed 0.5 ms.

The central part of the shower detector system which operated together with the EAS radio-emission registration facility in Summer 2016 is a carpet of 72 scintillation detectors placed nearly equidistantly in the nodes of a net with $4\,m \times 3\,m$ spatial step in the shape of an elongated rectangle with a total area of about $900\,m^2$. The EAS selection algorithm which was applied at that time ensures a 100% registration probability for showers with primary energy $E_0 > (5 - 7) \times 10^{14}$ eV whose axes pass through the central scintillation carpet; in every case of EAS registration a special control pulse signal, the shower trigger, was generated to synchronize any detector subsystems which were working simultaneously with the shower installation.

The positions of the detectors within the central part of the Tien Shan shower installation and the basic scintillation detectors are shown in Fig. 1. A detailed description of the whole detector complex and the data operation algorithms of the Tien Shan mountain station, as well as a report on the results of calibration measurements made there in 2014–2015, can be found in [6].

The zenith and azimuth angles of EAS arrivals are estimated by the mutual time delays between scintillation signals in two detector pairs placed symmetrically at opposite sides of the central carpet (these detectors are

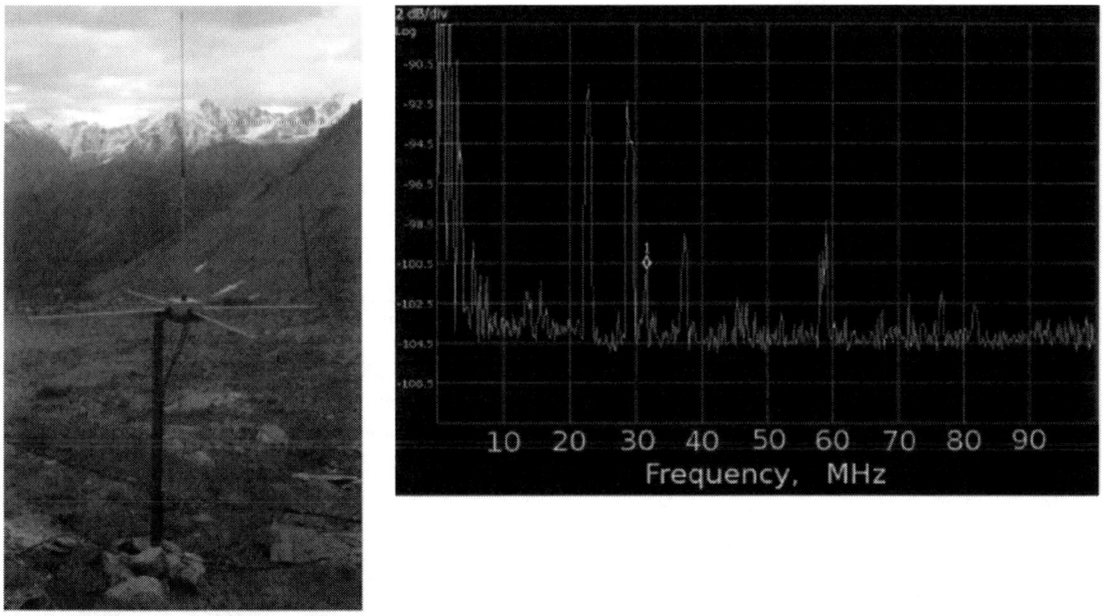

Figure 2. Three dipole antennae set installed at point A1 at the Tien Shan mountain station (left), and the spectrum of background radio-emission measured with the Agilent Technologies N1996A wave analyzer (right).

shown as black squares in Fig. 1). These delays are measured automatically for every registered EAS event using a 4-channel Tektronix TDS 2014 digital oscilloscope which operates with a 4 ns granularity for the time axis.

A shortcoming of the Tien Shan shower installation in its current state of Summer 2016 was the absence of any operational detectors in its periphery. This limited the range of registered events with satisfactorily defined shower characteristics for only EAS whose axes came within the central scintillation carpet. To overcome this, additional scintillation detectors are now being installed at various points at distances of 20, 40, and 70 m from the carpet's center.

During the test measurements searching for EAS connected radio-signals at Tien Shan three radio-antennae sets placed along the northern side of scintillation carpet were used, as shown by A1-A3 in Fig. 1. Three dipole antennae were placed in each of these points; one of them being set vertically, and other two-mutually perpendicularly in the horizontal plane with their axes oriented in parallel to meridional and latitudinal directions (see the left panel of Fig. 2). The RF-signals from these antennae after detection in 3-channel receiver devices designed by Karlsruhe Institute of Technology and placed immediately at the antennae positions were transmitted through matched 50 Ω cable lines to the data collection center.

The frequency spectrum of background radio-emission accepted by the antennae system after its installation at the operation place (with no connection to any external trigger signal) is shown in the right panel of Fig. 2. According to the position of the spectrum density peaks it is seen that the considered antennae system is sensitive mostly in the 25–85 MHz frequency range. A high level of electromagnetic interference was also found to be temporarily present at these frequencies, in particular during the night time, which practically cancels any attempt of EAS signal registration by masking it between intensive high frequency noise. The problem of interference control must be one of the most important in the definitive design of the EAS radio system at Tien Shan.

A fast ADC system consisting of a pair of 4-channel CAEN DT5720 electronic modules of Italian production [9] was used to register the detected signal from radio-antennae set in the experiment.

This system ensures digitizing the analogue input signals with a 4 ns time resolution and keeps in its internal memory continuous records of the time history of the input signal stored for some tens of microseconds as a minimum. Presently, the outputs of all three antennae dipoles of the A1 point, and two horizontal dipoles of the point A2 are connected to these ADC modules.

Following the arrival of an EAS trigger signal generated by the scintillation shower system, the whole data set kept within the internal ADC memory is transmitted to a special program of the main control computer, and the latter in turn sends this information to the general Tien Shan database. Simultaneously, the spatial distribution of the EAS particle density measured with the central scintillation carpet detector system are also loaded in the same database; these distributions can be used later for determining the parameters of the corresponding shower: the position of the shower core in the plane of the shower installation, zenith and azimuth directional angles of the shower axis, the parameters of shower age and size (the size has a physical meaning of the total number of particles in the shower. It is proportional to the primary energy of a cosmic ray particle, E_0, which caused the event, and can be used, with some assumptions, for its immediate estimation). All records in the Tien Shan database are supplied with a time stamp which marks the moment of EAS registration with an accuracy of

DOI: 10.1051/epjconf/201714511003

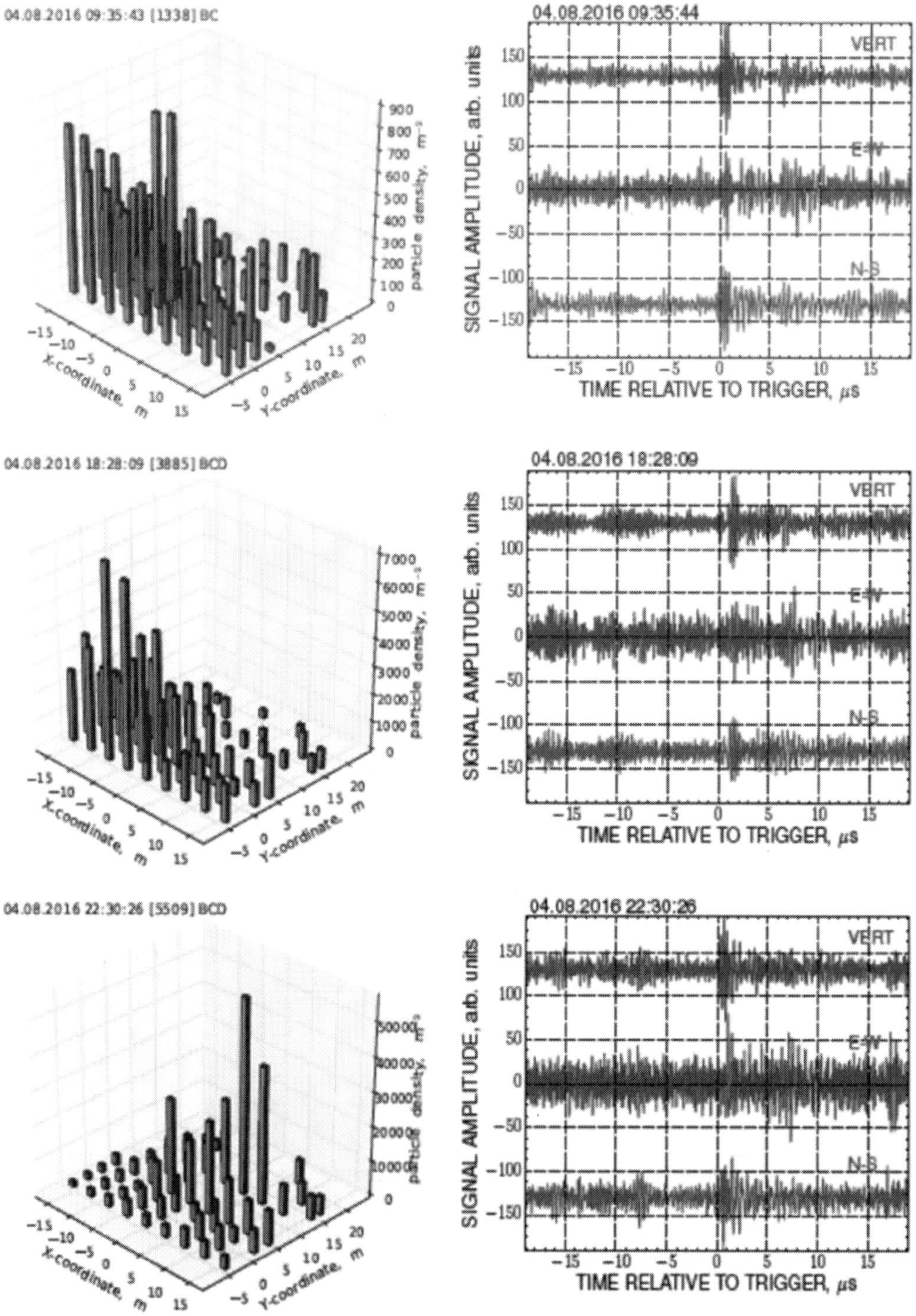

Figure 3. Sample events with a noticeable radio-signal from EAS. Left panels: 2D distribution of the density of charged shower particles in the plane of the central scintillation carpet; right panels: time series of the detected radio-signal amplitude from the vertical and two horizontal (East-West and North-South) antennae dipoles of the A1 point. Zero point of time axis in the right panel plots corresponds to the arrival time of the shower trigger.

1 s (in the UT time zone); the clock of the local data registration system is verified continuously according to external Internet and GPS time servers.

3. Experimental results

The test experiment on the simultaneous registration of the data from the shower and radio systems of the Tien Shan station was held during four days in August 2016, the total life-time of the installation's operation was 70 h. The purpose of this experiment was to clarify the real operational conditions for the registration of radio-signals at Tien Shan and to check the whole complex of hardware and software tools necessary for synchronous recording of the data from both detector system types. All records of radio-signal time series written in this period were looked over visually, and the signs of the presence of radio-emission signal above the usual fluctuation level in the $\pm 10\,\mu$s vicinity of shower trigger moment were searched for in every registered radio-event. The total amount of examined radio-records is about 30000, of which 400 candidate events were selected which have an indication of a noticeable radio pulse shortly after EAS arrival time. Characteristic examples of such event type are presented in the plots of Fig. 3.

For each of the events in Fig. 3 the spatial 2D distribution of the density of EAS charged particles is shown in the plots of the left column, while the right column plots demonstrate the time distributions of detected radio-signal intensity obtained from three dipoles of antennae at point A1 (antennae signals at point A2 are quite similar to those shown in the figure). Every one of the shown distributions is registered with a 4 ns temporal resolution and consists of 10000 sequential measurements of signal intensity which correspond to the full duration of the measured time series of $40\,\mu$s. All distributions are centered relative to the arrival time of the EAS trigger which came from the scintillation shower system; hence, the leading $20\,\mu$s in every time series corresponds to the 'prehistory' of the radio-signal which precedes the EAS time, and the next $20\,\mu$s describe the time behaviour of the radio-signal after the passage of the shower front through the antennae system.

A noticeable radio-signal pulse with amplitude significantly above the mean level of stochastic fluctuations is present in all events of Fig. 3, and this pulse always arrives with a constant time delay of 1–3 μs duration relative to the trigger. At the same time, it is seen from the particle density distributions in the left column that the axis of corresponding shower has evidently come somewhat northward of the central scintillation carpet, i.e. at the side of the radio-antennae location, at least in the events of 09:35:43 UT and 18:28:09 UT. Such an asymmetric disposition of EAS core is typical for the greater part of the events selected by the presence of a visible radio-signal just after the shower trigger, which is evidence that the observation of both signal types (i.e. an EAS and a radio-pulse delayed for a constant small time) in these events cannot simply be a random coincidence.

Because of the side location of EAS core in selected events relative to the central scintillation carpet and the absence of peripheral particle detectors in the operational state during the considered measurements run any strict determination of shower parameters in the presented events is impossible. If it is supposed that the shower axis of the 09:35:43 UT and 18:28:09 UT events is somewhere around the region of the radio-antennae sets, i.e. at a ~30–40 m distance from the center of the scintillation carpet, then according to the registered distributions of the particle density one can state that the primary EAS energy, E_0, in these events must be, correspondingly, at least $(2\text{–}3)\cdot 10^{16}$ eV and 10^{17} eV respectively (these estimations should be taken in the sense of lower limits). As for the 22:30:26 UT event, its maximum registered particle density of the order of 40000 particles/m^2 corresponds to a primary energy $E_0 > (3\text{–}5)\cdot 10^{17}$ eV. Considerable E_0 estimations which result, again, for the most part of the events selected by the presence of a radio-pulse in the vicinity of their shower front is another confirmation that this coincidence is non-random. Remembering that the threshold of the shower trigger during the measurements was about $E_0 > (5\text{–}7)\cdot 10^{14}$ eV, and the overwhelming part of registered shower events have, indeed, energies around 10^{14}–10^{15} eV (since the intensity of the primary cosmic ray flux decreases roughly with increasing energy), then the high E_0 values which result at once in all selected events cannot be fully accidental. It should also be noted that according to both modern experiments [4] and [5] the lower threshold of EAS radio-emission registration in 10–100 MHz diapason is of the order of $5\cdot 10^{16}$ eV which satisfactorily agrees with present estimations for E_0 in the events with a feasible radio-signal observation.

Besides the comparatively prominent radio-signal maximum within the closest 1–3 μs vicinity of EAS triggers frequently there is another secondary weak pulse which can often be seen in the time series of radio-intensity in selected candidate events, and usually follows the primary one at a fixed time of 5–7 μs after the trigger. Such a secondary pulse can be observed in the time distributions of Fig. 3, in particular in the 09:35:43 UT event. The characteristic 5–7 μs time delay of the secondary pulse corresponds to a linear distance of about 1.5–2 km, and this estimation, in its turn, to the distance to the nearest mountain slopes surrounding the Tien Shan station together with its antennae system sets. Hence, the secondary radio-pulse observed in time distributions of selected events can be interpreted as an echo signal resulting from the reflection of EAS radio-emission from the local relief details of Tien Shan station.

It is supposed that, in future, events with an appreciable EAS signal like the ones manually selected in the test experiment should serve as a prototype for the elaboration of an automatic recognition algorithm of the events with EAS radio-emission which is necessary for the final version of the corresponding software complex and any strict statistical investigations.

4. Conclusion

Setting up a radio-signal registration system mostly sensitive in the 25–85 MHz RF range is now in progress at the Tien Shan mountain cosmic ray station. This system is aimed for simultaneous operation with the EAS particle density and Cherenkov radiation detectors which are present at the station, and will permit a mutual calibration of all these independent methods of EAS investigation. Later on, with the use of an EAS radio-emission registration method, an enlargement of the energy range of primary cosmic ray investigation at Tien Shan up to $E_0 \sim 10^{19}$ is anticipated.

As a result of the preliminary test experiment, which was made using a newly installed dipole radio-antennae set, some candidate events were selected which demonstrate the presence of a noticeable radio-signal pulse in closest 1–3 μs vicinity of EAS arrival time. Specific features of particle density distribution in these events permit to state that most of them have a rather high primary energy $E_0 > (2\text{–}5) \cdot 10^{16}$ eV and a close location of their shower cores near to the radio-antennae set, so the time coincidence of the observed radio-pulse with the shower front arrival time in these events cannot be fully accidental. Hence, the radio-antennae system installed at Tien Shan together with a designed program complex for registration of its signal do indeed ensure an effective selection of radio-emission from EAS particles, and further work in this direction should be seriously considered.

References

[1] G.A. Askaryan, Excess negative charge of an electron-photon shower and its coherent radio emission, Soviet Phys. J.E.T.P. **14**, 441–443, 1962

[2] F.D. Kahn and I. Lerche. Radiation from cosmic ray air showers. Proceedings of the Royal Society of London. Series A. Mathematical and Physical Sciences **289**(1417), 206–213, 1, 1966

[3] H.R. Allan, Progress in Elementary Particle and Cosmic Ray Physics, page 169. North Holland Publ., Amsterdam (1971)

[4] D. Ardouin, A. Belletoile, D. Charrier, R. Dallier, L. Denis, P. Eschstruth, T. Gousset, F. Haddad, J. Lamblin, P. Lautridou, A. Lecacheux, D. Monnier-Ragaigne, A. Rahmani, and O. Ravel, Radio-detection signature of high-energy cosmic rays by the CODALEMA experiment. Nucl. Instrum. Methods A **555**(1–2), 148–163, 12, 2005

[5] A. Nigl, W.D. Apel, T. Arteaga, J.C. Asch, J. Auffenberg, F. Badea, L. Bhren, K. Bekk, M. Bertaina, et al., Frequency spectra of cosmic ray air shower radio emission measured with LOPES. Astron. Astrophys. **488**, 807–817, 12, 2008

[6] A.P. Chubenko, A.L. Shepetov, V.P. Antonova, R.U. Beisembayev, A.S. Borisov, O.D. Dalkarov, O.N. Kryakunova, K.M. Mukashev, R.A. Mukhamedshin, N.F. Nam, R.A. Nikolaevsky, V.P. Pavlyuchenko, V.V. Piscal, V.S. Puchkov, V.A. Ryabov, T. Kh. Sadykov, N.O. Saduev, N. M. Salikhov, S.B. Shaulov, A. V. Stepanov, N.G. Vildanov, L.I. Vildanova, M.I. Vildanova, N.N. Zastrozhnova, and V.V. Zhukov, New complex EAS installation of the Tien Shan mountain cosmic ray station. Nucl. Instrum. Methods A **832**, 158–178, 2016

[7] R.U. Beisembaev, Yu.N. Vavilov, M.I. Vildanova, N.G. Vildanov, O.D. Dalkarov, V.V. Zhukov, S.K. Machavariani, R.A. Nam, V.P. Pavlyuchenko, V.A. Ryabov, N.O. Saduev, T.H. Sadykov, A. V. Stepanov, and Zh.S. Takibaev, The first results obtained with the installation HORIZON-T. J. Phys.: Conf. Ser. **409**, 012127, 2013

[8] Analog Devices. 2.7 V to 5.25 V, micropower, 8-channel, 125 kSPS, 12-bit ADC in 16-lead TSSOP AD7888 Datasheet, `http://www.analog.com`, 2015

[9] CAEN. Mod. DT5720 4 channel 12 bit 250 MS/s Digitizer Manual Rev. 9, `http://www.caen.it/servlet/checkCaenManualFile?Id=10257/`, 2014

Horizon-T experiment status

Dmitriy Beznosko[1,a], Rashid Beisembaev[2], KanatBaigarin[1], Elena Beisembaeva[2], Oleg Dalkarov[2], Vladimi Ryabov[2], Turlan Sadykov[3], Sergei Shaulov[2], Alekse Stepanov[2], Marina Vildanova[2], Nikolay Vildanov[2], and Valeriy Zhukov[2]

[1] Nazarbayev University, Physics Department, Astana, KZ
[2] P. N. Lebedev Physical Institute of the Russian Academy of Sciences, Moscow, Russia
[3] LLP "Institute of Physics and Technology", Almaty, Kazakhstan

Abstract. Horizon-T is an innovative detector system constructed to study Extensive Air Showers (EAS) in the energy range above 10^{16} eV coming from a wide range of zenith angles (0°–85°). The system is located at the Tien Shan High-altitude Science Station of the Lebedev Physical Institute of the Russian Academy of Sciences at ~ 3340 meters above sea level. It consists of eight charged particle detection points separated by distances up to one kilometer as well as an optical detector subsystem to measure the Vavilov-Cherenkov light from the EAS.

The time resolution of charged particles and Vavilov-Cherenkov light photons passage of the detector system is a few ns. This level of resolution allows conducting a research of the atmospheric development of individual EAS. This report focuses on a general description of the detector system and the individual sub-systems providing an overview of the operations and latest results.

1. Introduction

The Tien Shan High-altitude Science Station (TSHSS), of the P. N. Lebedev Physical Institute of the Russian Academy of Sciences, is located at a height of ~3340 above sea level. The Horizon-T [1] (HT) detector system was constructed there in order to study both the spatial and temporal distribution of the charged particles within the EAS with a time resolution of few ns. This new method allows to collect more data from each detection point (particle density and time of passage) and increase the measurement precision of the EAS parameters and accurately measure the core position even if it falls outside of the detector system area.

HT consists of eight charged particle detection points separated by distances up to one kilometer as well as an optical detector subsystem to view the Vavilov-Cherenkov light from the EAS. The aerial view of the detector system is shown in Fig. 1.

2. HT detectors

The HT detector system consists of detectors of several different types: plastic scintillator [2] with MELTS [3] FEU49 model PMT, plastic scintillator with Hamamatsu [4] R7723 PMT assembly, glass with R7723 PMTs and a Vavilov-Cherenkov radiation detector with both FEU49 and Hamamatsu H6527 PMT assemblies.

2.1. Scintillator detectors

The Scintillator Detectors (SD) all use polystyrene-based square-shaped cast scintillator with 1 m^2 area and 5 cm thickness. The near periphery SD (detection points 1, 4, 5, 6 and 7) use R7723 PMT assembly as photo-detector, and the far periphery (points 2, 3 and 8) all use FEU49. This is due to the fact that long cables are used to connect these points to the Data Acquisition system (DAQ) which is located at detection point 1. The cable calibration [5] shows that signal widening becomes significant thus justifying the use of FEU49 as they have higher gain but wider initial pulse. Three SD, oriented perpendicular to each other in the x, y and z planes, are located at each detection point. The z-plane is parallel to the sky, the x-axis is directed north.

Each SD is calibrated using the signal from minimum ionizing particles (MIP) from the incoming cosmic ray flux. For that, an additional trigger detector consisting of FEU49 and a 15 cm diameter scintillator is spaced under each SD being tested. Double-coincidence schema is used. The resulting calibration gives the area of a single MIP signal as well as the width of the MIP signal pulse from each SD. Due to analysis specifics of the data, calibration is taken between the 0.1 and 0.9 of the pulse area. This also excludes the baseline noise effects. The area of a MIP signal for the z-plane SD from detection point 1 is shown in Fig. 2. The Landau curve fit gives the most probable value used for the calibration.

We define the front edge of a pulse as the time between 0.1 and 0.5 of the pulse area. With that definition, the pulse front of the SD with R7723 PMT is 5.9±1.1 ns as seen from Fig. 3. The total width is a time between 0.1 and 0.9 of the pulse area and is 18.2±4.2 ns. Both values are for the z-plane SD from detection point 1 but are typical for any SD with R7723 PMT. Here all uncertainties quoted are the widths of the distributions from the fit and not the errors on the parameter. The uniformity of the response across the SD area is also studied.

[a] e-mail: dmitriy.beznosko@nu.edu.kz

DOI: 10.1051/epjconf/201714511004

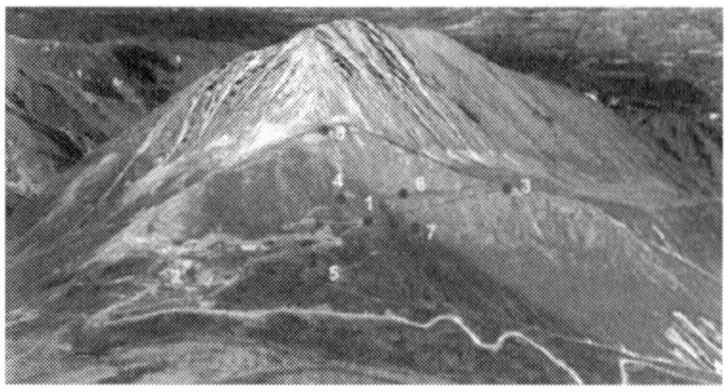

Figure 1. Aerial view of the Horizon-T detector system.

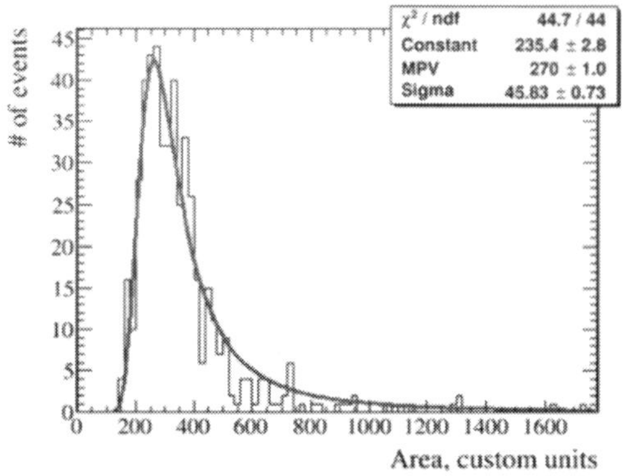

Figure 2. MIP area for SD.

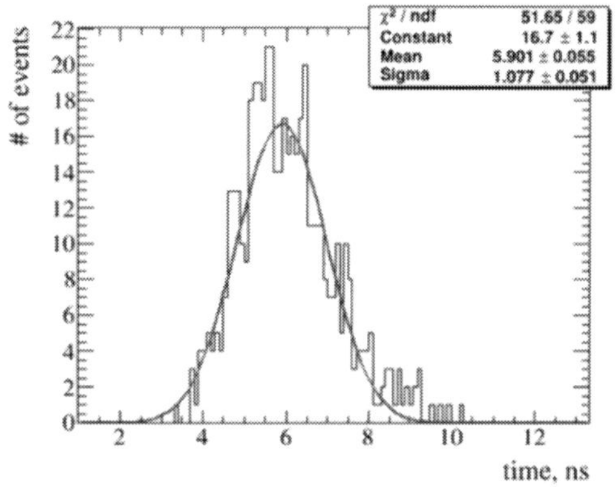

Figure 3. Front edge for SD detector pulse with R7723 PMT.

The SD with the FEU49 also get their MIP response calibrated using exactly the same procedure. From that process, we also get the following typical timings: pulse front 23.2±5.1 ns and total width of 102±29 ns (the pulse front and the total duration for the FEU49 itself are 15.7±0.5 ns and 38.9±1.4 ns). Data is for the z-plane SD from detection point 2 and includes the clearly seen influence of the long cables used to connect these points to the DAQ.

2.2. Glass detectors

At detection points 1, 4, 5, 6 and 7 there are also glass-based detectors (GD) together with the SD. Each GD uses 50 cm × 50 cm × 3 cm optical glass that is painted white using TiO_2 and R7723 PMT assembly. The reason that only the near periphery is equipped with these detectors is that the very narrow pulse they produce gets widened

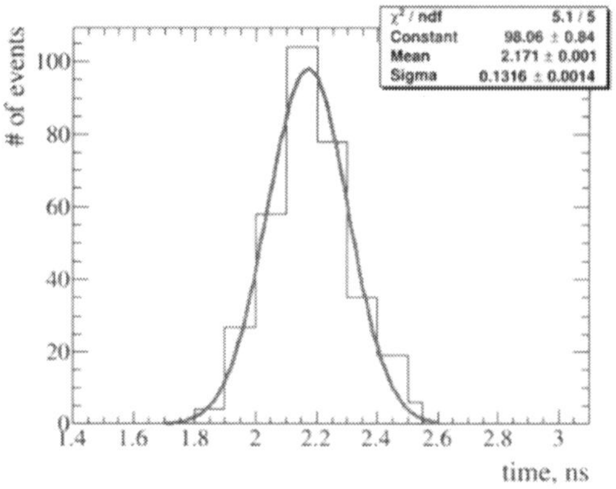

Figure 4. Front edge for GD detector pulse.

Figure 5. Vavilov-Cherenkov radiation detector.

by the long cable thus diminishing the usefulness of the GL. As measured at detection point 1, the pulse front is 2.17±0.13 ns (see Fig. 4) and the total width is 5.1±0.7 ns.

2.3. Vavilov-Cherenkov radiation detector

Next to detection point 1 there is a Vavilov-Cherenkov radiation detector (VCD), shown in Fig. 5. The VCD consists of three parabolic mirrors of 150 cm diameter and focal length of 65 cm. They are mounted on the rotating support allowing detection in zenith angle range of 0°–80° and in azimuthal angle range of 0°–360°.

A 15 cm diameter FEU49 and a 15 cm H6527 PMTs are located in the focal point of each mirror. The field of view of each mirror + PMT is then ∼ 3°.

3. Data acquisition system and trigger logic

The DAQ for the HT detector system currently consists of 3 analog to digital converter (ADC) units and in-house written custom software. The units used are the desktop

 DOI: 10.1051/epjconf/201714511004

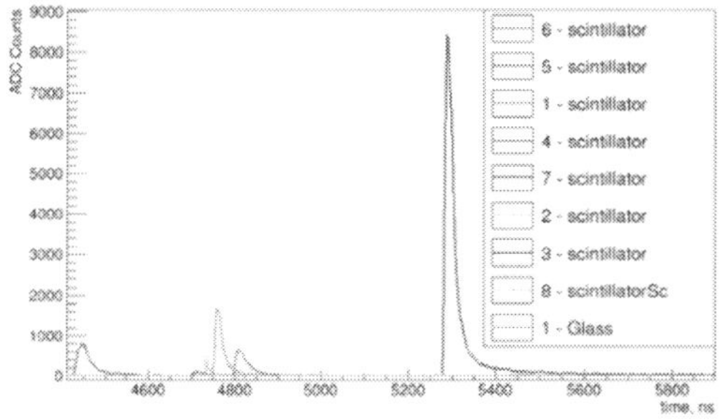

Figure 6. HT sample event.

CAEN [6] DT5730 ADC that do not require a separate crate for operation. They are each read by a computer via USB port.

The multi-threaded DAQ software takes advantage of the multiple CPUs in current computers. One thread is handling data communications, one is for data display, one saves the data as a compressed binary tree format, another controls the process for the failure – resistant operations of the DAQ. The graphical interface and file format are implemented using ROOT [7] framework. Both the internal ADC time stamp and the computer system time are used for best event timing precision.

One of the ADCs is designated as a main board and is used to form a hardware master trigger using double coincidence from detection points 5 and 6 (or 4 and 7). Then the signal is fed via fan-out unit into all three ADCs. The same trigger signal is used for the synchronization pulse that is recorded by the 8th channel of each ADC. The total channel capacity of the DAQ is then 21 channels (with 3 channels used for the sync pulse).

The hardware trigger used is quite relaxed, thus a number of empty events are collected. The offline trigger is applied to all files before any further analysis. Typically we require at least 4 detection points to have a signal in them at the level above the 7 widths of the noise floor, taken as the standard deviation over the first 200 points of each channel that never have any signal in them by design. The resultant rate of EAS registration intensity rate is $\sim$120 events/day at the HT detector system.

4. Sample data

A sample of a typical event detected by the near periphery is shown in Fig. 6. The raw data for each channel is shown, only cable length correction is introduced.

5. Conclusion

The Horizon-T detector system is designed for measuring the spatial and temporary structure of incoming EAS with axis in a wide range of zenith angles from 0°–80°. HT is located at an altitude of about 3340 meters above sea level centered at geographical coordinates of 43°02′ 49″ N and 76°56′43″ E. The overall detection rate of EAS starting with primary particle energy at $\sim 10^{16}$ eV is 120 events/day.

HT has eight charged particle detection points, separated by hundreds of meters. Each point has scintillator detectors, near periphery is also equipped with glass ones.

Detectors are read out by both FEU49 and Hamamatsu R7723 PMTs. Glass detectors with R7723 reach pulse front edge resolution of $\sim$2.2 ns.

The detector system also includes Vavilov-Cherenkov radiation detector that is made from mirrors with FEU49 and H6527 PMTs and has a field of view of each mirror + PMT of $\sim$13°. This detector allows measurements of light from EAS disks in the large range of zenith angles down to 80°.

References

[1] R.U. Beisembaev et al., `arXiv:1605.05179`, (2016)
[2] A Baitenov et al., `arXiv:1601.00086`, (2016)
[3] MELZ-FEU, 4922-y pr-d, 4c5, Zelenograd, g. Moskva, Russia, 124482 `http://www.meltz-feu.ru`
[4] Hamamatsu Photonics, 314–5 Shimokanzo, Toyooka-village, Iwatagun, Shizuoka-ken, 438–0193 Japan `http://www.hamamatsu.com`
[5] T Beremkulov et al., `arXiv:1608.04312`, (2016)
[6] CAEN S.p.A. Via della Vetraia, 11, 55049 Viareggio Lucca, Italy `http://www.caen.it`
[7] R. Brun and F. Rademakers, Nucl. Instrum. Meth. A **389**, 81–86 (1997)

Modern status of the Tien-Shan cosmic ray station

V.A. Ryabov[1,a], A.M. Almenova[2], V.P. Antonova[3], R.U. Beisembayev[1,5], S.P. Bezshapov[1], A.S. Borisov[1], A.P. Chubenko[1], O.D. Dalkarov[1], A.V. Gurevich[1], A.N. Karashtin[4], O.N. Kryakynova[3], G.G. Mitko[1], R.A. Mukhamedshin[10], A.M. Mukhashev[9], R.A. Nam[1,5], N.F. Nikolaevsky[3], V.P. Pavlyuchenko[1], V.V. Piscal[1,5], M.O. Ptitsyn[1], V.S. Puchkov[1], N.O. Saduev[8], T.Kh. Sadykov[2], N.M. Salikhov[3], S.B. Shaulov[1], A.L. Shepetov[1], Yu.V. Shlyugaev[6], A.V. Stepanov[1,5], W.M. Thu[7], L.I. Vildanova[1,5], M.I. Vildanova[1,5], N.N. Zastrozhnova[2], V.V. Zukhov[1,5], and K.P. Zybin[1]

[1] P.N. Lebedev Physical Institute of the Russian Academy of Sciences (FIAN), 119991, Leninskii pr., 53, Moscow, Russia
[2] Institute for Physics and Technology, Almaty, 050032, Ibragimova, 11, Kazakhstan
[3] Institute of Ionosphere, 050020, Kamenskoye plato, Almaty, Kazakhstan
[4] Research Radiophysics Institute, 603950, Bolshaya Pechyorskaya, str., 25/12a, Nizhny Novgorod, Russia
[5] Tien Shan Mountain Cosmic Ray Station, 050020, Mitina-str., 3, Almaty, Kazakhstan
[6] Institute of Applied Physics of RAS, 603950, Ulyanova str., 46, Nizhny Novgorod, Russia
[7] Moscow Institute of Physics and Technology (State University), 117303, Moscow, Russia
[8] Institute for Experimental and Theoretical Physics of al-Farabi Kazakh National University, al-Farabi pr., 71, Almaty, Kazakhstan
[9] Abay Kazakh National Pedagogic University, Dostyk pr., 13, Almaty, Kazakhstan
[10] Institute for Nuclear Research, 60-letiya Oktyabrya pr. 7a, 117312 Moscow, Russian Federation

Abstract. We present a description of the new complex installation for the study of extensive air showers which was created at the Tien Shan mountain cosmic ray station, as well as the results of the first measurements made there in 2015–2016. We also present new results on high-energy radiation observed during a thunderstorm.

1. Introduction

The Tien Shan cosmic ray station is situated in the mountains of Northern Tien Shan (Kazakhstan), 43° North, 75° East at an altitude of 3340 m above sea level. The average atmosphere depth at the height of the station location is 690 g/cm^2. At the station a multifunction complex for investigating the various components of extensive air showers has been created [1,2]. The new complex aims to investigate the following fundamental problems in astroparticle and atmospheric physics:

- High-precision measurements of the primary cosmic ray spectrum, mass composition, and distribution of the arrival directions in the range of $E_0 \sim 10^{14} - 10^{17}$ eV using a wide-spread scintillation shower system, a hybird ionization-neutron calorimeter, muon detectors, Cherenkov radiation detectors;
- Investigation of the parameters of high-energy nuclear interactions, especially shows in the region of EAS cores using a dense carpet of scintillation detectors with an enhanced up to $(1-2) \cdot 10^6$ dynamic range of amplitude measurements, in combination with an ionization calorimeter, saturation-free neutron detectors, X-ray films, underground neutron and muon installations;
- A search for strong interacting dark matter particles whose existance in the Universe is supposed to be in the form of exotic nuclei (e.g. strangelets) having a specific combination of low charge $Z \leq 10$ and large mass $A \geq 350$;
- Monitoring the intensity of $10^8 - 10^{11}$ eV cosmic rays of solar origin with application of the low- and high-energy neutron and gamma-radiation detectors; the control for "cosmic weather";
- Investigation of the CR influence on different environmental phenomena; lightning initiation in the atmosphere, intensity of the near-Earth radioactive background, anticipated seismic effects from the passage of powerful EAS, etc.

The Tien-Shan station is a unique place to examine the interconnection between cosmic rays and lightning discharges [3,4]. During a summer season (from the end of May until the beginning of September) thunderstorm clouds pass directly at the height of station. The uniqueness of the experimental conditions and the presence of modern specialized equipment at Tien-Shan provide new data on the processes under investigation. Simultaneous observations of gamma-ray, optical and radio emission, high-energy electrons, positrons, muons and neutrons at high altitudes became a new tool for studying the processes in the thunderstorm atmosphere [5–8].

2. Instrumentation

2.1. EAS Installation

The new experimental complex includes two systems for precision studies of the shower core, timing system

[a] e-mail: ryabov@x4u.lebedev.ru

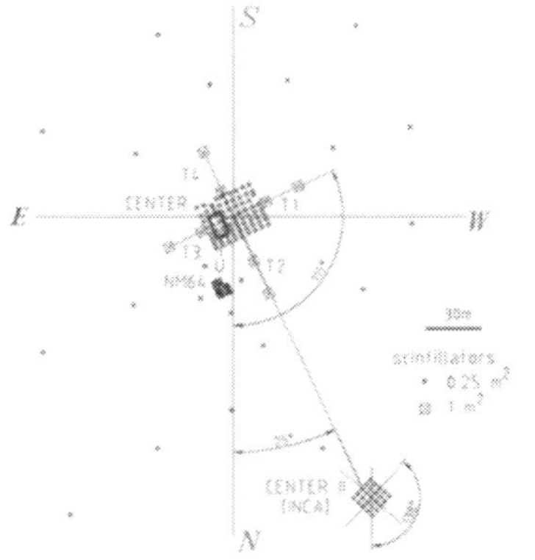

Figure 1. General plan of the Tien Shan EAS detector system.

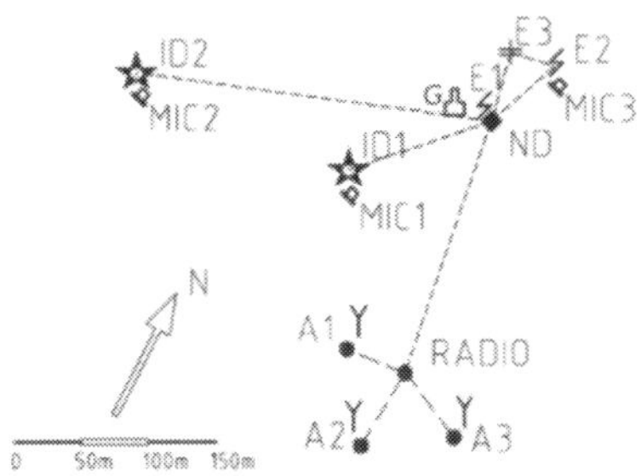

Figure 2. Schematic of the detector complex used in the present experiment. ID1 and ID2 — ionization detectors; G— scintillation detector of the low-energy gamma rays; ND — the neutron detector complex; A1-A3 — radio-antennae; E1-E3 — electric field sensors; MIC1-MIC3 — acoustic sensors of the thunder sound.

for determination of the EAS arrival directions, the underground system of muon and hadron detectors for registration of the penetrating components of the shower, ionization-neutron calorimeter and a neutron supermonitor. The layout of the new EAS detector complex on the territory of station is shown in Fig. 1.

The shower system is built around two points, Center and Center II, where the shower core detectors are placed: the dense scintillation "carpet" for registration of the EAS e/γ component, a set of timing scintillators T1T4 for determination of the EAS arrival directions according to relative time delays of shower front, the complex of muon and hadron detectors in the underground room, U, situated beneath the Center point; another scintillation carpet, and the big ionization calorimeter INCA in Center II. The NM64 type neutron supermonitor is also placed in the vicinity of the first detector center which is used as a detector of the EAS hadron component. A number of additional scintillation detectors should be spread nearly concentrically over the station territory around the two center cores for registration of the peripheral EAS particles.

At present, the system for registration of the electromagnetic shower component consists of ~ 100 detector points built on the basis of plastic scintillator plates with a sensitive area of $0.25\,\mathrm{m}^2$ and $1\,\mathrm{m}^2$, spread equidistantly over $\sim 10^4\,\mathrm{m}^2$ space. The dynamic range of scintillation amplitude measurements is currently about $(3-7)\cdot 10^4$, and there is a prospect of it being extended up to $\sim 10^6$. The direction of shower arrival is defined by signal delays from a number of the scintillators placed cross-wise at the periphery of the detector system.

To investigate the nuclear active shower components a multi-tier $55\,\mathrm{m}^2$ ionization-neutron calorimeter with a total absorber thickness of $\sim 1000\,g/\mathrm{cm}^2$, a typical spatial resolution of the order of 10 cm, and dynamic range of ionization measurement channel about $\sim 10^5$ was created. In addition, the use of saturation-free neutron detectors is anticipated for registration of the high- and low-energy hadron components in the region of the shower core. A complex of underground detectors is designed for the study of muonic and penetrative nuclear active components of the shower.

The full stack of data acquisition, detector calibration, and shower parameters restoration procedures is now completed.

2.2. Thunderstorm installation

The present disposition of the Thunderstorm complex at the Tien-Shan station is shown in Fig. 2.

The "Thunderstorm" complex now comprises the following facilities: an EAS registering system; a system of NaI scintillation detectors for registration of the gamma- and X-ray emission in the atmosphere; the multi-layer ionization detectors of energetic charged particles; a neutron super monitor for registration of the high-energy hadronic component of cosmic rays; a set of detectors for low-energy (thermal) neutron background; two independent radio systems; electrostatic detectors of the local electric field and its high frequency component.

3. Observational results

3.1. First EAS measurements

The integral spectra of the registered EAS over the number of charged particles N_e are shown in Fig. 3, where the spectrum obtained in the current experiment is compared with the ones measured at different time by the old shower detector system of the Tien Shan mountain station [9].

The approximation of the Aragats spectrum published in [10] is also shown on this plot. The newly obtained spectrum agrees satisfactorily with other ones, both in its absolute intensity and slope; its slight deviation from the correct pure power form in the range of $N_e \geq 2\cdot 10^6$ can be explained by the absence of peripheral shower detectors in our test measurements which resulted in a noticeable side effect with a relative rise of reconstructed N_e values when the lateral size of shower is comparable with geometrical size of the whole Center carpet.

As an illustration of the data newly obtained with the neutron calorimeter in the underground room, Fig. 4 shows the spectra of neutron events over the multiplicity of the registered evaporation neutrons M (which is connected with the total energy deposit E_h of the initial CR particles inside the detector unit).

The top pair of spectra shown in this plot were calculated for the events registered in either the upper or in the lower unit of the vertical neutron calorimeter under the control of the corresponding internal trigger which has been generated at the moment of the fast transient increase of neutron multiplicity inside the unit; generally, these spectra are similar to old ones published in [11]. Another

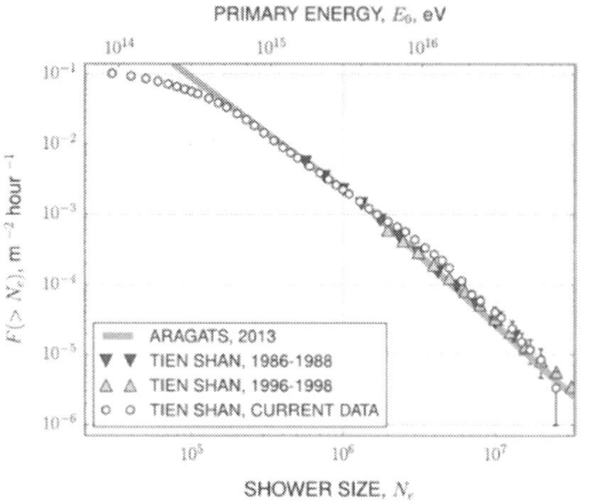

Figure 3. The integral EAS size spectrum obtained at the new shower installation (circles) compared with the spectra measured earlier at Tien Shan [9] (triangles), and Aragats [10] (gray line). For convenience, the upper axis represents the estimations of the corresponding primary energy E_0 which have been calculated as $E_0 = 3 \cdot 10^9 N_e$ eV.

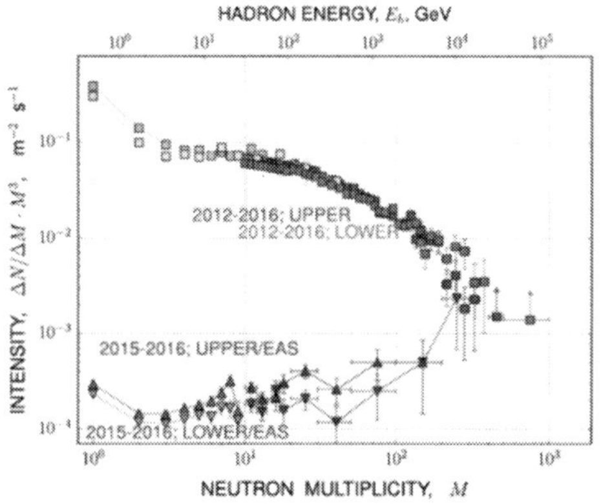

Figure 4. The neutron multiplicity spectra of events registered within the upper and lower units of the vertical underground calorimeter. For reference, the upper axis corresponds to estimations of the hadron energy deposit in the neutron detector unit which have been calculated according to the neutron monitor calibration formula taken from: $E_h = 0.49 \cdot M^{1.8}$ GeV.

spectra pair with much lower absolute intensity shown at the bottom of Fig. 4 corresponds to the events triggered by EAS passages at the time of the test exploitation of the new shower system in 2015–2016. With an increase of statistics, such data will be useful to clarify the nature of underground neutron events.

3.2. Observation of high-energy radiation during thunderstorms

The high-energy radiation observed during a thunderstorm is of two types [8]. One is a long-period emission lasting some tens of minutes or even hours. Another is a short one but very intense. It lasts less than a second, and usually the main part of this emission lasts even a few milliseconds. The emission of the first type is connected with thunderclouds. It is usually supposed to

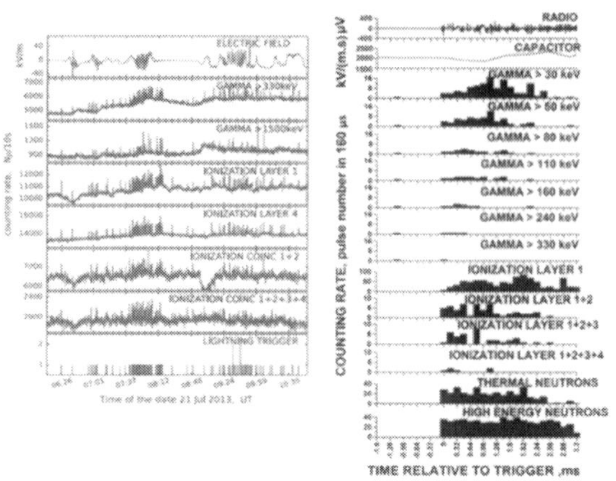

Figure 5. Left (a): monitoring mode with a time resolution of $10\,s$, Right (b): the triggering mode with a time resolution of $160\,\mu$s. In this figure we present the amplitude of radio emission on the antenna input, the strength of electric field and the counting rates in different types of particle detectors: gamma-rays for different energy thresholds, electrons for coincidences layers in respective energies ([1]−1 MeV, [1+2] −7.6 MeV, [1+2+3] −12 MeV, [1+2+3+4] −16 MeV), and also neutrons for thermal (~ 0.025 eV) and high energy range (≥ 1 MeV).

be determined mainly by the emanation of radioactive gases (Rn or Rn daughter). This long-period emission gives the main input into the high-energy emission of thunderstorms. In Fig. 5a it is shown that lightning could be diagnosed by its high-energy emission in the 10 s monitoring mode.

For a detailed study of penetrating emissions during thunderstorms, a high-resolution triggering method should be used. The wide complex of all emission type measurements in a $160\,\mu$s triggering mode was fulfilled (Fig. 5b).

The spectrum of lightning gamma emission was obtained. Together with gamma quanta, relativistic high-energy electrons are observed (Fig. 6). The energy spectrum of electrons is registered up to tens of MeV. The gamma and electron high-energy spectra are obtained simultaneously in multiple lightning events. The intensity of both electrons and gamma rays in lightning discharges prevail over the background emission by 1.5 to 2 orders of magnitude. Additionally, the lightning neutron emission was registered. It was shown to consist both of low-energy neutrons and of the intensive flow of energetic neutrons.

4. Conclusions

The new complex for investigation of the various components of extensive air showers has been created at the Tien Shan station. Test measurements carried out in 2014–2016 confirmed the efficient operation of the shower detector subsystems of the EAS installation. The results obtained so far demonstrate the reliable measurement of the EAS particle density and shower size spectrum within the energy range of $10^{14} - 10^{17}$ eV. High-resolution time series of neutron intensity, and the spectra of energy deposits in the hadron component were registered both on the surface of the Tien Shan station and underground by the neutron monitor like installations used for detecting

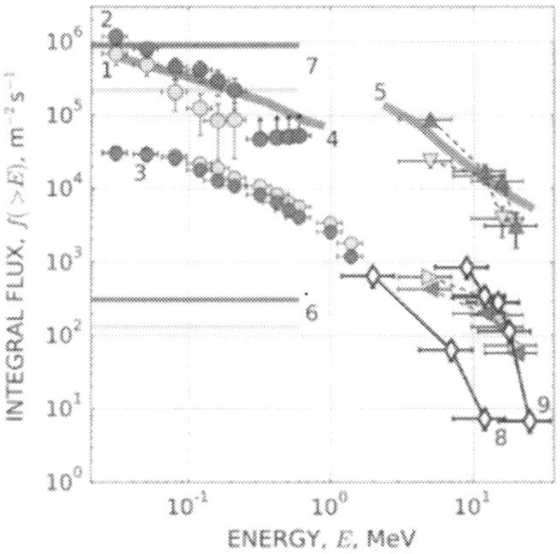

Figure 6. The energy spectra of gamma-radiation (circles) and accelerated electrons (triangles). 1 and 2 — two spectra for the transient radiation bursts around lightning trigger time in the high-resolution series for the two thunderstorm events; — the average spectra calculated over monitoring type measurements at Tien Shan in two thunderstorm days; 4 and 5 — correspondingly, the gamma-ray and electron spectra obtained in Geant4 simulation of an electron avalanche developing in the electric field of thunderclouds; 6 and 7 — the levels of neutron intensity corresponding to a quite time and just in the moment of transient burst on the two thunderstorm events; 8 and 9 — electron spectra registered in two thunderstorm events at Aragats [12].

neutron bearing cosmic ray particles. Similar neutron detector technique is planned to be widely applied in the operation of the big hybrid Tien Shan calorimeter INCA. The current development level of different detector subsystems at the Tien Shan station allows us expect new results in various fields of cosmic ray physics in the near future.

The measurements of all types of penetrating emissions were fulfilled during a thunderstorm at the Tien-Shan station. The measurements were made in both the monitoring mode and in the high-resolution trigger mode. We have demonstrated that the lightning high-energy emission could be directly observed in the 10 s monitoring mode. The results are confirmed by the simultaneous 160 μs triggering mode measurements. We also showed a significant difference in the magnitudes of penetrating radiation fluxes, depending on the time resolution of the recording apparatus. This integrated approach opens up a new area of research in atmospheric physics at high energies.

References

[1] A.S. Borisov, A.P. Chubenko, O.D. Dalkarov, et al., Proc. of Science, V. 30-July-2015, Article number 570 (2015)

[2] A.P. Chubenko, A.L. Shepetov, V.P. Antonova, et al., Nuclear Instruments and Methods in Physics Research A **832**, 158 (2016)

[3] A.V. Gurevich, A.N. Karashtin, V.A. Ryabov, et al., Phys.-Uspekhi **52**(7), 735 (2009)

[4] A.V. Gurevich, G.G. Mitko, V.P. Antonova, et al., Phys.Lett. A **373**, 3550 (2009)

[5] A.V. Gurevich, A.N. Karashtin, A.P. Chubenko, et al., Phys.Lett. A**325**, 389 (2004)

[6] A.V. Gurevich, V.P. Antonova, A.P. Chubenko, et al., Phys. Rev. Lett. **108**, 125001 (2012)

[7] A.V. Gurevich, V.P. Antonova, A.P. Chubenko, et al., Phys. Rev. Lett. **111**, 165001 (2013)

[8] A.V. Gurevich, A.M. Almenova, V. P. Antonova, et al., Phys. Rev. D **94**, 023003 (2016)

[9] V.P. Antonova, A.P. Chubenko, S.V. Kryukov, et al., J. Phys. G **28**(2), 251 (2002)

[10] P. Garyaka, R.M. Martirosov, S.O. Sokhoyan, et al., Bull. Natl. Acad. Sci. Armen. Phys. **48**, 79 (2013)

[11] A.P. Chubenko, A.L. Shepetov, L.I. Vildanova, et al., Bull. Lebedev Phys. Inst. **34**(4), 107 (2007)

[12] A. Chilingarian, B. Mailyan, and L. Vanyan, in 2nd International Cosmic Ray Workshop, Nor-Amberd, Armenia, 151 (2011)

The MoEDAL experiment at the LHC

James Pinfold[a]

Physics Department, University of Alberta, Edmonton, Alberta, Canada

Abstract. MoEDAL is a pioneering experiment designed to search for highly ionising messengers of new physics such as magnetic monopoles or massive (pseudo-)stable charged particles, that are predicted to exist in a plethora of models beyond the Standard Model. Its ground-breaking physics program defines a number of scenarios that yield potentially revolutionary insights into such foundational questions as: are there extra dimensions or new symmetries; what is the mechanism for the generation of mass; does magnetic charge exist; what is the nature of dark matter; and, how did the big-bang develop at the earliest times. MoEDAL's purpose is to meet such far-reaching challenges at the frontier of the field. In conclusion we will briefly report on current results; discuss plans to instal a new detector designed to search for very long-lived neutral particles as well as mini-charged particles; and, briefly delineate plans for an astroparticle extension of MoEDAL called Cosmic-MoEDAL.

1. Introduction

MoEDAL (Monopole and Exotics Detector at the LHC) [1–3], the 7th experiment at the Large Hadron Collider (LHC) [4], was approved by the CERN Research Board in 2010. It is designed to search for manifestations of new physics through highly-ionising particles in a manner complementary to ATLAS and CMS [5]. The most important motivation for the MoEDAL experiment is to pursue the quest for magnetic monopoles and dyons at LHC energies. Nonetheless the experiment is also designed to search for any massive, stable or long-lived, slow-moving particles [6, 7] with single or multiple electric charges arising in many scenarios of physics beyond the Standard Model (SM). A selection of the physics goals and their relevance to the MoEDAL experiment are described here and elsewhere [8]. For an extended and detailed account of the MoEDAL discovery potential, the reader is referred to the recently published MoEDAL Physics Review [9].

The structure of this paper is as follows. Section 2 provides a brief description of the MoEDAL detector. The physics reach of MoEDAL is discussed in Sect. 3, whilst Sect. 4 is dedicated to a discussion of recent results. Section 5 deals with the proposal to instal the MoEDAL detector for Penetrating Particles (MAPP) and a possible astroparticle extension of the MoEDAL-LHC project i.e. Cosmic-MoEDAL.

2. The MoEDAL detector

The MoEDAL detector [2] is deployed around the intersection region at Point 8 of the LHC in the LHCb experiment Vertex Locator (VELO) [10] cavern. A three-dimensional depiction of the MoEDAL experiment is presented in Fig. 1. It is a unique and largely passive LHC detector comprised of three sub-detector systems, with a fourth in the planning stage. MoEDAL bypasses the experimental difficulties encountered by the main LHC experiments in the detection of Highly Ionising Particle (HIP) messengers of new physics by using a passive plastic NTD technique to detect the ionisation trail of HIPs as well as a novel trapping array - called the MMT (Magnetic Monopole Trapper) for detecting HIPs that slow down and stop within its sensitive volume. Neither of these detector systems requires trigger or read-out electronics.

As the MoEDAL NTD stacks are less than 1-cm-thick there is little chance that a HIP will be absorbed. Also, NTDs provide a tried-and-tested and cost-effective method to accurately measure the track of a HIP and its effective charge. Importantly, MoEDAL's exposed HIP films will be directly calibrated in a heavy-ion beam. MoEDAL's roughly one tonne of MMT detectors insures that a small but significant fraction of the HIPs produced will be trapped for further study in the laboratory. MoEDAL's ability to retain a permanent record, and even capture new particles for further study - will make it an invaluable asset in the elucidation of any Terascale BSM scenario covered by its extensive physics repertoire. There are no SM particles that can produce such distinct signatures - thus, even the detection in MoEDAL of few HIP particle messengers of new physics would herald a discovery. The only 'real-time' sub-detector system is the TimePix2 pixel array that will be used to monitor low-energy highly ionising beam-related backgrounds.

2.1. The NTD detector sub-system

The Low Threshold (LT-NTD) array, part of the NTD sub-detector system, is the largest array of NTD detectors ever deployed at an accelerator. The NTD plastics employed are polyallyl-diglygol-carbonate (PADC) commonly known as CR39®a transparent rigid plastic and the polycarbonate Makrofol®. Each of the roughly 300 ($25 \times 25\,\text{cm}^2$) LT-NTD stacks is comprised of three sheets of CR39®and three of Makrofol as shown in Fig. 2. The CR39 layers

[a] e-mail: jpinfold@ualberta.ca

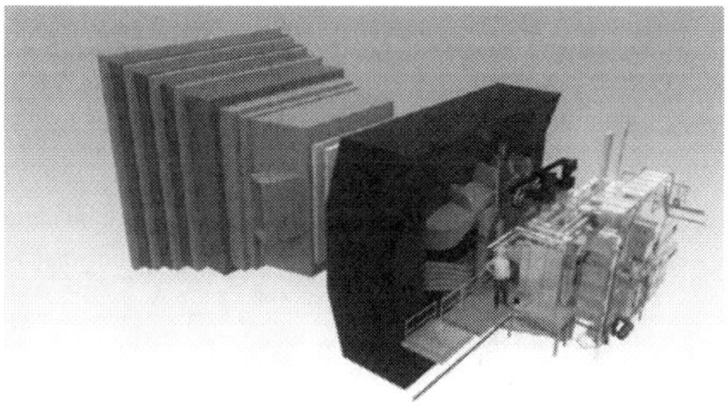

Figure 1. A schematic view of the MoEDAL detector.

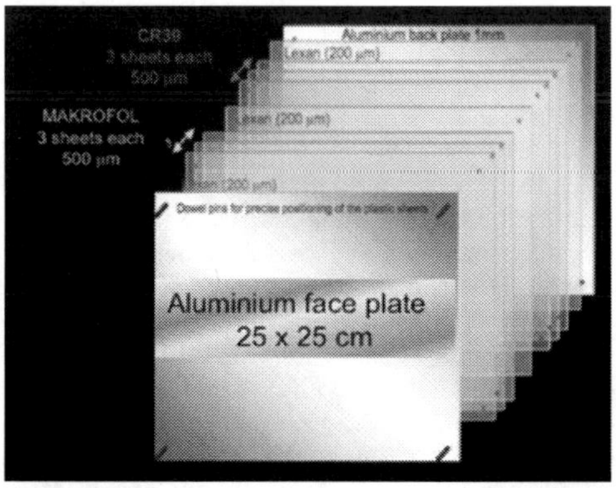

Figure 2. A MoEDAL stack comprised of CR39 and Markrofol NTD detectors.

in the LT-NTD array can detect particles with ionisation equivalent to five normal particles with charge resolution, better than 0.2 e, where e is the charge of an electron.

The TDR NTD array has been enhanced by the high charge catcher (HCC) sub-detector – with threshold Z/β of approximately 50 – comprised of stacks of three Makrofol plastic sheets in an Al foil envelope. These lightweight low-mass detector stacks can be deployed in previously inaccessible areas on and around LHCb's VELO (Vertex Locator) detector increasing the geometrical acceptance for magnetic monopoles to ~60%. Importantly, a plane of roughly 4 m^2 of HCC detectors – termed 'the shower curtain' – has been placed in the forward acceptance of the LHCb detector between the LHCb's RICH (Ring Imaging Cherenkov detector) and its first station of tracking detectors (TT) where the material budget is at a minimum. Although NTD technology was used in the past to search for monopoles [11], this is the first time at an accelerator that such a large area array of NTDs has been used to search for a wide range of magnetically and electrically charged HIPs.

The passage of a highly-ionising particle through the plastic detector is marked by an invisible damage zone along the trajectory. The damage zone is revealed as a cone-shaped etch-pit when the plastic detector is etched using a hot sodium hydroxide solution. Then the sheets of plastics are scanned looking for aligned etch pits in multiple sheets using "intelligent" computer controlled optical scanning microscopes. A HIP with ionising power greater than or equal to 5 times that of a relativistic charged SM particle will leave a characteristic set of at least 6 collinear etch pits in the 3 CR39 sheets in each

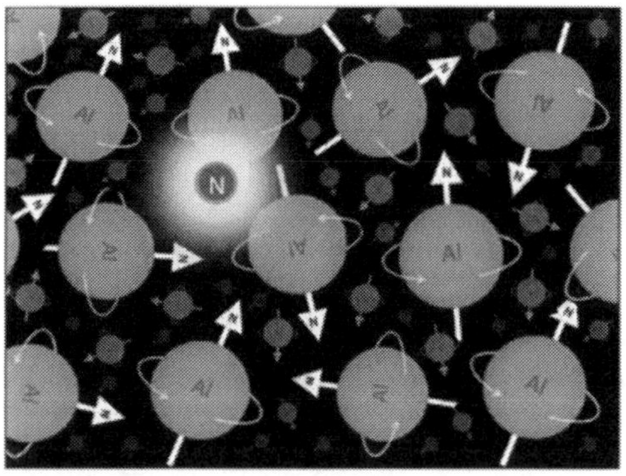

Figure 3. A monopole captured within an aluminium trapping element of the MoEDAL Trapping detector.

NTD stack. Magnetic monopoles, each with an ionising power that is thousands of times that of a SM particle will leave an unprecedented trail of 12 etch pits in a MoEDAL stack with 6 NTD sheets. These aligned etch pits, of size typically in the range 20–50 microns, accurately define a track that points towards to Interaction Point (IP). There is no known Standard Model background to this signal.

2.2. The trapping detector sub-system

The MMT is the newest sub-detector system to be added to the MoEDAL detector. The MMT detector consists of roughly 1 tonne of aluminium (Al) paramagnetic volumes placed at three points around the intersection point, IP8, that MoEDAL shares with the LHCb experiment. Al has an enhanced capability to trap monopoles due to its anomalously large nuclear magnetic moment. A fraction of the massive HIPs created will stop and be captured in the MMT detector as illustrated in Fig. 3. MoEDAL is the first experiment to use purpose-made trapping volumes to capture magnetic and electrically charged particles. The exposed monopole trapping volumes are monitored at the ETH Zurich SQUID facility for the presence of captured monopoles, a schematic description of the facility is given in Fig. 4. The use of SQUIDs to detect trapped magnetic charge has also been thoroughly tested in particle and astroparticle experiments where the search has been performed on 'found' or alternate use objects (such as beam-pipes) on a 'one-off' basis.

The signal for a magnetic monopole in the monopole trapping detectors at the ETH facility would be a sustained current – resulting from the passage of a monopole through the SQUID detector. Test solenoids are used to calibrate the response of the SQUID to a trapped monopole. Using test solenoids we found that the SQUID can detect magnetic charges as small as 0.1 g. After the SQUID scan has been performed it is envisaged that the trapping volumes will be sent to SNOLAB – 2 km underground – to be monitored for the decays of very long-lived electrically charged particles ($\tau > 10^7$ s). The MoEDAL search for very long-lived particles using a dedicated detector deployed deep underground has some advantages over searches carried out with the LHC GPEs. For example, MoEDAL has no trigger requirement and thus can perform relatively model-independent studies. Using long test solenoids to mimic a monopole we have

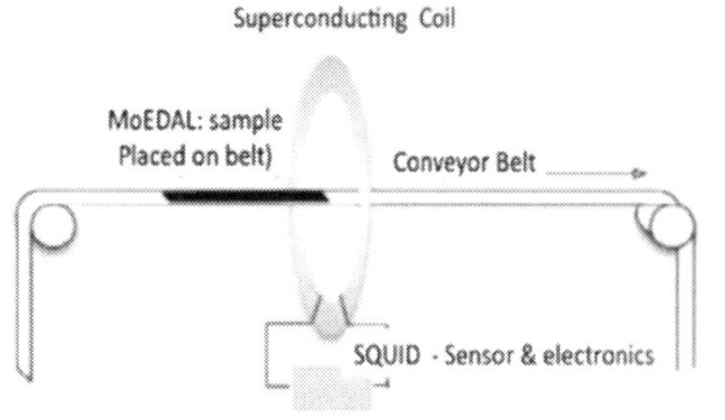

Figure 4. A schematic depiction of the use of a SQUID magnetometer to detect the presence of a trapped magnetic charge.

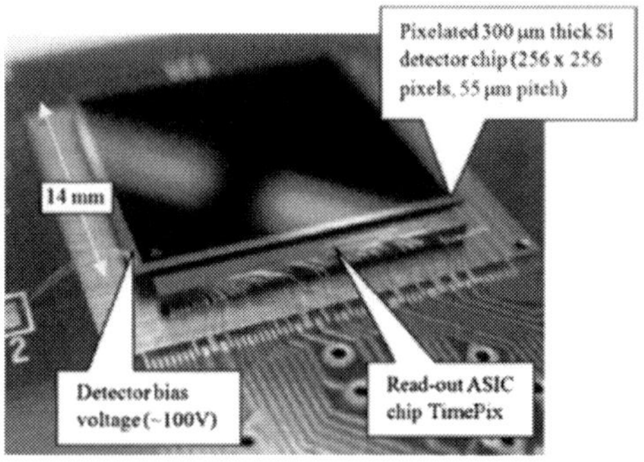

Figure 5. A photograph of a TimePix pixel device.

determined that the SQUID can detect trapped monopole with magnetic charge as small as 0.1 of a Dirac charge ($0.1g_D$).

2.3. The TimePix radiation monitoring system

An array of six TimePix2 pixel detectors are used to monitor the low energy highly-ionizing beam related backgrounds. Each pixel of the TimePix chip contains a preamplifier to enhance the signal, a discriminator to severely reduce electronic noise and a 4-bit DAC for threshold adjustment, and a 14-bit counter. MoEDAL uses the TimePix device's 'Time-over-Threshold' (ADC) and 'Time of Arrival' modes so that each pixel can supply an energy measurement. A photograph of a TimePix pixel chip is shown in Fig. 5. The TimePix detector is capable of providing a colour image of complete spallation events in its 300 micron thick silicon sensitive volume – with energy encoded in the colour.The TimePix sub-detector is read out via the web.

3. The physics reach of MoEDAL

As discussed above, the standard general-purpose Collider detectors at the LHC are not designed or optimised to detect massive slow-moving HIPs or mQPs. MoEDAL's new light on the high-energy frontier is provided by the use of massive HIPs or mQPs – for which there are no SM counterparts – as direct probes of pioneering new physics at the Terascale. Such an approach requires a new state-of-the-art in the quest for massive HIP messengers of beyond the SM physics that is provided by the custom-designed MoEDAL experiment.

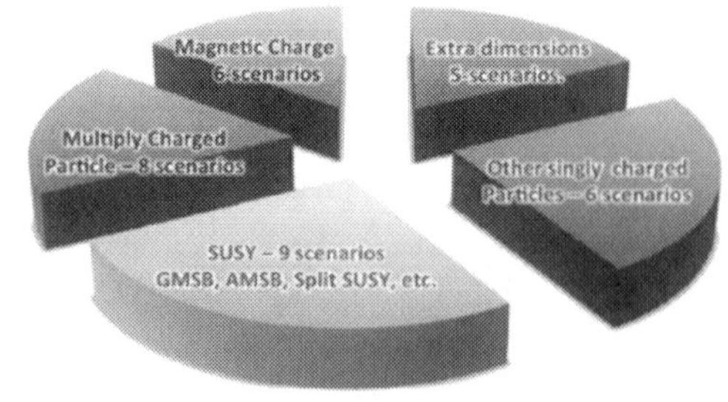

Figure 6. MoEDAL Physics Programme defined in Ref. [9] broken down by topic. We have identified 34 different existing possible discovery scenarios.

MoEDAL's physics programme [9] covers many fundamentally important BSM scenarios allowing MoEDAL to significantly expand the LHC's discovery horizon in a complementary way, as illustrated in the pie chart shown in Fig. 6. MoEDAL's Physics Coordinator and Theory Board oversee the physics programme in order to ensure that the physics objectives are maximised and met. The MoEDAL physics programme addresses fundamental questions such as: does magnetic charge exist, are there extra dimensions; are there new symmetries of nature; and, what is the nature of dark matter.

One of the main objectives for MoEDAL is the state-of-the-art search for the magnetic monopole, just as the search for the Higgs was the prime motivation for the LHC and its GPEs. The work of our Theory Board has renewed interest in the LHC search for a Terascale EW monopole [9, 12] arising from the SM. If discovered, this would be the first topological particle to be observed – with the utmost consequences for our understanding of the Universe. But MoEDAL is designed to do much more, as indicated in Fig. 6 by probing BSM physics by searching for anomalously electrically charged particles.

4. Recent MoEDAL results

The first MoEDAL results utilized a 160 kg of prototype MoEDAL trapping detector exposed to 8-TeV proton-proton collisions at the LHC, for an integrated luminosity of $0.75\,\text{fb}^{-1}$ during LHC's Run I. No magnetic charge exceeding 0.5 gD was detected in any of the exposed samples, allowing limits to be placed on monopole production in the mass range $100\,\text{GeV} \leq M_{monopole} \leq 3500\,\text{GeV}$ [13]. Model-independent cross-section limits have been presented in fiducial regions of monopole energy and direction for $1g_D \leq |g| \leq 6g_D$, and model-dependent cross-section limits are obtained for Drell-Yan (DY) pair production of spin-1/2 and spin-0 monopoles for $1g_D \leq |g| \leq 4g_D$. Under the assumption of Drell-Yan cross sections, mass limits are derived for $jgj = 2g_D$ and $|g| = 3g_D$ for the first time at the LHC, surpassing the previous result from the ATLAS Collaboration [14, 15] which placed limits only for monopoles with magnetic charge $|g| = 1g_D$.

The first search for magnetic monopole production in 13 TeV proton-proton collisions during LHC's Run-2 using the trapping technique have extended the previous results with 8 TeV data during LHC's Run-1. In this case a total of 222 kg of MoEDAL trapping detector samples was exposed in the forward region and

 DOI: 10.1051/epjconf/201714512002

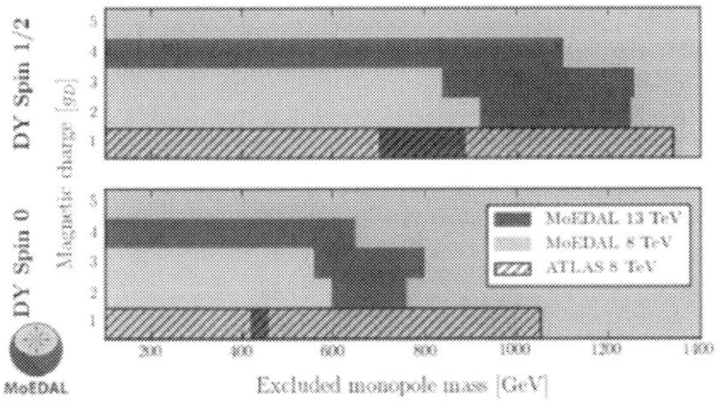

Figure 7. Monopole lower mass limits (95% confidence level) in models of spin-1/2 and spin-0 DY pair production in LHC pp collisions for monopole charges $|g|$ up to 4_D. These limits are based upon cross sections computed at leading order. These cross sections are only indicative since the monopole coupling to the photon is too large to allow for perturbative calculations. Previous results obtained in 8 TeV collisions are from Ref. [13] (MoEDAL prototype trapping detector) and Ref. [14,15] (ATLAS).

analysed by searching for induced persistent currents after passage through a superconducting magnetometer. Magnetic charges exceeding half the Dirac charge are excluded in all samples and limits are placed for the first time on the production of magnetic monopoles in 13 TeV pp collisions. This search probes mass ranges previously inaccessible to collider experiments for up to five times the Dirac charge. A summary of the limits obtained by both the above analyses are summarized in Fig. 7.

5. Future MoEDAL developments

A key aspect of the MoEDAL experiment is that it is sensitive to massive slow-moving and very HIPs, messengers of new physics for which the standard LHC GPE detectors are not optimised. MoEDAL is preparing a proposal to add a new sub-detector that is sensitive to particles with charge as small as a thousandth that of the electron - a mini-charged particle (mQP). The main LHC detectors are essentially blind to such particles. This addition to its detector system is consistent with MoEDAL's ethos of extending the physics reach of the LHC by searching for anomalously charged messengers of new physics in a way that is complementary to the existing capability provided by the main LHC detectors.

MoEDAL is proposing to deploy the MAPP (MoEDAL apparatus for detecting penetrating particles) in a tunnel shielded by some 30 m to 50 m of rock and concrete from the interaction point (IP8), as shown in Fig. 8. The purpose of the detector is to search for particles with fractional charge as small as one-thousandth the charge of an electron. This detector would also be sensitive to neutral particles from new physics scenarios via their interaction or decay in flight within the volume of the detector. The isolation of the detector means that the huge background from SM processes in the main detectors is largely absent.

The first apparatus specifically designed to detect mini-charged particles was the SLAC (Stanford Linear Accelerator Centre) 'beam dump' type detector, comprised of scintillator bars read out by photomultiplier tubes [16]. MoEDAL's new detector, shown in Fig. 9, and another apparatus proposed for deployment near to the CMS detector [17] also designed to search for minicharged

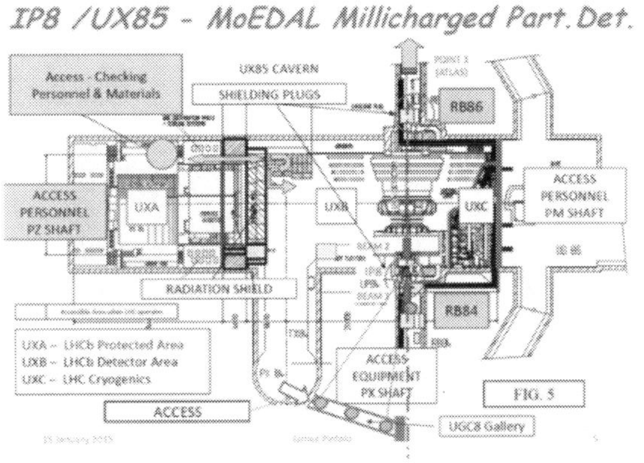

Figure 8. MoEDAL is preparing a proposal to install a mini-charged particle detector positioned in the forward region of IP8 some 30 m from the LHCb/MoEDAL intersection point (IP8).

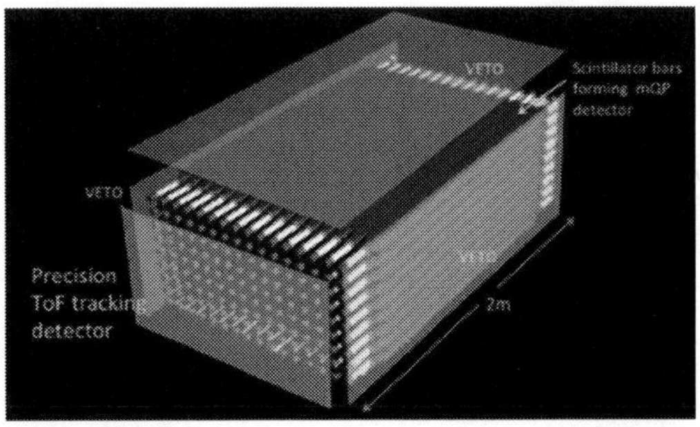

Figure 9. A depiction of a MAPP detector subunit. The final detector could contain up to four such units.

particles, both have a design that harks back to the original SLAC detector. In order to reduce backgrounds from natural radiation the photomultiplier tubes and scintillator detectors of the MoEDAL apparatus will be constructed from materials with low natural backgrounds currently utilised in the astroparticle physics arena. Its calibration system utilises neutral density filters to reduce the received light of high incident muons that manage to penetrate to the sheltered detector from the interaction point, in order to mimic the much lower light levels expected from particles with fractional charges.

5.1. Cosmic-MoEDAL

The MoEDAL collaboration is preparing an astroparticle extension to the MoEDAL-LHC experiment that will enable the search, for example, for magnetic monopoles to be extended from TeV scale at the LHC up to the Grand Unification (GUT) scale. In addition we propose to use the same detector technology for 'Cosmic-MoEDAL' as we use for MoEDAL-LHC. SLIM was the first experiment to use such an approach to extend the search for cosmic monopoles with masses from the GUT scale well below the GUT scale, with a high sensitivity. SLIM was necessarily deployed at high altitude at the Mt Chacaltaya lab. in Bolivia with an elevation of 5,400 m. However, SLIM's modest size (400 m^2) precluded it from the search for a flux of cosmic monopoles below the Parker Bound (an upper bound on the density of magnetic monopoles that

Figure 10. Artist's impression of the Cosmic-MoEDAL array deployed at an altitude of 5 km on Mt Chacaltaya.

is obtained from arguments based on the existence of a galactic magnetic field).

Cosmic-MoEDAL is proposed as a 50,000–100,000 m^2 of plastic NTDs (CR39) deployed at high altitude. Such an array would be able to take the search for cosmic monopoles with velocities $\beta \gtrsim 0.1$ from the TeV scale to the GUT scale for monopole fluxes well below the Parker Bound. Possible sites for Cosmic MoEDAL include: Chacaltaya (5 km) and Tenerife-Tiede (3 km). An artists impression of Cosmic-MoEDAL on Mt Chacaltaya is shown in Fig. 10.

6. Summary and conclusion

In 2015, the LHC restarted operations at the unparalleled energy of 6.5 TeV beam energy and the enormous collision rate of around a billion collisions per second. The LHC has been compared to a time machine enabling us to recreate the conditions that occurred when the Universe was only roughly 100 picoseconds old, providing an unprecedented laboratory for the study of the cosmology of the nascent Universe at the earliest times. In parallel, non-accelerator experiments are also exploring the Tera-universe via high-energy astrophysics.

The first deployment of the MoEDAL detector employing largely passive detector systems, tuned to the prospect of discovery physics, was at Point 8 on the LHC ring, in the Winter of 2014. As we have seen the novel MoEDAL detector, has a dual nature. First, it acts like a giant camera, comprised of NTDs – analysed offline by ultra fast scanning microscopes – sensitive only to new physics. Second, it is uniquely able to trap the particle messengers of physics beyond the SM for further study. MoEDAL's radiation environment is monitored by a state-of-the-art real-time TimePix pixel detector array. A new MoEDAL sub-detector, MAPP, sensitive to mini-charged particles and long-lived neutral particles, is being studied.

The first MoEDAL results using only a small part of their overall detector system have provided the world's best limits on multiply charge monopoles. In 2017 and beyond we expect to exploit MoEDAL's full detector repertoire to provide further world class limits on highly ionizing magnetically and electrically charged particles from a number of well predicated theoretical scenarios. In addition, MoEDAL has an aggressive plan to extend the physics reach of MoEDAL LHC with the MAPP detector and a possible astroparticle extension to MoEDAL, called Cosmic MoEDAL, that will be capable of taking the search for, for example, magnetic monopoles up to the GUT scale.

July 2012 the ATLAS and CMS experiments operating at the LHC announced the discovery of the Higgs boson. But, the LHC is only just getting started. Many fundamental questions still remain to be answered. Are there new symmetries of nature? Are there extra spatial dimensions? Is there a deeper substructure? Does magnetic charge exist? Why is gravity so weak compared to the other fundamental forces? What is the nature of dark matter? What was the physics of the earliest era of the universe? The list goes on. MoEDAL is designed to provide insights into such fundamental questions.

References

[1] For general information on the MoEDAL experiment, see: `http://moedal.web.cern.ch/`
[2] MoEDAL Collaboration, Technical Design Report of the MoEDAL Experiment CERN Preprint CERN-LHC-2009-006, MoEDAL-TDR-1.1 (2009), and references therein
[3] M. Fairbairn and J. L. Pinfold, published online by Contemporary Physics, `http://www.tandfonline.com/doi/full/10.1080/00107514.2016.1222649`
[4] L. Evans and P. Bryant, JINST **3**, S08001 (2008)
[5] A. De Roeck, A. Katre, P. Mermod, D. Milstead and T. Sloan, Eur. Phys. J. C **72**, 1985 (2012)
[6] M. Fairbairn, A. C. Kraan, D. A. Milstead, T. Sjostrand, P. Z. Skands and T. Sloan, Phys. Rept. **438**, 1 (2007)
[7] S. Burdin, M. Fairbairn, P. Mermod, D. Milstead, J. Pinfold, T. Sloan andW. Taylor, Phys. Rept. **582**, 1 (2015)
[8] V. A. Mitsou [MoEDAL Collaboration], EPJ Web Conf. **95** (2015) 04042
[9] B. Acharya et al. [MoEDAL Collaboration], Int. J. Mod. Phys. A **29**, 1430050 (2014)
[10] A. A. Alves, Jr. et al. [LHCb Collaboration], JINST **3**, S08005 (2008)
[11] K.A. Olive, K. Agashe, C. Amsler, M. Antonelli, J.-F. Arguin, D.M. Asner, et al., Particle Data Group, Magnetic monopole searches, Chin. Phys. C **38**, 090001 (2014)
[12] Y.M. Cho and D. Maison, Phys. Lett. B 391 (1997)
[13] B. Acharya et al. [MoEDAL Collaboration], JHEP **1608**, 067 (2016)
[14] G. Aad et al. [ATLAS Collaboration], Phys. Rev. Lett. **109**, 261803 (2012)
[15] G. Aad et al. [ATLAS Collaboration], Phys. Rev. D **93**(5), 052009 (2016)
[16] A.A. Prinz, et al., Phys. Rev. Lett. **81**, 1175 (1998)
[17] A. Haas, C.S. Hill, E. Izaguirre, and I. Yavin, Looking for millicharged particles with a new experiment at the LHC, Phys. Lett. B **746**, 117 (2015)

KASCADE-Grande: Composition studies in the view of the post-LHC hadronic interaction models

A. Haungs[1,a,b], W.D. Apel[1], J.C. Arteaga-Velázquez[2], K. Bekk[1], M. Bertaina[3], J. Blümer[1,4,14], H. Bozdog[1], I.M. Brancus[5], E. Cantoni[3,6,15], A. Chiavassa[3], F. Cossavella[4,16], K. Daumiller[1], V. de Souza[7], F. Di Pierro[3], P. Doll[1], R. Engel[1], D. Fuhrmann[8,17], A. Gherghel-Lascu[5], H.J. Gils[1], R. Glasstetter[8], C. Grupen[9], D. Heck[1], J.R. Hörandel[10], T. Huege[1], K.-H. Kampert[8], D. Kang[1], H.O. Klages[1], K. Link[4], P. Łuczak[11], H.J. Mathes[1], H.J. Mayer[1], J. Milke[1], B. Mitrica[5], C. Morello[6], J. Oehlschläger[1], S. Ostapchenko[12], T. Pierog[1], H. Rebel[1], M. Roth[1], H. Schieler[1], S. Schoo[1], F.G. Schröder[1], O. Sima[13], G. Toma[5], G.C. Trinchero[6], H. Ulrich[1], A. Weindl[1], J. Wochele[1], and J. Zabierowski[11]

[1] Institut für Kernphysik, KIT - Karlsruhe Institute of Technology, Germany
[2] Universidad Michoacana, Inst. Física y Matemáticas, Morelia, Mexico
[3] Dipartimento di Fisica, Università degli Studi di Torino, Italy
[4] Institut für Experimentelle Kernphysik, KIT - Karlsruhe Institute of Technology, Germany
[5] Horia Hulubei National Institute of Physics and Nuclear Engineering, Bucharest, Romania
[6] Osservatorio Astrofisico di Torino, INAF Torino, Italy
[7] Universidade São Paulo, Instituto de Física de São Carlos, Brasil
[8] Fachbereich Physik, Universität Wuppertal, Germany
[9] Department of Physics, Siegen University, Germany
[10] Dept. of Astrophysics, Radboud University Nijmegen, The Netherlands
[11] National Centre for Nuclear Research, Department of Astrophysics, Lodz, Poland
[12] Frankfurt Institute for Advanced Studies (FIAS), Frankfurt am Main, Germany
[13] Department of Physics, University of Bucharest, Bucharest, Romania
[14] now: Head of Division V at KIT - Karlsruhe Institute of Technology, Germany
[15] now at: Istituto Nazionale di Ricerca Metrologia, INRIM, Torino
[16] now at: DLR Oberpfaffenhofen, Germany
[17] now at: University of Duisburg-Essen, Duisburg, Germany

Abstract. The KASCADE-Grande experiment has significantly contributed to the current knowledge about the energy spectrum and composition of cosmic rays for energies between the knee and the ankle. Meanwhile, post-LHC versions of the hadronic interaction models are available and used to interpret the entire data set of KASCADE-Grande. In addition, a new, combined analysis of both arrays, KASCADE and Grande, was developed significantly increasing the accuracy of the shower observables. First results of the new analysis with the entire data set of the KASCADE-Grande experiment will be the focus of this contribution.

1. KASCADE-Grande

The Extensive Air Shower (EAS) experiment and cosmic-ray facility KASCADE finally stopped the active data acquisition end of 2012 after 20 years of measurements. KASCADE, its extension KASCADE-Grande as well as the radio test experiments LOPES and CROME are meanwhile fully decommissioned. The collaboration, however, continues the detailed analysis of the high-quality air-shower data. In addition, with KCDC, the KASCADE Cosmic-ray Data Center, we provide to the public the scientific data via a customized web page.

The multi-detector experiment KASCADE [1] (located in south-west Germany in the beautiful valley of the river Rhine at 49.1°N, 8.4°E, 110 m a.s.l.) started operation in 1993 and was extended to KASCADE-Grande in 2003. Main parts of the facility were the Grande [2] array spread over an area of $700 \times 700\,\mathrm{m}^2$, the original KASCADE array covering $200 \times 200\,\mathrm{m}^2$ with unshielded and shielded detectors, a large-size hadron calorimeter, and additional muon tracking devices. The radio antenna field LOPES [3,4] and the microwave experiment CROME [5,6] were also important components of the experimental set-up of KASCADE-Grande. The full facility was in operation until end of 2012.

The main goal of the measurements was the estimation of energy and mass of the primary particles in a wide energy range. The analysis is based on the combined investigation of the charged particle, the electron, and the muon components measured by the detector arrays of Grande and KASCADE. The general idea of the data analyses of the KASCADE and KASCADE-Grande experiments is the determination of the chemical composition in the primary energy range $10^{14} - 10^{18}$ eV by reconstructing individual mass group spectra. Structures observed in these individual

[a] e-mail: haungs@kit.edu
[b] Speaker for the KASCADE-Grande Collaboration.

 DOI: 10.1051/epjconf/201714513001

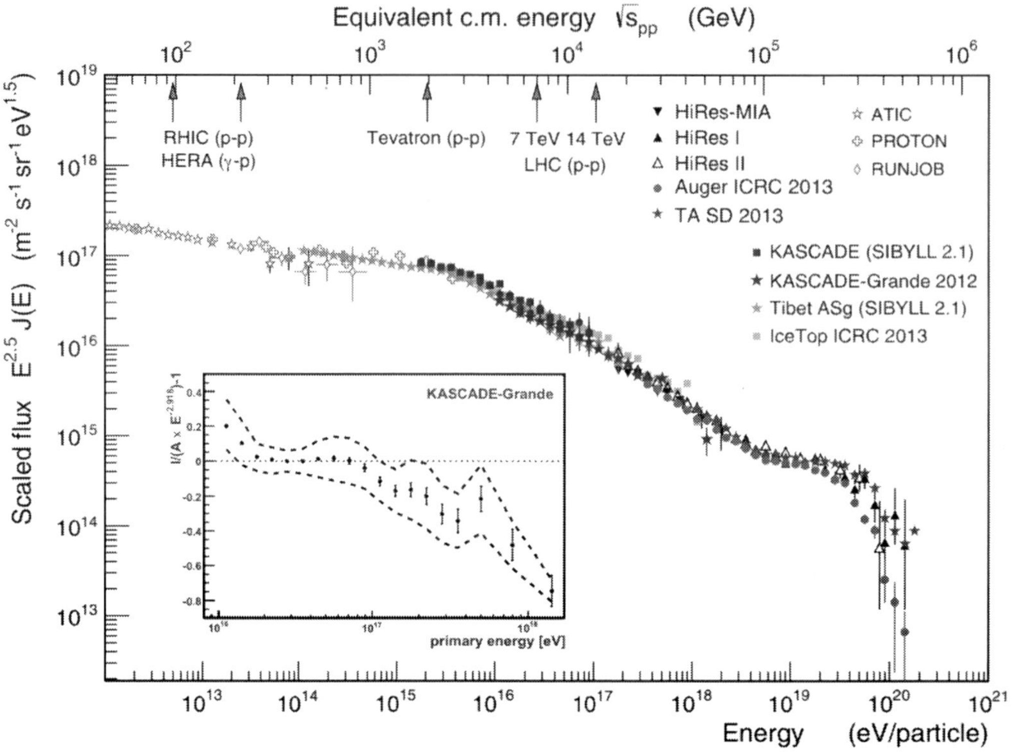

Figure 1. The all-particle energy spectrum [7] obtained with KASCADE and KASCADE-Grande (based on the QGSJet-II model and unfolded, i.e. corrected for the reconstruction uncertainties) . Shown are the spectra in comparison with results of other experiments. In addition, the corresponding interaction energy at accelerators are indicated. The inset displays the spectrum of KASCADE-Grande from 10^{16} eV to 10^{18} eV (from reference [8]) in a way to better see the spectral features.

spectra provide strong constraints to astrophysical models of origin and propagation of high-energy cosmic rays.

2. The all-particle energy spectrum

By using the hadronic interaction model QGSJet-II as a baseline, a composition independent all-particle energy spectrum of cosmic rays was reconstructed in the energy range from 10^{16} eV to 10^{18} eV from the data of the Grande extension (using the Grande array for charge particle measurements and the KASCADE muon array detectors to reconstruct the low energy muon component) within a total uncertainty in flux of 10–15% [8]. The spectrum is in the overlapping region in agreement with the earlier published spectrum by KASCADE [9] (10^{15} eV to 10^{17} eV). Significant structures are observed in the all-particle spectrum (Fig. 1): The justification of the 'knee' at a few times 10^{15} eV is given since many years. In addition, with Grande, there is now clear evidence that just above 10^{16} eV the spectrum shows a significant 'concave' behavior. A further feature in the spectrum is a small break, i.e. knee-like feature at around 10^{17} eV (see inset of Fig. 1). Found first by KASCADE-Grande this is meanwhile confirmed by other experiments (e.g., Tunka-133 [10] or IceTop [11]). This 'second knee' occurs at an energy where the rigidity dependent, i.e. charge dependent, knee of the iron component would be expected, if the 'first knee' is caused by light primaries. The concave part of the spectrum is then a consequence of knee-like features of the spectra of medium masses.

The co-location of LOPES with KASCADE-Grande as well as Tunka-Rex (the Tunka Radio extension) with Tunka-133 was used for an interesting cross-check of the absolute energy scale reconstructed by KASCADE-Grande: The radio technique enables the reconstruction of the primary energy around 100 PeV based on the amplitude of the radio signal. This amplitude can be measured absolutely, where the signal experiences no attenuation in the atmosphere. Due to a consistent amplitude calibration for Tunka-Rex and LOPES achieved by using the same reference source, the reconstruction is used to determine the air-shower energy on an absolute scale at both experiments. This was also used to compare the energy scales of the host experiments Tunka-133 and KASCADE-Grande. As a result, it was confirmed that the energy scales of cosmic-ray measurements by the independently calibrated experiments KASCADE-Grande and Tunka-133 are consistent with each other on the level of 10% [12].

3. Elemental composition of cosmic rays

Already in 2005 KASCADE proved [9, 13] that the knee is caused by a decrease of the light mass group of primary particles and not by medium and heavy primary particles. With KASCADE-Grande we investigated such individual mass group spectra also at higher primary energies [14, 15]. We performed all the simulations for this analysis with the air-shower simulation package CORSIKA [16]

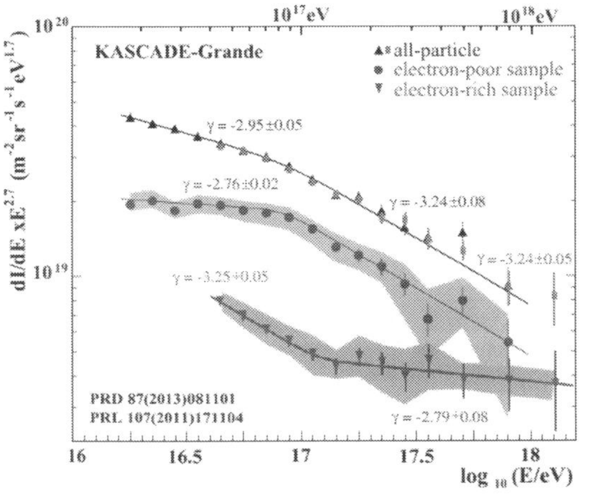

Figure 2. All-particle, electron-poor, and electron rich energy spectra from KASCADE-Grande [7]. References are given in the figure. All the results combined in this figure are based on the hadronic interaction models QGSJet-II-02 and FLUKA.

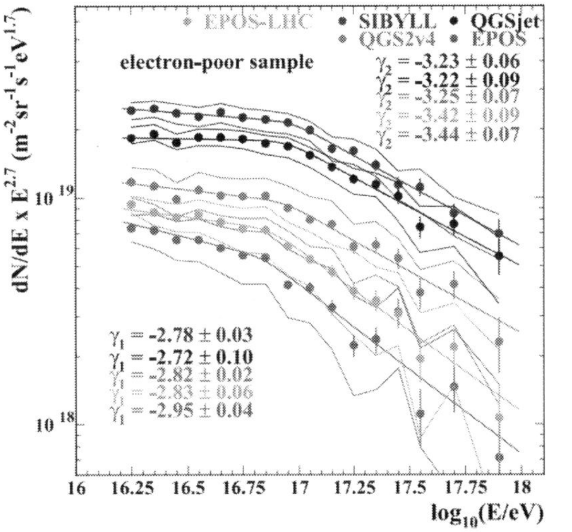

Figure 3. Reconstructed energy spectra of the heavy primary component for five hadronic interaction models. The error bars show the statistical uncertainties and the bands the systematic uncertainties; fits on the spectra and resulting slopes before and after the heavy knee are also indicated [19].

allowing simulations based on various hadronic interaction models. The application of this methodical approach to shower selection and separation in various mass groups was performed and cross-checked in different ways, where Fig. 2 shows the main results. In a first step we separated the Grande data in two mass groups only, i.e., in groups of heavy and light primary masses according to the electron-muon ratio in the air showers:

The reconstructed spectrum of the electron-poor events, i.e. the spectrum of heavy primaries, shows a distinct knee-like feature at about $8 \cdot 10^{16}$ eV with a statistical significance of 3.5 σ [14,17]. The change of the spectral slope is $\Delta\gamma = -0.48$ from $\gamma = -2.76 \pm 0.02$ to $\gamma = -3.24 \pm 0.05$ with the break position at $\log_{10}(E/eV) = 16.92 \pm 0.04$. The selection of heavy primaries enhances the knee-like feature that is already present in the all-particle spectrum. The analysis was repeated on the basis of different hadronic interaction models [18,19]. Despite the fact that the relative abundance of the heavy particles varies significantly dependent on the model in use, the spectral feature of this 'heavy' knee is visible in all the spectra. Figure 3 shows the results for the heavy primary spectra, but also for the light and all-particle spectra. It was found that the structures are similar for all models, but the relative abundances vary. This 'heavy' knee was confirmed by a detailed KASCADE-like unfolding analysis applied to the data of the Grande array [17]. This fits to a rapidity dependent knee scenario with the results of KASCADE [9], where a reduction of the light component beyond the first knee was observed. In addition, an ankle-like feature was observed in the spectrum of the electron-rich events (Fig. 2), i.e. light elements of the primary cosmic rays, at an energy of $10^{17.08 \pm 0.08}$ eV. At this energy, the spectral index changes by $\Delta\gamma = 0.46$ with a statistical significance of 5.8 σ [15].

4. Hadronic interaction models

Historically, a great achievement of KASCADE was the invention of the CORSIKA air-shower simulation tool [16], which meanwhile is used by all major experiments worldwide. Mainly data of the hadron calorimeter and additional muon counters at the central detector and the muon tunnel were used within KASCADE to iteratively test and improve the various versions of the hadronic interaction models optionally available in CORSIKA (see e.g., [13,20] and references therein).

Regarding the analysis of KASCADE and KASCADE-Grande data one has to conclude that all versions of hadronic interaction models of the newest generation provide a 'physical' result in terms of energy and composition of primary cosmic rays. Here, physical means here that the mean composition lies within the band spanned by primary proton and iron simulations. However, the absolute energy and, in particular, the mass scale varies significantly from model to model. In addition, it varies partly also within the models if different shower observables taken from the electromagnetic, muonic, or hadronic components are used. This ambiguity can not be resolved by looking at one single observable or experiment, only. One approach to disentangle this puzzle is the combined analysis of data from the KASCADE and the KASCADE-Grande experiments.

5. KASCADE and KASCADE-Grande combined analysis

A specific analysis focuses on combining the KASCADE and KASCADE-Grande arrays already at the shower reconstruction. The number of charged particles and the number of muons is the result of a fit to the lateral densities measured by both arrays. Therefore, events located in the Grande array gain additional 252 density measurements and events located in the KASCADE array gain 37 additional measurements compared to the individual fits to the lateral densities (see Fig. 4). The additional stations at larger distances allow us to reach for KASCADE located events higher energies and for Grande located events a higher accuracy.

DOI: 10.1051/epjconf/201714513001

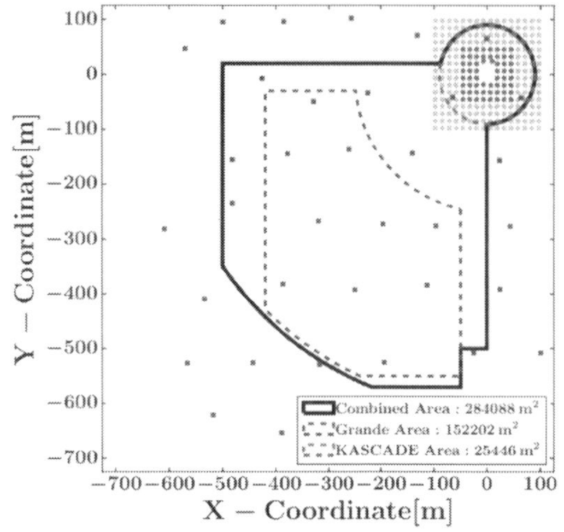

Figure 4. The stations of the KASCADE-Grande array (rectangles) are shown relative to the KASCADE array. In addition, the fiducial areas used for the two stand-alone reconstructions as well as the combined analysis are shown [21].

After converting the energy deposits to the number of particles at each detector using the methods of the stand-alone analyses but now for all detector stations of both arrays, the reconstruction procedure of the combined detector array follows the usual one: The number of particles corresponding to a certain energy deposited in a detector is calculated by dividing the total energy deposit by the effective energy deposited per single charged particle taking into account also the energy deposited by photons and the shower's radial dependent e/γ ratio. The electron size is obtained by fitting a NKG-like lateral density function to the measured particle densities. After this step, the shower size is the number of charged particles. The reconstruction of the number of muons is done simultaneously to the 192 muon detectors of the KASCADE array. Now, the muon Lateral Distribution Function (LDF) is included in the next iteration of the fitting procedure for the unabsorbed detectors resulting in the reconstruction of the shower size as the number of electrons. The reconstruction accuracies of the electron and muon shower size using the combined reconstruction, improves considerably compared to the stand alone analyses. The accuracy of the electron shower size at 10^5 electrons improves from about 6% to about 4 to 5% for the combined reconstruction, reaching an accuracy of better than 3% towards higher energies.

The improvement in the muon reconstruction due to the combined use of both detectors is rather small at the threshold of full efficiency, where the number of muons is reconstructed with an accuracy improving from about 25% to 17%. Therefore, the combined reconstruction reaches an improved accuracy and better uniformity over the entire range. The combination of the detectors right at the beginning of the reconstruction procedure results in much more accurately estimated observables. This is especially true for KASCADE-Grande, for which an improvement of several percent is achieved.

The new 2-dimensional shower size spectrum (Fig. 5) is then the object of further analysis. The distribution

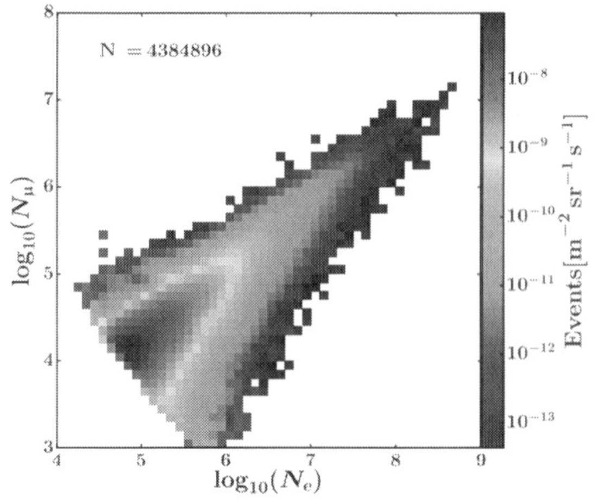

Figure 5. The two-dimensional shower size spectrum for KASCADE-Grande data obtained by a combined shower reconstruction of data from the KASCADE and the Grande arrays [21].

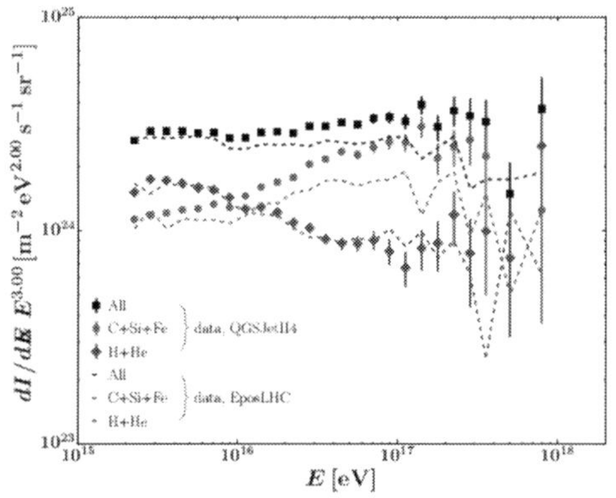

Figure 6. It is worth noting that all spectra obtained are labeled as 'RAW', which means that they have not been corrected for detector and reconstruction effects.

contains events from a larger fiducial area, are reconstructed with higher accuracy, and cover more than three orders of magnitude in primary energy. This plot contains 4,384,896 reconstructed events. The final goal of the analysis is to obtain from this event sample the all-particle as well as mass group spectra by one methodically consistent reconstruction procedure [21].

As a first step, the method from the Grande stand alone data analysis (k-parameter method, where k is basically the electron-muon ratio of the individual showers taking into account the incident zenith angle) was applied to the new reconstructed two-dimensional shower size distribution to obtain the all-particle energy spectrum, as well as the spectra of the heavy- and light-induced air showers. For the calibration of the energy and mass of the primary particle a set of simulations is needed which is obtained for each hadronic interaction model, separately. The same hadronic interaction model was always used to define the separation value (the cut in the parameter 'k') and to interpret the full data set.

Figure 6 shows the results of this analysis based on QGSJet-II-04 and EPOS-LHC for comparisons. The

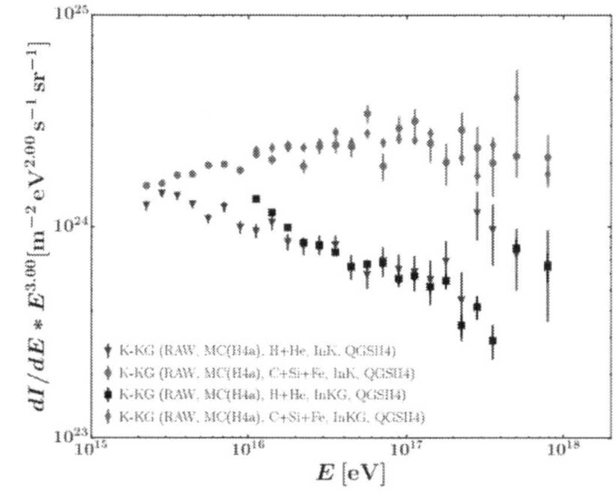

Figure 7. The reconstructed spectra for the heavy and light component are shown for data simulated using QGSJetII4 as the hadronic interaction model and H4a as the astrophysical model. The spectra have been obtained with the combined reconstruction, however, separately for the two data sets defined by the core being located in KASCADE and KASCADE-Grande, respectively. In both cases the calibration based on QGSJetII4 has been used.

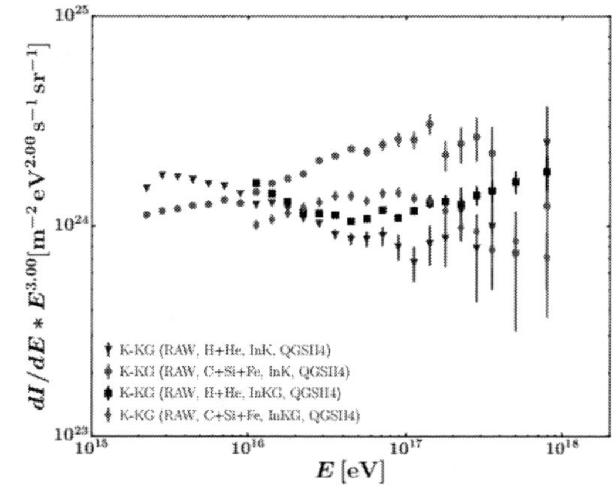

Figure 8. The reconstructed spectra for the heavy and light component are shown for measured data using calibrations based on QGSJetII4 as the hadronic interaction model. The spectra have been obtained with the combined reconstruction, however, separately for the two data sets defined by the core being located in KASCADE and KASCADE-Grande, respectively.

combined analysis now provides these spectra for an energy range of more than 3 orders of magnitude based on one unique coherent analysis. Considering the QGSJet-based results the spectra confirms the observation of all features found earlier, i.e. the light and heavy knees at approximately 3 and 100 PeV, respectively as well as the hardening at approximately 10 PeV and the light ankle at approx. 100 PeV. This is also true for the results based on EPOS-LHC (Fig. 6) and SIBYLL 2.3 (not shown). However, again, also for the post-LHC models the relative abundances of the light and heavy generated spectra differ quite significantly from model to model.

It is interesting to note, that comparing EPOS-LHC with QGSJet-II 04 the spectra of the light component agrees quite well, but not those of heavy primaries. This

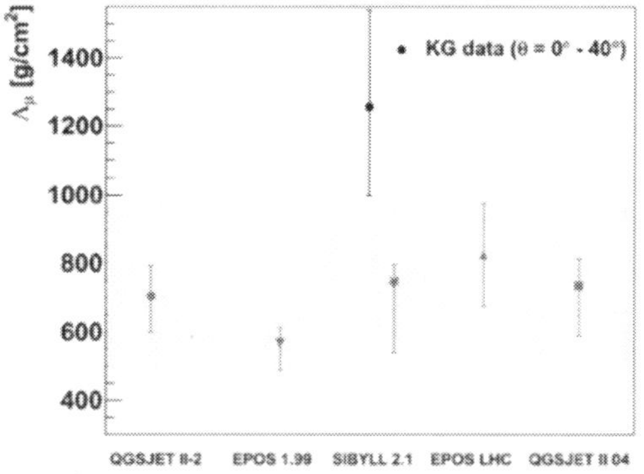

Figure 9. Muon attenuation lenghts for KASCADE-Grande in comparison with those obtained by detailed Monte-Carlo simulations based on various hadronic interaction models [27].

Figure 10. Logo of KCDC. The link to KCDC is: `https://kcdc.ikp.kit.edu`.

can be interpreted as a hint that proton-proton interactions are better described in the post-LHC models than nucleus-nucleus interactions. This is understandable as taking the LHC-data into account for the latter one an additional extrapolation or theoretical prediction is necessary to build the model.

To learn more about the source of the differences in interpreting the same data with different hadronic interaction models we deepened the analysis in the following way: For each interaction model separately a set of data was generated with the true, i.e. simulated spectra of the five elements H, He, C, Si, and Fe corresponding to the predictions of an astrophysical model which is relatively close to the obtained results, in this case the H4a [22] model provided by Tom Gaisser. In addition, the unique feature of KASCADE-Grande allows for cross-checking the results by dividing the data in one set containing events located within KASCADE (InK) and another set limited to events located in KASCADE-Grande (InKG). The advantage is that in one of the samples we obtain the muon number by measuring muons close to the shower core, for the other sample by measuring muons far (300–700 m distance) from the shower core. Other shower parameters are obtained with similar systematics in both samples.

The purpose of this procedure is to see how the three hadronic interaction models QGSJet-II-04 [23, 24], EPOS-LHC [25], and Sibyll 2.3 [26] compare to one another and to the data.

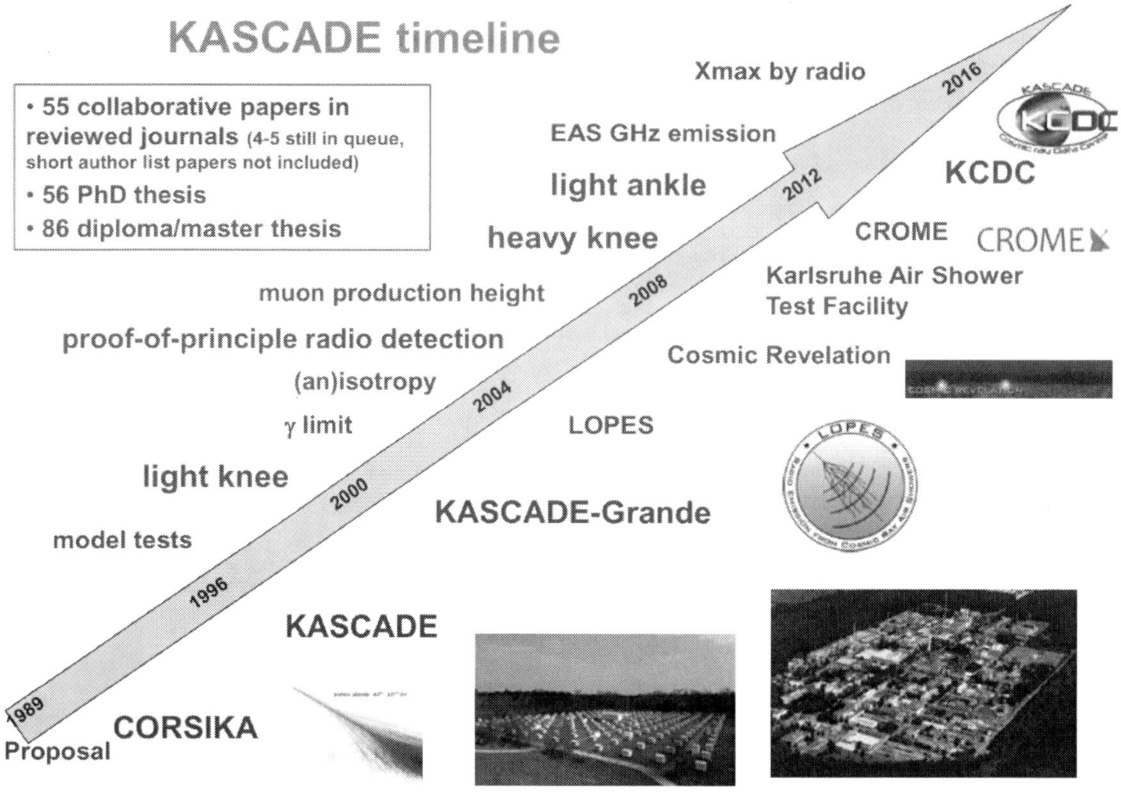

Figure 11. Timeline of the KASCADE(-Grande) facility. Shown are below the arrow of time in red the various hardware related parts of the facility and above the arrow in blue some highlights of physics results. 'Cosmic Revelation' was an art & science project at KASCADE in 2009 [29], which had a revival in summer 2016 (see `http://www.imachination.net/cosmicrevelation`).

If we now apply the full (including the detector response) simulated data to the above mentioned reconstruction we expect to get back the right answer, i.e. the input spectra for both event samples. This worked within the systematic uncertainties for all three studied interaction models (see as an example Fig. 7 for QGSJet-II 04) proving an internal consistency of the models, at least. Applying the reconstruction to the two measured samples, independently, the spectra show significant differences. As an example this is shown in case of the QGSJet-II-04 model (but we found similar differences for the other two models) in Fig. 8. It is obvious that a systematic deficiency appears for the model to describe the muon content of the showers consistently. There are either too few muons predicted in the center of the EAS or too few at larger distances (i.e. the lateral distribution of muons is predicted to be too flat). The presented investigation is neither sensitive to the total number of muons (this would slightly change the energy scale predicted by the model, only) nor to give a statement of which of these two cases is more probable. But, these are strong hints that there are deficiencies in describing the slope of the lateral distribution of muons in air showers.

Another analysis studies the attenuation of muons in the Earths' atmosphere [27]. This is especially interesting for cross-checking hadronic interaction models and confirms that there are deficiencies in describing the muon production in extensive air showers at a primary energy around 100 PeV. It is based on the idea of constant-intensity that assumes a flux of CRs that is isotropic in zenith angle. The number of muons measured for a certain flux is decreasing with increasing zenith angle. Therefore, the attenuation of muons in the atmosphere can be obtained (Fig. 9), where it is clear that none of the five models studied describes the observed attenuation length. The analysis was performed for the events reconstructed by the Grande stand-alone method for an event sample with a certain range of distance to the shower core and primary energy. A more detailed analysis of the muon attenuation will be published soon.

6. The KASCADE Cosmic-ray Data Center: KCDC

The KASCADE facility operated several large-area detectors for the measurement of EAS for more than 20 years. The major goal of KCDC [28] is the installation and establishment of a public data centre for high-energy astroparticle physics. In the research field of astroparticle physics, such a data release is a novelty, whereas the data publication in astronomy has been established for a long time. Therefore, there are no completed concepts, how the data can be treated/processed so that they are reasonably usable outside the collaboration. The project faces thereby open questions, e.g. how to ensure a consistent calibration, how to deal with data filtering and how to provide the data in a portable format as well as how a sustainable storage solution can be implemented. In addition, access rights and license policy play a major role and are considered.

Already with the first releases, KCDC fulfills three basic requirements needed to address the general public:
KCDC as a data provider: There is free and unlimited open access to KASCADE cosmic-ray data, where a selection of fully calibrated and reconstructed quantities per individual air shower is provided. The access has to rely on a reliable data source with a guaranteed data quality.
KCDC as an information platform: For a meaningful usage of KCDC, a detailed experiment description as well as sufficient meta information on the provided data is needed for any kind of data analysis. This is accompanied by a reasonable description of the physics background as well as tutorials, which are focused on a level for teachers and pupils.
KCDC as a long-term digital data archive: To constitute a sustainable piece of work, KCDC also serves as an archive of software and data for the collaboration as well as for the public.

KCDC has started to make the data from the KASCADE experiment available to the community, where the platform to external users is already released [28] and meanwhile several times updated. Interested colleagues are invited to visit KCDC under `https://kcdc.ikp.kit.edu` (Fig. 10).

7. Summary

In summary, after separating the KASCADE-Grande measured events into light and heavy components, knee-like features are identified in the spectrum of the light component by KASCADE and of the heavy component by Grande, as well as an ankle-like feature is observed in the spectrum of the light component. Whereas the 'heavy-knee' occurs at an energy where the rigidity dependent knee of the iron component is expected, the 'light-ankle' might indicate an early transition from galactic to an extragalactic origin of cosmic rays.

Whereas general structures observed in the all-particle spectrum as well as in the spectra of individual mass groups are meanwhile established and credible, the validity of the hadronic interaction models is proven to be still insufficient. KASCADE-Grande presented a new approach on comparing the three post-LHC hadronic interaction models, where in general the main message is the same for all of them: While the predictions of the hadronic interaction models are becoming more and more similar, larger discrepancies between the simulated data and measurements remain, which become obvious when comparing the results obtained for events located in KASCADE to events located in KASCADE-Grande. The main difference is observed in the muonic component of the measured air-showers. This implies that the underlying lateral density function as seen in simulations differs from the one implemented in nature.

The KASCADE-Grande collaboration has started to provide the entire data set to the public within the project KCDC, the KASCADE Cosmic Ray Data Center.

Figure 11 illustrates the timeline and some highlights of 27 years activity of the KASCADE(-Grande) facility.

The authors would like to thank the members of the engineering and technical staff of the KASCADE-Grande collaboration, who contributed to the success of the experiment. The KASCADE-Grande experiment is supported in Germany by the BMBF and by the 'Helmholtz Alliance for Astroparticle Physics - HAP' funded by the Initiative and Networking Fund of the Helmholtz Association, by the MIUR and INAF of Italy, the Polish Ministry of Science and Higher Education, and the Romanian Authority for Scientific Research UEFISCDI (PNII-IDEI grants 271/2011 and 17/2011). J.C.A.V. acknowledges the partial support of CONACYT (grant CB-2008/106717) and the German-Mexican bilateral collaboration grants (DAAD-CONACYT 2009–2012, 2015–2016).

References

[1] Antoni T et al. - KASCADE, NIM A **513**, 490 (2003)
[2] Apel W D et al. - KASCADE-Grande, NIM A **620**, 202 (2010)
[3] H. Falcke et al. - LOPES, Nature **435**, 313 (2005)
[4] F.G. Schröder et al. - LOPES, J. Phys. Conf. Ser. **632**, 012102 (2015)
[5] R. Smida et al. - CROME, EPJ Web Conf. **53**, 08010 (2013)
[6] R. Smida et al. - CROME, PRL **113**, 221101 (2014)
[7] A. Haungs et al. - KASCADE-Grande, ICRC 2015, The Netherlands, PoS(ICRC2015)278
[8] W.-D. Apel et al. - KASCADE-Grande, Astropart. Phys. **36**, 183 (2012)
[9] T. Antoni et al. - KASCADE, Astropart. Phys. **24**, 1 (2005)
[10] V.V. Prosin et al. - Tunka-133, NIM A **756**, 94 (2014)
[11] M.G. Aartsen et al. - IceCube, PRD **88**, 042004 (2013)
[12] Apel, W.D. et al. - Tunka-Rex and LOPES Collaborations, Physics Letters B **763**, 179 (2016)
[13] W.-D. Apel et al. - KASCADE-Grande, Astrop. Phys. **65**, 55 (2015)
[14] W.-D. Apel et al. - KASCADE-Grande, PRL **107**, 171104 (2011)
[15] W.-D. Apel et al. - KASCADE-Grande, PRD **87**, 081101(R) (2013)
[16] D. Heck et al., report Forschungszentrum Karlsruhe, FZKA **6019** (1998)
[17] W.-D. Apel et al. - KASCADE-Grande, Astropart. Phys. **47**, 54 (2013)
[18] W.-D. Apel et al. - KASCADE-Grande, Adv. Space Res. **53**, 1456 (2014)
[19] M. Bertaina et al. - KASCADE-Grande, ICRC 2015, The Netherlands, PoS(ICRC2015)359
[20] T. Antoni et al. - KASCADE, J. Phys. G **36**, 035201 (2009)
[21] S. Schoo et al. - KASCADE-Grande, ICRC 2015, The Netherlands, PoS(ICRC2015)263
[22] T. Gaisser, Astropart. Phys. **35**, 801 (2012)
[23] S. Ostapchenko, PRD **83**, 014018 (2011)
[24] S. Ostapchenko, EPJ web conf. **52**, 02001 (2013)
[25] T. Pierog et al., Phys. Rev. C **92**, 034906 (2015)
[26] R. Engel, ICRC 1999 **415**, 905 (1999)
[27] J.C. Arteaga et al. - KASCADE-Grande, ICRC 2015, The Netherlands, PoS(ICRC2015)314
[28] S. Schoo et al. - KASCADE-Grande, ICRC 2015, The Netherlands, PoS(ICRC2015)262
[29] T.O. Roth and A. Haungs, Transactions, Leonardo **42**, 3 (2009)

Multiple production up to $\sqrt{s} = 13\,\text{TeV}$ with the generator GHOST adapted to cosmic ray simulation

Jean-Noël Capdevielle[1,a], Zbigniew Plebaniak[2], Barbara Szabelska[2], and Jacek Szabelski[2]

[1] APC Université Paris Diderot, 10 rue A. Domon et V. Duquet, 75013 Paris and Académie des Sciences d'Outre Mer, 15 rue La Pérouse, 75116 Paris, France
[2] National Centre for Nuclear Research, Astrophysics Division, Cosmic Ray Laboratory, ul. 28 Pułku Strzelców Kaniowskich 69, 90-558 Łódź, Poland

Abstract. The model HDPM of CORSIKA has been updated and developed on the base of the recent measurements by LHCb, CMS, TOTEM... The new model GHOST involving a 4-source production reproduces correctly the pseudorapidity distributions of charged secondaries and has been testified with the data in the mid and forward rapidity region, especially in the complex case of TOTEM and also with the recent measurements of CMS up to $\sqrt{s} = 13$ TeV ($9.2 \cdot 10^{16}$ eV).

Special calculations have been devoted to the semi-inclusive data playing an important role in the cosmic ray simulation (fluctuations in the earliest collisions, individual cascades and remarkable events). Taking into account the violation of KNO scaling, the negative binomial distribution has been used pointing out an asymptotic behaviour of total charged multiplicities at primary energies exceeding 40 TeV ($7 \cdot 10^{17}$ eV). The interpretation of the most recent measurements suggests for EAS generated by primary protons a larger production of muons and a cascade maximum at higher altitude than previously assumed.

1. Introduction

In the sections that follow, we shall discuss the new improvements in cosmic ray simulation suggested by the most recent results of the LHC, especially the data of ATLAS, CMS, TOTEM, LHCb... up to $\sqrt{s} = 13$ TeV used as the largest energy now available to elaborate and testify the new Monte Carlo generator suitable for cosmic ray simulations.

Section 2 is devoted to the consequences of the violation of KNO scaling on the semi-inclusive data after considering the limits of the earliest scaling behaviour only valid now in small energy bands.

Section 3 is focused on the basic features of very high energy hadron interactions and guidelines useful for extrapolations of EAS simulations up to primay energies of 10^{20} eV. The 4-component model GHOST (Generator Hadronic Oversteering Secondary Treatment) reproduces general characteristics of p-p collisions up to $\sqrt{s} = 13$ TeV.

Section 4 evocates some measurements not explained in cosmic ray data above $\sqrt{s} = 1$ TeV, especially in high energy spectra of secondaries measured in EAS, spikes and coplanar emission in Gamma-ray families.

Observing that Extensive Air Showers (EAS) are random individual events, we emphasize here the interest of the semi inclusive data delivered by the LHC at ultra high energy.

[a] e-mail: capdev@apc.univ-paris7.fr

2. Consequences of the violation of KNO scaling and semi inclusive data

Each new step in energy reached by the colliders is an opportunity to evaluate the predictions of the interaction models. The simulation of the hadronic cascade in cosmic ray physics needs at least the parameters describing the different cross sections, the total multiplicity of secondaries, their energy and transverse momenta distributions, the energy dependence of those parameters and the statistic laws governing their fluctuations. The violation of Koba, Nielsen, Olesen (KNO) scaling observed in UA5 [1], for instance, introduced the negative binomial distribution well adapted to the prediction of multiplicities of secondaries in cosmic ray cascades.

2.1. KNO scaling in central region

Several attempts have been performed to introduce other scaling based on the Feynmann's scaling [2] describing the rapidity distribution as a central plateau with gaussian wings; the link with the pseudorapidity distribution was treating the central dip with dynamical considerations, whereas the extension from ISR energies up to SPS energies introduced both a rising plateau and a width increase derived from UA5 data on pseudorapidity distributions. The KNO scaling (concerning all the distribution) was rejected in 1983 following the contradiction between ISR and UA5 data; taking into account the wider distributions of multiplicity than KNO predicted also by the Dual Parton Model (DPM) [3] and the "fireball model" [1]. It was decided to explore the multiplicity distributions in limited pseudorapidity intervals.

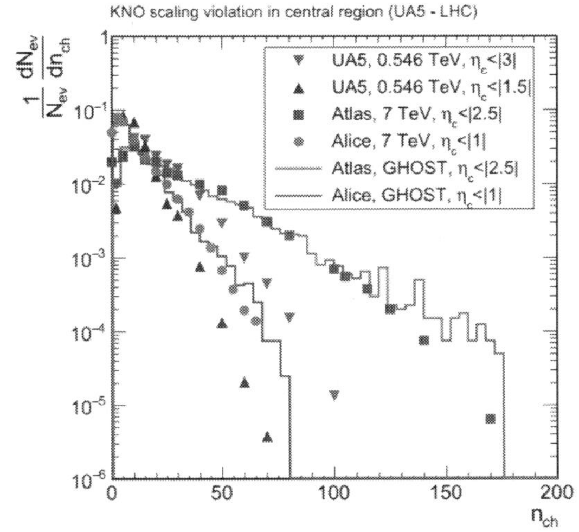

Figure 1. Charged multiplicities in central regions. Comparison between ATLAS, ALICE at $\sqrt{s} = 7$ TeV, respectively for $\eta \leq 2.5$ (squares), $\eta \leq 1.0$ (circles) and UA5 data, $\eta \leq 3$ (triangle down), $\eta \leq 1.5$ (triangle up). The histograms are the respective MC simulation results with GHOST for ATLAS and ALICE.

A tolerable scaling agreement was obtained in central region for $\eta \leq 1.3$ or $\eta \leq 1.5$ in the case of ISR, UA5 and UA1 for instance. However, several uncertainties remain. We have simulated with GHOST such situation for comparison between LHC data ($\sqrt{s} = 7$ TeV) and UA5 ($\sqrt{s} = 540$ GeV). Important discrepancies (Figure 1) of the distributions are distinguished between ATLAS and UA5 for $\eta \leq 2.5$ and $\eta \leq 3$. as well as in the case of ALICE and UA5 respectively for $\eta \leq 1.$ and $\eta \leq 1.5$ [4] [5]. Since LHC measurements of the average multiplicities $\bar{n_{ch}}$ are not yet very accurate, a strict comparison versus the KNO variable $z = n_{ch}/\bar{n_{ch}}$ cannot be shown, but there remains poor chance for this scaling in central region. Our histograms calculated with GHOST confirm the large discrepancies with UA5 data when the energy is rising up to $\sqrt{s} = 7$ TeV.

Similarly, the attempts made in the fragmentation region of the inclusive pseudorapidity distributions plotted in the beam frame were not conclusive; the measurements were obtained for ISR up to $\eta - y_{beam} = -0.75$ for 53 GeV in ISR, but only up to $\eta - y_{beam} = -2.4$ for 900 GeV in UA5 [6]. The genuine fragmention region cannot be reached because the range of forward directions at $|x_f| \geq 0.05$ is experimentally unavialable.

2.2. Empirical scaling and semi-inclusive data

Fortunately an "empirical scaling" was pointed out in the central region by UA5 as a relation between the normalized central density and the normalized multiplicity [6]. The ratio $\rho(0)/\bar{\rho}(0) = f(z)$ (instead of $f(z,s)$ is here independent of $\sqrt{s}$), $\rho(0)$ being the central density for a given charged multiplicity n, $\bar{\rho}(0)$ being the average central density (for a more simple legibility n, n_{ch}, N and N_{ch} are used interchangeable). The KNO variable $z = n/\bar{n}$ is the normalized multiplicity. The general dependence on z is shown in the Fig. 2. Following our adjustment, we

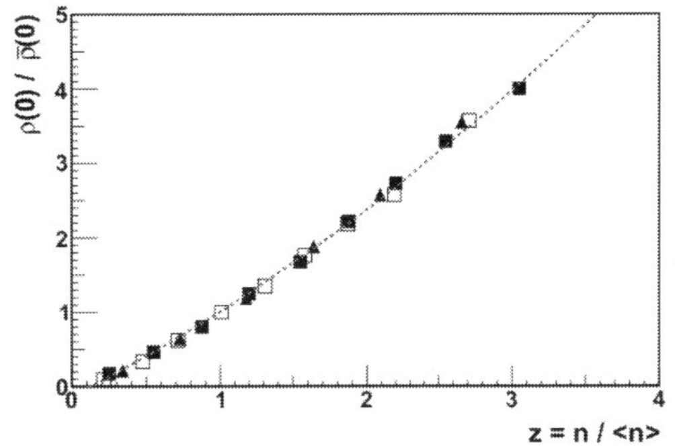

Figure 2. Empirical scaling function $\zeta = f(z)$ for $\sqrt{(s)} = 200$, 546 and 900 GeV for NSD collisions. In this figure $\bar{\rho}(0)$ is noted as $\rho(0)$ whereas $\rho(0)$ in text is specified as $\rho(0)_n$.

propose for $f(z)$ the numerical representation:

$$\zeta = f(z) = 0.11499\, z^2 + 1.0231\, z - 0.1319 \quad (1)$$

This kind of scaling is not yet confirmed in LHC data as the semi-inclusive measurements are performed only in small windows of pseudorapidity, but it remains present in models extrapolations at larger energies.

Following UA5 and our considerations in CORSIKA user's guide [7], the fluctuations of $< N_{ch} >$ for a fixed primary energy can be reproduced at very high energy by a more simple relation corresponding to the negative binomial distribution (NBD) at very high energy; this simple function [8] represents in terms of KNO reduced variables$z = \frac{n}{\bar{n}}$ the fluctuations as:

$$\Psi(z,k) = \bar{n} P_n = \frac{k^k}{\Gamma(k)} z^{k-1} e^{-kz} \quad (2)$$

$$k^{-1} = a + b\, ln\sqrt{s} \quad (3)$$

$$(a = -0.104, b = 0.058) \quad (4)$$

The values of the parameters correspond here to $\bar{n} = < N_{ch} >$ for the charged NSD component, P_n being the probability of an incoming multiplicity N_{ch}. The evolution of those fluctuations of N_{ch} is plotted in the Fig. 3 at ISR and UA5 energies, completed by the curve and the histogram illustrating the calculation with GHOST at $\sqrt{s} = 8$ TeV. The asymptotic form at very high energy (40 TeV) in the unfulfilled project of the Superconducting Super Collider (SSC 1986) is the most enlarged variation with the simple expression $\Psi(z) = 4z\ e^{-2z}$. The probability in p-p collision to receive randomly very small N_{ch} or very large N_{ch} increases at very high energy as shown in the Fig. 3.

Such contrast changes the statistical treatment of the earliest interactions in hadronic cascades at UHE and we underline the neighbourhood of the histogram at $\sqrt{s} = 8$ TeV with the asymptotic expression of the SSC. Thanks to the function ζ the semi-inclusive data can be governed by the integro-differential system:

$$\left.\frac{dN}{dy}\right|_{y=0} = m_r \left.\frac{dN}{d\eta}\right|_{\eta=0} = m_r\, \zeta\, \bar{\rho}(0) \quad (5)$$

$$\int \frac{dN}{dy} dy = z < n > \quad (6)$$

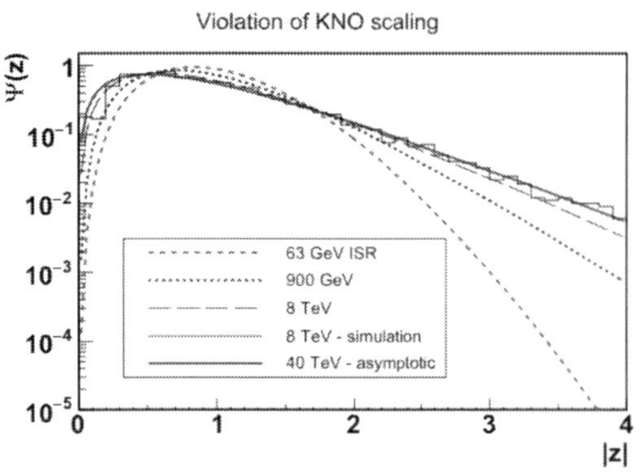

Figure 3. Fluctuations of total charged NSD multiplicity. The full line represents the asymptotic behaviour at $\sqrt{s} = 40$ TeV, the histogram and the dashed line correspond both to the calculation of GHOST and the function $\Psi(z, k)$ at $\sqrt{s} = 8$ TeV, whereas the dotted and dashed curves correspond respectively to $\sqrt{s}$ =900 GeV and $\sqrt{s} = 53$ GeV.

m_r is the ratio of central mean rapidity density and mean central pseudorapidity density derived from the "dip" existing in the centre of the pseudorapidity distribution, resulting from the mass m and the transverse mass m_T of the secondaries as $m_r = \sqrt{1 - m^2/m_T^2}$. In the case of the 4 gaussian generation (one pair of functions in each hemisphere), symmetrics around the center of mass,

$$\frac{dN}{dy} = \sum_{i=1}^{i=2} a_i(e^{-0.5u_i} + e^{-0.5v_i}) \tag{7}$$

$$u_i = \left(\frac{(y - y_i)}{\sigma_i}\right)^2, \quad v_i = \left(\frac{(y + y_i)}{\sigma_i}\right)^2 \tag{8}$$

it is possible to use the opportunity of the scaling of function ζ in the relation between the center y_i and the width σ_i of each gaussian function as

$$y_i = \sigma_i \left(2ln\left(\frac{z < n >}{\sqrt{(2\pi)}\,\zeta\,\sigma_i}\right)\right)^{0.5}$$

After introducing one proportion χ of the multiplicity distributed to the pair of gaussian centered in central region and in mid-rapidity region, it is possible to obtain with a minimal Monte Carlo generation of random deviates the total original rapidity distribution source of the distribution of pseudorapidity shown in the Fig. 4 for various range of multiplicity (respectively $\chi = 20\%$ and 80% or for the total NSD multiplicity in the Fig. 7). The couples rapidity-transverse momentum (y, p_t) are generated using the same p_t distribution as in CORSIKA [7]. The total average inelasticity of the NSD collisions calculated with GHOST is 0.71 indicating a larger proportion of the primary energy distributed to the secondaries. The generator GHOST has been used here to appreciate the elements provided by semi-inclusive measurements which will be distributed soon by the LHC.

The NSD behaviour is shown here in the Fig. 4, corresponding to the histogram of the Fig. 3 for $\sqrt{s}$ = 8 TeV.

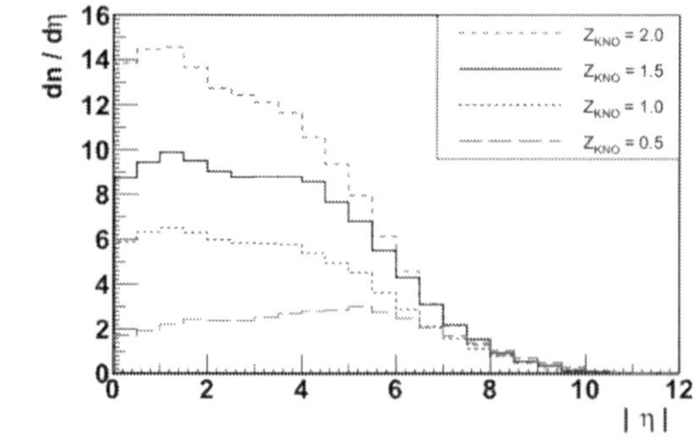

Figure 4. Semi-inclusive NSD pseudorapidity distribution, z = 0.5, 1., 1.5, 2. correspond respectively to relative intervals of N_{ch} [0.25–0.75], [0.75–1.25], [1.25–1.75], [1.75–2.25].

3. Basic features of VHE interactions

3.1. Theoretical models and phenomenological collision generators

High energy cosmic ray interactions are mainly understood with the "soft physics" data provided by the colliders. Therefore, the approach of microscopic models starts from the parton distribution functions, distinguished for valence quarks and diquarks, sea quarks and gluons. In the case of p-p collisions, there are common features between models used in accelerators or in cosmic rays simulations, respectively models such as Dual Parton Model, Phojet, Pythia on one side, models like QGSJet, Sybill on the other side [9].

The geometrical and statistical models such as the cluster model GENCL of UA5 [10] are more connected with the phenomenological features of the experimental data and built on analogy with quantum optics. Similarly, the Hybrid Dual Parton Model (HDPM) was inspired by the most simple double pomeron structure of DPM. It represents the pair of quark-diquark chains stretched between the valence quark and diquark of the projectile and the target by one pair of gaussian functions with a correct overlap to reproduce the expected rapidity distribution with the profile of one plateau with gaussian-shaped wings [11–13].

Unfortunately, in the interpretation of cosmic ray data, the approach has a limit due to the difficulties of measurement in the very forward region of rapidity and the determination of the properties of the leading cluster. In other words, the role of the pilot particle fundamental in hadronic cosmic ray cascades remains dependent on adequate assumptions.

All the models have been extended to p-A and A-A collisions in order to describe the collision of cosmic ray hadrons with air constituents and how the comparison with the LHC data becomes also possible.

3.2. Guidelines and extrapolations at ultra high energy

Conversely, there exist also models elaborated directly for collisions of heavy nuclei which can be tuned to describe p-p collisions; this is the case, for instance, of the non-equilibrium statistical relativistic diffusion model (RDM) based on a Fokker-Planck type transport equation [14]. Hence, the earliest and most simple data

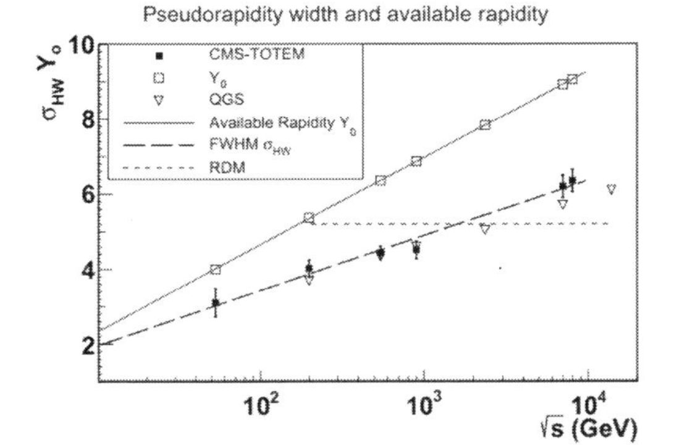

Figure 5. Energy dependence of the available rapidity Yo and of the FWHM, σ_{HW}, respectively, open squares, solid line and dashed line from GHOST, dark squares from CMS-TOTEM, triangles from QGSJET01 model, horizontal line from the relativistic diffusion model RDM [14].

produced by colliders and accelerators is the charged Non Single Diffraction (NSD) distribution. The characteristics of this distribution: average central rapidity density $\rho(0)$, total average multiplicity of charged secondaries n_{ch} (as far as it can be approached by the integration of the densities collected till the smallest angles of emission), the maximum and the full width at half maximum (FWHM), transverse momentum distribution, are the normal tests of the preliminary validity of the models recognized.

A next step of quality is reached with additive criteria adapted to the same pseudorapidity distribution data considered in small bands of multiplicity i.e. in a semi-inclusive analysis as it was shown in the Fig. 4.

3.3. Energy dependence of total charged multiplicity and central pseudorapidity densities at LHC

Up to now the average central pseudorapidity density has been measured up to $\sqrt{s} = 13$ TeV (only for the inelastic component as measured recently by CMS), the other components being limited at $\sqrt{s} = 8$ TeV. For instance the widening of the full pseudorapidity distribution at half maximum (FWHM) can be used as an estimator up to $\sqrt{s} = 8$ TeV thanks to TOTEM data giving a part of the pseudorapidity densities near the forward region, but this is not yet the case at $\sqrt{s} = 13$ TeV as shown in the Fig. 5.

We can observe also in the Fig. 5 the different behaviour of the full width at half maximum (FWHM), σ_{HW}, on the one hand, and the rise of the available rapidity (or rapidity Y_0 of the center of mass in the laboratory system) on the other hand. Both dependencies on s reflect a logarithmic increase

$$\sigma_{HW} = 0.3182\, ln(s) + 0.5 \qquad (9)$$

$$Y_0 = 0.502\, ln(s) + 0.3 \qquad (10)$$

We begin however to have new elements in the interval $\sqrt{s} = [1. - 13.]$ TeV useful for a lever arms of two guidelines in cosmic ray simulation, the average multiplicity dependence and average central pseudorapidity dependence to describe in better condition the cosmic ray data up to about 10^{17} eV. New extrapolations risked up to 10^{20} eV could be evaluated.

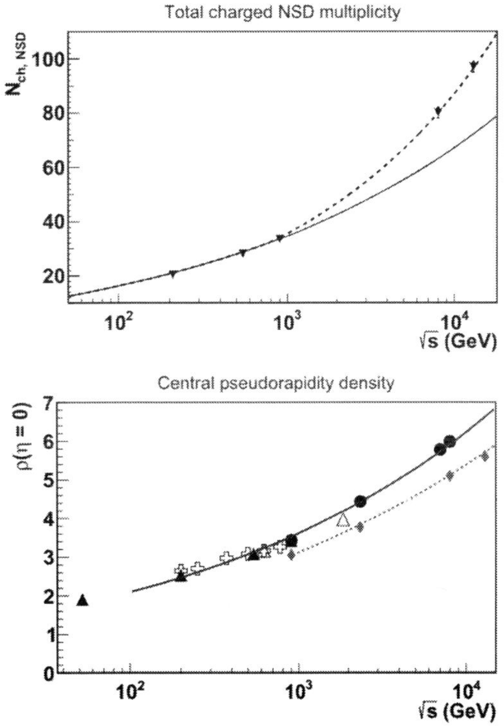

Figure 6. Figure 6a (upper) Total charged multiplicity (NSD) dependence on s for $E_0 = 10^{13}-10^{17}$ eV following UA5 (solid line) and the asymptotic tendency using GHOST (dashed line). 3 lower points are from UA5, the upper point at 13 TeV is estimated with GHOST. Figure 6b (bottom) Central pseudorapidity density dependence on s for $E_0 = 10^{13}-10^{17}$ eV (solid line for NSD component with one point from GHOST at $\sqrt{s} = 13$ TeV, dashed line with diamonds devoted to the inelastic component is aproximately $0.8695 \cdot \bar{\rho}(0)$ NSD).

The energy dependence in the Fig. 6 of the total mean charged multiplicity (NSD component) is shown hereafter adding to the $Sp\bar{p}S$ collider measurements [1] the data of CMS and TOTEM at $\sqrt{s} = 8$ TeV. An important enhancement is thus appearing when we take into account the multiplicity estimated using GHOST after integration of the density of both rapidity and pseudorapidity in the best agreement with the pseudorapidity measured at $\sqrt{s} = 8$ TeV by CMS and TOTEM [15]. We note a large discrepancy of the asymptotic behaviour (solid line) with the previous tendency (dashed line) expected from UA5 data [1]. The first estimation of the total average multiplicity $< N_{ch-UA5} >$ by a power law has to be updated as we suggest above $\sqrt{s} = 1$ TeV introducing another power law for $< N_{ch} >$ as follows:

$$< N_{ch-UA5} > = -7.0 + 7.2\, s^{0.127} \qquad (11)$$

$$< N_{ch} > = -0.74 + 2.59\, s^{0.191} \qquad (12)$$

There is also an important rise of multiplicity in the central region characterized by the average central pseudorapidity density $\bar{\rho}(0)$ with a power law dependence on s such as $0.708 s^{0.118}$. However we use hereafter the parabolic logarithmic fit of CMS [16], the different behaviour with

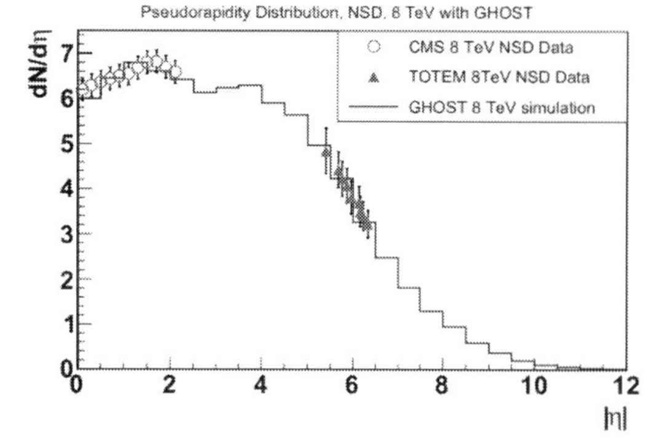

Figure 7. Average pseudorapidity density simulated with GHOST at $\sqrt{s} = 8\,\text{TeV}$ compared with CMS and TOTEM measurements taking into account respective limits of Pt.

the logarithmic dependence $\bar{\rho}_1(0)$ of UA5 replaced by $\bar{\rho}(0)$ as follows is shown in the Fig. 6b

$$\bar{\rho}(0) = 3.17 - 0.372 ln(s) + 0.0291 ln(s^2) \quad (13)$$

$$\bar{\rho}_1(0) = 0.24\, ln(s) + 0.1 \quad (14)$$

The dependence on energy evaluated in LHC up to $\sqrt{s} = 13\,\text{TeV}$ exhibits multiplicity enhancement in central rapidity region as well as in mid-rapidity region and a total charged mean multiplicity close to 100 particles can be expected at $\sqrt{s} = 15\,\text{TeV}$.

3.4. Average rapidity and pseudorapidity distributions up to $\sqrt{s}$ = 13 TeV

At present the charged density measurements at $\sqrt{s} =$ 8 TeV cover central and mid-rapidity region limited to 7 units of pseudorapidity. A comparison of a profile generated with the double gaussian generator GHOST is shown in the Fig. 7 and compared with the NSD measurements of CMS and TOTEM [15].

This result takes into account for both rapidity and pseudorapidity the generation of individual collisions containing for instance the semi-inclusive data shown in the Fig. 4. The simple integration of the simulated distribution in agreement with the data on average densities of pseudorapidity $< N_{ch} > = \int \frac{dN_{ch}}{d\eta}\, d\eta$ gives for $\sqrt{s} =$ 8 TeV an average total multiplicity $< N_{ch} > = 80 \pm 2$.

The results obtained by GHOST are also compared with the recent data concerning the inelastic pseudorapidity respectively at $\sqrt{s} = 8\,\text{TeV}$ and $\sqrt{s} = 13\,\text{TeV}$ [16] as shown in the Fig. 8. There are not measurements published up to now concerning the NSD component at $\sqrt{s} = 13\,\text{TeV}$. An attempt with GHOST generator (dotted histogram) suggests one possible pseudorapidity distribution among several tendencies; it is only one example taking account the enhancement of total multiplicity suggested by the Fig. 6, a measurement of central NSD pseudorapidity being necessary.

4. Cosmic ray data related with the LHC energy domain

A large part of cosmic ray remarkable events such as coplanar emission [17] or spikes [9] have been observed

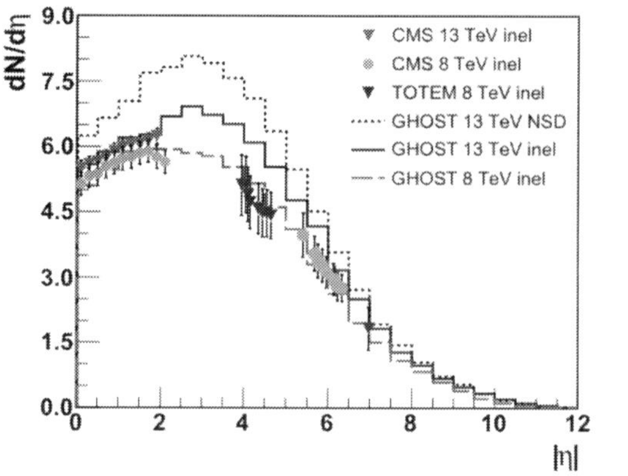

Figure 8. Average pseudorapidity density simulated with GHOST compared with CMS measurements for the inelastic data of CMS at $\sqrt{s} = 8\,\text{TeV}$ (lower histogram) and at $\sqrt{s} = 13\,\text{TeV}$ (middle histogram). The upper histogram (dotted) concerns the NSD data and is just a preliminary attempt.

in different conditions (mountain altitude or stratosphere). They appear mainly between 1 PeV and 100 PeV in laboratory frame being covered now by the LHC data up to $\sqrt{s} = 13\,\text{TeV}$. Very small emulsion X-ray chambers of $40 \times 50\,\text{cm}$ have collected data in the stratosphere with balloons (i.e. JACEE, siberian flights above 30 km altitude), Concorde supersonic flights at 17 km altitude and even with japanese air liners near 10 km altitude). In parallel Emulsion-X-ray chambers or and X-ray chambers were also installed above 4 km altitude in Pamir, Tibet with very large area in order to collect events at energies exceeding 10^{17} eV.

As expected more A-A collisions were collected with balloons, when more p-A collsions were collected during airplane flights with a statistics exhausted above $5 \cdot 10^{16}$ eV interesting with "spikes" in pseudorapidity distribution. This data contains events with local multiplicity fluctuations which were also observed at lower energy in UA5 data suggesting the examination of different hypothesis such as formation of quark-gluon plasma, but the Monte Carlo simulations of the CERN suggest also that the fluctuations occurred with a frequency similar to the data sample. New comparison of those statistics in LHC and in altitude flights appears necessary.

Similarly near and above $\sqrt{s} = 5\,\text{TeV}$, the unexpected alignment of secondaries has been observed in 2 independent stratospheric flights, in parallel with the collection of Pamir registrations. Additional simulations are now necessary to understand the coplanar emission in a new context; for instance diquark breaking of valence diquark could generate very large transverse momenta from the extreme string tension between quarks contained in each diquark. There could be also other investigations in relation with the ridges seen in p-p and p-A collisions at the LHC.

5. Conclusion

The 4 sources model GHOST is a very fast Monte Carlo collision generator able to reproduce both inelastic and NSD data up to $\sqrt{s} = 13\,\text{TeV}$. This circumstance allows a more refined analysis of cosmic ray data at energies above the "knee" at $3 - 5\ 10^{15}$ eV. The enhancement of multiplicity suggests in EAS simulation a larger abundance of muons and a maximum of development at higher altitude than with previous models. Altogether an abundance of heavy primaries lower than with previous models can be expected above the knee. Furthermore, the energy guide lines derived from the complete data of the LHC shall provide new elememts in the polemics on GZK cut-off.

The data collected at $\sqrt{s} = 7 - 8$ TeV is confirmed and will be completed soon for the NSD, inelastic components at $\sqrt{s} = 13$ TeV and hopefully for cosmic ray data in the leading cluster with LHCf. This could help the interpretation of cosmic ray anomalies in the "knee" region above 2 TeV. The high energy photons in EAS in Tian Shan and the gamma ray families above 5 TeV (near 10 PeV in laboratory frame) exhibit steeper integral energy spectra than at lower energies. Such signal could be in relation with an important part of energy lost in the leading cluster.

This work was performed thanks to the French-Polish Collaboration Agreement IN2P3-COPIN. Part of Z. Plebaniak work was supported by NCN grant UMO-2016/20/T/ST9/00589.

References

[1] G. Alner, K. Alpgard, P. Anderer, R. Ansorge, B. Asman, S. Berglund, K. Berkelman, D. Bertrand, C. Booth, C. Buffam et al., Physics Reports **154**, 247 (1987)
[2] R.P. Feynman, Phys. Rev. Lett. **23**, 1415 (1969)
[3] A. Capella, U. Sukhatme, C.I. Tan, J.T.T. Van, Physics Reports **236**, 225 (1994)
[4] G. Aad et al. (ATLAS), New J. Phys. **13**, 053033 (2011), 1012.5104
[5] K. Aamodt et al. (ALICE), Eur. Phys. J. **C68**, 89 (2010), 1004.3034
[6] G.J. Alner, R.E. Ansorge, B. Åsman, C.N. Booth, L. Burow, P. Carlson, R.S. DeWolf, A. Drees, B. Eckart, G. Ekspong et al., Zeitschrift für Physik C Particles and Fields **33**, 1 (1986)
[7] D. Heck, J. Knapp, J.N. Capdevielle, G. Schatz, T. Thouw, *CORSIKA: a Monte Carlo code to simulate extensive air showers* (1998)
[8] P. Carruthers, C. Shih, Physics Letters B **127**, 242 (1983)
[9] J.N. Capdevielle, Il Nuovo Cimento C **19**, 623 (1996)
[10] G. Alner, K. Alpgård, R. Ansorge, B. Åsman, C. Booth, L. Burow, P. Carlson, R. DeWolf, B. Eckart, G. Ekspong et al., Nuclear Physics B **291**, 445 (1987)
[11] J.N. Capdevielle, F. Cohen, B. Szabelska, J. Szabelski, Journal of Physics G: Nuclear and Particle Physics **36**, 075203 (2009)
[12] J.N. Capdevielle, Journal of Physics G: Nuclear Physics **14**, 503 (1988)
[13] J.N. Capdevielle, Journal of Physics G: Nuclear and Particle Physics **15**, 909 (1989)
[14] G. Wolschin, Physics Letters B **698**, 411 (2011)
[15] S. Chatrchyan et al. (CMS, TOTEM), Eur. Phys. J. **C74**, 3053 (2014), 1405.0722
[16] V. Khachatryan, A. Sirunyan, A. Tumasyan, W. Adam, E. Asilar, T. Bergauer, J. Brandstetter, E. Brondolin, M. Dragicevic, J. Er et al., Physics Letters B **751**, 143 (2015)
[17] J. Capdevielle, S. Slavatinsky, Nuclear Physics B - Proceedings Supplements **75**, 12 (1999)

LHC data and cosmic ray coplanarity at superhigh energies

R.A. Mukhamedshin [a]

Institute for Nuclear Research, Russian Academy of Sciences, Moscow, Russia

Abstract. A new phenomenological model FANSY 2.0 is designed, which makes it possible to simulate hadron interactions via traditional and coplanar generation of most energetic particles as well as to reproduce a lot of LHC (ALICE, ATLAS, CMS, TOTEM, LHCf) data. Features of the model are compared with LHC data. Problems of coplanarity are considered and a testing experiment is proposed.

1. Introduction

A number of models of hadron interactions with nuclei is concurrently applied in cosmic-ray experiments as none of them can explain the entire experimental data set of EAS features. Besides, a number of phenomena observed in mountain-based and stratospheric X-ray–emulsion chamber (XREC) experiments are not yet explained. One of these interesting phenomena is the so-called coplanarity of the most energetic cores of γ-ray–hadron families, i.e., groups of high-energy ($E \gtrsim n \cdot 1$ TeV) particles in relatively young EAS cores initiated by protons and nuclei of the primary cosmic radiation (PCR) (see Sect. 3).

The phenomenological model FANSY 1.0 was designed a few years ago, which helped to understand general features of coplanar events [1]. FANSY 1.0 gives reasonable results as compared with experimental data on intensity of γ-ray families, energy dependence of the muon and hadron spectra and so on [1].

However, FANSY 1.0 cannot properly reproduce LHCf data on high-X_F γ-rays e.g. [2], and neutrons, which are really important for cosmic ray experiments, as well as the transversal size of γ-ray families, e.g. [3].

To improve this situation, a new phenomenological model FANSY 2.0 is designed, which simulates traditional and coplanar particle generation and reproduce experimental data with a higher accuracy as compared with FANSY 1.0. FANSY 2.0 reproduces a number of ALICE, ATLAS, CMS, TOTEM data on charged, strange and charmed stable and resonance particle generation, LHCf data on high-X_F gamma-rays and neutrons.

This paper is devoted to pp interactions simulated at superhigh energies ($\sqrt{s} = 900$ GeV -13 TeV). Compatibility of FANSY 2.0 and LHC data will be discussed in more detail in another work [4].

This paper is organized as follows. Section 2 presents comparison of LHC data and simulated results. Coplanarity problem is briefly considered in Sect. 3. Final comments are given in Conclusion.

[a] e-mail: rauf_m@mail.ru

2. High energy interactions

2.1. Cross sections

Figure 1 shows FANSY 2.0's results (curve) and data of different experiments on the energy dependence of the inelastic pp cross section σ^{pp}_{inel} (top); single and double diffraction sections, σ^{pp}_{SD} and σ^{pp}_{DD} (bottom). Inelastic pp cross section data include results of the ATLAS [5] and CMS [6] Collaborations at $\sqrt{s} = 13$ TeV.

2.2. Low-(pseudo)rapidity data

Experimental LHC data are mainly related to low-(pseudo)rapidity particles. This kinematic range is not very important for consideration of cosmic-ray-initiated EAS's in the atmosphere. However, models, which cannot describe these data, will inevitably cause mistrust. Experimental and simulated data are shown in Sect. 2 with filled and empty symbols, respectively.

2.2.1. Charged particles

Figure 2 shows ATLAS $dN_{ch}/d\eta$ distributions of charged particles at $-2.1 < \eta < 2.1$, $n_{ch} \geq 2$, and $p_t > 0.1$ GeV/c in NSD pp interactions at $\sqrt{s} = 13$ TeV (circles) [7], 7 TeV (squares) and 900 GeV (diamonds) [8].

Figure 3 shows experimental $dN_{ch}/d\eta$ distributions of charged particles in pp NSD interactions at $\sqrt{s} = 7$ TeV of the CMS experiment [9] at $-2.5 < \eta < 2.5$ (left) and the LHCb experiment at $2.0 < \eta < 4.5$ (right) [10].

2.2.2. Strange particles

Figure 4 shows FANSY 2.0 and CMS $dN/d|y|$ distributions of K^0_s mesons in pp NSD interactions at $\sqrt{s} = 900$ GeV (lower squares) and $\sqrt{s} = 7$ TeV (upper triangles) [11].

Figure 5 shows FANSY 2.0 and LHCb experiment's $d\sigma/dy$ distributions of ϕ mesons with $0.6 < p_t < 5.0$ GeV/c at $\sqrt{s} = 7$ TeV (upper squares) [12].

2.2.3. Charmed particles

Among different aims, FANSY 2.0 is designed to study the "forward-physics" aspects of the generation of charmed

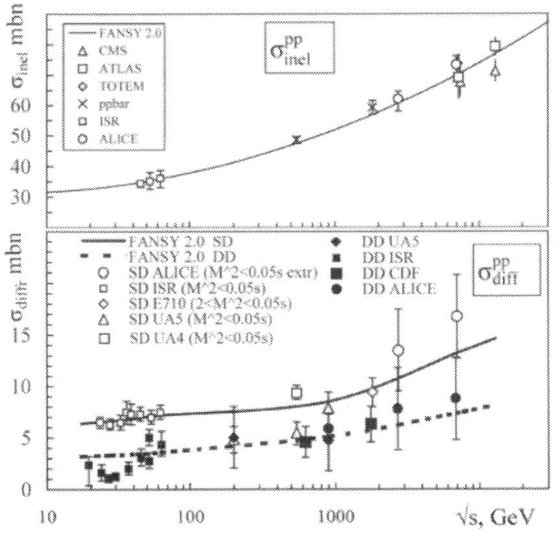

Figure 1. Experimental (points) and FANSY 2.0's (curve) energy dependencies of inelastic pp cross section (top); single and double diffraction sections (bottom).

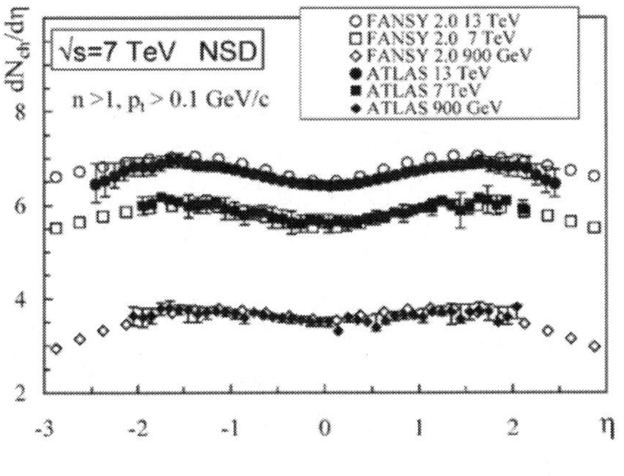

Figure 2. FANSY 2.0 and ATLAS $dN_{ch}/d\eta$ distributions of charged particles at $-2.1 < \eta < 2.1$, $n_{ch} \geq 2$, and $p_t > 0.1$ GeV/c in NSD pp interactions at $\sqrt{s} = 13$ TeV (circles), 7 TeV (squares) and 900 GeV (diamonds).

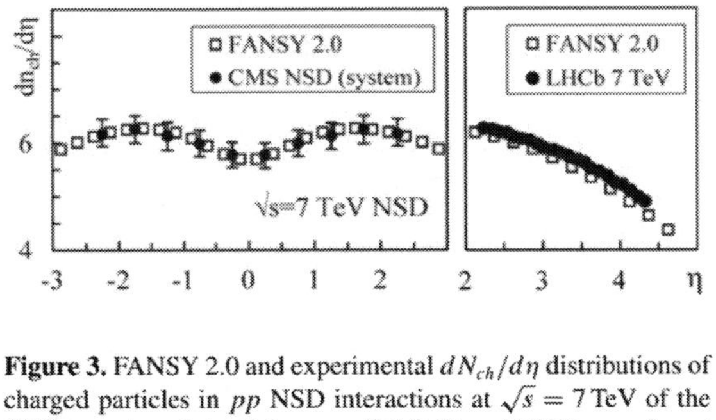

Figure 3. FANSY 2.0 and experimental $dN_{ch}/d\eta$ distributions of charged particles in pp NSD interactions at $\sqrt{s} = 7$ TeV of the CMS experiment at $-2.5 < \eta < 2.5$ (left) and LHCb experiment at $2.0 < \eta < 4.5$ (right).

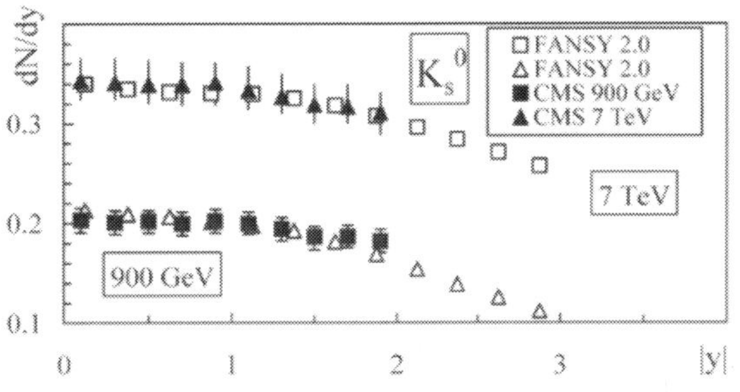

Figure 4. CMS $dN/d|y|$ distributions of K_s^0 mesons in pp NSD interactions at $\sqrt{s} = 900$ GeV (lower squares) and $\sqrt{s} = 7$ TeV (upper triangles).

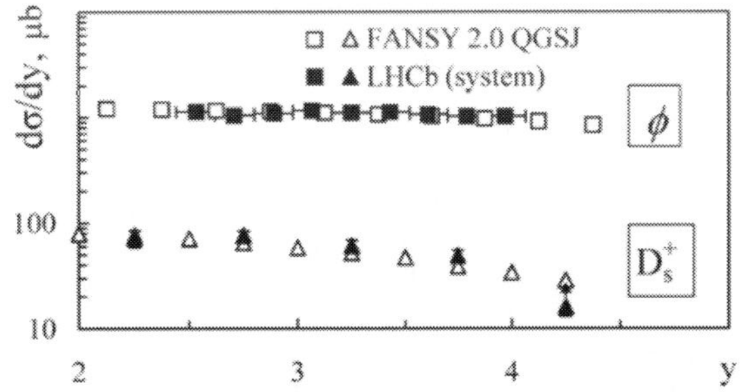

Figure 5. FANSY 2.0 and LHCb $d\sigma/dy$ distributions of ϕ mesons at $0.6 < p_t < 5.0$ GeV/c (upper squares); $D_s^{\pm}$ mesons at $1.0 < p_t < 8.0$ GeV/c (lower triangles) at $\sqrt{s} = 7$ TeV.

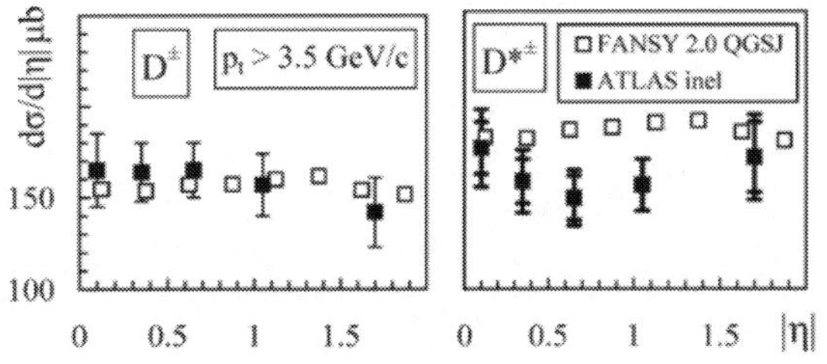

Figure 6. ATLAS and FANSY 2.0's $d\sigma/d|\eta|$ distribution of $D^{\pm}$ mesons (left) and $D^{*\pm}$ mesons (right) with $p_t > 3.5$ GeV/c.

particles. So these particles are considered below in more detail as compared with light-quark hadrons.

Figure 5 shows LHCb $d\sigma/dy$ distribution of $D_s^{\pm}$ mesons with $1.0 < p_t < 8.0$ GeV/c at $\sqrt{s} = 7$ TeV in pp interactions (lower triangles).

Figure 6 shows FANSY 2.0 and ATLAS $d\sigma/d|\eta|$ distributions of $D^{\pm}$ (left) and $D^{*\pm}$ (right) mesons with $p_t > 3.5$ GeV/c in pp interactions at $\sqrt{s} = 7$ TeV [13].

Figure 7 shows ALICE $d\sigma/dp_t$ distributions of D^+ (top,left), D^0 (top,right) mesons as well as strange D_s^+ (bottom,left), vector D^{*+} (bottom,right) mesons at $|y| < 0.5$ and $\sqrt{s} = 7$ TeV in pp interactions [14]. Experimental systematic uncertainties are approximately equal to statistical ones.

Figure 8 shows LHCb data on $d^2\sigma/dp_t dy$ distributions of D^+ mesons in five rapidity regions,

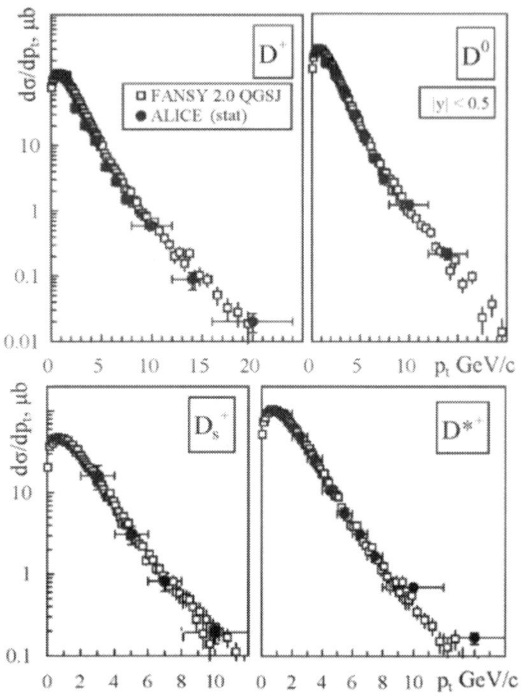

Figure 7. ALICE $d\sigma/dp_t$ distributions of D^+ (top, left), D^0 (top, right) mesons as well as strange D_s^+ (bottom,left), vector D^{*+} (bottom,right) mesons at $|y| < 0.5$.

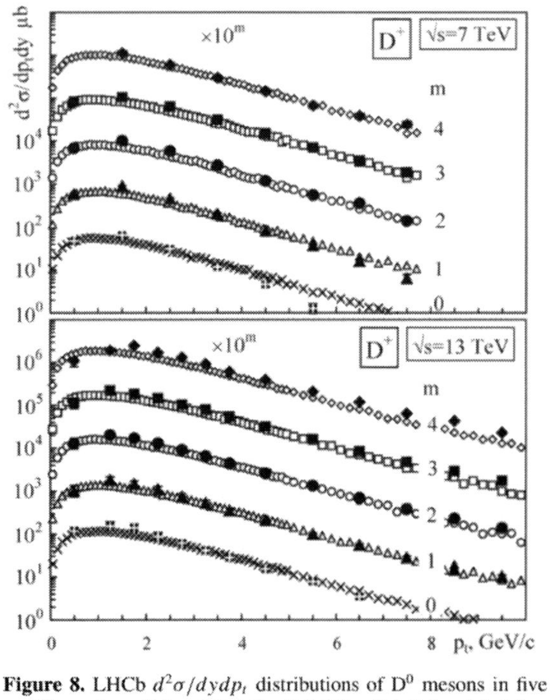

Figure 8. LHCb $d^2\sigma/dydp_t$ distributions of D^0 mesons in five rapidity regions ($2.0-2.5, 2.5-3.0, 3.0-3.5, 3.5-4.0, 4.0-4.5$) at $\sqrt{s} = 7$ TeV (top) and $\sqrt{s} = 13$ TeV (bottom) with filled diamonds, squares, circles, triangles, and crosses, respectively. Spectra are multiplied by 10^m, where $m = 4, 3, 2, 1, 0$ for the above-mentioned y ranges, respectively.

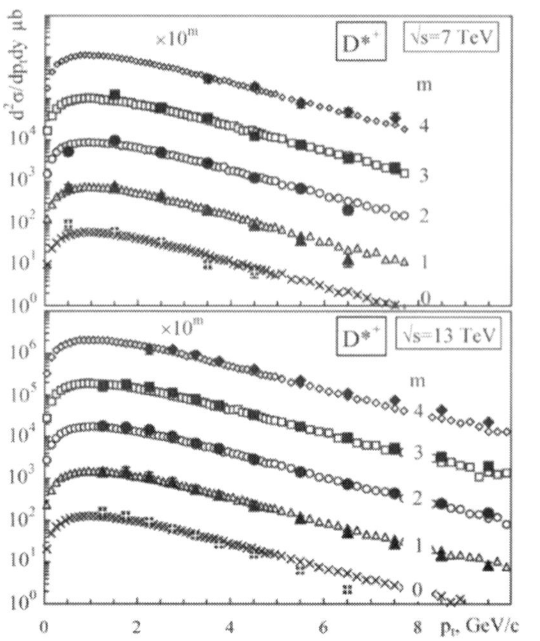

Figure 9. LHCb $d^2\sigma/dp_tdy$ distributions of D^{*+} mesons in five rapidity regions. Notations are the same as in Fig. 8.

Table 1. D^0, D^+ and D^{*+} charmed-meson generation cross sections (mbarn). ALICE data include statistical and systematical errors. FANSY 2.0 data include statistical errors.

Particles	p_t, GeV/c	ALICE $\sigma\pm$ stat. $\pm$ syst.	FANSY 2.0 $\sigma\pm$ stat.
D^0	1–16	$412\pm33\,^{+55}_{-140}$	457 ± 6
D^+	1–24	$198\pm24\,^{+42}_{-73}$	201 ± 4
D^{*+}	1–24	$203\pm23\,^{+30}_{-67}$	180 ± 2

namely, at $2.0-2.5$, $2.5-3.0$, $3.0-3.5$, $3.5-4.0$, and $4.0-4.5$ at $\sqrt{s} = 7$ TeV (top) and $\sqrt{s} = 13$ TeV (bottom) with diamonds, squares, circles, triangles, and crosses, respectively [15, 16]. Spectra are multiplied by 10^m, where $m = 4, 3, 2, 1, 0$ for the above-mentioned rapidity ranges, respectively.

Figures 9 and 10 show LHCb $d^2\sigma/dp_tdy$ distributions of D^{*+} and D_s^+ mesons, respectively [15, 16]. All the notations are the same as in Fig. 8.

Figure 11 shows data of the ALICE experiment on $d^2\sigma/dydp_t$ of ω^0 (squares) and ϕ mesons (triangles) at $2.5 < y < 4.5$ [17]. Comparison of cross sections of generation of these mesons gives information on the suppression of strange quark generation.

Figure 12 shows experimental $d\sigma/dp_t$ spectra of $D^\pm$ (left) and $D^{*\pm}$ (right) mesons at $|\eta| < 2.1$ by the ATLAS experiment at $\sqrt{s} = 7$ TeV.

Table 1 shows ALICE and FANSY 2.0 cross sections of D^0, D^+ and D^{*+} charmed-meson generation at $|\eta| < 0.5$ in wide p_t ranges.

Table 2 shows ALICE and FANSY 2.0 ratios of yields of charmed mesons, D_s^+/D^+ and D_s^+/D^0, as well as a ratio of prompt vector mesons to prompt vector+pseudoscalar mesons, $P_V = D^*/(D^{*+} + D^+)$, at $|\eta| < 0.5$.

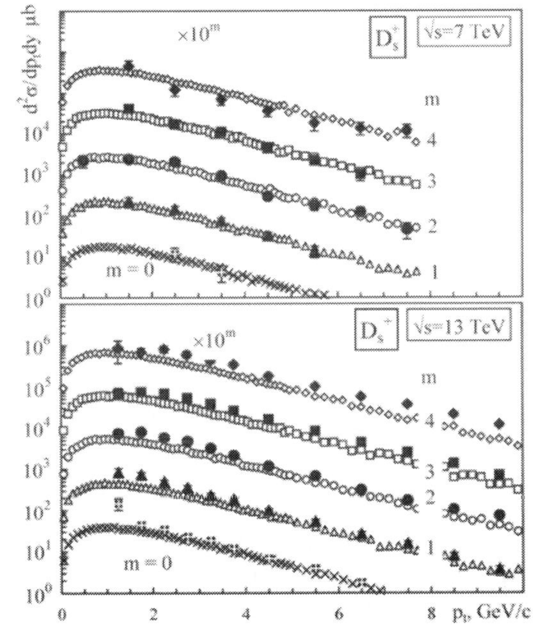

Figure 10. LHCb $d^2\sigma/dp_t dy$ distributions of D_s^+ mesons in five rapidity regions. Notations are the same as in Fig. 8.

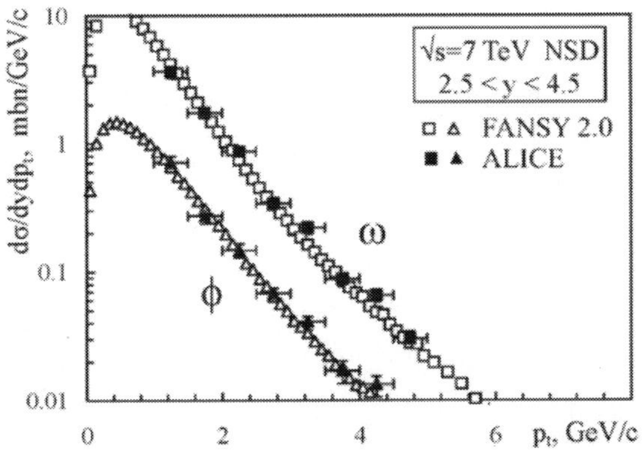

Figure 11. FANSY 2.0 and ALICE $d^2\sigma/dydp_t$ distributions of ω^0 (upper squares) and ϕ (lower triangles) mesons, respectively, at $2.5 < y < 4.5$ at $\sqrt{s} = 7$ TeV.

2.2.4. *High η and X_F data*

Experimental high η and X_F data are more interesting for cosmic ray experiments. Unfortunately, at the present time the specific design of colliders makes it possible to derive highest η and X_F data for low-p_t neutral particles only.

Data of the CMS+TOTEM experiment [18] at $\sqrt{s} =$ 8 TeV are of especial interest. Figure 13 shows "NSD-enhanced" $dn_{ch}/d\eta$ distributions for events selected at $\sqrt{s} = 8$ TeV under the following requirements. Number of detected charged particles $n_{ch} \geq 1$ at $-6.5 < \eta < -5.3$ **and** $5.3 < \eta < 6.5$ (a); "more-forward" data for events with $n_{ch} \geq 1$ at $-6.5 < \eta < -5.3$ **or** $5.3 < \eta < 6.5$ (b); "SD-enhanced" data derived for events with $n_{ch} \geq 1$ *only* at $-6.5 < \eta < -5.3$ **or only** $5.3 < \eta < 6.5$ (lower triangles) (c); "more-forward" data obtained with displaced interaction points for events with $n_{ch} \geq 1$ at $-7.0 < \eta < -6.0$ **or** $3.7 < \eta < 4.8$ (upper squares)(d).

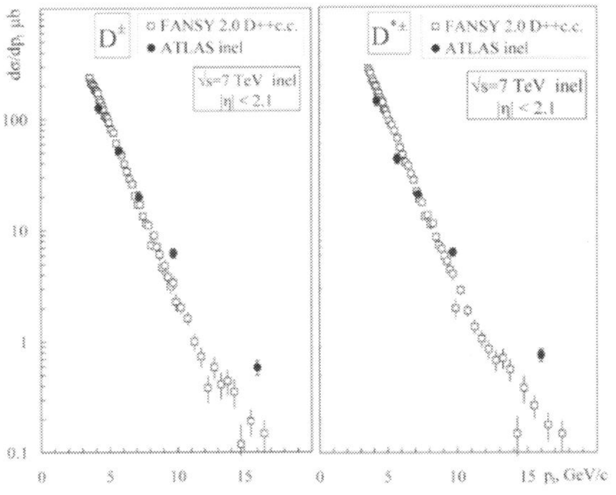

Figure 12. ATLAS $d\sigma/dp_t$ spectra of $D^{\pm}$ (left) and $D^{*\pm}$ (right) mesons at $|\eta| < 2.1$ and $\sqrt{s} = 7$ TeV.

Table 2. D_s^+/D^+, D_s^+/D^0 ratios and ratio of vector mesons to prompt vector+pseudoscalar ones, P_V, at $|\eta| < 0.5$. ALICE data include statistical and systematical errors. FANSY 2.0 results include statistical errors.

Particles	p_t, GeV/c	ALICE	FANSY 2.0
D_s^+/D^+	2–12	$0.36 \pm 0.11 \pm 0.12$	0.35 ± 0.01
D_s^+/D^0	2–12	$0.20 \pm 0.05 \pm 0.06$	0.15 ± 0.01
$P_V = D^{*+}/(D^{*+} + D^+)$	1–24	$0.59 \pm 0.06 \pm 0.08$	0.55 ± 0.01

Figure 14 shows FANSY 2.0 and the LHCf "neutron" ($n, \bar{n}$, K_s, L^0) $d\sigma/dx_F$ spectrum at $p_t < 0.11x_F$) at $\Delta\phi = 360°$ and $\sqrt{s} = 7$ TeV. Figure 15 shows LHCf and simulated "neutron" $d\sigma/dE$ energy spectra at $8.99 < \eta < 9.2$ (top) and at $\eta > 10.76$ (bottom).

Figure 16 shows FANSY 2.0 and LHCf γ-ray n_γ/N_{inel} energy spectra in the following pseudorapidity and azimuthal-angle ranges [2], $8.81 < \eta < 8.89$, $\Delta\phi = 20°$ (top) and $\eta > 10.94$, $\Delta\phi = 360°$ (bottom).

3. Coplanar particle generation

3.1. General view

A tendency to a coplanarity of the most energetic cores of γ-ray–hadron families observed was first found by the *Pamir* Collaboration [19–23] and confirmed later in [24–27]. The probability W_{tot}^{fluct} for the total set of these experimental results to be produced by cascade fluctuations is much lower than 10^{-10} [1,28]. This result illustrates that strong "forward-physics" interactions at superhigh energies are not well-described with the quark-gluon string model (QGSM) concept.

The phenomenon is related to hadron-nucleus interactions at $E_0 \gtrsim 10^{16}$ eV ($\sqrt{s} \gtrsim 4$ TeV) [29], characterized by a large cross section (comparable with σ_{inel}^{pp}) and was initially interpreted as a manifestation of large transverse momenta of the most energetic fragmentation-range particles ($X_{Lab} = E/E_0 \gtrsim 0.1$) [28].

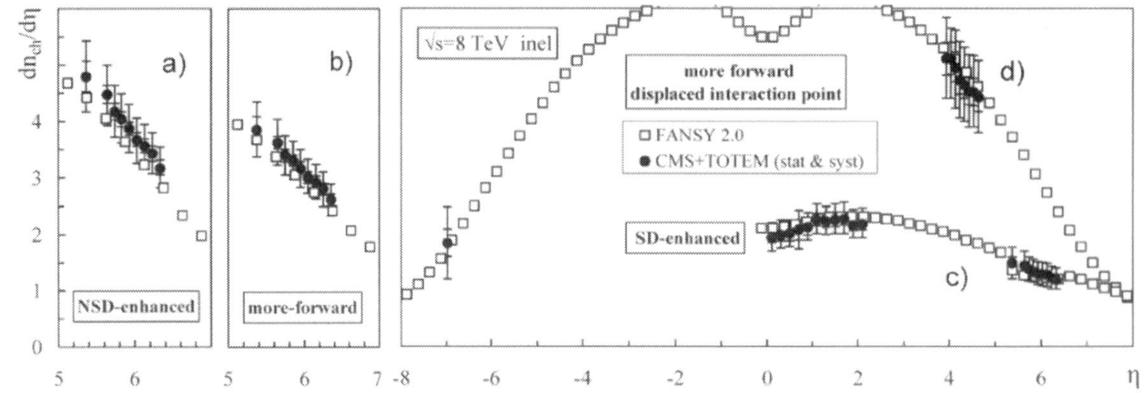

Figure 13. CMS+TOTEM and FANSY 2.0's $dn_{ch}/d\eta$ data for events selected at $\sqrt{s} = 8\,\text{TeV}$ under the following requirements. "NSD-enhanced" data: $n_{ch} \geq 1$ in the ranges $-6.5 < \eta < -5.3$ **and** $5.3 < \eta < 6.5$ (a); "more-forward" data: $n_{ch} \geq 1$ in the ranges $-6.5 < \eta < -5.3$ **or** $5.3 < \eta < 6.5$ (b); "SD-enhanced" data (lower triangles): $n_{ch} \geq 1$ **only** in the ranges $-6.5 < \eta < -5.3$ **or only** $5.3 < \eta < 6.5$ (c); "more-forward" data (upper squares) derived with displaced interaction points: $n_{ch} \geq 1$ in the ranges $-7.0 < \eta < -6.0$ **or** $3.7 < \eta < 4.8$ (d).

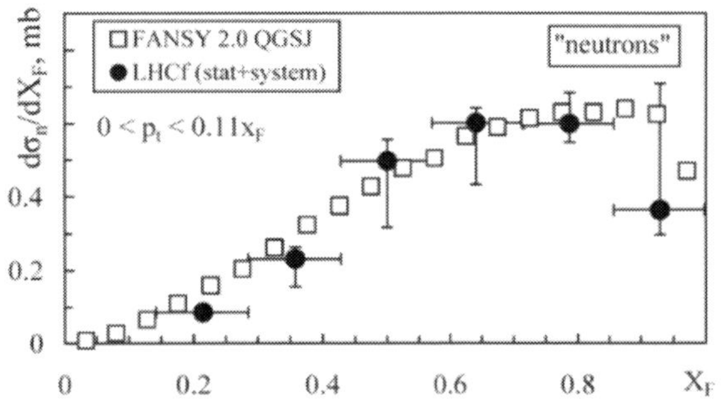

Figure 14. FANSY 2.0 and LHCf "neutron" $d\sigma/dx_F$ spectrum at $p_t < 0.11x_F$.

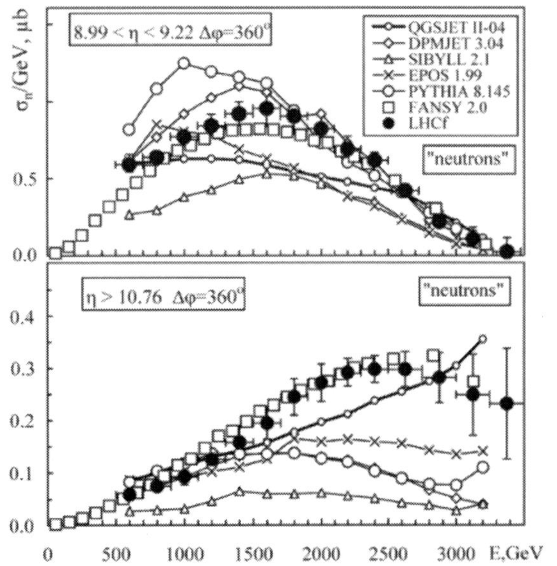

Figure 15. FANSY 2.0 and LHCf "neutron" $d\sigma/dE$ energy spectrum at $8.99 < \eta < 9.2$ (top) and at $\eta > 10.76$ (bottom), $\Delta\phi = 360°$ and $\sqrt{s} = 7$ TeV.

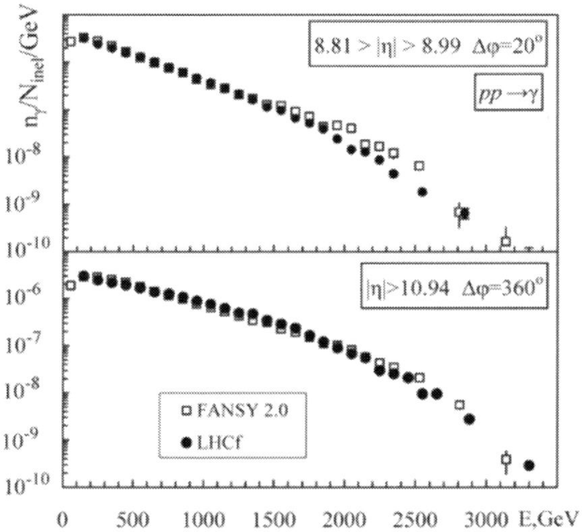

Figure 16. LHCf data and FANSY 2.0 results on γ-ray n_γ/N_{inel}/GeV energy spectra at $8.81 < \eta < 8.89$, $\Delta\phi = 20°$ (top) and $\eta > 10.94$, $\Delta\phi = 360°$ (bottom).

This phenomenon could be reproduced in the framework of two concepts as a result of a) conservation of the angular momentum of a relativistic fast-rotating quark-gluon string (QGS) stretched between colliding hadrons [31]; b) semihard double diffraction (SHDID) dissociation and appearance of coplanarity as a result of QGS tension inside the diffraction cluster between a semihardly scattered constituent quark and other spectator quarks of the projectile hadron and its following rupture [32] with a lower multiplicity and higher average energy of particles; c) the most extraordinary explanation assumes that this phenomenon could be described within the recently proposed hypothesis of "crystal world", with latticed and anisotropic spatial dimensions and decrease of dimension number with increasing energy [33].

In this work, only the first approach is considered.

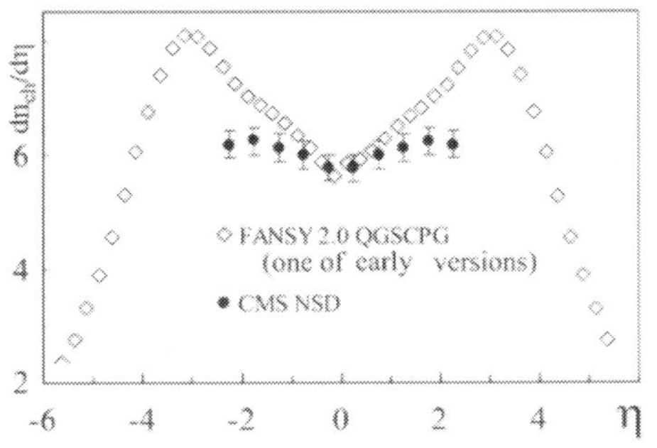

Figure 17. Peaks in $d\sigma/d\eta$ distribution originated by growth of transverse momenta of most energetic particles in primary-concept CPG versions.

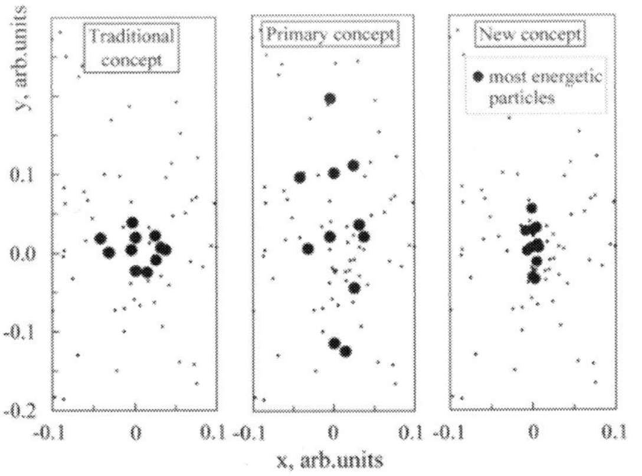

Figure 18. Tracks of particles on the target plane in cases of traditional interaction (left), coplanar interaction within the primary concept of increased p_t (middle), coplanar interaction within the new concept of traditional p_t (right). Most energetic particles are shown with large filled circles.

To study this problem, FANSY 2.0 QGSCPG version is designed [34], which simulates both traditional and coplanar particle generation. More details of comparison of LHC data with results of simulation with the above-described basic FANSY 2.0 QGSJ version are given in [34].

3.2. Coplanarity concepts

Simulation using tentative FANSY 2.0 QGSCPG versions demonstrated the following fundamental problem.

The originally exploited concept qualitatively explains the observed coplanarity of momenta of most energetic particles with assuming their *high transverse momenta* in a coplanarity plane. However, in this case a significant p_t growth suppresses $d\sigma/dy$ and $d\sigma/d\eta$ distributions of hadrons at highest $|y|$ and $|\eta|$ values and creates robust peaks at $2 \lesssim |\eta| \lesssim 4$ which are contrary to LHC data (Fig. 17). This was an unsolvable problem for all the QGSCPG versions based on the primary coplanarity concept. However, the coplanarity is observed in cosmic-ray experiments. Is it possible to reconcile this result with the LHC data?

Simulation has shown that a general agreement of LHC data and idea of coplanar generation becomes real only using a new concept of coplanarity origin, namely, some *decrease of particle's transverse momenta directed normally to the coplanarity plane* takes place so that the absolute p_t values *do no change*.

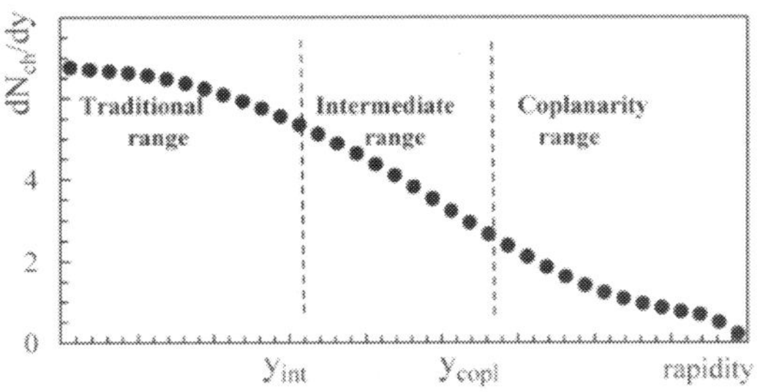

Figure 19. Qualitative dependence of the traditional, intermediate and coplanarity ranges on rapidity.

Figure 18 shows three examples of tracks of particles, generated in the same imaginary interaction, on a target plane, placed at some distance from the interaction point, in the cases of traditional QGSM-like interaction (left), primary-concept coplanar interaction with increased p_t (middle), and new-concept coplanar interaction with traditional p_t (right). Most energetic particles are shown with large black circles. The geometric scale is given in arbitrary units.

3.3. Coplanarity simulation

All simulated interaction characteristics (excluding azimuthal ones) simulated with FANSY 2.0 QGSJ and QGSCPG versions, are similar. The QGSJ and QGSCPG versions merge smoothly at $\sqrt{s} \lesssim 2$ TeV.

In the QGSCPG version all characteristics of particles are primarily simulated with the traditional way. If the summary energy of secondary particles is higher than a fixed value, transversal momenta of high-rapidity particles in the coplanarity range ($|y| > y_{copl}$), the algorithm turns $\overrightarrow{p}_t$ of each such particles towards the coplanarity plane. This plane is determined by the momenta of the interacting hadrons and $\overrightarrow{p}_t$ of leaders surviving after the collision. The trend of turning of transverse momenta of secondary particles to this plane is weakening at $y_{int} < |y| < y_{copl}$ (in the intermediate range) and disappears at $|y| < y_{int}$ (the traditional range). Here $|y_{max}| = \sqrt{s}/2/m$, $y_{copl} \approx y_{max} - \Delta y \approx 5 - 6$, $y_{int} \approx 2 - 3$, $\Delta y = 3 - 5$, m is particle mass.

Figure 19 shows a qualitative dependence of the traditional, intermediate and coplanarity ranges on rapidity.

3.4. On the search for coplanarity at the LHC

The CASTOR experiment seems to be promising in studying coplanarity (at least, in the framework of FANSY 2.0). Figure 20 shows a simplified CASTOR's cross section scheme and an example of detection of one coplanar interaction. The detector consists of 16 segments and is divided in the middle by a vertical slit. Particles are considered to be detected if their pseudorapidity values are in the range of $5.3 < \eta < 6.5$ and they do not fall into the vertical slit. Black circles in Fig. 20 show tracks of

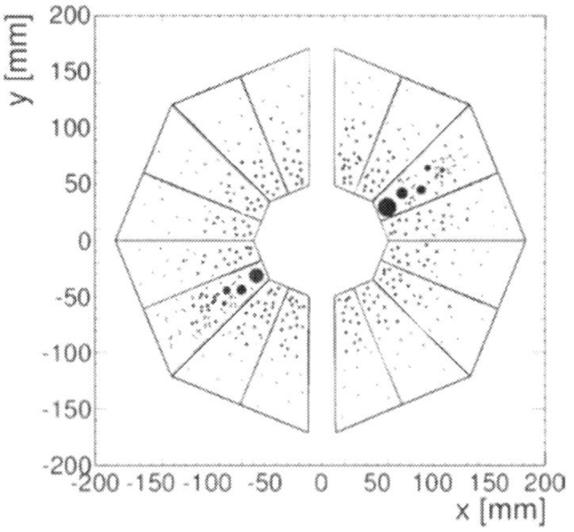

Figure 20. CASTOR's simplified cross section scheme. Black circles show tracks of particles. High-energy particles shown with large circles demonstrate a tendency to coplanarity.

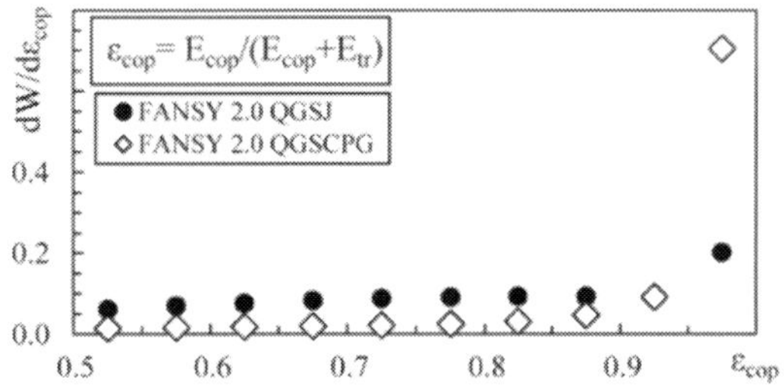

Figure 21. CASTOR's ε_{cop} distributions for FANSY 2.0 QGSJ and QGSCPG at $\sqrt{s} = 7$ TeV.

particles. The larger the circle size, the higher the energy of the particle. High-energy particles show a tendency to some coplanarity. Low-energy particles form a more or less azimuthally symmetric halo.

To analyze events, energy values "measured" in each of the segments, E_i ($E_i \geq 100$ GeV, i is the number of a segment), are used. Events with total energy $\sum E_i \geq$ 1 TeV "measured" in two or more segments are only analyzed. The first number is assigned to a segment with a maximum release of energy, E_{max}. Here and below, $E_1 = E_{max}$, $E_{cop} = E_1 + E_9$; $E_{tr} = E_5 + E_{13}$, i.e., it is the energy measured in 9th and 13th segments, perpendicular to the first segment. A simple parameter is applied, namely, $\varepsilon_{cop} = E_{cop}/(E_{cop} + E_{tr})$, which characterizes the event coplanarity degree. If $\varepsilon_{cop} = 1$, the degree of event coplanarity is maximum.

Figure 21 shows ε_{cop} distributions for FANSY 2.0 QGSJ and QGSCPG versions at $\sqrt{s} = 7$ TeV. Difference between model predictions becomes very large at $\varepsilon_{cop} \to 1$.

Conclusion

The FANSY 2.0 Monte Carlo code is designed to study superhigh-energy cosmic-ray "forward physics" interactions and includes traditional *QGSJ* QGSM-based version as well as a *QGSCPG* one which realizes a coplanar particle generation (CPG).

CPG process simulated with FANSY 2.0 QGSCPG does not contradict LHC data and could be tested in the CASTOR experiments. To reconcile experimental and simulated data, it is necessary to replace the primary concept of *growth* of transverse momenta of most energetic particles in the coplanarity plane with a new concept of *reduction* of transverse momenta directed normally to the coplanarity plane.

A version of the appearance of CPG processes in double-diffraction interactions only requires a separate consideration, as in this case the energy spectrum of secondary particles is more hard which is important in cosmic-ray experiments.

References

[1] R.A. Mukhamedshin, Eur. Phys. J. **C 60**, 345 (2009)
[2] O. Adriani, L. Bonechia, M. Bongia et al., Physics Letters **B 703**, 128–134 (2011)
[3] J. Kempa, R. Mukhamedshin, M. Tamada, J. Phys.: Conf. Ser. **409**, 012109 (2013)
[4] R.A. Mukhamedshin, (to be sent to Eur.Phys.J.**C**)
[5] The ATLAS Collaboration, `arXiv:1606.02625v2 [hep-ex]`
[6] The CMS Collaboration, `arXiv:1607.02033v1 [hep-ex]`
[7] The ATLAS Collaboration, `arXiv:1606.01133v1 [hep-ex]`
[8] The ATLAS Collaboration, New J. Phys. **13**, 053033 (2011)
[9] The CMS Collaboration, `arXiv:1005.3299v2 [hep-ex]`
[10] The LHCb Collaboration, Eur. Phys. J. **C 72**, 1947 (2011)
[11] The CMS Collaboration, J. High Energ. Phys. **05**, 064 (2011)
[12] Phys. Lett. **B 703** (2011) 267–273
[13] The ATLAS Collaboration. `arXiv:1512.02913v1 [hep-ex]`
[14] B. Abelev et al. (ALICE Collaboration). J. High Energ. Phys. **01**, 128 (2012)
[15] The LHCb Collaboration, Nucl. Phys. **B 871**, 1 (2013)
[16] The LHCb collaboration. R.Aaij, C. Abelln Beteta et al., J. High Energ. Phys. **03**, 159 (2016)
[17] The ALICE Collaboration, `ArXiv: 1112.5 [hep-ex]`
[18] `arXiv:1405.0722v1`
[19] A. Borisov et al. (Pamir Collaboration), in *Proc. 4th Int. Symp. on Very High Energy Cosmic Ray Interactions, Beijing* 4 (1986)
[20] I.P. Ivanenko et al., JETP Lett. **50**, 2125 (1992)
[21] V.V. Kopenkin et al., Phys. Rev. **D 52**, 2766 (1995)
[22] Pamir Collaboration, *Preprint* INP MSU no. 89-67/144 (1989)
[23] Borisov A.S. et al., Nucl. Phys. B (Proc. Suppl.) **97**, 118 (2001)
[24] L. Xue et al., in *Proc. 26th Int. Cosmic Ray Conf., Salt Lake City* **1**, 127 (1999)
[25] A.V. Apanasenko et al., in *Proc. 15th Int. Cosmic Ray Conf., Plovdiv* **7**, p. 220 (1977)
[26] A.K. Managadze et al., in *Proc. 27th Int. Cosmic Ray Conf., Hamburg* **1**, p. 1426 (2001)
[27] J.N. Capdevielle, J.Phys. **G 14**, 503 (1988)
[28] R.A. Mukhamedshin, J. High Energy Phys. JHEP **05**, 049 (2005)

[29] A.S. Borisov et al., Nucl. Phys. B (Proc. Suppl.) **75A**, 144 (1999)
[30] A.K. Managadze et al., in *Proc. 29th Int. Cosmic Ray Conf., Pune* **9**, 81 (2005); A.K. Managadze et al., Physics of Atomic Nuclei **70**(1), 184 (2007)
[31] T. Wibig, `arXiv:0003230 [hep-ph]`
[32] I.I. Royzen, Mod. Phys. Lett. **A9**(38), 3517 (1994)
[33] L.A. Anchordoqui, D.C. Dai, M. Fairbairnet et al., `arXiv:1003.5914 [hep-ph]`; L.A. Anchordoqui, D.C. Dai, H. Goldberg et al., `arXiv:1012.1870v2 [hep-ph]`
[34] R.A. Mukhamedshin (to be sent to Eur.Phys.J. C)

Extrapolation of proton-proton cross section to cosmic ray energies using geometrical model

Zbigniew Plebaniak[1,a] and Tadeusz Wibig[1,2,b]

[1] National Centre for Nuclear Research, Astrophysics Division, Cosmic Ray Laboratory, ul. 28 Pułku Strzelców Kaniowskich 69, 90-558 Łódź, Poland
[2] Faculty of Physics and Applied Informatics, University of Łódź, ul. Pomorska 149/153, 90-236 Łódź, Poland

Abstract. We present a new parameterization of the hadronic matter distribution in the geometrical picture of elastic scattering. Our model contains four free energy-dependent parameters that allows us to propose a better description of differential elastic cross section data measured in accelerator experiments from ISR to LHC energies. A simple linear dependence of all parameters on the logarithm of energy above $\sqrt{s} = 300$ GeV allows an extrapolation of elastic, inelastic and total proton-proton cross sections to ultra high energies seen in cosmic rays. Results of the extrapolation are in agreement with cross sections presented by ground-based cosmic ray experiments like AUGER or Telescope Array. Our parameterization also shows agreement with "BEL behaviour" of proton.

1. Introduction

The hadronic cross section is one of the most important parameter of models dedicated to describing Extensive Air Shower (EAS) development at ultra high energies. It is necessary for example to distinguish between EAS produced by lighter and heavier nuclei. It is also strongly connected with the multiplicity of produced particles. Unfortunately, due to QCD nature it is not possible to calculate this parameter from fundamental physical laws. This fact forces building complicated phenomenological models, based on accelerator measurements at lower energies, that allow for cross section extrapolations to energies observed in cosmic rays. Such models can also give information about the internal structure of colliding particles.

In this paper we concentrate on the problem of extrapolating hadronic cross section to cosmic ray energies. For this purpose we developed optical geometrical picture, assume energy dependent parameterization of the hadronic matter distribution in protons. The presented 4-parameter model describes the distribution of two hadronic matter fractions in protons. We received the energy dependence of parameters based on fits to the differential elastic cross section from accelerator data.

Elastic scattering of hadrons has been studied experimentally in a wide energy range. This work is based on data taken by ISR, SPS, Tevatron and LHC experiments. The energy of $\sqrt{s} = 8$ TeV for proton-proton collistions at TOTEM and ATLAS experiments is today the highest available energy useful for our considerations. Building our model we obtain the simple logarythmic dependence of four parameteres for higher energies, that allows for extrapolation to cosmic ray energies.

The material is organised as follows. In Sect. 2 we present the phenomenological description of the scattering process and eikonal representation of the elastic scattering amplitude. In Sect. 3 we describe the model assumptions. Section 4 is devoted to the results of fits and extrapolation of the cross section. Discussion and physical interpretation are the content of Sect. 5.

2. Scattering amplitudes and their eikonal representation

The differential elastic cross section is described by the elastic scattering amplitude $F(s, t)$ as follows:

$$\frac{d\,\sigma_{el}}{d|t|} = \pi\,|F(t)|^2 \tag{1}$$

We want to depict the absorption processes in a geometrical framework. The relation between interaction geometry and the momentum transfer space is defined with the Fourier transform with the help of the profile function $\Gamma(s, b)$ [1].

$$\begin{aligned} F(s,t) &= i\int_0^\infty J_0\left(b\sqrt{-t}\right)\,\Gamma(s,b)\,b\,db \\ &= i\int_0^\infty J_0\left(b\sqrt{-t}\right)\,\{1-\exp[-\chi(s,b)]\}\,b\,db, \end{aligned} \tag{2}$$

where $q^2 = -t$ is the four-momentum transfer, b - the impact parameter.

The Eikonal $\chi(s, b)$ may be expressed by:

$$\chi(s,b) = (\lambda(s) + i)\Omega(b,s) \tag{3}$$

[a] e-mail: `zp@zpk.u.lodz.pl`
[b] e-mail: `t.wibig@gmail.com`

The complex form of this function is defined by the energy-dependent parameter $\lambda(s)$ which represents the ratio of the real to the imaginary part of $\Omega(s, b)$:

$$\lambda(s) = \frac{\Re(\Omega(b, s))}{\Im(\Omega(b, s))}. \quad (4)$$

In our calculations, we used a slightly modified energy dependence of λ parameterized, e.g., by Menon in Refs. [2,3]. In Fig. 2 we present, related to λ, the calculated ratio of the real and the imaginary part of the scattering amplitude $\rho = \Re(F(\mathbf{b} = 0, s))/\Im(F(\mathbf{b} = 0, s))$, for central collisions (two dimensional impact parameter $\mathbf{b} = 0$), with accelerator data.

Using this formalism, cross sections can be described as follows [4]:

$$\sigma_{tot}(s) = 2\int \left[1 - \Re(e^{i\chi(b,s)})\right] d^2\mathbf{b} \quad (5)$$

$$\sigma_{el}(s) = \int \left[1 - e^{i\chi(b,s)}\right]^2 d^2\mathbf{b} \quad (6)$$

$$\sigma_{inel}(s) = \int 1 - \left[e^{i\chi(b,s)}\right]^2 d^2\mathbf{b} \quad (7)$$

The transmission coefficient $\Omega(b, s)$ which means the opaqueness of two colliding hadrons in impact parameter space, was introduced by Chou and Yang in 1968 [5], and can be described by the following convolution:

$$\Omega(b) = \Omega(\mathbf{b}) = iK_{pp} \iint_{-\infty}^{\infty} D(\mathbf{b} - \mathbf{b}')D(\mathbf{b}')\, d^2\mathbf{b}', \quad (8)$$

where $D(b)$ is a hadron profile function representing the integral of the density function $\rho(x, y, z)$ in the z direction as a collision axis:

$$D(\mathbf{b}) = \int_{-\infty}^{\infty} \rho(x, y, z)\, dz, \quad (9)$$

One of the simplest hadronic matter distributions $\rho(x, y, z)$ we can imagine is the one used in [6]. It represents the exponential decreasing of hadron matter density with radius:

$$\rho_h(\mathbf{r}) = \frac{m_h^3}{8\pi} e^{-m_h|\mathbf{r}|}. \quad (10)$$

Results of the scattering cross section obtained using this form shows agreement with accelerator data below and at the region of the first dip. Above this region, the deficit of higher transfer momenta ($|t| > 1\,\mathrm{GeV}^2$) can be observed. Currently, in the light of new LHC data at $\sqrt{s}$ of 7 and 8 TeV we tried to update the model modifying the distribution described by Eq. (10).

3. Model

To obtain a more satisfying description of the elastic scattering, we introduced the sum of two exponents with different slopes, m_1 and m_2, and two differrent normalization factors, c_1 and c_2, instead of one in Eq. (10).

Using Eq. (11), experimental data between 20 GeV and 8 TeV were fitted. Four parameters $m_{1,2}$ and $c_{1,2}$ were found for ten energy data sets starting from $\sqrt{s} = 20\,\mathrm{GeV}$, through five energies of ISR, SPS point at 546 GeV, Tevatron at 0.6 and 2 TeV and LHC data set measured at 7 and 8 TeV in centre of mass [7]. Results are shown in Fig. 1.

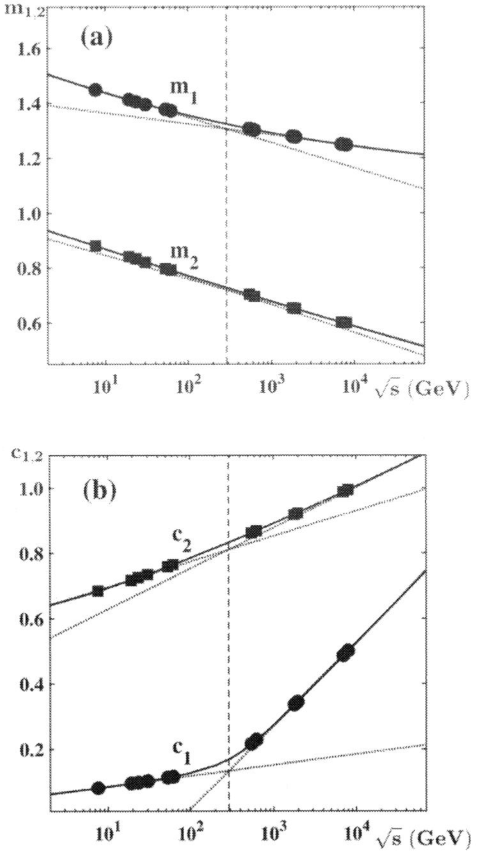

Figure 1. The values of adjusted parameters $m_{1,2}$ (a) and $c_{1,2}$ (b) are shown by the solid symbols. Dashed lines show the asymptotic linear energy dependencies (for high and low energies) of all parameters. Solid lines are the smooth overall energy dependencies.

$$\rho_h(\mathbf{r}) = \frac{1}{8\pi}\left(c_1 m_1^3 e^{-m_1|\mathbf{r}|} + c_2 m_2^3 e^{-m_2|\mathbf{r}|}\right). \quad (11)$$

The integral of this distribution may be expressed by analytic function as:

$$D(r(x, y)) = c_1\frac{m_1^2}{4\pi} m_1 r K_1(m_1 r) + c_2\frac{m_2^2}{4\pi} m_2 r K_1(m_2 r) \quad (12)$$

The first fraction, with higher slope m_1, represents the hard inner core of proton. Its opacity related to normalisation c_1 is going up much faster with energy over $\sqrt{s} = 300$ GeV. The second fraction, m_2, c_2, represents a softer and bigger hadron matter cloud. All four parameters indicate a change around 300 GeV in the center of mass energy which indicates that some new processes start at this point. The decrease of both slopes and the increase of both normalisation factors when the collision energy rise, confirm that the proton becomes Blacker, Edgier and Larger according to postulated BEL behaviour.

Table 1. The extrapolated cross sections in mb at higher energies.

Energy ($\sqrt{s}$)	8 TeV	14 TeV	24 TeV	30 TeV	57 TeV	95 TeV
Fagundes et al. [29]		108.6 ± 1.2				
Bourelly et al. [30]		103.63 ± 1.0				
Petrov et al. [31]		106.73				
Block, Halzen et al. [32]		107.30				
Islam et al. [33]		110.00				
Jenkovszky et al. [34]		111.00				
Block [35]					133.40 ± 1.6	
AKENO [28]	101±16	104 ± 26	124 ± 34			
Fly's Eye [26]				120 ± 15		
AUGER [18]					133.20 ± 13	
Telescope Array [22]						170.00 ± 50
LHC TOTEM [36]	102.9					
LHC ATLAS [37]	96.07					
presented model	**96.95**	**105.56**	**115.8**	**120.33**	**132.66**	**143.09**

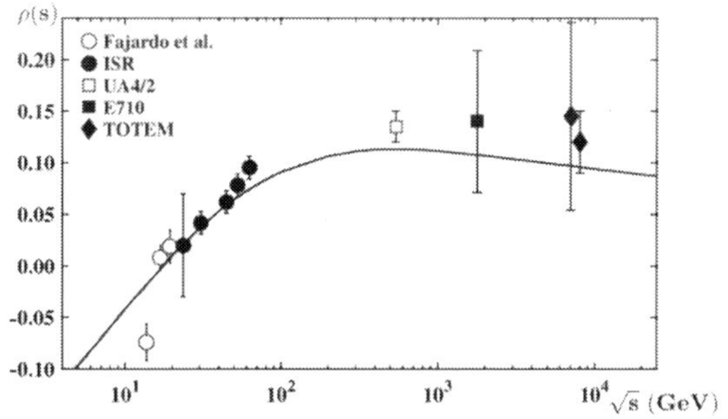

Figure 2. Ratio of the real and the imaginary part of the scattering amplitude $\rho = \Re(F(\mathbf{b}=0,s))/\Im(F(\mathbf{b}=0,s))$. The solid curve is the result of the parameterisation of $\lambda(s)$ from Ref. [2,3] modified slightly by us used in the present work. Points represent data from [8–12].

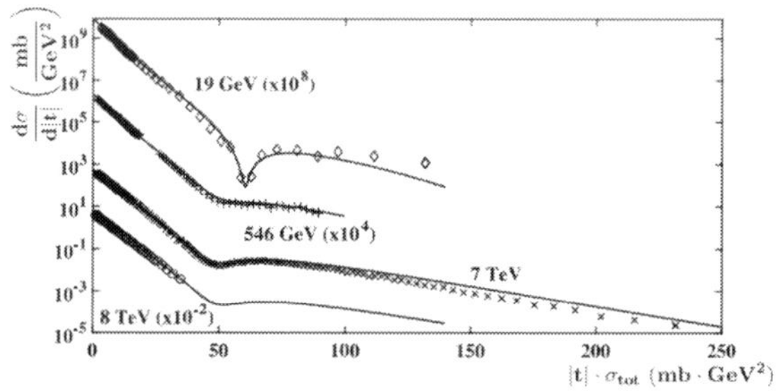

Figure 3. The differential elastic cross sections from our model for c.m.s energies of 19 GeV, 546 GeV and 7 TeV shown as a function of the ($|t| \times \sigma_{\text{tot}}$) compared with the measurements [13–17].

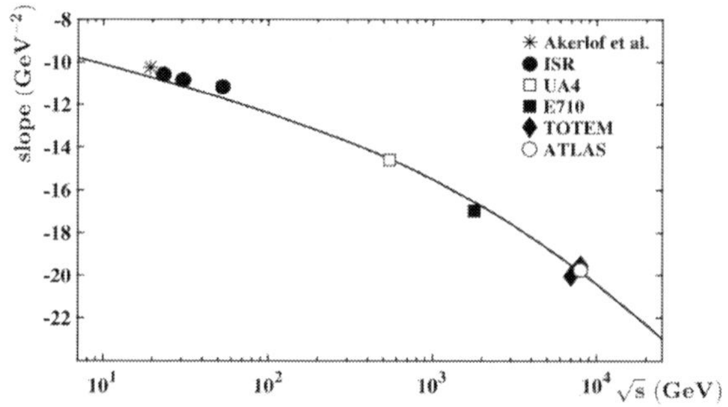

Figure 4. Calculated slopes of the elastic differential cross section as a function of interaction energy. Line shows our model predictions, points are fits to the experimental results from [13,16–21] in the range of $0.1 < |t| < 0.3\,\text{GeV}^2$.

4. Results

Using the presented model, we calculated the elastic, inelastic and total cross sections according to Eqs. (6), (7) and (5). Their compatibility with experimental data is presented in Fig. 5.

In Fig. 3 we plotted the calculated elastic scattering amplitudes related to the measured differential elastic cross sections at three energies. We can see an agreement with data over the first dip region which was a weakness of models prepared in past.

We used the calculated amplitudes such as in Fig. 3 to determine low $|t|$ slopes. In Fig. 4 results of our model are presented in good agreement with published data. Points are fits to the experimental results in the range of $0.1 < |t| < 0.3\,\text{GeV}^2$.

5. Discussion

We have developed a modified simple optical model of protonproton scattering with four model parameters, which allows us to describe the hadronic matter distribution of colliding protons. Values of all the parameters were adjusted to the elastic scattering data in the wide range of energies from fixed target experiments below the ISR to the 8 TeV energy of LHC $|t|$ distribution data. Using the eikonal parameterization the correct description of the total and elastic cross sections, the elastic slope and the differential elastic cross section at large values of momentum transfer have been found.

We found that for the low and the high energies there are asymptotic regimes which seem to be linear in the logarithmic scale of the available interaction energy $\sqrt{s}$. In agreement with "BEL" behaviour our model shows that the proton becomes blacker and larger as the energy increases.Thanks to the new parameterization of the hadronic matter distribution we obtain better agreement with scattering differential cross section accelerator data for higher momenta transfers for all available energies.

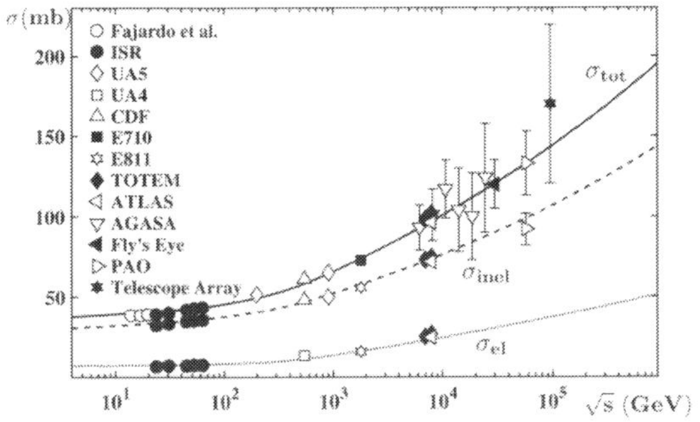

Figure 5. Values of the elastic, inelastic and total cross section calculated with our model as a function of the interaction energy compared with the measurements. The solid line represents total cross section, dashed inelastic and dotted line elastic cross section predictions. Points are experimental results from Refs. [8–10, 12, 18, 20, 22–28].

It constitutes a solid base for the extrapolation of the above scattering cross sections available from accelerator measurements at present, up to very high energy domains of cosmic rays and could help us to interpret the data from an astrophysical and a high energy physics point of view.

References

[1] V. Barone, E. Predazzi *High-energy particle diffraction* (Berlin: Springer, 2002), ISBN 3-540-42107-6
[2] D. Fagundes, M. Menon, G. Silva, The European Physical Journal C **71**, 1637 (2011)
[3] A.F. Martini, M.J. Menon, Phys. Rev. D **56**, 4338 (1997)
[4] T. Wibig, D. Sobczynska, Journal of Physics G: Nuclear and Particle Physics **24**, 2037 (1998)
[5] T.T. Chou, C.N. Yang, Phys. Rev. **170**, 1591 (1968)
[6] T. Wibig, Journal of Physics G: Nuclear and Particle Physics **39**, 085003 (2012)
[7] Z. Plebaniak, T. Wibig, Physics Letters B **761**, 469 (2016)
[8] U. Amaldi, K. Schubert, Nuclear Physics B **166**, 301 (1980)
[9] N.A. Amos et al. (E710 Collaboration), Phys. Rev. Lett. **68**, 2433 (1992)
[10] G. Antchev et al. (TOTEM Collaboration), Europhys. Lett. **101**, 21004 (2013)
[11] C. Augier et al. (UA4/2 Collaboration), Physics Letters B **316**, 448 (1993)
[12] L.A. Fajardo, R. Majka, J.N. Marx, P. Némethy, L. Rosselet, J. Sandweiss, A. Schiz, A.J. Slaughter, C. Ankenbrandt, M. Atac et al., Phys. Rev. D **24**, 46 (1981)
[13] C.W. Akerlof et al., Phys. Rev. **D14**, 2864 (1976)
[14] G. Antchev et al. (TOTEM Collaboration), EPL (Europhysics Letters) **95**, 41001 (2011)
[15] G. Antchev et al. (TOTEM Collaboration), EPL (Europhysics Letters) **101**, 21002 (2013)
[16] R. Battiston et al. (UA4 Collaboration), Phys. Lett. **B127**, 472 (1983), [141(1983)]
[17] M. Bozzo et al. (UA4 Collaboration), Phys. Lett. **B155**, 197 (1985)
[18] P. Abreu et al. (The Pierre Auger Collaboration), Phys. Rev. Lett. **109**, 062002 (2012)
[19] N.A. Amos et al. (E-710 Collaboration), Phys. Lett. **B247**, 127 (1990)
[20] G. Antchev et al. (TOTEM Collaboration), Phys. Rev. Lett. **111**, 012001 (2013)
[21] G. Antchev et al. (TOTEM Collaboration), Nucl. Phys. **B899**, 527 (2015), `1503.08111`
[22] R.U. Abbasi et al. (Telescope Array), Phys. Rev. **D92**, 032007 (2015), `1505.01860`
[23] F. Abe et al. (CDF Collaboration), Phys. Rev. D **50**, 5550 (1994)
[24] G.J. Alner et al. (UA5 Collaboration), Z. Phys. **C32**, 153 (1986)
[25] C. Avila et al. (E811 Collaboration), Phys. Lett. **B445**, 419 (1999)
[26] R.M. Baltrusaitis, G.L. Cassiday, J.W. Elbert, P.R. Gerhardy, S. Ko, E.C. Loh, Y. Mizumoto, P. Sokolsky, D. Steck, Phys. Rev. Lett. **52**, 1380 (1984)
[27] M. Bozzo et al. (UA4 Collaboration), Phys. Lett. **B147**, 392 (1984)
[28] C. Patrignani et al. (Particle Data Group), Chin. Phys. **C40**, 100001 (2016)
[29] D.A. Fagundes, M.J. Menon, P.V.R.G. Silva, Journal of Physics G: Nuclear and Particle Physics **40**, 065005 (2013)
[30] C. Bourrely, J.M. Myers, J. Soffer, T.T. Wu, Phys. Rev. D **85**, 096009 (2012)
[31] Petrov, V.A., Prokudin, A.V., Eur. Phys. J. C **23**, 135 (2002)
[32] M.M. Block, F. Halzen, Phys. Rev. **D83**, 077901 (2011), `1102.3163`
[33] M.M. Islam, R.J. Luddy, A.V. Prokudin, Modern Physics Letters A **18**, 743 (2003)
[34] L.L. Jenkovszky, A.I. Lengyel, D.I. Lontkovskyi, Int. J. Mod. Phys. **A26**, 4755 (2011), `1105.1202`
[35] M.M. Block, Phys. Rev. D **84**, 091501 (2011)
[36] G. Antchev et al. (TOTEM) (2016), `1610.00603`
[37] M. Aaboud, G. Aad, B. Abbott, J. Abdallah, O. Abdinov, B. Abeloos, R. Aben, O. AbouZeid, N. Abraham, H. Abramowicz et al. (ATLAS), Physics Letters B **761**, 158 (2016)

Extensive Air Showers with unusual structure

Dmitriy Beznosko[1,a], Rashid Beisembaev[2], Kanat Baigarin[1], Elena Beisembaeva[2], Oleg Dalkarov[2], Vladimir Ryabov[2], Turlan Sadykov[3], Sergei Shaulov[2], Aleksei Stepanov[2], Marina Vildanova[2], Nikolay Vildanov[2], and Valeriy Zhukov[2]

[1] Nazarbayev University, Physics Department, Astana, Kazakhstan
[2] P. N. Lebedev Physical Institute of the Russian Academy of Sciences, Moscow, Russia
[3] LLP "Institute of Physics and Technology", Almaty, Kazakhstan

Abstract. A total of 23500 Extensive Air Showers (EAS) with energies above $\sim 10^{16}$ eV have been detected during the ~3500 hours of the Horizon-T (HT) detectors system operations before Aug. 2016. Among these EAS, more than a thousand had an unusual spatial and temporary structure that showed pulses with several maxima (modals or modes) from several detection points of the HT at the same time. These modes are separated in time from each other starting from tens to thousands of ns. These EAS have been called multi-modal. Analysis shows that the multi-modal EAS that have been detected by Horizon-T have the following properties:

1. Multi-modal EAS have energy above $\sim 10^{17}$ eV.
2. Pulses with several modes are located at large distances from the EAS axis.

An overview of the collected data will be provided. General comments about the unusual structure of the multi-modal EAS will be presented.

1. Introduction

The Horizon-T [1] detector system is located at the Tien Shan High-altitude Science Station (TSHSS) of the Lebedev Physical Institute of the Russian Academy of Sciences at ~3340 meters above sea level. It consists of eight charged particle detection points, separated by distances up to one kilometer, as well as an optical detector subsystem to measure the Vavilov-Cherenkov light from the EAS. An aerial view of the detector system is shown in Fig. 1. The coordinates of each detection point with respect to the point 1 are given in Table 1.

HT is an innovative detector system constructed to study both the spatial and temporal distribution of the charged particles within the EAS in the energy range above 10^{16} eV coming from a wide range of zenith angles (0°–85°) with a time resolution of a few ns. This novel approach allows to collect two-fold the data from each detection point (particle density and time of passage) and increase the measurement precision of the EAS parameters such as measuring the core position when it falls outside of the detector system area. The HT detector system consists of detectors of several different types: plastic scintillator [2] with MELTS [3] FEU49 model PMT, plastic scintillator with Hamamatsu [4] R7723 PMT assembly, glass with R7723 PMTs and a Vavilov-Cherenkov radiation detector with both FEU49 and Hamamatsu H6527 PMT assemblies. All detection points connect to Data Acquisition (DAQ) via cables that are being calibrated separately [5].

During more than ~3500 hours of operation from the start of the first prototype up to August 2016 the HT detector system detected 23500 EAS with energy above $\sim 10^{16}$ eV – the threshold energy for HT. Among these EAS, more than a thousand had an unusual spatial and temporal structure that showed pulses with several maxima (also called modes) from several detection points of the HT at the same time. These modes were separated from each other by times from tens to thousands of ns. These EAS are called so called Multi-Modal events (MME). Initially attributed to delayed particles, they were first observed by J. Jelly and W. Whitehouse [6].

[a] e-mail: dmitriy.beznosko@nu.edu.kz

2. Standard EAS definition

As a solid starting point, the definition of the 'standard' EAS is required. For that, we adopt the output of the EAS simulation package CORSIKA [7]. The simulated EAS serves as a reference that is simple to compare to, and allows to demonstrate certain simple properties of the EAS disk at the observation level.

Firstly, every parent particle regardless of species and energy gives rise to a single uniform EAS without any breaks in the structure or large-scale fluctuations with statistical significance.

Secondly, the particle density drops smoothly with distance when moving away from the core. Symmetry is assumed here for a vertically incoming EAS disk. The charged particles distribution vs. the distance from the core is illustrated in Fig. 2. These simulation results are supported by the result of the experiment 'Hadron' published earlier [8].

Thirdly and finally, the thickness of the EAS disk increases with the distance from the core. Additional characteristics may be sought from the simulated EAS but

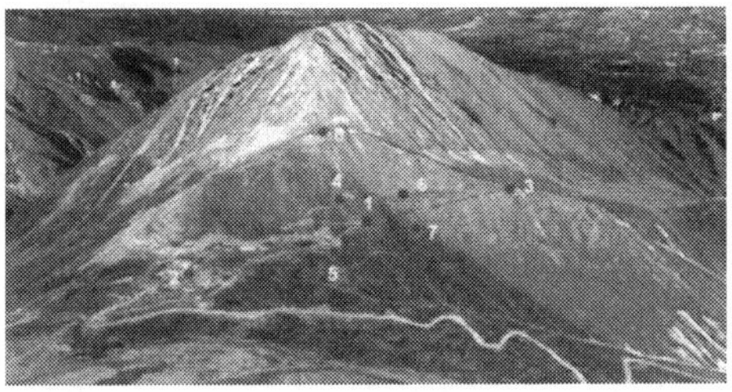

Figure 1. Aerial view of the Horizon-T detector system.

Table 1. HT detection points coordinates.

Station #	X, m	Y, m	Z, m	R, m
1	0	0	0	0
2	−445.9	−85.6	2.8	454.1
3	384.9	79.5	36.1	394.7
4	−55.0	−94.0	31.1	113.3
5	−142.4	36.9	−12.6	147.6
6	151.2	−17.9	31.3	155.4
7	88.6	178.4	−39.0	203
8	221.3	262	160.7	374

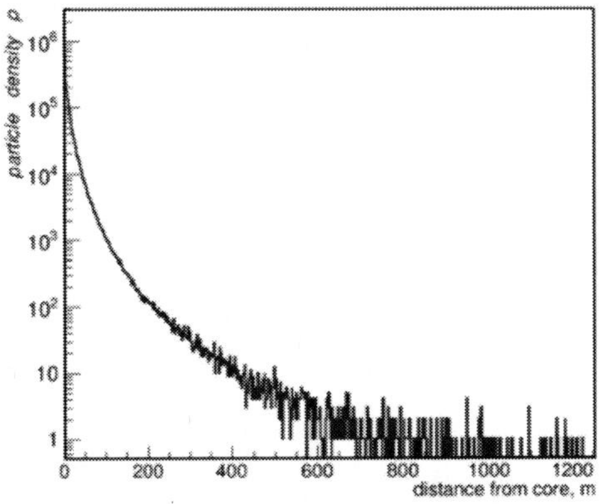

Figure 2. Charged particles density vs. distance from EAS core.

these three seem to be vital ones to both standard definition and the illustration of the unusual structure of the detected EAS. To illustrate the disk widening, the particles arrival time distribution over the 4 m^2 area at 50 and 100 meters is shown in Figs. 3 and 4 respectively.

The event display of the standard EAS as detected by the HT detector system is shown in Fig. 5. We can see that the EAS disk arrives from the side of the detector point 4 grazing it with its edge, then hits points 5, 1 and 6 almost simultaneously with the core passing further then hitting point 7 following the arrival direction.

3. EAS with unusual structure

Besides the initial observation of the MME in 1953 [6], other scientists were trying to study these events at different times. Initially, the effect was attributed by many of them to the birth in the EAS of a very massive particle with a low γ factor that would get delayed and then start a new shower of its own, thus causing the multi-modality of the detected signals.

During the 1960s to 1980s, MME events were studied by British and US physicists, such as Linsley [9],

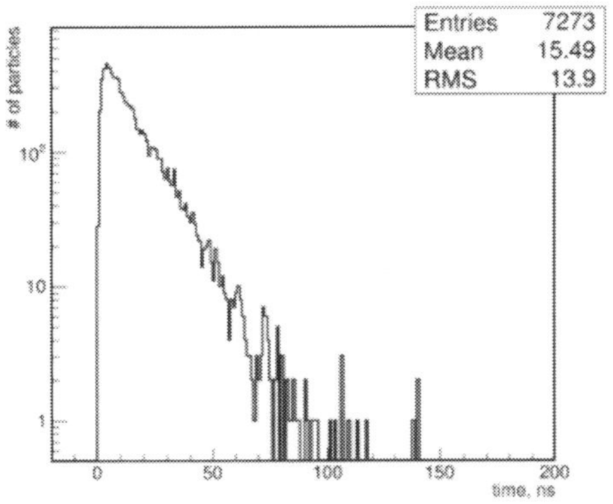

Figure 3. Charged particles arrival time at 50 m from EAS core.

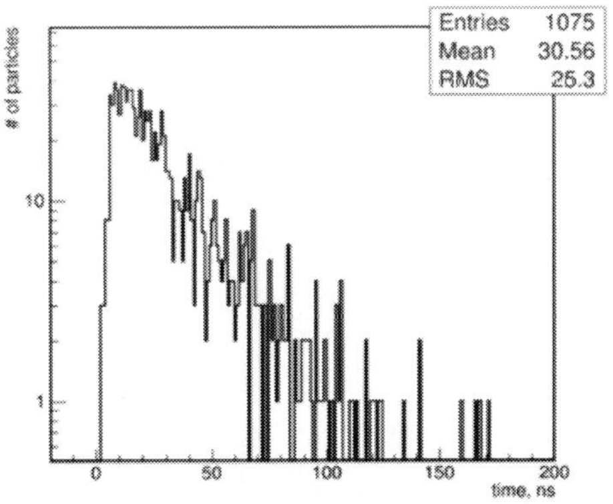

Figure 4. Charged particles arrival time at 100 m from EAS core.

Watson [10] and Yodh [11]. Since the 1970s, the delayed particle effect has been studied in Japan (Sakuyama [12], Inoue [13]). Starting from the 1980s, studies of MMEs were continued by Moscow State University [14] and at the Lebedev Physical Institute of the Russian Academy of Sciences [15,16]. In the 1990s, MME pulses were registered at the Yakutsk detector system [17], and in the 2000's they were detected by the "TUNKA" detector system [18].

Note that the delayed particle explanation, in order to support the time delays observed experimentally, requires frequent birth of particles with rest masses exceeding 10^{12} eV. Not only we don't know of such a particle, but, according to current gauge invariance theories, the probability of such a particle being born is inversely proportional to the square of the mass contradicting the frequency of observation. Thus, MME researchers have always admitted that they have extreme difficulty explaining the multi-modality using the current physics knowledge and that MME may indicate the existence of a new physical phenomena.

3.1. MME example at HT

Most MME events detected at HT seem to be parts of a larger event arriving far away from the detector system. However, there are several events recorded with the MME axis falling near one of the far periphery points, allowing the observation of the complex structure of such an event. An example of such an MME event is given here.

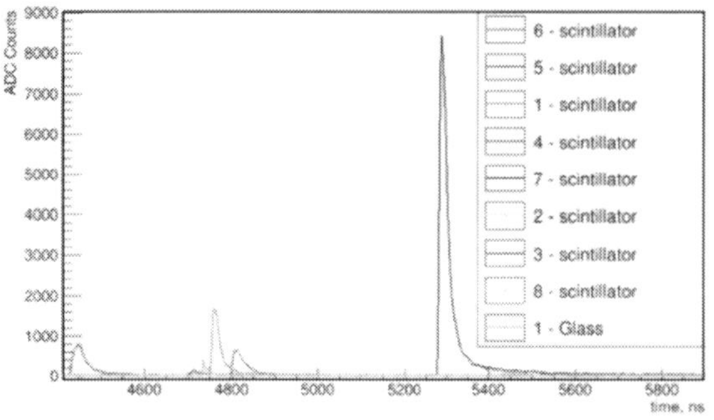

Figure 5. Sample 'standard' event as seen by the HT detector system.

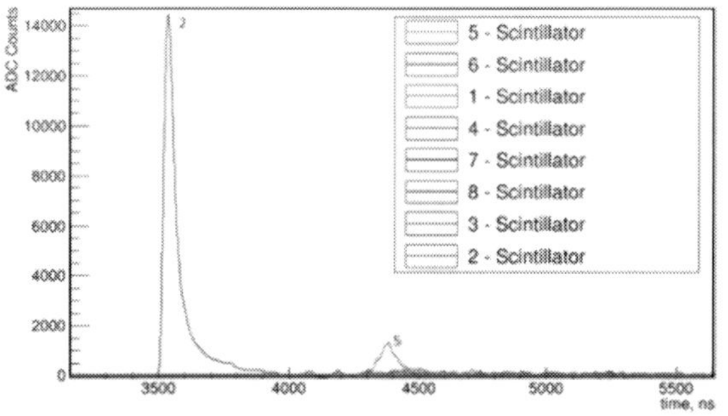

Figure 6. MME as seen by the HT detector system.

The event in Fig. 6 was detected on Jan. 26 2016 at 3:51 UTC. The event display shows the very large signal at detection point 2 and and the next pulse is visible from point 5, thus indicating that the MME axis has arrived somewhere between these points much closer to point 2. The amplitudes of pulses from other detection points is much smaller and are not visible in this figure. If we take the pulses with the highest amplitude and use them as for a standard EAS analysis, then the axis of this event is tilted at $\theta \sim 50°$ and the arrival direction is approximately along the line, connecting detection points 2 and 3, and deviating towards point 5. Note also that there is a pulse from point 2 resembling what is expected from a standard EAS, and the pulse from point 5 seems to start to deviate from that expected of a smooth pulse shape.

Pulses from detection points 1 and 7 only are shown in Fig. 7. Note that these points are both located about the same distance from the axis and further than point 5. Both have a complicated structure, with the signal from point 7 having the modes spread farther apart than for the point 1 signal.

The pulses from detection points 6 and 4 display full multi-modality (see Fig. 8). The peaks are separated by hundreds of ns and feature at least one of the pulses having a complex shape.

The pulses from the detection points 3 and 8 are shown in Fig. 9. These points are furthest from the axis of this event and display the features similar to points 4 and 6, except that the pulses are further separated by several hundreds of ns. The signal from point 8 has the same 3 distinct modes, however, point 3 has only 2 modes visible. It is also the furthest point from the axis of the event, and it is possible that some modes are too weak to be detected.

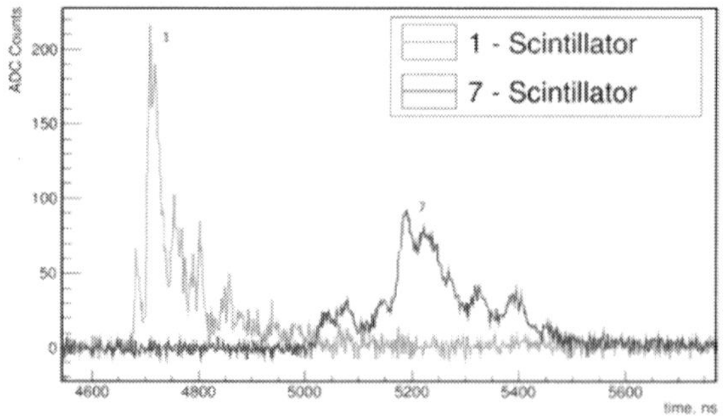

Figure 7. MME as seen by the HT detector system.

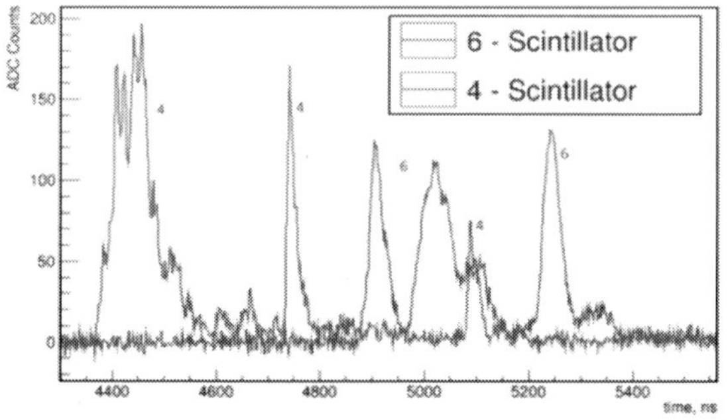

Figure 8. MME as seen by the HT detector system.

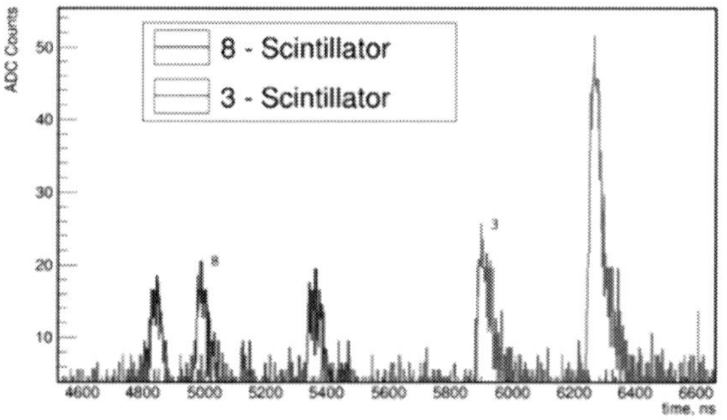

Figure 9. MME as seen by HT detector system.

Using the pulses from all detection points we can infer that this is a very high energy event as its radius is at least 1 km. There are several clean events like this that have been observed at HT and they all follow the same pattern - a large single pulse (that may have structure that is just too fine to be resolved), then the complicated structure pulses at intermediate distances and fully separated multi-peaked pulses at the farthest points from the event axis. It is also plausible that other MME events detected are just parts of the events that have the structure described above. Further studies and analysis of such unusual events are in progress.

4. Conclusion

A large number of MME events were detected at HT. The observations show a large variety of shapes and delays between the modes, thus discarding the hypothesis of delayed particles, because in this case all detection points would have shown a similar structure. There is a strong indication that all the variety of forms detected comes from the unified structure that was observed in several events with the axis close to the far periphery detection points, allowing observation of the event structure at different distances.

References

[1] R U Beisembaev et al., `arXiv:1605.05179`, (2016)
[2] A Baitenov et al., `arXiv:1601.00086`, (2016)
[3] MELZ-FEU, 4922-y pr-d, 4c5, Zelenograd, g. Moskva, Russia, 124482 `http://www.meltz-feu.ru`
[4] Hamamatsu Photonics, 314-5 Shimokanzo, Toyooka-village, Iwatagun, Shizuoka-ken, 438-0193 Japan `http://www.hamamatsu.com`
[5] T Beremkulov et al., `arXiv:1608.04312`, (2016)
[6] J.V. Jelly and W.J. Whitehouse, Proc. Phys. Soc. **A66**, 454 (1953)
[7] D. Heck et al., Forschungszentrum Karlsruhe Report FZKA **6019**
[8] Adamov D.S. et al., ICRC **460A**, 5 (1987)
[9] Linsley J. and Scarsi L., Phys. Rev. **128**, 2384 (1962)
[10] Baxter A.J., Watson A.A. and Wilson J.G., Proc. 9 ICRC **2**, 724 (1965)
[11] Yodh G.B. et al., Proc. 18 ICRC **11**, 264 (1983)
[12] Sakuyama H. et al., Proc. 16 ICRC **8**, 135 (1979)
[13] Inoue N. et al., Proc. 19 ICRC **7**, 316 (1985)
[14] Atrashevich V. B. et al., RAS News Bul. **58**(12), 98 (1994)
[15] Beisembaev R.U. and Vavilov Yu. N. et al., Proc. 24 ICRC. Roma **1**, 454 (1995)
[16] Beisembaev R.U. et al., Nuclear Physics T.72. **11**, 1–4 (2009)
[17] Glushkov A. B. et al., Letters to Journal of Experimental and Theoretical Physics **67**(6), 361 (1998)
[18] Prosin V. V. et al., RAS Bulletin **73**(5), 627 (2009)

A method of observing Cherenkov light at the Yakutsk EAS array

Lev Timofeev[a] and Anatoly Ivanov[b]

The Yu.G. Shafer Institute of Cosmophysical Research and Aeronomy, Russia

Abstract. A proposed new method for measuring the cherenkov light from extensive air showers (EAS) of cosmic rays (CR), which allows to determine not only the primary particle energy and angle of arrival, but also the parameters of the shower in the atmosphere – the maximum depth and "age". For measurements it is proposed to use Cherenkov light produced by EAS in a ground network of wide-angle telescopes which are separated from each other by a distance 100–300 m depending on the total number of telescopes operating in coincidence, acting autonomously, or includes a detector of the charged components, radio waves, etc. as part of the EAS. Results of such an array should be developed. The energy measurement and CR angle of the arrival data on the depth of the maximum and the associated mass of the primary particle generating by EAS. This is particularly important in the study of galactic cosmic ray for $E > 10^{14}$ eV, where currently there are no direct measurements of the maximum depth of the EAS.

1. Introduction

Investigation of Cherenkov light induced by cosmic rays (CRs) in the atmosphere began in the middle of the previous century in the UK and USSR. An exhaustive description of early developments is given in the book [1], while recent reviews of the whole area are given, e.g., in Refs. [2,3].

It was realized then that the angular and temporal structure of the Cherenkov light emitted by extensive air shower (EAS) can be used to infer the longitudinal development parameters of the shower [4–7]. the angular distribution of Cherenkov photons from EAS was calculated by V.I. Zatsepin [4] assuming it was determined mainly by that of the electrons in the shower. Then Fomin and Khristiansen proposed [5] to use the pulse shape of the Cherenkov signal, namely the pulse width, to indicate the shower maximum position, X_{max}, in the atmosphere. Experimental measurements of the Cherenkov signal pulse shape were made initially in Yakutsk and at Haverah Park [6,7]. The results were used not only to estimate X_{max} but attempts were made to evaluate cascade parameters of electrons at CR energies around 10^{17} eV [7,9]. A variety of detectors are used then, for instance, at the Tunka experiment forming an array of Cherenkov detectors functioning currently near Lake Baikal [10].

Our intention to develop a Cherenkov telescope functioning as a differential detector of EAS was motivated by the possibility to measure the depth of the cascade maximum and/or the shower age via angular and temporal distributions of the Cherenkov signal [8]. Combining X_{max} and the shower age with other characteristics measured with surface detectors of the EAS array, e.g. the energy and muon content, one is able to estimate the average mass composition of CRs. Experimental arguments in elucidating the origin of the knee and ankle in the CR spectrum will significantly strengthen due to the measurements of the angular and temporal distributions of the Cherenkov signal in the energy range above 10^{15} eV. Existing scenarios of CR acceleration in the sources are different in the expected model composition around the knee and in the transition region between galactic and extragalactic components [11]. So an accurate estimation of the average mass of the CR nuclei in addition to improved measurements of the sharpness of the knee and ankle should allow us to discriminate some scenarios. The PMT in such a prototype telescope is used as the observing camera and placed at the focal length in front of a spherical mirror see Fig. 1. Since the radius of the photocathode R2486 equals 3 cm, the mirror diameter was chosen to be 26 cm with a radius of curvature of 22.5 cm. Mirror parameters have been chosen from considerations to ensure as wide a field of view as possible, but so that possible distortions do not exceed the pixel size, 3.8 mm. As a result, the width of the field of view is $-14° \leq \alpha \leq 14°$.

2. Data acquisition system

The signals from each channel of the multi-anode PMTs continuously input to preamplifiers and onwards to the Analogue to Digital converters and stored in the buffer memory of a 32-channel industrial computer (Fig. 2). On the arrival of a trigger generated by the EAS array, an industrial computer saves the accumulated pre-history as a separate file labeled with the time mark and the assignment of a serial number.

We carried out the photomultiplier sensitivity check from plastic scintillator in the voltage range from 900 to 1300 volts and obtained a coordinate-dependent sensitivity

[a] e-mail: timofeevlev@ikfia.ysn.ru
[b] e-mail: aivanov@ikfia.ysn.ru

DOI: 10.1051/epjconf/201714514002

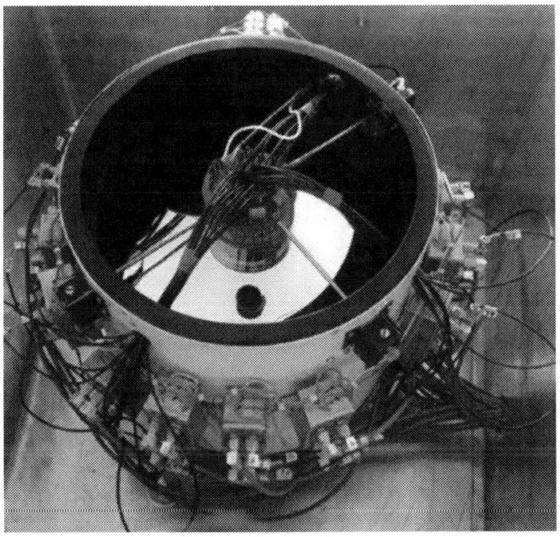

Figure 1. Wide field of view Cherenkov telescope prototype.

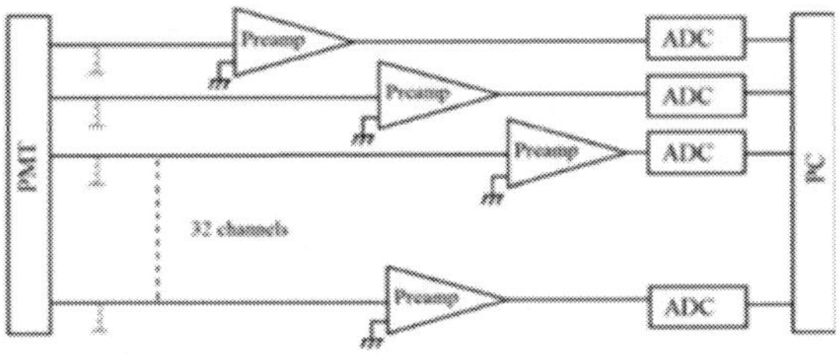

Figure 2. Data acquisition system.

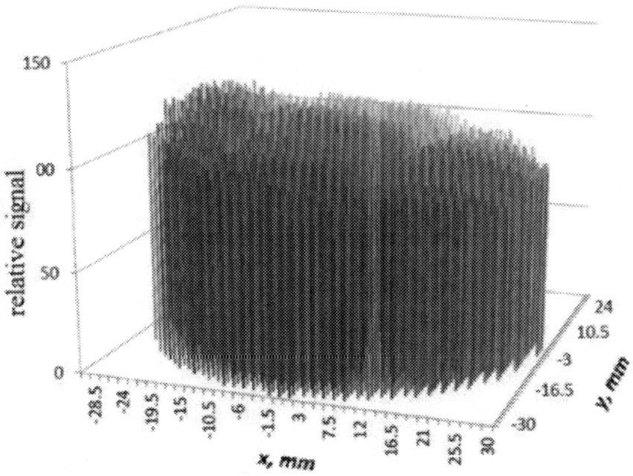

Figure 3. Normalized coordinate-dependent PMT sensitivity.

of the Photo Multiplier Tube from which a radial nonuniformity of sensitivity is seen in Fig. 3. We also measured the angular dependence of the telescope, the energy spectrum of telescope channels, and found that the Photo Multiplier Tube dark current is minimal at a voltage less than 1300 volts. Currently, the telescope is installed at the Yakutsk EAS array at station obscura-3 at a distance 500 meters from the center of the array, where the field test is run.

An example of the output signals from the multi anode PMT is shown in Fig. 4. The average number

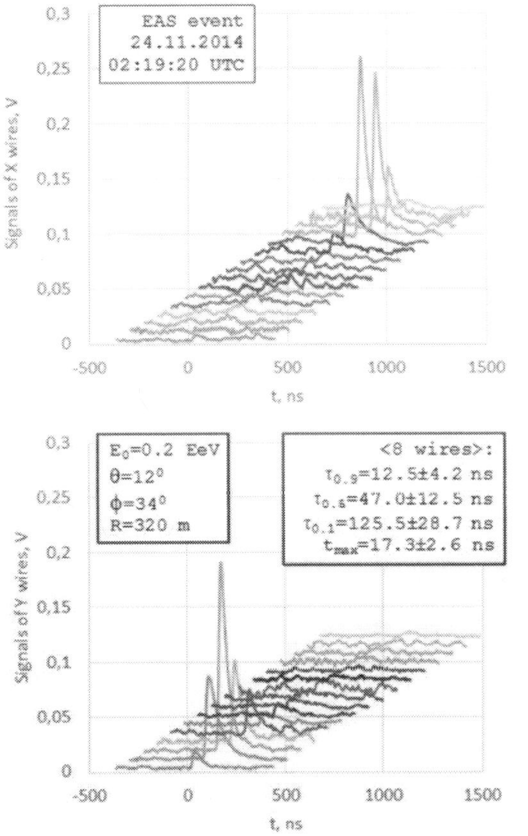

Figure 4. Signals of 32 wires from the multi-anode PMT at the focus of the telescope. The EAS parameters in a particular shower are estimated using data of the surface detectors. Pulse durations (τ) are measured at 0.1, 0.5, 0.9 levels of the pulse maximum V_{max}. The time difference (t_{max}) between V_{max} and 0.1 V_{max} of the rising signal is shown in the right panel.

of channels triggered in the event, where the signal is definitely exceeds the noise level is 17 ± 6. So we get a good accuracy of the pulse duration by means of several independent measurement channels.

3. Pulse duration as a function of the distance to the shower core

Here we focus on the pulse shape parameters of the Cherenkov signal from EAS. The main feature of the Cherenkov signal is its rise time with the shower core distance due to geometrical reasons. The data was accumulated during the observation period from 2012 to 2015, when 2000 coincident EAS events were detected. A sample of showers is selected in the energy interval from 10^{16} to 10^{19} eV. Additional cuts are applied to exclude showers with axes out of the array area and zenith angles more than sixty degrees. A specific cut is used for saturated signals where EAS events generate more Cherenkov light than the Cherenkov telescope can properly detect. These signals are also rejected. The number of EAS events surviving after cuts is 733 with the primary energy distribution shown in Fig. 5.

DOI: 10.1051/epjconf/201714514002

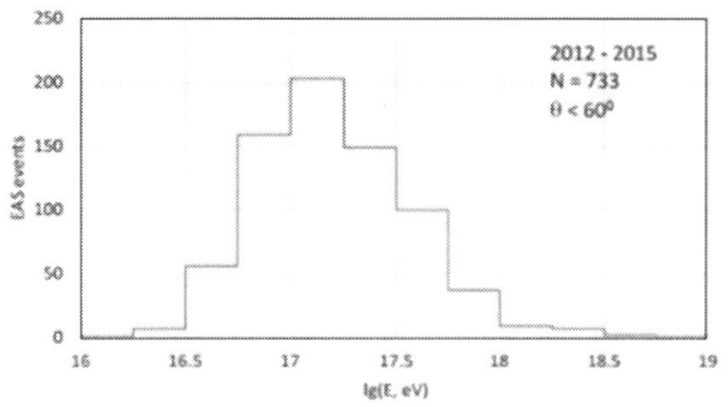

Figure 5. Energy distribution for EAS events detected by coincidence scintillation detectors of Yakutsk array and wide field of view Cherenkov telescope.

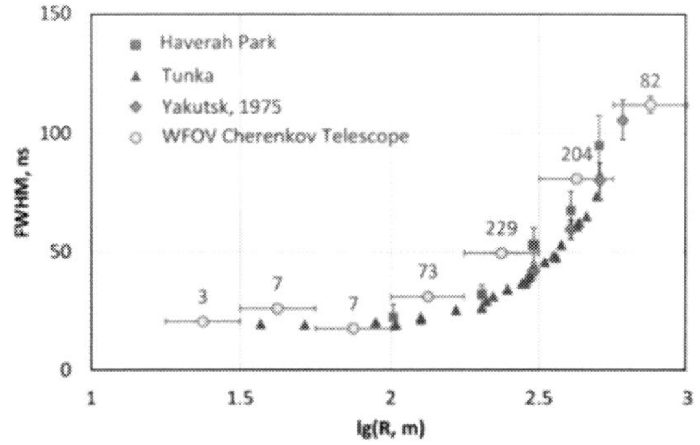

Figure 6. Full width at half-maximum of the Cherenkov signal as a function of the shower core distance, R, measured by EAS arrays.

The main feature of the signal is its rise time variation with the shower core distance due to geometrical reasons. The first measurements of the parameter were made in Yakutsk [12] and Haverah Park [13]. The most recent results are provided by the Tunka array [14]. Our measurement of the full width at half-maximum of the signal confirms the previous results (see Fig. 6). Event numbers are indicated above our data points. The energy ranges relevant are: Haverah Park (E) ~0.2 EeV; Yakutsk, 1975 E $\in$ (0.001, 1) EeV; Tunka E $\in$ (0.003, 0.03) EeV.

We analyzed the full width at half-maximum (FWHM) dependence on other shower parameters such as zenith and azimuth angles, and energy. No significant variation versus angles is found exceeding the instrumental errors. On the other hand, an indication of the energy dependence of the pulse width is found in our data. Application of the FWHM(R) function is possible in the shower core location procedure in the array plane. The method was pioneered by John Linsley [15] using charged particle signal width vs R. While in the Yakutsk array group the particle and Cherenkov photon density distribution functions are used to locate the shower core, the FWHM(R) function can be used additionally (or instead) to refine the core coordinates in the array plane. As an illustration, the signal width is shown as a function of energy in Fig. 7. A systematic variation of the width is visible, although the tendencies are contradictory in the shower core distance intervals. More data is needed, especially at distances to the shower core less than one hundred meters, to reveal the reliable energy dependence. Application of the FWHM(R)

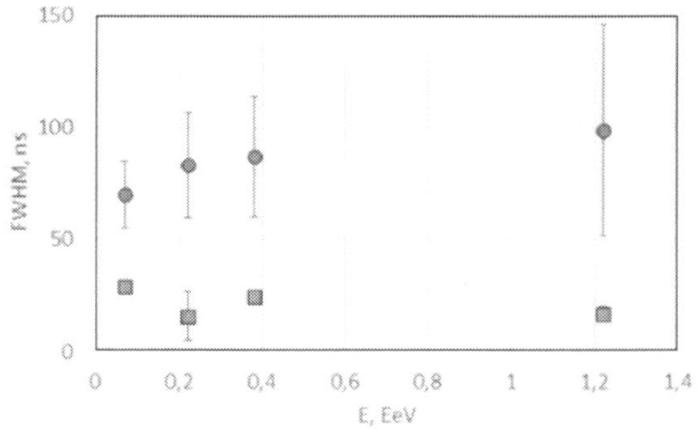

Figure 7. Energy dependence of Full Width at Half-Maximum of the Cherenkov signal.

function is possible in the shower core location procedure in the array plane at least additionally to the Cherenkov photon density distribution.

4. Method to estimate the distance to the shower maximum via cherenkov light measurements

Common methods, which rely on Cherenkov light detectors to find the height in the atmosphere, h_{max}, where the number of EAS particles reaches a maximum, are based on measurements of the lateral distribution of the photon density and pulse duration [16–18]. We propose another approach to evaluate h_{max} resulting from our measurements of the Cherenkov light induced by EAS. Knowing the time delay of the Cherenkov signal maximum in the detector relative to the shower axis crossing the array plane at distance, R, one is able to calculate the distance, h_θ, to the height in atmosphere, h_{max}, where the emission of Cherenkov photons reaches a maximum. Other shower parameters needed are the coordinates of the shower axis and the EAS arrival angles. In our case, these parameters can be provided by the set of synchronized surface detectors of the Yakutsk array. The time difference, Δt, is determined by triangles consisting of R, h_{max}, h_θ. More information about this method can be found in the article [19].

$$c\Delta t = \sqrt{h_\theta^2 + R^2 - 2Rh_\theta \sin\theta\cos\phi} - h_\theta,$$

where ϕ, θ are shower arrival angles; c = 0.3 m/ns. A solution is given by

$$h_\theta = \frac{0.5(R^2 - (c\Delta t)^2)}{c\Delta t + R\sin\theta\cos\phi}.$$

The solution of the equation is shown in Fig. 8. Rather large values of the derivative $dh_\theta/d\Delta t$ near the time difference threshold restrict the maximum possible locations within narrow limits due to moderate timing accuracy. On the other hand, relatively small heights can be resolved with reasonable timing resolution. We illustrate in Fig. 9 the maximum height location accuracy achievable with the different shower core and time measurement errors. The present accuracy for the Yakutsk array detectors is not sufficient to measure the distance to the shower maximum with reliable results. But the Yakutsk

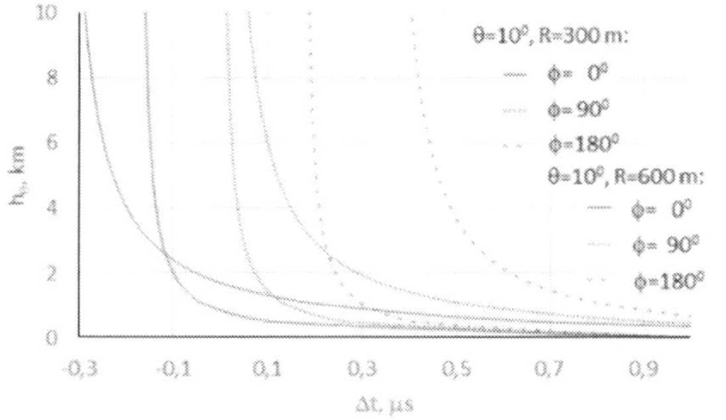

Figure 8. The height of Cherenkov light emission maximum as a function of time difference, Δt, between Cherenkov photons arriving at the detector and to the shower core in the array plane. The azimuth angle, ϕ, is between the detector and the shower axis projected on the array plane.

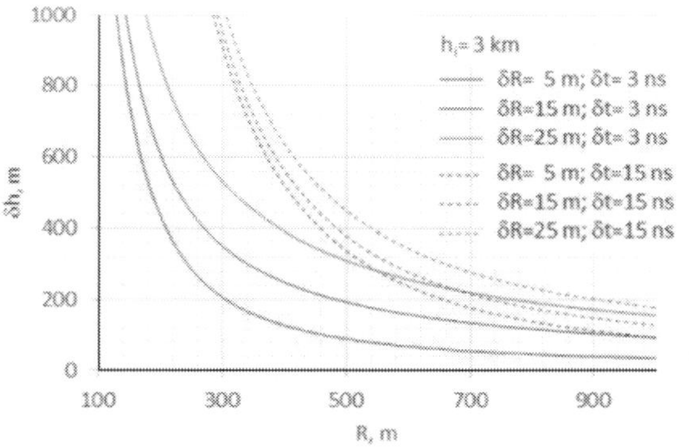

Figure 9. Estimation of the h_{max} reconstruction accuracy as a function of the shower core distance. Different EAS core location and timing errors are given.

array modernization program now in progress includes the best part of the improvements needed. Specifically, a target synchronization accuracy of detectors will equal 5 ns due to Gigabit Ethernet via optical fibers. New scintillation counters and Cherenkov light detectors will provide the core location accuracy of showers within 15–20 m. Accordingly, the modernized array will be able to measure the Cherenkov X_{max} to an accuracy of 3–4%, if the time resolution of the Cherenkov light detectors will be better than 5 ns.

5. Conclusions

As part of the modernization program we developed two new integral Cherenkov telescopes, equipping them with the latest electronics we achieved a significant improvement in the time resolution, and combined them into the Cherenkov light registration system with adjustable zenith angle, to get information about the longitudinal shower development see Fig. 10.

A new approach is explored in estimating the distance to the shower maximum based on the time difference between Cherenkov photons arriving at the detector and the shower core in the array plane. Fast optical detectors and a nanosecond-order synchronization system of the array detectors are essential in future implementations of the method. A Cherenkov telescope prototype obtained experimental data for the period from 19.10.2012 to 11.04.2016, on 203 clear moonless nights with about 1500 hours of observation, during the observation of the master load 30,000 times, of which the Cherenkov telescope gave a signal to the coincidence of about 1200 times. Measuring the width at half-maximum as function of distance from the shower core, (R), will help determine the shower core in the array plane. Further modernization of the array will enable to calculate the height of the shower maximum by the time delay.

Figure 10. Integral Cherenkov telescope with Fresnel lens.

The work is supported in part by RFBR (grants 11-02-00158, 11-02-12193, 12-02-10005, 12-02-31550) and the Russian Ministry of Education and Science (contracts 02.740.11.0248, 16.518.11.7075). Grant NOFMU academic mobility and Head of Sakha Republic Grant for young scientist 17/16.

References

[1] J.V. Jelley, *Cherenkov Radiation and its Applications* (Pergamon Press, 1958)
[2] A.S. Lidvansky, Radiation Physics and Chemistry **75**, 89 (2006)
[3] A.A. Watson, `astro-ph/1101.4535`
[4] V.I. Zatsepin, J. Exp. Theor. Phys. **47**, 689 (1964)
[5] Yu.A. Fomin and G.B. Khristiansen, Yad. Fiz. **14**, 642 (1971)
[6] N.N. Kalmykov et al., J. Exp. Theor. Phys. Lett. **21**, 66 (1975)
[7] R.T. Hammond et al., Nuovo Cim. **C1**, 315 (1978)
[8] M.N. Dyakonov et al., Nucl. Instrum. Meth. A **248**, 224 (1986)
[9] V.M. Grigorjev et al., J. Exp. Theor. Phys. Lett. **30**, 727 (1979)
[10] S.F. Berezhnev et al., Nucl. Instrum. Meth. A **692**, 98 (2012)
[11] E.G. Berezhko et al, Astropart. Phys. **36**, 31 (2012)
[12] N.N. Kalmykov et al., J. Exp. Theor. Phys. Lett. **21**, 66 (1975)

[13] R.T. Hammond et al., Nuovo Cim. **C1**, 315 (1978)
[14] V.V. Prosin et al., Nucl. Phys. B - Proc. Suppl. **190**, 247 (2009)
[15] J. Linsley, Proc. 19th ICRC **9**, 434 (1985)
[16] M. Nagano and A.A. Watson, Rev. Mod. Phys. **72**, 689 (2000) 689
[17] S.P. Knurenko and A.V. Sabourov, Astroph. Space Sci. Trans. **7**, 251 (2011)
[18] E.G. Berezhko, S.P. Knurenko and L.T. Ksenofontov, Astropart. Phys. **36**, 31 (2012)
[19] A.A. Ivanov for the Yakutsk array group, EPJ Web of Conf. **53**, 04003 (2013)

Higgs production enhancement in P-P collisions using Monte Carlo techniques at $\sqrt{s}$ =13 TeV

M.H.M. Soleiman[1,a], S.S. Abdel-Aziz[1], and M.S.E. Sobhi[1]

[1] Cairo University, Faculty of Science, Physics Department, Giza, Egypt
[2] CTP, British University in Egypt, Egypt

Abstract. A precise estimation of the amount of enhancement in Higgs boson production through pp collisions at ultra-relativistic energies throughout promotion of the gluon distribution function inside the protons before the collision is presented here. The study is based mainly on the available Monte Carlo event generators (*PYTHIA* 8.2.9, *SHERPA* 2.1.0) running on PCs and $CERNX$-Machine, respectively, and using the extended invariant mass technique. Generated samples of 1000 events from *PYTHIA* 8.2.9 and *SHERPA*, 2.1.0 at $\sqrt{s} = 13$ TeV are used in the investigation of the effect of replacing the parton distribution function (*PDF*) on the Higgs production enhancement. The *CTEQ66* and $MSRTW2004nlo$ parton distribution functions are used alternatively on *PYTHIA* 8.2.9 and *SHERPA* 2.1.0 event generators in companion with the effects of allowing initial state and final state radiations (*ISR* and *FSR*) to obtain evidence on the enhancement of the SM-Higgs production depending on the field theoretical model of SM. It is found that, the replacement of *PDFs* will lead to a significant change in the SM-Higgs production, and the effect of allowing or denying any of *ISR* or *FSR* is sound for the two event generators but may be unrealistic in *PHYTIA* 8.2.9.

1. Introduction

In 1964, a boson field was theorized by Robert Brout, Franois Englert, Peter Higgs, Gerald Guralnik, C. R. Hagen, and Tom Kibble called Higgs boson H^0 to explain how the elementary particles in the SM acquire their masses via the Higgs mechanism. In order to find a connection between the experiment and the theoretical models, a wave function had to be introduced to describe the hadrons in terms of its constituent partons using what is called the Factorization theory [1], which gives a prediction for the cross sections by factorizing out long-distance effects (which are not perturbatively calculable) from short-distance effects (which are perturbatively calculable) in a systematic fashion. The long distance effects are outlined into functions describing the distributions of partons in hadrons, where these functions are called the parton distribution functions (*PDF*) [2]. It can be measured experimentally by finding common functions that fit all the data concerning many observations [3]. Many analysis approaches have been introduced to study these functions for data from lepton deep inelastic scattering reactions (either through electron scattering or neutrino scattering [4]) by suggesting suitable parametrizations of the structure functions, characterizing the structure of the target nucleon to leptonic probes [5]. Other models are introduced to describe the evolution of the parton density distributions with the square of the exchanged momentum Q^2, taking into account their physical dependence on gluons, quarks, and antiquarks and deducing what is called the $DGLAP$ equation [6].

On July 31, 2012, *ATLAS* [7] and CMS [8] experiments at the Large Hadron Collider (LHC) at CERN announced the discovery of the Higgs boson with a mass of 125.09 ± 0.21 GeV, where the most significant channels for the detection are $H \rightarrow \gamma\gamma$ and $H \rightarrow ZZ \rightarrow 4l$, since all final state particles can be measured very precisely to reconstruct the invariant mass, m_H, with good identification resolution. This discovery confirmed the validity of the standard model for describing elementary particles and their interactions. A clear signal for the Higgs mass in the diphoton channel for the entire data set collected by the CMS experiment in PP collisions during 2011 and 2012 LHC running periods, gives the invariant mass of 124.7 ± 0.34 GeV with the significance of 5.7σ, where ggH process is expected using next-to-leading order (NLO) matrix elements [9].

The parton distribution functions multiply the parton-parton scattering cross section with the summation over all parton families to formulate the total scattering cross section of the hadron-hadron interactions. The parton structure of hadrons in QCD, drives a rapid increase in the proton-proton scattering cross section at high energies [10]. The possibility of enhancement of heavy boson production through hadron-hadron collision has been studied theoretically for $W^{\pm}$, Z bosons at the LHC [11]. The study assumed that the standard model parameters are fixed at standard values and the *PDFs* of *MRST99* and *CTEQ5* started at scale of $Q^2 = 1\,\text{GeV}^2$. The non-negligible variations in the production cross section have

[a] e-mail: mhm.soleiman@gmail.com; mhmsoleiman@cu.edu.eg

been predicted. The initial state radiation (*ISR*) and final state radiation (*FSR*) must be taken into consideration in constructing the Higgs invariant mass. The study of the effects of *ISR*/*FSR* on the reconstruction of heavy lepton (top quark) masses are studied in [12] using *PYTHIA* 8 and *POWHEG* 1.0 Monte Carlo event generators. *ISR* and *FSR* contribute to the systematic uncertainty of the estimation of the number of Higgs events in association with $W^{\pm}$, Z bosons by about 5.2% in the search for Higgs boson production [13]. *ISR*/*FSR* modeling is vital in the reconstruction of Jets and photons at CMS. While the inclusive analysis is useful for constraining $PDFs$, the more exclusive channel could probe details of *ISR* and *FSR* modeling [14]. Search for the Higgs boson in all the hadronic final state using the full CDF data set taking into account the effects of uncertainties from *ISR* and *FSR* is done in Ref. [15, 16].

In this work, we study the Higgs production in the gluon channel $gg \to H^0$ through the diphoton decay ($H^0 \to \gamma\gamma$) using different parton distribution functions ($PDFs$) available already in Monte Carlo event generators (*PYTHIA* 8.2.9 and *SHERPA* 2.1.0). The first PDF is named *CTEQ66* and expresses two pp collisions in excited proton states and the second is identified as $MSTRW2004nlo$ and describes pp collision in normal proton states.

1.1. The discovery channel $H^0 \to \gamma\gamma$

The decay mode $H^0 \to \gamma\gamma$ is very relevant to experimental searches for the Higgs boson. Although having a small branching ratio (about 0.23%) [17], it has a clean signature and lower background than the other decay modes. The energy and momentum of the photons can be measured very precisely and an accurate reconstruction of the invariant mass of the decayed particle could be obtained by a simple calculation algorithm.

In the $H^0 \to \gamma\gamma$ channel a search is performed for a narrow peak over a smoothly falling background in the invariant mass distribution of two high p_T photons. The decay mode can be well identified experimentally but the signal rate is small compared to the backgrounds both coming from two prompt photons (irreducible background) and from those in which one or more of the photons are due to decay products or misidentified particles in jets (reducible background). In order to optimize search sensitivity and separate the various Higgs production modes, *ATLAS* and CMS experiments split events into several mutually exclusive categories. Diphoton events containing a high p_T muon, electron, dijets, or missing energy consistent with the decay of a W or Z boson are tagged in the vector boson to Higgs (VH) production category, and those containing energetic dijets with a large mass and pseudorapidity difference are assigned to the vector boson fusion (VBF) production category and the remaining events (99% of the total) are considered in the gluon fusion production (gF) category. While the VH category is relatively pure, the VBF category has significant contamination from the gluon fusion process. *ATLAS* uses the diphoton transverse momentum orthogonal to the diphoton thrust axis in the transverse plane [17].

Table 1. The x and Q^2 ranges of the grids for two $PDFs$ and the order of the running of α_s at the value of M_Z.

PDF	x range	Q^2 range (GeV^2)	α_s	$\alpha_s(M_Z)$
$MSTW$	10^{-6}–1	1.00–10^9	NLO	0.12018
$CTEQ66$	10^{-8}–1	1.69–10^{10}	NLO	0.1180

1.2. *PDFs* available in monte-carlo event simulations

Most of the MonteCarlo generators offers several $PDFs$ with different orders; leading order (LO), next-to-leading order (NLO) and some next-to-next leading order ($NNLO$), where the higher orders $PDFs$ give more precise results of the cross sections as the partons inside the proton are allowed to have a total momentum larger than the momentum of the proton [18]. The effect of some of the next-to-leading-order corrections permits a large value of the gluon distribution function at small scale parameter x without compromising the quark distributions at large x. Two separate groups obtained full sets of parton distributions by fitting the structure function with data. The MC-adapted $PDFs$ released so far from the $MRST$ group with LO and NLO corrections [18], and from the $CTEQ$ group with $MC1$, $MC2$ and MCS [19], where the most basic features of each PDF are listed in Table 1.

- $MSTR$ – ***PDFs***: These $PDFs$ have been introduced by the $MSTW$ group. Many releases are available based on different fits from several experiments. The group produces $PDFs$ at LO, NLO and $NNLO$.
- $CTEQ$ – ***PDFs***: These $PDFs$ have been introduced by the $CTEQ$ group, adopting an approach, which is very similar to that of $MSTW$. The groups fit the majority of the available data, the recent significant update in widest use is *CTEQ66* [20]. This is slightly older than the $MSTW2008$ sets and is based on Tevatron data. They are available at NLO.

1.3. Diphoton reconstruction in MONTE CARLO analysis

In our analysis, we introduce the algorithms followed to reconstruct the Higgs invariant mass from photon candidates using two different Monte Carlo generators: *PYTHIA* 8.2.9 [21] and *SHERPA* 2.1.0 Monte Carlo generator [22] at $\sqrt{s} = 13$ TeV. The algorithms do not make any hypothesis as to whether the photons originating from the Higgs particle (i.e Assuming $m_H = 125$ GeV according to the latest updates from the LHC) or not, and without any selection criteria. The reconstruction process in any event is based on the collection of all possible photon candidates available in that event without any restrictions on the original parent particle.

The parton distribution functions ($PDFs$) available in both *PYTHIA* 8.2.9 and *SHERPA* 2.1.0 Monte Carlo generators are based on the calculations and fits done by the $MSTW$ group and the $CTEQ$ group. Our selections are $MRSTW2004nlo$ and *CTEQ66*. These $PDFs$ are already available officially through the programmatic library $LHAPDF$ [23].

As shown in Fig. 1 the diphoton decay mode is mediated by loop diagrams containing charged particles. The $topquark$ loop and the $W^{\pm}$ boson loop diagrams dominate the decay amplitude, though they contribute with

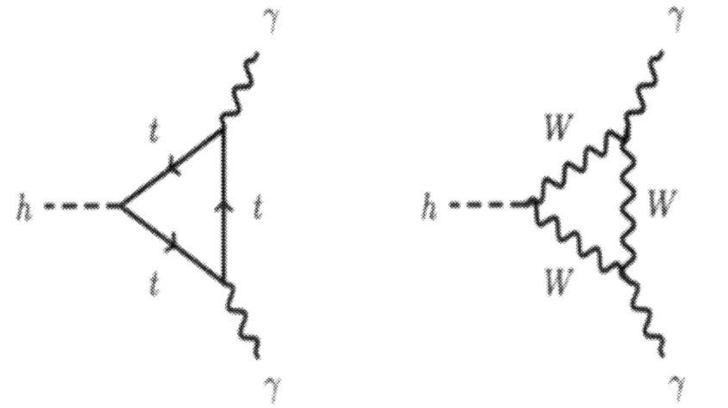

Figure 1. The Feynman diagram for Higgs decays into diphoton: (Left) $topquark$ loop (Right) $W^{\pm}$ boson loop.

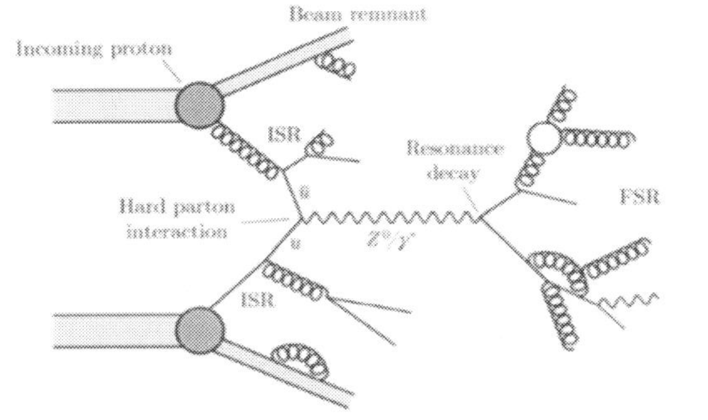

Figure 2. The Feynman diagram for Higgs decays into diphoton: (Left) $topquark$ loop (Right) $W^{\pm}$ boson loop.

opposite signs. In *PYTHIA* 8, the multiple interaction (MI) process $f\bar{f} \rightarrow W^{\pm}W^{\mp}$ (mainly $f = \tau, b, t, c$) have been chosen to be the interference process. This process goes through weak double Boson exchange.

In *SHERPA* 2.1.0, the shown process for Higgs production is the inclusive gluon fusion. The inclusive process is calculated at next-to-leading order (NLO) accuracy, including all interference effects between Higgs-boson production and the SM channel $gg \rightarrow \gamma\gamma$ contributions with all partonic processes are allowed. The corresponding matrix elements are taken from [24] and [25].

1.4. Effect of initial state radiations and final state radiations

The parton showers in Monte Carlo generators are subdivided into two types, initial-state radiation (*ISR*) and final-state radiation (*FSR*). Initial-state radiation (*ISR*) arises because the incoming charged particles can radiate before entering the hard scattering process. A constituent parton from each of the incoming hadrons starts at high energy with low virtuality and evolves to higher space-like virtuality by emitting partons and losing energy. The showering of these partons terminates when they collide to initiate the hard subprocess, and that is why *ISR* is considered to be complicated. It is ordered in increasing virtuality but since the simulation starts with the hard interaction, a corresponding backward-evolutions algorithm was formulated by Schumann and Krauss [26] and adopted in all shower models. The key point in backward evolution is that the evolution probability depends on the amount of partons that could have given rise to the one being evolved. As the collision energy increases, partons at smaller scale parameter x become involved and more particles can be produced in the *ISR* (see Fig. 2).

The Final-state radiation (*FSR*) starts after the hard collision as it takes the products of the collision and describes how they branch into more particles. After the *FSR* the hadronization starts, and the partons produced in the previous stages split up and combine into colorless hadrons than can survive and possibly reach a detector. As shown in Fig. 2, the *ISR* are competing processes and can modify the total energy by further *ISR* branching, while *FSR* redistributes the energy among the outgoing partons. That is why *ISR* can lead to changes in the incoming $PDFs$. In the case of *FSR*, a color dipole can stretch from a radiating parton to a beam remnant, leading to momentum shuffling between the beam and the parton. *ISR* can result in large x values taken from the beams, and leading to flavor changes in the $PDFs$.

2. Event generation and analysis

The aim of the study is to prove the effect of *PDF* on the Higgs production in pp collision at $\sqrt{s} = 13\,\text{TeV}$. The invariant mass plot for the Higgs decay into diphoton has been constructed using two different Monte Carlo event generators *PYTHIA* 8.2.9 and *SHERPA* 2.1.0. The diphoton invariant mass distribution is performed over the range $0 < m_{\gamma\gamma} < 300\,\text{GeV}$. Binned fits are used for the speed of computation, and the bin size chosen is 1200 MeV, which is sufficiently small compared to the mass resolution, so that no information is lost. The description of the expected peak of Higgs invariant mass $m_{\gamma\gamma}$ is executed for the invariant mass region $100 \leq m_{\gamma\gamma} \leq 150\,\text{GeV}$. The suitable fitting functions describing the Higgs peak, the background, and the total invariant mass distribution are given as follows:

$$f_1(x; p[1], p[2], p[3]) = (p[1] + p[2]x)e^{-p[3]x} \tag{1}$$

where $f_1(x; p[1], p[2], p[3])$ is exponential times polynomial function.

$$f_2(x; p[4], p[5], p[6]) = p[4]e^{-e^{-z}-z+1}; \quad z = (x - p[5])/p[6] \tag{2}$$

where $f_2(x; p[4], p[5], p[6])$ is non symmetric extreme peak function.

$$f_{total}(x; p[1]...p[6]) = f_1(x; p[1], p[2], p[3]) + f_2(x; p[4], p[5], p[6]) \tag{3}$$

where $f_{total}(x; p[1], \ldots, p[6])$ is the total fitting function.

These fitting functions are found to be the best functions describing the background and the Higgs peak in the invariant mass plot inside the $m_{\gamma\gamma}$ region from 100 to 150 GeV.

The invariant mass extended technique is:

1. Obtain the invariant mass plot of $m_{\gamma\gamma}$ for a sample of (say 50 events) of *PYTHIA* 8.2.9 and *SHERPA* 2.1.0 events and for two predetermined $PDFs$.
2. Fit the invariant mass spectrum in the region $100 \leq m_{\gamma\gamma} \leq 150\,\text{GeV}$ with the total fitting function f_{total} to obtain the six fitting parameters.

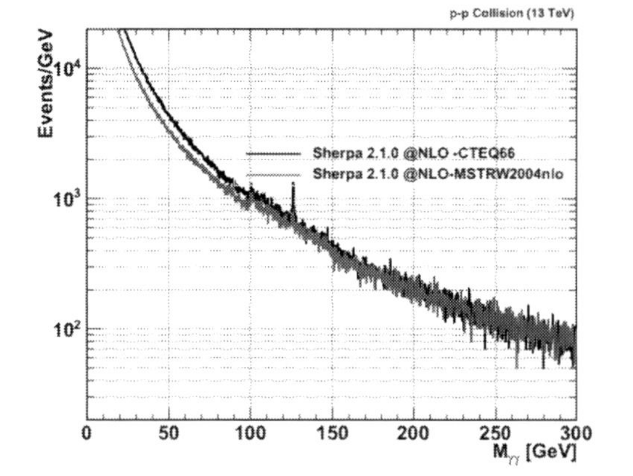

Figure 3. The diphoton invariant mass plot using *SHERPA* 2.0.1 for two *PDFs* (*CTEQ66* and $MRSTW2004nlo$).

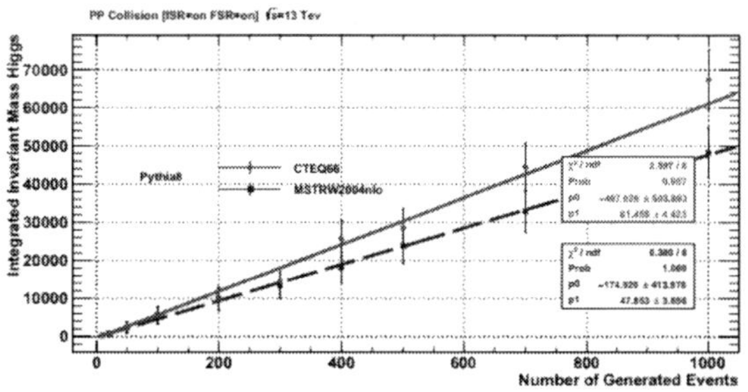

Figure 4. The number of generated events vs. the integrated invariant mass using *PYTHIA* 8.2.9. The error bars are scaled by factor 30.

3. Subtract the integration of the exponential times polynomial fitting function from the invariant mass spectrum in the region $100 \le m_{\gamma\gamma} \le 150\,\text{GeV}$ to obtain a quantity proportional to the generated number of Higgs.
4. Obtain a plot between the integrated number of generated Higgs bosons and the number of generated events.
5. Determine the slope of the line fitting the points on the plot from the previous step.

The slopes of the fitting lines in the obtained graphs are sensitive to the replacement of *PDFs*. We take the error bars to be calculated by the square root of the integrated number of Higgs bosons from the invariant mass plot using the above technique. The calculation, fittings, and plots are produced on $ROOT$ 5.34 environment.

3. Results and discussion

The diphoton invariant mass distributions are performed over the range $0 < m_{\gamma\gamma} < 300\,\text{GeV}$. Binned fits are used for speed of computation, and the bin size chosen to be 1200 MeV, is sufficiently small compared to the mass resolution that no information is lost (see Fig. 3).

Table 2, shows the fit parameters of this combined functions for 1000 events generated by *SHERPA* 2.1.0.

The invariant mass spectrum shown in Fig. 3, has been produced with different number of events (i.e. 1000, 700, 600, 50, 30, 30, 10), with the two different *PDFs*. The integration under the peak signal around 126 GeV has been estimated. Figures 4 and 5 show the relation between the number of generated events vs. the integrated invariant mass for *PYTHIA* 8.2.9 and *SHERPA* 2.1.0 respectively.

The two plots show a linearity between the number of generated events and the number of Higgs events calculated by the integration under the peak signal, where the slope of each straight line shows an improvement in case of *CTEQ66* than $MSTRW2004nlo$. The values for these straight line parameters are shown in Table 3, where $P[0]$ is the y-intercept while $p[1]$ is the slope of each line.

Due to the color-charged partonic initial states, every hard process at hadron colliders is accompanied by initial- and subsequent final-state radiation [8]. The results presented are obtained by keeping the initial state radiation off (see Figs. 6 and 7) and the final state radiation off (see Figs. 8 and 9).

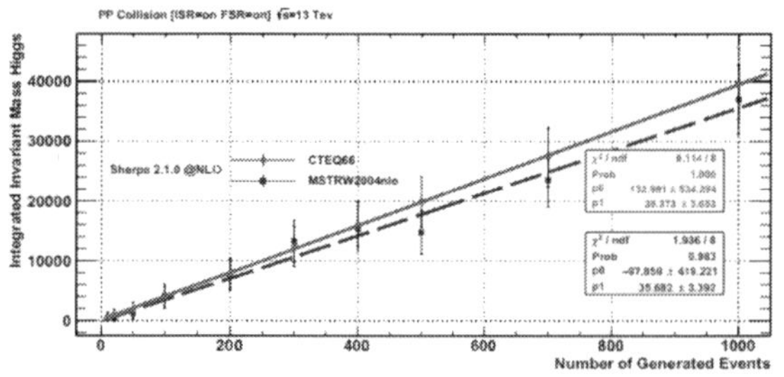

Figure 5. The number of generated events vs. the integrated invariant mass using *SHERPA* 2.1.0. The error bars are scaled by factor 30.

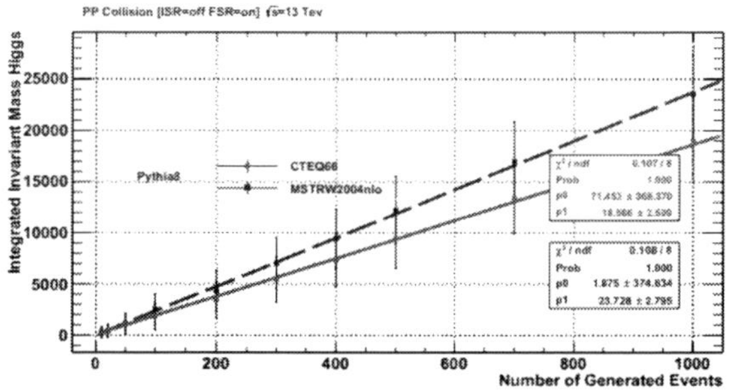

Figure 6. The number of generated events vs. the integrated invariant mass using *PYTHIA* 8.2.9 with *ISR* off. The error bars are scaled by factor 30.

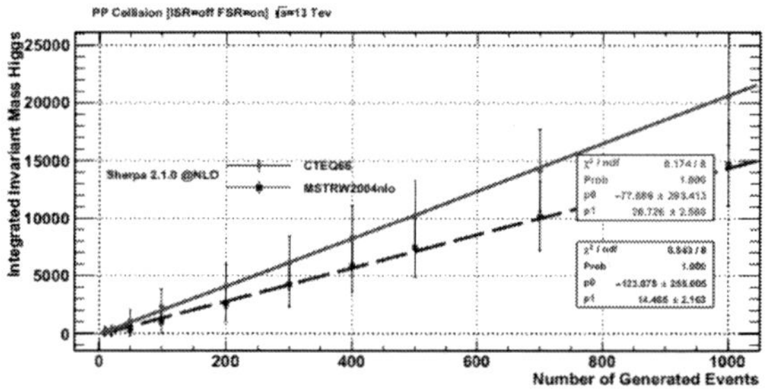

Figure 7. The number of generated events vs. the integrated invariant mass using *SHERPA* 2.1.0 with *ISR* off. The error bars are scaled by factor 30.

Table 2. The fit parameter for the diphoton invariant mass plot using *SHERPA* 2.1.0.

$P[1]$	$P[2]$	$P[3]$	$P[4]$	$P[5]$	$P[6]$
60	126	$4.2E-02$	$1.59993E+03$	$-3.2952E-03$	$1.42408E-02$

Table 3. The fit parameter for the integrated invariant mass for different number of events in the case of *ISR* on and *FSR* on.

PDF	**Monte Carlo generator**	$p[0]$	$p[1]$	$\chi^2/n.d.f$
CTEQ66	*PYTHIA* 8.2.9	-407.029 ± 503.893	61.455 ± 4.423	2.597/8
MSTRW2004nlo	*PYTHIA* 8.2.9	-174.920 ± 413.978	47.853 ± 3.856	0.380/8
CTEQ66	*SHERPA* 2.1.0	132.991 ± 534.294	39.373 ± 3.653	0.114/8
MSTRW2004nlo	*SHERPA* 2.1.0	-97.859 ± 419.221	35.682 ± 3.392	1.936/8

Table 4. The fit parameter for the integrated invariant mass for different number of events in the case of *ISR* off and *FSR* on.

PDF	**Monte Carlo generator**	$p[0]$	$p[1]$	$\chi^2/n.d.f$
CTEQ66	*PYTHIA* 8.2.9	71.453 ± 368.370	18.566 ± 2.509	0.107/8
MSTRW2004nlo	*PYTHIA* 8.2.9	1.875 ± 374.634	23.728 ± 2.795	0.108/8
CTEQ66	*SHERPA* 2.1.0	-77.689 ± 293.413	20.726 ± 2.560	0.174/8
MSTRW2004nlo	*SHERPA* 2.1.0	-123.075 ± 255.005	14.465 ± 2.163	0.543/8

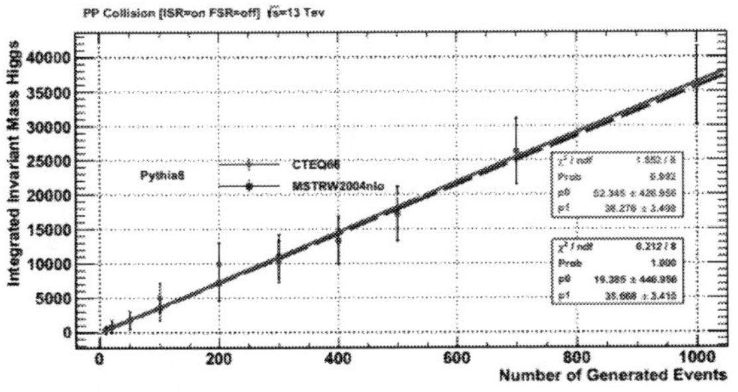

Figure 8. The number of generated events vs. the integrated invariant mass using *PYTHIA* 8.2.9 with *FSR* off. The error bars are scaled by factor 30.

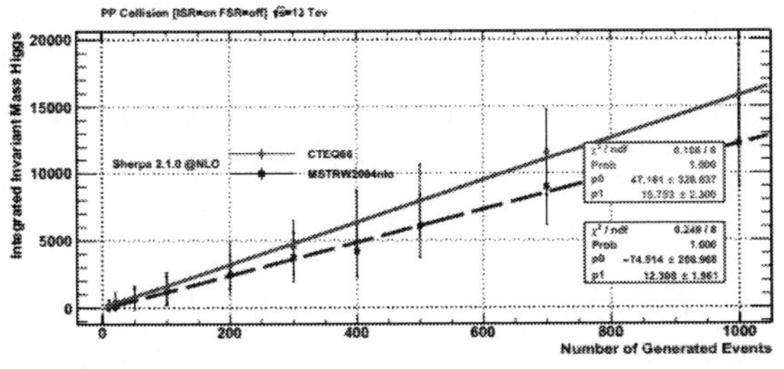

Figure 9. The number of generated events vs. the integrated invariant mass using *SHERPA* 2.1.0 with *FSR* off. The error bars are scaled by factor 30.

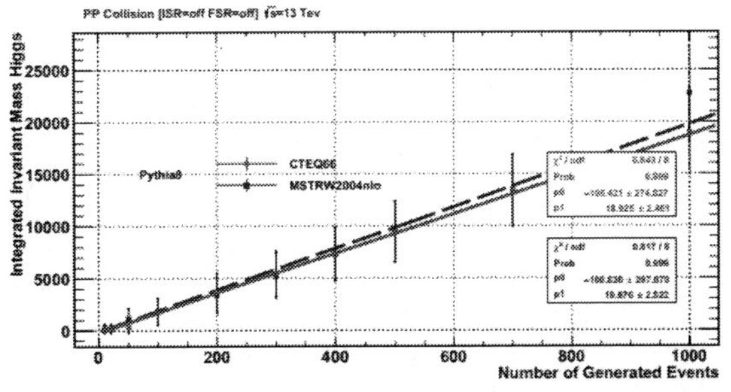

Figure 10. The number of generated events vs. the integrated invariant mass using *PYTHIA* 8.2.9 with *ISR* and *FSR* are off. The error bars are scaled by factor 30.

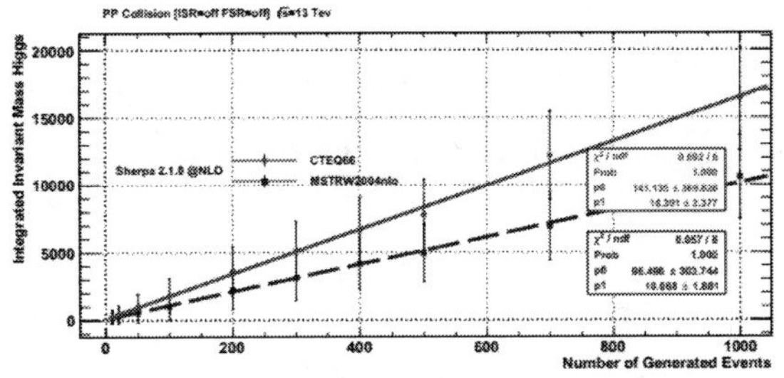

Figure 11. The number of generated events vs. the integrated invariant mass using *SHERPA* 2.1.0 with *ISR* and *FSR* are off. The error bars are scaled by factor 30.

As shown in Figs. 6 and 7, the predictions from *SHERPA* 2.1.0 do not match with the predictions given by *PYTHIA* 8.2.9 (See Table 4). The results obtained from *PYTHIA* 8.2.9 show an inversion in the Higgs integrated production, where *CTEQ66* is expected to be higher than *MSTRW2004nlo*. The inversion appears in Fig. 6 represents that, *PYTHIA* 8.2.9 modifies the *PDFs* before starting the hard process.

Also, in Figs. 8 and 9, the predictions from *SHERPA* 2.1.0 do not match with the predictions given by *PYTHIA* 8.2.9 (See Table 5) when the final state radiation is off. The two graphs are consistent in the ordering of the behaviors for the selected *PDFs*, but *FSR* is expected to be of minor effect on the Higgs production.

Table 5 gives the predictions from *SHERPA* 2.1.0 and *PYTHIA* 8.2.9 where *FSR* is off. In *PYTHIA* 8.2.9, *CTEQ66* and *MSTRW2004nlo* give approximately the same results, while in *SHERPA* an enhancement can be observed with *CTEQ66*.

The ideal case study is where both *ISR* and *FSR* are off. This case expresses the Higgs production enhancement including only the effects of replacing *PDFs*. Figures 10 and 11 show the results of the extended invariant mass technique obtained for *PYTHIA* 8.2.9 and *SHERPA* 2.1.0 where *ISR* and *FSR* are off. Table 6 gives the obtained

Table 5. The fit parameter for the integrated invariant mass for different number of events in the case of *ISR* on and *FSR* off.

PDF	**Monte Carlo generator**	$p[0]$	$p[1]$	$\chi^2/n.d.f$
CTEQ66	*PYTHIA* 8.2.9	752.345 ± 428.956	36.276 ± 3.409	$1.552/8$
MSTRW2004nlo	*PYTHIA* 8.2.9	19.385 ± 446.956	35.668 ± 3.410	$0.212/8$
CTEQ66	*SHERPA* 2.1.0	47.161 ± 328.837	15.753 ± 2.300	$0.108/8$
MSTRW2004nlo	*SHERPA* 2.1.0	-74.514 ± 208.968	12.308 ± 1.961	$0.249/8$

Table 6. The fit parameter for the integrated invariant mass for different number of events in the case of *ISR* off and *FSR* off.

PDF	**Monte Carlo generator**	$p[0]$	$p[1]$	$\chi^2/n.d.f$
CTEQ66	*PYTHIA* 8.2.9	-196.421 ± 274.027	18.925 ± 2.461	$0.843/8$
MSTRW2004nlo	*PYTHIA* 8.2.9	-100.636 ± 297.879	19.876 ± 2.522	$0.817/8$
CTEQ66	*SHERPA* 2.1.0	141.135 ± 369.626	16.391 ± 2.377	$0.092/8$
MSTRW2004nlo	*SHERPA* 2.1.0	96.496 ± 303.744	10.068 ± 1.881	$0.057/8$

fit parameters for this case study. In this last case, the inversion in the behavior is observed for the events generated by *PYTHIA* 8.2.9. The resolution between the two *PDFs* plots for *PYTHIA* 8.2.9 is very low in general, while it is clear in *SHERPA* 2.1.0.

4. Conclusions

In the present work we study the possibility of pre-evolution of parton distribution function with Q^2 to affect the production of Higgs bosons in the frame of SM. The sensitive method of integrated number of Higgs invariant mass is used to study the effect of replacing *PDFs* (*CTEQ66* for enhanced gluon state and $MSTRW2004nlo$ for ground state) in each of *PYTHIA* 8.2.9 and *SHERPA* 2.1.0 event generators.

The conclusions are given in the following:

1. The observation of the enhancement in the Higgs production is possible in the generated events from *PYTHIA* 8.2.9 and *SHERPA* 2.1.0. In the events where *ISR* and *FSR* are allowed, the replacement of *PDFs* shows significant differences in the Higgs production (The samples used in this study contain 1000 events for each event generator and for each *PDFs*).
2. The effect of replacing *PDFs* in *PYTHIA* 8.2.9 is more sound than in *SHERPA* 2.1.0 by a factor of three.
3. The expected behavior of the integrated number of Higgs invariant mass is accepted for all cases of *ISR* and *FSR* in *SHERPA* 2.1.0 while in *PYTHIA* 8.2.9, the absence of *ISR* leads to an inversion of the behavior. This shows that, *PYTHIA* 8.2.9 modifies the gluon distributions before executing the hard process.
4. In the case of removing the effect of *ISR* and *FSR*, there is an inversion in the behavior for the Higgs production of the two *PDFs* in *PYTHIA* 8.2.9 and a suppression in the Higgs production for both *PYTHIA* 8.2.9 and *SHERPA* 2.1.0 event generators with good resolution in *SHERPA* 2.1.0.

Thanks are due to Cairo University, the research committee for partial funding the symposium attendance and encouraging the publishing of this work.

References

[1] D. E. Soper, J. C. Collins, and G. Sterman, Adv. Ser. Direct. High Energy Phys. **5**, 91 (1989)
[2] D. A. Kosower, W. T. Giele, S. A. Keller, `arXiv:hep-ph/0104052v1`, 23 (2011)
[3] M. Dittmar, et al., `arXiv:hep-ph/0511119`, 160 (2005)
[4] Rein Dieter, et al., Annals Phys. **133**, 106 (1981)
[5] Ta. Pie Cheng., *Gauge Theory of Elementary Particle Physics*, (Oxford University Press, ISBN 0-19-851961-3, 1983)
[6] P. Kooijman, and J. Engelen, Progress in Particle and Nuclear Physics **41**, 1–47 (1998)
[7] ATLAS Collaboration (G. Aad, et al.), Phys. Rev. Lett. **716**, 1–29 (2012)
[8] CMS Collaboration (Chatrchyan, Serguei, et al.), Phys. Lett. B **716**, 30–61 (2012), `arXiv:[hep-ex]1207.7235`, `CMS-HIG-12-028`, `CERN-PH-EP-2012-220`
[9] CMS Collaboration (Khachatryan, Vardan, et al.) Eur. Phys. J. C **74**(10), 3076 (2014), `arXiv: [hep-ex] 1407.0558`, `CMS-HIG-13-001`, `CERN-PH-EP-2014-117`
[10] Loyal Durand, and Pi Hong, Phys. Rev. Lett. **58**, 303 (1987)
[11] Martin, D. Alan, et al., Eur. Phys. J. C **14**, 133–145 (2000), `hep-ph/9907231`, `DTP-99-64`, `OUTP-99-31-P`, `RAL-TR-1999-047`
[12] Valentina Tudorache, Mihai Cuciuc, Romanian Reports in Physics **64**(4), 945–956 (2012)
[13] CDF Collaboration, CDF/ANAL/EXOTIC/PUBLIC, 8957 (2007)
[14] M. Voutilainen (for CMS Collaboration), Presented at EDS Blois Conference, Proceedings, 14th Workshop on Elastic and Diffractive Scattering (EDS Blois Workshop) on Frontiers of QCD (December 15–21, 2011)
[15] CDF Collaboration, Phys. lett. B **717**, 1973 (2012)
[16] CDF Collaboration, Springer-Verlagh, JHEP **02**, 004 (2013), `ArXiv:1208.6445`
[17] LHC Higgs Cross Section Working Group (S. Heinemeyer), *Handbook of LHC Higgs Cross Sections: 3. Higgs properties: Report of the LHC Higgs Cross Section Working Group*, CERN, (2013)
[18] A. Sherstnev, and R. S. Thorne, `arXiv:[hep-ph]0807.2132v1`, (2008)
[19] H. L. Lai, J. Huston, et al., JHEP **04**, 089 (2007)

[20] Tomas Kasemets, *Inclusion of parton distribution functions in PYTHIA 8*, Master's thesis, Department of Astronomy and Theoretical Physics, Lund University (February 15, 2010)

[21] S. Mrenna, T. Sjöstrand, and P. Skands, Comput. Phys. Commun. **178**, 852–867 (2008), arXiv:[hep-ph] 0710.3820, CERN-LCGAPP-2007-04, LU-TP-07-28, FERMILAB-PUB-07-512-CD-T

[22] T. Gleisberg, et al., JHEP **0902**, 07 (2009), arXiv:[hep-ph] 0811.4622, FERMILAB-PUB-08-477-T, SLAC-PUB-13420, ZU-TH-17-08, DCPT-08-138, IPPP-08-69, EDINBURGH-2008-30, MCNET-08-14

[23] Andy Buckley, (LHAPDF collaboration), Eur. Phys. J. C**75**, 132 (2015), arXiv:1412.7420 [hep-ph] GLAS-PPE-2014-05, MCNET-14-29, IPPP-14-111, DCPT-14-222

[24] L. J. Dixon and Y. Li, Phys. Rev. Lett **111**, 111802 (2013), arXiv:[hep-ph] 1305.3854

[25] T. Sjöstrand, Phys. Lett. B**157**, 321-325 (1985), FERMILAB-PUB-85-023-T, Torbjörn Sjöstrand, et al.,Comput. Phys. Commun. **135**, 238-259 (2001), arXiv:[hep-ph] 0010017, LU-TP-00-30

[26] S. Schumann, and F. Krauss, JHEP **0803**, 038 (2008), arXiv:[hep-ph] 0709.1027

The study of initial conditions in collisions of light, intermediate and heavy nuclei

A.A. Loctionov[1,a], A.Kh. Arginova[1], A.Sh. Gaitinov[1], and T.N. Kvochkina[2]

[1] Institute of Physics and Technology, Almaty, Kazakhstan
[2] Institute of Nuclear Physics, Almaty, Kazakhstan

Abstract. The system size dependence for multiparticle processes has been recognized in both cosmic ray ("Stratosphere" collaboration) and at accelerator ("EMU" collaboration) experiments. The strong enhancement in multiplicity fluctuations for the most central light-light – (C, O, Ne) + (C/N/O) – collisions has been revealed at JINR-AGS-SPS energies. The sharp difference of light nuclear interactions are interpreted as the sign of intrinsic alpha-clustering in light nuclei.

1. Introduction

Understanding the initial stages of high energy nuclear interactions is one of the key problems for heavy ion physics. The importance of the new insight on the problems were pointed out in the discussions [1,2] just after QM 2014. To understand the problems of particle production in cosmic ray interactions the ever-increasing forward physics programs are considered at the LHC and RHIC colliders [3]. At accelerators the broad system size studies are being performed in the modern NA 61/SHINE experiments at the SPS as part of the program of a systematic search for the onset of deconfinement [4]. The new and far-reaching results have recently been obtained by ALICE in the comparative analysis of freeze-out radii in p-p, p-Pb and Pb-Pb interactions [5]. A better understanding of the initial state characteristics can be obtained from small system interactions. These problems were considered at QM 2014 [6], and at WWND 2016 [7] and more detailed at QM 2015 [8]. The new perspective fixed target experiment BM@N, as the 1-st stage in the heavy ion program at the NICA project [9], will be launched next year. In this work the system size dependence for multiparticle processes is considered at cosmic ray and accelerator experiments.

2. The difference of multiparticle processes generated by cosmic ray protons and light nuclei

The system size dependence for multiparticle processes generated by CR protons and light nuclei has been investigated on the data of the cosmic ray collaboration "Stratosphere" [10]. For this goal dynamical fluctuations of multiplicity and the transverse momentum spectra for γ-quanta were studied. In the well-known intermittency approach [11] the "genuine" dynamical fluctuations over the background of statistical fluctuations were revealed on the basis of the filtering properties of the scaled factorial moments. To overcome the "empty bin" effect we used the third averaging over the start point of the location of the original region D, where N_{step} is the number of small steps ($step/D \ll 1$) of the start position of the original region ,D, in the area of pionization.

$$\langle F_q \rangle = \frac{1}{M} \sum_m \frac{1}{N_{step}} \sum_{step} \frac{1}{N_{evt}} \sum_{evt} \frac{\langle n_m^{[q]} \rangle}{\langle n_m \rangle^q}$$

Three times F_q averaging has increased the value of the q-interval by at least two units in the comparison with the standard approach, i.e. up to rank q = 8. Quantitatively the patterns of the intermittency exponents $\phi(q)$, obtained on the high rank tails, clearly indicate two distinct regions of the q-behaviour for events produced by protons and light nuclei (Fig. 1). The additional

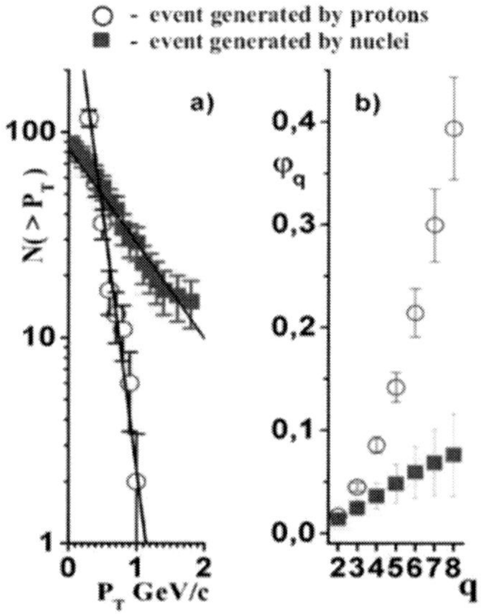

Figure 1. The distribution of transverse momentum and the intermittency exponent ϕ(q) behaviour for events produced by protons and nuclei.

[a] e-mail: loction@sci.kz

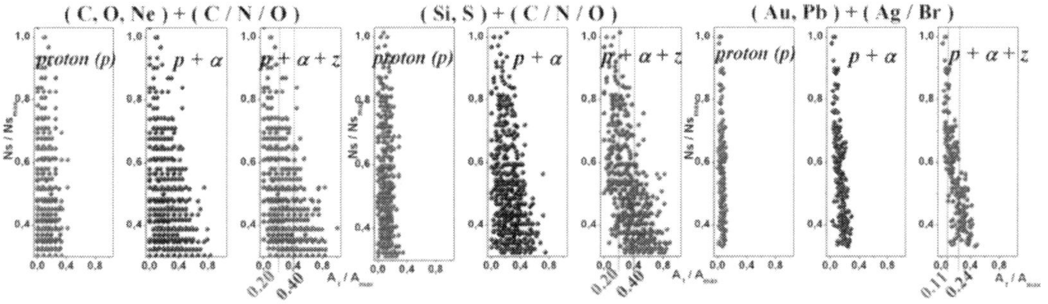

Figure 2. Three variants of the correlation plots $N_s/(N_s)_{max} - A_f/A_{max}$ for collisions of light-light, intermediate-light and heavy-heavy nuclei. In these figures different spectators take into account consecutively in the left (proton-spectators), middle (proton + alpha-particle-spectators) and right frame (proton + alpha + heavy mass fragment-spectators). All correlation panels show centrality intervals for the most central collisions in the top-left, whereas for the most peripheral ones in the bottom-right.

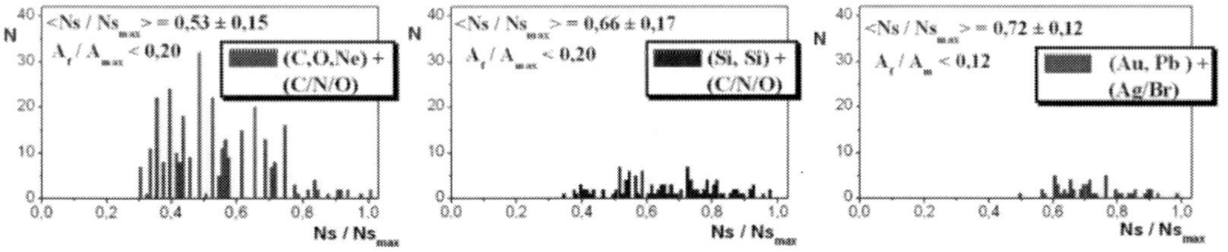

Figure 3. Multiplicity distributions for "central" interactions in light-light – (C, O, Ne) + (C/N/O), intermediate-light – (Si, S) + (C/N/O) and heavy-heavy – (Au, Pb) + (Ag/Br) collisions.

analyses of the transverse momentum spectra of all secondary γ-quanta in the soft region (up to 2 GeV/c) again demonstrated the two component patterns with large differences between proton/nuclei events: the inverse slope of the distribution $T_p \sim 0.2$ GeV/c, is essentially smaller than $T_A \sim 0.8$ GeV/c, (Fig. 1).

It is very important to compare our cosmic ray data with current collider experiments. However, there are some serious problems [12] to perform this analysis.

3. The system size dependence in nuclear collisions at accelerator energies

At accelerator energies the system size dependence for the initial stages of interactions has been investigated in the centrality selected collisions of light and heavy nuclei using the data of JINR-AGS-SPS fixed target emulsion experiments. The basic variables of the analyses are: the normalized multiplicity of the produced charged particles – $N_s/(N_s)_{max}$, and the normalized sum of the mass of all charged, – light and heavy, – fragment-spectators: proton-spectators, alpha-particle-spectators and heavy mass fragment-spectators – A_f/A_{max}. A_{max} corresponds to the mass of the projectile nucleus. $(N_s)_{max}$ corresponds to the maximal multiplicity in the experiment. The transfer from Z to A was performed on the basis of data for high energy cosmic ray nuclei. In order to research the multiparticle processes only events with high multiplicities $N_s > 1/3(N_s)_{max}$ were accepted.

The new solution for centrality determination has been proposed and realized for this goal [13], using correlations between the multiplicity and sum of all fast charged, light and heavy, – fragment-spectators A_f. In this approach the number of participant nucleons for projectile nucleus $N_{part}(projectile) = A_{max}\ A_f$. The model of quantitative Glauber nuclear geometry for the description of (near) right spherical nucleus collisions was not used.

From considering the light-light – (C, O, Ne) + (C/N/O), intermediate-light – (Si, S) + (C/N/O) and heavy-heavy – (Au, Pb) + (Ag/Br) nuclear collisions one can conclude that:

1. nonlinearity becomes stronger with the impact parameter increasing in both the multiparticle production processes and in the processes of projectile fragmentation,
2. there is a clear trend to the fluctuation enhancement with system size decreasing [13], (Fig. 2).

For more detailed comparative analyses of the "genuine production processes" we have to remove or, at least, minimize the uncontrolled fluctuations of the volume of the interaction. The projectile participants are adequately determined with our centrality estimator. In order to get rid of fluctuations in the target we have to consider the most central nuclear interactions [14]. With this goal we compare the multiplicity distributions for the "most central" interactions in light-light, intermediate-light and heavy-heavy collisions. The results have shown (Fig. 3) that the width of the normalized multiplicity $N_s/(N_s)_{max}$ distributions is inversely proportional to the volume of interacting systems. For:

- light-light interactions $\langle N_s/(N_s)_{max}\rangle = 0.53 \pm 0.15$,
- intermediate-light interactions $\langle N_s/(N_s)_{max}\rangle = 0.66 \pm 0.17$ and
- heavy-heavy interactions $\langle N_s/(N_s)_{max}\rangle = 0.72 \pm 0.12$.

It is a very significant trend with the connection of the system size behaviour for different measures of fluctuations [15]. On the other hand, it can provide a new insight into collision geometry for light nuclei, as the fluctuation increasing with system size decreasing can reflect the failure of the spherical nuclear pattern at the transition from heavy-heavy to light-light interactions. The obtained results can be interpreted as a clear sign of intrinsic alpha-clustering in the multiparticle processes for light nuclei.

It should be pointed out in this context, that interesting patterns, reflecting the alpha-cluster nature in the exited fireballs, have recently been computed and discussed for high energy asymmetrical light-heavy nuclear interactions [16].

4. Summary

The studies of proton and nucleus interactions at cosmic ray and accelerator energies have shown a clear system size dependence in the transverse momentum spectra for γ-quanta and multiplicity distributions for charged particles. The sharp difference in the multiplicity spectra for light nuclear interactions is interpreted as a sign of intrinsic alpha-clustering in light nuclei.

The work was supported by the project 0228/PCF-15 of Ministry of Education and Science of Kazakhstan Republic.

References

[1] F. Antinori et al., 24-26 May 2014, arXiv: 1409.2981
[2] U. Heinz et al., 13-15 Sep 2014, arXiv:1501.06477
[3] T. Sako, "RHICf forward experiment", ISVHECRI 2016, 22-27August, Moscow
[4] M. Mackowiak-Pavlowska, Nucl. Phys. A 00(2016) 1-5
[5] ALICE Collaboration, arXiv:1404.1194
[6] B. Schenke, R.Venugopolan, Nucl. Phys A **931** (2014) c1039–c1044
[7] Q. Xu, http://indico.cern.ch/event/435555/contributements/1236137/1815104/wwnd-QiaoXu_final.pptx
[8] C. Loizides, Nuclear Physics A **956** (2016)
[9] B. Kekelidze, http://www.gpad.ac.ru/info/contributions/kekelidze_11.04.2016.pdf
[10] A.Kh. Argynova, N.A. Dobrotin, T.N. Kvochkina et al., Proc. of 27-th Intern. CR Conf. 2001.-P.1477-1480, Hamburg, Copernicus Gesellschaft, 2001
[11] A. Bialas, R. Peschanski, Nucl. Phys B **273** (1986), 703; B **308** (1988), 857
[12] A.Kh. Argynova, N.I. Kochelev, T.N. Kvochkina et al., http://pos.sissa.it/archive/conferences/236/244/_244.pdf
[13] A.Kh. Argynova, A.Sh. Gaitinov, A.A. Loctionov, http://indico.cern.ch/event/219436/session/2/
[14] E. Bratkovskaya, et al., Journ. of Phys.: Conf. Ser. **623** (2015) 012005
[15] V.V.Begun, M. Gazdzicki et al., arXiv:0404056, 2004
[16] W. Broniowski, E. R. Arriola, Phys. Rev. Lett. **112**, 112501 (2014)

Detection of high energy electromagnetic and hadron components of air-shower cores in the new hybrid experiment "Pamir-XXI"

M. Tamada[1,a], N. Inoue[2], A. Misaki[2], and A. Ohsawa[3]

[1] Faculty of Science and Engineering, Kinki University, Higashi-Osaka, Japan
[2] Faculty of Science, Saitama University, Saitama, Japan
[3] Institute for Cosmic Ray Research, University of Tokyo, Kashiwa, Japan

Abstract. In the Chacaltaya hybrid experiment we have shown that the observed characteristics of the events accompanying atmospheric families (a bundle of high energy particles in the air-shower core) can not be well described by current simulations. The atmospheric families detected so far by emulsion chambers (sandwiches of X-ray films and lead plates) are key ingredients in the analysis. But the number of analyzed events with atmospheric family is still small due to the limited size of the experiment. Now a new very large hybrid experiment "PAMIR-XXI" is proposed to be constructed at the Pamirs. The notable feature of the experiment is to construct large hadron calorimeters at the center of air-shower arrays to study the air-shower core in detail. We study the possibility to analyze high energy air-shower cores in the "Pamir-XXI" experiment by using the burst density of scintillation detectors instead of using the family data of emulsion chambers. It is shown that the unusual characteristics of the events observed by the Chacaltaya hybrid experiment can be well seen in the hybrid experiment "PAMIR-XXI" too.

1. Introduction

In the Chacaltaya hybrid experiment[1–3] we have studied in detail high-energy air-shower cores by observing three components in the air-showers, i.e., air-shower size, burst-size (hadron-component) and accompanied family energy (high energy e, γ -component). We have shown that the characteristics of the observed events, in the energy region of primary particles $E_0 = 10^{15} \sim 10^{17}$ eV, can not be fully explained by simply changing the chemical composition of primary particles, though many experimental groups claim that the fraction of heavy primaries increases rapidly beyond the "knee" and that the change of the chemical composition of primary cosmic-rays is an origin of the "knee" and well explain the observed characteristics of the air-showers[4–7].

In the Chacaltaya hybrid experiment, the observed characteristics of each component itself look like they are well described by simulations, but if the events are divided into two groups, with family and without family, the simulations fail to describe the characteristics of both groups of data. Thus the information on the families is a key ingredient for understanding the characteristics of cosmic-ray interactions in the energy range of $10^{15} \sim 10^{17}$ eV. To date atmospheric families are detected by emulsion chambers, but the measurement and analysis of the family data using emulsion chambers is very time-consuming. Furthermore there are some difficulties in connecting the air-shower to the accompanied family, because the latter has no time information. So the number of observed events with atmospheric family is still very small and it is necessary to get at least ten times more family events to make the argument clear. It is preferable to use scintillation counters instead of X-ray films at a suitable depth of the lead plates to detect high-energy atmospheric families in future large-scale hybrid experiment. Here we study the possibility to analyze high energy air-shower cores by using scintillation detectors instead of using emulsion chambers for detecting high energy atmospheric families in the new comprehensive EAS experiment "Pamir-XXI" at the Pamir [8].

The experimental data detected by the Chacaltaya hybrid experiment are converted into "experimental data" expected to be detected in the "PAMIR-XXI" experiment through detailed detector simulations of both experiments and then the characteristics of the "expected experimental data" are compared with those of simulated data of air-shower cores.

2. Results of the hybrid experiment at Mt. Chacaltaya

2.1. Chacaltaya hybrid experiment

The air-shower array covers a circular area within a radius about 50 m by 35 plastic scintillation detectors to measure the lateral distribution of the electron density of the air-showers. In the center of the air-shower array, 32 blocks of emulsion chambers (0.25 m^2 each) are installed. Each block of the emulsion chamber consists of 30 lead plates each of 0.5 cm thickness and 14 sensitive layers of X-ray film which are inserted after every 1 cm lead. Burst detectors with 5 cm thick plastic scintillator are installed

[a] e-mail: msnbtamada@gmail.com

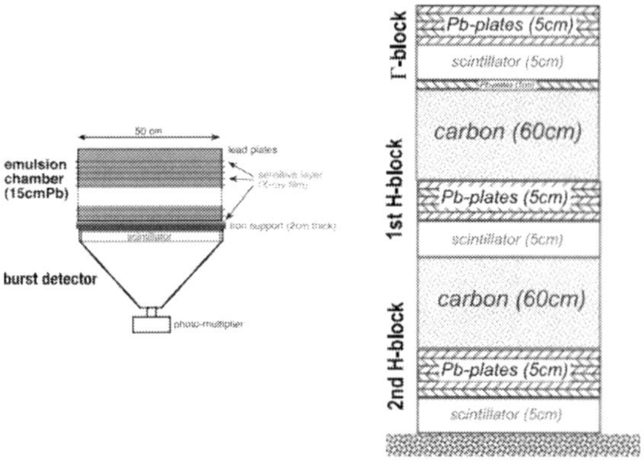

Figure 1. Structure of the hadron-calorimeter of the Chacaltaya hybrid experiment (left) and that assumed for the PAMIR-XXI experiment (right).

underneath the respective blocks of the emulsion chamber (see Fig. 1). A 2 cm thick iron support is inserted between the emulsion chamber and the hadron calorimeter. The total area of the emulsion chambers is 8 m^2. Details of the Chacaltaya hybrid experiment are described in Refs. [1,2].

In the Chacaltaya hybrid experiments, we can obtain the air-shower size, N_e, from the air-shower array data, burst-density (the number of charged particles in $50 \times 50\,\text{cm}^2$), n_b, which are closely connected to the hadron component in the air-shower, from the hadron calorimeter (burst detector) and the energy and geometrical position of individual high energy electromagnetic particles of atmospheric families from the emulsion chamber. Correlations between air-showers and accompanying families were studied by comparing experimental and simulated data.

2.2. Main results

We have analyzed 1,037 events with maximum burst density $n_b^{max} \geq 10^4$ among 32 blocks of burst detectors and air-shower size $N_e \geq 10^6$ in the $\sim 40\,\text{m}^2$ year exposure of hadron calorimeters. Among them 62 events are accompanied with a family of $\Sigma E_\gamma \geq 10$ TeV [3,9].

Figure 2 shows air-shower size, N_e, dependence of the maximum burst-density, n_b^{max}, of the event for all the events with $n_b^{max} \geq 10^4$ and for those which accompany an atmospheric family of $\Sigma E_\gamma \geq 10$ TeV. The experimental data are compared with those of simulated data by CORSIKA [10] using QGSJET01 [11]. The experimental data look like they are well described by the model calculations if we use all the events with $n_b^{max} \geq 10^4$, but if we pick only the events accompanied by a family, the simulations fail to describe the experimental data, i.e., n_b^{max} of the experimental data are systematically larger than that of simulations.

Figure 3 shows a correlation diagram between n_b^{max} and accompanying family energy ΣE_γ. The experimental data are compared with those of simulated data. As naturally expected, n_b^{max} is roughly proportional to the family energy in the simulated data irrespective of the primary particles though n_b^{max} of the events coming from iron-primaries is larger than that from proton-primaries. We can see much of the experimental data is located

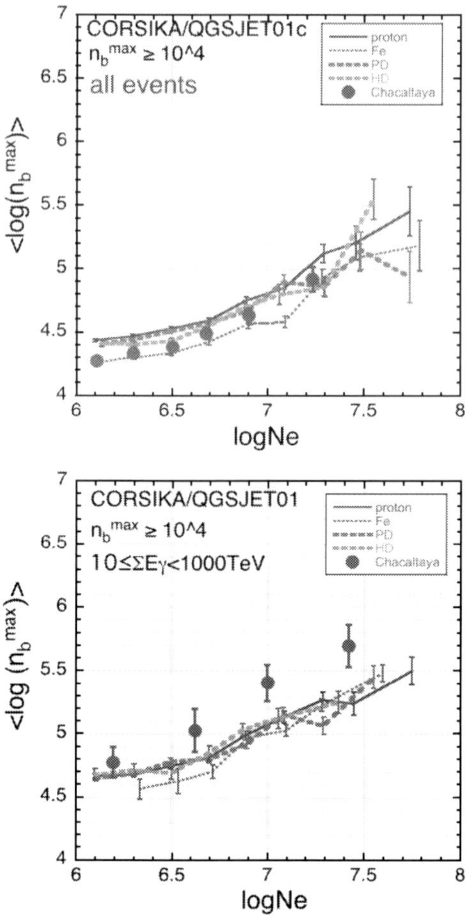

Figure 2. Shower size, N_e, dependence on maximum burst density, n_b^{max}, for the total observed events (upper) and for those accompanying family with energy $\Sigma E_\gamma \geq 10$ TeV (lower) in the Chacaltaya hybrid experiment. Lines are those for the simulated events by CORSIKA/QGSJET01 with different primary particles.

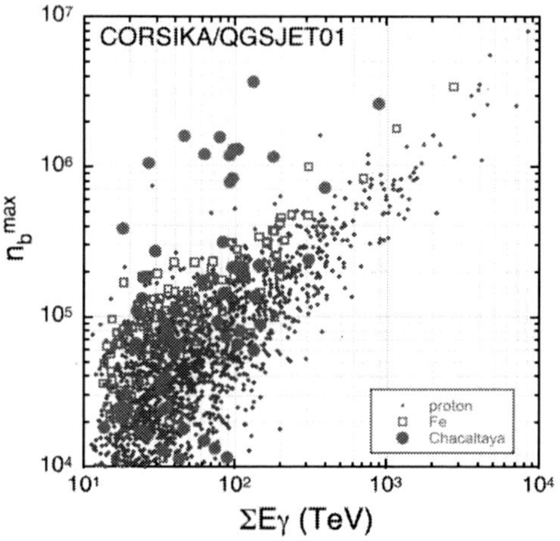

Figure 3. Correlation diagram between n_b^{max} and family energy ΣE_γ of the events with atmospheric families in air-showers of $10^6 \leq N_e < 10^8$. Large closed circles are those for the Chacaltaya data. Small crosses are for simulated data of proton primaries and squares for iron primaries.

far from the simulated events in the diagram, i.e., there exist events which have larger burst-density, $n_b^{max} >\sim 10^6$, though the accompanied family energy is not so large.

3. The project of "PAMIR-XXI" experiment

A new hybrid experiment "PAMIR-XXI" at the Pamir mountains (~4,300 m above sea level) is proposed to study the primary cosmic-ray spectrum and chemical composition in $10^{15} \sim 10^{18}$ eV as well as nuclear interactions at super high energy. The extraordinary characteristics of the experiment is to construct very large lead-carbon calorimeters (192 m^2 with thickness of $\sim 3.5\lambda_{int}$) to study in detail the air-shower core at the center of the dense EAS array ($80 \times 80\,\mathrm{m}^2$) surrounded by sparse EAS array ($1 \times 1\,\mathrm{km}^2$). Details of the experiment are described in Ref. [8].

4. Simulations for the "PAMIR-XXI" experiment

4.1. Air-showers

To generate extensive air-showers we use the CORSIKA simulation code (version 74005) [10] employing the QGSJET model (QGSJET01d) [11] for cosmic-ray nuclear interactions. Primary particles of $E_0 \geq 10^{15}$ eV are sampled respectively from the power law energy spectrum of integral power index −1.8, for pure protons and pure iron nuclei. The thinning energy in the calculations in air-showers is fixed to be 10 GeV. Shower sizes, N_e, at Chacaltaya and the Pamir are calculated by using EGS4. Here we set the low energy cut-off of the particles to be $E_{cut} = 0.3$ GeV for hadrons and muons and $E_{cut} = 0.001$ GeV for electromagnetic particles.

In order to compare with the Chacaltaya hybrid experiment, we assume 32 hadron calorimeters of the "Pamir-type", each of which has an area $50 \times 50\,\mathrm{cm}^2$, are located similar to those of the Chacaltaya hybrid experiment [1,2].

4.2. Calculation of the burst-density in Γ-block and hadron-block of the calorimeter

Here we consider calorimeters with two carbon layers of 60 cm thick (see Fig. 1) for the "Pamir-XXI" experiment[1]. The upper part, Γ-block, of the calorimeter which consists of the emulsion chamber of 6 cm Pb and scintillator of 5 cm thickness, mainly detect electromagnetic cascade showers of the e,γ-component in the air-shower core and the lower part, hadron-blocks each of which consists of an emulsion chamber of 5 cm Pb and scintillator of 5 cm thickness under a 60 cm thick carbon layer mainly detect hadronic showers produced by local nuclear interactions of the hadron component in the lead plates and carbon layer. We use GEANT4 code (version 4.10.0) [12] with the QGSP model for hadronic interactions to calculate the burst-density.

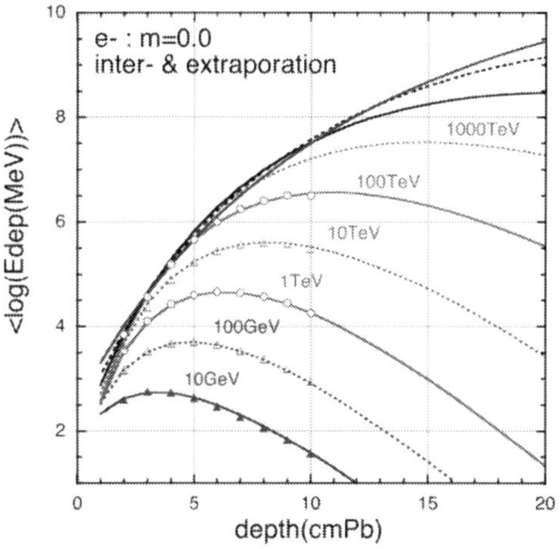

Figure 4. Average cascade transition on energy deposit in the 5 cm thick plastic scintillators for electrons with vertical incidence and various energies. Curves are those of approximated numerical functions based on the NKG function.

We calculate nuclear and electromagnetic cascades in the calorimeter and the energy deposit in the 5 cm thick plastic scintillator, located in the Γ-block and two hadron-blocks, is calculated for incident hadrons (pions, protons, neutrons and kaons) as well as for muons and e, γ with 5 different energies of 10 GeV, 100 GeV, 1 TeV, 10 TeV and 100 TeV, and 3 different zenith tangent of arrival direction, $\tan\theta = 0.0, 0.5, 1.0$.

For the Γ-block, the energy deposit is calculated assuming the 5 cm thick plastic scintillator is located at various depths, under 2 cm Pb, 3 cm Pb, and so on, in order to find the best position of the scintillation counter. Figure 4 shows examples of the average transition on energy deposit in the scintillator for electrons with vertical incidence. For the hadron-block, the energy deposit is calculated assuming the scintillator is located under 5 cm Pb. In Fig. 5 we show examples of distributions of the energy deposit in the scintillator of the 2nd hadron-block for pions with vertical incidence. The distribution of the energy deposit obtained by GEANT4 simulations are approximated by numerical functions. The distributions of the energy deposit obtained by sampling from these numerical functions which approximate GEANT4 results are shown in the figure. The distributions agree well with those obtained by full simulations using GEANT4.

In order to save computing time, the energy deposit is sampled from the average transition curves approximated based on the NKG function (see curves in Fig. 4) for the Γ-block and from the approximated numerical functions for the hadron-block (see curves in Fig. 5) for every particle incident upon the hadron calorimeter. This is converted into a particle number using average energy loss (= 10 MeV) of a single muon in the 5 cm thick plastic scintillator. Finally we get the burst-density, n_b, defined by the number of particles per 0.25 m^2, in each block of 32 hadron calorimeters[2].

[1] There are 3rd hadron-block with carbon layer and 4th hadron-block with thick lead layer under the 2nd hadron-block in the original design of the calorimeter of the "PAMIR-XXI" [8].

[2] High energy e,γ particles give energy deposits in the scintillation counter of the hadron block. The contribution from e,γ particles to the burst-density is less than ~5% in the 2nd

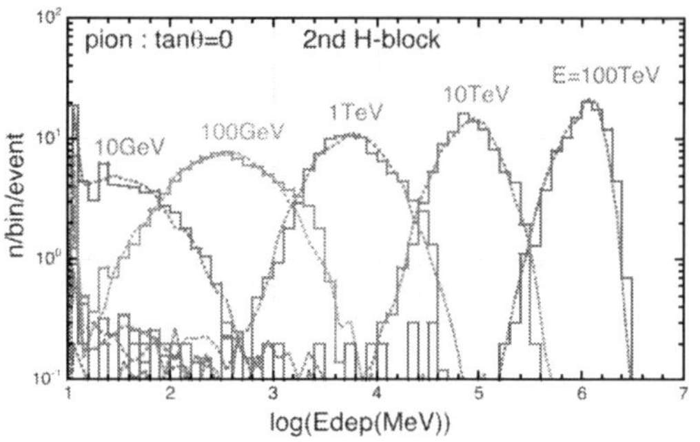

Figure 5. Distribution of the energy deposit, E_{dep}, in the 5 cm thick plastic scintillator located under 5 cm Pb in the 2nd hadron-block of the calorimeter for pions with vertical incidence. Histograms are results obtained by the GEANT4 simulation and dashed-curves are those obtained by sampling from approximated numerical functions.

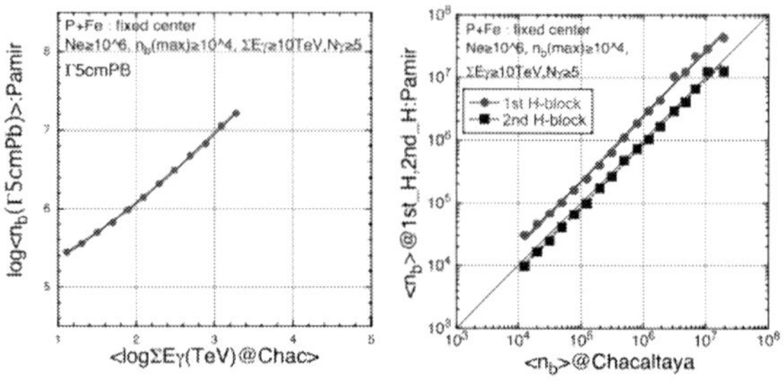

Figure 6. The average relationship between the family energy, ΣE_γ, given by the Chacaltaya hybrid chamber and burst density in the scintillator of the Γ-block, $n_b(\Gamma)$, of the "PAMIR-XXI" experiment (left), and that between the burst density given by the Chacaltaya burst detector, n_b(Chac) and that by the hadron-block, n_b(1st_H), and n_b(2nd_H), of the PAMIR-XXI detectors (right).

5. Expected results in "PAMIR-XXI"

5.1. Estimation of the expected experimental data

Let's consider how the events detected by the Chacaltaya hybrid experiment can be observed by the detectors of "PAMIR-XXI" experiment. Suppose the one and the same simulated air-shower event is detected by both the Chacaltaya hybrid experiment and by the"PAMIR-XXI" experiment. Through the detector simulations for both experiments[3], we can compare the family energy, ΣE_γ, observed by the Chacaltaya emulsion chamber and the burst density in the scintillator of the Γ-block, $n_b(\Gamma)$, of the "PAMIR-XXI" experiment. Then we get an average relationship between ΣE_γ at Chacaltaya and $n_b(\Gamma)$ at the Pamir. The relationship between ΣE_γ and the burst density under 5 cm Pb in the Γ-block, $n_b(\Gamma 5\,\mathrm{cm\ Pb})$, is shown in Fig. 6. In the same way as above, we can get an average relationship between the burst density given by the Chacaltaya burst detector, n_b(Chac) and that by the 1st hadron-block, n_b(1st_H), and also the relationship between the burst density, n_b(Chac), and that of the 2nd hadron-block, n_b(2nd_H), of the "PAMIR-XXI" detectors. The results are shown in Fig. 6. Using these relationships, we can transform the experimental data of the Chacaltaya hybrid experiment into "expected experimental data" for the "PAMIR-XXI" experiment. The fluctuation around the average relationships is rather large. Also some of the events possibly have no family of $\Sigma E_\gamma \geq 10$ TeV at Chacaltaya but give a large burst-density in the Γ-block at the Pamir experiment. These effects are not taken into account in the present analysis.

hadron-block but is 30 ∼ 40%, in average, in the 1st hadron-block.

[3] Details of detector simulations of the Chacaltaya hybrid experiment is described in Ref. [13].

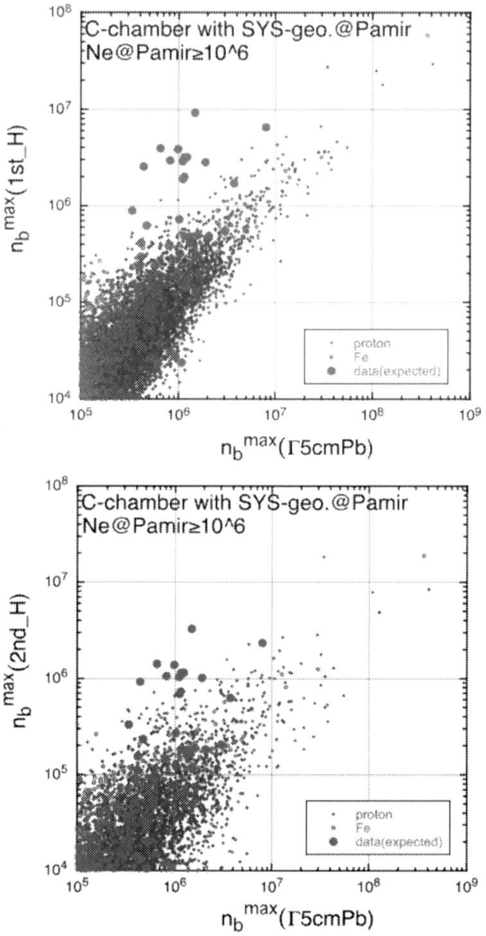

Figure 7. Correlation diagram between the maximum burst density at the 1st hadron-block, n_b^{max}(1st_H), and the burst density at 5 cm Pb of the corresponding Γ-block, $n_b^{max}(\Gamma 5\,\mathrm{cm\ Pb})$, (upper) and that between the maximum burst density at the 2nd hardon-block, n_b^{max}(2nd_H), and $n_b^{max}(\Gamma 5\,\mathrm{cm\ Pb})$ (lower). The large red circles are for the "expected experimental data" of "PAMIR-XXI". Small crosses are for the simulated data of proton primaries and small circles are for iron primaries.

5.2. Comparison of "expected experimental" data with simulations

Figure 7 shows a correlation diagram between the maximum burst density at the 1st hadron-block, n_b^{max}(1st_H),

DOI: 10.1051/epjconf/201714515001

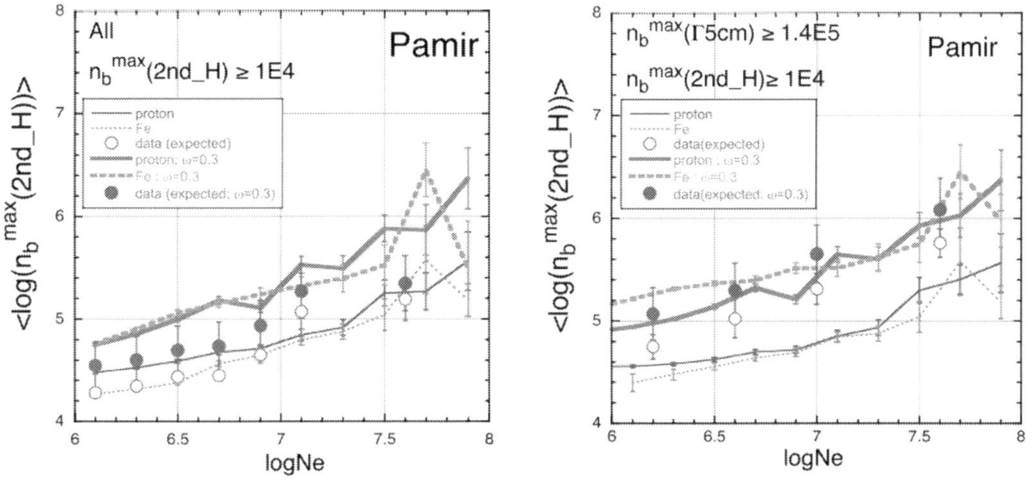

Figure 8. Shower size, N_e, dependence on the maximum burst density of the 2nd hadron-block, n_b^{max}(2nd_H), expected in the "PAMIR-XXI" for all events with $n_b^{max}(\text{2nd\_H}) \geq 10^4$ (left) and for events with a large burst density at the Γ-block, $n_b^{max}(\Gamma\ 5\,\text{cm Pb}) \geq 1.4 \times 10^5$ (right). Open circles are those for "expected experimental data". Thin lines are those for the events simulated by CORSIKA/QGSJET01 with proton (solid) and Fe (dashed) primaries. Closed circles and thick lines are those in the case of introducing fluctuations in hadron-Air interactions in simulations with a parameter of relative fluctuation $\omega = 0.3$.

and the burst density at 5 cm Pb in the corresponding Γ-block, $n_b^{max}(\Gamma 5\,\text{cm Pb})$[4], expected in the "PAMIR-XXI" experiment (see Fig. 3 for comparison with the Chacaltaya data). Also shown is that between the maximum burst density of the 2nd hadron-block, n_b^{max}(2nd_H), and $n_b^{max}(\Gamma 5\,\text{cm Pb})$ of Γ-block. We can see that a considerable number of "expected experimental" events accompany larger burst density in the hadron-block than do simulated events.

Figure 8 shows the shower-size, N_e, dependence on the maximum burst density of the 2nd hadron-block, n_b^{max}(2nd_H), in "PAMIR-XXI" for all events with $n_b^{max}(\text{2nd\_H}) \geq 10^4$ (upper) and for events with large burst density at the Γ-block, $n_b^{max}(\Gamma\ 5\,\text{cm Pb}) \geq 1.4 \times 10^5$ (lower), which correspond to events with family energy $\Sigma E_\gamma \geq 10\,\text{TeV}$ in the Chacaltaya hybrid experiment. If we analyze all the events together, the "expected experimental" data look like they are well described by the simulations, but if we pick up only the events with larger burst density in Γ-block, model calculations fail to describe the "expected experimental" data, i.e., the burst density in the hadron-block in those events is systematically larger than those of simulated events (see Fig. 2 for the comparison with the Chacaltaya data).

5.3. Effect of fluctuations in hadron-Air interaction

The possible existence of fluctuations in the cross-section of hadronic interaction is discussed in references [14, 15] and this fluctuation affects the attenuation of hadronic component in air-showers [15]. In the analysis of the Chacaltaya hybrid experiment, we have shown that the disagreement between experimental and simulation data becomes smaller when cross-section fluctuations are taken into account [3]. Here we again introduce the fluctuation of the cross-section in high-energy hadron-air interactions and study the effect on the correlation between the burst density in the Γ-block and in the hadron-block in the "PAMIR-XXI" experiment. We apply the same procedure used in paper [15] , that is, the cross-section, $\sigma(E)$, of hadron-air is sampled from the uniform distribution in the interval $[1 - \sqrt{3}\omega\sigma_0(E), 1 + \sqrt{3}\omega\sigma_0(E)]$. Here $\sigma_0(E)$ is the mean cross-section at energy, E, and ω is the relative fluctuation of the cross-section. We apply these modifications in the CORSIKA simulation code and calculate air-showers.

The results are shown in Fig. 8 for the case of $\omega = 0.3$. The fluctuation of the cross-section affects the attenuation of the hadron component, then it also affects the "expected experimental" data. In the figure the closed circles are for the "expected experimental" data and thick lines for the simulated data taking into account the fluctuation of the cross-section. As seen in the figure, the "expected experimental" data are well described by model calculations including cross-section fluctuation especially for events with larger burst densities in the Γ-block.

6. Summary and discussions

We have shown that the unusual characteristics observed by the Chacaltaya hybrid experiment, in which the emulsion chamber data is one of the key ingredient of the analysis, can also be well seen in the "PAMIR-XXI" hybrid experiment, in which scintillator counters are used instead of emulsion chambers. If the "PAMIR-XXI" experiment is carried out as planned, statistics of the events increase ~10 times larger than those detected in the Chacaltaya hybrid experiment after 2~3 years exposure. If the unusual characteristics, i.e., disagreement between experimental data and simulations, still remain in the "PAMIR-XXI" experiment, we must seriously consider the reason to explain it. As one of the possible reasons, we considered

[4] The results using the burst-density $n_b^{max}(\Gamma 4\,\text{cm Pb})$ or $n_b^{max}(\Gamma 6\,\text{cm Pb})$ are not much different from those using $n_b^{max}(\Gamma 5\,\text{cm Pb})$.

the effect of fluctuations in the cross-section of hadron-air interactions. It is found that the discrepancy between experimental and simulated data becomes smaller when the cross-section fluctuation is taken into account. Some changes of the characteristics of particle production may also cause the discrepancy between experimental and simulated data, e.g., recent results of the LHCf experiment show a more abundant neutron production rate to photon production than in the current model prediction in the very forward region in p-p collisions at $\sqrt{s} = 7\,\text{TeV}$ [16].

References

[1] N.Kawasumi et al., Phys. Rev. D **53** (1996) 3534
[2] C. Aguirre et al., Phys. Rev. D **62** (2000) 032003
[3] M. Tamada et al., Proc. 17th Int. Symp. on Very High Energy Cosmic Ray Interactions, Berlin (2012), EPJ Web of Conference **52** (2013) 07006
[4] Tibet ASγ Collaboration (M.Amenomori et al.), Phys. Rev. D **62** (2000) 112002-1, 072007-3
[5] Tibet ASγ Collaboration (M.Amenomori et al.), Phys. Lett. B **632** (2006) 58
[6] J. Blumer, R. Engel and J. R. Horandel, Prog. in Part. Nucl. Phys. **63** (2009) 58
[7] W.D. Apel et.al., Advances in Space Res. **53** (2014) 1456
[8] A.S. Borisov, K. Muminov, V. Galkin and V. Puchkov, Proc. 17th Int. Symp. on Very High Energy Cosmic Ray Interactions, Berlin (2012), EPJ Web of Conferences **52** (2013) 11002, Slide of the talk: `https://indico.desy.de/conferenceOtherViews.py?view=standard&confId=4594`
[9] J. Kempa, B. Pattison, E. Glaysz-Dziadus, L.W.Jones, R. Mukhamedshin, M. Tamada and Z. Wlodarczyk, Cent. Eur. J. Phys. **10** (2012) 723
[10] D. Heck, J. Knapp, J.N. Capdevielle, G. Schatz and T. Thouw, Report **FZKA 6019** (1998), Fortshungzentrum Karlsruhe; available from `https://www.ikp.kit.edu/corsika/70.php`
[11] N.N. Kalmykov, S.S. Ostapchenko and A.I. Pavlov Nucl. Phys. B, Proc. Suppl. **52B** (1997) 17
[12] S. Agostinlli et al. Geant4 collaboration, Nucl. Inst. and Meth. A **506** (2003) 250
[13] T. Yamasaki and M. Tamada, Proceedings of 32th ICRC, Beijing (2011) Vol.1, p. 216
[14] B. Blättel et al., Phys. Rev. D **47** (1993) 2761
[15] G. Wilk and Z. Wlodarczyk, Phys. Rev. D **50** (1994) 2318
[16] O. Adriani et al. (LHCf Collaboration), Phys. Lett. B **750** (2015) 360

A summary of recent results from the GRAPES-3 experiment

S.K. Gupta[1,2,a]

[1] Tata Institute of Fundamental Research, Homi Bhabha Road, Mumbai 400005, India
[2] The GRAPES-3 Experiment, Cosmic Ray Laboratory, Raj Bhavan, Ooty 643001, India (**On behalf of the GRAPES-3 Collaboration**)

Abstract. The GRAPES-3 experiment is a combination of a high density extensive air shower (EAS) array of nearly 400 plastic scintillator detectors, and a large 560 m^2 area tracking muon telescope with an energy threshold E_μ >1 GeV. GRAPES-3 has been operating continuously in Ooty, India since 2000. By accurately correcting for the effects of atmospheric pressure and temperature, the muon telescope provides a high precision directional survey of the galactic cosmic ray (GCR) intensity. This telescope has been used to observe the acceleration of muons during thunderstorm events. The recent discovery of a transient weakening of the Earth's magnetic shield through the detection of a GCR burst was the highlight of the GRAPES-3 results. We have an ongoing major expansion activity to further enhance the capability of the GRAPES-3 muon telescope by doubling its area.

1. Introduction

The GRAPES-3 (**G**amma **R**ay **A**stronomy at **P**eV **E**nergie**S**-phase **3**) is a high density extensive air shower (EAS) array which was designed for a precision study of the acceleration of charged particles in nature by using a number of phenomena as explained below. The GRAPES-3 experiment had a modest beginning with its first component comprising 256 plastic scintillator detectors (each 1 m^2 in area) in Ooty, India in 2000 as shown schematically in Fig. 1. The array has now been expanded to about 400 detectors which are deployed in a dense hexagonal configuration with an inter-detector distance of 8 m, in Ooty (2200 m altitude, 11.4°N latitude, 76.7°E longitude). About 25% of these detectors are at present instrumented with two fast photomultiplier tubes (PMTs) for the purpose of extending their dynamic range to $\sim 10^4$ particles m^{-2}. The scintillators, the signal processing electronics, and the data acquisition systems were all designed, and fabricated in-house to minimize costs, and to optimize performance [1].

The second component of the GRAPES-3 array contains a large area (560 m^2) tracking muon telescope [2] to measure the cosmic ray induced muon flux with high statistical precision. This high statistical precision has permitted us to probe the acceleration of muons in the atmospheric electric fields set up during major thunderstorm activity. These thunderstorm activities are also known to result in the generation of terrestrial flashes of γ-rays [3–5]. The muon telescope has also provided new information with very high sensitivity on solar flares, coronal mass ejections (CMEs), and Forbush decrease (Fd) events [6,7] besides serving as a sensitive probe of the interplanetary magnetic field, and the turbulence caused in it due to the propagation of CMEs [8,9].

It had been shown earlier by us that the primary cosmic ray spectrum, and composition can be extracted using the muon multiplicity, and the shower size distributions of the detected EAS [10]. By using the GRAPES-3 data, this technique was employed to determine the composition of the primary cosmic rays from 30 TeV to 1 PeV, along with a sizeable overlap with the direct measurements from balloon- and satellite-borne detectors. However, this interpretation of EAS data involved the extrapolation of hadronic interaction properties to energies beyond those available from accelerators. This necessitates modeling of high energy interactions assuming certain reasonable assumptions leading to several models of hadronic interactions which depend on the nature of underlying assumptions [11–16]. Therefore, the interpretation of the EAS data in terms of composition and energy spectrum gets coupled to the choice of the hadronic interaction models used. In GRAPES-3 data, we used the overlap in the energy with the direct measurements to identify the hadronic interaction models that had yielded composition and energy spectra consistent with the direct measurements in the overlap energy region from 300 TeV to 1 PeV [17].

2. Experimental details

To operate the GRAPES-3 EAS array at the lowest possible threshold energy, a simple 3-line coincidence of detectors is used to generate the Level-0 trigger, that acts as the fast GATE and START for the charge to digital (ADCs) and time to digital converters (TDCs), respectively. Since this trigger selects very small local showers and also larger showers whose cores land far from the area covered by the array, we require at least 12 out of the inner 127 detectors should have triggered their discriminators within 1 μs of the Level-0 trigger. This Level-1 trigger with an observed EAS rate of 15 Hz is used to record the charge (ADC) and the arrival time (TDC) of the pulses from each detector [1].

[a] e-mail: `gupta.crl@gmail.com`

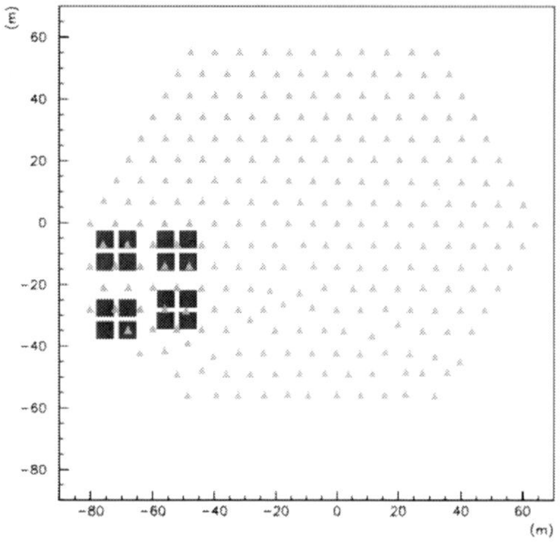

Figure 1. A schematic layout of the GRAPES-3 EAS array with 256 detectors (triangles). Sixteen squares represent a 35 m^2 area muon telescope with $E_\mu \geq 1$ GeV.

With the expansion of the array the trigger rate has now increased to 35 Hz. The pulse charge is later converted into equivalent number of minimum-ionizing particles (MIPS) using the most probable charge for a single measured MIP. The high sensitivity of scintillator detectors allows the atmospheric radon decay products to be routinely detected during every episode of rain in Ooty.

In Fig. 1 each square shown in the lower left-side, represents a 4-layer muon telescope module with an energy threshold of 1 GeV for vertical muons. Each layer contains 58 proportional counters (PRCs), each $600 \times 10 \times 10\,\mathrm{cm}^3$ in size. Thus, the sixteen modules that constitute a total area of 560 m^2 of the muon telescope are shown as sixteen squares in Fig. 1. The energy threshold of 1 GeV was achieved by placing a concrete absorber of thickness $550\,\mathrm{g\,cm}^{-2}$ above the bottom PRC layer. Thus, each module with a sensitive area 35 m^2 consists of 232 PRCs arranged in four layers of 58 PRCs each, with alternate layers placed in orthogonal directions. Two successive layers of PRCs are separated by a 15 cm thick concrete layer. This four-layer configuration allows a 3-D reconstruction of each muon track to an accuracy of $\sim 4°$. The sixteen telescope modules detect ≥ 1 GeV muons at a rate of $4 \times 10^9\,\mathrm{d}^{-1}$, allowing tiny variations due to atmospheric, solar, and other systematic effects to be precisely measured. Using the reconstructed muon direction the muons are segmented into nine solid-angle bins, labeled Vertical (V), North (N), South (S), East (E), West (W), South-East (SE), North-East (NE), South-West (SW), North-West (NW) covering a solid angle of 2.3 sr.

2.1. Muon module efficiency corrections

The daily rate of muons measured for a period of 100 days (1 Jan.–10 Apr. 2006) by modules 4 and 5 is shown in Fig. 2a, Fig. 2b, respectively. The rates are nearly the same, and display similar variation caused by the atmospheric and solar effects. However, the ratio of their rates shown in Fig. 2c displays a significantly smaller (<0.4%), yet a very systematic variation, which represents the ratio of the efficiencies of modules 4 and 5. This was caused by a gradual change in the detection efficiency of the PRCs which was modeled by using a fourth order polynomial in time. Since the GRAPES-3 muon telescope consists of sixteen modules, a total of 120 independent combinations of efficiency ratios could be formed using the corresponding fourth order polynomials (for sixteen modules) for daily data. The 120 series of 365 ratios for the year 2006 represent the ratios of their unique efficiency profiles of the corresponding modules. Since the magnitude of this variation is rather small (<1.0%), this problem could be readily solved after performing a Taylor series expansion, and by using the singular value decomposition (SVD) technique which yielded the five coefficients corresponding to each fourth order polynomial describing the time-dependent efficiencies of each of the sixteen modules. The details of this technique are described elsewhere [18]. The efficiency ratios for modules 4 and 5 obtained by SVD technique yielded a fit which described the data very well as is seen from Fig. 2c. When the muon rates for modules 4, and 5 were corrected by using the corresponding efficiency variations, the resultant ratios shown in Fig. 2d almost equal unity as is expected for an ideal correction.

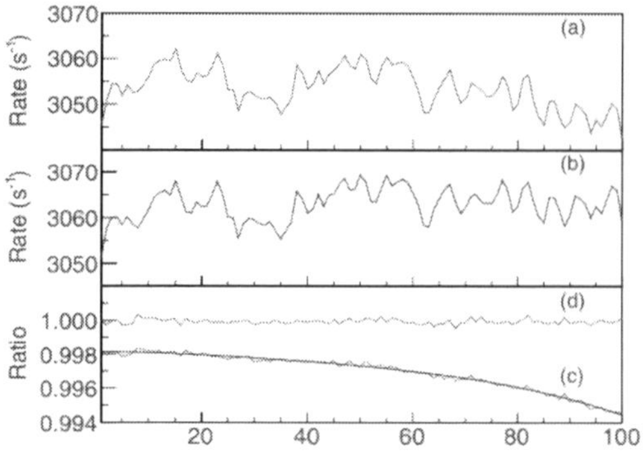

Figure 2. Muon rates for the telescope modules, (a) 4, (b) 5 for 100 days (1 Jan.–10 Apr. 2006. Ratio of the muon rates, (c) before efficiency correction, and a fourth order polynomial fit, (d) after the efficiency correction.

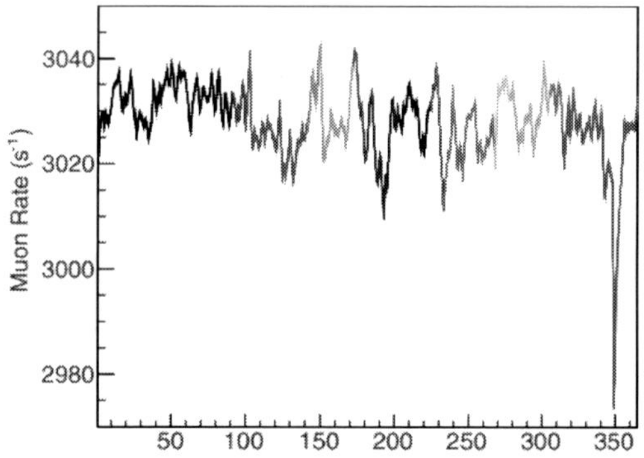

Figure 3. Muon rates for 364.1 d for sixteen telescope modules from 1 Jan.–30 Dec. 2006 after efficiency correction modeled by a fourth order polynomial.

In Fig. 3 the daily rates of all modules are shown for the whole year (2006) after correcting for efficiency variations using the fourth order polynomial correction. Except for an Fd event in December 2006, the muon rates displayed

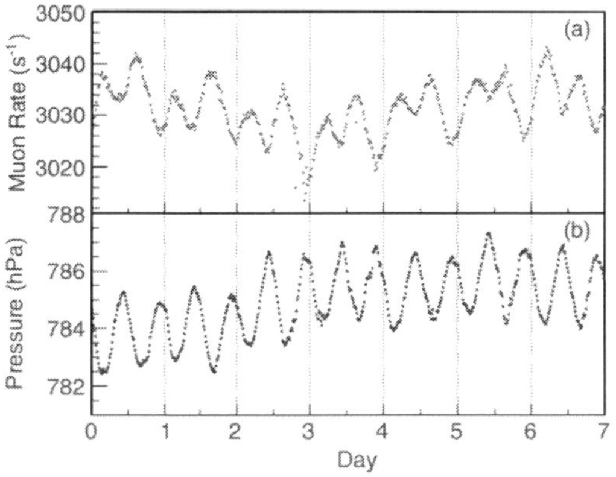

Figure 4. Variation of, (a) mean muon rate, (b) pressure (hPa) for the week 1 – 7 March 2006 [20].

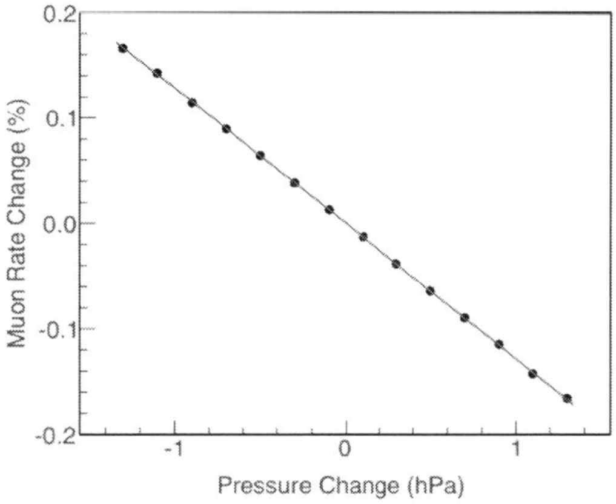

Figure 5. Dependence of muon rate on atmospheric pressure relative by IFFT [20].

a narrow variation of only 20 (3010–3040) counts for the rest of the year. Also, the sixteen rates are nearly indistinguishable from each other. The rms spread in the daily muon rates was only $0.32\,s^{-1}$ which was comparable to the statistical error of $0.26\,s^{-1}$. Thus, the efficiency correction was fairly effective, and the rms spread of $0.32\,s^{-1}$ in a rate of $3000\,s^{-1}$ implies a precision of 0.01% in the measured relative efficiencies.

2.2. Pressure correction

The muons detected by GRAPES-3 are produced by GCRs of energy ~100 GeV, and are modulated by the solar activity, and therefore, display a 24 h periodic variation due to the rotation of the Earth. However, at near-equatorial locations such as the GRAPES-3 [2] also show a 12 h periodic variation in addition to the 24 h variation of solar origin. The 12 h period is induced by the heating of the upper atmosphere which results in a non-linear response of the atmosphere with a dominant 12 h oscillation in pressure at near-equatorial locations (amplitude ~1 hPa) [19].

Since variations in atmospheric pressure change the total mass of the air column above the telescope, it causes a corresponding change in the rate of muons which was observed as an anti-correlation of the muon rate with pressure. The measured muon rate variation is a combined effect of the change induced by the pressure and the solar activity. A correction for the pressure variation can be implemented by using the pressure coefficient β which can be obtained from the measured muon rate. The pressure at Ooty displays a dominant 12 h periodic variation as shown in Fig. 4a for an interval of one week. This variation produces a synchronous response in the muon rate due to an inherent anti-correlation of these two variables. This anti-correlation was studied by obtaining the fast Fourier transform (FFT) of the two sets of data. Since the use of FFT requires an uninterrupted time series of 2^N intervals, where N is a positive integer, the muon rates were converted into an uninterrupted stream of $2^{17} = 131072$, 4 min intervals covering 364.01 d starting from 1 January 2006. The deviation in muon rate from its mean was calculated in percent, and an FFT performed on muon rate as well as on the pressure data. By using a narrow band filter centered at a period of 12 h the FFT power from the muon and pressure data were extracted. These power spectra permitted the segregation of the 12 h barometric effect in the muon data from all other phenomena. Next by taking the inverse fast Fourier transform (IFFT) of the two filtered data sets, the muon, and pressure data were converted back into time domain [20].

In Fig. 5 the IFFT muon data are plotted against corresponding pressure data. The error on each data point is invisible because of its small value. A linear fit shown by the solid line yielded a pressure coefficient β of $-0.128\%\ hPa^{-1}$ [20].

2.3. Temperature correction

Atmospheric temperature undergoes annual variations due to the Earth's orbital movement around the Sun. This causes an annual variation in the atmospheric density, resulting in a similar modulation in the measured muon rate. This annual modulation was first reported over six decades ago. A phenomenological framework was also sketched by defining the concept of an effective temperature 'T_{eff}'. T_{eff} can be calculated by using the thermal profile of the upper atmosphere. The significance of T_{eff} is that a real atmosphere behaves exactly as an isothermal atmosphere at a fixed temperature of T_{eff}. However, the calculation of T_{eff} requires as input the attenuation length, λ, of the hadron in the atmosphere [21]. Due to its near-equatorial location (11.4°N) the upper atmospheric temperature at Ooty shows a relatively small annual variation (~1 K). The GeV muons recorded by GRAPES-3 telescope are produced by the GCRs of average energy < 100 GeV. The mesons produced by the GCRs in the atmosphere have energies of 10s of GeV which decay in the upper atmosphere into GeV muons which are detected by the GRAPES-3 telescope. The energy loss of the muons along a direction θ is $\leq 2\sec(\theta)$ which is comparable to their energy. This results in the decay of a significant fraction of the muons before being detected by the GRAPES-3. When the temperature of the atmosphere decreases, the fraction of the muons decaying also decreases due to reduced length traveled due to thermal contraction of the atmosphere. This results in muons displaying an anti-correlation with the temperature which was measured by using the GRAPES-3 data collected over six years (2005–2010). The anti-correlation

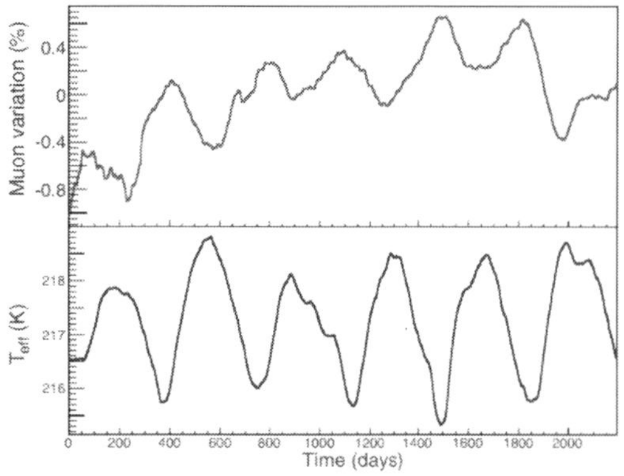

Figure 6. Variation of (a) muon rate, (b) T_{eff} (K) [22].

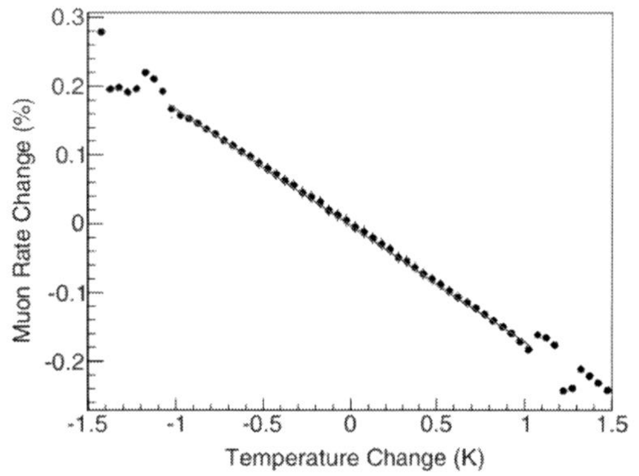

Figure 7. Variation in muon intensity as a function of ΔT relative to respective mean values [22].

between the annual variation in T_{eff} and the muon rate was used to measure the temperature coefficient α_T by the fast Fourier transform (FFT) technique described in [20].

During the six year interval (2005–2010), relatively faster variations caused by solar phenomena such as the Forbush decreases, the 27 d solar rotation etc. were removed by using an effective low-pass filter created by the running average over a window of 60 d. The resultant percent deviation in the muon rate, and T_{eff} are shown in Fig. 6a, and Fig. 6b, respectively. By using the FFT technique as outlined above in §2.2, and a narrow-band filter centered at a frequency of 1 per year the dependence of the muon rate on T_{eff} was obtained as shown in Fig. 7.

For an attenuation length $\lambda = 120\,\mathrm{g\,cm^{-2}}$ the value of the temperature coefficient, α_T, was found to be $(-0.169 \pm\ 0.001)\%\,\mathrm{K^{-1}}$. It was also found that the magnitude of α_T scaled with the assumed value of λ in the range $\lambda = 80$–$180\,\mathrm{g\,cm^{-2}}$. However, the actual magnitude of the correction to the muon rate was independent of the λ value used, indicating the intrinsic robustness of the result [22].

3. GRAPES-3 results

3.1. Atmospheric acceleration

It was observed that thunderstorm activity is accompanied by sudden changes in the rate of detected muons over a time scale of ~30 minutes. However, this phenomena is not observed in the muon rate in all directions, but is confined to a limited solid-angle. In Fig. 8 color coded reduction in the rate of muons in percent is shown along 49 directions constituting the GRAPES-3 field of view (FOV) of 2.3 sr. This event was observed during a massive thunderstorm on 30 September 2015 which lasted for 27 minutes. A significant deficit may be seen in NE, and SW directions.

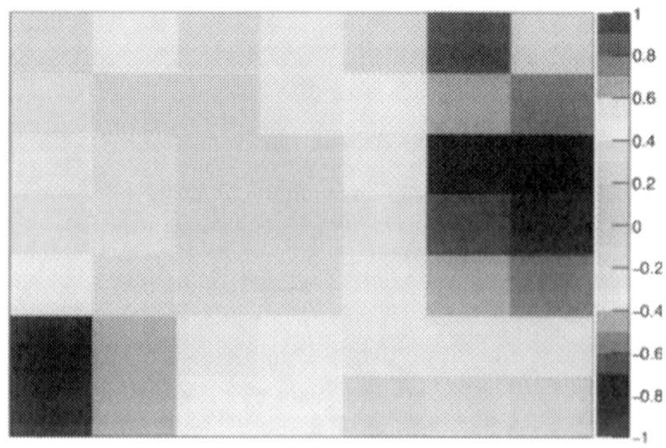

Figure 8. 7×7 panels display variation of muon rate along 49 directions on 30 September 2015 from 17:53 to 18:20 h (IST) during a thunderstorm. A significant deficit is seen in NE, and SW directions for 27 minutes.

During this event the reduction in the muon rate was accompanied by the presence of a large electric field that was recorded by the GRAPES-3 electric field monitors (EFMs) on ground. The polarity of the electric field indicated that the ground was at a large positive electric potential relative to the upper atmosphere. The variation in the rate of muons during 17:53 to 18:20 h Indian Standard Time (IST) on 30 September 2015 in the GRAPES-3 FOV is shown in Fig. 8 where a large reduction in the rate of muons is clearly seen in the NE and SW directions.

The GRAPES-3 EFMs are distributed widely, with a typical distance of ~3 km between them. During this event all four EFMs showed positive electric potential on ground which was concurrent with the deficit seen in the muon rate. It is well-known that the ratio of the μ^+ to μ^- is >1.0 at energies of a few GeV which is known to gradually increase at higher energies [23–25].

This phenomena can be understood when a cloud in the upper atmosphere is negatively charged. This would result in deceleration of μ^+ and acceleration of μ^-. Since their ratio is greater than one the rate of muons, which is a sum of both polarities, will show a net reduction. The GRAPES-3 muon telescope does not operate with magnetic field so it is not possible to distinguish the two types of muons. At present we are carrying out large-scale Monte Carlo simulations by using CORSIKA [26] to generate the muons in the atmosphere which are subsequently detected by the muon telescope. We had also modified CORSIKA to include the presence of electric potential, which leads to the acceleration of these muons. In Fig. 9 the percentage change in the muon rate is shown as a function of applied electric potential [27]. It is to be noted that when a positive electric potential relative to the ground is applied, the muon rate initially decreases slowly up to 1 GV, and thereafter more rapidly. However, when a negative potential is applied then the muon rate initially increases slowly up to 0.8 GV, and thereafter decreases

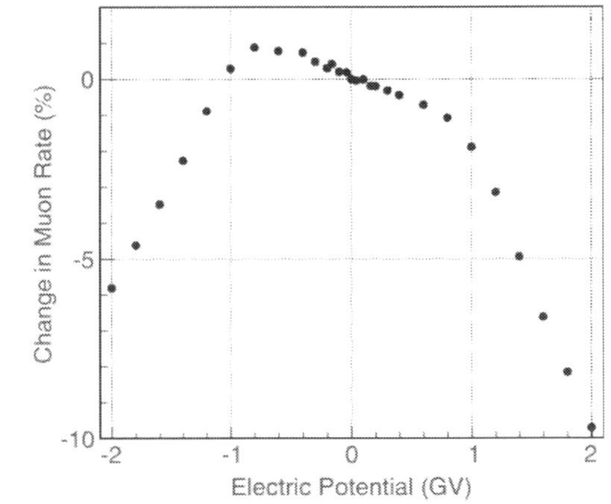

Figure 9. Variations in muon rate due to an applied electric potential in the upper atmosphere for intervals of 100 MV during a thunderstorm.

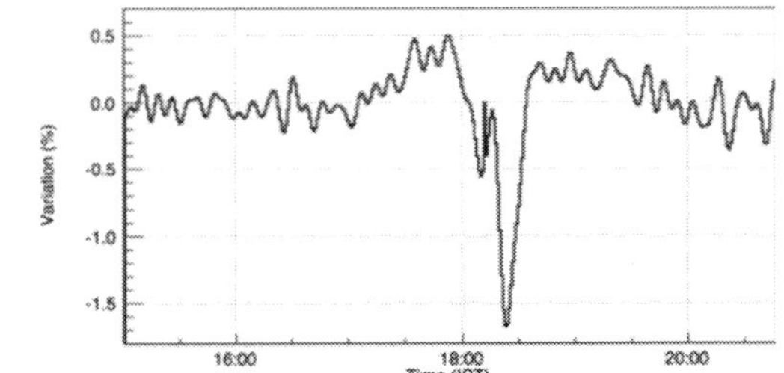

Figure 10. Percent variations in muon rate observed during the thunderstorm on 30 September 2015 during interval 15–21 h IST. An electric potential of 1.2 GV is required to produce observed maximum decrease of 1.7%.

rapidly. The muon rates for the nine NE, and five SW directions were combined, and are shown in Fig. 10 as a function of time for the 30 September 2015 thunderstorm event. The observed maximum decrease of 1.7% in the rate of muons shown in Fig. 10 requires an electric potential of nearly 1.2 GV to be present during this thunderstorm.

3.2. Detection of a GCR burst

The GCRs producing the low-energy (GeV) muons detected by the GRAPES-3 telescope are affected by the geomagnetic field, and the Monte Carlo simulations can accurately account for it. Depending on the location on Earth, the geomagnetic field causes a deflection of the incoming GCR in the surrounding space. Since the GCRs are overwhelmingly positive particles, the minimum energy per unit charge, also called the cutoff rigidity has a lower value from the west when compared to the east direction. This geomagnetic deflection has a significant effect on the muons detected by GRAPES-3 due to the location of Ooty close to the geomagnetic equator. To account for the deflection of the GCRs in the Earth's magnetic field, a model of the geomagnetic field, namely, "International Geomagnetic Reference Field 11th generation model (IGRF-2011)" [28] was employed.

The characteristics of the muons detected, such as the energy spectrum and the angular distribution are governed by the interaction properties of GCRs producing

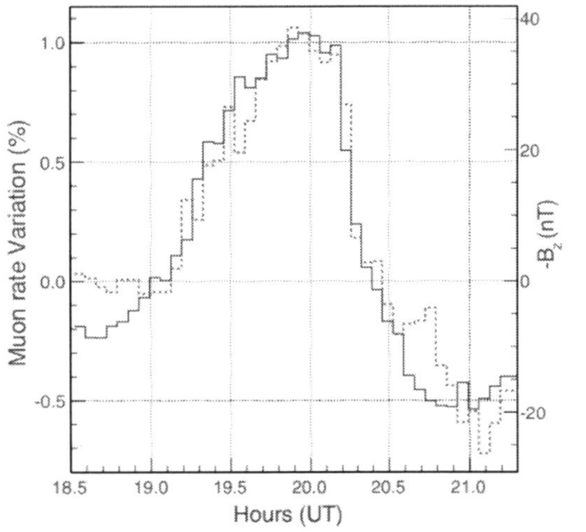

Figure 11. Muon-rate (solid-line), and -B_z (broken-line) on 22 June 2015, correlation coefficient $R = -0.94$ from [30].

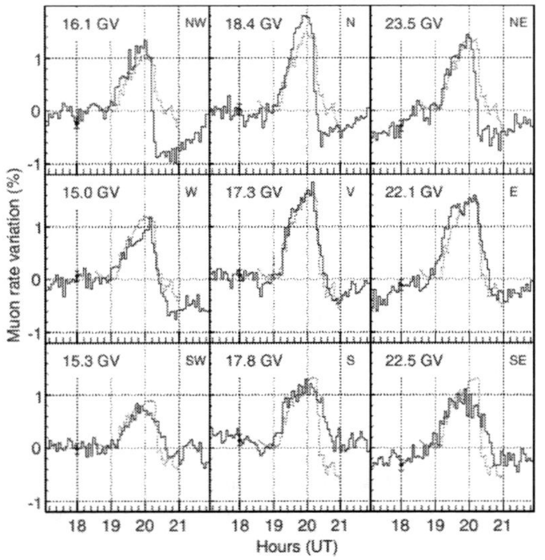

Figure 12. Muon-rate variations in nine directions observed by GRAPES-3 on 22 June 2015 shown by solid-line. Simulation results normalized to data by scaling the IMF 17 times shown by the broken-line. Cut-off rigidities (GV), and error bars are shown for each direction [30].

them. The angular distribution of the muons is a consequence of several effects which include attenuation in the atmospheric, the hadronic cross sections of parent particles, their production heights, particle multiplicities, and their pseudo rapidity distributions. These interaction properties are parametrized in a hadronic interaction generator based on specific assumptions which include extrapolation of various properties from accelerator based measurements.

Simulations were carried out using CORSIKA [26], and FLUKA [13] to describe low-energy, and SIBYLL [15] for high-energy hadronic interactions for protons as GCRs with a differential spectral slope of −2.7 near the top of the atmosphere. The trajectory of a proton (p) was simulated using back-tracing methods by launching

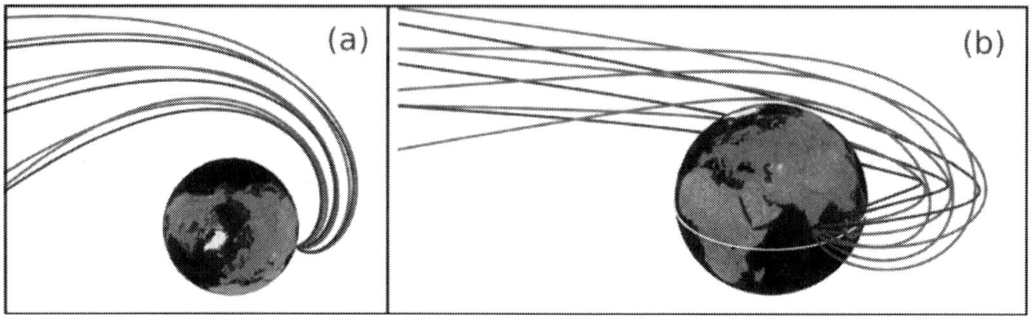

Figure 13. Trajectories of the GCRs near cutoff rigidity viewed from, (a) north pole, (b) equator. NW, N, NE shown in blue, W, V, E in green, SW, S, SE in red [30].

an anti-proton ($\bar{p}$) of the same energy in the same direction into space from the GRAPES-3 muon telescope. If the $\bar{p}$ trajectory escapes into space without striking the Earth, then a 'p' of the same energy would be able to reach the Earth from outer space from the same direction. The minimum energy of $\bar{p}$ escaping into space defines the cutoff rigidity of a p in that direction by the use of this "back tracing" method. Protons above the geomagnetic cutoff rigidity were subjected to the deflection in the geomagnetic field by using a cosmic ray trajectory program [29]. For this purpose, 10^9 GCR protons above 10 GeV were simulated. The muons above the 1 GeV threshold were tracked down to the location of GRAPES-3. Thereafter, each muon incident at a zenith angle θ was tested if it crossed the detection threshold of $\sec(\theta)$ GeV. The muons satisfying this cut were registered as detected. The resultant hits in the four layers of PRCs were histogrammed into nine bins, namely NE, E, SE, N, V, S, NW, W, SW as was done for real data. 10^9 incident GCR protons resulted in detection of 2×10^8 simulated muons.

The GRAPES-3 had relatively high cutoff rigidities (15–24 GV) along the nine directions covering a solid angle of 2.3 sr. On 22 June 2015 18:40 UT a coronal mass ejection (CME) reached the Earth and triggered a severe G4-class geomagnetic storm, hereafter 'storm'. Starting at 19:00 UT, the muon telescope registered a 2 h duration burst of GCRs that showed a strong 94% correlation with a spike in the interplanetary magnetic field (IMF) of 40 nT as shown in Fig. 11 [30].

Simulations carried out to explain the occurrence of this burst to be due to the change in the cutoff rigidities caused by the arrival of the IMF showed that a large (17×) compression of the IMF to 680 nT which led to the lowering of the cutoff rigidities could generate this burst as shown in Fig. 12. Here, the value of 680 nT represented a short-term change in the geomagnetic field around the Earth, averaged over seven times its volume [30].

Due to the lowering of the cutoff rigidities the GCRs were deflected from the day-side of the Earth to the night-side by 210° in longitude as shown in Fig. 13 from polar, and equatorial views. This offered a natural explanation of the night-time detection of the burst by the GRAPES-3 muon telescope. The simultaneous occurrence of the burst in all directions pointed to its origin close to the Earth possibly within the magnetosphere. Detection of this burst indicated a transient weakening of the Earth's magnetic shield. A study of these events may provide a better understanding of future solar induced storms that could have devastating impact on the modern technological infrastructure on the Earth [30].

As part of the expansion of the muon telescope, a total of 3780 large proportional counters (PRCs) each 600 × 10 × 10 cm^3 in size, are required. Out of these 2500 PRCs (65%) have already been fabricated and tested. About 20% of the total PRCs have been deployed in the field. The ongoing expansion activity for enhancing the sensitivity of the GRAPES-3 muon telescope by doubling its area is expected to be completed in a years time.

4. Summary

Measurements using the high efficiency tracking muon telescope in the GRAPES-3 EAS array have been used to observe the acceleration of muons in large-scale electric fields during powerful thunderstorms. By exploiting the high statistics GRAPES-3 muon data, the relative efficiencies of the sixteen modules of the GRAPES-3 muon telescope could be accurately determined. By the application of the fast Fourier transform technique corrections to the muon rate due to variations in the atmospheric pressure, and the atmospheric temperature could be accurately measured in a fairly robust manner. These corrections paved the way for the discovery of transient weakening of the Earth's magnetic shield during a severe G4-class geomagnetic storm. The ongoing major expansion activity for enhancing the sensitivity of the GRAPES-3 muon telescope by doubling its area is expected to be completed in a years time.

We thank the members of the GRAPES-3 collaboration who were instrumental in obtaining these exciting results discussed here.

References

[1] S.K. Gupta et al., Nucl. Instr. Meth. A **540**, 311 (2005)
[2] Y. Hayashi et al., Nucl. Instr. Meth. A **545**, 643 (2005)
[3] D.M. Smith, Science **307**, 1085 (2005)
[4] M.S. Briggs, AIP Conf. Proc. **1366**, 47 (2011)
[5] `https://science.nasa.gov/science-news/science-at-nasa/2014/31dec_tgfs`
[6] T. Nonaka et al., Phys. Rev. D **74**, 052003 (2006)
[7] P. Subramanian et al., Astron. Astrophys. **494**, 1107 (2009)

[8] K.P. Arunbabu et al., Astron. Astrophys. **555**, A139 (2013)
[9] K.P. Arunbabu et al., Astron. Astrophys. **580**, A41 (2015)
[10] S.K. Gupta et al., Phys. Rev D **68**, 052005 (2003)
[11] H. Fesefeldt, Report PITHA-85/02 RWTH Aachen (1985)
[12] H. Petersen, et al., Phys. Rev. C **78**, 044901 (2008)
[13] `http://www.fluka.org`
[14] S. Ostapchenko, Phys. Rev. D **83**, 014018 (2011)
[15] R.S. Fletcher et al., Phys. Rev. D **50**, 5710 (1994)
[16] T. Pierog and K. Werner, Nucl. Phys. B Proc. Suppl. **196**, 102 (2009)
[17] H. Tanaka et al., J. Phys. G: Nucl. Part. Phys. **39**, 025201 (2012)
[18] P.K. Mohanty, Ph.D. Thesis Unpublished (2014)
[19] R.S. Lindzen and S. Chapman, Space Sci. Rev. **10**, 3 (1969)
[20] P.K. Mohanty et al., Astropart. Phys. **79**, 23 (2016)
[21] P. Barrett et al., Rev. Mod. Phys. **24**, 133 (1952)
[22] K.P. Arunbabu et al., Under preparation (2017)
[23] S. Haino et al., Phys. Lett. B **594**, 35 (2004)
[24] P. Achard et al., Phys. Lett. B **598**, 15 (2004)
[25] P. Adamson et al., Phys. Rev. D **76**, 052003 (2007)
[26] D. Heck et al., CORSIKA: A Monte Carlo code to simulate EAS, **6019** (1998)
[27] B. Hariharan et al., Proc. of Science PoS (ICRC2015) 448 (2015)
[28] C.C. Finlay et al., Geophys. J. Int. **183**, 1216 (2010)
[29] D.F. Smart et al., Space Sci. Rev. **93**, 305 (2000)
[30] P.K. Mohanty et al., Phys. Rev. Lett. **117**, 171101 (2016)

Exotic geophysical phenomena observed in an environmental neutron flux study using EAS PRISMA detectors

Victor Alekseenko[1], Anastasia Bagrova[1,2], Shuwang Cui[3,a], Yayun He[3], Bingbing Li[3], Xinhua Ma[4], Egor Pozdnyakov[5], Oleg Shchegolev[1], Yuri Stenkin[1,2], and Vladimir Stepanov[1]

[1] Institute for Nuclear Research, Russian Academy of Sciences, Russia
[2] Nuclear Research National University "Moscow Engineering Physics Institute", Russia
[3] The College of Physics and Information Engineering, Hebei Normal University, China
[4] Key Laboratory of Particle Astrophysics, Institute of High Energy Physics, Chinese Academy of Sciences, China
[5] CJSC NPF "Luminofor", Russia

Abstract. Some exotic geophysical events are observed by a global net of electron-neutron detectors (en-detectors) developed in the framework of the PRISMA EAS project. Our en-detectors running both on the Earth's surface and underground are continuously measuring the environmental thermal neutron flux. Thermal neutrons are in equilibrium with media and are therefore sensitive to many geophysical phenomena, which are exotic for people studying ultra high-energy cosmic rays or carrying out low background experiments deep underground.

1. Introduction

The en-detectors were developed for Extensive Air Shower (EAS) experiments in the framework of the PRISMA project [1]. The main issue of the project is the possibility of the EAS array to record secondary thermal neutrons accompanying EAS. The project will be realized in a few years in collaboration with the LHAASO project as the PRISMA-LHAASO array. One of the PRISMA-LHAASO aims is to monitor the environmental thermal neutron flux to study variations of possible neutron background change. We have found some interesting phenomena in the neutron flux having geophysical or Sun-Earth origin. They are: rain and snow effect, Forbush effect, daily temperature variations, seasonal and moon tidal waves, barometric pumping effect, etc.

2. En-detector

A novel type of thermal neutron scintillation detector *(en-detector)* based on a well-known specialized solid granulated alloy, ^{6}LiF+ZnS (Ag), is used to detect thermal and epithermal neutrons. The thickness of the scintillator layer is $\sim 30\,\mathrm{mg/cm^2}$. The sensitive area of the scintillator surface is $0.75\,\mathrm{m^2}$. A schematic view of the en-detector and its performances can be found elsewhere [2]. The detection efficiency for thermal neutrons was found to be $\sim 20\%$. The efficiency displays a $\sim 1/v$ dependence on the neutron velocity v, typical for thin detectors. Due to the thin scintillator layer, relativistic charged particles, such as electrons or muons, produce very small signals below the threshold. This allows us to use the detector in a scaler mode to study the variations in the thermal neutron flux. Due to the existence of several scintillating time constants in ZnS(Ag) (from 40 ns to several μs), the pulse shape produced by heavy and slowly moving particles (alpha and triton emission after neutron capture on a ^{6}Li target) is longer than that produced by relativistic charged particles or by PMT's noise. Therefore, we exclude background signals applying a pulse shape selection method.

A distinctive feature of our data acquisition process is that all pulses from PMT are integrated with a time constant of 5 μs and then digitized by a FADC in a 100 μs time gate with a step of 50 ns. In other words, on-line pulse shape analysis and selection are used to count real neutron pulses and to reject, though still counting, the noise pulses. The technical details and some preliminary results can be found elsewhere [2]. The en-detector used in this work is located at a depth of $25\,\mathrm{hg/cm^2}$ in a shaft of muon detectors of the EAS-MSU array (Moscow State University) [3]. The underground room where the detector has been running for $\sim$ 4 years has no forced ventilation and has no protection against soil gas penetration. Experimental data were accumulated as a time series with 1 min steps. The detector has very stable long-term behavior, which is controlled through daily energy deposit spectra.

3. Overview of the observed "exotic" phenomena in neutron flux

As mentioned above, the en-detectors were developed for Extensive Air Shower studies in the framework of the PRISMA project. Studying environmental neutron backgrounds at different sites, we have observed some "exotic" effects. We have to emphasize here that the observed phenomena are exotic only for high-energy physicists but are well-known for geophysicists.

[a] e-mail: cui_shuwang@163.com

DOI: 10.1051/epjconf/201714515003

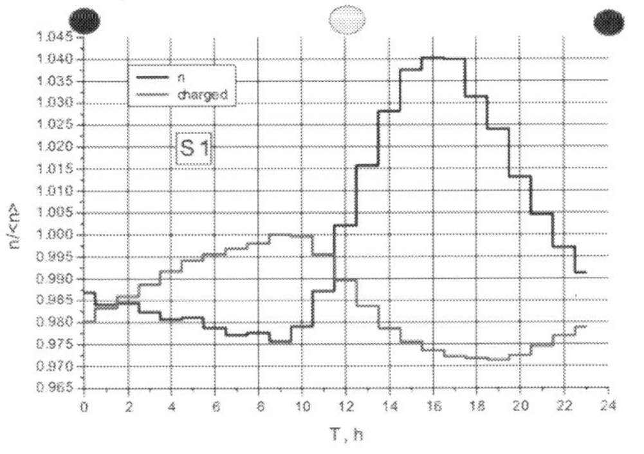

Figure 1. Diurnal wave S1 observed in PRISMA-YBJ en-detectors in local astronomical time. Circles shows sun culminations. No corrections were applied.

3.1. Diurnal wave

Four of our en-detectors have been running continuously in Tibet (Yangbajing, 4300 m a.s.l.) since August 2013 as the PRISMA-YBJ array [2] being a prototype of PRISMA. During this study a long data time series was accumulated and analyzed. Using the superimposed epoch method, the diurnal wave, S1, was obtained for both pulses selected as "neutron" and "charged" shown in Fig. 1. As one can see, the two types of events anti-correlate. This could be explained as follows:

i) Neutron wave is very close to the near surface air temperature wave because thermal neutrons are in equilibrium with the media and their mean velocity, $<v>$, is proportional to sqrt(T) while the flux, F, is $\sim C<v>$, where C is the neutron concentration in air;
ii) Pulses selected as "charged" are partially connected with decays of radon in air [3] and so they have not a temperature origin, and mostly follow the wave of radon exhalation from soil and its concentration in air.

3.2. Moon tidal waves

Using the superimposed epoch analysis for longer periods, we found some known lunar tidal periods, namely the lunar month (T = 29.5 d) and its 4th harmonic which is shown in Fig. 2. As one can see the wave is clearly observed with an amplitude of about 0.5%. In addition, a delay of ~1 day between the syzygies and the flux maximum in neutrons can be seen, while that for "charged" is ~1.5 day. This phenomenon can be explained by the influence of Moon tidal forces on the Earth's crust resulting in a change in soil radon diffusion and thus neutron production in the upper soil level.

3.3. Barometric pumping effect

A 2-day delayed anti-correlation was observed in the underground en-detector situated at a depth of 25 m of water equivalent in the mine of the EAS-MSU array [4]. Cosmic ray hadrons do not penetrate to this depth and thus the neutron barometric coefficient should be close to zero. But, when the air pressure monotonously decreases for many days, all underground gases including radon

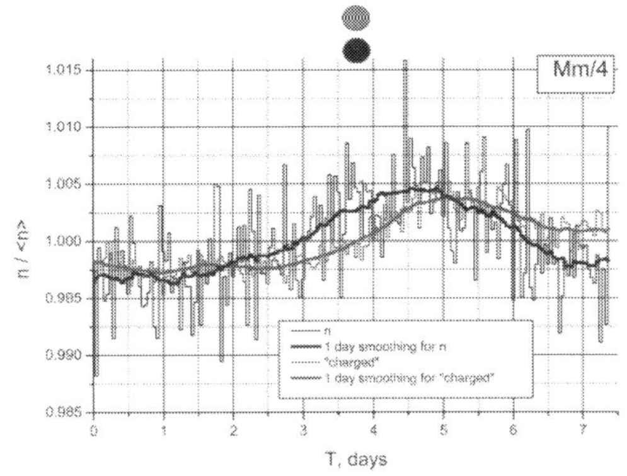

Figure 2. 4th harmonic of Moon month (T=7.38d) observed at Tibet level. Positions of moon syzygies are marked by circles.

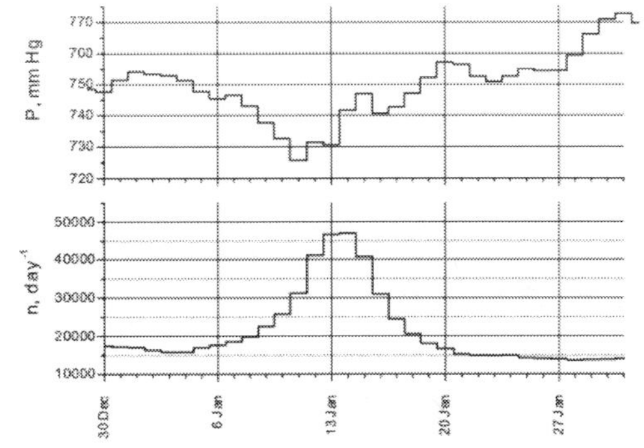

Figure 3. Anti-correlation between daily air pressure and normalized daily neutron counting rate in January, 2014.

(Rn-222) undergo the so called barometric pumping effect well known to geophysicists, when soil gases effectively diffuse from deeper to upper soil levels and thence to the atmosphere. We have found the barometric pumping effect exists not only for radon but also for thermal neutrons. The effect results in a spontaneous rise of the thermal neutron flux up to a factor of 3. After applying the 2-day delay to the pressure data, we estimated the correlation between barometric pressure and neutron counts at the period when the biggest neutron increase was observed (Jan. 4 to Jan. 23, 2014). Anti-correlation between the absolute 2-day delayed mean daily air pressure, P, and the normalized daily neutron counting rate, $n/<n>$, for this 19 day period is shown in Figs. 3, 4. As one can see, a significant anti-correlation is observed yielding the correlation coefficient equal to $R = -0.93$ and the regression coefficient equal to $\sim (6.4 \pm 0.6)\%/$ mm Hg. Therefore, the resultant negative barometric coefficient at this period is a factor of 6.4 larger than that for cosmic ray hadrons at the Earth's surface being equal to $\sim -1\%/$ mm Hg.

3.4. Response to Nepal's 2015 earthquake

During the strong Nepal earthquake of magnitude 7.8 in April 2015 the PRISMA-YBJ array was running at a distance of 450 km from the epicenter. It is known that radon could be a precursor of nearest earthquake due to its long (3.8 d) life-time when it can migrate to long distances in both air and in soil. Therefore, our method based on the recording of radon-due thermal neutrons makes it possible

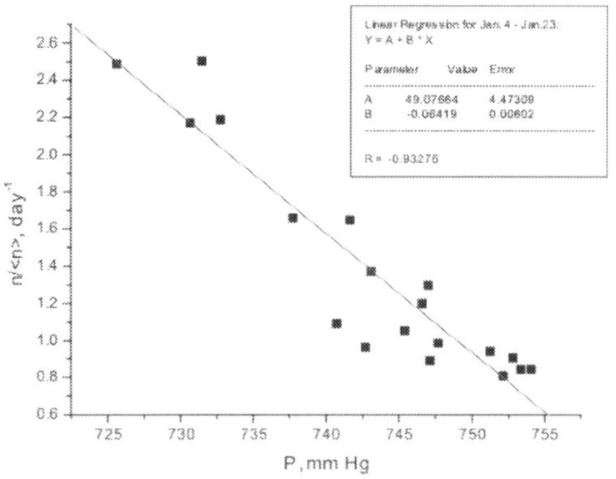

Figure 4. Anti-correlation plot between absolute 2-days delayed daily air pressure and normalized daily neutron counting rate for the dates Jan. 4–Jan. 23 2014.

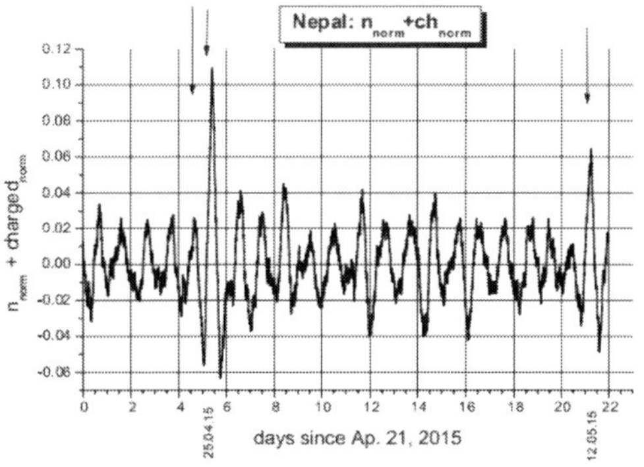

Figure 5. Response of the PRISMA-YBJ detectors to the Nepal earthquakes in April-May 2015.

to monitor soil radon concentration and thus to detect its change during earthquakes. For this purpose we use both of our counting rates: for thermal neutrons and for "charged". The latter was found to be sensitive to radon decays in air, while neutrons are sensitive to radon decays in soil [3].

Results are shown in Fig. 5 where the sum ($\sum$) of normalized "neutrons" and "charged" is plotted. We use such a function to emphasize the change in phase of the daily time series of neutrons. It was mentioned above (see Fig. 1) that normally these two branches have anti-correlated phases of daily waves. But, during the days with a nearby strong earthquake, the phases coincide, the two waves do not compensate each other and $\sum$ has a maximum seen in Fig. 1.

It is noticeable that the effect we saw appeared not on April 25 when the strongest earthquake occurred ($M = 7.8$), but only the next day (April 26) when an aftershock quake with $M = 6.7$ was observed. Two weeks later, another strong aftershock was observed and again we saw a significant peak of $\sim 6\%$ amplitude. We can add that such high amplitude peaks were not observed during the whole period of data taking.

4. Conclusion

Using the en-detectors developed for EAS studies of variations of the environmental neutron flux, we have observed some geophysical phenomena very unusual for high-energy cosmic ray studies. We hope that the use of nuclear physics methods applied to geophysics can give novel interesting and unexpected results.

This work was supported in Russia by RFBR (grants 14-02-00996, 16-32-00054 and 16-29-13067_ofi_m), RAS Presidium Program 'Fundamental properties of matter and astrophysics', and in China by NSFC (No. 10975046, No. 11375052).

References

[1] Yu.V. Stenkin, Nucl. Phys. B (Proc. Suppl.) **196**, 293–296 (2009)
[2] B. Bartoli et al., Astropart. Phys. **81**, 49–60 (2016)
[3] Yu.V. Stenkin et al. Pure and Appl. Geophys, in press; arXiv:1605.01283 [physics.geo-ph]
[4] S.N. Vernov et al. Proc. 16th Intern. Cosmic Ray Conf, **6**, 129–134 (1979)

EAS primary particle parameter estimation with the complex Pamir-XXI detector array

V.I. Galkin[1,2,a], A.S. Borisov[3,b], R. Bakhromzod[1,4], V.V. Batraev[1], S. Latipova[5], and A. Muqumov[1,4]

[1] Lomonosov Moscow State University, Faculty of Physics, Moscow, Russia
[2] Lomonosov Moscow State University, Skobeltsyn's Institute of Nuclear Physics, Moscow, Russia
[3] Lebedev Physical Institute of the Russian Academy of Sciences, Moscow, Russia
[4] Umarov Physical-Technical Institute of Academy of Sciences of Tajikistan, Dushanbe, Tajikistan
[5] Tajik National University, Faculty of Physics, Dushanbe, Tajikistan

Abstract. Some new developments in EAS methods made in the framework of the Pamir-XXI project are presented. First, primary energy and direction definition accuracies by a network of fast scintillators are considered, optimum network cell size is defined for 10 PeV showers. Second, the same accuracies for a network of fast optical (Cherenkov) detectors are considered for 30 TeV–10 PeV showers. Third, the possibilities of separation of EAS initiated by protons, nitrogen and iron nuclei of 1 and 10 PeV energies using a wide-angle Cherenkov telescope are discussed. Finally, the results of the extraction of 30–50 TeV gamma showers from the proton shower background with the same telescope are presented.

Presumably, our developments can help in the study of PCR mass composition and ultra high energy gamma ray astronomy in other projects.

1. Introduction

The Pamir-XXI project was originally planned as a multi-component study of EAS in a wide energy range 30 TeV–1 EeV bearing in mind the solution of astrophysical problems (detailed studies of PCR energy spectrum and mass composition, detection of γ-rays of point-like and diffuse sources above 30 TeV) and the problems of hadron interaction at super high energies including exotic phenomena in EAS cores.

At present there is no certainty in the terms of realizing the project, but we still continue developing the experimental procedures using detailed simulations of EAS characteristics. To reach the denoted aims a complex detector array is required, incorporating numerous detectors of different types [1].

In this paper we concentrate on the analysis and selection of optimal detector configurations for:

- Cherenkov γ-ray astronomy;
- PCR energy spectrum and mass composition studies.

2. Statement of the problem and the solution logic

In order to succeed in the above mentioned studies it is enough to measure primary particle energy E_0, type (mass), arrival direction $\vec{\theta}$ and core location $\vec{x}_{core}$ at the observation level. These parameters are usually considered to be equally demanding with regard to the amount of experimental data used while making the parameter estimates. Our analysis shows this is not the case as the primary type (mass) proves to be much more demanding.

That is why we insist that a detector array and data processing methods should be optimized with respect to the informatively most exacting problems, namely, the primary mass estimation and the γ-event selection.

For the measurements to be as consistent as possible one must consider the most information-rich component to be the reference component for all other measurables. Among the EAS components the Cherenkov light (CL) is definitely the most informative and makes it possible to solve all stated problems within the energy range mentioned even with a small duty cycle ($\leq$10%).

Taking into account these statements, one is forced, first of all, to construct an array of Cherenkov detectors optimized for the solution of primary type/mass estimation problem. The array should analyze both Cherenkov pulses (spatial-temporal distribution of CL) and Cherenkov [angular] images (CL spatial-angular distribution) and thus will include a network of widely spaced fast optical detectors and a set of a few optical telescopes.

A network of charged particle detectors will be used at $E_0 > 1$ PeV for evaluation of E_0, $\vec{\theta}$, $\vec{x}_{core}$ to help the central calorimeter and muon detectors (presumably trackers) to estimate the primary mass.

The method for the primary mass estimation by charged particle data (when finally established) can be tuned with the help of the respective Cherenkov light method.

[a] e-mail: v_i_galkin@mail.ru
[b] e-mail: asborisov55@mail.ru

3. Artificial events

3.1. Generation

Artificial events for the observation level 4250 m a.s.l. were generated with CORSIKA6.990/QGSJET01 [2] for a set of fixed E_0 from 30 TeV to 100 PeV and different primary types: p, He, N, S, Fe, γ. CL spatial-temporal and spatial-angular distributions and CORSIKA standard particle files were stored.

As for super high energy hadron interaction model used, one might object that QGSJET01 is not the one that best fits the EAS data pool. Our answer is: nobody knows what is the true hadron interaction model, although all the models incorporated in CORSIKA reproduce EAS development satisfactorily and definitely recreate relative changes of the EAS characteristics with E_0 and A (primary mass). This is what matters most of all when one looks for a function of measurable values that can help to distribute EAS events by A. When such a function is found one can correct it for the hadron interaction model and even do without any model using experimental distributions.

3.2. Processing

To process the spatial-temporal data of simulated events simple polynomial models were used for CL and particle lateral distributions LD(R) and fronts $\tau(R)$:

$$\mathrm{LD(R)} = \mathrm{a_0}/(1 + \mathrm{a_1 R} + \mathrm{a_2} R^2 + \mathrm{a_3} R^3),$$
$$\tau(R) = R(a_0 + a_1 R + a_2 R^2).$$

The shower direction definition is done simultaneously with the core location search by fitting the measured front arrival times by the front model and the measured number of photons (photoelectrons)/particles by the lateral distribution model. During this fit τ(R) and LD(R) parameters are also varied so that we obtain the best front and lateral distribution approximations for each event. LD(R) is further integrated within a circle of a certain radius to form an estimate of E_0.

Processing of CL angular images does not require any model functions, it comprises the location of a spot's long axis and integration over certain regions of field of view (FOV). The primary direction estimate by spatial-temporal data is used.

4. Optical part of Pamir-XXI detector array

The optical part of the detector array uses spatial-temporal and spatial-angular distributions of CL for the estimation of primary particle parameters and, thus, must incorporate two sets of detectors (Fig. 1).

4.1. Multiple use of event Cherenkov data

Event sample volumes vary from 200 for 30–100 TeV primaries to 30 for 100 PeV primaries but CL data of each event includes:

- a 600 × 600 × 302 histogram of spatial-temporal distribution covering a 600 m × 600 m carpet and ∼ 300 ns delay interval and

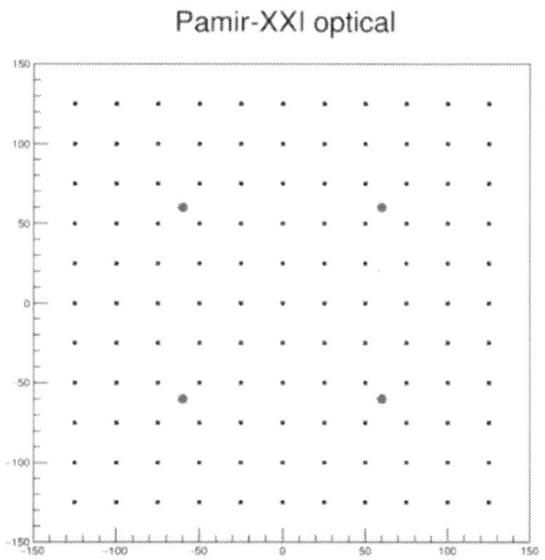

Figure 1. Optical detector array layout used in simulations: 1) rectangular network of 11 × 11 wide angle fast optical detectors of area ∼1 m^2 and aperture ∼1 sr set with 25 m step (squares), which must be able to determine the shower direction with an accuracy of at least 0.1°, core location with an accuracy of ∼1 m and the primary energy within 15% accuracy and also makes it possible to analyse the CL pulse shape; 2) optical image telescopes (circles) with mirrors of area ∼4 m^2, field of view diameter of about 30° and pixel diameter of ∼0.8° spaced by ∼100 m, which should enable the analysis of the CL angular distribution close to the shower core and be sensitive to the primary particle mass.

- a 250 × 250 × 108 × 108 histogram of spatial-angular distribution within a 500 m × 500 m square of observation level with every 2 m × 2 m bin imitating an imaging telescope of 27° × 27° field of view.

Thus, each event can be used many times by choosing different 1 m × 1 m bins of the 1st histogram as detectors of fast network and different 2 m × 2 m bins of the 2nd histogram as imaging telescopes.

4.2. CL spatial-temporal distribution processing results

Uncertainties of core location and primary direction determination by fast optical detector network meet the above set requirements (∼ 1 m and < 0.1°). The results for pure CL signal in photons and signal + night sky background (BG) ($10^8\,\mathrm{cm^{-2}\,sr^{-1}\,s^{-1}}$) in photoelectrons are almost the same for $E_0 \geq 1$ PeV. The situation changes for primary energies 30–100 TeV but the uncertainties still lie within the pre-set limits which gives hope to detect gamma-events in this energy range (Table 1).

The primary energy E_0 is estimated as an integral CL300 of CL lateral distribution (LD) over a circle of radius 300 m. Parameters of LD are obtained as a result of a joint fit of CL LD and front models to CL spatial-temporal data of individual EAS. Uncertainties of CL300 for $E_0 \geq 1$ PeV are indifferent to night sky BG and conversion to photoelectrons (Table 2). The situation is different for the 30–100 TeV range: mixing of CL and BG photons and their conversion to photoelectrons increase the uncertainties (Table 3). Introduction of a threshold

Table 1. Core location, m, and primary direction, °, uncertainties.

event/signal	core loc.mean	core loc.RMS	prim.dir.mean	prim.dir.RMS
60 TeV p, CL	1.6	0.93	0.044	0.033
60 TeV p, CL+BG	2.6	1.9	0.076	0.064
60 TeV p, CL+BG,>200phel	2.4	1.6	0.059	0.048
30 TeV γ, CL	1.0	0.59	0.030	0.020
30 TeV γ, CL+BG	1.8	1.3	0.058	0.052
30 TeV γ, CL+BG,>200phel	1.5	1.1	0.051	0.046

Table 2. CL300 relative fluctuations for $E_0 \geq 1$ PeV, %.

E_0, PeV	proton	nitrogen nucleus	iron nucleus
1	15	10	8
10	14	8	5

Table 3. CL300 relative fluctuations for $E_0 = 30$–100 TeV, %.

E_0, TeV, prim.patricle	CL photons	CL + BG, photoelectrons	CL + BG, >200 phel
60, proton	18	20	18
100, proton	13	16	16
30, γ	5	10	9
50, γ	2	9	8

(e.g., 200 photoelectrons) for each fast detector somewhat reduces the uncertainties.

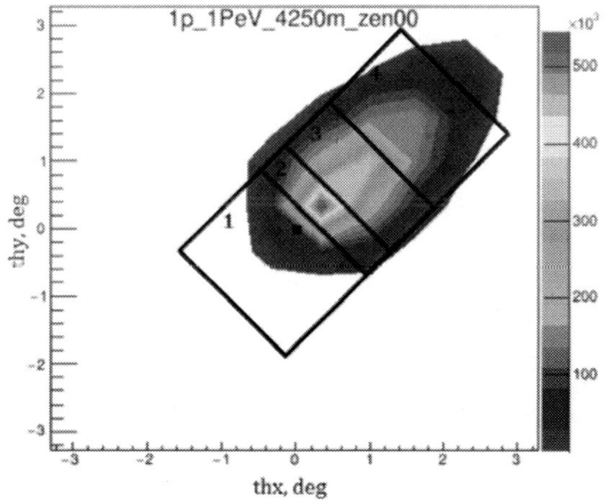

Figure 2. Shower image and rectangular areas of integration aligned along the spot long axis. Black square in the center of FOV marks the shower arrival direction. CL spot is sliced by rectangles 1,2,3,4 across its long axis. A stripe formed by the rectangles is 5° wide. The outer side of rectangle 1 is placed at $-1.5°$ with respect to the arrival direction. The image is integrated within the rectangles, integrals S_i form ratios r_{ij}. The widths of the rectangles are varied so that to find r_{ij} giving the best separation of images produced by different primary particles.

4.3. CL spatial-angular distribution processing results

The method for processing of CL angular images uses the spot shape parameters that provide maximum separation of showers initiated by primaries of different types. We calculate the integrals S_i, $i = 1, 2, 3, 4$ over rectangular segments of the FOV in the vicinity of the CL spot and use their ratios $r_{ij} = S_i/S_j$ as features characterizing the events. Maps for 0.75° × 0.75° pixel size are used because this resolution turned out to be the most effective in preliminary studies [1]. Figure 2 gives the geometry of the image processing.

Optimum selection criteria have been found for p-N and N-Fe pairs at 1 and 10 PeV for different core distances R. Generally, optimum r_{ij} depends on the pair, E_0 and R. In each case the borders between criterion class distributions are adjusted so that misclassification errors (e.g., $P\{p \to N\}$ and $P\{N \to p\}$) are equal. Thus, Table 4 shows only one value. Errors for pure CL signal are slightly lower than for CL+BG. Analysis shows that integration intervals along the spot long axis, corresponding to optimum r_{ij}, mostly depend on R and a universal feature $r = r(R)$ can be defined behaving almost optimally. Figure 3 presents integration limits for the universal feature as functions of R.

The same approach to CL images processing is capable of suppressing at least 99% of nuclear background in 30–100 TeV γ-ray detection. In this case borders between γ and proton classes are placed to leave only one proton in the γ domain. Best results for CL+BG turned into photoelectrons are shown in Table 5.

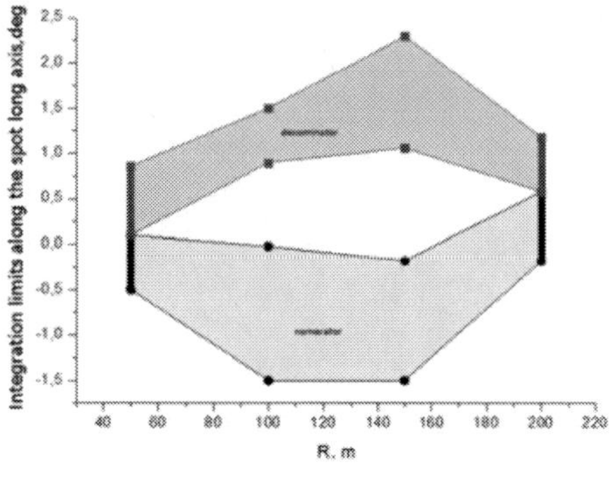

Figure 3. Integration limits for universal feature r vs. R. Lower area: numerator integral. Upper area: denominator integral.

5. Charged particle detector network

We consider a fragment of charged particle detector network of 5 × 5 detectors 1 m × 1 m each in order to optimize the spacing (grid step) and find the lower limits of uncertainties in the core location, arrival direction and primary energy estimation. Two versions of particle spatial-temporal distribution processing are considered:

a) according to CORSIKA particle data, the number of charged particles and the moment of particle front arrival in each detector are defined; using

Table 4. Misclassification probabilities for p-N and N-Fe pairs, E_0= 1, 10 PeV. All samples contain 240 images.

signal	criterion	E_0	1 PeV				10 PeV			
		R, m	50	100	150	200	50	100	150	200
CL,		p-N	0.05	0.10	0.15	0.17	0.09	0.13	0.17	0.21
photons	optimum	N-Fe	0.05	0.10	0.15	0.15	0.004	0.04	0.09	0.12
CL+BG,		p-N	0.05	0.11	0.15	0.17	0.09	0.13	0.17	0.20
photoelectrons	optimum	N-Fe	0.05	0.10	0.15	0.16	0.01	0.04	0.09	0.13
CL,		p-N	0.06	0.10	0.16	0.18	0.09	0.13	0.18	0.21
photons	universal	N-Fe	0.05	0.11	0.15	0.16	0.004	0.05	0.10	0.12
CL+BG,		p-N	0.05	0.11	0.15	0.17	0.09	0.13	0.17	0.20
photoelectrons	universal	N-Fe	0.05	0.11	0.17	0.16	0.01	0.05	0.10	0.13

Table 5. Probabilities $P\{p \to \gamma\}/P\{\gamma \to p\}$. All samples contain 400 images.

Pair	50 m	100 m
30 TeV γ – 60 TeV p	0.0025/0.345	0.0025/0.5125
50 TeV γ – 100 TeV p	0.0025/0.255	0.0025/0.545

these measured data the core location and arrival direction are estimated as well as particle lateral distribution function which is then integrated to give a E_0 estimate;

b) **all** CORSIKA particles within a detector are tracked through its body (2 mm of Al + 2 cm of polystyrene), by the deposited energies and centers of mass of ionization pulses the same event parameters as in a) are estimated.

10 PeV proton showers are used. In case a) 60 events are considered, in case b) 10 events. In both cases each event is used many (1600) times. Three grid steps are tried: 10, 15 and 20 m. The shower axis is kept within the detector network fragment.

Step 15 m shows the best overall results in both cases, uncertainties are very similar: core location ~1 m^2, arrival direction ~ 0.18°, primary energy ~ 12%. E_0 is estimated by the integral of particle LD within a circle of 100m radius in case a) or of ionization LD within the same radius as in case b).

6. Conclusions

1. The proposed optical part of Pamir-XXI is capable of achieving the required accuracies in core location (~1 m), arrival direction (better than 0.1°) and primary energy ($\lesssim$15%) within the target primary energy range 30 TeV–100 PeV. Using this detector array it is also possible to divide all primary nuclei into three groups (1–100 PeV) and reject not less than 99% of nuclear background events while selecting γ-events (30–100 TeV).
2. The charged particle network can estimate the core location within ~1 m accuracy, arrival direction with uncertainty ~0.2° and primary energy with uncertainty better than 15%. A preferable grid step is 15 m.
3. The question of whether and how this is possible to estimate the primary mass using charged particle detector data is still open. Probably, it is possible with muon trackers and/or central calorimeter with high spatial resolution.

References

[1] A. S. Borisov and V. I. Galkin, Journal of Physics: Conference Series, **409**, No.1, 012089 (2013)
[2] D. Heck and T. Pierog *Extensive Air Shower Simulation with CORSIKA: A User's Guide*, (Karlsruher Institut fur Technologie, 2011)

Observation of high-energy cosmic rays by very inclined muon bundles in the NEVOD-DECOR experiment

O. Saavedra[1,a], N.S. Barbashina[2], A.G. Bogdanov[2], D.V. Chernov[2], L.I. Dushkin[2], S.S. Khokhlov[2], V.A. Khomyakov[2], V.V. Kindin[2], R.P. Kokoulin[2b], K.G. Kompaniets[2], G. Mannocchi[3], A.A. Petrukhin[2], V.V. Shutenko[2], G. Trinchero[3], I.I. Yashin[2], K.O. Yurin[2], and E.A. Zadeba[2]

[1] Dipartimento di Fisica dell' Universita di Torino, Italy
[2] National Research Nuclear University MEPhI (Moscow Engineering Physics Institute), Moscow, Russia
[3] Osservatorio Astrofisico di Torino – INAF, Torino, Italy

Abstract. The Russian-Italian NEVOD-DECOR experiment on measurements of the local muon density spectra at various zenith angles gave the possibility to obtain important information on the primary cosmic ray flux and interaction characteristics in a wide energy range from 10^{15} to more than 10^{18} eV. At large zenith angles and high muon densities, a considerable excess of muon bundles has been found in comparison with expectation. In this paper, an update of these investigations is presented and some new results obtained by the collaboration are discussed.

1. Introduction

The first cooperation agreement between MEPhI and Torino group (Institute of Cosmo-geophysics, Torino, and Torino University) was signed in June 1992. The main purpose of the cooperation was the creation of the coordinate detector DECOR around the Cherenkov water calorimeter NEVOD [1] for investigations of multi-particle events in cosmic rays at large zenith angles. According to the agreement, streamer tube chambers, front-end electronics and other equipment used earlier in the NUSEX experiment [2] in the Mt. Blanc tunnel were delivered to Moscow. The first supermodules of the detector were put into operation in 1998. In 2001, the construction of the DECOR (eight supermodules with total area about 70 m^2) was completed [3]. Long-term experiments (about 20,000 hours net operation time) with the NEVOD-DECOR complex were conducted in 2002-2007. The main objects of the investigations were the characteristics of multi-muon events (muon bundles) in a wide range of zenith angles, from about 30° to the horizon. Unique experimental data on the zenith-angular dependence of muon bundle intensity with particle multiplicities from 3 to about 100 were obtained. A novel approach to the analysis of data on muon bundles based on the phenomenology of the local muon density spectra (LMDS) at various zenith angles was developed [4], that allowed to extract information about the primary cosmic ray flux and interaction characteristics in a record wide energy range from 10^{15} to more than 10^{18} eV. In 2012, after a deep modernization of the measuring systems of the NEVOD calorimeter [5], the experiments with the complex were resumed, and the available experimental statistics of muon bundles has been increased nearly twice. In this paper, a brief overview of the earlier DECOR data and some new results of the analysis of the combined experimental statistics are presented.

2. Muon bundles in the coordinate-tracking detector

A general layout of the NEVOD-DECOR is shown in Fig. 1. The side part of the coordinate-tracking detector consists of 8 supermodules (SM) and is deployed in the galleries of the NEVOD building, from three sides of the water tank with inner volume $9 \times 9 \times 26$ m^3 and 0.5 m thick concrete walls. Two pairs of SMs are located in the opposite short galleries, and four SMs occupy a long gallery. Each SM has an effective area of 8.4 m^2 and consists of eight planes of streamer tube chambers with resistive cathode coating hung vertically with 6 cm spacing between them. The chambers are operated in a limited streamer mode with a continuous flow of (Ar + CO_2 + n-pentane) gas mixture. The length of the chambers is 3.5 m, inner tube cross-section is 9×9 mm^2. The planes of the chambers are equipped with a two-dimensional system of external readout strips (256 horizontal X- and 256 vertical Y-strips in each plane with 1.0 and 1.2 cm pitch, respectively). The spatial and angular accuracy of the reconstruction of the tracks crossing the SM is about 1 cm and better than 1°, correspondingly. As a first-level trigger condition, a 4-fold coincidence of signals from the planes of the SM is used.

Initially, the coordinate-tracking detector also included four horizontal SMs deployed on movable platforms on the roof of the NEVOD water tank (see Fig. 1). In 2005, the upper supermodules were re-assembled, equipped with a newly developed fast electronics, and a wide-aperture muon hodoscope URAGAN [6] with an area

[a] e-mail: saavedra@to.infn.it
[b] e-mail: rpkokoulin@mephi.ru

DOI: 10.1051/epjconf/201714516001

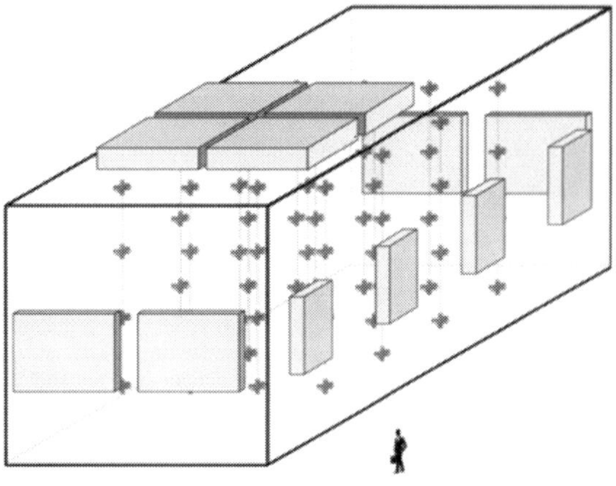

Figure 1. General layout of the NEVOD-DECOR complex.

of about 45 m^2 was created on the basis of the upper part of the DECOR. The URAGAN ensures the detection and reconstruction in real-time mode of about 6,000 particle tracks per second, and is now being used for the solution of various tasks related with cosmic ray muon flux variations.

In 2002–2007, a long series of measurements with NEVOD-DECOR complex were conducted. Net operation time with a full configuration of the side DECOR (8 SMs) amounted to 19,922 h. The set of the hardware event selection conditions based on the coincidences of the signals from different parts of the coordinate detector and signals formed by the NEVOD triggering system was kept unchanged during the measurements. Selection of muon bundles from DECOR data is based on the fact that tracks of muons produced in the atmosphere are nearly parallel. The selection procedure included several stages: trigger condition of the coincidence of the signals from at least three DECOR supermodules; off-line track reconstruction and soft program selection of muon bundle candidates containing a required minimal number of quasi-parallel tracks; final event classification and track counting by the operators. Muon bundles leave a very bright signature in the coordinate detector (see Fig. 2), and cannot be mixed with other types of events. A more detailed description of experimental data treatment is given in [7].

Since the procedure of track counting by the operators is rather laborious, the data were analyzed part by part, for different ranges of muon multiplicity and zenith angles. In all, about 40 thousand events were selected from the data accumulated in 2002–2007 measurement series. Phenomenological distributions of muon bundles in particle multiplicity for several zenith angle intervals and in zenith angle for different minimal numbers of muons are presented in Fig. 3. As is seen from the plots, the obtained data on muon bundles cover about six orders of magnitude in the event intensity.

3. Local muon density spectra at various zenith angles

At large zenith angles, EAS are detected at the ground as a practically pure muon component. Due to a large distance from the shower generation point to observation level, transverse dimensions of EAS in muons rapidly increase with zenith angle, the effect being additionally enhanced by the particle deflection in the geomagnetic field. Therefore, a muon detector with sizes of the order of tens meters may be considered as a point-like probe. In an individual event, the local density of EAS muons at the observation point may be estimated as the ratio of muon bundle multiplicity to the detector area for a given direction.

Contribution to the events with a certain muon density give showers with different primary energies, detected at different (random) distances from the axis. However, due to a fast decrease of the primary cosmic ray (PCR) intensity with the increase of energy, the effective primary energy band appears relatively narrow (see Fig. 4). It is important to mark that at different zenith angles the events with a fixed muon density are formed by primary particles with substantially different energies. At that, the event collection area is determined not by the detector size, but by the dimensions of the EAS in the muon component which near the horizon reach several kilometers. Two last circumstances allow one to explore a very wide range of PCR energies in a single experiment with a relatively small detector by means of measuring LMDS at various zenith angles.

A general approach to the analysis of DECOR data on muon bundles included the following steps: iterative reconstruction of local muon density spectra in a detector-independent form on the basis of the measured distributions of muon bundle characteristics, taking into account Poisson fluctuations of the number of muons that hit the setup, triggering and selection conditions, detector efficiencies, etc.; calculation of the expected LMDS for various assumptions about the primary spectrum, mass composition and hadron interaction models using two-dimensional muon lateral distribution functions (LDF) simulated by means of the CORSIKA code [8]; comparison of data with calculation results.

In Fig. 5, experimental LMDS reconstructed from DECOR data are compared with CORSIKA-based calculations for 4 different zenith angles (35°, 50°, 65°, and 78°). As a model of the primary flux for the comparison of data and calculations, a power type energy spectrum with differential slope $(\gamma + 1) = 2.7$ below the knee energy (4 PeV) steepening to $(\gamma + 1) = 3.1$ above the knee was assumed; absolute normalization was chosen according to the bulk of experimental data around the knee presented in the review [9]; two limiting cases of the mass composition (only protons and only iron nuclei) were considered. The curves in the figure correspond to two widely used hadron interaction models SIBYLL2.1 and QGSJET01 (dashed and solid curves, respectively). Arrows in the upper part of each frame indicate the values of effective (mean logarithmic) primary particle energies.

As seen from the figures, the experimental LMDS is in reasonable agreement with calculations for moderate zenith angles ($\theta = 35°$, upper left frame in Fig. 5) at energies around the knee, including the absolute normalization. The increase of the LMDS slope related with the knee in the primary spectrum is clearly seen. This result may be considered as an indirect check of

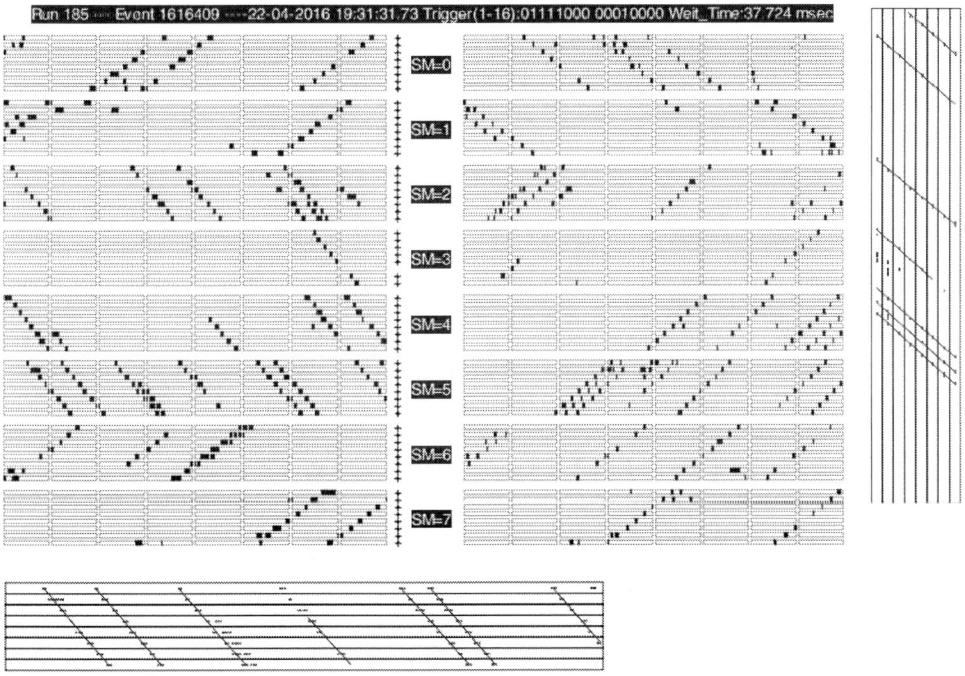

Figure 2. Example of detection of muon bundle event in DECOR. On the right and at the bottom, enlarged views of the response of SM ♯ 5 are shown.

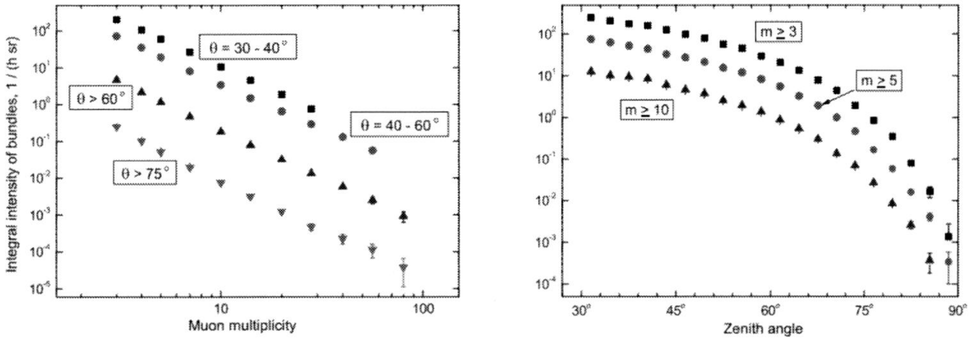

Figure 3. Integral distributions of muon bundles in the particle multiplicity for several ranges of zenith angle (left) and zenith angle distributions of the bundles with different minimal number of muons (right). Data of 2002–2007 [7].

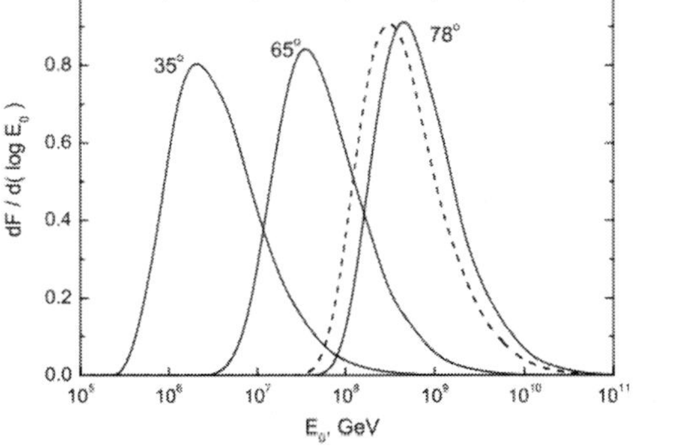

Figure 4. Distribution of primary cosmic ray particle energies contributing to events with a fixed muon density (D = 0.2 muons/m^2) at different zenith angles [7]. Solid curves correspond to calculations for primary protons, dashed curve for iron nuclei.

the energy scale calibration. At effective primary energies $E_0 > 10^{16}$ eV (zenith angles 50° and 65°), a progressive increase of LMDS in comparison with calculations is observed, which can be interpreted as a trend to a heavier mass composition, though an alternative interpretation related with increasing deficit of muons in simulated EAS is also possible. Near 10^{17} eV, a hint for a second knee was found: partial fits of the data at 65° below and above this energy show an increase of the integral LMDS slope, β, by 0.20 ± 0.09. This result is in agreement with slope estimates derived from the data at other zenith angles (thin lines in the figures, the intervals of $E_0 = 10^{16} - 10^{17}$ eV for 50° and $E_0 > 10^{17}$ eV for 78°). Finally, data points at high muon densities and the largest zenith angle (78°), which correspond to primary energies around 10^{18} eV, lie near the upper edge of the calculation uncertainty band, and even somewhat exceed the expectations for a pure iron flux.

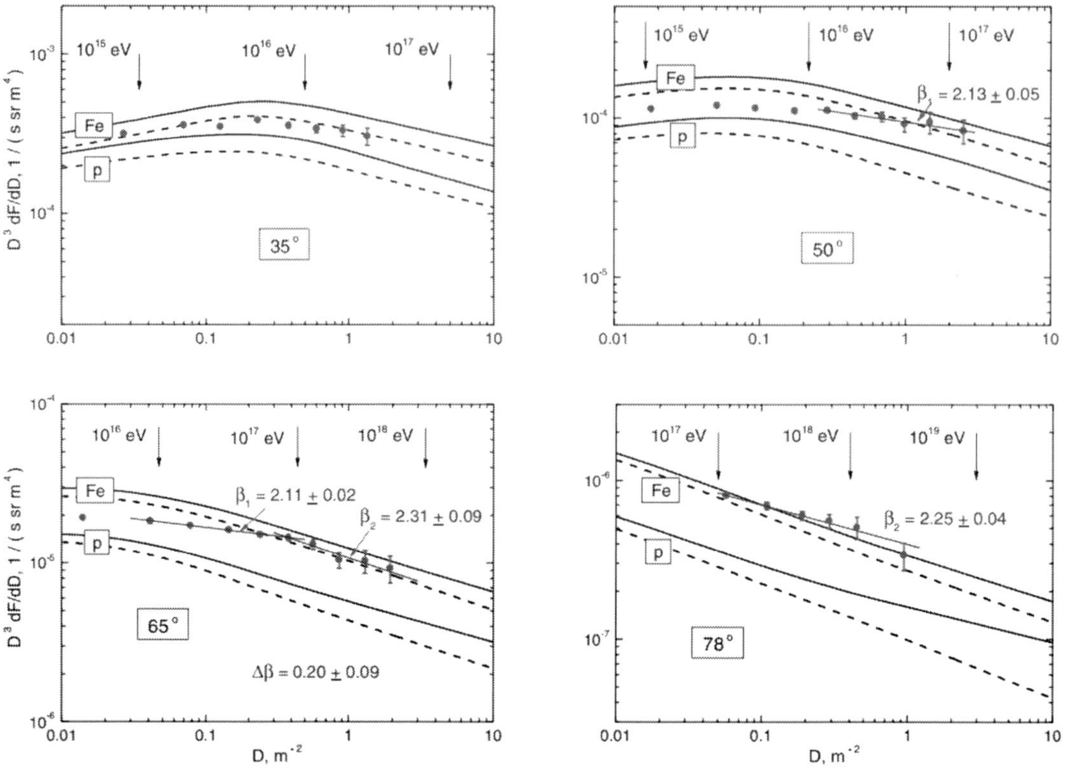

Figure 5. Measured (points) and calculated differential local muon density spectra for 4 zenith angles [7, 10]. Thin lines represent partial power fits of the data between 10^{16} and 10^{17} eV (integral spectrum slope β_1), and above 10^{17} primary energy (β_2).

4. Excess of muon bundles at ultra-high energies

Under certain assumptions about primary mass composition and hadron interaction model, LMDS may be converted into estimates of the primary cosmic ray spectrum [7] and directly compared with other measurements of the primary energy spectrum. Such a procedure was applied to muon bundle samples with high muon multiplicities ($m \geq 10$) and two zenith angle intervals: $\theta \geq 75°$ and $\theta \geq 80°$ (395 and 49 events, respectively). These data correspond to mean logarithmic primary energies around 10^{18} eV and higher. The conversion was performed with two limiting assumptions on primary composition (pure protons and pure iron nuclei), using five different hadron interaction models available in CORSIKA: SIBYLL2.1, QGSJET01, QGSJET-II, EPOS1.61 and EPOS1.99. Thus, for each of the two data samples, 10 estimates of the differential primary energy spectrum were derived [7, 10].

In Fig. 6, primary intensity estimates obtained from DECOR data are compared with results of measurements of the PCR energy spectrum in the EeV energy range. Though a reasonable agreement in absolute normalization with AGASA data [11] is observed, it is seen from the figure that none of the examined interaction models allows us to match the data on muon bundles at large zenith angles with PCR spectrum measurements performed by means of the fluorescence method in HiRes [12], Auger [13] and Telescope Array [14] experiments, even for extremely heavy mass composition. This contradiction becomes even more significant, if we take into account that both HiRes and Auger data on the X_{max} distribution of EAS favor a light (predominantly proton) primary composition near 10^{18} eV [15, 16]. If one assumes only proton primaries, the observed intensity of muon bundles will be about 2–3 times higher than the expectation. Recently, based on the analysis of the surface detector data on the muon content in inclined EAS, the Auger collaboration came to the same conclusion that "none of the current shower models, neither for proton or iron primaries, are able to predict as many muons as are observed" [17]. Later, the inconsistence muon abundance in EAS at ultra-high energies has got the name "muon problem" or "muon puzzle".

5. Observation of the second knee in the EAS muon component

As mentioned above, after the modernization of electronic systems of the Cherenkov water detector, the combined operation of NEVOD and DECOR was resumed in 2012. The main task of the experiment now being conducted is the measurement of the energy deposit of muon bundles in the NEVOD calorimeter [18]. Meanwhile, the available statistics of muon bundle events is continuously increasing, and by February 2016 the total net operation time of DECOR exceeded 41,000 h. The differential local muon density spectra for three zenith angles (57°, 64° and 72°) reconstructed on the basis of the combined statistics of 2002–2007 and 2012–2016 (for 3 zenith angle intervals, 18187, 17540 and 2920 bundles with muon multiplicity $m \geq 5$) are shown in Fig. 7. Arrows in the figure indicate

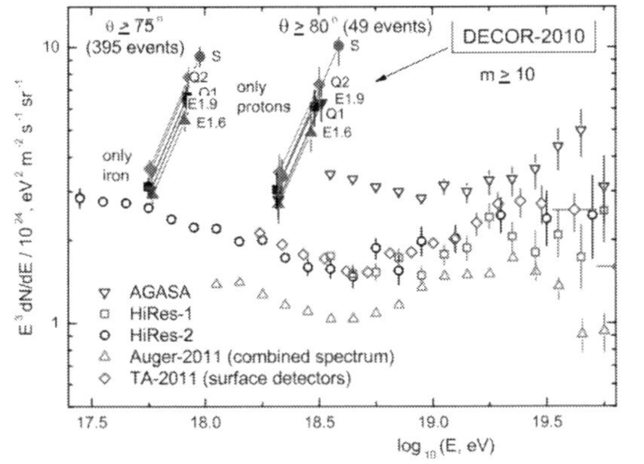

Figure 6. Comparison of the differential primary energy spectrum estimates derived from DECOR data on muon bundles (dark points) with the results of EAS measurements in AGASA [11], HiRes [12], Auger [13] and TA [14] experiments. Labels near the points indicate hadron interaction models mentioned in the text.

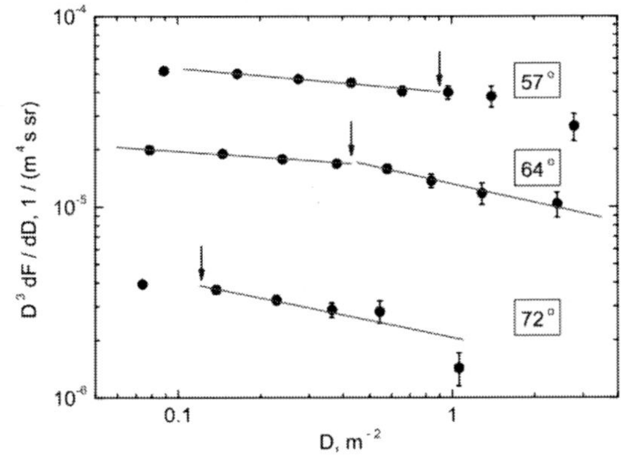

Figure 7. The LMDS reconstructed on the basis of combined DECOR data of 2002–2016 for three different zenith angles and piece-wise power function fits of the data (lines). Arrows indicate the position of the effective 10^{17} eV primary particle energy for different angles.

the positions of the effective 10^{17} eV primary particle energy for different zenith angles. Thin lines represent the power function fits of the spectra in two ranges of the primary energy: below and above 10^{17} eV. As seen from the figure, there are two well separated spectrum slopes in the two energy regions. Combining statistically independent results of the fits for different zenith angles, we obtain $\beta_1 = 2.107 \pm 0.014$ below 10^{17} eV and $\beta_2 = 2.303 \pm 0.054$ above this energy, or $\Delta\beta = 0.196 \pm 0.056$. This is the first statistically significant confirmation of the presence of the second knee measured with muons (at about 3.5σ level).

6. Conclusions

During long-term measurements conducted with the NEVOD-DECOR experimental complex, unique data on inclined muon bundles detected at ground level have been accumulated. Application of a novel approach to EAS studies based on the phenomenology of local muon density spectra at various zenith angles gave the possibility to obtain important information on the primary cosmic ray flux and interaction characteristics in the energy range from 10^{15} to 10^{18} eV in the frame of a single experiment with a relatively small setup. Comparison of the measurement results with CORSIKA-based simulations showed the increase of the LMDS slope at PeV energies of primaries (corresponding to the first knee), and a trend to a heavier primary mass composition at energies above the knee. At large zenith angles and high muon densities (corresponding to primary energies around 10^{18} eV) a considerable excess of muon bundle events in comparison with the expectation based on independent estimates of the primary spectrum and widely used hadron interaction models has been found. The presence of the second knee (the steepening of the spectrum around 10^{17} eV) in multi-muon event intensity has been confirmed with 3.5σ significance.

Experimental investigations are conducted at the Experimental Complex NEVOD (MEPhI) with support from the Ministry of Education and Science of the Russian Federation (contract RFMEFI59114X0002, MEPhI Academic Excellence Project, and government task). Simulations were performed using the resources of the MEPhI high-performance computing center.

References

[1] V.M. Aynutdinov et al., Astrophys. Space Sci. **258**, 105 (1998)
[2] G. Battistoni et al., NIM A **245**, 277 (1986)
[3] N.S. Barbashina et al., Instruments and Experim. Techniques **43**, 743 (2000)
[4] N.S. Barbashina et al., Nucl. Phys. B (Proc. Suppl.) **165**, 317 (2007)
[5] S.S. Khokhlov et al., Astrophys. Space Sci. Trans. **7**, 271 (2011)
[6] N.S. Barbashina et al., Instruments and Experim. Techniques **51**, 180 (2008)
[7] A.G. Bogdanov et al., Phys. Atom. Nucl. **73**, 1852 (2010)
[8] D. Heck et al., Forschungszentrum Karlsruhe Report FZKA 6019 (1998)
[9] S. Eidelman et al. (Particle Data Group), Phys. Lett. B **592**, 228 (2004)
[10] O. Saavedra et al., J. Phys.: Conf. Ser. **409**, 012009 (2013)
[11] M. Takeda et al., Astropart. Phys. **19**, 447 (2003)
[12] R.U. Abbasi et al., Phys. Rev. Lett. **100**, 101101 (2008)
[13] J. Abraham et al., Phys. Lett. B **685**, 239 (2010)
[14] D. Ikeda et al., Proc. 32nd ICRC (Beijing), vol. 2, 238 (2011)
[15] R.U. Abbasi et al., Phys. Rev. Lett. **104**, 161101 (2010)
[16] J. Abraham et al., Phys. Rev. Lett. **104**, 091101 (2010)
[17] G. Rodrigues et al., Proc. 32nd ICRC (Beijing), vol. 2, 95 (2011)
[18] A.G. Bogdanov et al., EPJ Web Conf. **99**, 06004 (2015)

Measurements of the energy deposit of multi-muon events in the Cherenkov water calorimeter

R.P. Kokoulin[1,a], N.S. Barbashina[1], A.G. Bogdanov[1], D.V. Chernov[1], L.I. Dushkin[1], S.S. Khokhlov[1], V.A. Khomyakov[1], V.V. Kindin[1], K.G. Kompaniets[1], G. Mannocchi[2], A.A. Petrukhin[1], O. Saavedra[3], V.V. Shutenko[1], G. Trinchero[2], I.I. Yashin[1], and E.A. Yurina[1]

[1] National Research Nuclear University MEPhI (Moscow Engineering Physics Institute), 115409 Moscow, Russia
[2] Osservatorio Astrofisico di Torino – INAF, 10025 Torino, Italy
[3] Dipartimento di Fisica dell' Universita di Torino, 10125 Torino, Italy

Abstract. To solve the problem of the excess of multi-muon events observed in several cosmic ray experiments at ultra-high energies (so-called "muon puzzle"), an analysis of the energy characteristics of the muon component of extensive air showers (EAS) is required. A possible approach to such investigations is the measurement of the energy deposit of muon bundles in the detector, which provides information on the mean muon energy. In the experiment now being conducted at the NEVOD-DECOR complex, the local muon density and EAS arrival direction are determined according to the data of the coordinate-tracking detector DECOR, whereas the energy deposit is measured by means of the Cherenkov water calorimeter NEVOD. Results of the measurements of the energy deposit of inclined muon bundles based on the data accumulated during more than 23,200 h observations in 2012 – 2016 are presented and compared with simulations performed by means of the CORSIKA code.

1. Introduction

Unexpectedly high rates of high-multiplicity muon bundle events were found in cosmic ray experiments conducted at the LEP detectors ALEPH [1] and DELPHI [2]. At that time, it was interpreted as an indication for a heavy mass composition at very high primary particle energies. However, the analysis of the DECOR data [3,4] on inclined muon bundles with a high multiplicity at large zenith angles showed that the measured event intensity cannot be explained in the frame of the existing hadron interaction models even under the assumption of an extremely heavy (pure iron nuclei) mass composition at primary energies around 10^{18} eV. Later, studies of the muon content in inclined extensive air showers (EAS) in the energy range 10^{19} eV at the Pierre Auger Observatory [5,6] confirmed the excessive amount of muons: considerably more muons in simulations were needed to match the data and simulations. Now, the inconsistant muon abundance in EAS at ultra-high energies is often referred to as the "muon problem" or "muon puzzle".

It is natural to assume that to explain the observed excess of multi-muon events the existing models of the development of the nuclear cascade in the atmosphere need a revision. A key to the solution of this problem may be found by investigating the energy characteristics of the muon component of EAS and their variations with primary energy [7]. A possible approach to such investigations is the measurement of the energy deposit of muon bundles in the detector material. The specific energy loss of muons in the region above hundreds GeV almost linearly increases with muon energy, and the appearance of an excessive flux of high-energy muons should change the dependence of the average muon bundle energy deposit on the energy of the primary particle. An experiment to investigate the energy deposit of inclined muon bundles was started at the experimental complex NEVOD in 2012. The complex includes a Cherenkov water calorimeter [8,9] and a coordinate-tracking detector [10] to ensure a reliable identification of multi-muon events. Detection of muon bundles in a wide range of multiplicities and zenith angles provides a possibility to explore a very wide interval of primary energies in the framework of a single experiment. First results of the measurements of muon bundle energy deposit were reported at the preceding Symposium [11]. For the moment, the available experimental statistics have been increased about 2.4 times and are now based on 23,200 h of net observation time. In the paper, updated results of the NEVOD-DECOR experiment on muon bundle energy deposit measurements are presented and compared with simulations performed using the CORSIKA code [12].

2. Experimental data

The Cherenkov water detector (CWD) NEVOD [8] has an inner volume of 9 × 9 × 26 cubic meters and is equipped with a spatial lattice of quasi-spherical measuring modules (QSM). Each QSM includes six FEU-200 photomultipliers with flat 15 cm diameter cathodes directed along the axes of the orthogonal coordinate system. In total, the measuring system contains 91 QSMs (546 PMTs), arranged in vertical strings (3 or 4 QSMs in each) with distances 2.5 m along the main axis of the water tank

[a] e-mail: rpkokoulin@mephi.ru

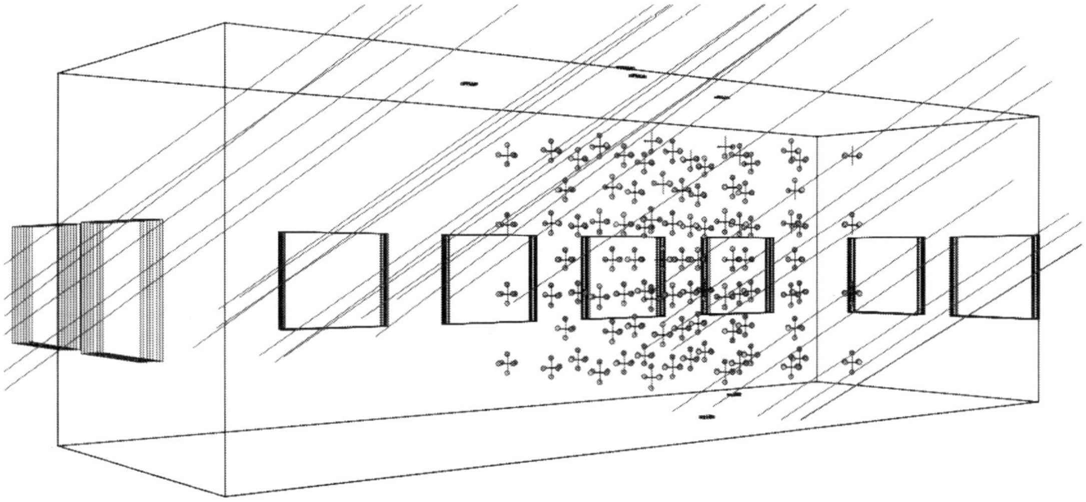

Figure 1. An example of a muon bundle event detected in the NEVOD-DECOR complex. Thin lines represent muon tracks reconstructed from DECOR data; small circles indicate hit photomultipliers of the CWD; big rectangles around the CWD volume are DECOR SMs; small dark rectangles at the top and the bottom represent hit counters of the CTS.

and 2.0 m across it and over the depth. After a recent modernization [9], the electronic systems of the detector ensure measurements of the signals from each PMT in a wide dynamic range from 1 to 10^5 photoelectrons. The setup also includes a calibration telescope system (CTS); 40 plastic scintillation counters each with area $20 \times 40\,\mathrm{cm}^2$ are located on the roof of the water tank, above the QSM lattice, and 40 counters on the tank bottom. The coordinate-tracking detector DECOR [10] with a total area about 70 m^2 includes eight supermodules (SMs) deployed in the galleries of the NEVOD building from three sides of the water tank. The sensitive area of each supermodule is 8.4 m^2. Supermodules consist of 8 vertical planes of plastic streamer tube chambers with resistive cathode coating and a two-dimensional system of external readout strips. The spacing between neighboring planes is 6 cm. The spatial and angular accuracy of muon track reconstruction in the SM is better than 1 cm and 1°, respectively. A general layout of the setup with an example of a muon bundle event detection is shown in Fig. 1.

In the present analysis, data of three measurement series conducted during a four-year long period between May 2012 and May 2016 are used (measurements are now being continued). The total live time of the observations amounted to 23,215 h. In these data, 39542 muon bundle events with muon multiplicities $m \geq 5$ and zenith angles $\theta \geq 55°$ have been found. In addition, from the initial part of the accumulated experimental material (3253 h live time), muon bundles arriving at moderate zenith angles were selected: 15084 events with $m \geq 5$ and $40° \leq \theta < 55°$. The events were selected in two 60°-wide sectors of the azimuth angle where most of the DECOR supermodules (six of eight) were screened with the NEVOD water volume; data of only these shielded SMs were used for muon track counting. The average muon threshold energy for such selection conditions is about 2 GeV.

As a measure of the muon bundle energy deposit in the NEVOD, the sum, Σ, of the signals of all hit PMTs of the water calorimeter is used (in units of photoelectrons, ph.e.). It is assumed that the total yield of Cherenkov photons is proportional to the total energy loss of muons in the detector material (including the energies of secondary particles produced by muons and cascades from them). The local density of muons D (m^{-2}) in the event is estimated on the basis of the number of the detected muons and effective area S_{det} of six DECOR SMs for the measured direction of muon bundle arrival. A preliminary analysis showed that, as might be expected, to a first approximation the total energy deposit in CWD is proportional to the local muon density. Therefore, further we consider the specific energy deposit Σ/D, i.e. the Cherenkov calorimeter response normalized to the local muon density estimate in the event.

3. Results and discussion

Results of the measurement of zenith-angular dependence of the average specific energy deposit of muon bundles are presented in Fig. 2. At moderate zenith angles (less than 55°), a residual contribution of electromagnetic and hadron EAS components to the response of the unscreened water calorimeter located at the surface is observed. This contribution decreases nearly exponentially with the slant depth of the atmosphere [11], and at large zenith angles practically only muons and secondary relative to them particles remain in the bundle events. Arrows in the lower part of the figure indicate typical energies (calculated mean-logarithmic values [4]) of primary particles contributing to muon bundle formation at different zenith angles. As seen from the figure, the experimental data clearly demonstrate the increase of the specific energy deposit with zenith angle.

The curves in Fig. 2 represent the expected angular dependence of the specific energy deposit obtained on the basis of simulation of EAS muon component using the CORSIKA code [12] for protons and iron nuclei as primary particles. In simulations, we have used the

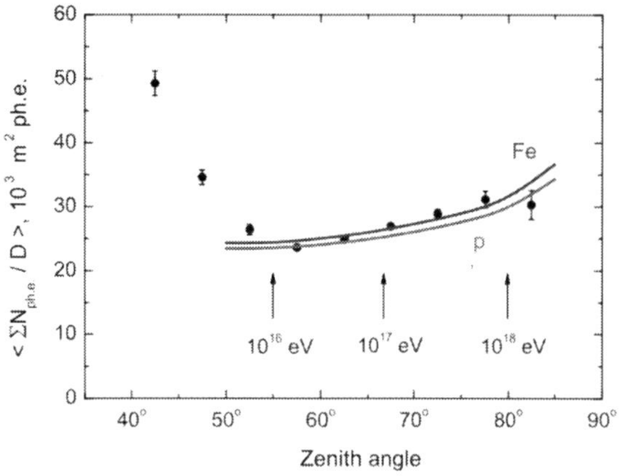

Figure 2. Dependence of the average specific energy deposit of muon bundles on the zenith angle. The points correspond to the present data. The curves represent results of CORSIKA-based calculations for primary protons and iron nuclei (labels 'p' and 'Fe', respectively).

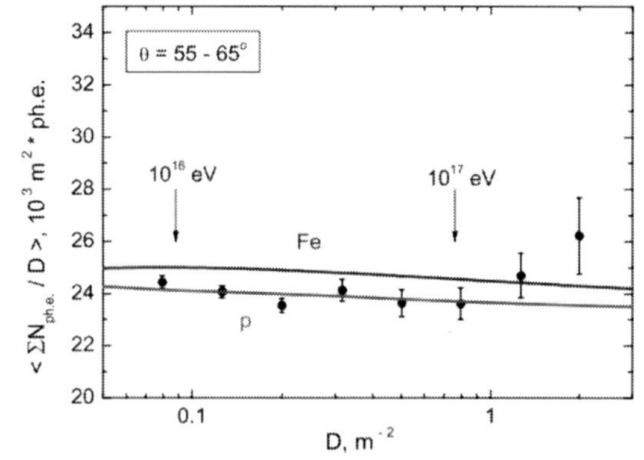

Figure 3. Dependence of the average specific energy deposit for muon bundles with arrival zenith angles $55° \leq \theta < 65°$ on the local muon density. Notations are the same as in Fig. 2.

combination of hadron interaction models SIBYLL2.3 [13] and FLUKA [14] for hadrons with energies above and below 80 GeV, respectively. As calculations show, the value of the mean energy of muons comprising the bundles rapidly increases with zenith angle and reaches approximately 500 GeV at $\theta = 85°$. The calculated dependence of the average specific energy deposit was obtained in the following way. First, for every zenith angle the average specific muon energy loss $\langle dE/dX \rangle$ was calculated using the energy loss tables [15]. Then, the aforementioned assumption that the total Cherenkov light yield in the calorimeter is proportional to the muon energy loss was done. Finally, the absolute calibration was derived from the normalization of the calculated curve for primary protons to measurement results in the zenith angle interval $55° - 65°$. As a whole, the measured dependence of the average specific energy deposit on the zenith angle confirms the increase of the mean muon energy in the bundles and is in reasonable agreement with expectation. However, at angles $\theta > 70°$ (at effective primary energies more than about 10^{17} eV) some indication for an excess compared to calculations is seen, even for a heavy primary nuclei assumption.

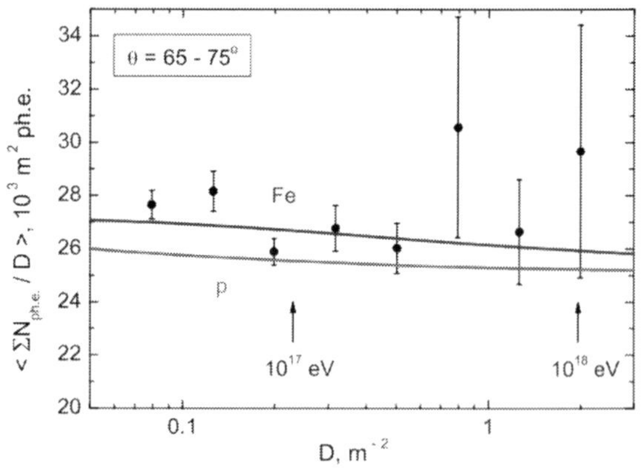

Figure 4. The same as in Fig. 3, but for zenith angle interval $65° \leq \theta < 75°$.

In Fig. 3, the measured average specific energy deposit of muon bundles detected in zenith angle interval $55° \leq \theta < 65°$ is plotted as a function of the local muon density. In fact, such consideration for a fixed interval of zenith angles allows to follow possible changes of the detector response (and correspondingly, of the mean muon energy in the bundles) with the increase of the primary energy. The curves in the figure are obtained on the basis of simulation for primary protons and iron nuclei for a fixed zenith angle $\theta = 65°$; the same normalization coefficient as in Fig. 2 has been applied. Simulation results demonstrate a trend to a slow decrease of the mean muon energy for increasing primary energy; on the contrary, the data give a hint (though not yet statistically significant) for an increase of the specific energy deposit at primary energies exceeding 10^{17} eV.

Of course, for a more definite conclusion about an excessive amount of high-energy muons in the bundles an increase of experimental statistics and careful analysis of possible systematic distortions are required. One of the possible systematic effects may be an underestimation of the number of muons in the bundles at high muon densities because of track masking due to a finite detector resolution. This difficulty may be overcome if we analyze the sample of events corresponding to larger zenith angles. For example, if we shift the minimal zenith angle from 55° to 65°, then approximately 3 times lower muon densities will correspond to the same primary energies, and the loss of tracks at effective energies around 10^{17} eV will be negligible. Simultaneously, a residual effect of non-muon components on the CWD response, if it exists, will be practically eliminated due to a significant increase of the slant depth of the atmosphere. In Fig. 4, the same dependence as in Fig. 3, but for zenith angle interval $65° \leq \theta < 75°$, is presented. Though statistical errors for the moment are large, an evidence for an increase of the average specific energy deposit of muon bundles with increasing primary energy from 10^{17} to 10^{18} eV is noticed.

4. Conclusion

Measurements of the energy characteristics of muon bundles in inclined EAS generated by primary cosmic ray particles in the energy range $10^{16} - 10^{18}$ eV are being conducted at the NEVOD-DECOR complex. The preliminary results are in reasonable agreement with CORSIKA-based simulations of EAS muon component, though some indication for an excess in the measured muon bundle energy deposit compared to calculations in the region of primary energies above 10^{17} eV is seen. This deviation may be an evidence for an inclusion of a new mechanism of generation of high energy muons at ultra-high energies of primary particles.

The work is being performed at the Experimental Complex NEVOD with support from the Ministry of Education and Science of the Russian Federation (contract RFMEFI59114X0002, MEPhI Academic Excellence Project, and government task). Simulations were performed using the resources of the MEPhI high-performance computing center.

References

[1] V. Avati et al., Astropart. Phys. **19**, 513 (2003)
[2] J. Abdallah et al., Astropart. Phys. **28**, 273 (2007)
[3] R.P. Kokoulin et al., Nucl. Phys. B (Proc. Suppl.) **196**, 106 (2009)
[4] A.G. Bogdanov et al., Phys. Atom. Nuclei **73**, 1852 (2010)
[5] L. Nellen (for Pierre Auger Collaboration), J. Phys.: Conf. Ser. **409**, 012107 (2013)
[6] A. Aab et al., Phys. Rev. D **91**, 032003 (2015)
[7] A.A. Petrukhin, NIM A **742**, 228 (2014)
[8] V.M. Aynutdinov et al., Astrophys. Space Sci. **258**, 105 (1998)
[9] S.S. Khokhlov et al., Astrophys. Space Sci. Trans. **7**, 271 (2011)
[10] N.S. Barbashina et al., Instrum. Experim. Techniques **43**, 743 (2000)
[11] A.G. Bogdanov et al., EPJ Web of Conf. **99**, 06004 (2015)
[12] D. Heck et al., Forschungszentrum Karlsruhe Report, FZKA 6019 (1998)
[13] F. Riehn et al., Proc. 34th Int. Cosmic Ray Conf., Hague, PoS (ICRC2015) 558 (2015)
[14] A. Ferrari et al., Report CERN-2005-10 (2005)
[15] D.E. Groom et al., Atom. Data and Nucl. Data Tables **78**, 183 (2001)

Testing the QGSJET01 and QGSJETII-04 models with the help of atmospheric muons

Leonid G. Dedenko[1,2,a], Anton V. Lukyashin[3,4,b], Tatiana M. Roganova[2,c], and Galina F. Fedorova[2,d]

[1] Faculty of Physics M.V. Lomonosov Moscow State University, Leninskie Gory, 119991 Moscow, Russia
[2] Skobeltsyn Institute of Nuclear Physics, Lomonosov Moscow State University, 119234 Moscow, Russia
[3] State Scientific Center Institute for Theoretical and Experimental Physics, National Research Center "Kurchatov Institute", 117218 Moscow, Russia
[4] National Research Nuclear University (MEPhI) Moscow Engineering Physics Institute, 115409 Moscow, Russia

Abstract. More accurate original calculations of the atmospheric vertical muon energy spectra at energies $10^2 - 10^5$ GeV have been carried out in terms of the QGSJET01 and QGSJETII-04 models. The Gaisser-Honda approximations of the measured energy spectra of primary protons, helium and nitrogen nuclei have been used. The CORSIKA package has been used to simulate cascades in the standard atmosphere induced by different primary particles with various fixed energies E. Statistics of simulated cascades for secondary particles with energies $(0.01 - 1) \cdot E$ was increased up to 10^6. It has been shown that predictions of the QGSJET01 and QGSJETII-04 models for these muon fluxes are below the data of the classical experiments L3 + Cosmic, MACRO and LVD by factors of ~ 1.7-2 at energies above 10^2 GeV. It has been concluded that these tested models underestimate the production of the most energetic secondary particles, namely, π-mesons and K-mesons, in interactions of primary protons and other primary nuclei with nuclei in the atmosphere by the same factors.

1. Introduction

Extensive air showers (EAS) are used as a tool to understand the origin and composition of cosmic rays, their possible sources and the transport of cosmic particles in various magnetic fields on their way to the Earth at very high energies. All features of the energy spectrum, arrival directions and composition of the primary cosmic particles should be determined through an analysis of the EAS data. These data as some signals in the surface and underground detectors are usually interpreted in terms of various models of hadronic interactions [1–8]. Such interpretation may be not obligatory correct. As an example, the energy of showers calculated in terms of the QGSJETII-03 [2] model with the help of the surface detectors signals at the Telescope Array (TA) [9] happened to be 1.27 times lager than such energy estimated with the help of the fluorescence light. To ensure that results of such an interpretation are as accurate as possible these models should be thoroughly tested. Usually these models are tested with the help of the accelerator data at small values ($\sim$0) of the pseudorapidity, η, where most of secondary particles are produced [10–12]. However, calculations have shown that the maximal energy flow carried by secondary particles occurs at much larger values ($\sim$8$-$10) of the pseudorapidity [13]. Let us also note that the longitudinal development of EAS depends strongly on the rate of the projectile particle energy fragmentation. So, it is of primary importance to verify the production of the most energetic secondary particles simulated in terms of these models. The atmospheric muon flux also depends strongly on such a production. So, it is a valuable suggestion to test various models of hadronic interactions by comparing model predictions of these muon fluxes with data. There are many beautiful data to be compared with model predictions. We select the classical experiments L3+Cosmic [14], MACRO [15] and LVD [16] and elaborate the smooth approximation of these muon data in the energy interval of $10^2 - 10^5$ GeV. To make model predictions of the muon flux we have suggested to simulate EAS induced by primary protons, helium and nitrogen nuclei with different fixed energies in the atmosphere with the help of the CORSIKA package [17] and calculate the muon energy spectrum in each individual shower. Results of these simulations for every type of primary particles multiplied by intensities of these primary particles should be integrated on the energy of these primary particles. Thus, we also need some expressions for the energy spectra of various primary particles. Indeed, there are results of many measurements (e.g. ATIC-2 [18], CREAM [19], RUNJOB [20], AMS02 [21,22], PAMELA [23]) of the fluxes of the primary cosmic nuclei. We will use the Gaisser-Honda approximations [24] for energy spectra of the various primary nuclei. Thus, with the help of any interaction models [1–8] and the package CORSIKA [17] and data on fluxes of the primary cosmic nuclei [18–23] or any approximations of the data, one can predict the energy spectra of atmospheric vertical high energy muons at sea level. These predictions can be compared with data observed by the L3+Cosmic [14], MACRO [15]

[a] e-mail: ddn@dec1.sinp.msu.ru
[b] e-mail: lukyashin.anton@physics.msu.ru
[c] e-mail: rogatm@yandex.ru
[d] e-mail: fdr@dec1.sinp.msu.ru

and LVD [16] collaborations at energies above 100 GeV. Finally, some conclusions can be drawn about the validity of various models.

In fact, some low energies have been tested with the FLUKA [25] package in such a way. We are sorry that our testing of some models in [26–28] are not correct. We do apologize for our mistake in input data for the atmosphere!

In this paper models QGSJET01 [1] and QGSJETII-04 [3] have been tested. A comparison of muon data observed in references [14–16] with the results of simulations allows us to draw a conclusion about the most energetic secondary particle production described by these models.

2. Method

The hadronic interaction models QGSJET01 and QGSJETII-04 have been tested with the help of the atmospheric muon data. To estimate the energy spectra $D(E_\mu)$ of atmospheric vertical muons in the energy range $10^2 - 10^5$ GeV we need to know the energy spectra dI_p/dE, dI_{He}/dE and dI_N/dE of the primary protons, He and N nuclei within the energy interval $10^2 - 10^7$ GeV and the energy spectra of vertical muons $S_p(E_\mu, E)$, $S_{He}(E_\mu, E)$ and $S_N(E_\mu, E)$ calculated from the primary protons, helium and nitrogen nuclei with various fixed energies, E, in terms of the QGSJET01 and QGSJETII-04 hadronic interaction models in the same energy range of $10^2 - 10^5$ GeV. For a comparison we need data on the vertical muon flux. The smooth approximation of the atmospheric muon data observed by the collaborations L3+Cosmic, MACRO and LVD had been used for comparison with results of simulations.

Functions $S_p(E_\mu, E)$, $S_{He}(E_\mu, E)$, $S_N(E_\mu, E)$ are the differential energy spectra of muons in showers induced by primary protons, helium and nitrogen nuclei with fixed values of energy, E. These spectra were calculated for 24 values of the energy E of the primary protons. The energy distributions of muons induced by primary helium and nitrogen nuclei $S_{He}(E_\mu, E)$, $S_N(E_\mu, E)$ have been also calculated as a combined distributions of protons based on the hypothesis of superposition [29]. As results coincide with simulations in terms of superposition, we will use this hypothesis for all nuclei.

The energy spectra of the primary particles are important ingredients of simulations. As the energy per nucleon is of importance, only the energy spectra of the primary protons, He and N nuclei should be taken into account. We had used approximations (1) for $(dI_p/dE)_{GH}$, $(dI_{He}/dE)_{GH}$ and $(dI_N/dE)_{GH}$ suggested by Gaisser and Honda (GH).

We have taken into account a possible change of primary spectrum above the "knee". At energies above $E_1 = 3 \cdot 10^6$ GeV for primary protons and above $E_2 = 6 \cdot 10^6$ GeV for primary He and N nuclei we used modified GH approximations (2) of the energy spectra of primary particles. The values of parameters for the Gaisser-Honda approximation are listed in Table 1. The approximation parameters: α, b and c are dimensionless units. Parameter K has dimensions $[1/(GeV \cdot m^2 \cdot s \cdot sr)]$. E_k is the kinetic energy per nucleon in GeV units.

$$\frac{dN}{dE_k} = K \cdot \left(E_k + b \cdot exp\left(-c \cdot \sqrt{E_k}\right)\right)^{-\alpha} \quad (1)$$

Table 1. Parameters for the Gaisser-Honda approximation.

Nuclei	A	α	K	b	c
H	1	2,74	14900	2,15	0,21
He	4	2,64	600	1,25	0,14
N	14	2,6	33,2	0,97	0,01

Table 2. Parameters for simulations: proton energy E and threshold energy E_{th} in GeV and statistics N.

E	E_{th}	N	E	E_{th}	N
113,3	100	10^6	$3,162 \cdot 10^4$	100	10^6
177,8	100	10^6	$5,623 \cdot 10^4$	100	10^6
237,1	100	10^6	$1 \cdot 10^5$	100	10^5
316,2	100	10^6	$1,778 \cdot 10^5$	100	10^5
421,7	100	10^6	$3,162 \cdot 10^5$	100	10^5
562,3	100	10^6	$5,623 \cdot 10^5$	100	10^5
749,8	100	10^6	$1 \cdot 10^6$	100	10^4
$1 \cdot 10^3$	100	10^6	$1,778 \cdot 10^6$	100	10^4
$1,778 \cdot 10^3$	100	10^6	$3,162 \cdot 10^6$	100	10^4
$3,162 \cdot 10^3$	100	10^6	$5,623 \cdot 10^6$	100	10^4
$5,623 \cdot 10^3$	100	10^6	$1 \cdot 10^7$	100	10^3
$1 \cdot 10^4$	100	10^6	$1,778 \cdot 10^7$	100	10^3
$1,778 \cdot 10^4$	100	10^6	$3,162 \cdot 10^7$	100	10^3
$1 \cdot 10^5$	10^3	10^6	$1 \cdot 10^6$	10^4	10^6
$1,778 \cdot 10^5$	10^3	10^6	$1,778 \cdot 10^6$	10^4	10^6
$3,162 \cdot 10^5$	10^3	10^6	$3,162 \cdot 10^6$	10^4	10^6
$5,623 \cdot 10^5$	10^3	10^6	$5,623 \cdot 10^6$	10^4	10^6

$$\frac{dN}{dE_k} = K \cdot \left(E_k + b \cdot exp\left(-c \cdot \sqrt{E_k}\right)\right)^{-\alpha} \cdot \sqrt{\frac{E_i}{E_k}} \quad (2)$$

The CORSIKA 7.4 package had been used to simulate the second important ingredients - the energy spectra $S_p(E_\mu, E)$, $S_{He}(E_\mu, E)$ and $S_N(E_\mu, E)$ of vertical muons in showers induced by primary protons, helium and nitrogen nuclei for various fixed energies, E, in terms of the QGSJET01 and QGSJETII-04 hadronic interaction models in the energy range of $10^2 - 10^5$ GeV with statistics 10^6 events for the most energetic muons. Table 2 shows fixed energies, E, of the primary protons and statistics of simulated showers.

The results of these calculations in the energy range $10^2 - 10^7$ GeV were interpolated for 100 values of energies E with equal intervals in decimal logarithmic scale. The energy interval $10^2 - 10^5$ GeV of muons was divided into 60 equal bins also in a decimal logarithmic scale. So, the width of the bin was equal to $h = \lg(E_{\mu,(i+1)}/E_{\mu,i}) = 0,05$. Let us note that average muon energies for the 1-st, 21-st and 41-st bins which we will use later are equal to 105.9, $1.059 \cdot 10^3$ and $1.059 \cdot 10^4$ GeV respectively. In fact simulations for He and N nuclei have been carried out only for energies 10^4 and 10^6 GeV to test the hypothesis of superposition. As results of simulations for the primary nuclei showed a good agreement with this hypothesis we used this hypothesis to estimate the flux of the nucleons from the primary helium and nitrogen nuclei.

The energy spectra $D_p(E_\mu)$, $D_{He}(E_\mu)$ and $D_N(E_\mu)$ of muons for the primary protons, He and N nuclei are calculated as integrals of the products of functions $S_p(E_\mu, E)$, $S_{He}(E_\mu, E)$ and $S_N(E_\mu, E)$ with corresponding intensities dI_p/dE, dI_{He}/dE and dI_N/dE of the primary protons, on the energy, E, of primary particles.

DOI: 10.1051/epjconf/201714516003

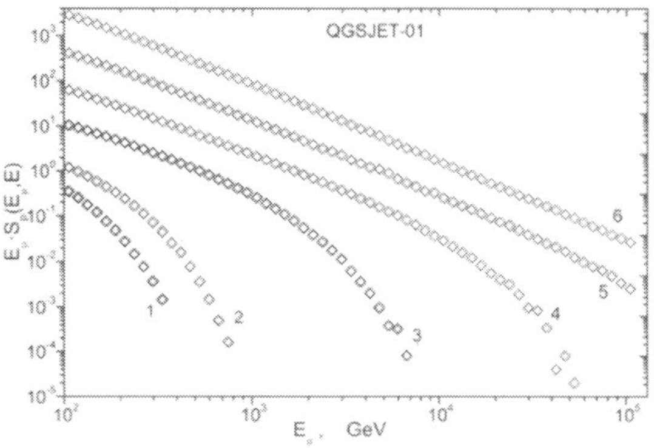

Figure 1. The energy spectra of muons in showers induced by primary protons with various fixed energies E (QGSJET01 model): 1–5 · 10^2; 2–10^3; 3–10^4; 4–10^5; 5–10^6; 6–10^7 GeV.

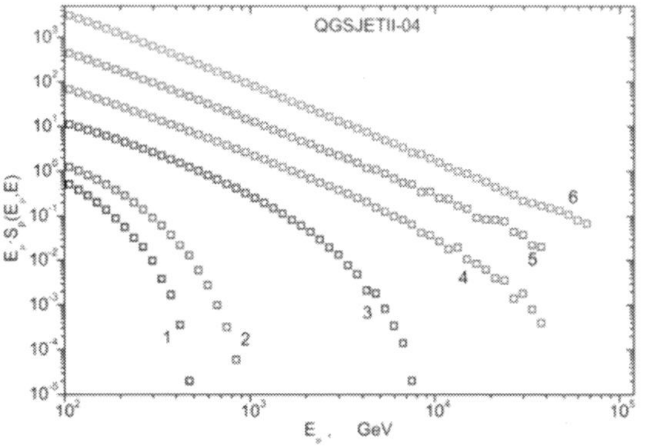

Figure 2. The energy spectra of muons in showers induced by primary protons with various fixed energies E (QGSJETII-04 model): 1–5 · 10^2; 2–10^3; 3–10^4; 4–10^5; 5–10^6; 6–10^7 GeV.

$$D_p(E_\mu) = \int \left(\frac{dI_p}{dE}\right) \cdot S_p(E_\mu, E) \cdot dE \qquad (3)$$

$$D_{He}(E_\mu) = \int \left(\frac{dI_{He}}{dE}\right) \cdot S_{He}(E_\mu, E) \cdot dE \qquad (4)$$

$$D_N(E_\mu) = \int \left(\frac{dI_N}{dE}\right) \cdot S_N(E_\mu, E) \cdot dE \qquad (5)$$

The resulting energy spectrum of atmospheric muons is the sum of partial energy spectra of muons produced by primary protons, helium and nitrogen nuclei

$$D(E_\mu) = D_p(E_\mu) + D_{He}(E_\mu) + D_N(E_\mu). \qquad (6)$$

3. Results of simulations

The energy spectra $S_p(E_\mu, E)$ of the atmospheric vertical muons simulated for various fixed energies, E, of primary protons in terms of the QGSJET01 and QGSJETII-04 hadronic interaction models are shown in Figs. 1 and 2 respectively. It is seen that statistics of $\sim 10^6$ at the higher energy end of the spectra is not enough.

Table 3. Average number of muons with energies above the threshold E_{th} in showers induced by primary protons with energies E.

	$E = 10^5$ GeV		
Model	Paper E_{th}	100 GeV	1000 GeV
QGSJET01	[30]	21,0	0,605
QGSJET01	This work	20,9	0,593
QGSJETII-04	This work	22,4	0,613
	$E = 10^6$ GeV		
Model	Paper E_{th}	100 GeV	1000 GeV
QGSJET01	[30]	132,3	3,612
QGSJET01	This work	132,8	3,490
QGSJETII-04	This work	141,3	3,583

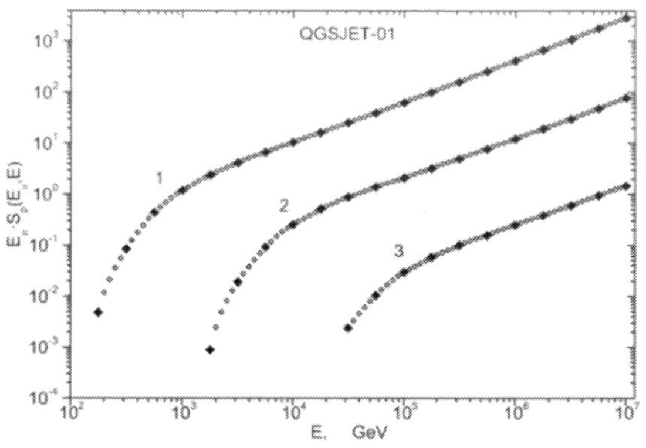

Figure 3. Dependence of $E_\mu \cdot S_p(E_\mu, E)$ on E for 3 muon bins. 1–$E_\mu = 105,9$ GeV; 2–$E_\mu = 1,059 \cdot 10^3$ GeV; $E_\mu = 1,059 \cdot 10^4$ GeV. ♦– direct calculation, ◇–interpolated results.

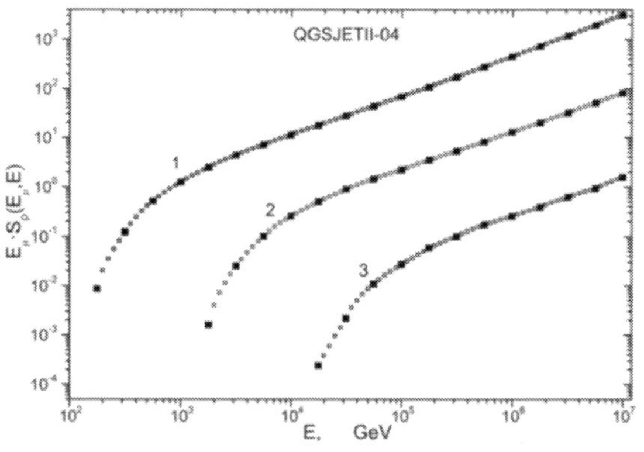

Figure 4. Dependence of $E_\mu \cdot S_p(E_\mu, E)$ on E for 3 muon bins. 1–$E_\mu = 105,9$ GeV; 2–$E_\mu = 1,059 \cdot 10^3$ GeV; $E_\mu = 1,059 \cdot 10^4$ GeV. ■–direct calculation, □–interpolated results.

Table 3 displays the total number of muons with energies above 10^2 and 10^3 GeV in showers induced by primary protons with energies 10^5 and 10^6 GeV estimated in terms of the QGSJET01 and QGSJETII-04 hadronic interaction models in our simulations and in reference [30]. A very reasonable agreement is seen.

Results of interpolation on energy, E, for the 1-st, 21-st and 41-st bins of muon energy spectra $S_p(E_\mu, E)$ as open symbols and direct results as solid symbols are displayed in Figs. 3 and 4 for the QGSJET01 and QGSJETII-04 models respectively. Good agreement is also seen.

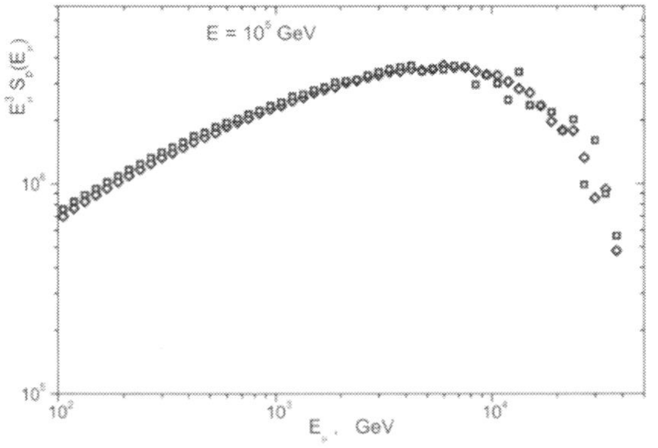

Figure 5. The energy spectra of muons in showers induced by primary protons with fixed energy E = 10^5 GeV. ◊ - QGSJET01, □ - QGSJETII-04.

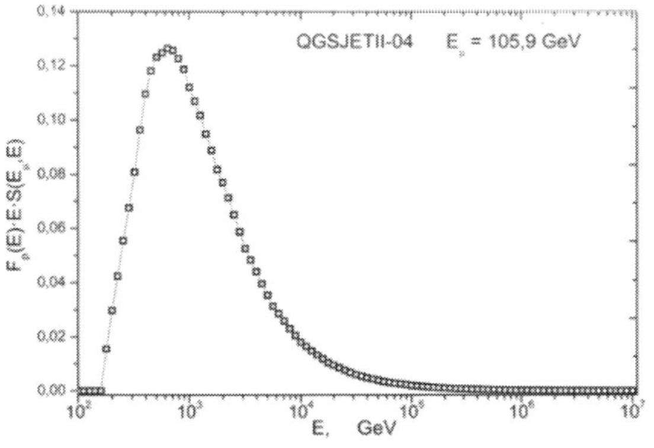

Figure 6. Contributions of the primary protons with various energies E into the 1-st bin of the muon spectra.

The next Fig. 5 demonstrates a comparison of the muon energy spectra $S_p(E_\mu, E)$ calculated in terms of the QGSJET01 model as open diamonds and the QGSJETII-04 model as solid squares for a fixed energy $E = 10^5$ GeV of the primary protons. The results for the QGSJETII-04 model are 10–15% above the QGSJET01 model values.

Figures 6, 7 and 8 demonstrate distributions of the primary nucleon energy which contribute much to the 1-st, 21-st and 41-st bins of muon energy spectra respectively for the QGSJETII-04 model. It is possible to note that nearly two orders of energy, E, are of importance for any fixed energy, E_μ, of muons. The maximal contributions occur at energies $\sim 7 \cdot 10^2$ GeV (1-st bin), $\sim 7 \cdot 10^3$ GeV (21-st bin) and $\sim 6,4 \cdot 10^4$ GeV (41-st bin). The results of comparison of calculations in terms of the QGSJET01 and QGSJETII-04 models spectra $D(E_\mu)$ with the smooth approximation of data [14–16] are illustrated in Fig. 9. It should be noted that at the above energies ~100 GeV both the simulated spectra and data are steepened. It is because the decay constant, B, for the charged mesons is equal to $B \sim 100$ GeV and the probability of decay for charged mesons with higher energies is decreasing.

It is seen that the calculated spectra are 2 times below the data for the case of the QGSJET01 model and 1.7 times below the data for the QGSJETII-04 model. Figure 10

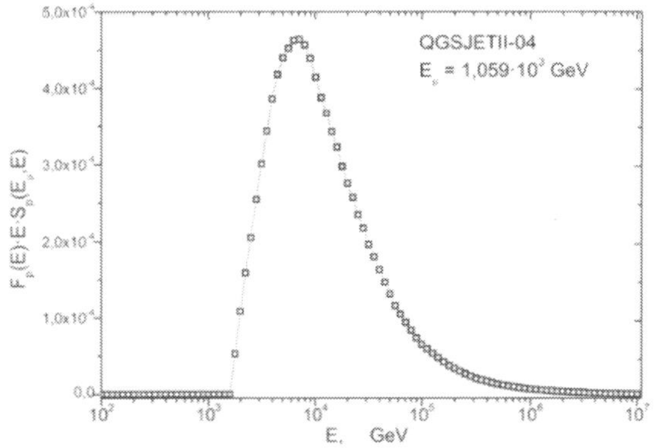

Figure 7. Contributions of the primary protons with various energies E in to the 21-st bin of the muon spectra.

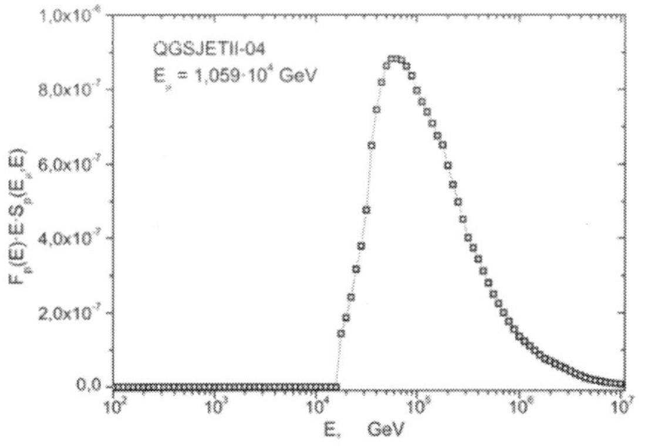

Figure 8. Contributions of primary protons with various energies E in to the 41-st bin of the muon spectra.

shows ratios MC/DATA for the two models. For the case of the QGSJET01 model this ratio decrease from 55% at $E = 10^2$ GeV to 50% at $E = 10^5$ GeV while in case of the QGSJETII-04 model this ratio decreases from 65% to 55% for the same values of muon energy E. Figure 11 shows various contributions to the total muon energy spectrum $D(E_\mu)$ from the primary protons, He and N nuclei. These contributions are decreasing from 90 to 85% for protons and are increasing from 9 to 14% for He nuclei and from 0,5 to 1% for N nuclei. These fractions are not exactly reliable because the true fraction of nuclei is unknown.

4. Conclusion

Muons which contributes much to the muon energy spectra are produced in decays of the most energetic π-mesons and K-mesons generated in the first interactions of the primary particles with nuclei in the atmosphere. As the calculated vertical muon energy spectra for the QGSJET01 and QGSJETII-04 models are 1.7 times and 2 times respectively below the data we can conclude that the production of the most energetic π-mesons and K-mesons in these models is considerably suppressed. This suppression may induce smaller values of signals in the surface scintillation detectors and will result in larger values of the calculated energy estimates. So, the

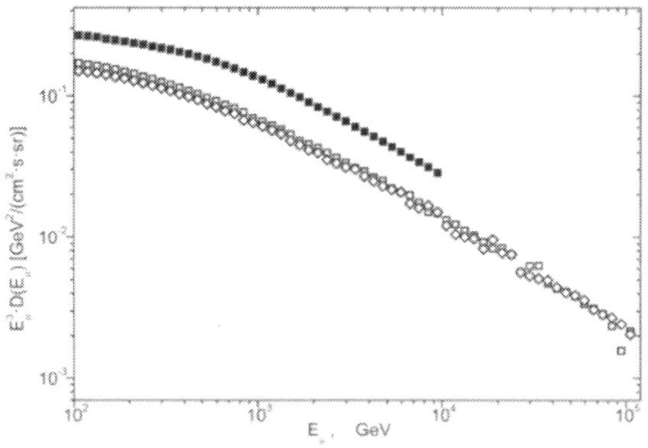

Figure 9. The energy spectra of near vertical muons. ◊–QGSJET01, □–QGSJETII-04, and (■)–experimental data (L3+Cosmic, MACRO, LVD).

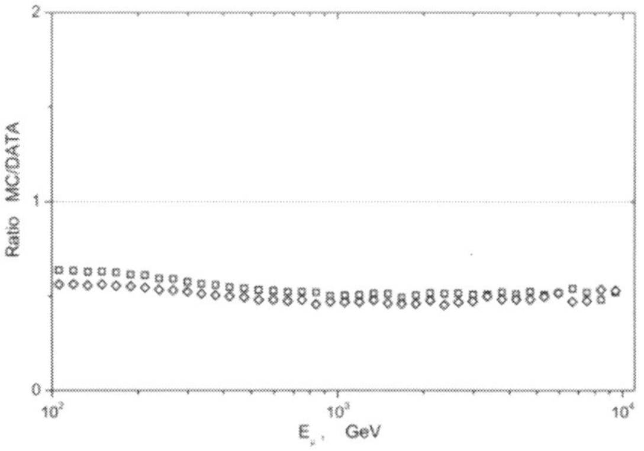

Figure 10. The ratio MC/DATA: ◊–QGSJET01, □–QGSJETII-04.

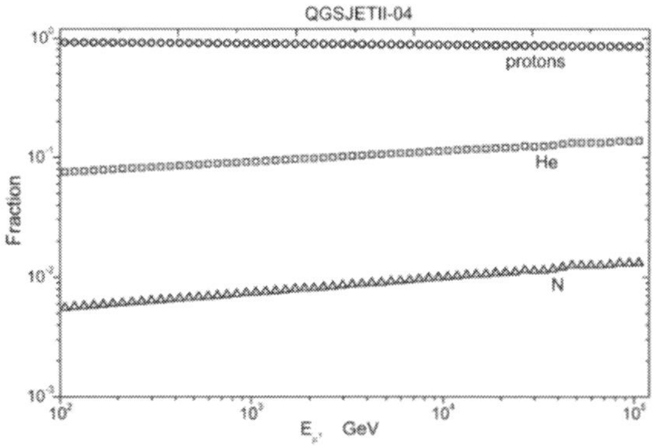

Figure 11. Relative contribution of protons, He and N nuclei into the muon spectrum.

coefficient 1.27 used by the TA collaboration [9] to decrease the energy estimates of showers calculated on the base of signals in the scintillation detectors may be understood as a result of this suppression. The increased intensity of the primary particle flux observed at the Yakutsk array at super high energies [31] may also be a result of smaller values of the calculated signals in surface scintillation detectors.

Authors thanks N. N. Kalmykov, pointing to alternative computing with other results and A. A. Lagutin for important assistance in the verification of the results for the QGSJET01 model.

References

[1] Kalmikov N. N., Ostapchenko S. S., Phys. Atom. Nucl. **56**, 346 (1993)

[2] Ostapchenko S. S., Phys. Rev. D. **74**, 014026; hep-ph/0505259 (2006)

[3] Ostapchenko S. S., Phys. Rev. D. **83**, 014018 (2011)

[4] Ahn E.-J., Engel R., Gaisser T. K., et al., Phys. Rev. D. **80**, 094003 (2009)

[5] Werner K., Liu F. M. and Pierog T., Phys. Rev. C. **74**, 044902; hep-ph/0506232 (2006)

[6] Pierog T. and Werner K., Nucl. Phys. Proc. Suppl. **196**, 102; arXiv:0905.1198 (2009)

[7] Ranft J., Phys. Rev. D. **51**, 64 (1995) hep-ph/9911213 and hep-ph/9911232 (1999)

[8] Werner K., Phys. Rep. **232**, 87 (1993)

[9] Abu-Zayyad T., Aida R., Allen M., et.al., (Telescope Array Collaboration) Astrophys. J. Lett. **768:L1** 5 (2013)

[10] Pierog T., EPJ Web of Conf. **99**, 09002 (2015)

[11] Ostaphcenko S. S., Progr. of Theor. Phys. Suppl., **193**, 204 (2012)

[12] D'Enterria D., Engel R., Pierog T., et. al., Astropart. Phys. **35**, 98 (2011)

[13] Engel R., Rebel H., Acta Phys. Polonica B. **35**, 321 (2004)

[14] Achard P., et al. (The L3 Collab.) Phys. Lett. B. **598**, 15–32, hep-ex/0408114v1 (2004)

[15] Ambrosio M., Antolini R., Auriemma G., et al., (The MACRO Collab.), Phys. Rev. D. **52**, 3793 (1995)

[16] Aglietta M., Alpat B., Alieva E. D., et al., (The LVD Collab.), Phys. Rev. D. **58**, 092005, hep-ex/9806001v1 (1998)

[17] Heck D., Knapp J., Capdevielle J.-N., et al., Report FZKA 6019 Forschungszentrum Karlsruhe, (1998)

[18] Panov A. D., Adams J. H., Ahn H. S. et al., Bull. Russ. Acad. Sci. Phys. **71**, 494 (2007); Bull. Russ. Acad. Sci. Phys. **73**, 564 (2009)

[19] Ahn H. S. (for the CREAM Collab.), Astrophys. J. Lett. **L 89**, 714 (2010)

[20] Derbina V. A., Galkin V. I. and Hareyama M. (for the RUNJOB Collab.), Astrophys. J. **L 41**, 628 (2005)

[21] Aguilar M., Aisa D., Alpat B., et. al., (AMS Collab.) Phys. Rev. Lett. **114**, 171103 (2015)

[22] Aguilar M., Aisa D., Alpat B., et. al., (AMS Collab.) Phys. Rev. Lett. **115**, 211101 (2015)

[23] Adriani O., Barbarino G. C., Bazilevskaya G. A., et. al., (PAMELA Collab.) Advances in Space Research **51**, 219-226 (2013)

[24] Gaisser T. K. and Honda M., Ann. Rev. Nucl. Part. Sci. **52**, 153–199 (2002)

[25] Battistoni G., Ferrari A., Muraro S. and Sala P. R., (Proc. Suppl.) Nucl. Phys. B. **168**, 286 (2007)
[26] Dedenko L. G., Roganova T. M., Fedorova G. F., JETP Lett. **100**, 223 (2014)
[27] Dedenko L. G., Roganova T. M., Fedorova G. F., Phys. Atom. Nucl. **78**, page numbers (2015)
[28] Dedenko L. G., Lukyashin A. V., Fedorova G. F. and Roganova T. M., EPJ Web of Conferences **99**, 10003 (2015)
[29] Dedenko L.G., Zatsepin G.T., Proceedings of the 6-th ICRC, Moscow, **II**, Extensive air showers and cascades process, 201–208 (1960)
[30] Lagutin A. A., Tyumentsev A. G., Yushkov A. V., J. Phys. G. **30**, 573–596, `hep-ph/0402070Journal` (2004)
[31] Glushkov A. V., Egorova V. P., Ivanov A. A., et al., Proceedings ot the 28th ICRC, Tsukuba **1**, 393 (2003)

DOI: 10.1051/epjconf/201714517001

Investigation of EAS cores

S.B. Shaulov[1,6,a], P.F. Beyl[2], R.U. Beysembaev[2], E.A. Beysembaeva[2], S.P. Bezshapov[1], A.S. Borisov[1], K.V. Cherdyntceva[1], M.M. Chernyavsky[1], A.P. Chubenko[1], O.D. Dalkarov[1,7], V.G. Denisova[1], A.D. Erlykin[1], N.V. Kabanova[1], E.A. Kanevskaya[1], K.A. Kotelnikov[1], A.E. Morozov[1], R.A. Mukhamedshin[5], R.A. Nam[2], N.M. Nesterova[1], N.M. Nikolskaya[1], V.P. Pavluchenko[1], V.V. Piskal[2], V.S. Puchkov[1], S.E. Pyatovsky[1], V.A. Ryabov[1,7], T.Kh. Sadykov[4], A.L. Schepetov[1], M.D. Smirnova[1], A.V. Stepanov[2], A.V. Uryson[1], Yu.N. Vavilov[1], N.G. Vildanov[2], L.I. Vildanova[2], I.S. Zayarnaya[1], J.K. Zhanceitova[3], and V.V. Zhukov[2]

[1] P.N.Lebedev Physical Institute of the Russian Academy of Sciences, Moscow, Russia
[2] Tien-Shan High Mountain Science Station, Almaty, Kazakhstan
[3] Baykonurov Dgezkazgan University, Kazakhstan
[4] Institute for Physics and Technology, Almaty, Kazakhstan
[5] Institute for Nuclear Research of the Russian Academy of Sciences, Moscow, Russia
[6] Moscow Institute of Physics and Technology (State University),MIPT (SU), Moscow, Russia
[7] National Research Nuclear University MEPhI (Moscow Engineering Physics Institute), Moscow, Russia

Abstract. The development of nuclear-electromagnetic cascade models in air in the late forties have shown informational content of the study of cores of extensive air showers (EAS). These investigations were the main goal in different experiments which were carried out over many years by a variety of methods. Outcomes of such investigations obtained in the *HADRON* experiment using an X-ray emulsion chamber (XREC) as a core detector are considered. The N_e spectrum of EAS associated with γ-ray families, spectra of γ-rays (hadrons) in EAS cores and the N_e dependence of the muon number, $\langle N_\mu \rangle$, in EAS with γ-ray families are obtained for the first time at energies of 10^{15}–10^{17} eV with this method. A number of new effects were observed, namely, an abnormal scaling violation in hadron spectra which are fundamentally different from model predictions, an excess of muon number in EAS associated with γ-ray families, and the penetrating component in EAS cores. It is supposed that the abnormal behavior of γ-ray spectra and N_e dependence of the muon number are explained by the emergence of a penetrating component in the 1st PCR spectrum 'knee' range. Nuclear and astrophysical explanations of the origin of the penetrating component are discussed. The necessity of considering the contribution of a single close cosmic-ray source to explain the PCR spectrum in the knee range is noted.

1. Introduction

The power spectrum of the primary cosmic radiation (PCR) has two specific ranges, namely, the so-called knee at energies of 10^{15}–10^{18} eV and the range of energies higher than 10^{19} eV where cut off of the PCR spectrum is supposed. The analysis of data submitted in this paper belongs to the knee range.

The PCR energy spectrum has two rather sharp breaks in this range, namely, at energies of 3 and 70 PeV. The origins of these peculiarities of the spectrum are still not clear though the main break at 3 PeV was found in 1958 [1]; evidence in favor of the second one was first observed in the nineties [2].

The almost only experimental way of solving the problem consists in determining the PCR mass composition in the knee range. Characteristics of EAS cores at mountain altitudes is most sensitive to the composition of PCR nuclei. The inclination the power spectrum of hadrons, $E_h^{-\kappa}$, at TeV energies sharply increases with the transition from primary protons to iron nuclei. In parallel, the power slope index, κ, changes from $\kappa \simeq 1$ to $\kappa \simeq 2$. The only detector allowing to carry out such measurements is the X-ray emulsion chamber (XREC) [3] due to its high spatial resolution (~ 100 microns) and its ability to measure the energies of the most energetic single hadrons in EAS cores.

The main problem of these measurements is the timing of the XREC data, that do not have a time reference, with data of the EAS-array electronic detectors. The solution of this problem [4] has permitted to carry out such measurements in the *HADRON* experiment.

The creation of a hybrid installation has become possible as a result of the development of EAS techniques developed in previous experiments. In the following section a short history of the development is given for example in the CR research in the P.N. Lebedev Physical Institute (LPI, or FIAN).

2. History

There has been more than 80 years of CR investigations in FIAN. A purposeful EAS study by the FIAN group started in the forties with the construction of the high-mountain cosmic-ray station in the Pamirs initiated by D.V.

[a] e-mail: ser101@inbox.ru

Scobeltsyn and S.I. Vavilov [5]. A series of observational operations with Geiger counters, a cloud chamber and calorimeter have been made. N.A.Dobrotin played a major role in expanding the work at the Pamir Station. The development of an EAS model or model of nuclear-electromagnetic cascades (NEC) in the atmosphere by G.T. Zatsepin became the major result of these operations [6,7].

EAS registration by ground-based installations up to now is virtually the only method to study PCR particles at energies higher than $\sim 10^{14}$ eV. The NEC model is a foundation of this method. Only the NEC model permits to determine such important characteristics as the primary energy and type of the primary nucleus.

The PCR nuclear composition consists of five groups of nuclei: p, He, CNO, Si and Fe. The fraction of each nuclear group is well-known at energies below $\sim 10^{14}$ eV from direct measurements carried out on board satellites.

At higher energies direct data is absent, and it is possible to determine the PCR composition by means of EAS study only. The PCR spectrum is measured up to a maximum energy $\sim 10^{20}$ eV but the problem of PCR composition has not yet been solved. It is a difficult problem so new observational methods are required to progress further.

In accordance with NEC models, the main information on the type of the primary nucleus is concentrated in the spectra of the most energetic hadrons located in the central area of EAS (EAS core). The size of this area is about one meter and is basically formed by nuclear interactions. The EAS periphery (hundreds meters) is mainly formed by low-energy electromagnetic processes and hadrons, and consists mainly of muons, electrons, positrons and γ-rays. These EAS components are important and used in experiment for different purposes.

The EAS periphery permits to determine the energy of the primary particle, initiating the EAS, by using the total number of charged particles (traditionally denoted N_e or N_{ch}). The transition from N_e to E_0 is performed by means of the EAS model.

The core area mainly gives information about the type of primary nucleus. Certainly such a division is conditional and was made with the aim to emphasize that the main difference in EAS installations relates to the type of core detector.

3. EAS experiment

A cloud chamber, spark chambers, calorimeters and XRECs were used as core detectors at Tien Shan in different years [8–11].

The big ionization calorimeter (BIC) at the Tien Shan station had a thickness of lead of 1000 g/cm^2 and an area of 36 m^2. The PCR spectrum, the sharp form of its break, the connection between the energy in the core and total number of electrons N_e have been obtained in the EAS experiment using BIC. These outcomes were obtained with active assistance of scientists from Bulgaria, Hungary and Czechoslovakia. Determination of the energy of the break (the knee) in the PCR spectrum ($3 \cdot 10^{15}$ eV) has permitted to calibrate measurements of EAS energies. Furthermore an estimation of the PCR composition in the range of the spectrum break has been obtained by studying muon number fluctuations [12].

Besides, the substantial growth of the hadron absorption length (a so-called "long-flying component") [13] was found in the region of the spectrum break. The hypothesis of abnormal magnification of charm generation cross-section in a narrow forward cone of interaction was proposed to explain the long-flying component [14].

The installation consisting of a large number of spark chambers placed underground has operated for the study of pairs and groups of muons.

The accuracy of determining the arrival direction of particles (~ 3 mrad) has permitted to detect pairs of penetrating particles (possibly muons), created above the installation by hadrons with energies of ~ 10 TeV [15].

Measurements of the anisotropy of PCR arrival directions in the energy range $E_0 = 5 \cdot 10^{13} - 5 \cdot 10^{14}$ eV was carried out in the Russian-Hungarian experiment, CLARA [16].

A target experiment with a calorimeter and cloud chamber placed in a magnetic field was run to determine the properties of nuclear interactions. The formation of a heavy cluster in nuclear interactions, so-called fireball, has been developed at energies above 100 GeV. This model is used till now in accelerator experiments.

Research for PCR at higher energies demanded an increase in the area of the installation. For this purpose the Wilson chamber was replaced by a big two-gap spark chamber of area 6 m^2, the largest one at that time [9]. As before, the installation included a calorimeter and an XREC installed under the spark chamber. Scientists of Łodz University (Poland) actively participated in this installation.

The sizes of the cloud and spark chambers were apparently insufficient for investigations in the range of the PCR spectrum "knee" ($E_0 \sim 3 \cdot 10^{15}$ eV). Installations with areas of hundreds square meters were required for this purpose. In response to a request of the Academy of Sciences the Soviet Government supported such a project: a big X-ray emulsion chamber of area of 1000 m^2 was assembled at the Pamirs and started to operate in 1975.

4. Experiment *Pamir*

The use of X-ray films under layers of lead for detecting particles was proposed in 1958 by I.D. Rapoport [17] and implemented by Yu.A. Smorodin in experiments with XREC exposed on board aircraft and at mountain altitudes [11] as well as by I.V. Rakobolskaya in underground XREC experiments for muon registration [18]. These experiments demonstrated the prospects of the new method. A scheme of XREC for detecting γ-rays and hadrons is the *Pamir* experiment is shown in Fig. 1.

The XREC is a passive detector of high energy particles. Hadrons and γ-rays with energies of a few TeV generate narrow electron-photon cascades with a typical size of $\sim 100\,\mu$ m in lead.

Gamma-rays with TeV energies are generated mainly through π^0-decay ($\pi^0 \to 2\gamma$) and create electromagnetic cascades in the upper lead block of the hadron XREC. The cascades create dark spots on X-ray films visible to the naked eye. Hadrons ($\pi^{\pm}$ mainly) interact with the carbon (gum) target and generate γ-rays (π^0) which are detected in the lower block of XREC.

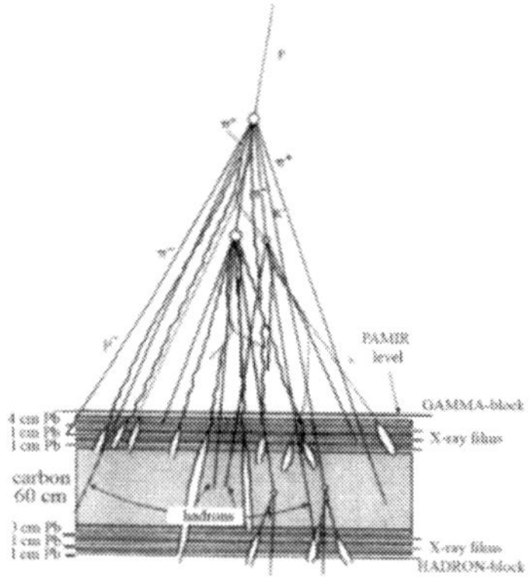

Figure 1. Scheme of XREC for detecting γ-rays and hadrons at the Pamirs.

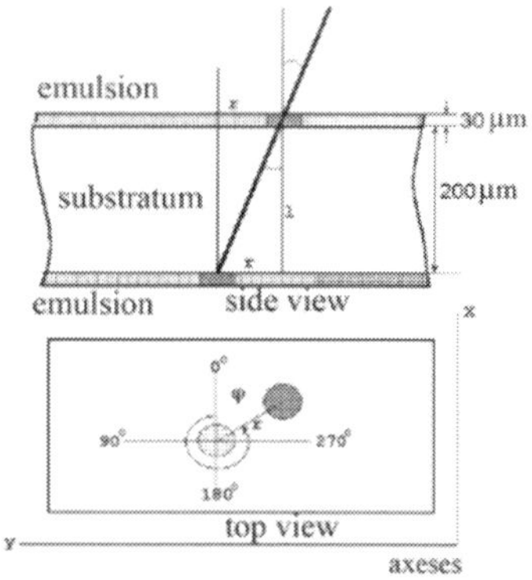

Figure 2. Scheme of angle measurements for cascades in XREC.

Measurements of the darkness of the dark spots allow the measurement of the energy and arrival angles of cascades. The energy is determined by means of a photometer and calculation procedure. Two emulsion layers on both sides of the plastic film substrate, permit to measure the shift between spots and determine the arrival angles of cascades. The scheme of angle determination in XREC is shown in Fig. 2.

The high threshold of γ-ray registration ($E_\gamma \gtrsim 1$ TeV) in XREC automatically selects the most energetic particles in EAS cores. The determination of arrival angles allows us to combine genetically bound cascades in gamma-ray – hadron families and study their energy and lateral distributions. These properties make the XREC a unique measuring tool for the study of EAS cores.

An international cooperation was established to carry out the *Pamir* experiment. It included more than a hundred research staff from Russia, Tajikistan, Kazakhstan, Armenia, Georgia, Poland, Japan and Brazil. The largest XRECs were assembled every year by efforts of the *Pamir* experiment in the Ak-Arhar gorge at an altitude of 4370 m a.s.l. (596 g/cm^2).

The *Pamir* cooperation has developed a detailed research technique for XREC usage. Some of the Russian X-ray films were exposed at Mt.Chacaltaya in Bolivia (540 g/cm^2) [3] and on the contrary Japanese films were exposed at the Pamir experiment for cross calibration of the two techniques. As a result, some supporting evidence for the long-flying component was derived [19] and a

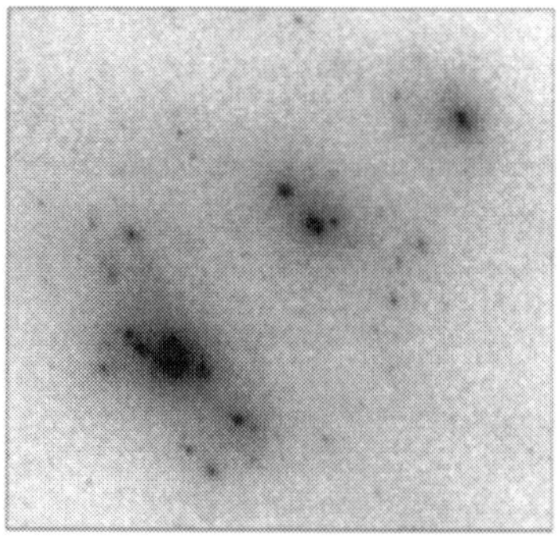

Figure 3. Aligned subgroups of high-energy particles detected in XREC.

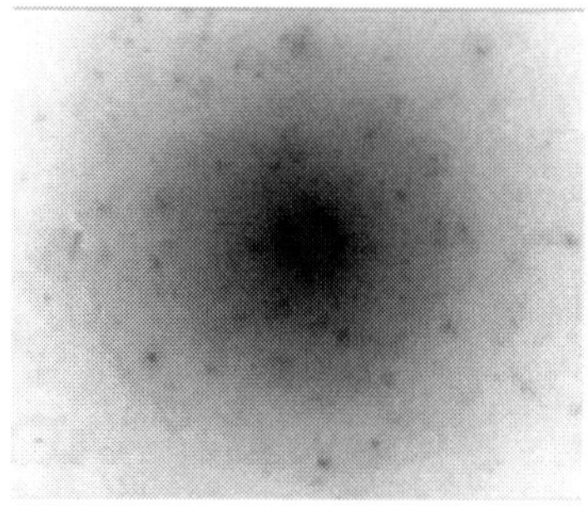

Figure 4. The halo (spot of a continuous darkening in the center of γ-ray family) with a typical size of ~ 1 cm.

number of new effects were observed. Interpretation of these results is prolonged till now. In particular, examples of interesting events found in the *Pamir* XRECs are shown in Figs. 3 and 4 which demonstrate an unusual alignment of cascade tracks along a straight line in the target plane (Fig. 3), and a so-called halo — a large dark spot at the center of the EAS core with a typical size of ~ 1 cm (Fig. 4) [20,21].

These outcomes are unique as similar effects are not observed in collider experiments. Two explanations are possible. Events of such a type can be generated by protons or nuclei but fall into the forward-cone range of the interaction, which is not visible in collider experiments, or these events have an astrophysical origin and are generated by unknown particles of a new type which are present in the primary cosmic radiation.

In the early '80s it became clear that the EAS and XREC methods separately exhausted their possibilities. On the one hand, the calorimeter technique has insufficient spatial resolution for a detailed study of the fine structure of cores. On the other hand, the XREC method does not give information on the primary particle energy. The time has come to combine these two methods.

The hybrid *HADRON* experiment at Tien Shan [22] has combined these two methods and continued to study EAS cores in a more appropriate way.

5. Experiment *HADRON*

The central part of the *HADRON* installation is schematically represented in Fig. 5.

This configuration makes it possible to determine the energy of PCR particles, the lateral distribution function

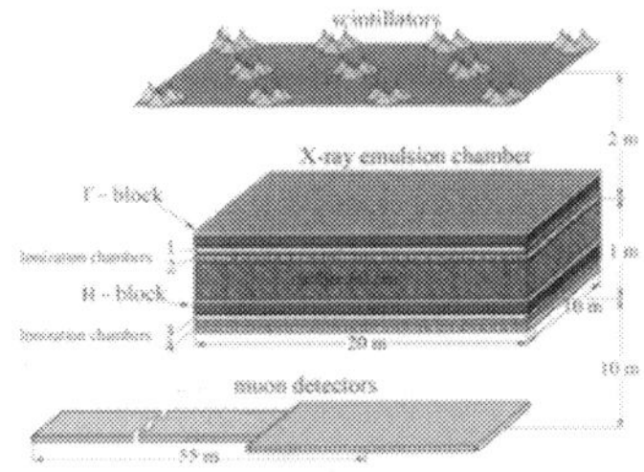

Figure 5. Scheme of the central part of the *HADRON* installation (top to bottom): ground-based scintillation detectors, XREC and ionization chambers, underground muon detector.

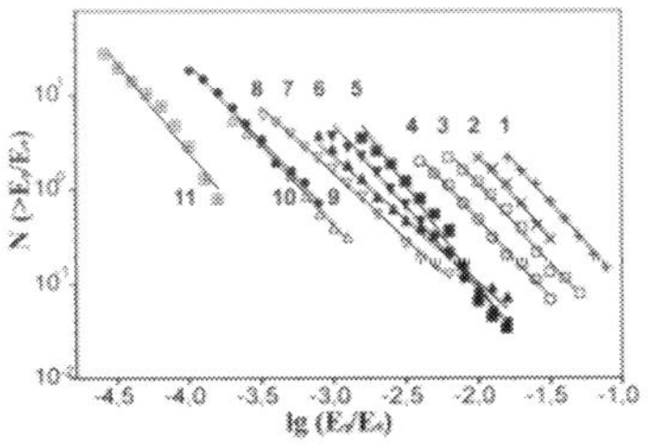

Figure 6. The experimental spectra of γ-rays energies in different intervals $N_e(E_0)$ around the break.

(LDF) of electrons and the number of muons in EAS. However, its most important part is the emulsion chamber used as a detector of the most energetic EAS-core particles. XREC makes it possible to extract unique information about the spectra of the most energetic hadrons due to the extremely high spatial resolution of XREC ($\sim 100\,\mu$m) and ability to measure energies of individual hadrons.

Extensive information has been obtained from measurements of EAS properties. In particular, valuable information has been derived from the analysis of EASs associated with γ-rays and/or high-energy hadrons.

EAS and XREC data were combined in this experiment, using procedures different from traditional experiments. Our group has been a pioneer in the application of a careful statistical method of combination [4]. As usual this procedure was performed by comparing the places and angles of arrival of an EAS and the associated γ-ray family. However, it was shown [4] that using only these criteria are insufficient to provide a rather small background, i.e., a low contribution of mistakenly combined events. The equation for the formation of background shows that a background value of 10–20% can be obtained only by including several additional conditions.

- The EAS installation should work not less than 90% of calendar time during XREC exposure time (one year).
- the position of EAS axis should be determined with an accuracy not worse 10–20 cm.
- Additional EAS selection has be done to decrease the number of EAS candidates in comparison to events in XREC. For example, one can require that a local burst in the ionization chamber with maximum ionization is above 1 TeV.

The first two requirements presented considerable challenges in the operation of the *HADRON* installation. Subsystems of *HADRON* have been doubled to support

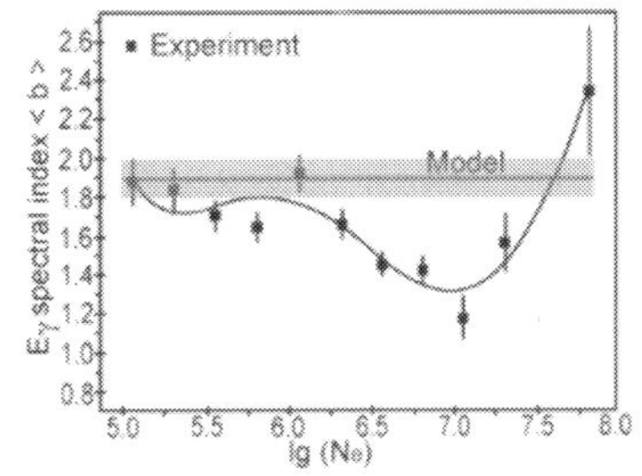

Figure 7. The N_e dependence of spectral index $\langle b\rangle$ of the energy spectra of γ-rays, E_γ^{-b}.

90% effectiveness in the case of unexpected faulty operation. Electronics of individual channels has been executed so that it could be quickly replaced. Communication circuits had photon-coupled decouplings for protection from thunderstorm and electrical noise in the conditions of high-mountainous operation and so on. The position of the EAS axis was determined by means of two layers of crosswise located ionization chambers that provided the 20-cm resolution.

The experiment *HADRON* was originally planned with a view to a direct measurement of the hadron spectra in the knee energy range. Usage of XREC allowed us to select proton showers among all EASs because EASs associated with γ-ray families must be mainly generated by primary protons as predicted by all the NEC models. On the whole, for this reason different models predicted a similar result: spectra in different intervals of primary energy have close slopes, i.e., an approximate scaling has to be observed. It is a consequence of the small efficiency of heavy nuclei to generate high energy hadrons (γ-rays). So the weighting of PCR composition behind the knee should not be affected since γ-ray–hadron families should be mainly generated by protons. Therefore any decrease of the proton fraction should only affect the intensity of events but not the form of spectra (declination). So scaling (or its weak violation) was expected because of nuclear and QCD effects associated with softer spectra of hadrons with increasing primary energy. However, expectations are not met [23].

5.1. Abnormal scaling violation in γ-rays spectra

The experimental spectra of γ-ray energies in different intervals of $N_e(E_0)$ are shown in Fig. 6. The numbers in the figure designate the numbers of logarithmic intervals in the range $\lg N_e = 5.0 - 7.25$ through $\Delta \lg N_e = 0.25$.

Figure 6 indicates that the declination of the spectra varies immediately behind the break of the EAS spectrum at $\lg N_e = 6.1$, and such a correlation is unlikely to be accidental. A N_e dependence of the spectral index $\langle b\rangle$ of the energy spectra of γ-rays, E_γ^{-b}, is shown in Fig. 7. It illustrates that a strong scaling violation takes place in contradiction with model predictions and this violation has an abnormal fashion. Declinations of the spectra are diminishing, i.e. the contribution of the most energetic hadrons increases and the spectra become harder. The regular character of $b(N_e)$ dependence is one more demonstration of a reality of observable scaling violation. Ya. Zeldovich spoke: "If experiment confirms the theory, it is pleasant. If it does not confirm, it is interesting".

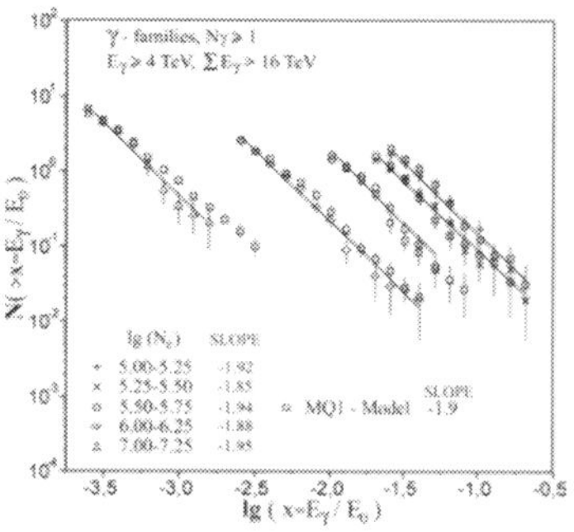

Figure 8. Observational and theoretical energy spectra of γ-rays in N_e intervals where they approximately agree.

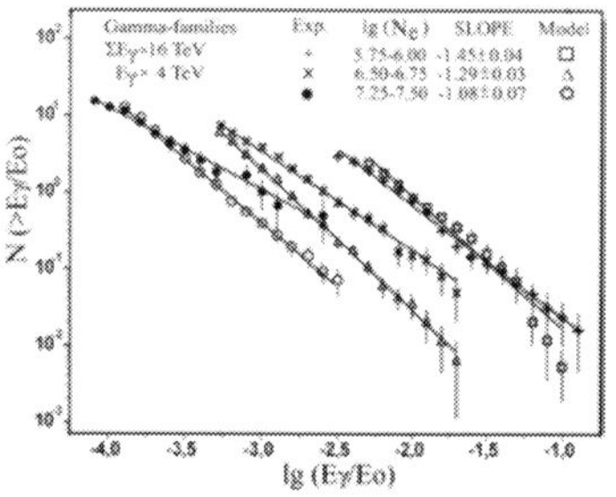

Figure 9. Observational and theoretical energy spectra of γ-rays in N_e intervals where they differ.

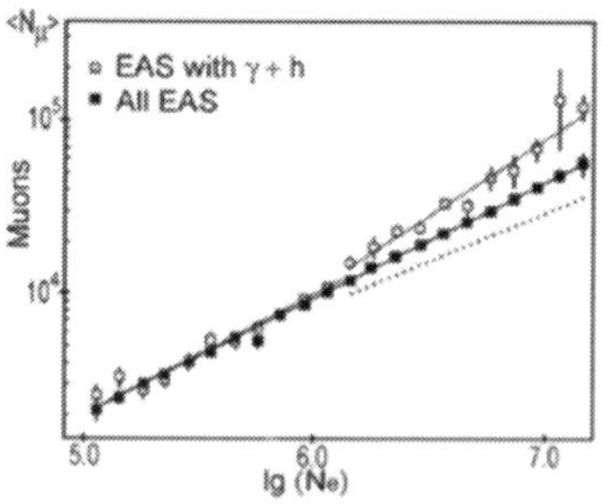

Figure 10. Experimental N_e dependence on the average muon number $\langle N_\mu(N_e)\rangle$: full squares —for all EAS (an inclination index $\kappa = 0.7$), empty squares —for EAS associated with γ-ray families ($\kappa = 0.9$). The dotted line is an expected dependence for EAS with γ-ray families.

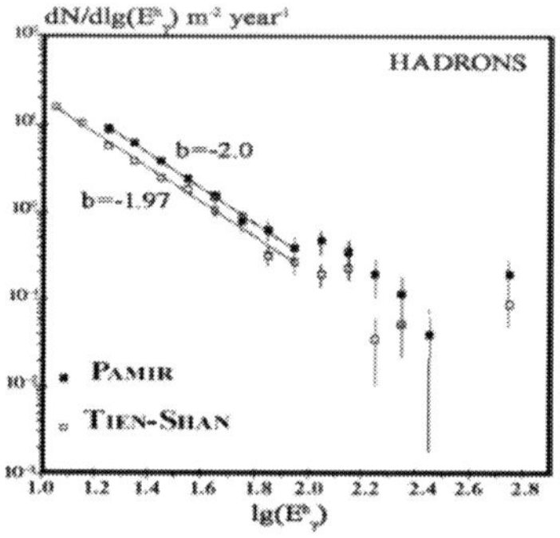

Figure 11. Comparison of hadron spectra observed with XREC at levels of the Pamirs (594 g/cm^2) and Tien Shan (690 g/cm^2).

A comparison between experimental and theoretical spectra of E_γ in Figs. 8 and 9 shows that there are N_e intervals where experimental and model spectra are in good agreement and where they differ considerably in slopes. Energies E_γ of γ-rays are divided on primary energy $E_0 \simeq 0.002 N_e$ to represent spectra in terms of the parameter $x = E_\gamma / E_0$, similar to the Feynman variable.

The results shown in Fig. 8 mean that the model correctly describes, in general terms, the experimental data. The sharp contrast between experimental and model spectra in Fig. 9 (difference exceeds 5σ) means that the break of CR spectra may be directly associated with some abnormal modification of CR nuclear composition.

Two explanatory variants of an experimental minimal value $b \simeq 1.2$ are possible: only light nuclei (protons) are present in the range behind the break, or an unusual new CR component arises here, which effectively generates γ-rays (hadrons) with hard spectra. Both explanations are exotic, but the variant with light nuclei is obviously contrary to muon data.

5.2. Excess of muons in EAS with γ-ray families

The N_e dependences of the average muon number $\langle N_\mu(N_e)\rangle$ for all EAS and EAS with γ-ray families are shown in Fig. 10.

All models predict that EAS with γ-rays are mainly generated by light nuclei. Therefore N_μ for them should be less than average values for all EAS. This prediction is arbitrarily shown by the dotted line in Fig. 10.

The observational data appears above the average values! A prominent feature of nuclear cascades is the following, namely, light nuclei (protons mainly) generate γ-rays of high energies, but the number of muons in their EAS have to be relatively low.

On the contrary, EAS generated by heavy nuclei should contain many more muons, but virtually do not contain high-energy γ-rays. Therefore EAS with abnormal rigid γ-ray spectra and, simultaneously, with a maximal number of muons cannot be explained by traditional nuclear-cascade models. It remains to speculate that the area of the break is shaped by showers of unknown origin.

This unexpected and very important result may signal the existence of a new kind of CR component, which does not consist of nuclei [24,25], or nuclear cascade developments change cardinally in such a way that cascades generated by Fe nuclei become more penetrating then proton-initiated cascades.

5.3. The penetrating component

The comparison of hadron spectra observed with XREC at levels of the Pamirs (594 g/cm^2) and Tien Shan (690 g/cm^2) gives an unusually large value of the absorption length: $\lambda_{att} = 240 \pm 41$ g/cm^2. The spectra are shown in Fig. 11.

Figure 12 shows the result of comparing energy spectra of γ-rays in γ-ray families with observed energies $\sum E_\gamma \geq 100$ TeV.

The maximum difference of E_γ spectra for levels of the Pamirs and Tien Shan does not exceed 2 but has to be $\sim$2.5. This difference disappears with increasing E_γ.

 DOI: 10.1051/epjconf/201714517001

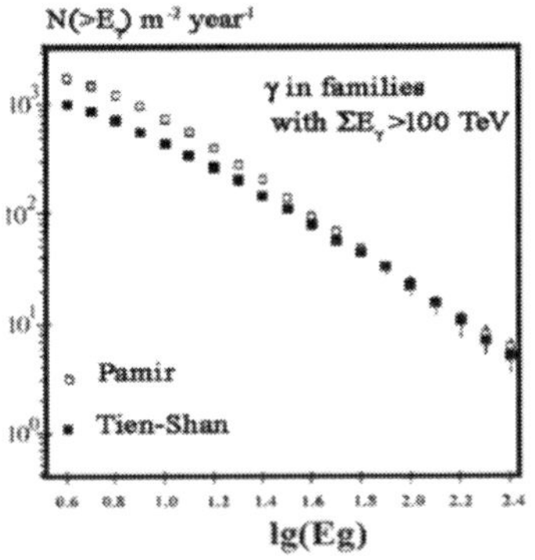

Figure 12. Energy spectra of γ-rays in γ-ray families with observed energies $\sum E_\gamma \geq 100$ TeV.

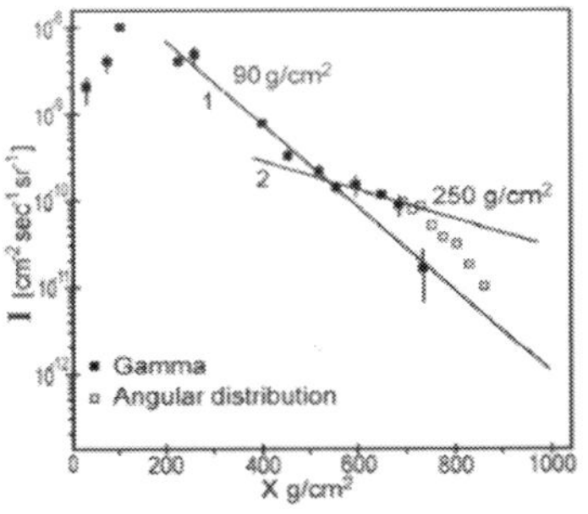

Figure 13. The absorbtion curve in an atmosphere of an electromagnetic component with $n_\gamma \geq 1$ TeV, $\sum E_\gamma \geq 5$ TeV.

The presence of a new penetrating component in CR proves to be true by the shape of the absorption curve of the electromagnetic component in the atmosphere detected by XREC shown in Fig. 13 [22]. These are results collected from different experiments using XRECs made on board balloons, airplanes and at different mountain heights of ground-based XREC [26]. Absorption is described by two exponents: with absorption length $\lambda_{att} \simeq 90\,\mathrm{g/cm^2}$ (1) (as predicted by traditional nuclear-cascade models) and with a much larger value $\lambda_{att} \simeq 250\,\mathrm{g/cm^2}$ (2) produced by some unknown component.

5.4. PCR spectrum and model of a close single source

One more important object of examination is the PCR energy spectrum. The spectrum measured in the *HADRON* [2] and GAMMA [28] experiments as well as in TA [29] and HiRes [30] experiments at higher energies are shown in Fig. 14. The scheme of shaping the PCR spectrum considered by a model of a single close source [31–34] is shown in Fig. 15. All spectra are multiplied by E_0^3 to emphasize features of the PCR spectrum.

Data of the γ-ray spectra (Fig. 7) and form of the PCR spectrum come into conflict with diffusion and accelerating models of the PCR spectrum in the area of the first knee. The model of a single close source may be added to correct this situation. It permits to represent the scheme of the PCR spectrum as the sum of a few components. The smooth spectrum 1 in Fig. 15 is explained by the radiation of many PCR sources in the Galaxy. Its declination varies

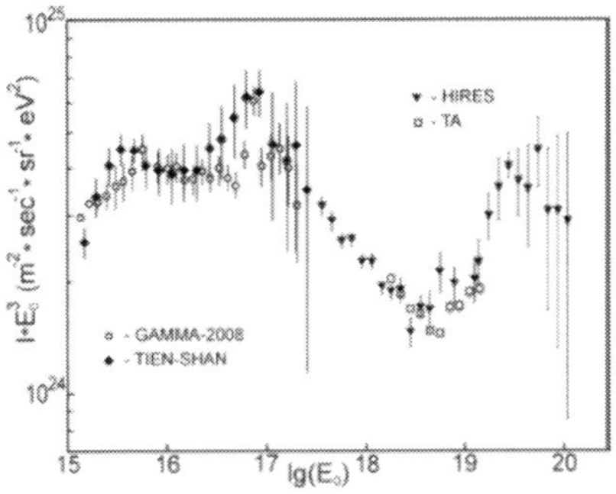

Figure 14. The PCR energy spectrum measured in *HADRON*, GAMMA, TA and HiRes experiments.

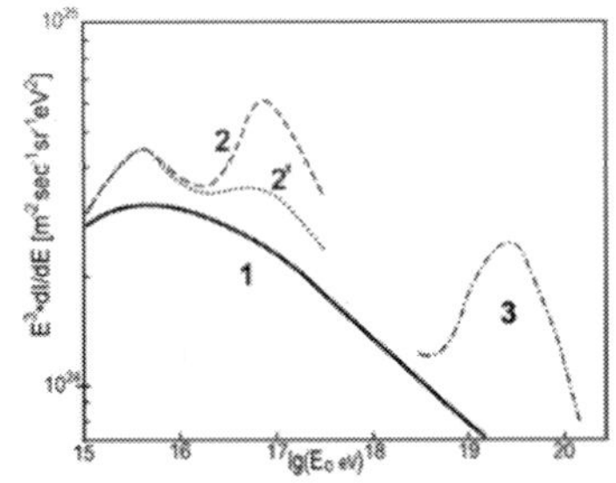

Figure 15. The scheme of the PCR spectrum formation in the model of a single close sours.

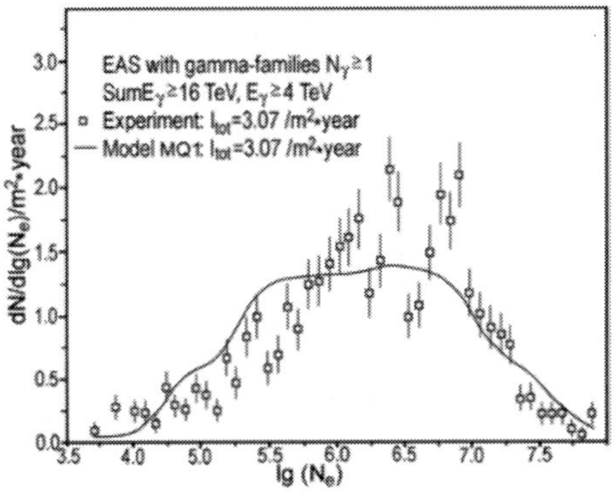

Figure 16. The N_e spectrum of EAS associated with γ-ray families.

continuously according to diffusion models and/or the accelerating model. The spectrum 2, which is limited by local area of energies, is shaped by a single close source. We have not yet touched on how part 3 of the PCR spectrum in Fig. 15 is shaped because it is far beyond the capabilities of our installation.

The form of spectrum 2 is represented with two variants 2 and 2′ because the bump at an energy of $\sim 10^{17}$ eV is not observed in experiments performed at sea level.

Another argument in favor of a model of a close source follows from the shape of the experimental spectrum of EAS with γ-ray families. The experimentally obtained spectrum and a spectrum calculated using the MQ1n model (solid line) [27] is shown in Fig. 16.

The spectrum of EAS associated with γ-ray families has been detected in the experiment *HADRON* for the first time. It demonstrates that EAS+ γ-ray family spectrum has a thin structure which consists of two–three peaks. The peaks are very narrow and are also located close to each other. It is possible only if the fluctuations are small enough and may be understood if there is only a single source.

What can we say about the possible position of the single close source? At first a spherical layer 200–400 parsec in thickness has been defined around the sun system where this source may be located [31]. Besides, an original method to determine the direction to the source was suggested by V.P. Pavluchenko. This direction was determined by means of data of the GAMMA experiment [28]. These two results show that source may be located in the Vela constellation.

6. Conclusions

- EASs contain a penetrating component.
- The presence of a penetrating component can explain the anomalous violation of scaling in the spectra of γ-rays in γ-ray families associated with EASs.
- To explain the excess of muons in EASs associated with γ-ray families, we should assume a fundamental change in the characteristics of nuclear cascades in the atmosphere.
- Experimental data confirm the model of a single close source.

The origin of the penetrating component may be associated with an unusual experimental effect of the growth of the proton elastic cross section found by I.M. Dremin from the comparison data of UNK (Serpukhov) and the LHC and presented at ISVHECRI 2016.

Depending on the size of this growth, features of nuclear cascades can change dramatically. An increasing absorption length should obviously lead to the emergence of a penetrating component and to some increase of γ-ray energies at the level of observation (anomalous scaling violation). More difficult to explain is the excess of muons. In this case it is necessary to assume that due to the large number of nucleons iron nuclei generates γ-ray families with hard spectra more efficiently than a single proton. In any case, this effect is of great interest from the point of view of possible explanations of the observed effects. The situation may be clarified by model calculations of nuclear cascades with an increase in the elastic cross section of nucleons.

References

[1] Khristiansen G.B., Kulikov G.V. About the extensive air showers particles number spectrum, JETPh, 1958, **35**, 3(9), p. 635–640 (in Russian)

[2] S.B. Shaulov, EAS data at the mountain level and a shape of the CR spectrum beyond the break., Nuclear Physics B (Proc. Suppl.) **122** (2003); `arXiv:1610.07427v1 [astro-ph.HE]` 24 Oct. 2016

[3] PAMIR Collaboration, Investigation of nuclear interactions in the area of energies of 10^{14}–10^{17} eV by EC method in cosmic rays (experiment "Pamir"), Trudy FIAN, 1984, **154**, p. 3–141 (in Russian)

[4] S.B. Shaulov, Statistical criteria for the combination events in EXC and EAS (Exp."Hadron"). Preprint FIAN, 1987, N 245, 3–12

[5] N.A. Dobrotin. Short history of the first phase of expedition jobs on the cosmic rays in P.N.Lebedev Physical institute AS of the USSR, Moscow 1998 (in Russian)

[6] G.T. Zatsepin, Nuclear cascade process and its role in development of the EAS, DAN USSR, 1949, **67**, p. 993 (in Russian)

[7] N.A. Dobrotin, G.T. Zatsepin, I.L. Rosental et al., Extenciv air showers of cosmic rays, Uspekhy Phys. Nauk, 1953, **49**, p. 186 (in Russian)

[8] Guseva D.D. et al., New installation..., Izv. AN USSR, ser. phis., 1966, **30**, p. 1574–1576 (in Russian)

[9] Kotelnikov K.A., Shaulov S.B., Zvonkov Ju.E., Two-gap spark chamber with area 6 m^2, PTE, 1977, N5, p. 66–67 (in Russian)

[10] T.P. Amineva et al., Installation for EAS study..., Trudy FIAN, 1970, **46**, p. 157–176, Moscow, Nauka (in Russian)

[11] Baradzey L.T., Kanevskaya E.A., Smorodin Ju.A., Procedure of measuring of energy..., Trudy FIAN, 1970, **46**, p. 200–211 (in Russian)

[12] Nikolsky S.I., Aseikin V.S., Benko D., Danilova T.V., Kabanova N.V., Stamenov J.N., Fluctuations of numbers of μ-meson and depandance of number of mesons on number of particles in showers at level of mountains, Izv. AN USSR, Ser. fiz. **34**, N9, p. 1955–1957, 01.1970 (in Russian)

[13] Yakovlev V.I., Long flying component: recent data and interpretation, Int. Symp. on VHE CRI. Ann Arbor. USA, 1992, p. 154

[14] Dremin I.M. et al., Monte Carlo simulations of long-flying cascades in cosmic rays and leading charm at SSC, Int. Symp. on VHE CRI. Ann Arbor, USA, 1992, p. 534

[15] Basarov E.V., Beisembaev R.U., Besshapov S.P., Vavilov Yu.N., Vildanova L.I. Muon pairs observed on Tien Shan arrangement. Proc. 17-th ICRC, Paris, **7**, p. 48–51; Besshapov S.P., Vavilov Yu. N. The possible role of heavy long living hadrons in muon pair creation by EAS hadrons in dense substance. Proc. 18-th ICRC, Bangalore, 1983, **5**, p. 106–109

[16] D.G. Benko, G. Erdos, S.I. Nikolsky, N.M. Nesterova, V.A. Romakhin, A.P. Chubenko, A.L. Shchepetov, A. Somogyi, A. Varga, Search for sources of primary cosmic rays at energies above 0.1 PeV at the Tien Shan, Nuclear Physics B (Proc. Suppl.), 2008, 175–176, p. 541–543, XIV ISVHECRI, 2006, Weihai, China

[17] Rapoport I.D., Photographic method of dense showers detection for the charged particles, JETF **34**, 1958, p. 998–1000 (in Russian)

[18] Rakobolskaya I.V., monograph, The investigation of muons of super high energies, 1975, Izd. NAUKA, p. 3–210 (in Russian)

[19] Hlytchieva V.S., Sveshnikova L.G. et al. Anomalies of hadron attenuation at large depth of lead. // Proc. of 6 ISVHECRI. Tarbes. France. 1990, p. 184; Arisawa T. et al. Observation of attenuation behavior of hadrons in extremely high energy cosmic ray interactions: New hadronic state?, Nucl. Phys. B **424** (1994) 241–287

[20] A.S. Borisov, Z.M. Guseva, V.G. Denisova et al., Coplanar particles emission in the process of

the multiple births at high energies (experiment "Pamir"), Izv RAN, ser. fiz., 1994, **58**, N12, p. 10–12 (in Russian)

[21] V.V. Arabkin, V.A. Borodkin, M.D. Smirnova, Yu.A. Smorodin, S.B.Shaulov. The haloes spectra of super families at the pressures 585 g/sm^2 and 690 g/sm^2 and nucleons absorption path at the energy about 10^{16} eV. 6-th ISVHECRI, Tarbe 1990, 257–260; Pamir Collaboration. Experimental researches of the halo at superfamilies. Izv. AN USSR. Ser. fiz., 1982, **46**, 9, 1790–1792 (in Russian)

[22] S.B. Shaulov, Cosmic rays about strange quark stars, monography, ISBN: 978-3-659-77433-1, LAP LAMBERT Academic Pablishin, Deutchland, 2015, p. 1–129

[23] Arabkin V.V., Cherdyntceva K.V. et al, The abnormal behaviour of γ-ray families characteristics at primary energy 10^{16} eV, Izv. AN USSR, ser. fiz. **55**, N4, p. 674–677, 1991 (in Russian)

[24] S.B. Shaulov, S.P. Bezshapov, Looking for Strange Quark Matter in Cosmic Rays, Proc. of XVII ISVHECRI 2012, European Physical Journal, Web of Conferences **52**, 04010, p. 1–8 (2013)

[25] S.B. Shaulov, Hypothesis of strange quark matter in cosmic rays, Preprint FIAN, N19, p. 3–42, 2012, Moscow

[26] S.B. Shaulov et al., A peculiarity of high energy gamma-rays absorption in the atmosphere (EC data), Nuclear Physics B (Proc. Suppl.) 122 (2003)

[27] A.M. Dunaevsky, M.V. Zimin, Study of the detection of gamma-hadron families in carbon chamber, BULLETIN de la société des sciences et des lettres de ŁÓDŹ, série: recherches sur les déformations, vol. XVI, 1994

[28] V.P. Pavluchenko and Collaboration GAMMA, Bulletin of the Lebedev Physics Institute, 2014, **40**(5), p. 55

[29] D. Ikedal for the Telescope Array collaboration*: Results from the Telescope Array Experiment, Astrophys. Space Sci. Trans. **7**, 257–263, doi: 10.5194/astra-7-257-2011, 2011

[30] R.U. Abbasi, T. Abu-Zayyad, M. Allen et al., First Observation of the Greisen-Zatsepin-Kuzmin Suppression, Phys. Rev. Lett. **100**, 101101, doi: 10.1103/PhysRevLett.100.101101, 2008a; `http://arxiv.org/abs/astro-ph/0703099v2` 15 Feb. 2008

[31] Erlykin A.D., Wolfendale A.W., A single source of cosmic rays in the range 10^{15}–10^{16} eV, J. Phys. G: Nucl. Part. Phys. **23** (1997) 979–989

[32] Erlykin A.D., Wolfendale A.W., High energy cosmic ray mass spectroscopy I. Status and prospects., Astroparticle Physics **7** (1997) 1–13

[33] A.D. Erlykin, A.W. Wolfendale, High energy cosmic ray mass spectroscopy II: Masses in the range $10^{14} - 10^{17}$ eV., Astroparticle Physics **7** (1997) 203–211

[34] A.D. Erlykin, M. Lipski, A.W. Wolfendale, High energy cosmic ray spectroscopy. IV. The evidence from direct observations at lower energies and directional anisotropies, Astroparticle Physics **8** (1998) 283–292

The VHE anomaly in blazar spectra and related phenomena

Timur Dzhatdoev[1,a], Emil Khalikov[1,b], Anna Kircheva[2], and Alexander Lyukshin[2]

[1] Federal State Budget Educational Institution of Higher Education M.V. Lomonosov Moscow State University, Skobeltsyn Institute of Nuclear Physics (SINP MSU), 1(2), Leninskie gory, GSP-1, Moscow 119991, Russian Federation
[2] Federal State Budget Educational Institution of Higher Education M.V. Lomonosov Moscow State University, Department of Physics, 1(2), Leninskie gory, GSP-1, Moscow 119991, Russian Federation

Abstract. Most of the recent research on extragalactic γ-ray propagation focused on the study of the $\gamma\gamma \rightarrow e^+e^-$ absorption process ("absorption-only model"). Starting from a possible anomaly at very high energies (VHE, $E > 100\,\mathrm{GeV}$), we briefly review several existing deviations from this model. The exotic interpretation of the VHE anomaly is not supported by the recent works. On the other hand, the process of intergalactic electromagnetic cascade development naturally explains these effects. We discuss phenomenology of intergalactic cascades and the main spectral signatures of the electromagnetic cascade model. We also briefly consider the hadronic cascade model; it also may explain the data, but requires low strength of magnetic field around the source of primary protons or nuclei.

1. Introduction

Observations made with imaging Cherenkov telescopes in the VHE range provide an opportunity to test extragalactic gamma-ray propagation models (e.g. [1,2]). The majority of registered extragalactic VHE γ-ray sources are blazars — active galactic nuclei that are likely to have narrowly collimated jets (with opening angles about $1°$–$5°$) pointing towards the observer.

VHE γ-rays are subject to the $\gamma\gamma \rightarrow e^+e^-$ absorption process. Some works found that the observed intensity in the optical depth region $\tau_{\gamma\gamma} > 2$ is too high to be explained in the conventional framework of the "absorption-only model" that only takes into account absorption of primary photons and their adiabatic losses [2,3]. The statistical significance of the anomaly claimed in [2] amounts to 4.2 σ. The minimal energy at which the anomaly reveals itself shows a good correlation with the redshift of the source, indicating that the anomaly is not connected with intrinsic properties of the source but is the signature of the incompleteness of the absorption-only model.

The existence of the anomaly was initially interpreted as evidence for γ–axion-like particle ($\gamma \rightarrow ALP$) oscillations [2]. This model is briefly considered in Sect. 2 of this paper. However, there are some other and less exotic ways to account for this anomaly. These explanations include questioning the validity of the theoretical EBL models invoked to explain observations. While the change of the EBL model could decrease the statistical significance of the anomaly at high energies, it certainly does not explain the whole observation set (see Sect. 3). The anomaly can also be explained in the framework of the electromagnetic (Sect. 4) or hadronic (Sect. 5) intergalactic cascade models.

[a] e-mail: timur1606@gmail.com
[b] e-mail: nanti93@mail.ru

2. Axion-like particles model

Axion-like particles (ALPs) are light bosons with zero spin that are a generalization of the axion — a particle that served as a natural solution to the CP-symmetry violation problem [4–6]. Both particles are characterized by the two-photon coupling. However, unlike axions, ALPs do not have a connection between their two main parameters: their mass, m_a, and the two-photon coupling constant, $g_{a\gamma}$. Photons can oscillate into ALPs and back into photons in a similar way as neutrinos do. Indeed, if the γ–ALP mixing does occur, then the primary γ-rays can convert to ALPs near the source, thus avoiding absorption on the way to the observer, and then reconvert back into photons near the observer. These extra photons may account for the excess in intensity in the observable spectrum.

The upper limit on $g_{a\gamma}$, that was established in [7], is shown in Fig. 1 as the brown area. Assuming that the VHE anomaly in blazar spectra is due to photon-ALP mixing, the lower limit on $g_{a\gamma}$ (depending on m_a) was first put in [8] (light blue area in Fig. 1). Besides the aforementioned one, the γ-ALP mixing phenomenon has another distinct and readily identifiable signature that is shown in Fig. 2 for the case of blazar 1ES 0229+200 (redshift $z = 0.14$). Together with the spectral energy distribution (SED = E^2dN/dE) obtained with the VERITAS Cherenkov telescope [9], Fig. 2 presents several model fits to this SED (γ-ALP mixing effects were taken into account according to [10], see [11] for more details). γ-ALP mixing is effective above a certain critical energy, therefore a step-like irregularity in the shape of the observed spectrum is expected. The drop in intensity is usually about 1/3 of its original value (this result is valid when two photon polarization states attain equipartition with one polarization state of ALP above the critical energy). However, this signature of γ-ALP mixing was not found [12], and the scenario in which γ-ALP oscillation

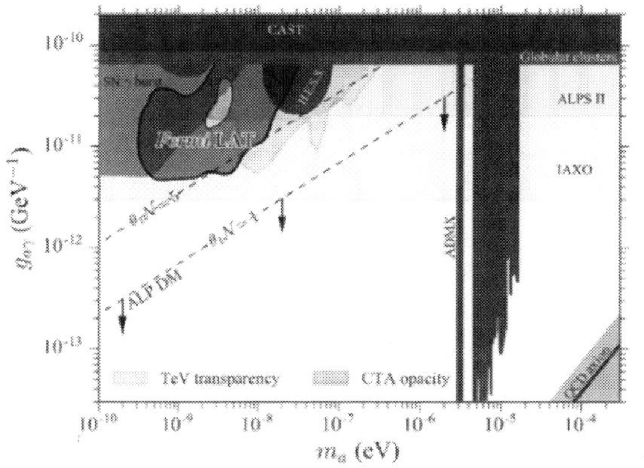

Figure 1. Some constraints on the γ-ALP mixing parameters $g_{a\gamma}$ and m_a (a figure from [12]).

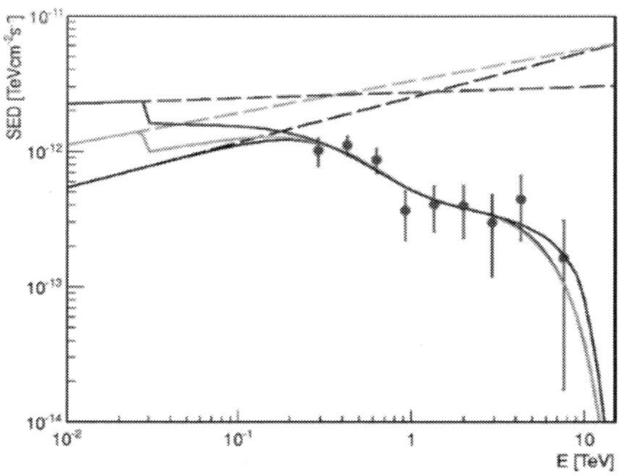

Figure 2. Spectral signatures of γ-axion-like particle (ALP) mixing (a figure from [11]). Black curve denotes absorption-only model spectrum, green — relatively weak γ-ALP mixing, blue — relatively strong γ-ALP mixing. Red circles with bars denote the VERITAS measurements together with uncertainties [9].

can modify the γ-ray opacity of the Universe was strongly constrained (the corresponding excluded region of the γ-ALP mixing parameters is shown as gray area in the upper-left part of Fig. 1).

3. Electromagnetic cascade in the expanding magnetized Universe

In this work we also consider some other explanations of the anomaly, that do not require any new physics, but nevertheless go beyond the absorption-only model, introducing qualitatively new observable effects. In these models the primary particle (a γ-ray [11,13,14] or a proton/nucleus [15–18]) creates secondary photons and electrons which form the intergalactic electromagnetic cascade. If the extragalactic magnetic field (EGMF) is weak enough, then cascade electrons are not strongly deflected or delayed by it, and thus can contribute to the observable spectrum of a point-like source.

Electromagnetic cascade develops on a composite photon target consisting of a dense, but rather low-energy cosmic microwave background (CMB) and comparatively

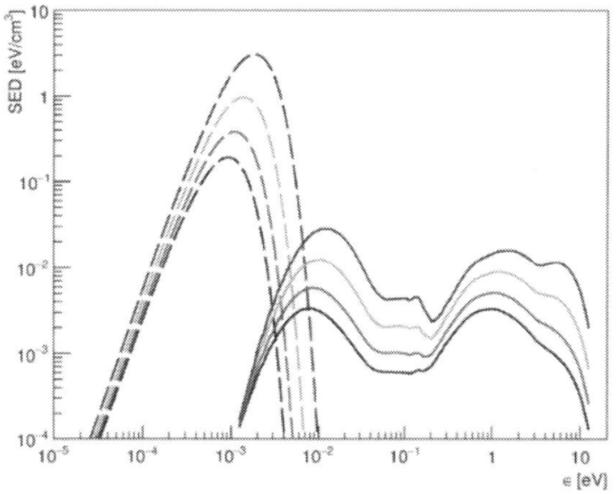

Figure 3. SED of low-energy photon field of the Universe: the EBL (solid curves) and the CMB (dashed curves). Black curves: $z = 0$, red — $z = 0.186$, green — $z = 0.5$, blue — $z = 1.0$.

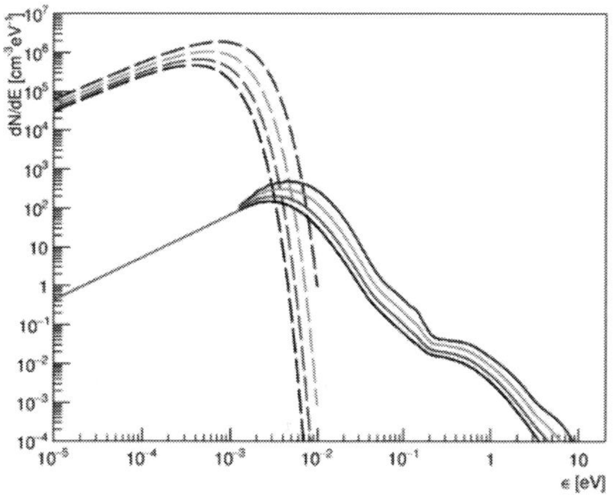

Figure 4. Spectral density of the EBL and CMB for the same values of z as in Fig. 3. Magenta line shows an additional low-energy component that may be added to the model density of the EBL.

dilute extragalactic background light (EBL) which consists of IR, visible and UV photons. SEDs of these photon fields are shown in Fig. 3 for various redshift values assuming the EBL model of [19]. Starlight is the primary source of visible and UV EBL photons; the radiation of heated dust mainly forms the IR region of the EBL spectrum. For the case of moderate redshift $z < 1$, primary γ-rays with energy $E_0 < 50\,\text{TeV}$ produce pairs mainly on the EBL photons as the threshold energy of target photon for the case of the $\gamma\gamma \rightarrow e^+e^-$ process is about $\epsilon_{Thr} > 0.25\,\text{eV}/(E_0/1\,[\text{TeV}])$. On the other hand, most of the inverse Compton (IC) interaction acts occur on the CMB photons due to their high concentration and the absence of energy threshold for this process. Indeed, this issue is clear from Fig. 4 where the spectral density of the CMB and EBL vs. energy is shown.

Following [20], we calculated interaction rates $R(E, z)$ for γ-ray (R_γ) and e (R_e) on the CMB and EBL for the energy range $1\,\text{GeV}–10\,\text{PeV} = 10^{16}\,\text{eV}$ (Fig. 5). For $E > 100\,\text{GeV}$ $R_e(E/(1+z), z)/R_e(E, 0) = (1+z)^3$

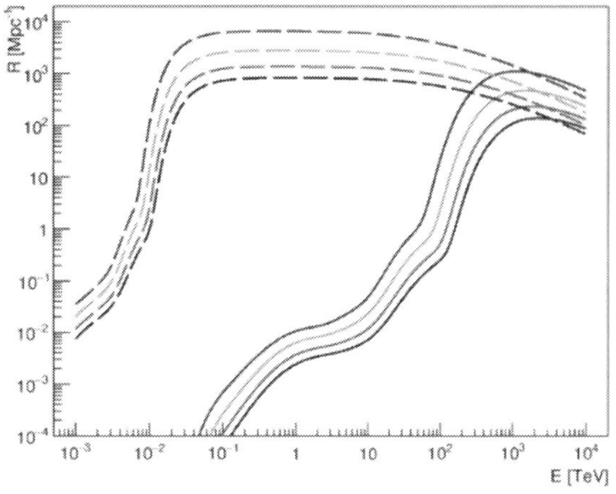

Figure 5. Interaction rates for γ-ray and electron for the same values of z as in Fig. 3.

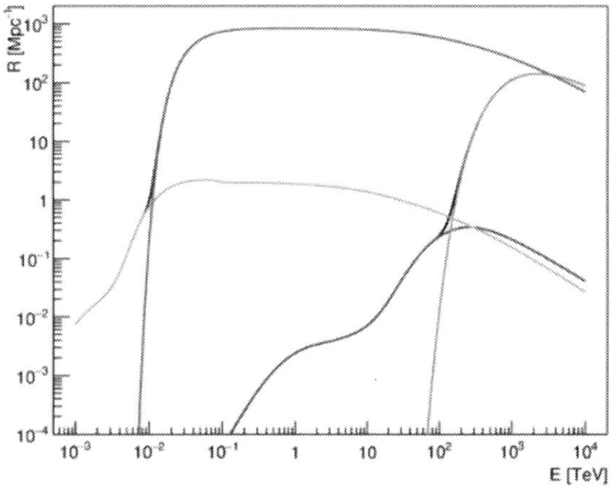

Figure 6. Interaction rate for γ-ray and electron for $z = 0$ decomposed on different components. Black line — total R_γ, red — $R_\gamma(EBL)$, green — $R_\gamma(CMB)$, blue — total R_e, cyan — $R_e(EBL)$, magenta — $R_e(CMB)$. Parameter E_{thr} [20] was set to 3 MeV.

with good precision, reflecting the CMB evolution with z; this relation also approximately holds for the case of γ-ray and $E > 140$ TeV; for γ-ray and $E < 100$ TeV $R_\gamma(E/(1+z), z)/R_\gamma(E, 0) \approx (1+z)^{p_{EBL}}$ with $p_{EBL} \approx$ 1.2–2.0 (depending on energy) due to the continuing stellar activity at $z < 1$.

Figure 6 shows total interaction rates at $z = 0$ for both γ-ray and e, as well as the partial contributions of the EBL and CMB to R_γ and R_e. A small suppression in R_e on the EBL at $E > 70$ GeV is induced by an abrupt cutoff of the model EBL spectrum at $\epsilon < 10^{-3}$ eV. We show a comparison between $R_e(EBL)$ calculated with additional component denoted as magenta line in Fig. 4 and without such a component (see Fig. 7). It is evident that without the artificial cutoff of the EBL spectral density at low energies the suppression in $R_e(EBL)$ is absent. While this spectral feature is clearly irrelevant for the final result on the observable spectrum of cascade γ-rays (its contribution to the total rate is ~0.1 %), it may appear of use while testing precision numerical Monte Carlo (MC) codes aimed at intergalactic cascade simulations.

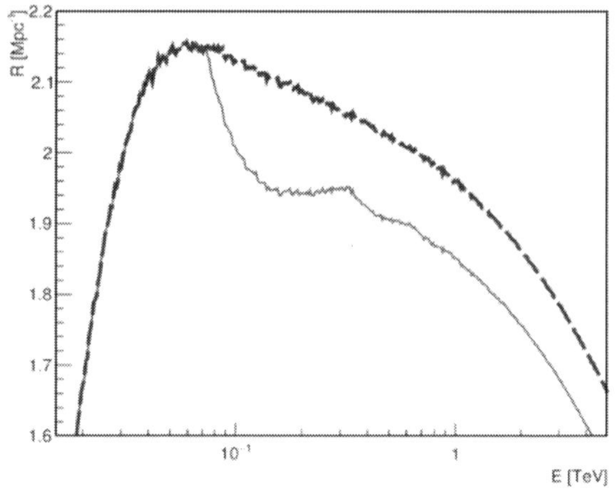

Figure 7. Interaction rate $R_e(EBL)$ for $z = 0$ (fragment) with (dashed thick blue curve) and without (solid red curve) additional component.

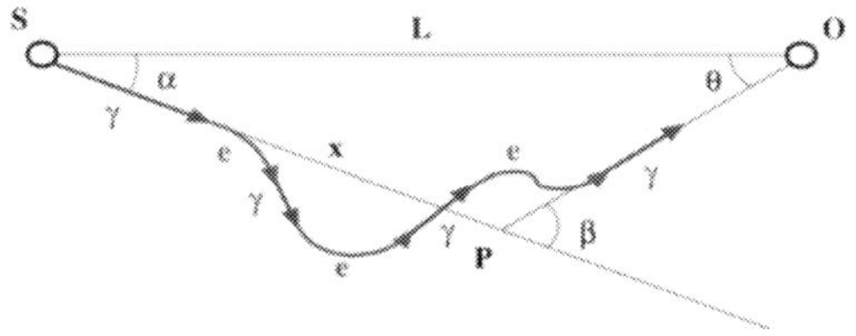

Figure 8. A geometrical sketch of intergalactic electromagnetic cascade development (a figure from [23]). S denotes source, O - observer, L is the distance from the source to the observer; β is the total deflection angle of the last electron in the cascade chain, α — the emission angle, θ — the observation angle.

The deflection of electrons in EGMF may impress certain spectral and timing signatures on the observable spectrum and time distribution of γ-rays (e.g. [21,22]). Using an approximate geometrical scheme (from [23], see Fig. 8), [20] derived the following expression for the time delay of photons caused by the deflection of electrons in the EGMF (c is the speed of light):

$$\Delta t \simeq (x(1 + \sin\alpha/\sin\theta) - L)/c. \quad (1)$$

Applying the small angle approximation ($\theta \simeq \sin\theta$; $\cos\theta \simeq 1 - \theta^2/2$) and the basic formula of trigonometry ($\sin\alpha = \sin(\beta - \theta) = \sin\beta\cos\theta - \cos\beta\sin\theta$) one can obtain that:

$$\Delta t \simeq \frac{x}{2c}\left(1 - \frac{x}{L}\right)\sin^2\beta. \quad (2)$$

Finally, we note that in the VHE energy range the additional broadening of cascade angular distribution in pair production and IC interaction acts and associated time delay are usually negligible (e.g. [24]).

4. Electromagnetic cascade model

Using the publicly available code ELMAG 2.02 [20], we calculated a typical observable spectrum of a blazar with

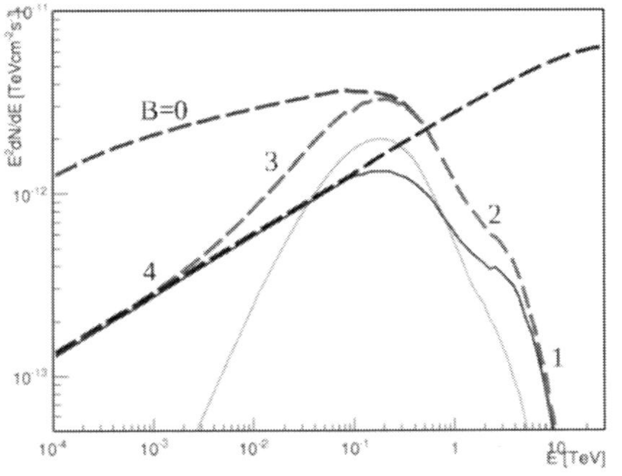

Figure 9. Signatures of the electromagnetic cascade model ($z = 0.186$) (a scheme from [25]). Green solid line denotes the cascade component, black solid line - primary absorbed component, black dashed line - primary intrinsic component, blue dashed line - observed spectrum with $B = 0$, red dashed line - observed spectrum for the case of non-zero EGMF.

a hard intrinsic spectrum in the TeV energy region (see Fig. 9). In this picture, several spectral signatures [25] are clearly seen, namely: 1) a high-energy cutoff, 2) an ankle, which is due to the intersection of the primary component of γ-rays absorbed on the EBL and the cascade component, 3) a possible low-energy cutoff due to delay of cascade electrons ("magnetic cutoff"), 4) a "second ankle" at comparatively low energy where the primary component again starts to dominate over the cascade one. The first signature, the high-energy cutoff, is very similar to the case of the absorption-only model and thus is not specific for the electromagnetic cascade model. Provided that the cascade component is not entirely suppressed, the high-energy ankle is present in the spectrum irrespectively to the EGMF strength. On the other hand, the magnetic cutoff and the second ankle are connected with non-zero EGMF strength.

There are many indications that the cascade component indeed contributes to the observed spectra of some blazars at energies $E < 300$ GeV. For instance, [26] reported the observation of a hard spectrum of Mkn 501 in the energy range 20–200 GeV. Figure 10 is a graph from this paper that shows the observed spectrum displaying two signatures of the electromagnetic cascade model (except the high-energy cutoff): a prominent magnetic cutoff and a possible "second ankle".

Other indication is the result of [28], which hints at the larger number of blazars with comparatively hard spectra in the energy range 10–300 GeV towards the directions to the Large Scale Structure (LSS) voids. These papers are discussed more thoroughly in [11]. In [29] an evidence for the existence of halos around various blazars was reported. The authors of this paper analyzed the spectra of 50 blazars. They divided them into two groups: BL Lac objects (24 sources) and Flat Spectrum Radio Quasars (26 sources). The first group had comparatively hard observed spectra and redshifts $z < 0.5$. The second group sources could have any redshift and rather soft spectra. The analysis of these spectra showed that some blazars with relatively hard spectra could not be considered point

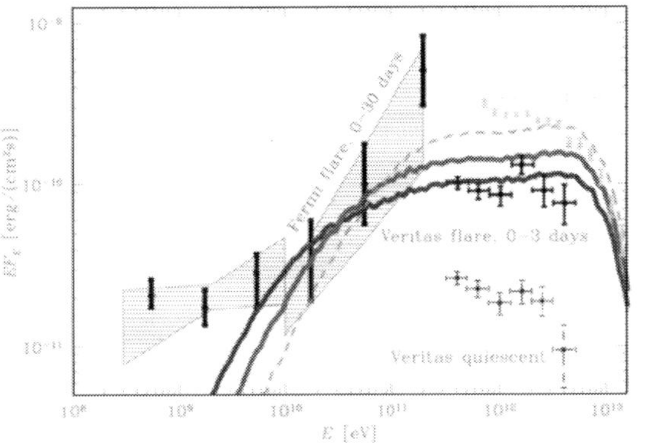

Figure 10. Fermi LAT and VERITAS spectrum of Mrk 501 (2009 flare, a figure from [26]). Dots with errors denote the VERITAS measurements [27] of the spectrum during the first 3 days of the flare (solid lines) and in the quiescent state (dashed lines). Model curves show cascade spectra for $B = 10^{-17}\,G$ (green dashed), $3 \cdot 10^{-17}$ G (red), $B = 10^{-16}$ G (blue). The EGMF correlation length is 1 Mpc.

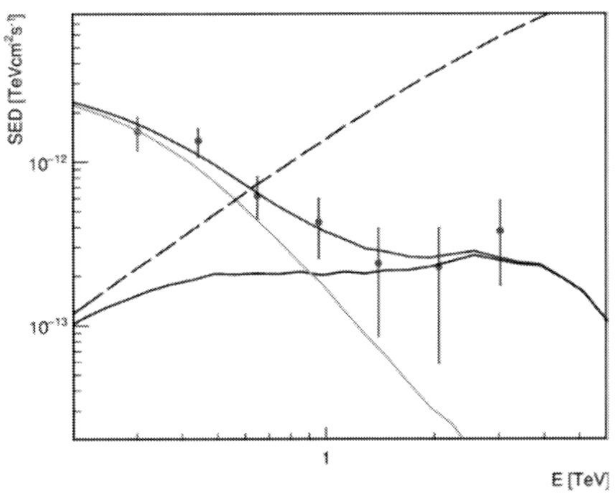

Figure 11. Electromagnetic cascade model fit for 1ES 0347–121 for the case of $B = 0$ (a figure from [11]).

sources. This result was interpreted in [29] as evidence for the deflection of cascade electrons and positrons in the EGMF with the strength in the range of 10^{-17}–10^{-15} G.

However, all these indications can be explained by other phenomena that do not make use of intergalactic cascade development process. For instance, the authors of [26] proposed some alternative explanations to their findings in the paper itself. The significance of the effect observed in [28] is below 3 σ. As for [29], the result could in principle be explained by the scattering of electrons and positrons accelerated in the source by a stronger local magnetic field ($B \sim 10^{-12}$–10^{-7} G), in which case quasi-isotropic halos may form around the source [30].

Let us consider the explanation of the anomaly [2,3] in the framework of the electromagnetic cascade model (more details are available in [11]). Figure 11 presents a fit to the observed SED of blazar 1ES 0347–121 ($z = 0.188$) assuming this model. Figure 12 shows the so-called "flux boost factor" — the ratio of spectra for the case of the electromagnetic cascade model and the

DOI: 10.1051/epjconf/201714517002

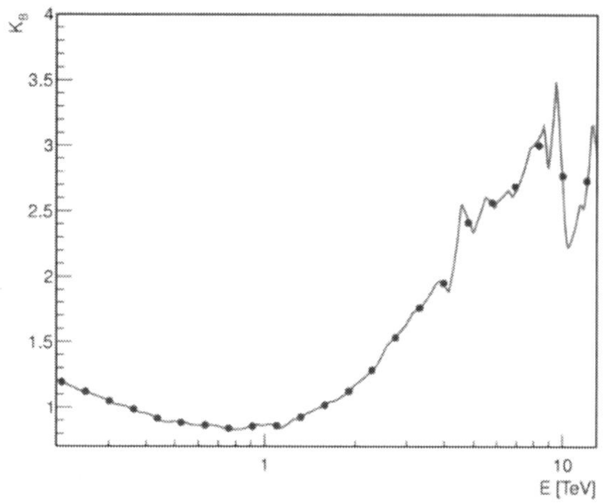

Figure 12. Flux boost factor vs. energy for 1ES 0347–121 and the model with voidiness = 1 (a figure from [11]).

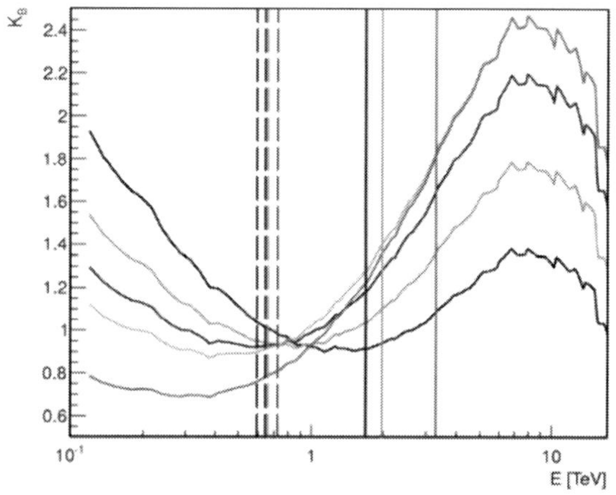

Figure 13. Flux boost factor vs. energy for 1ES0229+200 and models with different voidiness K_V (a figure from [11]): black curve — $K_V = 1$, green curve — 0.6, blue — 0.4, cyan — 0.3, magenta — 0.2.

absorption-only model. At E =2–10 TeV the value of K_B clearly exceeds unity, therefore the electromagnetic cascade model allows to explain the observed excess in intensity in this energy region, with respect to the absorption-only model. Finally, Fig. 13 presents several curves of the boost factor vs. energy labeled by different values of the fraction of space filled with voids — the so-called "voidiness" parameter [28]. It is interesting that at comparatively low values of voidiness (0.2–0.3) the values of K_B are much larger at high energies than for the case of voidiness = 1.

5. Hadronic cascade model

In hadronic cascade models, the primary particle is either a proton or a nucleus. They produce secondary particles on CMB and EBL via photopion processes and pair production; these secondary particles then initiate cascades the same way as they do in the electromagnetic cascade model. The graph of a typical observable SED of the hadronic cascade model for the case of zero EGMF and

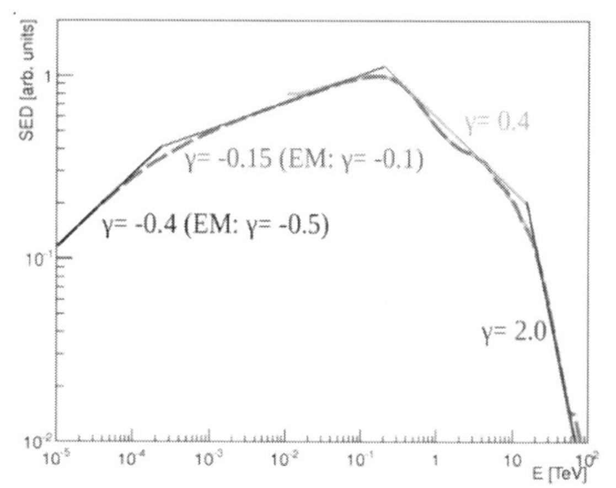

Figure 14. Spectral signatures of the hadronic cascade model (primary proton energy $E_{p0} = 3 \cdot 10^{19}$ eV = 30 EeV, $B = 0$) (a figure from [11]). Straight solid lines denote different power indices that are present in the overall spectrum.

primary proton is shown in Fig. 14. We approximated the spectrum with a broken poly-gonato power-law and compared the resulting power indices with the case of a purely electromagnetic cascade taken from [31]. Calculations of [31] apply to the case of the "universal regime", which usually sets for the case of sufficiently high primary energy ($E_0 > 100$ TeV) and sufficiently high redshift of the source.

The low-energy ($E < 200$ GeV) indices for the hadronic model are nearly the same as for the electromagnetic cascade model (the latter are shown in parentheses, the former — outside). However, at the higher energy the cutoff is not as marked for the case of the hadronic model. This fact makes it possible to effectively discriminate between EM and hadronic models and to identify the primary particle for the case of considered models. Cascade photons from nearby interactions of primary particles may form a significant excess over intensity expected for the case of the absorption-only model; therefore, the hadronic cascade model is, in principle, able to explain the VHE anomaly in blazar spectra [16, 17].

In [11] we performed a detailed calculation of a large grid of hadronic cascade models with different parameters for the case of blazar 1ES 0229+200, assuming the emission model of [32]. It appeared that this model is plausible only for the case of comparatively low value of magnetic field circumventing the central emitting object, $B < 100$ nG. Otherwise, the scattering of protons in this magnetic field significantly broadens the jet angular profile, thus weakening the total observable intensity in the VHE energy range much below the observed values.

6. Conclusions

The first statistically significant ($> 3\sigma$) deviation from the absorption-only model was the VHE anomaly, that was initially interpreted as an indication for the $\gamma \rightarrow ALP$ oscillations. However, this interpretation is not supported by recent research; in particular, another important signature of $\gamma - ALP$ mixing — the spectral irregularity at low energies — was not found. Therefore, we considered two alternative models that could explain such an effect, namely, the electromagnetic and hadronic cascade models.

We briefly reviewed the phenomenology of electromagnetic cascade development on the CMB/EBL and discussed the main spectral signatures of a cascade component in the observable spectrum. Several works support the presence of the cascade component in the spectra of some blazars. As we have shown, a characteristic "ankle" signature at the intersection of the primary and secondary components may qualitatively explain the VHE anomaly. Finally, we briefly discussed the hadronic cascade model that may also explain the anomaly, but demands a rather low strength of magnetic field circumventing the source.

To conclude, we have shown that several known deviations from the absorption-only model are naturally explained in the framework of the electromagnetic cascade model.

The work was supported by the Russian Foundation for Basic Research (grant 16-32-00823). Figures 3–7 were produced during the visit of T.D. to the Tokyo University, Institute for Cosmic Ray Research (ICRR). For this part of work, T.D. acknowledges the support of the Students and Researchers Exchange Program in Sciences (STEPS) and the hospitality of the Tokyo University ICRR.

References

[1] Stecker, F.W. & de Jager, O.C., ApJ **415**, 71 (1993)
[2] Horns, D. & Meyer, M., JCAP **033** (2012)
[3] Horns, D., preprint `astro-ph/1602.07499` (2016)
[4] Peccei, R.D. and Quinn, H.R., Phys. Rev. Lett. **38**, 1440 (1977)
[5] Weinberg, S., Phys. Rev. Lett. **40**, 223 (1978)
[6] Kim, J., Phys. Rev. Lett. **43**, 103 (1979)
[7] Andriamonje, S. et al. (CAST Collaboration), JCAP **04**, 010 (2007)
[8] Meyer, M. et al., Phys. Rev. D **87**, 035027 (2013)
[9] Aliu, E. et al., ApJ **782**, 13 (2014)
[10] Sanchez-Conde, M.A. et al., Phys. Rev. D **79**, 123511 (2009)
[11] Dzhatdoev, T., Khalikov, E., Kircheva, A., Lyukshin A. (Preprint `astro.ph/1609.01013`) (2016)
[12] Ajello M. et al., Phys. Rev. Lett. **116**, 161101 (2016)
[13] Aharonian, F. et al., A&A **349**, 11 (1999)
[14] Aharonian, F. et al., A&A **384**, 834 (2002)
[15] Uryson, A., JETP **86**, 213 (1998)
[16] Essey, W. & Kusenko, A., APh **33**, 81 (2010)
[17] Essey, W. et al., Phys. Rev. Lett. **104**, 141102 (2010)
[18] Essey, W. et al., ApJ **731**, 51, (2011)
[19] Gilmore, R. et al., MNRAS **422**, 3189 (2012)
[20] Kachelriess, M. et al., Comp. Phys. Comm. **183**, 1036 (2012)
[21] Neronov A., Vovk I., Science **328**, 73 (2010)
[22] Dermer, C.D. et al., ApJ Lett. **733**, L2 (2011)
[23] Dolag, K., Kachelriess, M., Ostapchenko, S., & Tom'as, R., ApJ **703**, 1078 (2009)
[24] Dzhatdoev, T.A., J. Phys. Conf. Ser. **632**, 012035 (2015)
[25] Dzhatdoev, T.A. et al., Talk at the Russian Cosmic Ray Conference, Dubna (2016)
[26] Neronov, A. et al., A&A **541**, A31 (2012)
[27] Abdo A., et al., ApJ **727**, 129 (2011)
[28] Furniss, A. et al., MNRAS **446**, 2267 (2015)
[29] Chen, W. et al., Phys. Rev. Lett. **115**, 211103 (2015)
[30] Aharonian, F. A. et al , ApJ **423**, L5 (1994)
[31] Berezinsky, V. & Kalashev, O., Phys. Rev. D **94**, 023007 (2016)
[32] Tavecchio, F., MNRAS **438**, 3255 (2014)

Requests of the cosmic ray community to accelerator high energy physicists

Lawrence W. Jones[a]

University of Michigan, Ann Arbor, Michigan, USA

Abstract. The requests of Cosmic Ray physicists to Accelerator physicists are summarized.

Introduction, current situation and future

This was a topic of discussion at the last three ISVHECRI meetings at Fermilab, in Berlin and at CERN respectively. Indeed, the cosmic ray request is still relevant. The issue is the understanding of cosmic ray air showers.

At and above PeV energies, most cosmic ray data is based on ground-level observations of air showers, which are the results of the interactions of primary cosmic rays with air nuclei. The observables, on which the energy and particle nature of the primary cosmic ray are based, are mostly electrons and muons, plus some hadrons. Of course, at these PeV energies ($> 10^{15}$ eV), there is no direct accelerator data on the collision of protons with air nuclei. However, colliding beam data, where a proton of 7.5 TeV collides with a nucleus of oxygen (O^{16}) of about 60 TeV (i.e., the same magnetic rigidity as a 7.5 TeV proton) would produce a final state distribution of observable particles which would be directly observed and studied. Such a collision would be equivalent to a proton of about a PeV colliding with a stationary oxygen nucleus.

The LHC has had runs with collisions of lead nuclei with other lead nuclei, and also with collisions of protons with lead nuclei, so colliding beams of heavy nuclei or collisions between protons and other nuclei are altogether practical. What is needed is an agreement with the leadership of at least one of the major collider experiments to collaborate in a proton-oxygen (or proton-nitrogen) colliding beam program. Probably only a week of data would be needed, to produce enough data to fine-tune the existing cosmic ray air-shower models. Perhaps some runs with collisions between light nuclei (e.g., He, Be, C, etc.) would also be useful. What is most needed is a close communication between active and involved members of the cosmic ray community (e.g., participants in the understanding of AUGER data) and members of at least one of the LHC experiments; ATLAS or CMS, for example.

It is hoped that agreement can be found to include a proton-oxygen run in the next program devoted to heavy-ion collisions.

[a] e-mail: `lwjones@umich.edu`

Open issues in hadronic interactions for air showers

Tanguy Pierog[a]

Karlsruhe Institute of Technology (KIT), IKP, 76021 Karlsruhe, Germany

Abstract. In detailed air shower simulations, the uncertainty in the prediction of shower observables for different primary particles and energies is currently dominated by differences between hadronic interaction models. With the results of the first run of the LHC, the difference between post-LHC model predictions has been reduced to the same level as experimental uncertainties of cosmic ray experiments. At the same time new types of air shower observables, like the muon production depth, have been measured, adding new constraints on hadronic models. Currently no model is able to consistently reproduce all mass composition measurements possible within the Pierre Auger Observatory for instance. Comparing the different models, and with LHC and cosmic ray data, we will show that the remaining open issues in hadronic interactions in air shower development are now in the pion-air interactions and in nuclear effects.

1. Introduction

Knowing the elemental composition of cosmic ray particles arriving at Earth is of crucial importance to understand the production and propagation of cosmic rays. Unfortunately, cosmic rays can be measured only indirectly above an energy of 10^{14} eV, through the cascades of secondary particles, called extensive air showers (EAS), that they produce in the atmosphere (for a recent review, see [1]). Only by simulating the generation of EAS and comparing the predictions with measurements can one draw conclusions on the primary mass composition of the arriving particles [2]. With the operation of modern large-scale experiments, the reliability of air shower simulations has become the source of the largest systematic uncertainty in the interpretation of cosmic-ray data [3–9]. While the electroweak interaction processes are reasonably well understood, modeling of hadronic multi-particle production is subject to large theoretical uncertainties that are, moreover, difficult to estimate [10–12].

The Large Hadron Collider (LHC) at the CERN laboratory allows us to access, for the first time, the energy region above the cosmic ray spectral knee in the laboratory with about 10^{17} eV in the laboratory frame. Therefore an analysis of inclusive particle data taken at the LHC is particularly interesting for constraining existing hadronic interaction models and for testing possible new mechanisms of hadron production [13]. The first published data from LHC experiments have mostly been taken with detectors covering the central phase space region in pseudorapidity ($|\eta| < 2.5$). This region is most easily accessible in collider experiments and is also the region of the highest rapidity-density of produced particles. The first data have been compared to cosmic ray models in [14]. On the other hand, since the number of particles in an air shower is roughly proportional to the energy of the primary particle, the most energetic outgoing particles of an interaction, emitted in the very forward region of a collider experiment – such as in diffractive interactions – are the most important ones for understanding air showers. For the first time at the LHC, collider experiments include a large variety of forward detectors to study forward particle and energy spectra which could have a direct impact on air shower development [15]. These latest measurements are not yet taken into account in the available hadronic interactions models, but are very important to understand the open issues in these models and for their future development.

At the same time, a new generation of hybrid cosmic ray detectors such as the Pierre Auger Observatory [7] (surface and fluorescence detectors), the IceCube/IceTop experiments [16,17] (low energy muons at the surface and high energy muons deep underground) or the KASCADE/KASCADE Grande collaboration [18,19] (muons of different energies and at different distances) gives access to various precise measurements of the mean logarithmic mass of cosmic rays within the same experiment. By definition the mean logarithmic mass should be independent of the measurement technique. If the physics is well described by a given hadronic model, the masses obtained from different observables should be consistent. This constraint is much stronger than the traditional test limiting the results to the range between proton and iron induced showers. This is now satisfied in most of the cases, but none of the current models is able to give a fully consistent picture of the different observable within a given experiment [20–22].

In this paper, we will discuss the remaining open issues in the hadronic model predictions after LHC data and their consequences on air shower observables. In the first section, we will describe the latest hadronic interaction models and then compare their results for the observables important for the air shower development in the second section. The main source of remaining uncertainties will be identified. Using detailed Monte Carlo simulations done with CONEX [23], the new predictions for X_{max} and for the number of muons will be presented. Finally

[a] e-mail: tanguy.pierog@kit.edu

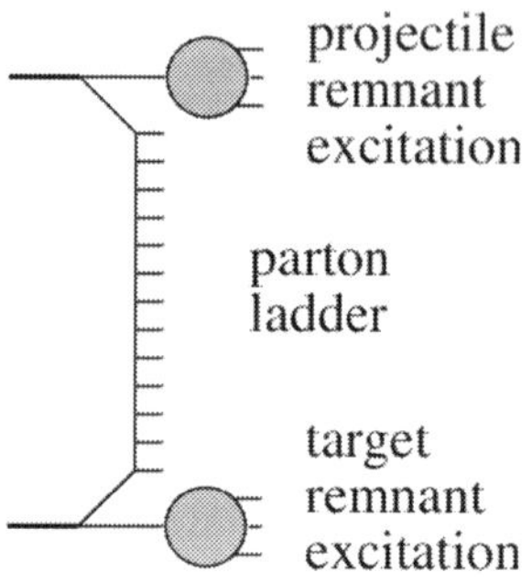

Figure 1. Elementary parton-parton scattering: the hard scattering in the middle is preceded by parton emissions attached to remnants. The remnants are an important source of particle production even at intermediate energies (~100 GeV cms).

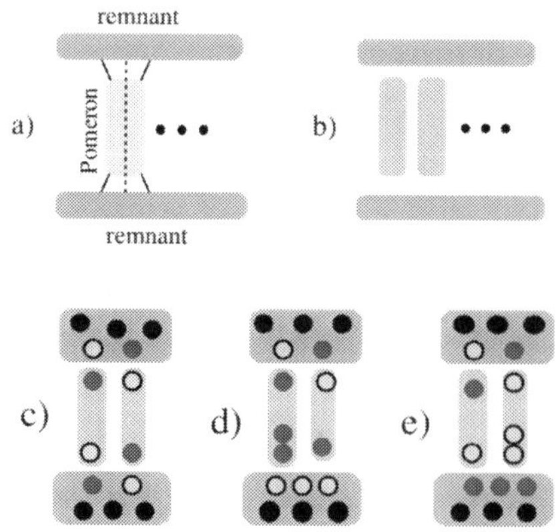

Figure 2. a) Each cut Pomeron is regarded as two strings b). c) The most simple and frequent collision configuration has two remnants and only one cut Pomeron represented by two $q - \bar{q}$ strings. d) One of the $\bar{q}$ string ends can be replaced by a qq string end. e) With the same probability, one of the q string ends can be replaced by a $\overline{qq}$ string end.

we will take the example of the muon production depth (MPD) measured by the Pierre Auger Observatory [20] to see how air shower measurements can constrain hadronic interaction physics and can be used to solve the remaining open issues.

2. Hadronic interaction models

There are several hadronic interaction models commonly used to simulate air showers. Here we will focus on the three high energy models which were updated to take into account LHC data at 7 TeV: QGSJETII-03 [24,25] changed into QGSJETII-04 [26], EPOS 1.99 [27,28] replaced by EPOS LHC (v3400) [29], and Sibyll 2.1 [30–32] updated to Sibyll 2.3 [33] all available since CORSIKA v7.5600 [34]. There is no major change in these models but in addition to some technical improvements, some parameters were changed to reproduce TOTEM [35] cross sections. These are based on Gribov-Regge multiple scattering, perturbative QCD and string fragmentation.

2.1. EPOS model

EPOS LHC is a minimum bias monte-carlo hadronic generator used for both heavy ion interactions and cosmic ray air shower simulations. It is based on EPOS 1.99 retuned to reproduce LHC data on a higher precision level. As with most of high energy hadron-hadron interaction models, it is based on the simple parton model which can be seen as an exchange of a "parton ladder" between the two hadrons.

In EPOS, the term "parton ladder" is actually meant to contain two parts [36]: the hard one, as discussed above, and a soft one, which is a purely phenomenological object, parameterized in Regge pole fashion. This is the so called Pomeron used as elementary parton-parton interaction in EPOS.

In addition to the parton ladder, there is another source of particle production: the two off-shell remnants, see Fig. 1. We showed in Ref. [37] that this "three object picture" can solve the "multi-strange baryon problem" of conventional high energy models, see Ref. [38].

Hence EPOS is a consistent quantum mechanical multiple scattering approach based on partons and strings [36], where cross sections and the particle production are calculated consistently, taking into account energy conservation in both cases (unlike other models where energy conservation is not considered for cross section calculations [39]). The main consequence of this energy sharing process is that the number of Pomerons generated event-by-event do not follow a simple Poissonian distribution. As a consequence it is much less likely to produce events with a very large number of Pomerons (large multiplicity) compared to the standard Gribov-Regge approach like in QGSJETII. Nuclear effects related to Cronin transverse momentum broadening, parton saturation, and screening have been introduced into EPOS [27]. Furthermore, high density effects leading to collective behavior in heavy ion collisions are also taken into account [40].

Within a Monte Carlo, first the collision configuration is determined: i.e., the number of each type of Pomerons exchanged between the projectile and target is fixed and the initial energy is shared between the Pomerons and the two remnants. Then particle production is accounted from two kinds of sources, remnant decay and cut Pomeron. A Pomeron may be regarded as a two-layer (soft) parton ladder attached to projectile and target remnants through its two legs. Each leg is a color singlet, of type $q\bar{q}$, qqq or $\overline{qqq}$ from the sea, and then each cut Pomeron is regarded as two strings, Cf. Fig. 2a) and b).

It is a natural idea to take quarks and anti-quarks from the sea as string ends for soft Pomerons in EPOS, because an arbitrary number of Pomerons may be involved.

Thus, besides the three valence quarks, each remnant has additional quarks and anti-quarks to compensate the flavors of the string ends, as shown in Fig. 2-c. According to its number of quarks and anti-quarks, to the phase space, and to an excitation probability, a remnant decays into mesons and/or (anti)baryons [37]. Furthermore, this process leads to a baryon stopping phenomenon in which the baryon number can be transferred from the remnant to the string ends (for instance in 2-d, depending on the process, the $3\bar{q} + 3q$ can be seen as 3 mesons or a baryon-antibaryon pair).

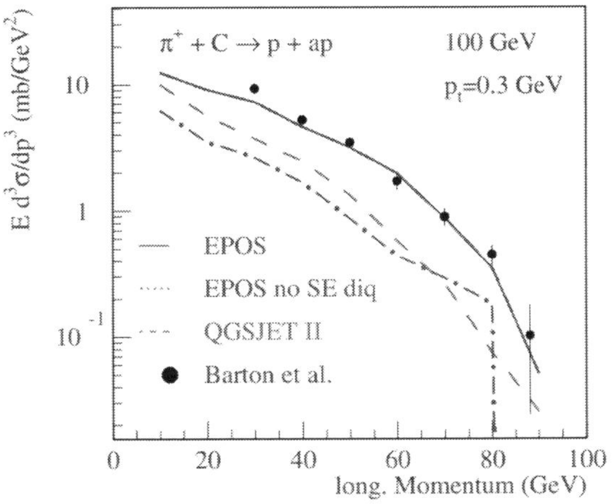

Figure 3. Model comparison: longitudinal momentum distributions of pion carbon collisions at 100 GeV from EPOS with (full) or without (dashed-dotted) string-end diquarks and QGSJETII (dashed) compared to data [41].

In the case of a meson projectile, this kind of diquark pair production at the string ends leads to an increase of the (anti)baryon production in the forward production in agreement with low energy pion-nucleus data [41] as shown Fig. 3. Comparing to the QGSJETII model which does not have diquark as string ends or using only $q\bar{q}$ as string end in EPOS, we can clearly see that this process is needed to reproduce experimental data. As a consequence it contributes to the larger number of muons in air shower simulations with EPOS.

Energy momentum sharing and remnant treatment are the key points of the model concerning air shower simulations because they directly influence the multiplicity and the inelasticity of the model.

2.2. QGSJETII model

QGSJETII-04 [26,42] model is a minimum bias hadronic interaction model optimized for air shower simulations. It has a minimum set of parameters to reduce the uncertainty due to the extrapolation to high energy and as a consequence has a less detailed description of the final stage of hadronic interactions (no final state effect, no rare particle production, etc.) which limit the data sets to which it can be compared. It is the current last evolution of the series of models based on Quark-Gluon and Strings model with Jet [43].

The elementary interaction used in QGSJETII is a semi-hard Pomeron very similar to the one used in EPOS. But to take into account non-linear effects at high energy, a very complex resummation scheme has been developed to take into account any type of Pomeron-Pomeron interactions resulting in a net fan diagram [42] which can be used for both cross-section and particle production. But then only the standard Gribov-Regge type of calculation can be used, meaning that energy sharing is not taken into account in cross-section calculation and when the total number of multiple scattering is calculated.

Like in EPOS, QGSJETII has some remnant from the projectile and target but in a simplified scheme which does not allow more than one quark exchange with the string ends. In particular diquarks are not allowed as string ends which gives quite different forward baryon production compared to EPOS as shown Fig. 3.

2.3. Sibyll model

Like QGSJETII, the Sibyll model is a minimum bias hadronic interaction model optimized for air shower simulations but with a different approach. The cross-section calculation is based on the same Gribov-Regge calculation without energy sharing and diffraction is now done with a two state approach like in QGSJETII. However the eikonal is based on the minijet model with an energy dependent p_t cutoff [44–47] and the string hadronization is using the Lund model [48] producing all type of particles.

The new Sibyll includes the same simplified remnant treatment as in QGSJETII. This allows for instance the increase of leading ρ^0 in $\pi-$p/A interactions. Compared to Sibyll 2.1 the new version has an improved production of baryon-anti-baryon pairs, in particular from the minijet (hard) particle production, and a phenomenological model for the production of charm particles which is important for the production of high energetic muons and neutrinos.

Unlike the two other models, Sibyll uses the semi-superposition model to treat the nuclear interactions meaning that for a nuclear projectile the result is the sum of a given number of nucleon-nucleus interactions without additional nuclear effects. Another main difference from the other models is a true scaling of forward particle production as discussed in [49].

So, even if the three models are based on very similar approaches (parton ladder and Gribov-Regge based multiple interactions), the detailed treatments of energy sharing, non-linear effects, nuclear effects and remnant production lead to different extrapolations in both proton and pion interactions and thus for air shower observables as shown in the next section.

3. Model comparison

A toy-model, as described in [50], only gives a very much over-simplified account of air shower physics. However, the model allows us to qualitatively understand the dependence of many air shower observables on the characteristics of hadronic particle production. Accordingly the parameters of hadron production which are most important for air shower development are the cross section (or mean free path), the multiplicity of secondary particles of high energy, the elasticity and the production ratio of neutral to charged particles. Until the start of the LHC, these parameters were not well constrained by particle production measurements at accelerators. As a consequence, depending on the assumptions of how to extrapolate existing accelerator data, the predictions of hadronic interaction models were very different [51]. We will show that the extrapolation to high energy is not really the issue anymore.

3.1. Cross section

As shown in [50], the cross section is very important for the development of air showers and in particular for the depth of shower maximum. As a consequence, the number

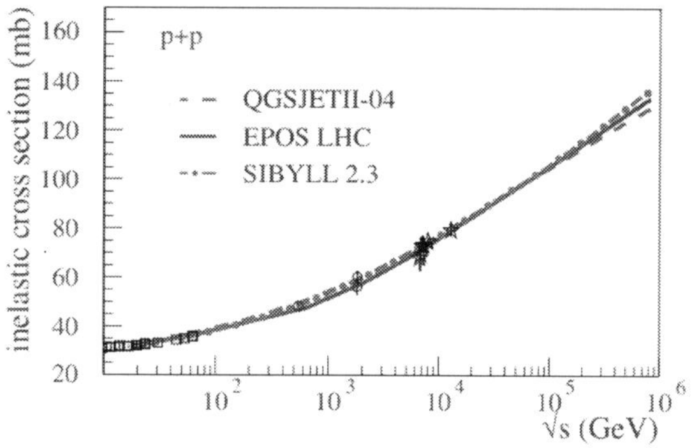

Figure 4. Inelastic p-p cross sections calculated with EPOS LHC (full line), QGSJETII-04 (dashed line), and Sibyll 2.3 (dashed-dotted line). Points are data from [53] and the stars are the LHC measurements [54–58].

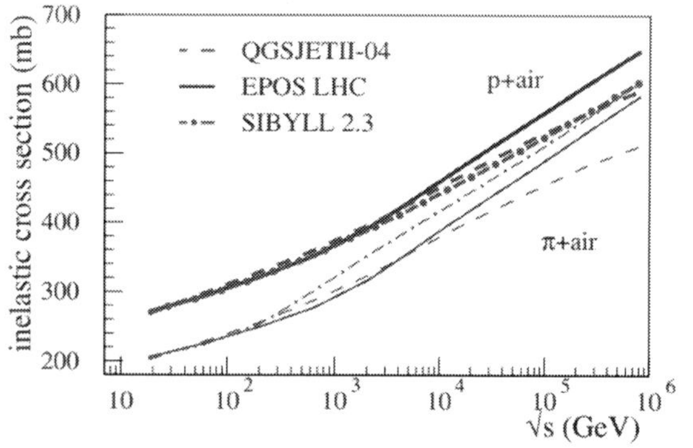

Figure 5. Inelastic p-air (thick lines) and π-air (thin lines) cross sections calculated with EPOS LHC (full line), QGSJETII-04 (dashed line), and Sibyll 2.3 (dashed-dotted line).

of electromagnetic particles at ground level is strongly correlated to this observable (if the shower maximum is closer to ground, the number of particles is higher).

The inelastic cross section of proton-proton scattering is usually used as an input to fix basic parameters in all hadronic interaction models. Therefore it is very well described by all the models up to the LHC, where data exist [52]. As shown on Fig. 4, thanks to the measurements at the LHC, even the extrapolations up to the highest energy are now very similar. In all the figures EPOS LHC is represented by a full (blue) line, QGSJETII-04 by a dashed (red) line and Sibyll 2.3 by a dashed-dotted (green) line.

However, using these models to predict proton-air and pion-air inelastic cross-sections as shown on Fig. 5, one can notice that significant differences appear which will have direct consequences on air shower development. Not only do the evolutions diverge at high energy, but for Sibyll 2.3 the relative behavior of the proton and pion-air cross-section is different from the other models (faster increase of the pion-air cross-section).

3.2. Multiplicity

According to [50], the multiplicity plays a similar kind of role as the cross section, but with a weaker dependence (log). On the other hand the predictions from the models have larger differences for the multiplicity compared to the cross section. As shown in Fig. 6 (left-hand side) and in a more general way in [59], the average multiplicity is well reproduced by all the models up to 1 TeV and even up to 13 TeV for EPOS LHC and QGSJETII-04 [60] and a difference appears between these two models only at the highest energy (beyond 100 TeV). However, in the case of a nuclear target the slope of the rise of the multiplicity as function of the energy is different for all three models leading to a difference of about 20–30% at the highest energies in p or π-air interactions (Fig. 6 right-hand side). This effect is small compared to the pre-LHC era [51] but can change the elongation rate of the air shower maximum development. Here again Sibyll 2.3 has a different behavior than the other models with a smaller slope and the same multiplicity for p or π-air interactions while other models have about 10% difference.

On Fig. 7 left-hand side it can be seen on the pseudorapidity distribution of charged particles at 7 TeV that even if the central density of particles is well reproduced by all models, the width of the distribution is too narrow in the case of Sibyll 2.3 which leads to a reduced total multiplicity as seen on Fig. 6. Furthermore on the right-hand side of Fig. 7, we can observe that the fluctuations are very similar for QGSJETII-04 and EPOS LHC but again Sibyll 2.3 seems to have problems to reproduce the shape of the distribution. This can be important for the fluctuations of the air shower maximum.

Finally by comparing the multiplicity of p-air, helium-air and iron-air interactions we can check that the difference between models increases due to nuclear effects. On Fig. 8 it is clear that from a maximum difference of about 30% for p-air interactions, one can reach more than a factor of 3 in the case of Fe-air. This is of course important for the simulation of air showers with a primary mass heavier than protons.

So, for both cross section and multiplicity, when the models are constrained by LHC data up to 7 TeV, the extrapolation to the highest energy in p-p are very similar but differences remain in nuclear and pion interactions because of the lack of data at high energy and with light ions (only heavy ion data available from RHIC and LHC at high energy).

3.3. Diffraction and elasticity

Another important observable determining air shower development is the elasticity [50] defined as the largest energy fraction carried by a secondary particle (the leading particle). The model predictions are shown on Fig. 9 for p-p, π-air and p-air (as inelasticity=1-elasticity) as a function of center of mass energy. Sibyll 2.3 has the largest elasticity and it is probably related to the fact that the multiplicity is lower (less energy taken from the leading particle). In the cases of EPOS LHC and QGSJETII-04 the difference is smaller for an air target compared to p-p interactions. This opposite behavior compared to the other observables can be explained by the fact that this quantity is very difficult to measure in collider experiments since the latter cannot cover 100% of the phase space. As a consequence there are only indirect constraints on the different contributions to the elasticity leading to a larger uncertainty in the models.

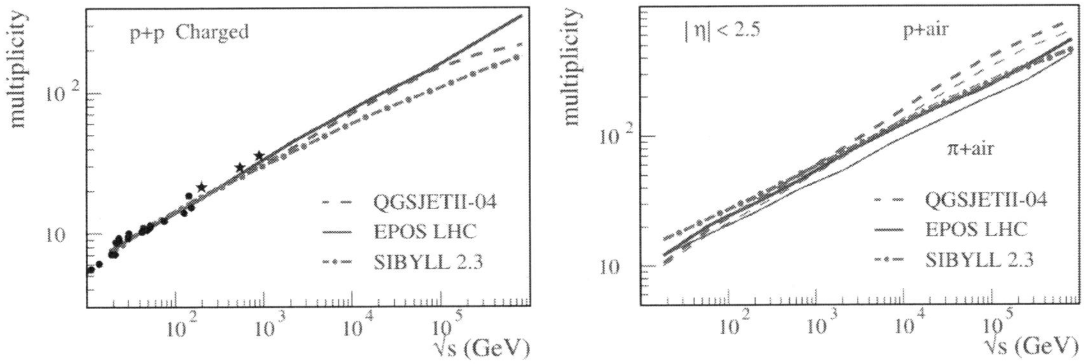

Figure 6. Averaged multiplicity (left-hand side) and multiplicity for $|\eta| < 2.5$ for p (thick lines) and π-air (thin lines) collisions (right-hand side) as a function of center of mass energy. Simulations are done with EPOS LHC (full line), QGSJETII-04 (dashed line) and Sibyll 2.3 (dashed-dotted line). Points are data from old experiments [53].

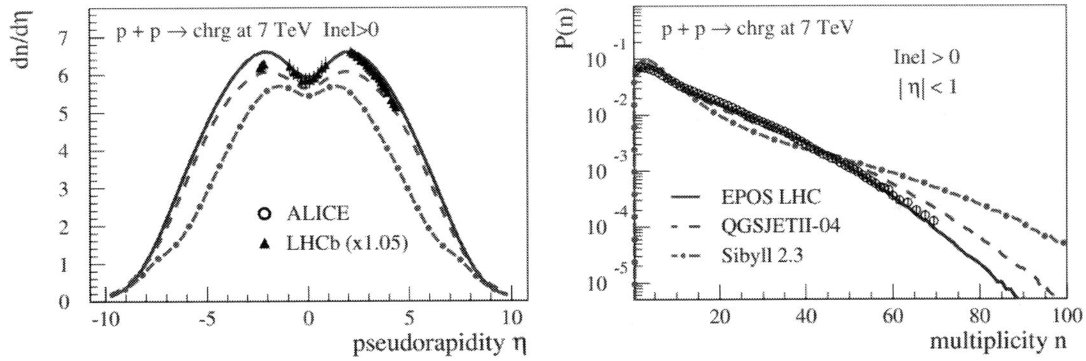

Figure 7. Pseudorapidity distribution $dN/d\eta$ for events with at least one charged particle with $|\eta| < 1$ (left-hand side) and corresponding multiplicity distribution (right-hand side) for p-p interactions at 7 TeV. Simulations with EPOS LHC (full line), QGSJETII-04 (dashed line) and Sibyll 2.3 (dashed-dotted line) are compared to data points from the ALICE and LHCb experiment (rescaled by 5% to take into account the different trigger in ALICE) [61,62].

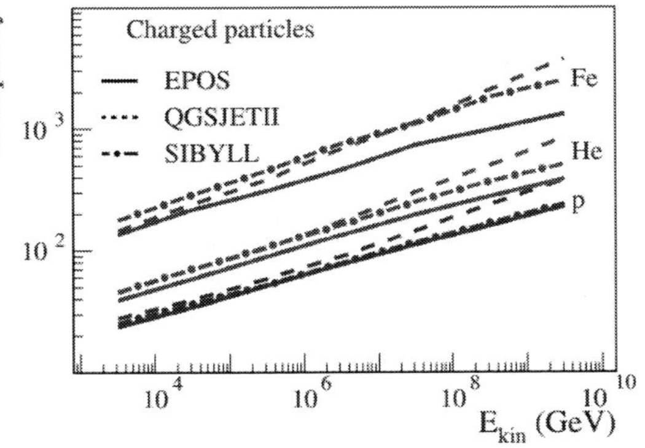

Figure 8. Averaged multiplicity for (from bottom to top) proton (blue lines), helium (red lines) and iron (purple lines) collisions with air as a function of center of mass energy. Simulations are done with EPOS LHC (full line), QGSJETII-04 (dashed line) and Sibyll 2.3 (dashed-dotted line).

One contribution is the diffractive dissociation. Indeed diffractive events are producing the largest elasticities and are important for air shower development, not only for the position of the shower maximum but also for the muon production [63]. At the LHC various measurements related to diffraction are now available [55,64–68]. Due to the difficulties of measuring very forward particles, the compatibility between the results is not as good as it is for the mid-rapidity measurements. This leads to some uncertainties in air shower simulations at a level of $10\,g/cm^2$ [69]. Nevertheless the difference between models seems to be even larger as illustrated on Fig. 10. The rapidity gap (range in pseudorapidity without particle detection in triggered events) cross-section measurement is poorly described by the model while it is directly related to the elasticity in general and diffraction in particular (the large rapidity gaps come from single diffractive events). For instance, the large probability for Sibyll 2.3 to produce rapidity gap around 2 to 4 is a direct consequence of the too narrow pseudorapidity distribution and implies a large elasticity.

It is interesting to notice that hadronic interaction models used for air shower simulations underestimate the diffraction dissociation as observed at LHC (as on Fig. 10 for large rapidity gaps). In fact in EPOS LHC this was improved, leading to a contradiction with air shower data as shown in Sect. 4.3.

3.4. Baryon and resonance production

Another important observable for EAS is the number of muons reaching the ground. It has been shown in [70] that the production of particles which are not π^0 (for

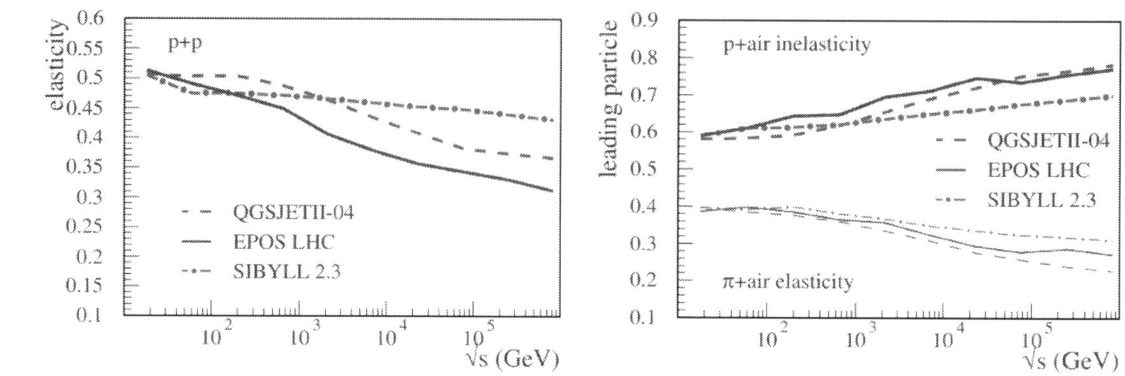

Figure 9. Elasticity (energy fraction of the leading particle) for p-p interactions (left-hand side) and for π-air (thin lines on the right-hand side) and inelasticity (1-elasticity) for p-air (thick lines on the right-hand side) as a function of center of mass energy. Simulations are done with EPOS LHC (full line), QGSJETII-04 (dashed line) and Sibyll 2.3 (dashed-dotted line).

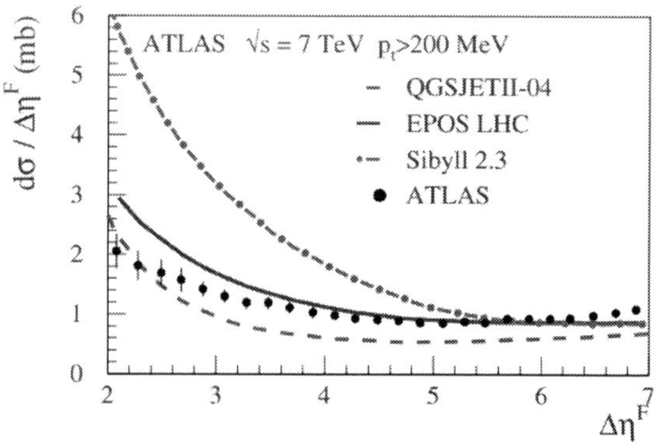

Figure 10. ATLAS measurement of the pseudorapidity gap $\Delta\eta F$ for particles with $p_{t,cut} > 200$ MeV in minimum bias events at 7 TeV [68] compared to EPOS LHC (solid line), QGSJETII-04 (dashed line) and Sibyll 2.3 (dash-dotted line) simulations.

instance baryon-antibaryon pairs or ρ^0 resonance) plays an important role in the muon production rate especially if we take into account the leading particle effect [71].

Recent measurements by NA61 [72] show that the ρ^0 production in π-C interactions seems to be underestimated by a relatively large amount (from 20% to 100%) potentially leading to a large increase of muon production [73]. On the other hand in [74] it is demonstrated that increasing the muon production by increasing the forward baryon pair production like in EPOS leads to a very deep muon production which seems to be in contradiction with data (see Sect. 4.3). Indeed from [75] it can be concluded that the excess of protons seen in [41] (Fig. 3) is not due to newly produce baryons but is due to some baryon stopping (protons from the nuclear target). As a consequence this effect does not lead to an increase of muon production by energy transfer as in EPOS LHC. Both results imply a change in the hadronic interactions models with strong implications on muon production in air showers as shown in the next section.

4. EAS simulations

4.1. Depth of maximum shower development

As shown in Fig. 11, the mean depth of shower maximum, X_{max}, for proton and iron induced showers simulated with CONEX is different for EPOS LHC, QGSJETII-04 and Sibyll 2.3 as a direct consequence of the differences shown in Sect. 3. However the elongation rate (the slope of the mean X_{max} as a function of the primary energy) is almost the same for all models since the difference between models is now much lower than it was in the past [51]. The difference between the models is a constant shift of about $+/-20\,g/cm^2$ around the value given by EPOS LHC. From the results shown in Sect. 3 it is likely that Sibyll 2.3 predicts too large values of the mean X_{max} since the multiplicity is already too low and the elasticity too high at the LHC.

Nevertheless the very similar elongation rate is very important for the study of the primary cosmic ray composition. If the models converge to a similar elongation rate, it will allow us to have a more precise idea on possible changes in composition at the "ankle", for instance, where the Pierre Auger Observatory measures a break in the elongation rate of the data [76].

In fact, further studies using the fluctuations of X_{max} around the mean can be used to test model consistency. Indeed both mean X_{max} and X_{max} fluctuations depend on the mass composition and since fluctuations are less dependent on the details of hadronic interactions (superposition model [50]) than the mean value, it can be checked that the composition corresponding to a given mean X_{max} is consistent with the observed fluctuations. In [76] the Pierre Auger collaboration shows that while it is possible to describe the observed data with EPOS LHC, QGSJETII-04 is in tension with data at a 1 sigma level (mean X_{max} too shallow by $\sim 15\,g/cm^2$).

4.2. Muons at ground level

Concerning the number of muons at ground level (for 40° inclined showers at a height of 1500 m), the difference between QGSJETII-04, EPOS LHC and Sibyll 2.3 is relatively small. We can see on Fig. 12 that model predictions differ only by about 10%. The studies by the Pierre Auger Observatory show that the absolute number of muons observed in vertical shower differ from the model predictions by $1.33 \pm 0.13 \pm 0.09$ [22] in the best case. This is a 2 sigma effect and in case of inclined showers the effect is less than 2 sigmas too [21]. Taking into account the ρ^0 measurement as explained in Sect. 3.4, it is not

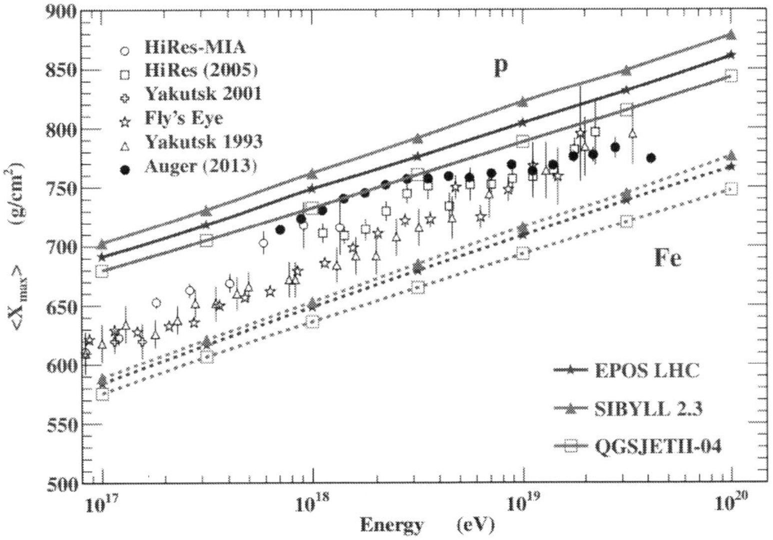

Figure 11. Mean X_{max} for proton and iron induced showers as a function of the primary energy. Predictions of different high-energy hadronic interaction models, full lines for proton and dashed lines for iron with full triangles for Sibyll 2.3, open squares for QGSJETII-04 and full stars for EPOS LHC, are compared to data. Refs. to the data can be found in [1] and [76].

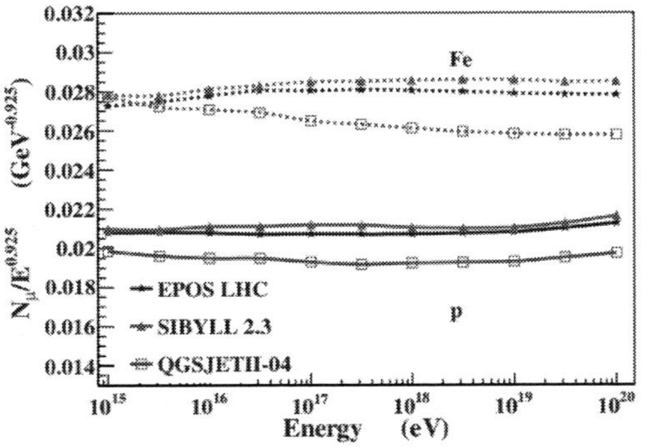

Figure 12. Mean number of muons at ground divided by the primary energy to the power 0.925 for proton and iron induced showers as a function of the primary energy. Predictions of different high-energy hadronic interaction models: full lines for proton and dashed lines for iron with full stars for EPOS LHC, open squares for QGSJETII-04, and full triangles for the results of Sibyll 2.3.

unlikely that the next generation of hadronic interaction models can reproduce the absolute number of muons, at least for vertical showers.

Even if the number of muons is much more similar now for all recent hadronic interactions models, and not so different compared to the data, there is still a large uncertainty related to the energy spectrum of the produced muons. This is an important factor for the attenuation length of the muons in the atmosphere [77] and for the muons at ground level in general [78]. As a consequence, one of the most sensitive measurement of how muons are produced in an air shower is the muon production depth and this is, in fact, not well reproduced by the current models.

4.3. Muon production depth (MPD)

We have seen in the previous section how LHC data could improve the description of EAS using updated hadronic interaction models. In fact, in one particular case, the update of EPOS leads to inconsistent results: the muon production depth measured by the Pierre Auger Observatory [20]. As shown on Fig. 13 the mean logarithmic mass $\langle \ln A \rangle$ calculated from $\langle X^{\mu}_{max} \rangle$ is incompatible with the one extracted from $\langle X_{max} \rangle$ and even out of the range defined by the proton and iron primary mass when EPOS LHC is used for the simulation. With QGSJETII-04 the resulting $\langle \ln A \rangle$ from $\langle X^{\mu}_{max} \rangle$ is below the iron line but not consistent with the one from $\langle X_{max} \rangle$. In a previous analysis [79], EPOS 1.99 was giving results lighter than iron, so the important shift observed in the MPD simulated with EPOS LHC can partially be explained by the change in elasticity due to the corrections in diffractive interactions needed to reproduce the rapidity gap distributions measured by the ATLAS collaboration [68]. We can see on Fig. 10 that EPOS LHC gives reasonable results while QGSJETII-04 is too low.

The change of the parameters needed to describe the rapidity gap correctly (the diffractive cross-section and the diffractive mass distribution) affected both proton and pion interactions because the same parameters were used for both type of projectile. While the change of diffraction and thus of elasticity in proton interaction has very little impact on $\langle X^{\mu}_{max} \rangle$, it appears that MPD are extremely sensitive to the elasticity of pion interactions. This can be understood by the fact that muons are produced at the end of the hadronic cascade after many generations of mainly pion-air interactions. As a consequence of this cumulative effect, even a small increase of only about 10% of the elasticity of pion-air interactions can lead to large shift in $\langle X^{\mu}_{max} \rangle$.

To check this hypothesis, the diffractive cross-section for pion interactions has been reduced in EPOS LHC to

DOI: 10.1051/epjconf/201714518002

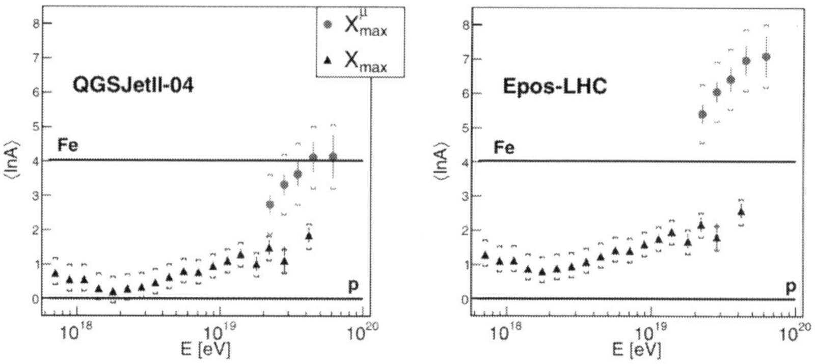

Figure 13. Mean logarithmic mass from $\langle X^{\mu}_{\rm max}\rangle$ (red) and from $\langle X_{\rm max}\rangle$ (black) as a function of energy from [20] using the QGSJETII-04 (left-hand side) or the EPOS LHC (right-hand side) hadronic interaction model.

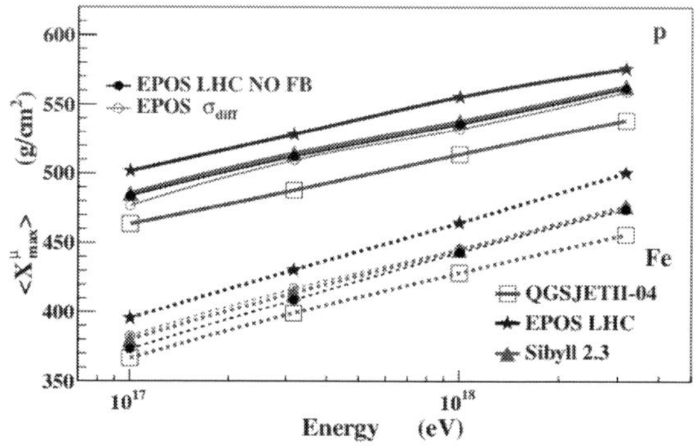

Figure 14. $\langle X^{\mu}_{\rm max}\rangle$ as a function of energy for EPOS LHC (triangles), EPOS 1.99 (stars) and EPOS LHC $\sigma_{\rm diff}$ with a reduced elasticity in pion-air interactions (open circles).

get a reduction of about 10% of the elasticity of the pion-air interactions. As a result $\langle X^{\mu}_{\rm max}\rangle$ is reduced by about 20 g/cm^2 as shown on Fig. 14 with the pink line with open circles. The diffraction has not been changed for proton interactions to keep full compatibility with LHC data and then the change in $\langle X_{\rm max}\rangle$ is limited to less than 5–10 g/cm^2. Another consequence is the increase of the number of muons at ground level by few percents.

Such a small change is compatible with all pion-nucleus data that are available at low energy and thus these two versions of EPOS cannot be discriminated from accelerator data. But the effect on MPD is so strong that data from the Pierre Auger Observatory can be used to constrain diffraction in pion interactions to get consistent results between the mean logarithmic mass which can be extracted from $\langle X^{\mu}_{\rm max}\rangle$ and the one deduced from $\langle X_{\rm max}\rangle$ which has very little dependence on pion hadronic interaction [74]. From EAS development we can thus say that elasticity of pion-air interactions should be lower than the elasticity of proton-air interactions.

The second factor explaining the large shift in MPD was identified in [74] as the too large production of forward baryons in pion interactions (which was indeed extended from low energy only in EPOS 1.99 to all energies in EPOS LHC to improve model consistency). Simply suppressing the production of diquark in string ends and thus the forward baryon pair production as in Fig. 3, the resulting $\langle X^{\mu}_{\rm max}\rangle$ is shown on Fig. 14 as a thin black line (on top of Sibyll 2.3 predictions) which is again about 20 g/cm^2 lower than the original EPOS LHC predictions. The electromagnetic $\langle X_{\rm max}\rangle$ is increased by less than 5 g/cm^2 by the change of forward baryon production (more energy in the π^0). The muons production is reduced at the level of QGSJETII-04.

Since these two effects are cumulative, changing both diffraction and forward baryon production leads to a value of $\langle X^{\mu}_{\rm max}\rangle$ very similar to the one from QGSJETII-04 and thus compatible with Auger $X^{\mu}_{\rm max}$ [20] with a $\langle X_{\rm max}\rangle$ reduced by only 5 g/cm^2 still compatible with the Auger $X_{\rm max}$ [76]. The decrease of muon production could be compensated by a larger ρ^0 production which is in fact related to the increase of diffractive dissociation in pion-nucleus interactions. This is the way to follow for future model development.

5. Summary

In [74] the uncertainty in the first proton(nucleus)-air interaction has been identified as the source of 70% of the uncertainty in the simulated $\langle X_{\rm max}\rangle$. The remaining 30% being linked to the pion-air interactions. Concerning the muon production, 90% is coming from the pion interactions and only 10% from the first interaction. In Sect. 3 we have shown that for the first interaction the uncertainty is not in the basic p-p interaction any more, very well constrained by LHC data, but by the nuclear effects which can not be tested properly with current model and data combination (data with heavy ions only at high energy and only EPOS LHC can treat heavy ion collisions properly). These nuclear effects, being important for both the air target and in the case of heavier primaries, are the main source of the systematic shift in X_{max} of about 20 g/cm^2 around EPOS LHC predictions. This uncertainty is comparable to the experimental uncertainty in the measurement of X_{max} and the elongation rate is now the same for all models for a constant composition. As a consequence the interpretation of the data using a post-LHC model will be more reliable, especially concerning the possible change in mass composition with energy as summarized in [80].

To further reduce these uncertainties and improve the description of air showers by hadronic interaction models, in particular the observables based on muons, it is crucial to improve the description of pion-nucleus interactions in general and the diffractive dissociation in particular which is likely to be different than in proton interactions. Forthcoming studies of diffraction at the LHC including with a nuclear target [15,81] will reduce the model uncertainty for the first interaction at its minimum. To further improve the models it is important to take into account that air shower measurements, such as the muon production depth, can also give very strong constraints on hadronic interactions in particular for pion interactions [74] for which cumulative effects due to the hadronic cascade can be observed. This should give qualitative input to improve the models which can then be quantitatively tested against past and future NA61 measurements for instance [82].

To conclude, we can say that LHC data contribute a lot to reducing the uncertainties in air shower simulations, providing better tools to analyze cosmic ray data. The differences between the hadronic models have been reduced but one should keep in mind that there are still uncertainties in the models themselves which have to be better quantified and transferred to the calculation of the systematic errors in EAS analysis. Consistency of different EAS observables can and should be used to test the hadronic interaction models. The open issues are now mainly in the treatment of pion interactions which have a direct influence on the geometry and energy of the muons in air showers. The next generation of models taking into account more detailed LHC data and what has been learned from the MPD study should significantly improve their description of air showers.

The author would like to thank the Pierre Auger Collaboration and Sergey Ostapchenko for useful discussions.

References

[1] J. Blümer, R. Engel, J.R. Hörandel, Prog. Part. Nucl. Phys. **63**, 293 (2009), 0904.0725
[2] J. Knapp, D. Heck, S.J. Sciutto, M.T. Dova, M. Risse, Astropart. Phys. **19**, 77 (2003), astro-ph/0206414
[3] T. Antoni et al. (KASCADE), Astropart. Phys. **16**, 245 (2002), astro-ph/0102443
[4] T. Antoni et al. (KASCADE), Astropart. Phys. **24**, 1 (2005), astro-ph/0505413
[5] M. Amenomori et al. (Tibet ASγ), Phys. Lett. **B632**, 58 (2006), astro-ph/0511469
[6] T. Abu-Zayyad et al. (HiRes-MIA), Phys. Rev. Lett. **84**, 4276 (2000), astro-ph/9911144
[7] J. Abraham (Pierre Auger Collaboration), Phys. Rev. Lett. **104**, 091101 (2010), 1002.0699
[8] A. Aab et al. (Pierre Auger), Phys. Lett. **B762**, 288 (2016), 1609.08567
[9] R.U. Abbasi et al. (HiRes), Phys. Rev. Lett. **104**, 161101 (2010), 0910.4184
[10] J. Knapp, D. Heck, G. Schatz (1996), in Wissenschaftliche Berichte FZKA 5828, Forschungszentrum Karlsruhe
[11] M. Zha, J. Knapp, S. Ostapchenko, Proc. of 28th Int. Cosmic Ray Conf., Tsukuba p. 515 (2003)
[12] R. Ulrich, R. Engel, M. Unger (2010), 1010.4310
[13] B. Alessandro, D. Bergman, M. Bongi, A. Bunyatyan, L. Cazon et al. (2011), 1101.1852
[14] D. d'Enterria, R. Engel, T. Pierog, S. Ostapchenko, K. Werner, Astropart.Phys. **35**, 98 (2011), 1101.5596
[15] K. Akiba et al. (LHC Forward Physics Working Group), J. Phys. **G43**, 110201 (2016), 1611.05079
[16] M.G. Aartsen et al. (IceCube), Astropart. Phys. **78**, 1 (2016), 1506.07981
[17] R. Abbasi et al. (IceCube), Nucl. Instrum. Meth. **A700**, 188 (2013), 1207.6326
[18] T. Antoni et al. (KASCADE), Nucl. Instrum. Meth. **A513**, 490 (2003)
[19] W.D. Apel et al. (KASCADE Grande), Phys. Rev. Lett. **107**, 171104 (2011), 1107.5885
[20] A. Aab et al. (Pierre Auger), Phys. Rev. **D90**, 012012 (2014), [Erratum: Phys. Rev. **D92**(1), 019903 (2015)], 1407.5919
[21] A. Aab et al. (Pierre Auger), Phys. Rev. **D91**, 032003 (2015), [Erratum: Phys. Rev. **D91**(5), 059901 (2015)], 1408.1421
[22] A. Aab et al. (Pierre Auger), Phys. Rev. Lett. **117**, 192001 (2016), 1610.08509
[23] T. Bergmann et al., Astropart. Phys. **26**, 420 (2007), astro-ph/0606564
[24] S. Ostapchenko, Phys. Rev. **D74**, 014026 (2006), hep-ph/0505259
[25] S. Ostapchenko, Phys. Lett. **B636**, 40 (2006), hep-ph/0602139
[26] S. Ostapchenko, Phys. Rev. **D83**, 014018 (2011), 1010.1869
[27] K. Werner, F.M. Liu, T. Pierog, Phys. Rev. **C74**, 044902 (2006), hep-ph/0506232
[28] T. Pierog, K. Werner, Nucl. Phys. Proc. Suppl. **196**, 102 (2009), 0905.1198
[29] T. Pierog, I. Karpenko, J.M. Katzy, E. Yatsenko, K. Werner, Phys. Rev. **C92**, 034906 (2015), 1306.0121
[30] J. Engel, T.K. Gaisser, T. Stanev, P. Lipari, Phys. Rev. **D46**, 5013 (1992)
[31] R. Engel, T.K. Gaisser, T. Stanev, P. Lipari, Proc. of 26th Int. Cosmic Ray Conf., Salt Lake City **1**, 415 (1999)
[32] E.J. Ahn, R. Engel, T.K. Gaisser, P. Lipari, T. Stanev, Phys. Rev. D **80**, 094003 (2009), 0906.4113
[33] F. Riehn, R. Engel, A. Fedynitch, T.K. Gaisser, T. Stanev (2015), 1510.00568
[34] D. Heck, J. Knapp, J. Capdevielle, G. Schatz, T. Thouw (https://web.ikp.kit.edu/corsika/), Wissenschaftliche Berichte, Forschungszentrum Karlsruhe **FZKA 6019** (1998)
[35] T. Csörgö et al. (TOTEM Collaboration), Prog. Theor. Phys. Suppl. **193**, 180 (2012), 1204.5689

[36] H.J. Drescher, M. Hladik, S. Ostapchenko, T. Pierog, K. Werner, Phys. Rept. **350**, 93 (2001), hep-ph/0007198
[37] F. Liu, J. Aichelin, M. Bleicher, H. Drescher, S. Ostapchenko et al., Phys. Rev. **D67**, 034011 (2003), hep-ph/0202008
[38] M. Bleicher, F. Liu, A. Keranen, J. Aichelin, S. Bass et al., Phys. Rev. Lett. **88**, 202501 (2002), hep-ph/0111187
[39] M. Hladik, H.J. Drescher, S. Ostapchenko, T. Pierog, K. Werner, Phys. Rev. Lett. **86**, 3506 (2001), hep-ph/0102194
[40] K. Werner, Phys. Rev. Lett. **98**, 152301 (2007), 0704.1270
[41] D.S. Barton et al., Phys. Rev. **D27**, 2580 (1983)
[42] S. Ostapchenko, Phys. Rev. **D81**, 114028 (2010), 1003.0196
[43] N.N. Kalmykov, S.S. Ostapchenko, A.I. Pavlov, Bull. Russ. Acad. Sci. Phys. **58**, 1966 (1994)
[44] T.K. Gaisser, F. Halzen, Phys. Rev. Lett. **54**, 1754 (1985)
[45] G. Pancheri, Y.N. Srivastava, Phys. Lett. **B182**, 199 (1986)
[46] A. Capella, J. Trân Thanh Vân, J. Kwieciński, Phys. Rev. Lett. **58**, 2015 (1987)
[47] L. Durand, P. Hong, Phys. Rev. Lett. **58**, 303 (1987)
[48] T. Sjostrand, Int. J. Mod. Phys. **A3**, 751 (1988)
[49] S. Ostapchenko, M. Bleicher, T. Pierog, K. Werner, Phys. Rev. **D94**, 114026 (2016), 1608.07791
[50] J. Matthews, Astropart. Phys. **22**, 387 (2005)
[51] R. Engel, D. Heck, T. Pierog, Ann. Rev. Nucl. Part. Sci. **61**, 467 (2011)
[52] H. Jung et al. (2009), 0903.3861
[53] C. Caso et al. (Particle Data Group), Eur. Phys. J. **C3**, 1 (1998)
[54] G. Antchev et al. (TOTEM), Europhys. Lett. **101**, 21004 (2013)
[55] B. Abelev et al. (ALICE), Eur. Phys. J. **C73**, 2456 (2013), 1208.4968
[56] G. Aad et al. (ATLAS), Nature Commun. **2**, 463 (2011), 1104.0326
[57] G. Antchev et al. (TOTEM), Phys. Rev. Lett. **111**, 012001 (2013)
[58] M. Aaboud et al. (ATLAS), Phys. Rev. Lett. **117**, 182002 (2016), 1606.02625
[59] D. d'Enterria, T. Pierog, JHEP **08**, 170 (2016), 1604.08536
[60] M. Aaboud et al. (ATLAS), Eur. Phys. J. **C76**, 502 (2016), 1606.01133
[61] K. Aamodt et al. (ALICE Collaboration), Eur. Phys. J. **C68**, 345 (2010), 1004.3514
[62] R. Aaij et al. (LHCb collaboration), Eur. Phys. J. **C74**, 2888 (2014), 1402.4430
[63] R. Luna, A. Zepeda, C.A. Garcia Canal, S.J. Sciutto, Phys. Rev. **D70**, 114034 (2004), hep-ph/0408303
[64] V. Khachatryan et al. (CMS), Phys. Rev. **D92**, 012003 (2015), 1503.08689
[65] G. Antchev et al. (TOTEM), Phys. Rev. Lett. **111**, 262001 (2013), 1308.6722
[66] M. Taševský (ATLAS), AIP Conf. Proc. **1654**, 040002 (2015)
[67] M. Meissner (LHCb), *Results of Soft-Diffraction at LHCb*, in *Proceedings, 15th conference on Elastic and Diffractive scattering (EDS Blois 2013)* (2013), 1310.2192
[68] G. Aad et al. (ATLAS Collaboration), Eur. Phys. J. **C72**, 1926 (2012), 1201.2808
[69] S. Ostapchenko, Phys. Rev. **D89**, 074009 (2014), 1402.5084
[70] T. Pierog, K. Werner, Phys. Rev. Lett. **101**, 171101 (2008), astro-ph/0611311
[71] S. Ostapchenko, EPJ Web Conf. **52**, 02001 (2013)
[72] A. Herve (NA61), PoS **ICRC2015**, 330 (2016), 1509.06586
[73] F. Riehn, Acta Phys. Polon. Supp. **8**, 993 (2015)
[74] S. Ostapchenko, M. Bleicher, Phys. Rev. **D93**, 051501 (2016), 1601.06567
[75] G.I. Veres, Ph.D. thesis, Eötvös Loránd University, Budapest (2001)
[76] A. Aab et al. (Pierre Auger), Phys. Rev. **D90**, 122005 (2014), 1409.4809
[77] J.C. Arteaga-Velazquez et al. (KASCADE Grande), *Study of the muon content of very high-energy EAS measured with the KASCADE-Grande observatory*, in *Proceedings, 33rd International Cosmic Ray Conference (ICRC2013): Rio de Janeiro, Brazil, July 2–9, 2013* (2013), p. 0772, 1308.3202
[78] J. Espadanal, L. Cazon, R. Conceição, Astropart. Phys. **86**, 32 (2017), 1607.06760
[79] P. Abreu et al. (Pierre Auger Collaboration), Proceeding of the 32nd ICRC, Beijing, China (2011), 1107.4804
[80] K.H. Kampert, M. Unger, Astropart.Phys. **35**, 660 (2012), 1201.0018
[81] Y. Itow (LHCf), PoS **ICRC2015**, 259 (2016)
[82] A. Häsler (NA61/SHINE), *Hadron Production Measurements in NA61/SHINE for the Precise T2K Neutrino Flux Prediction and the Cosmic ray Physics Program*, in *Proceedings, 16th Lomonosov Conference on Elementary Particle Physics: Particle Physics at the Year of Centenary of Bruno Pontecorvo: Moscow, Russia, August 22–28, 2013* (2015), pp. 327–330

Challenges for cosmic-ray experiments

Thomas Gaisser[a]

Bartol Research Institute and Department of Physics and Astronomy, University of Delaware, Newark, DE 19716, USA

Abstract. This paper is a commentary on presentations at ISVHECRI 2016 related to cosmic-rays, gamma-rays and neutrinos. Its goal is to highlight the unanswered questions raised during the conference about the sources of these cosmic particles and the relations among them.

1. Introduction

A current theme of astro-particle physics is "multi-messenger astronomy," which emphasizes use of gamma-rays and neutrinos to address still unanswered questions on the origin of high-energy cosmic rays. With this theme in mind, I review the presentations on cosmic-ray spectra and composition, gamma-ray astronomy and neutrino astronomy presented at ISVHECRI 2016. The role of hadronic interactions is addressed in the review paper of Tanguy Pierog [1].

2. Some cosmic-ray questions

Figure 1 gives a global view of the cosmic-ray spectrum.The abundant elements of the primary cosmic-ray spectrum are measured accurately to energies higher than a TeV per nucleon with spectrometers in space [2–4]. Calorimetric measurements with balloon-borne detectors [5,6] extend direct measurements to higher energy but with somewhat less precision. This means that we have good coverage of the composition with direct measurements up to about 100 TeV energy per nucleus. Indirect measurements with large detectors on the surface are needed for the higher energy cosmic rays. There are several questions of current interest associated with the various features in the energy spectrum:

- What is the composition in the knee region and how does it connect with direct measurements at lower energy?
- What is the cause of the hardening of the spectrum around 20 PeV?
- Where is the transition from Galactic to extra-galactic cosmic rays and how is it related to composition around the ankle?
- What is responsible for the apparent end of the spectrum around 100 EeV?
- Does the difference between the Auger and the TA spectrum in the cutoff region show that the cosmic-ray spectrum is different in different regions of the sky?

[a] e-mail: `gaisser@bartol.udel.edu`

2.1. The knee region

Most air shower measurements have a threshold around a PeV, while direct measurements extend only to $\sim$100 TeV. An exception is the ARGO-YBJ RPC carpet detector at high altitude in Tibet, which has the potential to cover the gap between direct and indirect measurements [8]. TAIGA-HISCORE is also starting to measure the spectrum down to $\sim$300 TeV [9]. ARGO-YBJ have reported measurements of both the all-particle spectrum and the spectrum of the light (p + He) component [10]. The measurements cover the energy range from $\sim$10 TeV to the knee region. While the measurement of the all-particle spectrum agrees with several other EAS measurements through the knee region, the light component appears to steepen starting around 700 TeV [8,10]. In contrast, KASCADE [11,12] shows the proton steepening above a PeV. In his presentation, DiSciascio compares the ARGJO-YBJ result with the Hörandel parameterization [13] of p + He, which also steepens at higher energy. The IceCube/IceTop composition analysis [15] starts around 3 PeV, too high to provide insight on this question.

Indirect measurements of the composition with EAS detectors are difficult, and the ARGO-YBJ result points to an ambiguity that needs to be resolved. A practical aspect is its implication for the atmospheric neutrino flux at high energy relevant for IceCube. Calculation of the flux of atmospheric neutrinos depends on the spectrum of nucleons as a function of energy per nucleon, which is dominated by protons and helium. If the proton and helium components steepen at 700 GV, then there should be a compensating increase in heavier nuclei to keep the all-particle spectrum constant. The sketch in Fig. 2 illustrates the effect, which would likely be a suppression of the flux of nucleons in a range around a PeV that arises if the all-particle spectrum is dominated by heavy nuclei in this region. This in turn would significantly reduce the flux of muons and muon-neutrinos around 100 TeV.

2.2. From the knee to the ankle

Figure 3 (left) compares measurements of the spectrum by KASCADE-Grande [12,16], TUNKA-133 [17,18] and IceTop [15,19,20]. The solid line shows a spectrum with a constant differential index of -3. The data show clear structure between the knee and the ankle, with a

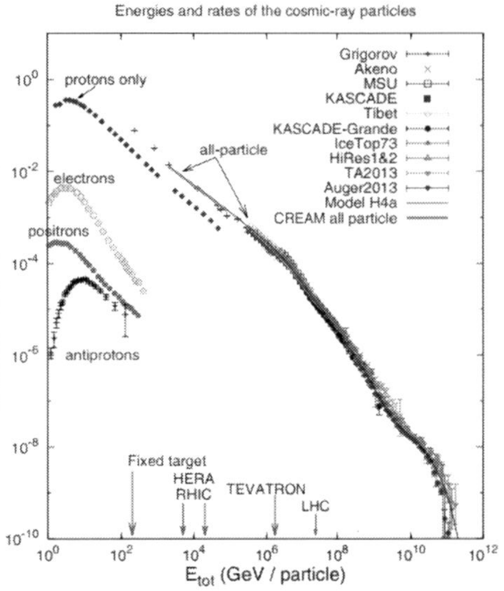

Figure 1. Overview of the energy spectra of various components of the cosmic radiation (Fig. 2.1 of [7]).

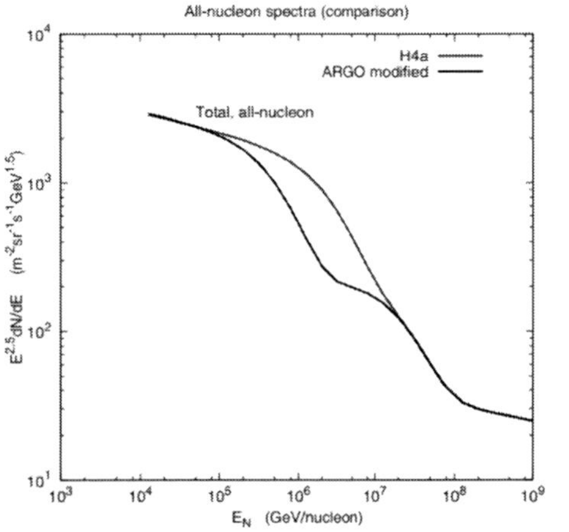

Figure 2. The spectrum of nucleons for the H4a model [14] compared with a modified version in which the cutoff rigidities for p and HE are reduced to 700 GV and the all-particle spectrum is restored by increasing the contribution of the CNO and Fe groups.

hardening around 2×10^{16} eV and a second knee above 10^{17} eV. The KASCADE-Grande analysis uses the fraction of muons to separate the spectrum into light and heavy components [12,21]. The data suggest that the sub-dominant light component increases relative to the heavy component as energy increases toward the PeV region, as shown in Fig. 3 (right). A possible interpretation is that the increase of the light component reflects the population of cosmic rays from extragalactic sources, while the steeper heavy component is the end of the Galactic population.

2.3. The highest energy cosmic rays

Measurements of the cosmic-ray spectrum to the highest energy were presented at ISVHECRI 2016 from both Telescope Array (TA) [22] and Auger [23]. They are in excellent agreement with each other through the ankle region within their systematic uncertainties in energy. However, after shifting the Auger energy assignment up by 8% (or the TA spectrum down by a similar amount) the TA spectrum remains somewhat higher than the Auger spectrum above 10 EeV.

The question of composition of the highest energy cosmic rays has long been an important unresolved issue. Both TA and Auger find a large fraction of protons in the EeV range, above which the interpretations differed, with Auger preferring heavier and TA lighter composition. Inferences about composition are based on both the mean depth of shower maximum as a function of energy and on fluctuations in depth of maximum in each energy bin, and they depend on the hadronic interaction model used to make the interpretation. The TA presentation [22] includes a plot of mean depth of maximum for both experiments obtained by the joint composition working group that includes members of both experiments. The results of the two experiments are not inconsistent with each other. In comparison with the interaction model QGSJETII-03 the depth of maximum measurements are between protons and iron, but closer to the proton limit. Thus at present the composition at the highest energy remains an open question.

Composition from 1–100 EeV is the key to what is one of the most important open questions in cosmic-ray physics, namely, the cause of the apparent cutoff in the spectrum at 100 EeV. There are two possibilities. If protons dominate at high energy, the natural interpretation would be the GZK process [24,25], energy loss to photo-pion production during propagation in the cosmic microwave background radiation. The other possibility is that the accelerators are reaching their maximum rigidity, as suggested by the Hillas plot [26]. The Auger presentation [23] illustrates the different energy-dependent compositions that characterize each of these possibilities [27]. The GZK explanation requires mostly protons at the highest energy while in the Hillas case an increasing fraction of heavy nuclei would be expected as the cutoff in rigidity affects protons first. In both cases effects of nuclear fragmentation during propagation must accounted for in addition to the source composition.

Both TA and Auger have initiated upgrades aimed at understanding the composition and the related question of anisotropy of the highest energy cosmic rays. TA × 4 will expand the surface detector with two large scintillator arrays (grid spacing 2.08 km) adjacent to the present array. One of the main aims is to get a better understanding of the hotspot observed in the TA data [28], which persists after seven years of observation.

The upgrade of the Pierre Auger Observatory [29] includes installation of a scintillator on top of each tank, improved electronics and underground muon detectors. With improved ability to separate the muon signal from the electromagnetic signal in the surface detector, the upgraded array surface array will have sensitivity to primary composition with 100% duty cycle. The goal is to determine the origin of the flux suppression at the highest energy and to determine the feasibility of astronomy with cosmic-ray protons.

2.4. Cosmic-ray overview

Using its low energy extension (TALE), the TA group has presented a combined spectrum that extends all the way

 DOI: 10.1051/epjconf/201714518003

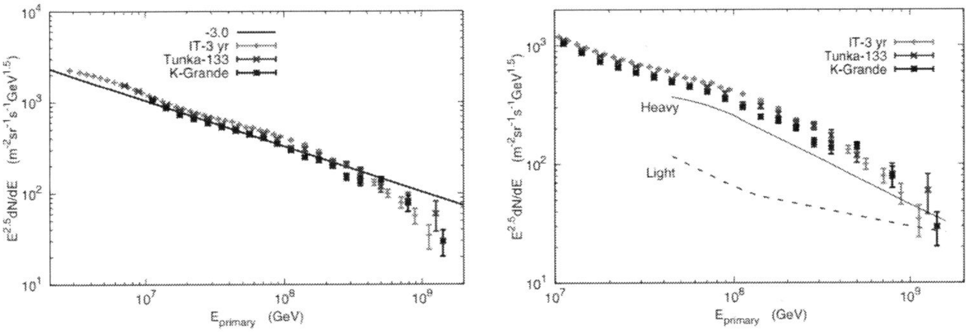

Figure 3. Measurements of the spectrum between the knee and the ankle. Left: compared to a constant E^{-3} differential spectrum (Fig. 17.2 of [7]) and Right: showing the heavy and light fractions as measured by KASCADE-Grande [21] separately (Fig. 17.3 of [7]).

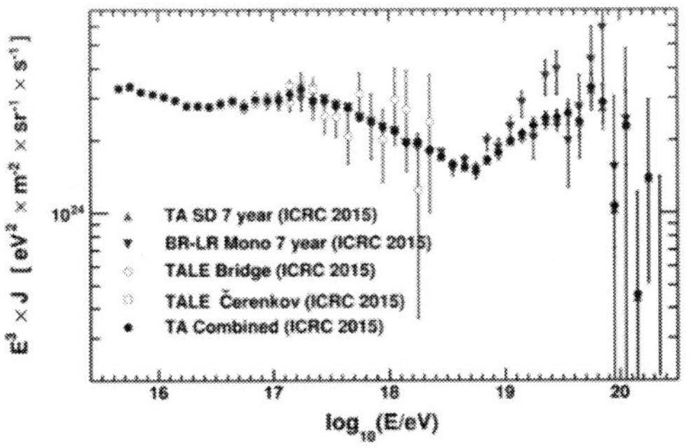

Figure 4. The combined spectrum of TA presented at this conference [22]) and shown here by kind permission of John Matthews.

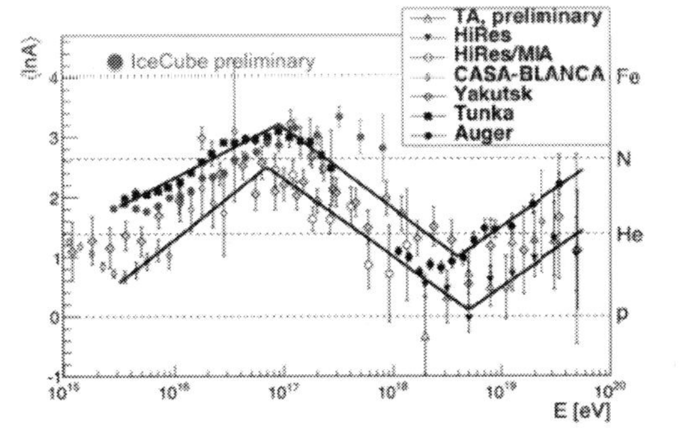

Figure 5. Mean value of the natural logarithm of the primary mass as a function of primary air shower energy. The plot is from the paper of Kampert & Unger [30]), and the solid lines indicate the ranged of data summarized there. Preliminary composition data from IceCube [31] are interpolated in the same way and added to the Kampert & Unger plot in red.

down to 5 PeV (Fig. 4). The data are consistent with those shown in Fig. 3 in the overlap region. The hardening of the spectrum at 2×10^{16} eV and the second knee just above 10^{17} eV are clearly visible, as well as the ankle and the suppression above 10^{20} eV.

The composition also shows an energy dependence with features that may be correlated with the spectrum, as illustrated in Fig. 5 [30]. The plot is made by interpolating the value of each measured depth of maximum between theoretical curves for protons and iron on a plot of X_{max} vs energy (in this case using SIBYLL 2.1 [32]). The increasing fraction of heavy nuclei up to 10^{17} eV, for example, can be interpreted naturally as a rigidity-dependent effect as Galactic accelerators reach their maximum energy. This is followed by an increase in the proton fraction that reaches a maximum (minimum in $\langle \ell n(A) \rangle$) at the ankle. The preliminary IceCube result [19,31] is shown in red in Fig. 5. It agrees with the trend of the data up to 10^{17} eV, but remains at a high level above that energy, although with large uncertainties.

In his presentation Ptuskin [33] reviewed the standard model of cosmic-ray origin, with Galactic supernovae of different types producing the cosmic-ray spectrum observed at Earth up to $\sim 10^{17}$ eV and an extra-galactic component dominating above 10^{18} eV. Analysis of low-energy iron isotopes [34] shows that there must have been at least two nearby supernova remnants within one or two million years of each other. The general picture of supernovae in the disk of the Galaxy exploding randomly and each accelerating cosmic rays for a period of time short compared to the propagation time is explored in Ref. [35]. The contribution of a particular supernova to the cosmic-ray flux currently observed at Earth will be determined by the energy-dependent residence time of cosmic-rays in the Galaxy coupled with time of the explosion and its distance from Earth. Such considerations also have implications for the observed anistropy of cosmic rays [36], which will reflect to some extent the particular history of supernova explosions in the Milky Way.

3. Multi-messenger astronomy

Photons and neutrinos, being electrically neutral, propagate from their sources in straight lines, unlike charged cosmic rays, which are bent in Galactic and cosmic magnetic fields. Therefore, to the extent that cosmic-rays interact in or near their sources, gamma-rays and neutrinos should be good probes of cosmic-ray sources. Neutrinos have the extra advantage of not being affected by interactions in transit (and the corresponding disadvantage of being difficult to detect). High-energy gamma-rays can be produced electromagnetically, for example by inverse Compton scattering by electrons, as well as from decay of π^0 produced in hadronic interactions. Neutrinos can only be produced by hadronic interactions. Just as atmospheric neutrinos are produced by interactions of cosmic rays in the atmosphere, any high-energy neutrino

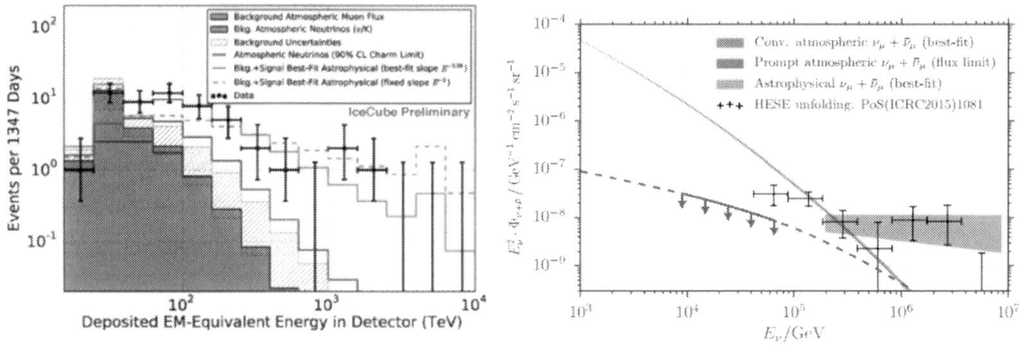

Figure 6. Spectrum of high energy neutrinos from IceCube. Left: 4-year HESE analysis [37]; Right: 6-year upward muon analysis [42].

of extraterrestrial origin will have been produced by the interaction of a proton (or less likely a nucleus) somewhere in the cosmos. The interaction may be with gas or by photo-pion production, but in both cases gamma-rays would be produced through the corresponding neutral pion channel. (The possibility of γ-rays and neutrinos from dark matter is not considered here.)

3.1. High-energy neutrinos in IceCube

Figure 6 shows the observation by IceCube of a hard spectrum of astrophysical neutrinos extending above the steeply falling background of atmospheric neutrinos. The signal is seen consistently in two channels. The high-energy starting event (HESE) analysis [38] requires the event to start well inside the instrumented volume of IceCube in such a way that the outer layers of optical modules exclude atmospheric muons from above, as well as atmospheric neutrinos of sufficiently high energy to be accompanied by a muon [39,40]. The astrophysical signal can be identified to lower energy by increasing the veto region and correspondingly reducing the size of the fiducial volume [41]. In the upward muon analysis [42], the Earth is used as a shield to exclude atmospheric muons. The signal then consists mainly of neutrino-induced muons. The astrophysical component is distinguished only by its hard spectrum extending to high energy.

Accompanying each of the neutrino analyses is a list of the highest energy events with their locations in the sky (right ascension and declination). In the case of the HESE analysis, the event type is given as cascade (ν_e and most ν_τ) or track (ν_μ). The four-year sample includes 54 events with deposited energy of 30 TeV and higher, of which fewer than twenty are background. For the upward muon analysis, all events are tracks, and 29 events with estimated muon energy $E_\nu > 200$ TeV are singled out as being the most likely to be of astrophysical origin. Sky maps are made for both cases, but no source of high significance emerges. The most energetic event in the upward muon sample has a deposited energy in the detector of 2.6 ± 0.3 PeV and an estimated muon energy of 4.5 PeV. The probability distribution for the parent neutrino energy peaks just below 10 PeV [42].

The spectral index that characterizes the astrophysical flux depends on the energy range of the analysis, and the possibility of a break and or a cutoff in the spectrum is not ruled out. A combined maximum-likelihood analysis of the high-energy astrophysical neutrino flux measured with several IceCube analyses gives a value of the energy flux of all neutrino flavors of

$$E_\nu^2 \frac{dN_\nu}{dE_\nu} \approx 2 \times 10^{-8}\ \mathrm{GeV\,cm^{-2}sr^{-1}\,s^{-1}} \tag{1}$$

at $E_\nu = 1$ PeV assuming the spectrum is a single power law [43]. The fitted differential spectral index, is -2.5 ± 0.09. It is interesting to note from Fig. 1 that the energy flux of cosmic rays has the same value at the ankle, which may be interpreted as indicating that the energy flux of the extragalactic population of cosmic rays is somehow related to the astrophysical neutrino flux, as suggested by Waxman and Bahcall [44,45].

In this context, it is interesting to note that detection of cosmogenic neutrinos from photo-pion production by interactions of ultra-high-energy cosmic rays with the CMB could resolve the origin of the cutoff of the cosmic-ray spectrum at 100 EeV. A search by IceCube for events of extremely high energy (EHE) with seven years of data [46] finds no neutrinos with energy $> 10^7$ GeV. This result places significant constraints on some models of cosmogenic neutrinos. Because neutrinos produced by interactions of cosmic-rays with energy-per-nucleon $> 10^9$ GeV would produce neutrinos with $E_\nu > 10^7$ GeV, the result also disfavors some models of astrophysical neutrinos associated with sources of ultra-high-energy cosmic rays.

3.2. Extragalactic source candidates

The tables of high-energy neutrinos [38,42] constitute a small fraction of the ~100,000 neutrinos per year reconstructed in IceCube. Most of these are atmospheric neutrinos. Nevertheless, the most sensitive searches for point sources of astrophysical neutrinos use the larger data samples that extend down to the TeV energy range. Maximum likelihood techniques that take account of the angular resolution and the deposited energy of each event are used to look for point sources of neutrinos emerging from the smooth foreground of atmospheric neutrinos. An all-sky analysis produces a sky map of significances. In addition, a list of sources selected, largely on the basis of gamma-ray observations, to be likely neutrino sources is investigated. The most sensitive search so far [47] uses seven years of data and finds no source of high significance in the all-sky analysis and places upper limits on the targeted sources.

The relation between a total observed intensity from all directions and the number of events to be expected from nearby sources is addressed in general in [48] and in the context of the IceCube discovery in [49]. Upper limits on extragalactic sources, together with an assumption about the cosmological evolution of the source class, lead to a lower limit on the density of sources. A recent analysis [50] of the total astrophysical intensity of neutrinos and the current upper limits of IceCube suggests that the density of extra-galactic sources should be $> 10^{-7}\,(\mathrm{Mpc})^{-3}$. Blazars and Gamma-Ray Bursts (GRB) are among the rare, luminous sources that are being constrained by this line of argument [51]. Moreover, there are experimental upper limits from IceCube on the contribution to the IceCube signal of both blazars [52] and GRBs [53].

Starburst galaxies are relatively abundant and have been proposed as a likely source of neutrinos [54]. The idea is that, because of the greater rate of supernova explosions and the turbulence they generate, there would be a higher intensity of cosmic rays produced and they would be confined in the host galaxy sufficiently long that most protons would interact (calorimetric limit). Since the target for the neutrino production is gas in the galaxy, the spectrum of neutrinos and gamma-rays would extend down to low energy. Fermi observations of the diffuse gamma background [55] have been used to constrain the starburst model [56,57]. In order to avoid producing more diffuse gamma-radiation than observed, the spectra of cosmic-ray induced neutrinos and photons must have a hard spectrum (differential spectral index $-\alpha$ with $\alpha < 2.2$ [56]. On the other hand, since most of the isotropic gamma-radiation comes from blazars [58], it would be surprising if there were not also neutrinos at some level [59].

Another possible neutrino source, discussed at this conference by Ptuskin [33], is Type IIn supernovae in external galaxies [60]. These are rare core-collapse supernovae that accelerate protons into the dense wind of the progenitor star. The environment of the progenitor wind is sufficiently dense that the SN expansion slows down on a time scale of 30 years. Like the starburst scenario, the target for neutrino (and gamma-ray) production is gas, so the spectra extend to low energy and the Fermi isotropic gamma limit applies (so the accelerated spectrum of protons must be hard). A maximum energy of 10^{17} eV/proton is found, which is sufficient to produce neutrinos up to $\sim PeV$ (but not beyond). Assuming a rate of 1% of all core collapse supernovae, a diffuse spectrum in rough agreement with the IceCube spectrum is found. Moreover, two tracklike events are in the direction of known Type IIn supernovae. HESE event 47 with 74 TeV deposited energy [37] is 1.35° from SNIIn 2005bx and event 11 of the upward muon sample with an estimated energy of 240 TeV [42] is 0.3° from SNIIn 2005jq.

3.3. Neutrinos and γ-rays from the Milky Way

Several of the events in the high-energy samples come from near the Galactic plane and could be of local origin in the Milky Way [61]. In fact a local component is expected because both neutrinos and photons are produced by cosmic-ray interactions of cosmic rays with interstellar gas during their diffusive propagation in the Galaxy. In his paper on neutrinos, Stecker [62] pointed out that these neutrinos would have the same spectrum as that of the primary cosmic rays ($\alpha \approx 2.7$). At high energy the spectrum of atmospheric neutrinos is one power steeper, so the Galactic neutrinos should become relatively more prominent as energy increases. In addition, the cosmic-ray spectrum in the central region of the Galaxy could have a harder spectrum than observed locally at Earth [63].

The Galactic diffuse γ emission is measured up to $\approx 100\,\mathrm{GeV}$ by Fermi [64]. It is concentrated along the Galactic ridge and is mostly from π^0-decay. The ARGO-YBJ measurements of the diffuse γ-emission up to a TeV are consistent with an extension of the Fermi measurement [65]. The corresponding spectrum of $\nu_\mu + \bar{\nu}_\mu$ should have the same shape and a magnitude of $2/3$ that of the gamma-rays at production and $1/3$ after oscillations [7]. In the presentation at this conference [8] the high intensity of gamma-rays from the region of the Cygnus Cocoon [66] was also described. This too should have a counterpart in neutrinos. The ANTARES neutrino detector has placed a limit [67] on the intensity of multi-TeV neutrinos from the Galactic plane that is approaching the higher prediction of Ref. [63].

Using gamma-ray data from Fermi-LAT as a template, predictions for Galactic neutrinos in the TeV to PeV range are examined in detail in Ref. [68]. The conclusion is that less than 10% of the high-energy events are of Galactic origin. Limits are also placed on the Galactic contributions in the TeV range.

4. Summary and outlook

Understanding the origin of ultra-high energy cosmic rays and whether their sources also produce neutrinos remains a key question [46]. It is possible that the sources of the astrophysical neutrinos seen by IceCube are not the sources of UHECR. The current status of neutrinos and cosmic-rays observed by IceCube is the subject of a recent review [69].

To improve the ability to identify the sources of high-energy astrophysical neutrinos, IceCube has started a real-time alert system [70]. Its goal is to maximize the ability to find electromagnetic counterparts of astrophysical neutrinos seen in IceCube. There are several types of alerts. Single track-like events (with good pointing and high energy) in the HESE and EHE samples generate alerts. Alerts for follow-up by optical, X-ray and γ-ray detectors trigger on multiple events from a single point in the sky in a limited time interval.

The discovery of high-energy astrophysical neutrinos motivates efforts to accumulate more data with new and larger detectors. Construction of KM3NeT is starting, with the deployment of test lines at the Italian site, Capo Passero [71] and a Letter of Intent [72]. The proposal is to build a high-energy unit called ARCA at Capo Passero and a densely instrumented array called ORCA at the French site near Marseille [71]. The first cluster of an expansion of Baikal to a Gigaton Volume Detector (Baikal-GVD) [73] was installed in 2015. Plans for a next generation IceCube-Gen2 are underway [74,75]. IceCube-Gen2 will include a densely instrumented subarray for neutrino physics [76].

The high-energy expansion of IceCube-Gen2 aims to increase the target volume by an order of magnitude. To reach the much larger volumes needed to detect a large number of cosmogenic neutrinos, the radio technique is being pursued. The first detectors of ARIANNA [77] and of ARA [78] prototypes are already in operation in Antarctica. A review of the current status of high-energy neutrino astronomy that covers current and future detectors and the science they will address is in preparation [79].

I thank John Matthews for permission to use his composite spectrum of Telescope Array and TALE (Fig. 4), and I thank Javier Gonzalez for adding the IceCube data points in Fig. 5. I thank Christopher Wiebusch for helpful comments on a draft version of this paper, and I acknowledge support from the U.S. National Science Foundation (PHY-1505990).

References

[1] T. Pierog, These Proceedings
[2] O. Adriani et al. [PAMELA Collaboration], "PAMELA Measurements of Cosmic-ray Proton and Helium Spectra," Science **332**, 69 (2011) [arXiv:1103.4055 [astro-ph.HE]]
[3] M. Aguilar et al. [AMS Collaboration], "Precision Measurement of the Proton Flux in Primary Cosmic Rays from Rigidity 1 GV to 1.8 TV with the Alpha Magnetic Spectrometer on the International Space Station," Phys. Rev. Lett. **114**, 171103 (2015)
[4] M. Aguilar et al. [AMS Collaboration], "Precision Measurement of the Helium Flux in Primary Cosmic Rays of Rigidities 1.9 GV to 3 TV with the Alpha Magnetic Spectrometer on the International Space Station," Phys. Rev. Lett. **115**(21), 211101 (2015)
[5] A. D. Panov et al., "Energy Spectra of Abundant Nuclei of Primary Cosmic Rays from the Data of ATIC-2 Experiment: Final Results," Bull. Russ. Acad. Sci. Phys. **73**, 564 (2009) [arXiv:1101.3246 [astro-ph.HE]]
[6] H. S. Ahn et al., "Energy spectra of cosmic-ray nuclei at high energies," Astrophys. J. **707**, 593 (2009) [arXiv:0911.1889 [astro-ph.HE]]
[7] Thomas K. Gaisser, Ralph Engel and Elisa Resconi, *Cosmic Rays and Particle Physics* (Cambridge University Press, 2016)
[8] G. Di Sciascio, this conference
[9] L. Kuzmichev, this conference
[10] B. Bartoli et al. [ARGO-YBJ and LHAASO Collaborations], "Knee of the cosmic hydrogen and helium spectrum below 1 PeV measured by ARGO-YBJ and a Cherenkov telescope of LHAASO," Phys. Rev. D **92**(9), 092005 (2015) [arXiv:1502.03164 [astro-ph.HE]]
[11] T. Antoni et al. [KASCADE Collaboration], "KASCADE measurements of energy spectra for elemental groups of cosmic rays: Results and open problems," Astropart. Phys. **24**, 1 (2005) [astro-ph/0505413]
[12] A. Haungs, this conference
[13] J. R. Hoerandel, "On the knee in the energy spectrum of cosmic rays," Astropart. Phys. **19**, 193 (2003) [astro-ph/0210453]
[14] T. K. Gaisser, "Spectrum of cosmic-ray nucleons, kaon production, and the atmospheric muon charge ratio," Astropart. Phys. **35**, 801 (2012) [arXiv:1111.6675 [astro-ph.HE]]
[15] M. G. Aartsen et al. [IceCube Collaboration], "The IceCube Neutrino Observatory – Contributions to ICRC 2015 Part III: Cosmic Rays," arXiv:1510.05225 [astro-ph.HE], pp. 37-44
[16] W. D. Apel et al., "The spectrum of high-energy cosmic rays measured with KASCADE-Grande," Astropart. Phys. **36**, 183 (2012)
[17] V. V. Prosin, this conference
[18] V. V. Prosin et al., "Tunka-133: Results of 3 year operation," Nucl. Instrum. Meth. A **756**, 94 (2014)
[19] H. Dembinski, this conference
[20] M. G. Aartsen et al. [IceCube Collaboration], "Measurement of the cosmic ray energy spectrum with IceTop-73," Phys. Rev. D **88**(4), 042004 (2013) [arXiv:1307.3795 [astro-ph.HE]]
[21] W. D. Apel et al., "Ankle-like Feature in the Energy Spectrum of Light Elements of Cosmic Rays Observed with KASCADE-Grande," Phys. Rev. D **87**, 081101 (2013) [arXiv:1304.7114 [astro-ph.HE]]
[22] J. Matthews, this conference
[23] D. Martello, this conference
[24] K. Greisen, "End to the cosmic ray spectrum?," Phys. Rev. Lett. **16**, 748 (1966)
[25] G. T. Zatsepin and V. A. Kuzmin, "Upper limit of the spectrum of cosmic rays," JETP Lett. **4**, 78 (1966) [Pisma Zh. Eksp. Teor. Fiz. **4**, 114 (1966)]
[26] A. M. Hillas, "The Origin of Ultrahigh-Energy Cosmic Rays," Ann. Rev. Astron. Astrophys. **22**, 425 (1984)
[27] A. di Matteo et al. in "The Pierre Auger observatory: Contributions to the 34th international Cosmic aAy Conference (ICRC 2015)," (arXiv:1509.03732) p. 103–110
[28] R. U. Abbasi et al. [Telescope Array Collaboration], "Indications of Intermediate-Scale Anisotropy of Cosmic Rays with Energy Greater Than 57 EeV in the Northern Sky Measured with the Surface Detector of the Telescope Array Experiment," Astrophys. J. **790**, L21 (2014) [arXiv:1404.5890 [astro-ph.HE]]
[29] A. Aab et al. [Pierre Auger Collaboration], "The Pierre Auger Observatory Upgrade - Preliminary Design Report," arXiv:1604.03637 [astro-ph.IM]
[30] K. H. Kampert and M. Unger, "Measurements of the Cosmic Ray Composition with Air Shower Experiments," Astropart. Phys. **35**, 660 (2012) [arXiv:1201.0018 [astro-ph.HE]]
[31] K. Rawlins [IceCube Collaboration], J. Phys. Conf. Ser. **718**(5), 052033 (2016). doi: 10.1088/1742-6596/718/5/052033
[32] E. J. Ahn, R. Engel, T. K. Gaisser, P. Lipari, T. Stanev, "Cosmic ray interaction event generator SIBYLL 2.1," Phys. Rev. D **80**, 094003 (2009) [arXiv:0906.4113 [hep-ph]]
[33] V. Ptuskin, this conference
[34] T. Hams et al., "Elemental Abundances of Ultra-Heavy GCRs measured by SuperTIGER and ACE-CRIS and the Origin of Galactic Cosmic Rays," PoS ICRC **2015**, 038 (2016)
[35] P. Blasi and E. Amato, "Diffusive propagation of cosmic rays from supernova remnants in the Galaxy. I: spectrum and chemical composition," JCAP **1201**, 010 (2012) [arXiv:1105.4521 [astro-ph.HE]]

[36] P. Blasi and E. Amato, "Diffusive propagation of cosmic rays from supernova remnants in the Galaxy. II: anisotropy," JCAP **1201**, 011 (2012) [arXiv:1105.4529 [astro-ph.HE]]
[37] M. G. Aartsen et al. [IceCube Collaboration], "The IceCube Neutrino Observatory - Contributions to ICRC 2015 Part II: Atmospheric and Astrophysical Diffuse Neutrino Searches of All Flavors," arXiv:1510.05223 [astro-ph.HE]
[38] M. G. Aartsen et al. [IceCube Collaboration], "Observation of High-Energy Astrophysical Neutrinos in Three Years of IceCube Data," Phys. Rev. Lett. **113**, 101101 (2014) [arXiv:1405.5303 [astro-ph.HE]]
[39] S. Schonert, T. K. Gaisser, E. Resconi and O. Schulz, "Vetoing atmospheric neutrinos in a high energy neutrino telescope," Phys. Rev. D **79**, 043009 (2009) [arXiv:0812.4308 [astro-ph]]
[40] T. K. Gaisser, K. Jero, A. Karle, J. van Santen, "Generalized self-veto probability for atmospheric neutrinos," Phys. Rev. D **90**(2), 023009 (2014) [arXiv:1405.0525 [astro-ph.HE]]
[41] M. G. Aartsen et al. [IceCube Collaboration], "Atmospheric and astrophysical neutrinos above 1 TeV interacting in IceCube," Phys. Rev. D **91**(2), 022001 (2015) [arXiv:1410.1749 [astro-ph.HE]]
[42] M. G. Aartsen et al. [IceCube Collaboration], "Observation and Characterization of a Cosmic Muon Neutrino Flux from the Northern Hemisphere using six years of IceCube data," Astrophys. J. **833**(1), 3 (2016) [arXiv:1607.08006 [astro-ph.HE]]
[43] M. G. Aartsen et al. [IceCube Collaboration], "A combined maximum-likelihood analysis of the high-energy astrophysical neutrino flux measured with IceCube," Astrophys. J. **809**(1), 98 (2015) [arXiv:1507.03991 [astro-ph.HE]]
[44] E. Waxman, J. N. Bahcall, "High-energy neutrinos from astrophysical sources: An Upper bound," Phys. Rev. D **59**, 023002 (1999) [hep-ph/9807282]
[45] J. N. Bahcall, E. Waxman, "High-energy astrophysical neutrinos: The Upper bound is robust," Phys. Rev. D **64**, 023002 (2001) [hep-ph/9902383]
[46] M. G. Aartsen et al. [IceCube Collaboration], "Constraints on Ultrahigh-Energy Cosmic-Ray Sources from a Search for Neutrinos above 10 PeV with IceCube," Phys. Rev. Lett. **117**(24), 241101 (2016) [arXiv:1607.05886 [astro-ph.HE]]
[47] M. G. Aartsen et al. [IceCube Collaboration], "All-sky search for time-integrated neutrino emission from astrophysical sources with 7 years of IceCube data," Astrophys. J. **835**(2), 151 (2017) [arXiv:1609.04981 [astro-ph.HE]]
[48] P. Lipari, "Proton and Neutrino Extragalactic Astronomy," Phys. Rev. D **78**, 083011 (2008) [arXiv:0808.0344 [astro-ph]]
[49] M. Ahlers and F. Halzen, "Pinpointing Extragalactic Neutrino Sources in Light of Recent IceCube Observations," Phys. Rev. D **90**(4), 043005 (2014) [arXiv:1406.2160 [astro-ph.HE]]
[50] K. Murase, E. Waxman, "Constraining High-Energy Cosmic Neutrino Sources: Implications and Prospects," Phys. Rev. D **94**(10), 103006 (2016) [arXiv:1607.01601 [astro-ph.HE]]
[51] M. Kowalski, "Status of High-Energy Neutrino Astronomy," J. Phys. Conf. Ser. **632**(1), 012039 (2015) [arXiv:1411.4385 [astro-ph.HE]]
[52] M. G. Aartsen et al. [IceCube Collaboration], "The contribution of Fermi-2LAC blazars to the diffuse TeV-PeV neutrino flux," Astrophys. J. **835**(1), 45 (2017) [arXiv:1611.03874 [astro-ph.HE]]
[53] M. G. Aartsen et al. [IceCube Collaboration], "An All-Sky Search for Three Flavors of Neutrinos from Gamma-Ray Bursts with the IceCube Neutrino Observatory," Astrophys. J. **824**(2), 115 (2016) [arXiv:1601.06484 [astro-ph.HE]]
[54] A. Loeb, E. Waxman, "The Cumulative background of high energy neutrinos from starburst galaxies," JCAP **0605**, 003 (2006) [astro-ph/0601695]
[55] M. Ackermann et al. [Fermi-LAT Collaboration], "The spectrum of isotropic diffuse gamma-ray emission between 100 MeV and 820 GeV," Astrophys. J. **799**, 86 (2015) [arXiv:1410.3696 [astro-ph.HE]]
[56] K. Murase, M. Ahlers and B. C. Lacki, "Testing the Hadronuclear Origin of PeV Neutrinos Observed with IceCube," Phys. Rev. D **88**(12), 121301 (2013) [arXiv:1306.3417 [astro-ph.HE]]
[57] K. Bechtol, M. Ahlers, M. Di Mauro, M. Ajello, J. Vandenbroucke, "Evidence against star-forming galaxies as the dominant source of IceCube neutrinos,"arXiv:1511.00688 [astro-ph.HE]
[58] M. Ackermann et al. [Fermi-LAT Collaboration], "Resolving the Extragalactic γ-Ray Background above 50 GeV with the Fermi Large Area Telescope," Phys. Rev. Lett. **116**(15), 151105 (2016) [arXiv:1511.00693 [astro-ph.CO]]
[59] E. Resconi, S. Coenders, P. Padovani, P. Giommi, L. Caccianiga, "Connecting blazars with ultra high energy cosmic rays and astrophysical neutrinos," arXiv:1611.06022 [astro-ph.HE]
[60] V. N. Zirakashvili and V. S. Ptuskin, "Type IIn supernovae as sources of high energy astrophysical neutrinos," Astropart. Phys. **78**, 28 (2016) [arXiv:1510.08387 [astro-ph.HE]]
[61] A. Neronov and D. V. Semikoz, "Evidence the Galactic contribution to the IceCube astrophysical neutrino flux," Astropart. Phys. **75**, 60 (2016) [arXiv:1509.03522 [astro-ph.HE]]
[62] F. Stecker, "Diffuse fluxes of cosmic high-energy neutrinos," Astrophys. J. **228**, 919 (1979)
[63] D. Gaggero, D. Grasso, A. Marinelli, A. Urbano and M. Valli, "The gamma-ray and neutrino sky: A consistent picture of Fermi-LAT, Milagro, and IceCube results," Astrophys. J. **815**, no. 2, L25 (2015) [arXiv:1504.00227 [astro-ph.HE]]
[64] M. Ackermann et al. [Fermi-LAT Collaboration], "Fermi-LAT Observations of the Diffuse Gamma-Ray Emission: Implications for Cosmic Rays and the Interstellar Medium," Astrophys. J. **750**, 3 (2012) [arXiv:1202.4039 [astro-ph.HE]]
[65] B. Bartoli et al. [ARGO-YBJ Collaboration], "Study of the Diffuse Gamma-ray Emission From the Galactic Plane With ARGO-YBJ," Astrophys. J. **806**, 20 [arXiv:1507.06758 [astro-ph.IM]]
[66] M. Ackermann et al., "A Cocoon of Freshly Accelerated Cosmic Rays Detected by Fermi in

the Cygnus Superbubble," Science **334**(6059), 1103 (2011)

[67] S. Adrian-Martinez et al. [ANTARES Collaboration], "Constraints on the neutrino emission from the Galactic Ridge with the ANTARES telescope," Phys. Lett. B **760**, 143 (2016) [`arXiv:1602.03036 [astro-ph.HE]`]

[68] M. Ahlers, Y. Bai, V. Barger and R. Lu, "Galactic neutrinos in the TeV to PeV range," Phys. Rev. D **93**(1), 013009 (2016) [`arXiv:1505.03156 [hep-ph]`]

[69] M. G. Aartsen et al. [IceCube Collaboration], "Neutrinos and Cosmic Rays Observed by IceCube," `arXiv:1701.03731 [astro-ph.HE]`

[70] M. G. Aartsen et al. [IceCube Collaboration], "The IceCube Realtime Alert System," `arXiv:1612.06028 [astro-ph.HE]`

[71] J. Brunner, for the KM3NeT Collaboration, presentation at RICAP 2016

[72] S. Adrian-Martinez et al. [KM3Net Collaboration], "Letter of intent for KM3NeT 2.0," J. Phys. G **43**(8), 084001 (2016) [`arXiv:1601.07459 [astro-ph.IM]`]

[73] A. D. Avrorin et al. [BAIKAL Collaboration], "The prototyping/early construction phase of the BAIKAL-GVD project," Nucl. Instrum. Meth. A **742**, 82 (2014) [`arXiv:1308.1833 [astro-ph.IM]`]

[74] M. G. Aartsen et al. [IceCube Collaboration], "IceCube-Gen2: A Vision for the Future of Neutrino Astronomy in Antarctica," `arXiv:1412.5106 [astro-ph.HE]`

[75] M. G. Aartsen et al. [IceCube Collaboration], "IceCube-Gen2 – The Next Generation Neutrino Observatory at the South Pole: Contributions to ICRC 2015," `arXiv:1510.05228 [astro-ph.IM]`

[76] M. G. Aartsen et al. [IceCube Collaboration], "PINGU: A Vision for Neutrino and Particle Physics at the South Pole," `arXiv:1607.02671 [hep-ex]`

[77] S. W. Barwick et al., "Radio detection of air showers with the ARIANNA experiment on the Ross Ice Shelf," `arXiv:1612.04473 [astro-ph.IM]`

[78] P. Allison et al. [ARA Collaboration], "Performance of two Askaryan Radio Array stations and first results in the search for ultrahigh energy neutrinos," Phys. Rev. D **93**(8), 082003 (2016) [`arXiv:1507.08991 [astro-ph.HE]`]

[79] *Neutrino Astronomy: current status, future prospects* (ed. Thomas Gaisser & Albrecht Karle, World Scientific 2017)

Inelastic proton-air cross section growth from 0.2 TeV to 10 PeV according to TIEN SHAN experimental cosmic ray data

Comparison with various models

N. Nesterova[a]

P.N. Lebedev Physical Institute, Russian Academy of Sciences, Moscow, Russia

Abstract. Conclusions are made about the increase with energy of the inelastic proton air cross section σ_{p-air} from 0.2 TeV (accelerator experiments with fixed targets) to 10 PeV (cosmic rays). Experimental data from the Tien Shan complex array on various components (mainly hadron, Cherenkov light, and electron) of extensive air showers at 0.5–10 PeV of primary cosmic rays are analyzed. They were compared with many results of different calculated models of cosmic ray interactions in the atmosphere. The analysis showed that the rise conforms to (7–9)% per one order of energy. These data correspond better to the QGSJET-II-04 version of the interaction model based on the recent LHC results. This model predicts better the slower rise of the cross-section than previous versions of QGSJET-II and some other models.

1. Introduction

The purpose of this work is to find the law of the rise of σ_{p-air} – the inelastic protonair cross section - with increasing primary cosmic ray (PCR) energy. It was done using the Tien Shan complex array experimental data at PCR energies, $E_0 = 0.5 - 10$ PeV, and their extrapolation from 0.2 TeV (accelerator experiments with fixed targets). Our initial publications are to be found in references [1–5]. Experimental data of various components (mainly hadrons, electrons and Cherenkov light in the atmosphere) of extensive air showers (EAS) initiated by PCRs were compared with many different former and modern models. Comparisons are also shown with some other experimental data up to Very High Energies (VHE).

2. Experimental results and comparisons with different models

The complex Tien Shan array (43.04 N, 76.93 E, $P =$ 685 g·cm^{-2}) comprises different EAS detectors: hadrons (the ionization calorimeter), electrons (scintillation and GM counters), muons (underground GM counters) and detectors of atmospheric Cherenkov light [5]. EAS were classified according to the total number of electrons N_e ($N_e \sim E_0$) at the Tien Shan level.

Firstly the Cherenkov light Q (photon·m^{-2}) lateral distributions at $R = 50 - 250$ m from the axis of EAS at PCR $E_0 = 1 - 10$ PeV based on experiments at Tien Shan [6] (Fig. 1) and the former "Pamir" [7] arrays were compared with model calculations. After that, experimental EAS "cascade curves" of electrons (N$_e$ as a function of the atmosphere depth P (g·cm^{-2}) at $E_0 > 2$ PeV (at a constant EAS intensity) were received for comparing with calculations [8] (Fig. 2). In these models various theoretical rates of increase of the inelastic proton–air cross section σ_{p-air} were used: 0%, 7%, and 10% per one order of E_0 from $\sigma_0(0.2\,\text{TeV}) = 265$ mb (accelerator data). Conclusions were made from these experiments that the σ_{p-air} rise is (7–9)% per one order of energy. Our inference conflicts with earlier conclusions of many other experimental groups and even with some models.

The main conclusions were made on the basis of the analysis of EAS hadron energy spectra at hadron energies E$_h$ > 1 TeV of EAS in various intervals of electron number $N_e(E_0)$. The special procedure of the processing for separation of each individual hadron was described in [9].

Many simulations showed that the number of hadrons $N_h(E_h = 1 - 5\,\text{TeV})$ at $E_0 = (0.5\text{–}10)$ PeV is practically independent of PCR mass composition, but is sensitive to some interaction parameters, especially to σ_{p-air} and the inelasticity coefficient, K_{inel}.

Initially [3,4] experimental results on $N_h(E_h > 1\,\text{TeV}, N_e)$ were compared with former calculation models [10–14] for different σ_{p-air} values at the inelasticity coefficient $K_{inel} = var$. Data on $N_h(E_h > 1\,\text{TeV}, N_e)$ indicate that the rise of σ_{inel}^{p-air} is 7–9% and $\sigma_{inel}^{p-air}(1\,\text{PeV}) =$ (350 ± 15) mb.

After that we compared the experimental hadrons energy spectra with the modern CORSIKA + QGSJET models for the same $N_e(E_0)$. Values of $N_e(E_0)$ were received in the special calculations. Spectra for different primary nuclei (p, He, O) were examined using the QGSJET-01 model. Energy spectra of hadrons at Tien Shan for PCR $E_0 = 1$ PeV, 3 PeV, 10 PeV are shown in Fig. 3. Calculations show that the number of hadrons per shower, N_h/N_{EAS}, at $E_h = (1\text{-}5)$ TeV is practically independent of the PCR mass composition. So, the number of hadrons in the experiment exceeds the number in this version of the QGSJET-01 model. Experimental data confirms the correctness of our conclusions up to 10 PeV PCR energies.

Then we compared experimental spectra with data of the QGSJET-II-03 model version for primary protons

[a] e-mail: `nester@sci.lebedev.ru`

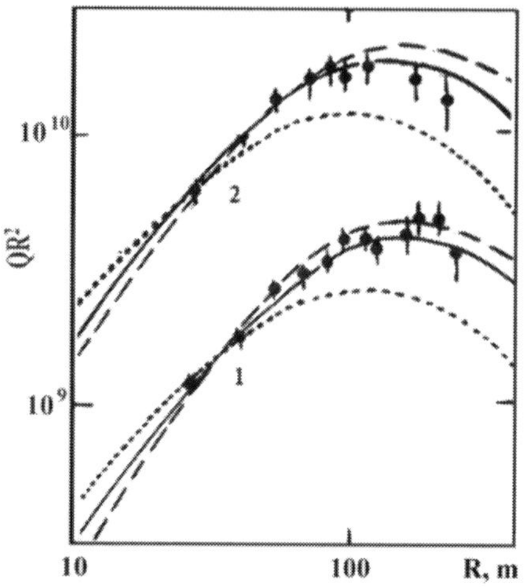

Figure 1. Cherenkov light lateral distributions at Tien Shan. 1. $E_0 = 2\,\mathrm{PeV}$, 2. $E_0 = 9\,\mathrm{PeV}$. Dashed line –0%, solid line –7%, dotted –10% σ_{inel}^{p-air} rise per one order of E_0.

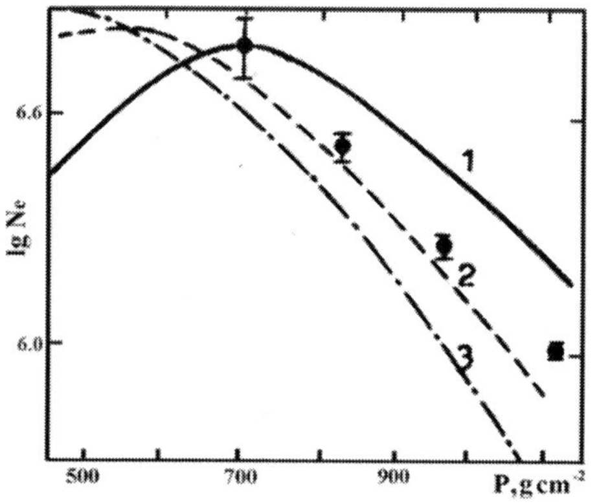

Figure 2. N_e vs. the atmospheric depth P (g·cm^{-2}) at $E_0 > 2\,\mathrm{PeV}$ for a constant intensity of EAS. 1. 0%, 2. 7%, 3. 10% – σ_{inel}^{p-air} rise per order of E_0.

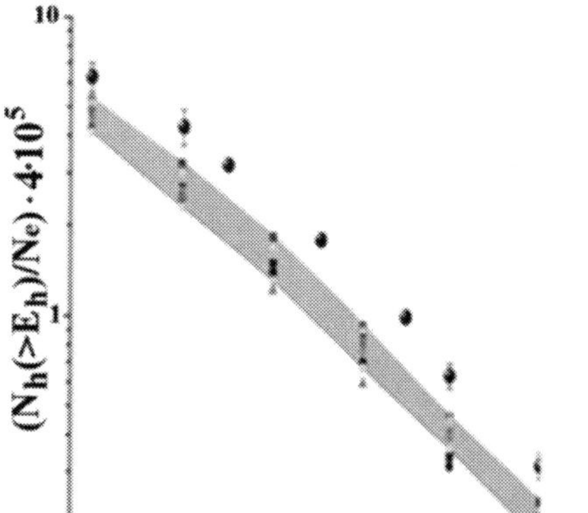

Figure 3. Hadron energy spectra at Tien Shan. The number of hadrons per shower (normalized to $N_e = 4 \cdot 10^5 (E_0 \approx 1\,\mathrm{PeV})$. $E_0 = 1\,\mathrm{PeV}$ (black), 3 PeV(grey), 10 PeV (light grey), from primary nucleus: P (squares), He (circles), O (triangles) – shaded area – QGSJET-I model. Experiment: black circles.

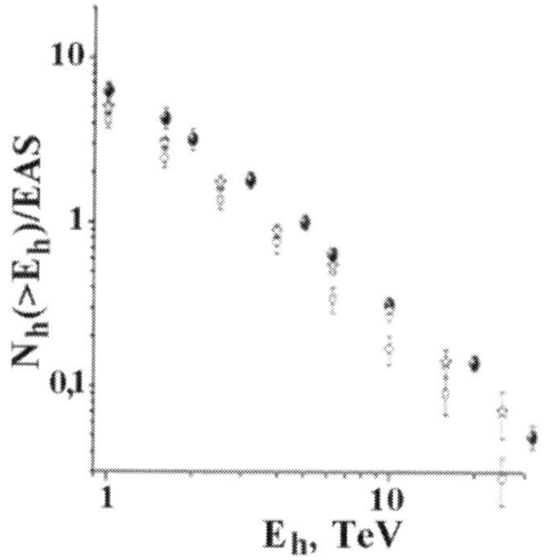

Figure 4. Hadron energy spectra at Tien Shan. The number of hadrons per shower for primary protons $E_0 = 10^{15}$ eV ($N_e = 4 \cdot 10^5$): CORSIKA+QGSJET-I – white squares, CORSIKA+ QGSJET-II white star, Experiment: black circles.

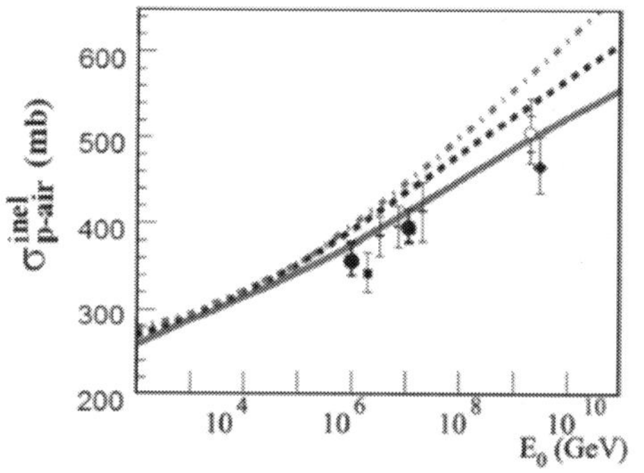

Figure 5. Proton-air cross section σ_{inel}^{p-air} vs PCR primary energy E_0. Versions: QGSJET-II-4 (solid line), QGSJET-II-03 (dashed), SIBYLL (dot-dashed). Notation: Tien Shan data -two black circles, EAS Top- black square, Yakutsk-dashes, Age- white circles, Diamonds-HiRes.

(Fig. 4). Results of the comparison with QGSJET-II-3, indicated that the number of hadrons in the experiment outnumbers the number in this model too. The difference between the experiment and the QGSJET- II-03 model was smaller in comparison with the QGSJET-I. It can explain smaller values of K_{inel} and $\sigma_{inel}^{\pi-air}$ in QGSJET II-03.

It is necessary to decrease σ_{inel}^{p-air} in these versions of QGSJET models. In these initial versions σ_{inel}^{p-air} (1 PeV) $\approx$ 385 mb and σ_{inel}^{p-air} rise is about 11% per one order of E_0. These values are somewhat more than our estimates.

The new version of the QGSJET-II model (QGSJET-II-04) was presented at the 32nd ICRC [15]. Changes of model were based on an analysis of LHC data on soft multi-particle production. In the new version the rise of σ_{inel}^{p-air} is about (8–9)% per one order of E_0. This rise is slower than in previous versions of QGSJET-II and corresponds better to our experimental data. Data of calculations and our experimental result as well as many other experimental data of recent years at $E_0 > 0.1\,\mathrm{PeV}$ are shown in Fig. 5.

3. Conclusion

Our analysis based on Tien Shan experimental results on EAS produced by PCRs at 0.5–10 PeV shows a slow rise of the inelastic protonair cross section σ_{inel}^{p-air} with increasing PCR energy [1–4]. Conclusions are based on a comparison of different models with experimental data on hadron energy spectra as well as EAS Cherenkov light lateral distributions in the atmosphere and electron cascade curves: $N_e(P)$. This rise of σ_{inel}^{p-air} corresponds to (7–9)% (not more than 10%) per one order of energy magnitude from 0.2 TeV (accelerators with fixed targets) to 10 PeV (EAS).

This corresponds to σ_{inel}^{p-air} (1 PeV) $\approx$ 350 mb and σ_{inel}^{p-air}(10 PeV) $\approx$ 380 mb.

The main conclusions were made on the basis of the method of analysis of EAS hadron energy spectra at $E_h > 1$ TeV. This value has the advantage that it is almost independent of mass composition of PCR at $E_0 = 0.5 -$ 10 PeV in agreement with QGSJET models and other former models. If conclusions based on experimental data are correct, the dissipation of the PCR energy in air is less than predicted by models such as CORSIKA+ QGSJET-I, old QGSJET-II (-01, -02, -03), SIBYLL and some previous models. In our recent work [3,4] the conclusion was made that it would be desirable to decrease σ_{inel}^{p-air} in QGSJET-I and old QGSJET-II models. The new version of QGSJET-II model (QGSJET-II-04) [15] corresponds better to our data and for some other data (HiRes, PAO, Ulrich et al. [16], Yakutsk, Knurenko et al. [17], EAS-TOP, Aglietta et al. [18]).

The author thanks colleagues in the development and operation of the Tien Shan array and help in the process of the work and L.G. Sveshnikova for the calculation.

References

[1] Chubenko A.P., Dubovoy A.G., Nesterova N.M. et al., Proc.18th ICRC (1983) **6**, p. 163
[2] Danilova T.V, Dubovy A.G., Krutikova N.P. et al. Proc.20th ICRC, Moscow (1987) **6**, p. 47
[3] Chubenko A.P., Dubovy A.G., Mukhamedshin R.A. et al., LG, 2009, Proc. 31st ICRC (Lodz) isrc0222
[4] Nesterova N.M., Sveshnikova L.G., 2009, Bull. RAS **73**, 612
[5] D.S. Adamov., V.V. Arabkin, K.V. Barkalov et al., Proc. 20-th ICRC, (1987) HE.6, 144
[6] Afanasiev B.N., Nesterova N.M. and Khein L.A., BLPH (1987) **3**, 42
[7] A.E. Chudakov, N.M. Nesterova. Nuovo Chimento N.2 Supplimento. (1955) **8**, Ser. X, p. 606
[8] Barkalov K.N., Dubovy A.G., Dyatlov P.A., BLPH (1987) **12**, 25
[9] N. Nesterova, Journal of Physics. Conference (2013) **409**, 012066
[10] Kalmykov N., Khristiansen G.B. and Motova M., Proc. 19th ICRC (La Jolla) (1985), **7**, 44
[11] Dunaevsky A.M. and Emelyanov Y.A., Proc. 17th ICRC (Paris) (1981), HE 5
[12] Ivanenko I.P., Kanevsky B.L. and Roganova T.P., Sov. J. Nucl. Fiz. (1979) **29**(3), p. 694
[13] Kempa J., Nuovo Cimento (1976) **31**(4), p. 568
[14] Fedorova G.F., Mukhamedshin R.A., Bull. Soc. Sci. Letter (Lodz) (1994) SER. Rech. Def., XVI, 137
[15] Ostapchenko S. Proc. 32nd ICRC (Beijing) (2011), HE 1.3, 681
[16] Pierre Auger Observatory Collaboration. Phys. Lett. B (2010) **685**, 239–246
[17] P. Knurenko, A.V. Sabourov. EPJ Web of Conferences (2013) **53**, 07006
[18] M. Aglietta et al., Phys. Rev. D (2009) **79**, 032004

DOI: 10.1051/epjconf/201714519002

The STM32 microcontroller based pulse intensity registration system for the neutron monitor

Alexander Shepetov[1,2,a], Alexander Chubenko[1], Olga Kryakunova[3], Nikolay Nikolayevsky[3], Nazyf Salikhov[3], and Victor Yanke[4]

[1] P. N. Lebedev Physical Institute of Russian Academy of Sciences (FIAN), 119234, Leninsky pr., 53, Moscow, Russia
[2] Tien Shan Mountain Cosmic Ray Station, 480020, Mitina str 3b, Almaty, Kazakhstan
[3] Institute of Ionosphere, 050020, Kamenskoye plato, Almaty, Kazakhstan
[4] P. N. Pushkov Institute of Terrestrial Magnetism, Ionosphere and Radio Wave Propagation of Russian Academy of Sciences (IZMIRAN), 142190 Troitsk, Russia

Abstract. We present the outlines of a new microcontroller based data acquisition system which is aimed for reliable operation in a typical cosmic ray particle registration experiment. The system supports connection of up to 16 input signals and ensures the following operation functionality: (1) stable monitoring of the intensity of a digital pulse signal, or digitization of a continuous potential level with a low time resolution (typically, about 1 s–100 s); (2) registration of uninterruptable high-resolution (up to 5–10 μs) time series of the input signal intensity; (3) synchronization of registered time series with both external (physical) or local (program-based) trigger signal; (4) possibility of an on-the-fly change of the whole configuration of informational system (both the combination and type of input signals, time resolution and sum duration of the time series measurements, trigger logic, etc.) immediately in operation time through convenient communication by plain text messages in dialog mode. In particular, the considered system is applied now for a long-term, high precision measurement of the counting rate of neutron signals at the NM64 type neutron supermonitor of the Tien Shan mountain cosmic ray station, with a real-time representation of the whole collected dataset in a WWW database.

1. Introduction

The data acquisition system newly created at the Tien Shan mountain cosmic ray station is made on the basis of a STM32F407 type microcontroller unit (MCU) [1] with the main purpose to ensure the stable registration of particle detector signals during a prolonged time period, in combination with flexible and simple control over the measurement process immediately in the time of its operation. The system is aimed to register the intensity of electric pulse signals which can arrive from various kinds of cosmic ray detectors; to the real time analysis of measurement results in an automatic regime; and to store these data for permanent keeping in a globally accessible database with a web oriented interface. Originally, the system was designed for operation with output pulses of neutron-sensitive proportional counters used in informational channels of the Tien Shan 18NM64 neutron supermonitor [2], but it can be applied in other detector set-ups of a similar kind. The system supports long-term, high-precision registration of the intensity of signal pulses on its inputs: it can operate continuously with an uninterruptable duty cycle of up to some years, and permits to measure intensity variations of input signal with relative accuracy of up to 0.05–0.1%. After being switched on, the system works fully autonomously, and does not require any special service from the side of qualified personnel.

The basic functionality of the considered system consists of continuous measurement of the counting rate separately for each neutron detector of the monitor, of automatic analysis of the monitor operation stability, and of regular sending of accepted results into some external database. The typical duration of the pulse number counting time in these *monitoring* type measurements is of the order of 1–100 s. (In the world wide net of the cosmic ray variations registration that the Tien Shan neutron supermonitor belongs to, a 1 minute periodicity of intensity measurements is historically accepted).

Besides the registration of the input pulses which come immediately from the neutron monitor, the data acquisition system ensures the detection of the so called neutron multiplicity events, i.e., arrival cases of a number of neutron signals within a short time (typically, of millisecond order). These events are caused by the momentary generation of a large number of evaporation neutrons inside the monitor as the interaction result of an energetic cosmic ray hadron. Produced neutrons are living and diffusing within the monitor during a rather considerable time being partially registered by monitor counters. The number of detected neutron pulses (called *multiplicity*), is proportional to the amount of generated neutrons, and the latter, in turn, depends on the energy of the primary cosmic ray interaction, since the average number of evaporation neutrons is a power function of this

[a] e-mail: ashep@tien-shan.org

energy. Hence, registration of multiplicity events gives an opportunity to study the variations of the energy spectrum of cosmic radiation [3].

The control program of the microcontroller unit forms the multiplicity events internally by analyzing the current flow of input pulses in real time in accordance with (arbitrary) rules which can be built into its embedded program code. For compatibility with previous data series, 6 types of multiplicity signal are currently generated at the Tien Shan monitor for the cases where 1, 2, 3, 4-5, 6-7, 8-10 neutron pulses were accepted from six counters of a standard NM64 type supermonitor unit within a $800\,\mu s$ long gate time. If needed, the algorithm of the multiplicity signal formation can be easily changed just in measurement time through re-programming of the MCU.

When using the considered system for the data acquisition from some specialized cosmic ray particle detectors it can be necessary to register coincidences between the signals which come to inputs of various informational channels. (The prototype installation of this kind is a particle telescope like set-up consisting of a number of spaced scintillators separated by an absorber which permits to estimate the energy and direction of penetrative cosmic ray particles—muons [4]). In the same manner as for multiplicity events, the microcontroller program permits to solve this task in parallel to basic intensity measurements by forming the coincidence signal internally from the current flow of input pulses. At that, it is possible to change promptly any parameters of the operation algorithm (combination of input signals to form coincidence from, the resolution time, etc.) immediately at the time of measurements through easy MCU re-programming. The existing program version ensures the registration of pairwise coincidences between two opposite scintillation counter layers, with 8 detectors in each layer, and with arbitrary combination of detector pairs for the muon telescope of IZMIRAN.

Both the formation of multiplicity events and tracing of coincidence signals are realized within the MCU in parallel and simultaneously with the basic pulse counting process.

One of the directions of the high energy cosmic ray investigation held at the Tien Shan mountain station is connected with rather rare events when the core of a 10^{15}–10^{17} eV extensive air shower (EAS) passes in the immediate vicinity (0–5 m) to the neutron monitor. Typically for these events is the instantaneous generation of an extremely large number (up to some thousands) of evaporation neutrons as an interaction result of energetic EAS core hadrons with monitor substances; later on, these neutrons are quickly thermalized and diffuse in the monitor during some milliseconds [5]. The MCU acquisition program permits to trace the events of such a type in the current signal flow, and to precisely register the time behavior of neutron intensity with 10–$30\,\mu s$ resolution over the time of some milliseconds uninterruptedly, which functionality, again, goes on in parallel and independently on the basic counting rate measurements. Switching on the tracing of EAS events, the configuration of registered temporal distributions (the list of informational channels which participate in EAS events formation, the resolution time, sum duration of distribution, etc.) can be set immediately during the measurements by corresponding commands sent to the MCU.

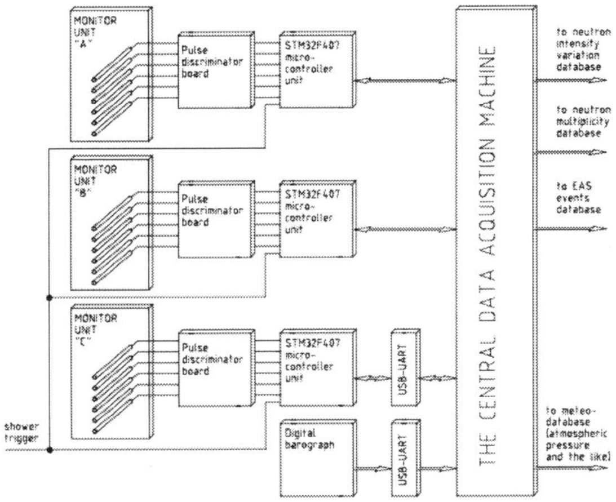

Figure 1. Block scheme of the cosmic ray intensity data acquisition system used at the Tien Shan 18NM64 neutron supermonitor.

Besides EAS events formed internally by the microcontroller itself, the high resolution time registration can be forced at any arbitrary moment by the arrival of a special control signal—the trigger, which can be elaborated outside the system, and comes as an electric pulse to a special input pin of the microcontroller. Such a synchronization type can be useful, for example, for the registration of the signals from some outside neutron and gamma-ray detectors which should operate simultaneously and in parallel with the neutron monitor. The outer trigger synchronization mode can be switched on or off, again, by demand at any arbitrary time moment.

2. Hardware design

A further description of the MCU based data acquisition system is made on the example of its variant which has remained in continuous operation at the 18NM64 neutron supermonitor of the Tien Shan mountain cosmic ray station since the beginning of 2015. The general block scheme of this system is shown in Fig. 1.

The Tien Shan neutron supermonitor consists of three separate units, each of which has the dimensions of $2 \times 3\,m^2$ and includes six big, $\varnothing 0.15\,m \times 2\,m$ proportional neutron counters [6,7]. Output pulses with typical amplitude of 5–20 mV and duration varying in the limits of 0.5–$2\,\mu s$ are taken from the anode wire of every counter and transmitted through a transistor amplifier into a short shielded coaxial cable which connects this counter with the corresponding informational channel on the data acquisition board. This board is located in the immediate vicinity of a monitor unit, such that the length of the connection cables does not exceed 0.5–2 m, and includes a set of six pulse discriminators necessary to properly form the neutron signals, as well as a processor module with an STM32F407 type MCU installed. The pulse signals from the discriminator outputs, which have a constant ~ 3 V amplitude and a fixed $1\,\mu s$ time duration, come to the MCU input pins for further operation. The high voltage of the neutron counters feeding, transmission coefficients of transistor amplifiers, and discriminator thresholds are individually tuned in such a way to ensure for every

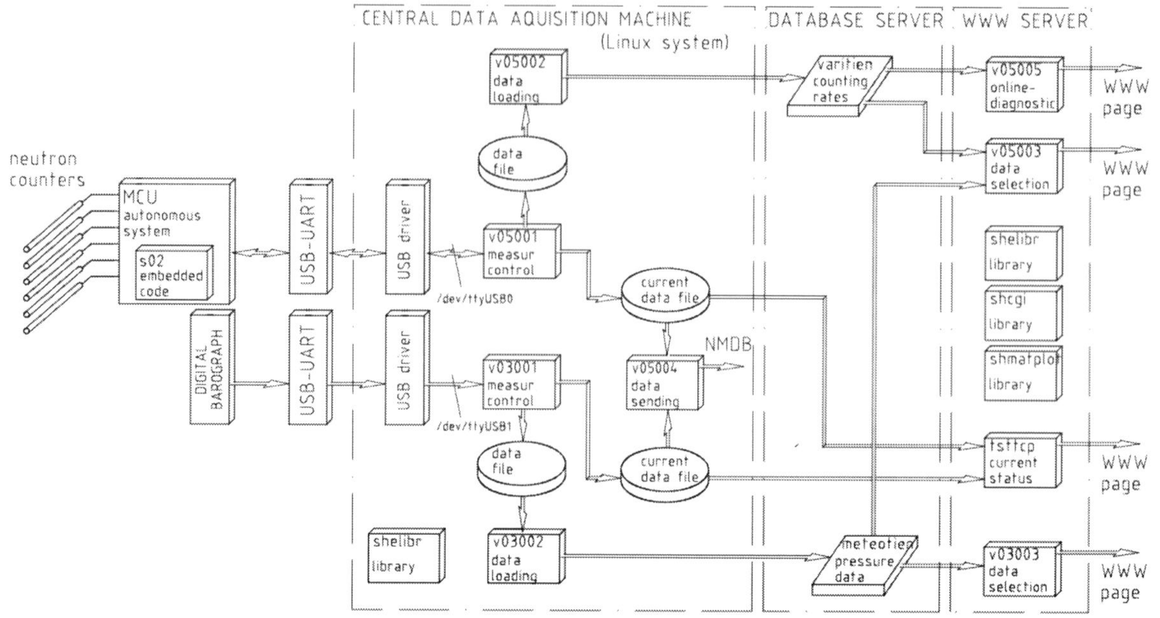

Figure 2. General schematic of the Tien Shan neutron supermonitor control program complex.

neutron counter its operation in the beginning part of the plateau of its counting characteristic.

Physically, each discriminator and MCU carrier board comprises a separate device—an autonomous data registration system. Presently, there are three such systems installed at the Tien Shan neutron supermonitor, each one serving its corresponding monitor unit.

All three microprocessor systems operate under the control of one and the same embedded program code which provides complete internal functionality of the pulse signal operation described in the previous section. Besides this, the embedded MCU code ensures the full duplex dialog-like interaction between the autonomous measurement control system and the central data acquisition machine. The latter is generally responsible for sampling the incoming neutron intensity data, their primary operation, and loading for further use in a permanent database, as well as for synchronous launching of the measurement process on all autonomous systems and continuous steady control over their current status during the whole operation time.

By means of either the universal asynchronous receiver-transmitter (UART) module, or an USB node both of which are in-built into the STM32F407 microprocessor, the autonomous registration system can interact with the central control computer via a serial interface line. Using an UART based variant is preferential in the case where the length of the communication cable to the autonomous system must be above 1.5–2 m (which is the case of monitor unit *C* in Fig. 1), so the use of additional transistor amplifiers of the interface signal with a further UART↔USB converter at the side of central computer is desirable; practically, distances of up to 30–50 m between the disposition of a detector system and control machine center can be achieved in this way without significant difficulty. The whole information exchange goes in the form of plain text messages, so there is no need for any specialized communication software, and any commonly used serial interface program is sufficient enough in the simplest testing case to send the commands to the autonomous system, and check back its reaction.

3. Software complex

The block scheme of the complete program complex presently used at the data acquisition system of the Tien Shan neutron supermonitor is shown in Fig. 2. As seen in this figure, the program set consists of the following main parts.

Every autonomous data registration board to which the outputs of the neutron counter channels are immediately connected operates under the control of a specialized embedded program *s02* which plays the role of the MCU operation system, and ensures its proper reaction on asynchronous program interrupts caused by the arrival of input pulse signals. For compatibility with other use, the version of the embedded program presently applied in the microprocessor registration systems of the Tien Shan supermonitor supports the connection of up to 16 separate signals to every autonomous board, though only 6 of them are now in use by each MCU. The embedded program *s02* is created on the basis of the free ARM architecture MCU programming library *libopencm3* [8] which provides the proper low-level access to the whole built-in periphery of the STM32F407 microprocessor box. The program is written in C programming language and converted into internal MCU machine code with the aid of a free open source toolchain from the *gcc-eabi* project [9].

The central computer that all autonomous measurement boards interact with works under the control of the free operation system Linux, which gives the possibility of both initiation of specialized programs to accept the data from autonomous registration systems, and of necessary network interaction with outer host machines: some database server to input the measurement results, the network time protocol (*nntp*) server to keep

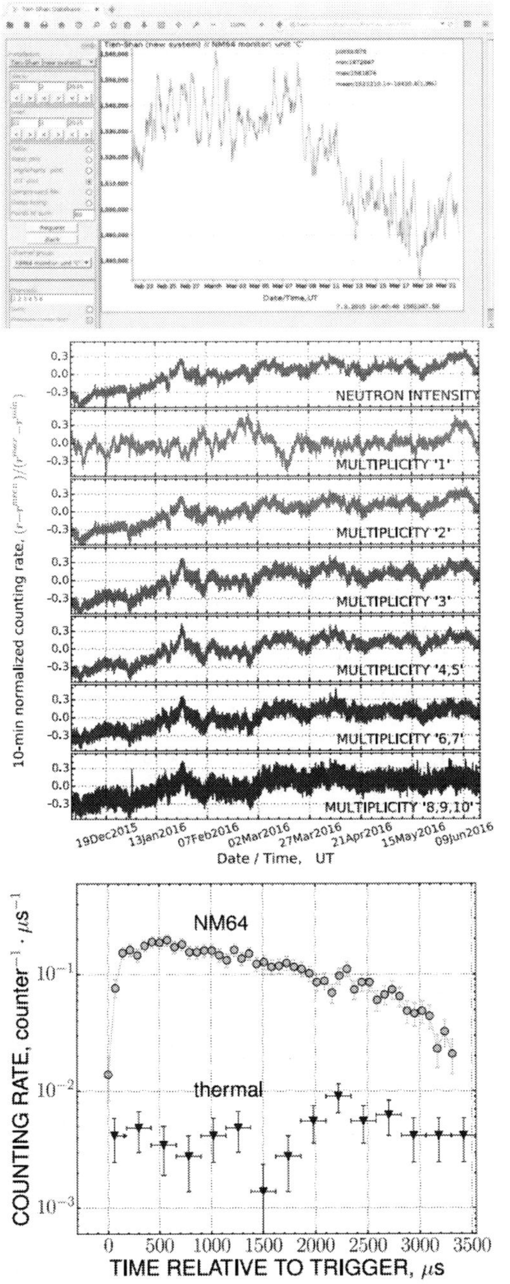

Figure 3. A sample of cosmic ray data from the Tien Shan neutron supermonitor. Top panel: representation of the current neutron counting rate measurements on the WWW site page in graphic form. Middle panel: the half-year long uninterruptable record of the neutron counting rate, and the intensity of neutron multiplicity events taken from archive database and presented with 10 min time resolution; all data are corrected for variations of the atmospheric pressure. Bottom panel: high-resolution (70 μs) time distributions of pulse intensity in the neutron counters of a NM64 supermonitor unit, and in a nearby thermal neutron detector by the close passage of an EAS core.

the constant synchronization of internal computer time, and an *http* server to present the current status of running measurements for remote observation through the web.

Normally, there is a specialized supervising program, *v05*, which supports operative control over the measurement process from the side of the central machine. It is this program which sends the commands for configuration of the autonomous system at the beginning of the measurements, keeps a steady check over its operation afterwards, accepts the results from the autonomous system, and saves these data temporarily in a local disk file. Simultaneously, there are just three separate *v05001* program instances constantly operating on the central machine, each of which controls the measurements of its own autonomous system. Periodically, according to timer signals another *v05002* type process is initialized which sends the whole information stored locally to the outer database server.

Another two programs, *v03001/v03002*, operate in an analogous fashion, and are aimed to track the atmospheric pressure which is necessary for proper correction of the raw neutron counting rate data, and is continuously provided by a digital barograph.

The general database of the Tien Shan mountain cosmic ray station is used as a permanent storage for information which comes from all experiments currently active at the station. At the present time the database functionality is supported by the *PostgreSQL* server program [10], and an access to different types of data kept there is provided through specialized script programs launched from the common WWW page of the Tien Shan station [11]. Clicking on the corresponding reference in this page the WWW interface programs *v05003* and *v03003* can be initiated which select the necessary information on the cosmic ray intensity measured by neutron counters of the Tien Shan supermonitor, and present it on the page in graphical or textual forms. A sample of such an output is shown in Fig. 3.

Besides the storage of collected neutron monitor information in the local database of the Tien Shan station, the *v05/v03* programs can re-send these data immediately during the measurements to the outer European Real-time Neutron Monitor Database (NMDB) [12]. The special program script *v05004*, which is initiated automatically after completion of every minutely neutron counting rate measurement, is responsible for proper communication with this remote server, and for timely data transmission.

High level program components of the *v05/v03* data manipulation complex are written in the interpreted programming language Python to ensure good portability between various operation environments which could be met at different server machines, as well as simplicity and transparency of program code from the programmer's point of view. For their operation all these programs use a number of auxiliary procedures defined in a separate library module *shelibr*; besides, the programs responsible for data representation on a WWW page utilize specific means of web server communication from the *shecgi* module, and *shematplot* convenience envelope over the free Python language graphic library *matplotlib* [13].

4. Conclusion

During a 2 year long period of continuous exploitation of the STM32F407 MCU based data acquisition system at the Tien Shan neutron supermonitor it has demonstrated good operation reliability, and high stability of precision measurements of the input pulse intensity, as well as sufficient convenience for practical use in cosmic ray experiments. Some typical results of this work are illustrated by the plots in Fig. 3.

The use of a cheap but sufficiently powerful modern MCU in the development of cosmic ray experimental set-ups opens a way for its further modernization and for a noticeable increase in the comprehension of the resulting measurement data without the need of a considerable redesign of any existing devices. Hence, an easy replacement of the input pulse counting module in the above described microcontroller driver program *s02* to an analogous one which can interact with the built-in analog-to-digital converter (ADC) node of MCU, leaving alone all other functional modules of this program (those of the USB and UART interface support, data storage facility, outer command interpreter, etc.) permits to apply the same autonomous registration board in the role of a fast multichannel ADC system. This option is applied at the Tien Shan cosmic ray station for developing various data sampling devices primarily intended for the detection of continuous electric signals (the radio- and optic emission sensors, electric field measurements, micropulsation of atmospheric pressure, and the like). By this, the feature of a dialog-based textual control over the whole configuration parameters set of the *s02* driver program is very convenient for a fast and flexible tuning of the data registration process to specific needs of every particular detector.

It should be noted that the use of a compact microprocessor technique with serial communication through a USB interface line and low power consumption allows the creation of portable data registration set-ups for various mobile autonomous detector systems (such as the distant EAS particle detectors [14], remote sensors of the optic and gamma-ray radiation from thunderclouds [15], seismological stations [16], and the like), which are supposed to be used widely at the Tien Shan cosmic ray detector complex.

References

[1] STMicroelectronics, *DM00037051 STM32F405xx, STM32F407xx Datasheet – production data*, `http://www.st.com/` (2015)

[2] A.G. Zusmanovich, O.N. Kryakunova, A.L. Shepetov, Adv. Space Res. **44**, 1194 (2009)

[3] R.A. Nobles, E.B. Hughes, C.J. Wolfson, Space Phys. **74**, 6459–6470 (1969)

[4] L.I. Dorman, *Cosmic rays in the Earths atmosphere and underground* (Springer Science & Business Media, 2013)

[5] V.P. Antonova, A.P. Chubenko, S.V. Kryukov, N.M. Nesterova, A.L. Shepetov, V.V. Piscal, L.I. Vildanova, G.B. Zhdanov, J. Phys. G **28**, 251 (2002)

[6] H. Carmichael, C.J. Hatton, Can. J. Phys. **42**, 2443 (1964)

[7] C.J. Hatton, *Progress in elementary particle and cosmic ray physics* (North-Holland Publishing Company, 1971), Vol. 10, ISBN: 9780720414103

[8] *The libopencm3 project. Main Page*, `http://libopencm3.org/` (2016)

[9] *GCC ARM Embedded*, `https://launchpad.net/gcc-arm-embedded/` (2015)

[10] The PostgreSQL Global Development Group, *PostgreSQL 9.5.0 Documentation* `http://www.postgresql.org/docs/manuals/` (2015)

[11] *The Tien-Shan Mountain Station's Database*, `http://www.tien-shan.org/she/` (2006–2016)

[12] *NMDB: Real-time database for high resolution neutron monitor measurements*, `http://www.nmdb.eu/` (2015)

[13] J. Hunter, D. Dale, E. Firing, M. Droettboom, *Matplotlib Release 1.4.0* `http://matplotlib.org/` (2014)

[14] A.P. Chubenko, A.L. Shepetov, V.P. Antonova, R.U. Beisembayev, A.S. Borisov, O. Dalkarov, O.N. Kryakunova, K.M. Mukashev, R.A. Mukhamedshin, R.A. Nam et al., Nucl. Instrum. Methods A **832**, 158 (2016)

[15] A. Gurevich, A. Almenova, V. Antonova, A. Chubenko, A. Karashtin, O. Kryakunova, V. Lutsenko, G. Mitko, V. Piscal, M. Ptitsyn et al., Phys. Rev. D **94**, 023003 (2016)

[16] L.I. Vil'danova, G.A. Gusev, V.V. Zhukov, G.I. Merzon, G.G. Mitko, A.S. Naumov, V.A. Ryabov, A.V. Stepanov, V.A. Chechin, A.P. Chubenko et al., Bull. Lebedev Phys. Inst. **40**, 74 (2013)

DOI: 10.1051/epjconf/201714519003

Shell-type SNRs as sources of cosmic rays

V.G. Sinitsyna[a], M.S. Andreeva, K.A. Balygin, S.S. Borisov, I.A. Ivanov, A.M. Kirichenko, A.I. Klimov, I.P. Kozhukhova, R.M. Mirzafatikhov, N.I. Moseiko, I.E. Ostashev, A.I. Palamarchuk, V.Y. Sinitsyna, and I.G. Volokh

P. N. Lebedev Physical Institute, Russian Academy of Science, Russia

Abstract. Investigations of VHE gamma-ray sources by any methods, including mirror Cherenkov telescopes, touch on the problem of the cosmic ray origin and, accordingly, the role of the Galaxy in their generation. SHALON observations have yielded results on Galactic supernova remnants (SNR) of different ages. Among them are: the shell-type SNRs Tycho's SNR (1572y), Cas A (1680y), IC 443 (age $\sim (3 \div 30) \times 10^3$ y), Cygni SNR (age $\sim (5 \div 7) \times 10^3$ y), G166.0 + 4.3 (age $\sim 24 \times 10^3$ y) and the classical nova GK Per (Nova 1901). Observation results are presented for each of the SNRs with spectral energy distributions by SHALON in comparison with other experiment data and images by SHALON together with data from X-rays by Chandra and radio-data by CGPS. The collected experimental data have confirmed the prediction of the theory about the hadronic generation mechanism of very high energy 800 GeV–100 TeV gamma-rays in Tycho's SNR, Cas A and IC443. For the first time, unique data on GK Per (Nova1901) TeV gamma-ray emission were obtained with the SHALON experiment. The X-ray data shows that the nova remnant of GK Per could be a younger remnant that will resemble older SNRs like IC 443 which interact with molecular clouds. GK Per is supposed to be a candidate for TeV gamma-ray emission due to accelerated particles in the reverse shock region.

1. Introduction

SHALON observations have yielded the results of two types of Galactic supernova remnants (SNR) of different ages. Among them are the shell-type SNRs: Tycho's SNR, Cas A, IC 443, γCygni SNR, G166.0 + 4.3 as well as the classical nova GK Per that is in the earliest SNR evolutionary stage. For each of the SNRs the observation results are presented in Table 1 and Fig. 1 with the spectral energy distribution by SHALON (△ and ▲) compared with other experimental data and images by SHALON (color scale) in comparison with data from X-ray and radio-data. Also the theoretical predictions are shown. The collected experimental data can help to solve the cosmic ray origin.

2. GK Per (Nova 1901)

Nova Persei 1901 (GK Per) is one of the most extensively studied classical nova shells over the entire electromagnetic spectrum. The nova remnant is detected at radio energies with the Very Large Array as a source of nonthermal, polarized radio emission [1]. These observations show the existence of a shocked interstellar material. The X-ray shell around GK Per was first discovered with the ROSAT experiment and later was observed by the Chandra telescope [2]. The X-ray emission of the same electron population has been detected with Chandra as an extension from radio wavelengths. The detection of X-rays from the supernova remnant shell which are primarily due to bremsstrahlung of shock accelerated relativistic electrons, supposed the detection of γ-ray emission originated from π°-decay, secondary pp-interactions [3] as well as a possible contribution emission produced via Inverse Compton scattering [3]. Chandra X-ray data shows that the nova remnant of GK Per could be a younger remnant that will resemble older SNRs like IC 443 ($(3 \div 30) \times 10^3$ y.) interacted with molecular clouds.

In accordance with a program on long-term studies of metagalactic γ-ray sources, fifteen-year-long observations of the central galaxy in the Perseus cluster, NGC 1275, are being carried out in the SHALON experiment [5,6]. During the observations of NGC 1275 the SHALON field of view contains the source of nonthermal radio and X-ray emission GK Per (Nova 1901) of classical nova type as it is located $\sim 3°$ North from NGC 1275. So due to the large telescopic field of view $> 8°$ the observations of NGC 1275 are naturally followed by the tracing of GK Per.

GK Per as a source accompanying NGC 1275 was observed with the SHALON telescope in the period from 1996 to 2012 for a total of 111 hours during clear moonless nights at zenith angles from 5° to 35°. The observations were performed using the standard (for SHALON) technique of obtaining information about the cosmic-ray background and gamma-ray-initiated showers in the same observing session. The SHALON method of selecting γ-ray showers from background cosmic-ray showers allows a 99.92% rejection of the background showers [6]. The gamma-ray source associated with GK Per was detected above 2 TeV by SHALON with a statistical significance 9.2σ determined by the Li&Ma method [4]. The signal significance for this SNR is less then one for the source with a similar flux and spectrum

[a] e-mail: `sinits@sci.lebedev.ru`

Table 1. The catalogue of galactic γ-ray sources observed by SHALON with parameters for spectrum fitting in the form of a power low with exponential cutoff $F(>E) \propto E^{k_\gamma} \times exp(-E/E_{cutoff})$.

Sources	Observable flux[a]	k_γ	E_{cutoff}, TeV	Distance, kpc	Type
Crab Nebula	(2.12 ± 0.12)	-1.36 ± 0.09	19.0 ± 2.0	2.0	PWN
Geminga	(0.48 ± 0.07)	-0.39 ± 0.05	5.4 ± 1.0	0.25	PSR or PWN
3C 58	(0.56 ± 0.15)	-1.33 ± 0.12	–	3.2	PWN
G54.1+0.3	(0.97 ± 0.35)	-1.43 ± 0.14	–	6.2	PWN
Tycho's SNR	(0.52 ± 0.04)	-0.93 ± 0.09	35.0 ± 5.0	3.1–3.3	Shell-type SNR
Cas A	(0.64 ± 0.10)	-0.91 ± 0.11	10.3 ± 2.5	3.1	Shell-type SNR
IC 443	(1.63 ± 0.23)	-1.94 ± 0.16	–	1.5	Shell-type SNR
γCygni SNR	(1.27 ± 0.11)	-0.93 ± 0.09	20.1 ± 4.2	1.5	Shell-type SNR
G166.0+4.3	(1.49 ± 0.25)	-1.95 ± 0.44	–	4.5	Shell-type SNR
GK Per	(0.31 ± 0.14)	-1.90 ± 0.36	–	0.46	Classical Nova
Cyg X-3	(0.68 ± 0.04)	-1.15 ± 0.08	75.0 ± 10.2	10.0	HMX Binary
4U 2129+47	(0.19 ± 0.06)	-0.42 ± 0.12	10.0 ± 3.0	6.0	LMX Binary
Her X-1	(0.45 ± 0.18)	–	–	6.6	LMX Binary
M57	(0.30 ± 0.17)	–	–	0.7	Planetary nebula

[a] Integral flux at energy > 800 GeV in units of 10^{-12} cm^{-2} s^{-1}.

index obtained in the same observation hours because of a smaller collection field of view relative to the standard procedure of the SHALON experiment [5,6]. Corrections for the effective field of view were made to calculate the source flux and energy spectrum. The average integral flux at energies above 2 TeV for GK Per is $I_{GK Per} = (2,9 \pm 1,3) \times 10^{-13}$ cm^{-2} s^{-1} (Fig. 1). Taking into account the proximity to a nearby source NGC 1275, we made the observation data procession first associated with NGC 1275 and then with GK Per. We found that 1% of showers are common for both sources. After a detailed analysis of the arrival direction and angular distance of these showers [6], less than 0.5% of NGC 1275 showers were recognized to be GK Per showers. This didn't change the average flux of NGC 1275 form [6]. The energy spectrum of gamma-rays in the observed energy region from 2 to 15 TeV is well described by the power law $F(E > 2\,\mathrm{TeV}) \propto E^{k_\gamma}$, with $k_\gamma = -1.90 \pm 0.35$. The spectral energy distribution, detailed images of GK Per are shown in Fig. 1 and [5]. Two TeV γ-ray emission regions were revealed: the main one coinciding with the position of the central source of GK Per and the weak emission of the shell, that is also observed in X-rays by Chandra [2] (Fig. 1, right, red lines).

3. Cas A supernova remnant (1680 year)

Cas A is a youngest of historical supernova remnants in our Galaxy. The supernova explosion that gave rise to Cas A occurred around 1680. Cas A was observed by SHALON during a 74 hour period of 2010–2014 [7,8]. The γ-ray source associated with Cas A was detected above 800 GeV at a level of 18.1σ (the γ-ray flux and spectrum details see [8] and Table 1). Figure 1, left presents the spectral energy distribution of the γ-ray emission from Cas A by SHALON (▲) in comparison with theoretical predictions [10,11] and other experimental data (see refs in [7,8]). Figure 1 right presents the Chandra X-ray image of Cas A (red lines) [9] in comparison with the TeV structure in the energy range of 0,8–30 TeV seen by SHALON.

The γ-ray emission in Cas A could be produced via IC scattering and by accelerated cosmic ray hadrons through interaction with the interstellar gas and then π°- decay. Solid lines in Fig. 1 show the very high energy γ-ray spectra of hadronic origin [10,11]. It was shown in [10] that the leptonic model with B = 0.3 mG predicts a 5–8 times lower γ-ray flux than observed; the model with B = 0.12 mG, which can broadly explain the observed GeV flux predicts the TeV spectrum with a cut-off energy about 10 TeV. The detection of γ-ray emission at 5–30 TeV by SHALON and the hard spectrum below 1 TeV would favor the π°-decay origin of the γ-rays in Cas A.

4. Tycho's Supernova Remnant (1572 year)

Tycho's SNR originated from the Ia type supernova which exploded in 1572. The supersonic expansion of the stellar debris visible in Tycho's SNR by Chandra has created two X-ray emitting shock waves: moving outward into the interstellar gas, and another moving back into the debris. The character of such a displacement of the shock and the contact discontinuity surfaces makes cosmic ray acceleration at the supernova shock very efficient.

In the observations of 1996 a new galactic source was detected by SHALON [12] in TeV energies with 17σ determined by [4]. This object was identified with Tycho's SNR. Tycho's SNR was also confirmed with VERITAS in observations of 2008–2010. The γ-ray emission from Tycho's SNR was detected with Fermi LAT in the range 400 MeV–100 GeV. The expected flux of γ-quanta from π°-decay, extends up to >30 TeV, while the flux of γ-rays originating from IC scattering has a sharp cut-off above a few TeV, so the detection of γ-rays with energies up to 80 TeV by SHALON [7] (Fig. 1) is an evidence of their hadronic origin. Also, the information on parameters of Tycho's SNR such as the distance (3.1–3.3 kpc) and interstellar medium density was obtained from the SHALON data within the nonlinear kinetic model [13]. The same parameters have been obtained in [14] calculations of structures visible by Chandra at X-ray energies (see red contours in Fig. 1, right).

5. γCygni SNR (age ~ $(5 \div 7) \times 10^3$ years)

γCygni SNR is a shell-type supernova remnant, its shell-like features are known in radio- and X-ray energy

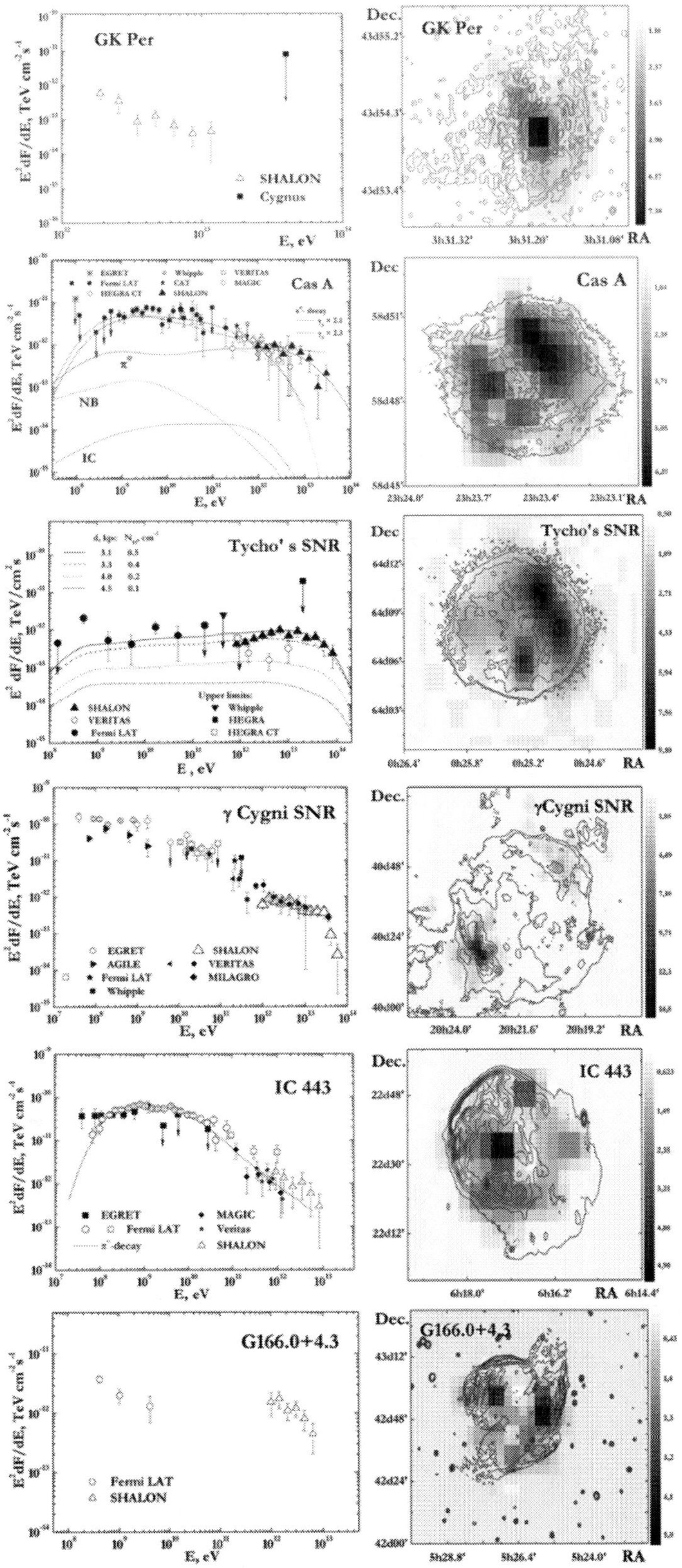

Figure 1. Spectral energy distributions (left) and images (right) of shell-type SNRs by SHALON compared with other experiments.

regions [15]. γCygni SNR is older than Cas A and Tycho's SNR, its age is estimated as 5000–7000 years [15] and is supposed to be in an early phase of adiabatic expansion.

During the observations of Cyg X-3 (since 1995) the SHALON field of view contains the γCygni SNR as it is located in the Cygnus Region at $\sim 2°$ SW from Cyg X-3. So due to the large telescopic field of view ($\sim 8°$) the observations of Cyg X-3 are naturally followed by observations of this SNR. As a result, γCygni SNR was detected above 800 GeV by SHALON (see [8,16]). TeV γ-ray emission regions correlate with the NW and SE parts of the shell visible in radio energies by CGPS (red contours in Fig. 1).

6. IC 443 SNR (age $\sim (3 \div 30) \times 10^3$ years)

IC 443 is a shell-type SNR with a complex shape consisting of two half-shells with different radii. IC 443 is one of the best candidates for investigating the connection among SNRs, molecular clouds and high- and very high energy γ-ray sources. The close placement of the dense shocked molecular clouds and detected GeV-TeV γ-ray emission [18] suggests that IC 443 can be considered as a candidate for a hadronic cosmic-ray source.

IC 443 was detected by SHALON with an integral flux above 0.8 TeV (see Table 1, [8,17] and Fig. 1) and a statistical significance of 9.7σ [4]. The favored scenario in which the γ-rays of 100 MeV–10 TeV energies are emitted in the shell of the IC443 SNR is π°-decay which are produced in the interactions of cosmic rays with the interstellar gas [18]. The solid lines in Fig. 1 show the γ-ray spectra of hadronic origin. Inverse Compton scattering can not explain the observed IC 443 γ-ray emission as there is no bright source of seed photons in the region of IC 443. The analysis of arrival directions of γ-rays with energies 800 GeV–10 TeV reveal the correlation of TeV γ-ray emission maxima with MeV-GeV emission observed by Fermi LAT [18]; also TeV γ-ray emission of South and South-West parts of IC 443 shell correlated with the position of the swept out dense molecular cloud (see red contours by CGPS in Fig. 1, right).

7. G166.0+4.3 (age $\sim 24 \times 10^3$ years)

G166.0 + 4.3 (VRO 42.05.01) is a supernova remnant with unusual morphology, containing a shell of circular component in the northeast intersected by a larger bowl-shaped component in the southwest (the wing), expanding into a low density medium [19]. There are important differences in the spatial distribution of the emission viewed in X-ray and radio energies. The radio image has an outstanding edge-brightened shape in the shell and wing component whereas the X-ray emission is not edge-brightened, but is dominated by a bright spot in the wing component [19]. As such remnant morphology can be results from the shock encountering a density discontinuity in the interstellar medium, G166.0 + 4.3 became a candidate for the investigation of particle acceleration in SNR shocks at high- and very high energies.

G166.0 + 4.3 was observed with the SHALON telescope during a period of 20.1 hours in 2015–2016. The gamma-ray source associated with the G166.0 + 4.3 SNR was detected above 800 GeV with a statistical significance of 6.1σ with a gamma-quantum flux above 0.8 TeV of $(1.49 \pm 0.25) \times 10^{-12}\,\mathrm{cm}^{-2}\,\mathrm{s}^{-1}$. The energy spectrum of gamma-rays in the observed energy region from 0.8 TeV to 7 TeV is well described by the power law $I(> E_\gamma) = (1.49 \pm 0.25) \times 10^{-12} \times E_\gamma^{-1.95\pm0.44}$. For each of the SNRs the observation results are presented in Table 1 and Fig. 1. The spectral energy distribution of G166.0 + 4.3 by SHALON together with Fermi LAT [20] data is presented in Fig. 1, left. Figure 1 right presents radio contours of G166.0 + 4.3 in comparison with the TeV structure in the energy range of 0,8–7 TeV by SHALON. TeV γ-ray emission regions correlate with the NE and SW parts of components visible in the radio energies by CGPS (red contours in Fig. 1, right). But TeV emission is more significant near the west, where the extended wing component and maximum of X-ray emission viewed by ROSAT [19] are located.

8. Conclusion

The observation results of Galactic shell-types supernova remnants at different evolution stages for GK Per (Nova 1901), Cas A, Tycho's SNR, γCygni SNR, IC 443 and G166.0 + 4.3 by SHALON mirror Cherenkov telescope are presented. The TeV γ-ray emission of classical nova GK Per, that could be a shell-type supernova remnant in an early evolution stage, was detected for the first time by SHALON as well as TeV γ-rays from the GK Per shell. The experimental data have confirmed the prediction of the theory about the hadronic generation mechanism of very high energy γ-rays in Tycho's SNR, Cas A and IC443.

References

[1] S.P. Reynolds, R.A. Chevalier, Astrophys. J. **281**, L33 (1984)
[2] S. Balman, Astrophys. J. **627**, 933 (2005)
[3] E. G. Berezhko, H. J. Völk, Astropart. Phys. **7**, 183 (1997)
[4] T.-P. Li and Y.-Q. Ma, Astrophys. J. **272**, 317 (1983)
[5] V.G. Sinitsyna and V.Yu. Sinitsyna, Bull. of the Lebedev Phys. Inst. **42**(6), 169 (2015)
[6] V.G. Sinitsyna and V.Yu. Sinitsyna, Astron. Lett. **40**(2–3), 75 (2014)
[7] V.G. Sinitsyna and V.Yu. Sinitsyna, Astron. Lett. **37**(9), 621 (2011)
[8] V.G. Sinitsyna and V.Yu. Sinitsyna *"Astroparticle, Particle, Space Physics and Detectors Physics Applications"* (Proc. 14th ICATPP, 2013) ed. S. Giani, C. Leroy, P.-G. Rancoita, L. Price and R. Ruchti (Singapore: World Scientific 2014), vol. 8, p. 135
[9] F.D. Seward, W.H. Tucker, R.A. Fesen, Astrophys. J. **652**, 1277 (2006)
[10] A. A. Abdo et al., Astrophys. J. **710**, L92 (2010)
[11] E.G. Berezhko, G. Pühlhofer, H.J. Völk, Astron. & Astrophys **400**, 971 (2003)
[12] V.G. Sinitsyna, et al. Nuclear Physics B (Proc.Suppl.) **97**, 215 (2001)
[13] E.G. Berezhko, H.J. Völk, L.T. Ksenofontov, Astron. & Astrophys. **483**, 529 (2008)

[14] D. Kosenko, S.I. Blinnikov, J. Vink, Astron. & Astrophys. **532**, 114 (2011)
[15] T.A. Lozinskaya, V.V. Pravdikova, A.V. Finogenov, Astron. Lett. **26**, 77 (2000)
[16] V.G. Sinitsyna and V.Yu. Sinitsyna, Bull. of the Lebedev Phys. Inst. **40**(5), 113 (2013)
[17] V.G. Sinitsyna and V.Yu. Sinitsyna, J. Phys.: Conf. Ser. **409**, 012110 (2013)
[18] A.A. Abdo, et al., Astrophys. J. **712**, 459 (2010)
[19] D.N. Burrows, Zh. Guo, Astrophys. J. **421**(1), L19 (1994)
[20] M. Araya, MNRAS **434**, 2202 (2013)

EPJ Web of Conferences **145**, 19004 (2017) DOI: 10.1051/epjconf/201714519004
ISVHECRI 2016

Very high energy emission sources beyond the Galaxy

V.G. Sinitsyna[a], M.S. Andreeva, K.A. Balygin, S.S. Borisov, I.A. Ivanov, A.M. Kirichenko, A.I. Klimov, I.P. Kozhukhova, R.M. Mirzafatikho, N.I. Moseiko, I.E. Ostashev, A.I. Palamarchuk, V.Y. Sinitsyna, and I.G. Volokh

P. N. Lebedev Physical Institute, Russian Academy of Science, Russia

Abstract. Active Galactic Nuclei (AGN) are considered as potential extragalactic sources of very and ultra high energy cosmic rays. According to theoretical predictions cosmic ray acceleration can take place at the shock created by the expanding cocoons around active galactic nuclei as well as at AGN jets. The measurements of AGN TeV spectra, the variability time scale of TeV emission can provide essential information on the dynamics of AGN jets, the localization of acceleration region and an estimation of its size. SHALON observations yielded data on extragalactic sources of different AGN types in the energy range of 800 GeV–100 TeV. The data from SHALON observations are compared with those from other experiments at high and very high energies.

1. Introduction

High and very high energy γ-rays have come to play an important role in the study of Active Galactic Nuclei (AGN). Many AGNs have been detected from MeV to TeV energies. The detection of any flares from AGNs in VHE gamma-rays and the determination of the time-scales of the flux variability in different wavelengths, including TeV energies, helps to clarify the location of the emission and limits the size of the emitting region.

The long term SHALON observations yielded data on AGNs of different types in the energy range of 800 GeV - 100 TeV (Table 1). Some of these AGNs were discovered at TeV energies by SHALON. The results on fluxes, spectral energy distributions and images are shown here and in [1–7]. All SHALON results are compared with data from experiments at high and very high energies (see Figs. 1,2,3,4).

2. Mkn 421 ($z = 0.031$), Mkn 501 ($z = 0.034$) and Mkn 180 ($z = 0.046$)

The observation results of known BL Lacs Mkn 421, Mkn 501 and Mkn 180 are presented (see Fig. 2, [7]). The images and spectral energy distributions at energies above 800 GeV are obtained for each of the sources.

A number of flare periods in different wavelengths including VHE γ-rays were found. For example, the last flaring state of Mkn 501 at TeV energies was detected by SHALON between March and June 2009. A flux increase was detected on 23–24 April and 23–25 May with an average flux of $(3.41 \pm 0.70) \times 10^{-12}\,\mathrm{cm^{-2}\,s^{-1}}$. The increase of the Mkn 501 flux in the 2009 SHALON data correlates with one at a lower energy range of $>$ 2 GeV by Fermi LAT and has a time lag with the flare at 0.2–2.0 GeV [8]. Also, this flare was detected at TeV energies by VERITAS, MAGIC and Whipple [8].

Mkn 180 was observed by SHALON in 2007, 2009–2014, for a total of 71 hours. After the standard analysis, a clear excess corresponding to an 18.5σ effect [9] was detected. The Mkn 180 spectrum fitting parameters are shown in Table 1.

Extreme variability in different wavelengths, including VHE γ-rays, on time-scales from minutes to years is the most distinctive feature of BL Lac objects, but the variations in total optical brightness are small. Also, no evidence for any significant flux variability was found during the full observation period of Mkn180 by SHALON.

3. 3C 382 ($z = 0.0578$)

3C 382 is a typical Broad Line Radio Galaxy (BLRG), essentially identical to a quasar except for the optical luminosity. 3C382 is a well-known X-ray source. Its radio structure shows several quasar-like features. The core is very bright and there is a narrow jet with a compact hotspot and with a lobe opposite the jet. In contrast to blazars, the jets in BLRGs do not point directly toward the observer, and the relativistic beaming effects and the related jet dominance are only moderate, so it becames possible to investigate both the accretion disk and the jet. Our observations are mainly aimed at searches for TeV γ-ray emission from 3C 382.

3C 382 was detected by SHALON at TeV energies [4] (in observations during the years 2009–2016, 46.3 hours in total) with a significance [9] of 9.5σ. The integral γ-ray flux above 0.8 TeV was estimated to be $(0.91 \pm 0.14) \times 10^{-12}\,\mathrm{cm^{-2}\,s^{-1}}$. The energy spectrum of γ-rays at energies above 800 GeV is described by a power law with index -1.08 ± 0.11. Correlations of the TeV flux with MeV-GeV energy fluxes were not found as the observations of 3C 382 at high energies with Fermi LAT [10] only gave an upper limit in the range 0.1–100 GeV (Fig. 2).

[a] e-mail: sinits@sci.lebedev.ru

Table 1. The catalogue of extragalactic γ-ray sources observed by SHALON with parameters for spectrum fitting in the form of a power low with exponential cutoff $F(>E) \propto E^{k_\gamma} \times exp(-E/E_{cutoff})$.

Sources	Observable flux[a]	k_γ	E_{cutoff}, TeV	Distance, z	Type
NGC 1275	(0.78 ± 0.05)	-2.18 ± 0.12	32 ± 7	0.018	Seyfert
SN2006 gy	(3.71 ± 0.65)	-3.10 ± 0.30	4.4 ± 1.9	0.019	SN
IC 310	(0.89 ± 0.09)	-0.84 ± 0.10	13 ± 3.9	0.019	RG
Mkn 421	(0.63 ± 0.05)	-1.51 ± 0.18	10 ± 3	0.031	BL Lac
Mkn 501	(0.86 ± 0.06)	-1.48 ± 0.15	11 ± 3	0.034	BL Lac
Mkn 180	(0.65 ± 0.09)	-1.84 ± 0.15	7.3 ± 2.2	0.046	BL Lac
3c382	(0.91 ± 0.14)	-1.05 ± 0.11	21 ± 7.0	0.0578	BLRG
4C+31.63	(0.73 ± 0.16)	-1.13 ± 0.16	10.4 ± 3.2	0.295	FSRQ
OJ 287	(0.26 ± 0.07)	-1.14 ± 0.11	9.5 ± 1.2	0.306	BL Lac
3C 454.3	(0.43 ± 0.07)	-0.52 ± 0.12	6.2 ± 1.0	0.859	FSRQ
4C+55.17	(0.90 ± 0.16)	-1.40 ± 0.15	5.4 ± 2.1	0.896	FSRQ
PKS 1441+25	(0.52 ± 0.23)	–	–	0.939	FSRQ
1739+522	(0.49 ± 0.05)	-0.50 ± 0.18	6.1 ± 1.2	1.375	FSRQ
B2 0242+43	(0.58 ± 0.20)	–	–	2.243	FSRQ
B2 0743+25	(0.37 ± 0.16)	–	–	2.979	FSRQ

[a] Integral flux at energy > 800 GeV in units of 10^{-12} cm^{-2} s^{-1}.

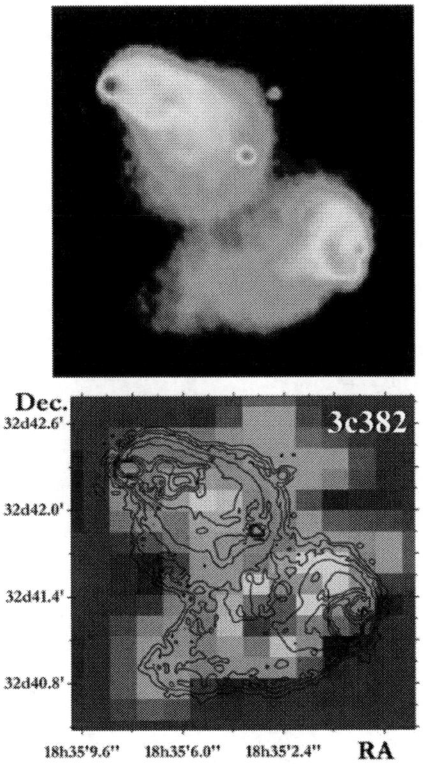

Figure 1. Top: VLA radio image of of 3c382. Bottom: SHALON image of 3c382 at energies 800 GeV −35 TeV. The contours represent the source's radio structure from VLA.

Figure 1 presents a VLA 1477 MHz image of 3C 382. It shows a core-lobe structure with hot-spots at the end of the North jet and at the South. The TeV image of 3C 382 is correlated with the structures visible in radio energies by VLA [11]. The structure viewed in TeV γ-rays by SHALON (Fig. 1) can be the result of heating or blowing the interstellar medium by jets similar to one observed in the NGC 1275 Seyfert galaxy. No significant emission was found from the position of core core, which is clearly seen in the radio-image.

4. OJ 287 (z = 0.306)

OJ 287 is a low-frequency peaked BL Lac object. It is one of the most studied blazars whose spectrum has been well measured through radio [12,13] to X-ray bands [14,15] and for optical studies. The most outstanding characteristic of OJ 287 is its 12 year period, which has been found in the optical [14] and also confirmed in X-rays. OJ 287 is supposed to be a binary black hole system in which a secondary black hole passes the accretion disk of the primary black hole and produces two impact flashes per period. The spectral energy distributions of blazars consist of two broad peaks. The first, lower frequency peak, is due to the synchrotron emissions of relativistic electrons in the jet. It is supposed that the second, higher frequency peak, is due to Inverse Compton emission of the same electrons [16]. GeV emission from OJ 287 has been detected by EGRET [17] and Fermi LAT [18]. It has also been suggested to be a TeV source.

OJ 287 was observed by SHALON in 1999, 2000, 2008–2010, for a total of 47.3 hours, at zenith angles ranging from 22° to 34°. The observations of 1999 and 2000 do not reveal a γ-ray flux from the position of OJ 287, but only an upper limit of $< 1.1 \times 10^{-13}$ cm^{-2} s^{-1}. In the observations of 2008, 2009 and 2010 (31.2 hours) a weak γ-ray flux was detected [3]. An excess corresponding to a 6.9σ effect was observed. The measured integral energy spectrum of OJ 287 can be well described by a power law with index -1.43 ± 0.18. The observed integral flux above 800 GeV is $(0.26 \pm 0.07) \times 10^{-12}$ cm^{-2} s^{-1}. The image of γ-ray emission from OJ 287 by SHALON telescope is shown in Fig. 2.

A flux increase over the detected average flux was found on 14, 15 November and 4, 5 December 2010. The average value of the increase is $(0.63 \pm 0.15) \times 10^{-12}$ cm^{-2} s^{-1} (statistical significance of 6.2σ [9]) with a softer energy spectrum with a power law of index -1.96 ± 0.16. A high flux was also detected on 16

DOI: 10.1051/epjconf/201714519004

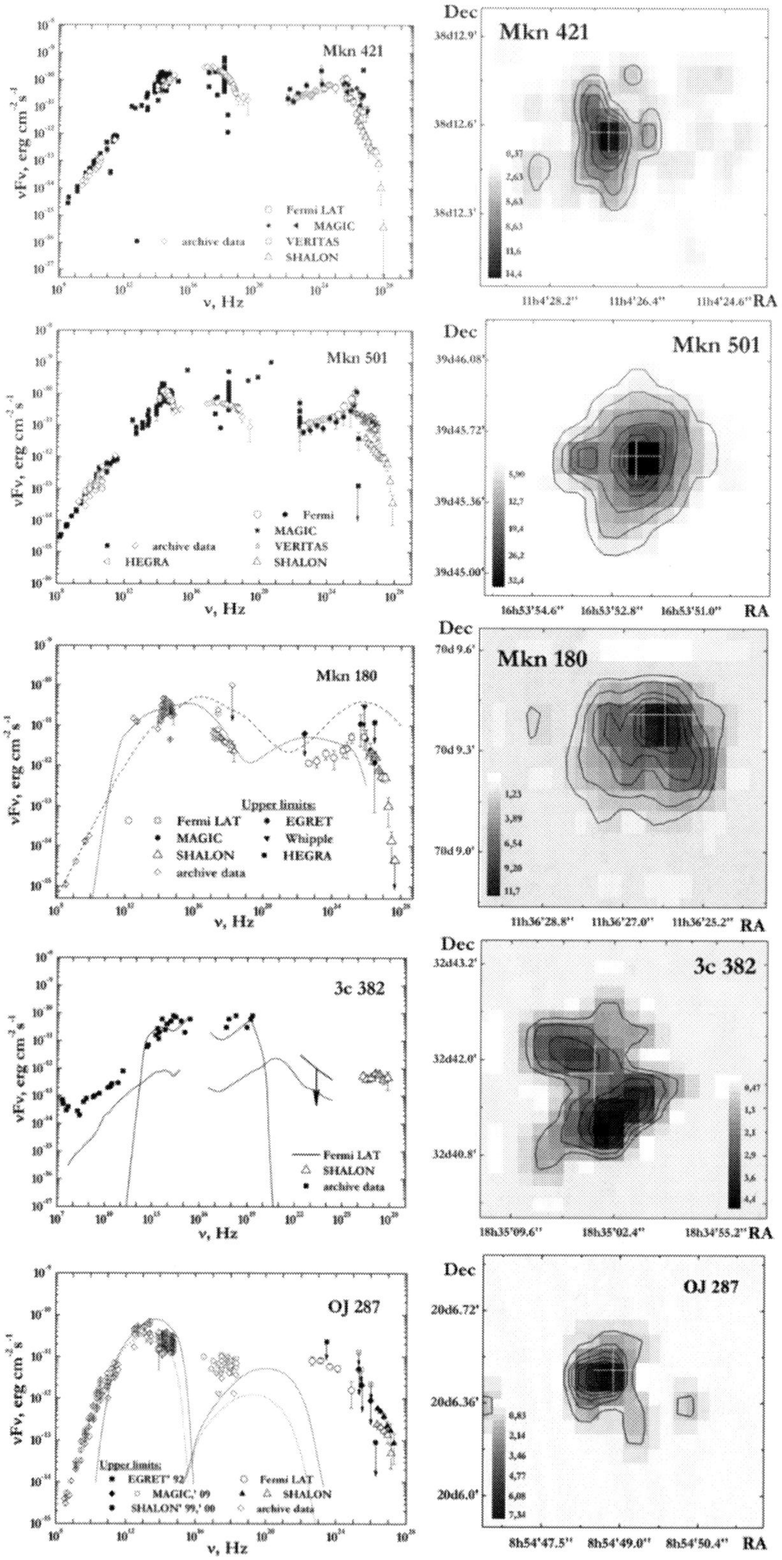

Figure 2. Spectral energy distributions (left) and images (right) of AGNs by SHALON compared with other experiments.

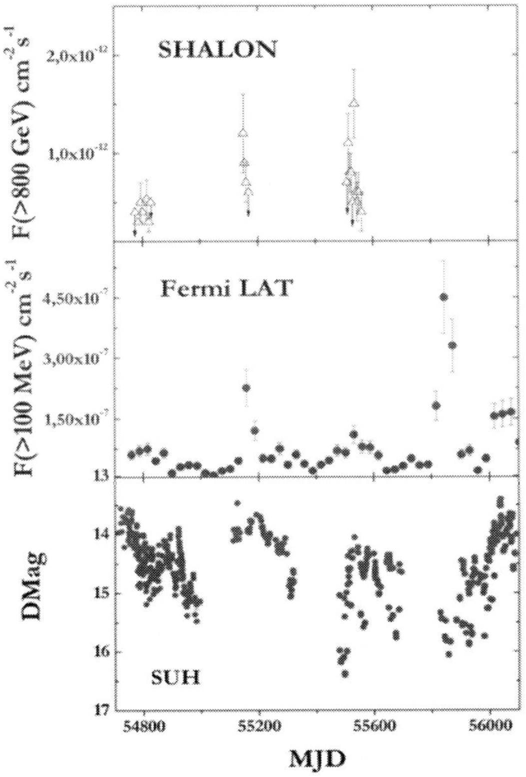

Figure 3. Comparison of the OJ 287 light curve observed by SHALON in the energy range 800 GeV −35 TeV with data in the MeV-GeV energies by Fermi LAT and in R-band.

November 2009. The duration of the flux increase in December 2010 was 2 days with a peak intensity $(1.16 \pm 0.51) \times 10^{-12}\,\mathrm{cm}^{-2}\,\mathrm{s}^{-1}$. No duration of intervals of flux increase were found in 2009, because the observations were interrupted due to weather conditions.

Figure 2 presents the spectral energy distribution of the γ-ray emission from OJ 287 by SHALON in comparison with other experiments' data – MAGIC [19], EGRET [17], and with theoretical predictions from [18] and [19]. The $\triangle$ at TeV energies in Fig. 2 are the SHALON spectrum of OJ287; an upper limit at > 0.8 TeV corresponds to SHALON observations in 1999, 2000. The $\blacktriangle$ present the γ-ray spectrum during the increased flux period of 2010. OJ 287 is the weakest extragalactic source observed by SHALON.

To reveal possible correlations of the emissions in various energy ranges, including those at high and very high energies, we compared the OJ287 γ-ray fluxes observed by SHALON in the periods when the observations were simultaneous with those by the Fermi LAT experiment (see Fig. 3) and data in R-band (Suhora Astronomical Observatory) [20]. A slight local flux increase can be seen in the period of November 2010 in the Fermi LAT data [21] and at R-band, which corresponds to the above-mentioned γ-ray flux increase observed by SHALON in 2010. The increase of OJ 287 flux in the 2009 SHALON data correlates with flares at MeV-GeV energies and R-band (see Fig. 3).

In [22] it is shown that the GeV flares are produced due to the interaction of moving plasma blobs with a standing shock. The flare correlations can indicate that the TeV emission observed by SHALON is a very high energy population of particles formed in the spectrum viewed by Fermi LAT, which are produced by the synchrotron self-Compton process at the jet away from the central core [22].

5. 4C+31.63 ($z = 0.295$)

To collect data for FSRQ type, sources at a large distance range from $z < 0.3$ to extremely high z sources from the first and second Fermi LAT AGN catalogue are observed by SHALON.

4C + 31.63 is a FSRQ type source at redshift $z = 0.295$ and was observed by SHALON in the period 2012–2015 for a total of 45.7 hours. γ-ray emission from the position of 4C + 31.63 was detected above 800 GeV [1] with a 9.7σ effect. Energy spectrum parameters are shown in Table 1. 4C + 31.63 was previously detected at high energies with Fermi LAT. The image of 4C + 31.63 in TeV γ-rays by the SHALON telescope is shown in Fig. 4. No high activity of 4C + 31.63 in TeV energies was found.

6. 3C 454.3 ($z = 0.859$)

3C 454.3 is a known flat-spectrum radio quasar at redshift $z = 0.859$. It has shown very high activity over the last fifteen years and displayed variability at all wavelengths. 3C 454.3 was detected at high energies by the EGRET telescope, with an average photon index of $\Gamma = 2.2$. 3C 454.3 was detected with Fermi LAT at high energies in 2008 [23]. This object has been particularly bright in the γ-ray band covered by AGILE and Fermi LAT.

In 1998 3C 454.3 was detected by SHALON at TeV energies [5] and was observed by SHALON in the period from 1998 to 2013, for a total of 69.2 hours, at zenith angles ranging from 18° to 35°. After the standard analysis, a clear excess corresponding to a 14.9σ effect [9] was determined. The observed energy spectrum parameters and integral γ-ray flux are presented in Table 1. The TeV γ-ray emission map is presented in Fig. 4 right. The spectral energy distribution of 3C 454.3 and limits on the intrinsic spectrum at TeV energies de-absorbed using extragalactic background light shapes (see [6]) is presented in Fig. 4 left with lines. 3C 454.3 has been detected with Fermi LAT at energies 200 MeV–300 GeV [23]. The spectrum by Fermi LAT is fitted with a broken power-law with photon indices $\Gamma_{low} = 2.27 \pm 0.3$, $\Gamma_{high} = 3.5 \pm 0.05$ with an average flux of $\sim 3 \times 10^{-6}\,\mathrm{cm}^{-2}\,\mathrm{s}^{-1}$, for energies > 100 MeV. Additional identification of detected sources with 3C 454.3 is performed because of the correlation of high level TeV fluxes with its flares at lower energies.

3C 454.3 shows a significant flux variability in the different energy ranges including high and very high energies. A number of flaring periods in observations of 3C 454.3 were detected by SHALON since 1998. A period of high level TeV flux was detected in 2006 from 20 to 28 September. A factor two increase of the TeV flux was detected on 15 November 2009. The flare of 2009 detected by Fermi LAT [24] has a ~10 day time lag with one at TeV energies. The last significant flaring state of 3C454.3 at TeV energies was detected in the SHALON observational

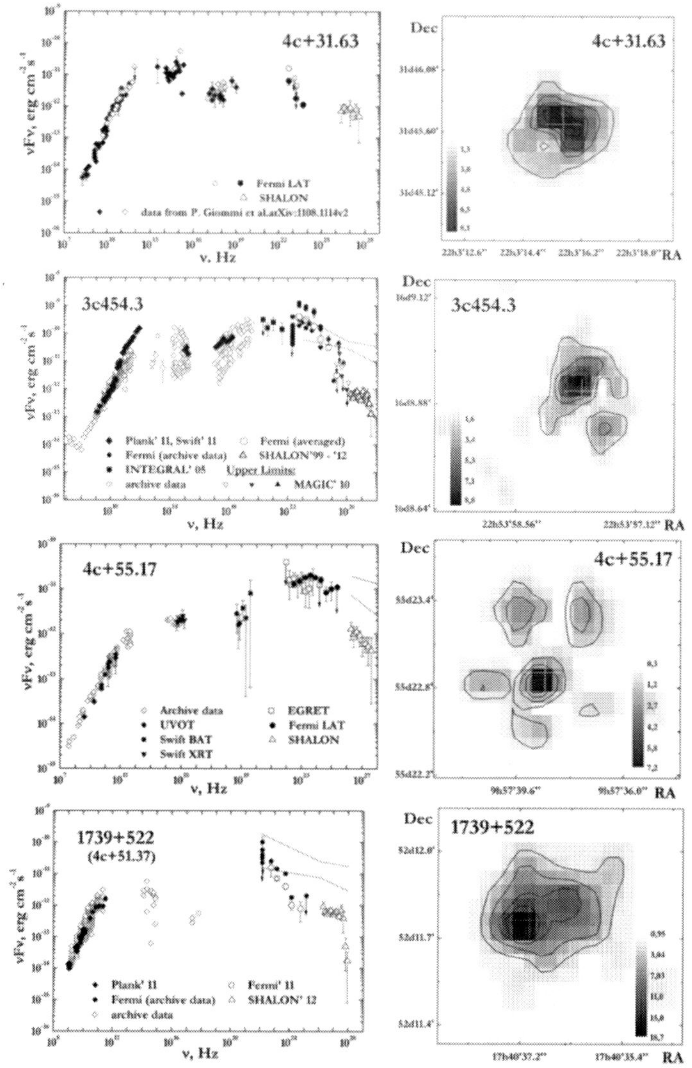

Figure 4. Spectral energy distributions (left) and images (right) of the γ-ray emission from AGNi observed by SHALON.

period of Nov.–Dec. 2010. The flare of TeV flux detected by SHALON on 10 Nov. 2010 correlates with a flux increase observed by Fermi LAT at the phase of the plateau and flare in R-band. The flux increase was detected on 2–4 Dec. with flux of $(3.41 \pm 0.70) \times 10^{-12}\,\mathrm{cm}^{-2}\,\mathrm{s}^{-1}$. This increase correlates with flares observed by Fermi LAT [25] and corresponds to the phase of increasing flux of the post-flare.

Very high energy flux variation time-scales observed by SHALON are similar to ones observed by Fermi LAT [24]. The short flaring times suggest parsec-scale size regions for the origin of TeV and MeV-GeV emission [24].

7. 4C + 55.17 (z = 0.896)

4C + 55.17 is a radio-loud active galaxy classified as a FSRQ. This object was previously detected at high energies with EGRET and Fermi LAT with an unusually hard MeV-GeV γ-ray spectrum [26]. 4C + 55.17 was observed by SHALON in the period from 2012–2016 for a total of 61.2 hours [1]. After the standard analysis, a clear excess corresponding to an 8.4σ effect [9] was determined with an integral flux and the observed energy spectrum shown in Table 1. Figure 4 shows the spectral energy distribution of 4C + 55.17, $\triangle$ represent the data from the SHALON telescope, and the line limited area at TeV energies is the intrinsic spectrum in the source corrected on the extragalactic background light attenuation [6]. The image of 4C + 55.17 obtained by SHALON at TeV energies is in Fig. 4.

It was shown in [26] and [27] that 4C + 55.17 demonstrates morphological, spectral properties of FSRQ-type blazars. The variability and flaring activity in different wavelengths including MeV-GeV energies and TeV γ-rays is the most distinctive feature of blazar-type objects. But no evidence for flux variability was found during the full period of observations of 4C + 55.17 by SHALON at TeV energies. Also, no evidence of high energy flux variability was found by Fermi LAT [26]. The detection of steady high and very high energy emission in 4C + 55.17 together with low variability in radio [28] points out an absence of features providing high activity at all wavelengths.

8. 1739 + 522 (4C + 51.37) (z = 1.375)

One more remote metagalactic γ-source of the FSRQ type was detected by SHALON in 1999 and is being intensively studied since then [1,5]. This object was identified with 1739 + 522 (4C + 51.37). It was observed by SHALON in the period from 1999, 2003–2013 for a total of 119.8 hours. The average value of the integral flux is $(0.49 \pm 0.05) \times 10^{-12}\,\mathrm{cm}^{-2}\,\mathrm{s}^{-1}$ at $E > 0.8\,\mathrm{TeV}$, with 15.1σ [9]. The energy spectrum parameters are shown in Table 1. The image of 4C + 51.37 at TeV energies by SHALON is presented in Fig. 4; Δ on the spectral energy distribution (fig. 4) represent the data from the SHALON telescope. Lines at TeV energies limit the range of the intrinsic spectrum in the source corrected on the extragalactic background light attenuation [6].

1739 + 522 is a high polarization radio quasar. It is detected by EGRET with an average γ-flux in the range $\sim$30 MeV to 50 GeV about $2 \times 10^{-8}\,\mathrm{cm}^{-2}\,\mathrm{s}^{-1}$ with a photon index of $\Gamma \sim 2.2$ [29]. The 100 MeV–100 GeV γ-ray emission from 1739 + 522 was detected by Fermi LAT with average flux of $F(> 1\,\mathrm{GeV}) = (2.5 \pm 0.2) \times 10^{-9}\,\mathrm{cm}^{-2}\,\mathrm{s}^{-1}$ and power-law with photon index $\Gamma = 2.5 \pm 0.04$ [30].

Two periods of TeV flux increase were found since the first detection of 1739+522. In observations from 8 to 20 June 2004 the flux increase from the level of the average value up to $(1.93 \pm 0.60) \times 10^{-12}\,\mathrm{cm}^{-2}\,\mathrm{s}^{-1}$ at $E > 0.8\,\mathrm{TeV}$ in the peak on 12 June 2004 was recorded in two night observations. A similar high flux increase was detected on 12 May 2005, but no duration of flux increase were found, because the observations were interrupted due to weather conditions.

9. B2 0242 + 43 and B2 0743 + 25 (z > 2)

It was discussed in [31] and later, that the γ-ray spectra of distant sources at 10s of GeV to TeV energies begin to be attenuated by the still poorly known EBL photon field due to the $\gamma + \gamma \rightarrow e^+e^-$ process. So, to estimate the cosmic γ-ray horizon two extremely high-redshifted sources from the first and second Fermi LAT AGN catalogue were observed by SHALON in the autumn-winter period of 2014.

B2 0242 + 43, with z = 2.243, is a radio-loud active galaxy classified as a FSRQ. It was observed by SHALON in 2014, for a total of 14 hours, at zenith angles from 16° to 33°. After the standard analysis, an excess corresponding to a 4.8σ effect [9] was determined with an integral flux of $(0.58 \pm 0.20) \times 10^{-12}\,\mathrm{cm}^{-2}\,\mathrm{s}^{-1}$. The unexpected high-level flux increase up to twice that detected on average was observed at the end of November. The area around B2 0242 + 43 of 20′ size in radius doesn't contain any candidate for TeV-emission nor any GRBs, supernova explosions or other powerful sources over the wide wavelength range variable sources that could be sources of the flares detected in the observation period. The list of nearby sources includes two weak radio-objects and one X-ray source weren't confirmed after their detection [32,33].

B2 0743 + 25, at redshift z = 2.979, is an FSRQ from the Fermi LAT catalogues. It was observed by SHALON in the period from 2014 to 2016 on clear moonless nights, for a total of 20 hours, at zenith angles ranging from 15° to 32°. The integral γ-ray flux of $I(> 0.8\,\mathrm{TeV}) = (0.37 \pm 0.16) \times 10^{-12}\,\mathrm{cm}^{-2}\,\mathrm{s}^{-1}$ was determined with a significance of 5.1σ [9]. An analysis of the region of 20′ in radius around B2 0743 + 25 was performed in order to find any close powerful sources that could be resolved as this distant quasar [33,34]. Among the nearby sources no candidate for TeV-emission were found as four quasar objects, known only due to identification by V B U magnitude, were not recognized in radio, X-ray or MeV-GeV energies [34].

Further observations and investigations of B2 0242 + 43, B2 0743 + 25 and the area around these objects are continuing to establish a reliable source of TeV γ-ray emission.

10. Conclusion

The results presented here are a brief summary of long-term observations of AGNs by the SHALON telescope such as spectral energy distributions and images. A number of variability periods at different wavelengths including VHE γ-rays were found. Short flaring times of very high energy emission from 3C 454.3 and 1739 + 522 are found and it suggests parsec scale size regions of TeV gamma-ray origin in these sources. In addition, all data obtained in SHALON observations are compared with those from other experiments at high and very high energies.

References

[1] V.G. Sinitsyna and V.Yu. Sinitsyna *"Astroparticle, Particle, Space Physics and Detectors Physics Applications"* (Proc. 14th ICATPP, 2013) ed. S. Giani, C. Leroy, P.-G. Rancoita, L. Price and R. Ruchti (Singapore: World Scientific 2014) vol. 8, p. 3; *ibid.* p. 20
[2] V.G. Sinitsyna,V.Yu. Sinitsyna, Astron. Lett. **40** (2–3), 75 (2014)
[3] V.G. Sinitsyna et al., In Proc. of 32nd ICRC, Beijing, China, 2011; vol. 8/11, p. 26
[4] V.G. Sinitsyna, V.Yu. Sinitsyna EPJ Web of Conferences **52**, 10005 (2013)
[5] V.G. Sinitsyna et al., Int. J. Mod. Phys. A **29**, 7023, 7026, 7029 (2005)
[6] V.G. Sinitsyna et al., J. Phys.: Conf. Ser. **409**, 012113 (2013)
[7] V.G. Sinitsyna et al., J. Phys.: Conf. Ser. **409**, 012112 (2013)
[8] A.A. Abdo et al., Astrophys. J. **727**, 129 (2011)
[9] T.-P. Li and Y.-Q. Ma, Astrophys. J. **272**, 317 (1983)
[10] J. Kataoka et al. Astrophys. J. **2011**, **740**, 29
[11] J.P. Leahy, R.A. Perley Astrophys. J. **102**, 537 (1991)
[12] E. Idesawa et al. Publ. Astron. Soc. Jpn. **49**, 631 (1997)
[13] H. Seta et al. Publ. Astron. Soc. Jpn. **61**, 1011 (2009)
[14] E. Valtaoja et al., Astrophys. J. **531**, 744 (2000)
[15] C.E. Tateyama et al. Astrophys. J. **608**, 149 (2004)
[16] G. Ghisellini et al. Mon. Not. R. Astron. Soc. **301**, 451 (1998)
[17] R.C. Hartman et al., ApJS **123**, 79 (1999)

[18] A.A. Abdo et al., Astrophys. J. **700**, 597 (2009); *ibid.* **715**, 429 132 (2010)
[19] M. Hayashida et al., In Proc. of 31st ICRC, Lodz, Poland, 2009
[20] P. Pihajoki et al. Astrophys. J. **764**, 5 (2013)
[21] M. Ackermann et al. Astrophys. J. **810**, 14 (2015)
[22] I. Agudo et al. Astrophys. J. Lett. **726**, L13 (2011)
[23] A.A. Abdo et al., Astrophys. J. **699**, 817 (2009)
[24] M. Ackermann et al. Astrophys. J. **721**, 1383 (2010)
[25] A.A. Abdo et al. Astrophys. J. **733**, L26 (2011)
[26] W. McConville et al. Astrophys. J. **738**, 148 (2011)
[27] A.P. Marscher et al. Astrophys. J. **577**, 85 (2002)
[28] P. Augusto et al. MNRAS **368**, 1411 (2006)
[29] R. Mukherjee et al. Astrophys. J. **490**, 116 (1997)
[30] P.L. Nolan et al. ApJS **199**, 31 (2012)
[31] A.I. Nikishov Sov. Phys. JETP **14**, 393 (1962)
[32] I.I.K. Pauliny-Toth et al., Astron. J. **77**, 265 (1972)
[33] W. Voges et al. Astron. Astrophys. **349**, 389 (1999)
[34] M.-P. Veron-Cetty, P. Veron Astron. Astrophys. **518**, A10-10 (2010)

Primary gamma ray selection in a hybrid timing/imaging Cherenkov array

E.B. Postnikov[1,a], A.A. Grinyuk[2], L.A. Kuzmichev[1], and L.G. Sveshnikova[1]

[1] Lomonosov Moscow State University Skobeltsyn Institute of Nuclear Physics (MSU SINP), 1(2) Leninskie gory, GSP-1, 119991 Moscow, Russia
[2] Joint Institute for Nuclear Research, 6 Joliot-Curie, 141980 Dubna, Moscow region, Russia

Abstract. This work is a methodical study on hybrid reconstruction techniques for hybrid imaging/timing Cherenkov observations. This type of hybrid array is to be realized at the gamma-observatory TAIGA intended for very high energy gamma-ray astronomy (> 30 TeV). It aims at combining the cost-effective timing-array technique with imaging telescopes. Hybrid operation of both of these techniques can lead to a relatively cheap way of development of a large area array. The joint approach of gamma event selection was investigated on both types of simulated data: the image parameters from the telescopes, and the shower parameters reconstructed from the timing array. The optimal set of imaging parameters and shower parameters to be combined is revealed. The cosmic ray background suppression factor depending on distance and energy is calculated. The optimal selection technique leads to cosmic ray background suppression of about 2 orders of magnitude on distances up to 450 m for energies greater than 50 TeV.

1. Introduction

Of several methods to detect primary gamma rays on the Earth's surface, an imaging technique is the most popular because of the high efficiency of cosmic ray background reduction [1]. Realization of this technique requires expensive and complicated installations called IACT (Imaging Air Cherenkov Telescopes) [2]. They collect Cherenkov light from extensive air showers by a set of mirrors and reflect it onto an array of photomultipliers called a camera in such a way that a 2-dimensional image of the event is formed (Fig. 1). The properties of an image are used to discriminate gamma ray induced showers against showers from the cosmic ray background.

The modern version of IACT implementation is stereoscopy [3], requiring a system of two or more imaging telescopes located at a distance of $\sim$100 m from each other [4]. The reason for using stereoscopy is the ability to determine the core position of a gamma ray shower, as well as the shower arrival direction, because different telescopes detect images of the same shower at different angles.

Another technique to detect primary gamma rays is measuring the arrival time of the Cherenkov light front along with the local intensity. This method requires a set of wide angle timing optical stations and for the first time was implemented in such experiments as THEMISTOCLE [5] and AIROBICC [6]. That kind of technique is much more cost-effective, the installations have a wide angle of view, and arrays of detectors can easily be developed on a large area. A significant disadvantage of this approach is a poor rate of cosmic ray background rejection. Some efforts [7] have been made by the authors of the present work to assess the ability of gamma ray selection based on differences in the form of Cherenkov light lateral distribution. The research revealed that such a selection can only be done for events with a large number of triggered detectors, but anyway the selection quality is still much weaker than that of the IACT technique.

All these reasons were taken into account when developing the gamma-ray observatory TAIGA (Tunka Advanced Instrument for cosmic ray physics and Gamma-ray Astronomy [8]). TAIGA combines the hybrid detection of air shower Cherenkov radiation by the wide angle timing array TAIGA-HiSCORE and the narrow angle imaging telescopes TAIGA-IACT [8]. At present, the first prototype of the telescope is being deployed in the Tunka valley (Fig. 1). The telescope mirrors have an area of $\sim$10 m^2 and a focal length of 4.75 m, the camera consists of 560 photomultipliers with a total field of view 9.72° (0.36° per pixel).

Preliminary results of simulation confirm that the joint operation of the HiSCORE array and an IACT could be a very cost-effective way to increase the energy range of gamma-ray astronomy [9].

This work is a methodical study on hybrid reconstruction techniques to be used within the TAIGA experiment.

2. Monte Carlo simulation

Simulated data of both types were used: the image parameters from the IACT, and the shower parameters reconstructed from the data obtained with the timing array.

The IACT data was simulated in a few consecutive steps. First, the shower development in the air was simulated by CORSIKA [10] to obtain the Cherenkov photon distribution on the observation level. Second, Cherenkov photons were traced through the optical

[a] e-mail: `evgeny.post@gmail.com`

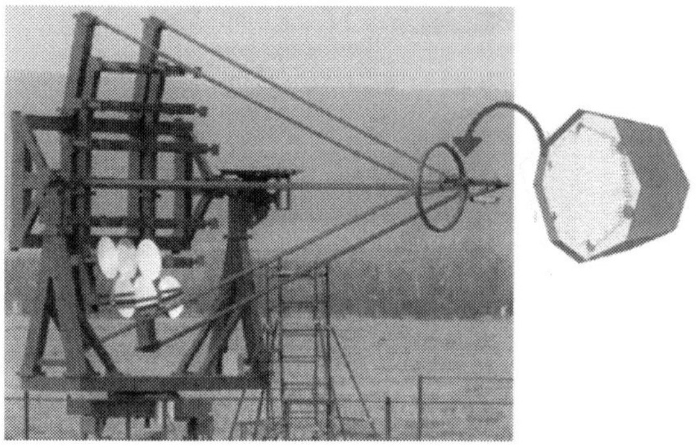

Figure 1. An IACT in the deployment phase (left panel) and a camera (right panel).

system of the IACT. Third, conversion from photons to photoelectrons in every photomultiplier of a camera was carried out using wavelength-dependent atmospheric absorption and photo-cathode sensitivity. Then the signal was sorted into 40 ns time bins and a night sky background added. Finally, a Cherenkov image was reproduced as a distribution of the number of photoelectrons in 560 photomultipliers.

The timing array data provides the arrival direction (with an accuracy of 0.1°–0.4° depending on energy), shower core position with 10 m accuracy, energy resolution with ~15% [8] accuracy. As a first step the monoenergetic data banks with primary particle energies from 20 TeV to 500 TeV were examined.

Gamma ray showers direction was fixed at the point source, whereas the background (proton) shower arrival direction was randomized around the gamma ray shower direction within a cone of 0.4° corresponding to the arrival direction resolution. This imitates the first selection step of showers in the cone of 0.4° based only on the timing array data.

3. Methods

3.1. Accuracy assessment

Data analysis technique for the IACT is usually based on the "Hillas analysis" [11]. For our task this method was modified. For each shower from the data bank various features of the image were calculated, and for each set of these features the selection quality factor, Q, was estimated as an accuracy indicator of our method.

The selection quality factor shows how much a significance, S_0, of the statistical hypothesis that the events detected by the installation do not belong to the background, can be improved after the event selection:

$$S = S_0 \times Q \tag{1}$$

In the framework of Poisson statistics, the significance has a simple form:

$$S = N_\gamma / \sqrt{N_{bckgr}}, \tag{2}$$

and thereby the selection quality factor is:

$$Q = \epsilon_\gamma / \sqrt{\epsilon_{bckgr}}, \tag{3}$$

where ϵ_γ and ϵ_{bckgr} are the fractions of gamma ray events and background events respectively after the selection stage. Optimal algorithms (and optimal parameter sets) were found in a process of maximizing the Q factor value under the condition:

$$\epsilon_\gamma \geq 0.5 \tag{4}$$

The fraction of background events after the selection corresponding to the Q value is:

$$\epsilon_{bckgr} = (\epsilon_\gamma / Q)^2 \tag{5}$$

3.2. Image cleaning

For every event the night sky background light was simulated N times to obtain more precise results, and the selection quality factor was averaged over all realizations. This multiple recalculation of background proved necessary because of strong fluctuations of the Q factor caused by random distortion of an image [12] (distributions of Q were plotted in [12]).

To perform image cleaning from random distortion, a few types of filtering were examined and the best algorithm as well as the optimal filtering parameters were chosen based on the Q maximization criterion [12]. Optimal filtering parameters in an image cleaning procedure are energy-dependent and distance-dependent, and their influence on Q is sufficient [12]. All the results below were obtained after optimal image filtering.

4. Results

4.1. Image parameters

The optimal set of parameters to be combined in a selection algorithm was revealed. The optimal imaging parameters from the IACT are:

- azwidth [11], or "azimuthal width", that is the "r.m.s. image width relative to an axis which joins the source to the centroid of the image" [11]. The telescope is supposed to point at the gamma ray source, so the "source" on the image plane is simply the geometrical center of the camera;
- core-azwidth, or "core-azimuthal width", suggested in the present article. This is the r.m.s. image width relative to an axis which joins the image centroid to the shower core position.

4.2. Optimal cuts

For every sample of events with fixed energy and interval of distance to the core position we have found the optimal cuts, corresponding to a maximal value of $Q(R, E)$. In Fig. 2 we present the obtained optimal azwidth cut depending on energy and distance to the shower core.

While the optimal cuts depend on energy and distance, one can conclude that the parameters determined by the timing array (primary energy, shower core position, shower arrival direction) are very essential for gamma ray selection.

An alternative way to take account of this dependence is by normalizing the Hillas parameters (image width etc.) depending on shower parameters. The Hillas parameters after normalization are called "scaled" ("scaled width" etc.) [13]. They are used in stereoscopy (Sect. 1), where the core distance is obtained as an intersection of axes

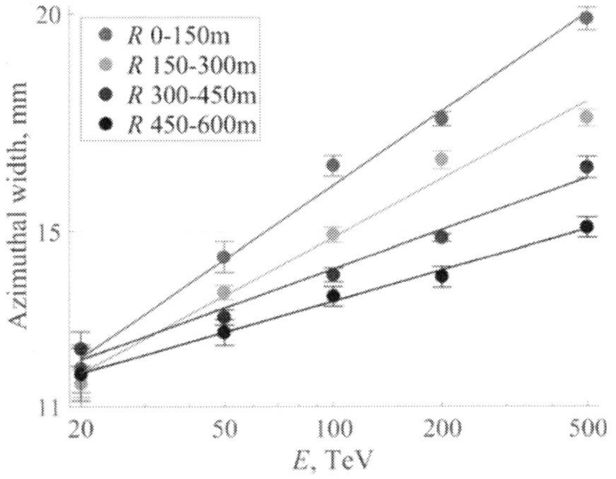

Figure 2. Selection cut depending on energy and distance.

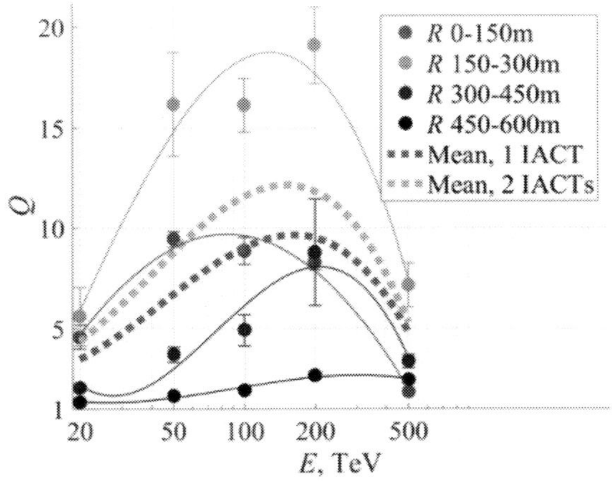

Figure 3. Q factor depending on energy and distance. The dashed lines are averaged over the whole installation area with one or two IACTs.

of multiple images. In a previous analysis of TAIGA simulation we also followed this approach ("hybrid scaled width" [9]), however, in the present work the scaling procedure is applied to the selection cuts. An additional difference of the present approach is scaling the cuts depending on both the core distance and primary energy from the timing array of TAIGA. Our previous steps [9] followed standard IACT "scaled cuts" analysis using total number of photoelectrons in an image ("image size") instead of the energy.

4.3. Optimal Q factor

Fig. 3 demonstrates the results of Q factor computation for a set of energies and core distance intervals. As can be seen, the cosmic ray background suppression about 100 times ($Q{\sim}5$) is to be achieved on a distance up to 450 m for an energy greater than 50 TeV.

In Fig. 3 we also present the mean value of $Q(E)$ averaged over the distance between the shower core and the nearest IACT. The upper curve $Q(R)$ represents the variant of two IACTs located at a distance 320 m apart from each other, the lower one – one IACT. A sensitivity

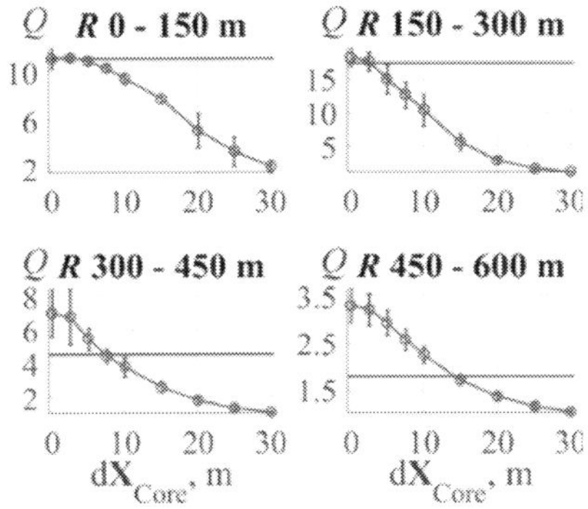

Figure 4. Q factor depending on the accuracy of core position determination. Different panels correspond to different distance intervals. Blue points – using core-azwidth parameter of this work, red line – using azwidth parameter [11].

of the gamma-ray observatory depends on the efficiency of background rejection as ${\sim}Q^{-1}(E)$. Therefore, from Fig. 3 the sensitivity improvement due to adding IACTs can be estimated.

4.4. Q dependence on core distance

As follows from the Q factor dependence on the core distance (Fig. 3), we can successfully use only one single IACT for efficiently selecting gamma ray showers up to ${\sim}450$ m in a hybrid installation. This situation differs from a stand-alone IACT, when the distance ${\sim}100$–150 m is a limit [4]. Therefore, the distance between two or more IACTs (if any) as parts of a hybrid installation can be significantly greater than in a stereoscopic system of IACTs (Section 1). In particular, the expected location of the second IACT in TAIGA is supposed to be ${\sim}320$ m apart from the first one.

4.5. New parameter

For the case of a core-azwidth parameter the optimal cut depends also on the accuracy of the shower core reconstruction $d\mathbf{X}_{Core}$. As a result, the quality factor Q also depends on $d\mathbf{X}_{Core}$. In Fig. 4 we present this dependence for different distance intervals. The meaning of the figure is that the new parameter proved more efficient than the standard parameter azwidth [11] only for a distance greater than ${\sim}450$ m. This result was obtained given the core position accuracy ${\sim}10$–15 m. However, the work aiming at the accuracy improvement is now in progress [14].

4.6. Possibility to use the IACT without the data from the timing array

A possibility to use the IACT without the data from the timing array (energy and distance to the core position) was also investigated in the energy interval 20–500 TeV to check the influence of the hybrid technique components on the selection quality. The Q factor obtained for this case is presented in Fig. 5. The figure demonstrates a major influence of timing array data on the selection quality at high energy.

Therefore, only the hybrid technique can lead to a proper functioning of a single IACT in the conditions of

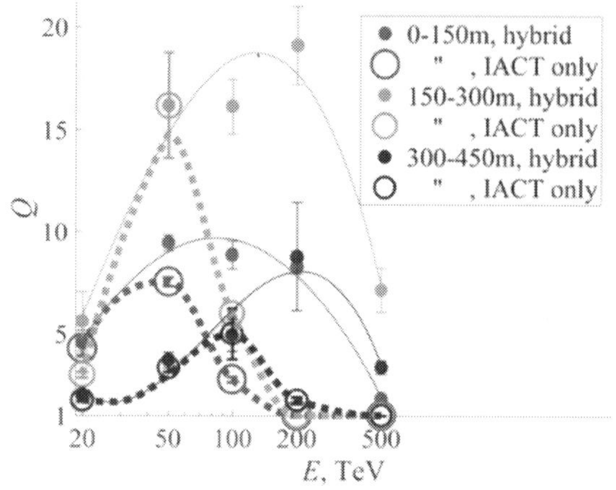

Figure 5. Q factor decrease in the absence of timing array data. Solid lines and filled circles: timing array data is present (hybrid technique), dashed lines and open circles: timing array data is absent (stand-alone IACT).

our research (the energy range 20–500 TeV and distances up to 600 m). However, in a hybrid installation this separate use of an IACT could be implemented at low energy (less than ~30 TeV), because in that case the shower core reconstruction and energy determination in a timing array of TAIGA is not available [8].

5. Discussion

Unlike stereosystems of IACT, a hybrid installation uses images only to select gamma rays, but not to solve other tasks. Therefore, in contrast to standard IACT systems, an image analysis in a hybrid technique can satisfy the only criterion of Q maximization. This helped us to perform full optimization and achieve very high selection quality as well as identify optimal values of cuts for selection and image cleaning different from those of IACT systems.

6. Conclusions

We performed a simulation-based methodical study on hybrid reconstruction techniques of both types of Cherenkov detectors: IACT + timing array. As a result we found an optimal set of imaging and timing parameters to be combined, as well as optimal procedures of image cleaning and gamma event selection. The cosmic ray background suppression about 100 times (Q factor ~5) is to be achieved for a distance up to 450 m for energies greater than 50 TeV. A separate use of the IACT (without the timing array) allows for only a suppression one ninth as weak as that of the joint operation (Q factor value one third as small) at 100 TeV. The investigation confirms in a quantitative way the efficiency of combining the timing-array technique with IACTs in the TAIGA gamma-ray observatory. All the results were obtained for the case that has never been investigated in other experiments.

The study was supported by the Russian Foundation for Basic Research, project no. 16-29-13035.

References

[1] F. Aharonian, *Very High Energy Cosmic Gamma Radiation. A Crucial Window on the Extreme Universe* (World Scientific, 2004)
[2] T.C. Weekes, M.F. Cawley, D.J. Fegan et al., ApJ **342**, 379 (1989)
[3] A. Kohnle, F. Aharonian, A. Akhperdzhanian et al., Astropart. Phys. **5**, 119 (1996)
[4] A.M. Hillas, Space Sci. Rev. **6**, 17 (1996)
[5] P. Baillon, L. Behr, S. Danagoulian et al., Astropart. Phys. **1**, 341 (1993)
[6] A. Karle, M. Merck, R. Plaga et al., Astropart. Phys. **3**, 321 (1995)
[7] A.Sh.M. Elshoukrofy, E.B. Postnikov, L.G. Sveshnikova, J. Phys.: Conf. Ser. (to be published, 2017)
[8] N. Budnev, I. Astapov, P. Bezyazeekov et al., J. Phys.: Conf. Ser. **718**, 052006 (2016)
[9] M. Kunnas et al., J. Phys.: Conf. Ser. **632**, 012040 (2015)
[10] D. Heck, J. Knapp, J.N. Capdevielle et al., *Report FZKA 6019* (Forschungszentrum Karlsruhe, 1998)
[11] A.M. Hillas, *Proc. 19th ICRC*, **3**, 445 (NASA Conf. Publ., La Jolla, 1985)
[12] E.B. Postnikov, A.A. Grinyuk, L.A. Kuzmichev et al., Bull. Russ. Acad. Sci. Phys. **4** (to be published, 2017)
[13] A. Daum et al., Astropart. Phys. **8**, 1 (1997)
[14] A.Sh.M. Elshoukrofy, E.B. Postnikov, E.E. Korosteleva et al., these proceedings

Parametric analysis of cherenkov light LDF from EAS for high energy gamma rays and nuclei: Ways of practical application

A.Sh.M. Elshoukrofy[1,2,a], E.B. Postnikov[1,b], E.E. Korosteleva[1], L.G. Sveshnikova[1,c], and H.A. Motaweh[2]

[1] Lomonosov Moscow State University Skobeltsyn Institute of Nuclear Physics (MSU SINP), 1(2) Leninskie gory, GSP-1, 119991 Moscow, Russia
[2] Faculty of Science, Damanhour University, El-Gomhouria St., 22516 Damanhour, El Beheria, Egypt

Abstract. In this paper we propose a 'knee-like' approximation of the lateral distribution of the Cherenkov light from extensive air showers in the energy range 30–3000 TeV and study a possibility of its practical application in high energy ground-based gamma-ray astronomy experiments (in particular, in TAIGA-HiSCORE). The approximation has a very good accuracy for individual showers and can be easily simplified for practical application in the HiSCORE wide angle timing array in the condition of a limited number of triggered stations.

1. Introduction

Ground-based gamma-ray astronomy has developed very fast since the discovery of the first TeV gamma-ray source, the Crab Nebula, in 1989 [1] by the Whipple collaboration. This experiment was based on the imaging air Cherenkov technique, IACT [2]. The main idea of this technique is the use of a telescope for collecting the Cherenkov light produced in the atmosphere by very energetic charged particles from extensive air showers (EAS). The competing method of Cherenkov light registration for gamma-ray astronomy is a wide angle timing array, which was first realized in THEMISTOCLE [3] and AIROBICC [4]. At the present time the gamma-ray timing array concept HiSCORE [5,6] (High Sensitivity Cosmic Origin Explorer) is realized in the Tunka valley in Siberia as part of the TAIGA observatory [7]. This is an array of wide-angle non-imaging Cherenkov light detectors spaced about hundred meters apart from each other. The wide angle technique assumes that the air shower front timing is measured by counting the number of Cherenkov photons, Q, emitted by secondary air shower particles and subsequently registered in every i-th station of the array at time, t, at a distance, R, from the shower core $Q_i(R,t)$. The analysis of $Q_i(R,t)$ gives the possibility to reconstruct all the parameters of the primary particle needed for gamma-ray astronomy: shower arrival direction, core position, energy and type of particles [8,9].

The critical point of the reconstruction procedure is the right choice of the Cherenkov light lateral distribution function (LDF) suitable for every kind of approximation. Up to now various approximations were only proposed for either energies greater than 1000 TeV [10,11] (non imaging technique) or up to 10 TeV for the imaging telescopes. In the HiSCORE experiment the method of event reconstruction was primarily developed for the Tunka-133 wide angle Cherenkov array [12] at energies above the PeV-range. A special piecewise fitting function [9,12,13] (referred below as the 'Tunka fit'), composed of 4 pieces, was designed to fit both smooth and sharp LDF. The total number of parameters of all constituent functions was reduced to two: a and bxy, the first one is a normalization factor, the second one characterizes the steepness of the whole LDF. The Tunka-133 approximation describes all the LDF for primary cosmic rays in the energy interval 10^{15}–10^{18} eV, for which it was elaborated. Nonetheless, our analysis reveals that the LDF of gamma rays, especially of ones incident at large zenith angles, very often cannot be reproduced. In [14] we proposed to use a simple 'knee-like' approximation of the Cherenkov light distribution, and have tested the quality of these approximations for gamma rays and hadrons in the energy region of interest, 30–3000 TeV. In this paper we mainly study the possibility of applying these functions to real experimental conditions of the HiSCORE experiment.

2. Event parameters reconstruction in the HiSCORE experiment

The HiSCORE array now includes 28 stations spaced with a step of 106 m and covering an area of 0.25 km^2. At present the main steps of event reconstruction are the following [9]:

1) Selection of events with more than 4–5 triggered stations (the stations, where the Cherenkov light pulse exceeds the night sky light background).
2) Rough estimation of zenith and azimuth angles of arriving particles (θ,ϕ) using the measured time of delay in every station with a plane approximation of the time front.
3) Reconstruction of the shower core position (X_0,Y_0) by minimizing the difference between the experimental $Q_i(R_i)$ and the 'Tunka fit', where R_i is

[a] e-mail: abeershehatamahmoud@yahoo.com
[b] e-mail: evgeny.post@gmail.com
[c] e-mail: tfl10@mail.ru

the distance between the i-th station and the core position (X_0, Y_0).

4) Re-estimation of θ and ϕ with the known core position (X_0, Y_0) using a more appropriate cone-like fitting function for the time front. It gives significant improvement of the directional resolution from $d\theta \sim 1°–2°$ to $d\theta \sim 0.1°–0.4°$, because the accuracy of the arrival direction reconstruction depends linearly on that of the shower core position. This point is very important for improving the noise/background ratio of a signal because a background flux depends on the observational solid angle as $\sim d\theta^2$.
5) Approximation of $Q_i(R_i)$ by the 'Tunka fit' function and estimation of the density of photons at a distance 200 m from the shower core, Q_{200}, and corresponding energy estimation by the formula:

$$E = aQ_{200}^{0.94} + b \qquad (1)$$

(a and b are constants obtained from simulations).
6) Parametric analysis of fitting functions for the estimation of the nature of the primary particle.

As is seen, the LDF fitting functions are used on 4 different steps: shower core reconstruction (3), arrival direction estimation (4), energy estimation (5), and primary particle type identification (6).

3. Knee-like approximation of the lateral distribution function

For parameterization of the lateral distribution function, $Q(R)$, of simulated Cherenkov light we used the function that we called a 'knee-like approximation', which was used earlier by J. Horandel [15] to describe the knee in the cosmic ray spectrum. It was a function of energy at that time, but we make it a function of the distance, R, from the shower axis. It depends on five parameters C, γ_1, γ_2, R_0, and α:

$$F_{appr.} = CR^{\gamma_1}\left(1 + \left(\frac{R}{R_0}\right)^{\alpha}\right)^{\frac{\gamma_2}{\alpha}} \qquad (2)$$

In (2) R is the distance from the shower axis, R_0 is the knee position, γ_1 is the slope of the LDF before the knee, $(\gamma_2+\gamma_1)$ is the slope of the LDF after the knee, and parameter α characterizes the sharpness of the knee. We applied that fitting function to simulated data to check the quality of the new approximation and used the HiSCORE bank of simulated events (from CORSIKA [16] code), where LDF functions are simulated with a space step of 5 m, for different types of primary particles such as protons, gammas, He, C, Fe in the wide energy interval 30–3000 TeV and zenith angle range 0°–50°. Below we call them 'true LDF. We showed [14] that the new approximation gives the possibility to fit the whole diversity of individual LDF for different nuclei and gamma rays at shower core distances of 20–500 m in the energy interval 30–3000 TeV with a very good accuracy. Parameters of the approximation depend on the energy and the type of primary particle and allow us to separate proton and gamma ray induced showers (details will be published in [17]).

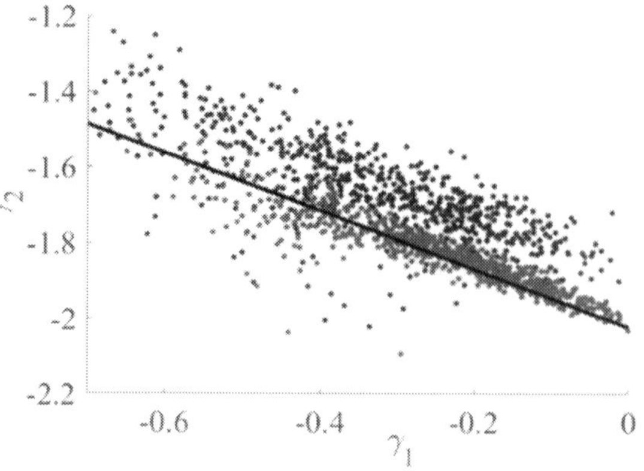

Figure 1. Correlation between γ_1 and γ_2 parameters in the fitting function (2) for gamma rays (red points) and protons (blue points) at 100 TeV.

However, the 5-parameter fitting function may not be suitable for the real conditions of registration, when we select events with a small number of triggered stations (4–5) to decrease the energy threshold, especially for the task with two additional unknown parameters (X-core, Y-core). Our study shows that the two-parameter fitting has a distinctive advantage when we work with a small number of triggered detectors; however, it reduces the diversity of LDF to a fixed set of curves. Therefore, different steps of reconstruction require fitting with different number of parameters.

To decrease the number of fitting parameters correlations between parameters were investigated. One of the examples of a strong correlation between γ_1 and γ_2 is presented in Fig. 1. A number of other correlations, including very important correlations of γ_1, R_0, and α with a depth of shower maximum, was also established. Using correlations of different parameters we developed four and three parameter versions of the knee-like fit. The 4-parameter version includes the least squares estimation of only 4 fit parameters: C, γ_1, R_0, and α, while the fifth parameter, γ_2, is to be obtained from γ_1 by an equation:

$$\gamma_2 = -0.77\gamma_1 - 2 \qquad (3)$$

The 3-parameter version sets α to the mean value for the given class of events.

4. Shower core reconstruction

We applied our 3– and 4-parameter fitting function to the step 3 of the HiSCORE method of event parameters reconstruction. In Fig. 2 we present, for illustration, an individual event from a 100 TeV gamma ray detected by 6 stations. In the top panel the stations location is shown and three versions of core position are depicted: the true one, the one obtained by the center of mass technique, and the one found by our method. In the bottom panel the 'true' LDF for this event is plotted together with the 3-parameter fitting function. For this event a simultaneous reconstruction of the core position and the shape of the true LDF looks quite satisfactory.

Below we present the statistical analysis of a core position resolution. For that purpose instead of the lateral

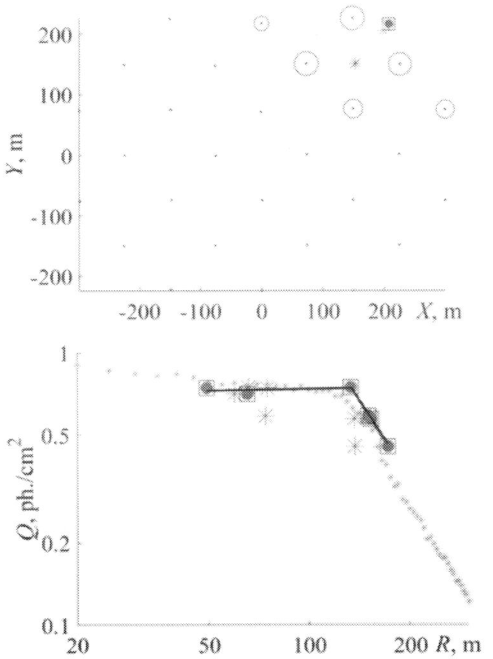

Figure 2. Example of a 100 TeV gamma ray event detected in the HiSCORE array (simulation). Top panel: grid points – 28 stations of the HiSCORE array, blue circles – hit detectors with diameter proportional to the value of Q_i in this detector, green star – true core, red star – weighted core, purple circle – core, obtained with 3-parameter fitting, square — core, obtained with 4-parameter fitting. Bottom panel: $Q(R)$, where R is calculated from different core positions: red stars – from the weighted core, purple circles – from the core, obtained with 3-parameter fitting, squares – from the core, obtained with 4-parameter fitting. Black line – 4-parameter fit, purple line – 3-parameter fit. Green circles – 'true' LDF.

distribution function, $Q(R)$, we applied the fit to the so called 'amplitude-distance function', $A(R)$ [9], which is the maximal value of the Cherenkov pulse rather than the total number of photons in the pulse. This variable is more robust and less disturbed by the noise, so it leads to better results in the core position estimation. Besides, the use of A instead of Q allows us to compare both techniques, the 'Tunka fit' and knee-like approximation, on the same kind of data, because the former is applied to $A(R)$ [9].

As we noted in Sect. 3, the 2-parameter method was found to give the best accuracy for the small number of hit detectors. Therefore, we calculated the averaged distance between the fitted core and the true core position for every number of hit detectors using this technique. The two parameters to be fitted to $A(R)$ by the least squares method are C and R_0. The other two parameters, γ_1 and γ_2, are obtained as functions of R_0 using the correlations between them, which were derived from the simulation (Fig. 3). The values of γ_1 and γ_2 as well as the correlation between them differ from those for $Q(R)$.

For the best results, the fit parameter R_0 is looked for only within the limits (constrained minimization of the mean squared error of the fit), and the limits are made

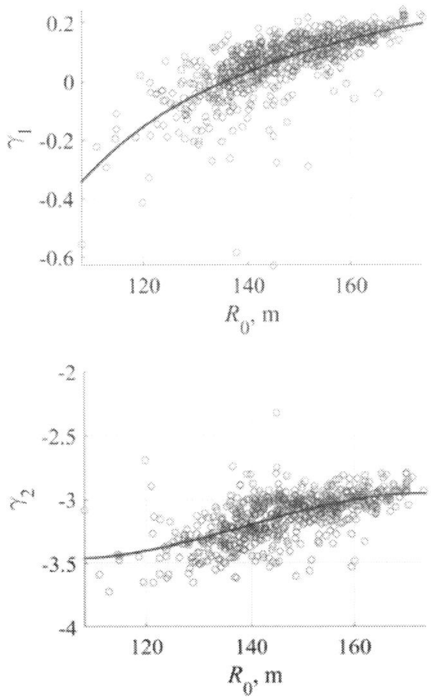

Figure 3. Correlation between parameters in the fit (2) of the amplitude-distance function: γ_1 and R_0 (top panel) and γ_2 and R_0 (bottom panel). Gamma rays at 30–3000 TeV and 28–42°.

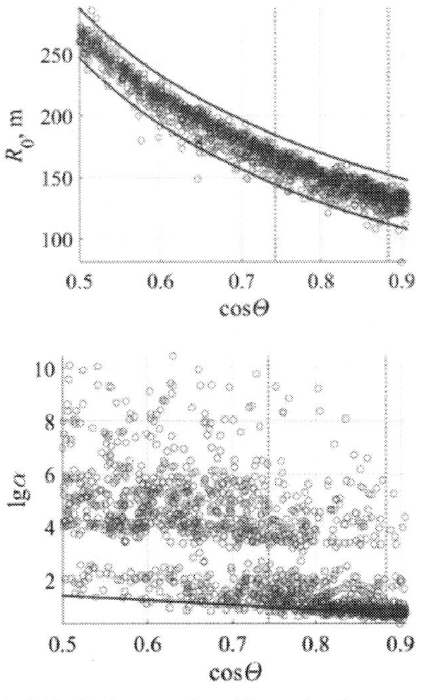

Figure 4. R_0 limits (top panel) and lgα value (bottom panel) for use in the 2-parameter knee-like fit of $A(R)$ for Gamma rays at 30–3000 TeV and 25°–60°. The dashed vertical lines are to select the angle interval 28–42° for the Crab Nebula observations.

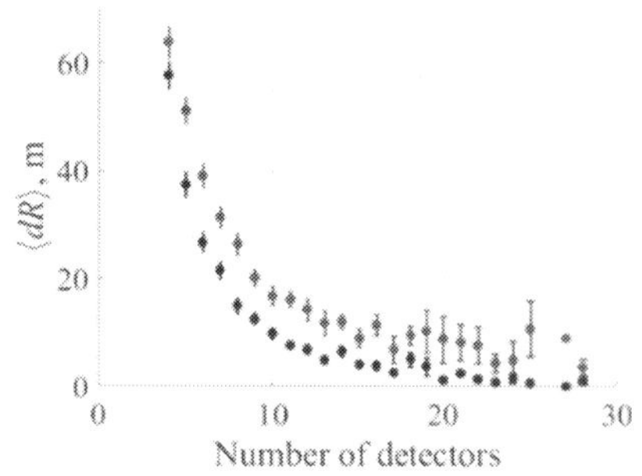

Figure 5. The averaged distance between the core position estimate and the true one for gamma events at (30–3000) TeV and 28°–42°, corresponding to the real conditions of the HiSCORE observation of the Crab Nebula signal. Blue points – knee-like fit of this article, red points – Tunka fit.

dependent on the shower angle, since the rough estimation of the angle is already performed on the previous stage (step 2 of Sect. 2). These limits are derived from the simulations and shown in Fig. 4 as well as the values of the knee sharpness parameter, α, which were also found to be slightly dependent on the shower angle.

Finally, in Fig. 5 we plot the mean accuracy $\langle dR \rangle$ of the core position determined by the 2-parameter version of the knee-like fit, which gives better accuracy for a small number of hit detectors. The figure also contains the results of the 'Tunka fit'. The averaging was performed for simulated gamma rays and not for cosmic rays (protons or nuclei), whereas the 'Tunka' fit was originally intended for cosmic ray analysis, and that is the reason for the difference in the accuracy of both techniques.

5. Conclusion

We proposed and studied a 'knee-like' 5-parameter approximation of lateral distributions of Cherenkov light emitted by extensive air showers. The knee-like approximation is capable of describing the whole diversity of individual lateral distribution functions for different nuclei and gamma rays at the shower core distance 20–500 m in the energy interval (30–3000) TeV with very good accuracy. Using correlations between different parameters of the knee-like fit we can decrease the number of parameters to 4, 3, or 2. Good quality of these versions of the knee-like fit was tested on simulated data in conditions of the real experimental setup of the TAIGA-HiSCORE project. An algorithm of a shower core reconstruction based on this fit was developed and tested on the same data.

This research confirms an ability of the knee-like fit technique to work with real data and its potential in improving the accuracy of event reconstruction in wide angle non-imaging experiments.

The study was supported by the Russian Foundation for Basic Research, project no. 16-29-13035.

References

[1] T.C. Weekes, M.F. Cawley, D.J. Fegan et al., ApJ **342**, 379–395 (1989)
[2] F. Aharonian, *Very High Energy Cosmic Gamma Radiation. A Crucial Window on the Extreme Universe* (World Scientific, 2004)
[3] P. Baillon, L. Behr, S. Danagoulian et al., Astropart. Phys. **1**, 341 (1993)
[4] A. Karle, M. Merck, R. Plaga et al., Astropart. Phys. **3**, 321 (1995)
[5] M. Tluczykont, D. Hampf, D. Horns, L.A. Kuzmichev et al., Astropart. Phys. **56**, 42 (2014)
[6] S.F. Berezhnev, N.M. Budnev, M.Buker et al., Bull. Russ. Acad. Sci. Phys. **79 3**, 348 (2015)
[7] N.M. Budnev, I.I. Astapov, A.G. Bogdanov et al. (TAIGA collaboration), JINST **9**, 09021 (2014)
[8] D. Hampf, M. Tluczykont, D. Horns, NIMA **712**, 137 (2013)
[9] V.V. Prosin, S.F. Berezhnev, N.M. Budnev et al. (TAIGA collaboration), NIMP A **756**, 94 (2014)
[10] A.A. Al-Rubaiee, Y. Al-Douri, U. Hashim, J. Astrophysics **2014**, 492814 (2014)
[11] A. Mishev, ISRN High Energy Phys. **2012**, 906358 (2012)
[12] E. Korosteleva, L. Kuzmichev, V. Prosin et al., *Proc. 28th ICRC*, 89 (Universal Academy Press, 2003)
[13] V.V. Prosin, S.F. Berezhnev, N.M. Budnev et al. (TAIGA collaboration), EPJ Web Conf. **99**, 04002 (2015)
[14] A.Sh.M. Elshoukrofy, E.B. Postnikov, E.E. Korosteleva, L.G. Sveshnikova, H.A. Motaweh, Bull. Russ. Acad. Sci. Phys. **4** (to be published, 2017)
[15] J. Hoerandel, Astropart. Phys. **199**, 193 (2003)
[16] D. Heck, J. Knapp, J.N. Capdevielle et al., *Report FZKA 6019* (Forschungszentrum Karlsruhe, 1998)
[17] A.Sh.M. Elshoukrofy, E.B. Postnikov, L.G. Sveshnikova, J. Phys.: Conf. Ser. (to be published, 2017)

Cosmic ray contribution in extragalactic background gamma-ray emission at an energy of 0.1 TeV

Anna Uryson[a]

Lebedev Physical Institute RAS, 119991 Moscow, Russia

Abstract. We estimate the intensity of gamma-quanta at the energy 0.1 TeV that is generated in extragalactic space in interactions of ultra-high energy cosmic rays with the cosmic background emission. Energy of 0.1 TeV is chosen because the Universe is mostly transparent for these quanta. In the paper three types of cosmic ray sources are analyzed: objects with red shifts up to $z = 1.1$ having monoenergetic particle spectra, $E = 10^{21}$ eV; the same objects with exponential particle spectra; objects with red shifts $0 < z \leq 0.0092$ i.e. located at distances less than ≈ 50 Mpc, also with exponential particle spectra. It is found that the cosmic ray contribution in extragalactic background emission at 0.1 TeV ranges from $f \ll 10^{-4}$ to $f \approx 0.1$ depending on the source characteristics. Thus the cosmic ray contribution in extragalactic background emission can be used for studying cosmic ray sources.

1. Introduction

The main problem of ultra-high energy (UHE) cosmic rays (CRs) is where they come from. Point UHECR sources seem to be appropriate but they are not discovered.

Information about UHECR origin is usually obtained from the particle energy spectrum. In space, particles lose energy in interactions with the cosmic microwave background. This results in a lack of particles at $E > 10^{20}$ eV at the Earth and in spectrum suppression (GZK-effect: [1,2]) if UHECRs come from distances of more than ≈ 100 Mpc. If particles arrive from nearer distances the UHECR spectrum has no GZK-suppression. Experimental data obtained with CR giant arrays – Pierre Auger Observatory and Telescope Array – indicate that the UHECR energy spectrum is suppressed. However, there may be another reason for spectrum suppression besides the GZK-effect: particles are accelerated to the maximum energy of approximately 10^{21} eV due to conditions in possible UHECR sources [3,4]. As sources are still unknown, the acceleration energy limit cannot be ruled out. So the reason for the spectrum suppression is not clear.

In addition to GZK-suppression there is another effect of UHECR interactions with the cosmic microwave background. Particles initiate electromagnetic cascades in extragalactic space [5,6]. We calculate the intensity of cascade quanta at the energy 0.1 TeV and find that it depends on the source characteristics: remoteness and initial CR energy spectra. We chose this energy because the Universe is almost transparent for such quanta [7], so their intensity is maximal.

2. Extragalactic cascades

In extragalactic space UHECRs interact with the microwave background: $p + \gamma_{rel} \longrightarrow p + \pi^0$, $p + \gamma_{rel} \longrightarrow n + \pi^+$. The cross section of reactions depends on the energies of the interacting particles. It has a peak of approximately 0.44 mb at photon energy $\epsilon^* \approx 340$ GeV in the proton rest system [8]. Thus the CR free path depends on the particle energy. For CRs at energies of about $4 \cdot 10^{19}$ eV their paths are ~ 100 Mpc, whereas particles at energies approximately 10^{20} eV have paths of only ~ 10 Mpc [9]. The energy transfer to pions is determined by the inelasticity of the interaction that is also given in [9].

Particles cascading is described in [5,6]. Generated pions decay via reactions $\pi^0 \longrightarrow 2\gamma$, $\pi^+ \longrightarrow \mu^+ + \nu$, $\mu^+ \longrightarrow e^+ + \nu + \overline{\nu}$, producing electrons and quanta which give rise to cascades.

The cascade develops through electron and quanta reactions with background emission including not only relic emission but also radio photons [10]. These reactions are pair production (PP) $\gamma + \gamma_b \longrightarrow e^+ + e^-$ and inverse Compton scattering (ICS) $e + \gamma_b \longrightarrow e' + \gamma'$ (here index b denotes background photons, index $'$ denotes scattered particles).

PP takes place provided that the photon energy is above the threshold energy

$$E_t = (mc^2)^2/\epsilon_b, \tag{1}$$

where mc^2 is the electron mass, $mc^2 = 0.511$ MeV, and ϵ_b is the energy of the background photon.

Cross sections of the PP and ICS are known [11]. For PP it is

$$\begin{aligned}\sigma_{\gamma\gamma} &\approx 3/8\sigma_T a^2/\{[2 + 2a^2 - a^4]\ln(a^{-1} + (a^{-2} - 1)^{1/2}) \\ &\quad -(1 - a^2)^{1/2}(1 + a^2)\},\end{aligned} \tag{2}$$

where $a = mc^2/E_{ec}$, E_{ec} is the photon energy in the centre-of-mass system,

$$E_{ec} = [E_e\epsilon_b(1 - \cos\psi)]^{1/2}, \tag{3}$$

[a] e-mail: `uryson@sci.lebedev.ru`

and ψ is the angle between photon impulses in the laboratory system. The cross section of the ICS at $E_e > E_t$ equals

$$\sigma_{IC} \approx 3/8\sigma_T(mc^2)^2/(E_e\epsilon_b)\{\ln[2E_e\epsilon_b(mc^2)^2] + 0.5\} \quad (4)$$

and equals the Thomson cross section, $\sigma_T \approx 6.65 \cdot 10^{-25}\,\text{cm}^2$, at $E_e < E_t$.

For electrons or quanta at lower energies, $E_e < E_t$, there is no PP, and comparatively soft quanta are produced in ICS, at mean energies

$$E_\gamma = 4/3\epsilon_b(E_e/mc^2)^2. \quad (5)$$

In this case the electron loses energy fairly slowly.

Minor interactions and the influence of magnetic fields in extragalactic space are discussed in [12].

Cosmic particles move in the expanding Universe and lose energy adiabatically. Passing the distance from a point with red shift z_i to a point with red shift z_{i+1} the particle energy decrease (ΔE_{ad}) equals:

$$(\Delta E_{ad}) = E(z_{i+1} - z_i)/(1 + z_i).$$

3. The model

Model assumptions concern three points: CR sources, background emission in extragalactic space, and the extragalactic magnetic field.

We assume that CR sources are point. There is no contradiction with results [13].

Next we assume that UHECR sources are active galactic nuclei: Blue Lacertae objects (Bl Lacs) that are powerful and remote at distances of hundreds Mpc (with red shifts up to $z = 1.1$), and Seyfert nuclei located in the radius of about 50 Mpc (with red shifts $0 < z \leq 0.0092$) which are comparatively low-power. These possible UHECR sources were analyzed in [14–17].

We obtained spatial distributions of sources using data on object red shifts from the catalogue [18]. The distributions are presented in [19].

We suppose that at $z > 0.1$ distance r is related to red shift z as

$$r = 2/3cH^{-1}(1 - (1 + z)^{-3/2})\,(\text{Mpc}) \quad (6)$$

(in accordance with the Einstein-de Sitter model with the parameter $\Omega = 1$), and as

$$r = cz/H\,(\text{Mpc}) \quad (7)$$

at lower z.

We assume a simple case, when CRs are accelerated in sources by either electrical fields, or by shock waves. In the first case the initial CR spectrum is monoenergetic, in the second it is exponential [19]. To describe UHECR energy spectra measured at the Earth the particle spectrum in Bl Lacs is either monoenergetic with energy $E = 10^{21}$ eV or exponential with the index in the range 2–3; in Seyfert nuclei the CR spectrum is exponential with the index in the range 2–3. Particles in sources are accelerated to the energy not higher than 10^{21} eV [3,4].

Hereinafter Bl Lacs are called remote sources, Seyferts are called near ones.

Thus in total we consider three types of UHECR sources: remote ones with a particle monoenergy spectrum, remote sources with the CR exponential spectrum, and near ones with the particle exponential spectrum.

We assume also that CRs consist of protons.

In the model, background emission is treated in the following way. The microwave background emission has Planck distribution in energy with a mean value $\epsilon_r = 6.7 \cdot 10^{-4}$ eV, the mean photon density is $n_r = 400\,\text{cm}^{-3}$. The high-energy "tail" consists of photons at the energy $\epsilon_t = 1 \cdot 10^{-3}$ eV, their mean density is $n_t = 42\,\text{cm}^{-3}$. The threshold energy (1) of PP and ICS equals $E_t \approx 10^{14}$ eV. Microwave background anisotropy has no influence on the cascade and is not considered.

The spectrum of the extragalactic background radiation was obtained experimentally in [20] and later theoretically in [21] allowing for the evolution of radio sources. We use results [21]: the radio background exists at the energy $\epsilon_{rad} \approx 4 \cdot 10^{-10}$ eV, so the threshold energy (1) equals $E_t \approx 6 \cdot 10^{20}$ eV.

We neglect electron losses due to synchrotron radiation in extragalactic magnetic fields.

4. Calculations

First we determine the number of gamma-quanta at the energy $E_\gamma \approx 10^{14}$ eV produced by a CR proton propagating from the source to the Galaxy by the following scheme.

Distances from UHECR sources are obtained in accordance with their spatial distribution: a source red shift, z_s, is generated by Monte-Carlo techniques, and the distance to the source is calculated. The proton mean free path in interactions with background emission being $< L >= 1/(n_{rel}\sigma_\gamma)$, the proton free path L is generated by the Monte-Carlo method.

At the point of interaction proton energy losses $\Delta(E_{ad})$ are calculated due to the expansion of the Universe. Then the energy of the CR proton after interaction is obtained. Both the cross-section of interaction and energy transfer are calculated using their dependences on ϵ^* [9]. The procedure is repeated until the proton reaches the Galaxy either its energy decreases to $E < 4 \cdot 10^{19}$ eV (because protons at this energy have paths of several hundred Mpc).

In π^0–decays quanta are produced at energy $E_\gamma = 0.5E_{\pi^0}$. In PP according to Eqs. (1–3), the angle ψ between impulses of a cascade photon and a background one is generated in the observer's frame of reference, then the photon energy E_{ec} in the centre of mass system is calculated. Next we obtain values of the cross section, $\sigma_{\gamma\gamma}$, and the free path of a cascade quanta $\lambda_{\gamma\gamma} = 1/(n_b\sigma_{\gamma\gamma})$. We chose between a radio photon and a relic one that photon whose path is shorter. Then the threshold energy E_t is derived according to Eq. (1). We suppose that one of the particles in the electron-positron pair has the energy E_t and the other one is at the energy $(E_\gamma - E_t)$. The ICS is considered in the same way using Eqs. (4, 5). The electron is treated until its energy decreases to 10^{14} eV (as at lower energies it loses energy fairly slowly scattering soft quanta).

Using this scheme we obtain the total number of gamma-quanta, N_γ, at energy $E \approx 10^{14}$ eV produced by one proton passing from a source to the Galaxy. The procedure is repeated for 1000 sources emitting one proton. Then the cascade-mean number of gamma-quanta at the energy $E \approx 10^{14}$ eV is derived.

A gamma-quantum at $E_\gamma \approx 10^{14}$ eV interacts with extragalactic background light (EBL) until its energy decreases to $E_{\gamma 1} \approx 0.1$ TeV. (The Universe is transparent for gamma-ray emission at this energy [7].) We suppose that in interactions with EBL a quantum at $E_\gamma \approx 100$ TeV produces $M_\gamma = 10^3$ quanta at the energy $E_{\gamma 1} = 0.1$ TeV, $M_\gamma = E_\gamma / E_{\gamma 1}$. As a result a CR proton propagating from the source to the Galaxy produces $N_\gamma M_\gamma$ gamma-quanta at the energy $E_{\gamma 1} \approx 0.1$ TeV.

5. Results

In the model of remote sources with the CR monoenergetic initial spectrum, $E = 10^{21}$ eV, protons on their way to the Galaxy give rise to $N_\gamma \approx 10^6$ gamma-quanta at the energy $E_\gamma \approx 100$ TeV, which in turn produce $M_\gamma = 10^9$ quanta at $E_{\gamma 1} = 0.1$ TeV.

Sources with a CR exponential initial energy spectra result in smaller values of N_γ. Protons from remote sources initiate $N_\gamma \approx 1.5 \cdot 10^3$ quanta at $E_\gamma \approx 100$ TeV along their way to the Galaxy, and produce $M_\gamma = 1.5 \cdot 10^6$ quanta at $E_{\gamma 1} = 0.1$ TeV. Near sources (which are Seyfert nuclei) result in $N_\gamma \approx 0$, $M_\gamma \approx 0$. These values of N_γ do not depend on the exponential index of the initial energy spectrum.

The intensity of gamma-ray emission produced in extragalactic cascades $I_\gamma(\geq 0.1\,\text{TeV})$ is estimated in the following way.

$$I_\gamma(\geq 0.1\,\text{TeV}) = M_\gamma I_{CR}(E > 4 \cdot 10^{19}\,\text{eV}). \quad (8)$$

Approximations of the CR integral intensity are [22]:

$$I_{CR}(> E) = 1 \cdot (E\,(\text{GeV}))^{-1.7}\,(\text{cm}^2\text{s sr})^{-1} \quad (9)$$

at $10\,\text{GeV} < E < 3 \cdot 10^6\,\text{GeV}$;

$$I_{CR}(> E) = 3 \cdot 10^{-10}(E(\text{GeV})/10^6)^{-2.1}\,(\text{cm}^2\,\text{s sr})^{-1} \quad (10)$$

at $E > 3 \cdot 10^6$ GeV. Using these approximations we get

$$I_{CR}(E > 4 \cdot 10^{19}\,\text{eV}) \approx 6 \cdot 10^{-20}\,(\text{cm}^2\text{s sr})^{-1}. \quad (11)$$

In our models the intensity of cascade gamma-ray emission is following remote and powerful sources with initial monoenergetic energy spectra, $E = 10^{21}$ eV:

$$I_\gamma(E \geq 0.1\text{TeV}) \approx 6 \cdot 10^{-11}\,(\text{cm}^2\text{s sr})^{-1} \quad (12)$$

remote sources with initial exponential energy spectra:

$$I_\gamma(E \geq 0.1\,\text{TeV}) \approx 6 \cdot 10^{-14}\,(\text{cm}^2\text{s sr})^{-1}, \quad (13)$$

near sources with the initial exponential energy spectra:

$$I_\gamma(E \geq 0.1\,\text{TeV}) \ll 6 \cdot 10^{-14}\,(\text{cm}^2\text{s sr})^{-1}. \quad (14)$$

Fermi LAT gives data on the intensity of extragalactic background gamma-ray emission [23]:

$$I_{EGB}(> 0.1\,\text{TeV}) \approx 7 \cdot 10^{-10}\,(\text{cm}^2\text{s sr})^{-1}. \quad (15)$$

Thus the fraction of cascade quanta in the intensity of extragalactic background gamma-ray emission depends on CR source parameters. At the energy $E_{\gamma 1} \approx 0.1$ TeV the fraction ranges from $f \ll 10^{-4}$ to $f \approx 0.1$.

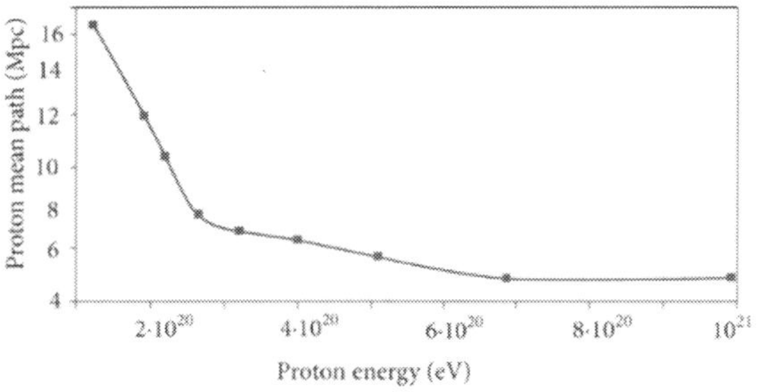

Figure 1. Proton mean path vs. proton energy in extragalactic space.

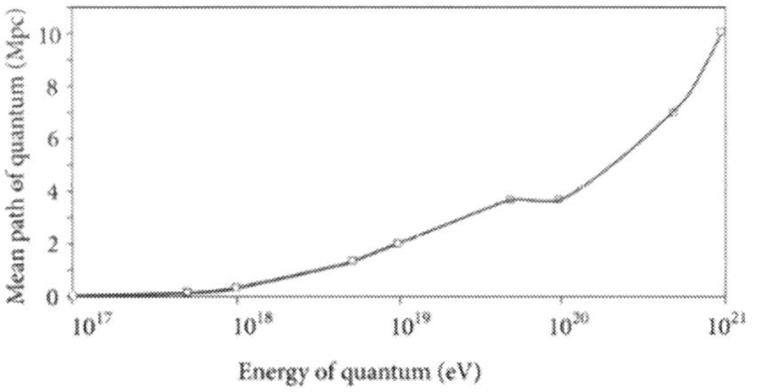

Figure 2. Quantum mean path vs. its energy in PP in extragalactic space.

6. Discussion

The variety in the number of cascade gamma-quanta N_γ arises from two reasons: resonance behavior of the cross section of $p - \gamma_{rel}$ - interaction and the initial CR spectrum. The emitted proton interacts with the microwave background radiation until its energy decreases to $4 \cdot 10^{19}$ eV. Then its free path increases to several hundred Mpc and the proton has a poor chance of interacting with background emission along the rest of its way to the Earth. As a result a proton from sources with a monoenergetic energy spectrum, the initial energy being 10^{21} eV, interacts with the cosmic microwave background about 10 times, initiating approximately 10 electromagnetic cascades. In sources with a CR initial exponential energy spectra the majority of protons are emitted at energies about $5 \cdot 10^{19}$ eV regardless of the exponential index of the spectrum. Having long free paths, protons interact with background emission 1–2 times only, thus giving rise to 1–2 cascades. In addition, distances from near sources are too short for cascade development. The proton mean path as a function of the proton energy is shown in Fig. 1.

The quantum mean path in PP as a function of quantum energy is shown in Fig. 2, and the electron mean path in ICS versus electron energy is shown in Fig. 3.

Paths do not exceed 10 Mpc for all quantum/electron energies. Quantum/electron propagation has no influence on the variety in the number, N_γ, in the models considered.

7. Conclusion

We estimated the intensity of gamma-ray emission at the energy $E_\gamma = 0.1$ TeV, produced by CR protons propagating in extragalactic space.

We analyzed three types of possible UHECR sources: remote and powerful sources located at hundreds Mpc with

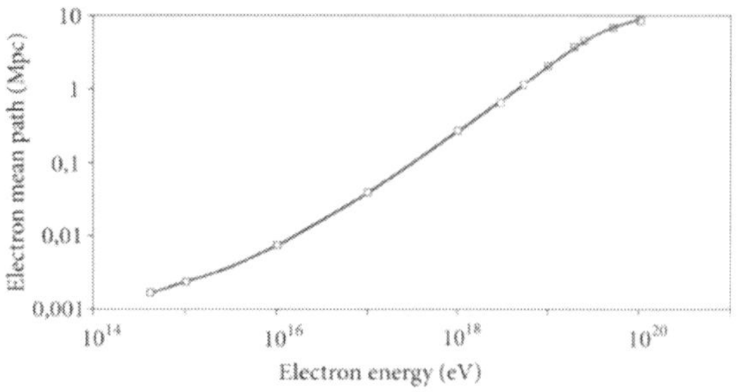

Figure 3. Electron mean path vs. its energy in ICS in extragalactic space.

red shifts up to $z = 1.1$, having an initial monoenergetic spectrum, $E = 10^{21}$ eV; remote sources but with an initial exponential energy spectra, and finally near sources located in the radius 50 Mpc, with red shifts $0 < z \leq 0.0092$, also having exponential spectra. Source parameters remoteness and initial energy spectrum – were chosen following [19].

We found that the cascade gamma-ray contribution in extragalactic background emission at energy $E_\gamma = 0.1$ TeV ranges from $f \ll 10^{-4}$ to $f \approx 0.1$ depending on the CR source parameters. In the estimation we used the Fermi LAT data [23].

References

[1] K. Greisen, Phys. Rev. Lett. **16**, 748 (1966)
[2] G.T. Zatsepin, V.A. Kuzmin, JETP. Lett. **4**, 78 (1966)
[3] M.V. Medvedev, Phys. Rev. **E67**, 045401 (2003)
[4] A.V. Uryson, Astron. Lett. **30**, 816 (2004)
[5] S. Hayakawa, Prog. Theor. Phys. **37**, 594 (1966)
[6] O. Prilutsky, I. L. Rozental, Acta Phys. Hung. Suppl. 1, **29**, 51 (1970)
[7] E. Dwek, F. Krennrich, Astroparticle Physics **43**, 112 (2013)
[8] Particle Data Group, Phys. Rev. **D 69**, 269 (2004)
[9] F. W. Stecker, Phys. Rev. Lett. **21**, 1016 (1968)
[10] V.S. Berezinsky, Physics of Atomic Nuclei **11**, 399 (1970)
[11] V.L. Ginzburg, *Theoretical Physics and Astrophysics* (Oxford, Pergamon Press, 1979)
[12] A.V. Uryson, In Proc. 34 ICRC, the Hague (2015)
[13] T. Abu-Zayyad et al., PAO and TA: Joint Contributions. In Proc. 33 ICRC, Beijing (2013)
[14] N. S. Kardashev, MNRAS **522**, 205 (1995)
[15] A.V. Uryson, JETP Letters **64**, 71 (1996)
[16] A.V. Uryson, Astron. Rep. **78**, 686 (2001)
[17] D.S. Gorbunov et. al., ApJ **577**, L93 (2002)
[18] M.P. Veron-Cetty, P. Veron, Astron. Astrophys. **412**, 399 (2003)
[19] A.V. Uryson, Physics of Particles and Nuclei. **37**, 347 (2006)
[20] T. A. Clark, L.W. Brown, J. K. Alexander, Nature **228**, 847 (1970)
[21] R. J. Protheroe, P. L. Biermann, Astropart. Phys. **6**, 45 (1996); 7, **181(E)** (1997)
[22] V.S. Berezinsky et al. Ed: V.L. Ginzburg, *Astrophysics of Cosmic Rays* (North-Holland,1990)
[23] M. Ackermann, M. Ajello, A. Albert et al. ApJ **799**, 1 (2015)

Gamma-ray families with halos: Main characteristics and possibilities of using them to estimate the p+He fraction in the mass composition of cosmic rays at energies 1–100 PeV

A.S. Borisov[1], V.G. Denisova[1], Z.M. Guseva[1], E.A. Kanevskaya[1], S.E. Pyatovsky[1,a], M.G. Kogan[1], A.E. Morozov[1], R.A. Mukhamedshin[2], V.S. Puchkov[1], S.B. Shaulov[1], G.P. Shoziyoev[1], and M.D. Smirnova[1]

[1] P.N. Lebedev Physical Institute of the Russian Academy of Sciences, Moscow, Russia
[2] Institute for Nuclear Research, Russian Academy of Sciences, Moscow, Russia

Abstract. Characteristics of γ-ray families with halos (XREC, Pamir) and data of experiments with EAS are analyzed to estimate the proton and helium (p+He) fractions in the primary cosmic radiation at $E_0 = 1$–100 PeV. It is shown that at energies $E_0 \sim 1$–100 PeV the fraction of p+He remains significant, namely, the fraction of p+He is near 40% at $E_0 = 10$ PeV.

1. XREC (Pamir) experiment

The X-ray emulsion chamber (XREC) (Pamir) (Fig. 1) represents a solid-state track chamber installed at the Pamirs at an altitude of 4400 m a.s.l. The air above the chamber is a thick target (600 g/cm^2) where nuclear-electromagnetic cascades (NEC) induced by particles of the primary cosmic radiation (PCR) take place. Single particles of the high-energy electromagnetic component (γ and $e^{\pm}$ called below for brevity "γ-rays") of the air cascades initiate electron-photon cascades (EPhC) in the upper lead plates of the chamber, assembled in a stack and called Γ-block. These EPhCs are recorded as dark spots by X-ray films placed between and under lead plates if the energy of incident γ-rays and electrons/positrons is high enough ($E_0 \geq 4$ TeV).

Processing of the experimental data was performed using the MC0 [1] and QGSJ models. The MC0 model well describes the XREC experimental data and corresponds to LHC data in the low energy region. The MC0 model designed for simulation of hadron-hadron and hadron-nucleus interactions in Extensive Air Showers (EAS) reproduces rather well the characteristics of the XREC experimental data and high-x_F LHC data.

Analysis of XREC (Pamir) experiment results using MC0 and QGSJ models shows that $\geq 90\%$ of the registered events are formed by the p+He component.

2. γ-ray families with halos in XREC (Pamir)

In γ-ray families with energies $\Sigma E_\gamma \geq 500$ TeV, γ-ray-initiated EPhCs overlap, creating regions with high optical density in the centers of these γ-ray families (Fig. 2). As follows from Fig. 3, simulations explain the nature of γ-ray families with halos in the framework of the traditional interaction models (MC0 and QGSJ).

To study the problem of halos, γ-ray families were simulated in the range of PCR particle energies $E_0 = 0.3$–3000 PeV under the following selection criteria:

- minimum energy of γ-ray families is 100 TeV;
- minimum number of particles in γ-ray families is $N_{chmin} = 3$;
- PCR particles are presented by nine groups of nuclei, namely, p, He, Li, C, O, Mg, Si, V, Fe;
- the observation level is 594 g/cm^2 for XREC (the Pamirs) and 695 g/cm^2 for EAS+XREC HADRON experiment (the Tian Shan);
- the threshold energy is 100 GeV for γ-rays and stable hadrons arriving on the top of the XREC from the air;
- the zenith angle varies from 0 to 0.9 radian.

Simulations of halos were previously done using simulated γ-ray families with the use of a Monte-Carlo program which applies a "chessboard" algorithm. A preliminary selection was made which was consistent with the criteria for processing experimental XREC data:

- only $e^{\pm}$ and γ-rays are selected;
- γ-ray families with energy $\Sigma E_\gamma \geq 100$ TeV and average distance of particles from the center R ≤ 15 cm are used.

The "chessboard" algorithm of the halos calculation applies Spatial Distribution Functions (SDF) of $e^{\pm}$ and γ-rays. The SDFs are calculated by the model simulation for XREC (Pamir) with the configuration shown in Fig. 1. The SDF considers the distance of a γ-ray (R, cm) from the cascade core, the depth of the cascade development (T, rad.unit), $E_{e^\pm}$ (MeV) and the probability that the first interaction will occur at a certain depth in XREC.

Figure 2 demonstrates examples of a scanned image of the experimental "FIANIT" halo event (left) and calculated halo formed by the primary proton ($E_p = 420$ PeV) (right) with close halo areas, S_{halo}, of 1020 mm^2 and 2100 mm^2, respectively.

[a] e-mail: `sep@lebedev.ru`

DOI: 10.1051/epjconf/201714519008

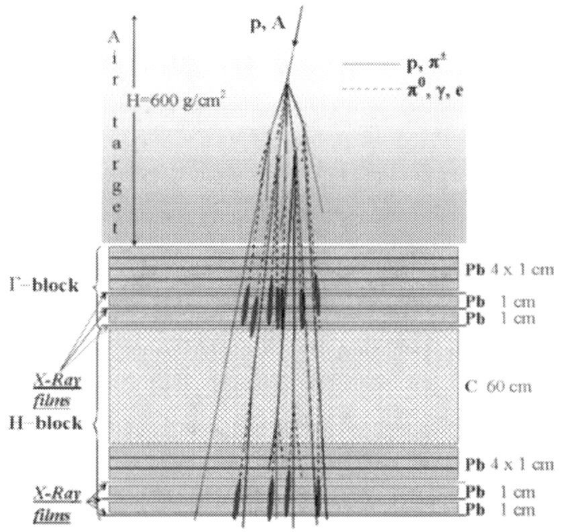

Figure 1. Layout of the Pamir experiment setup.

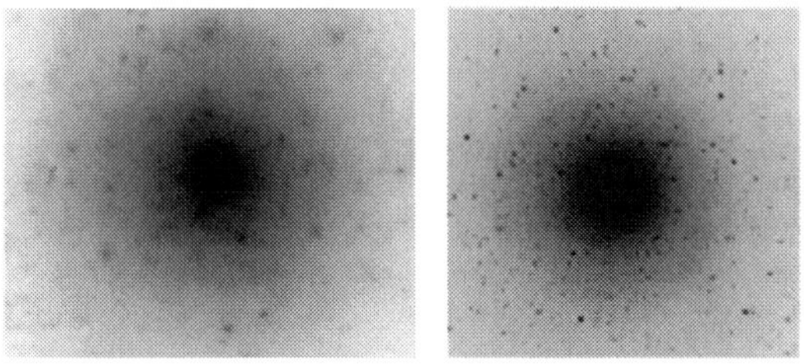

Figure 2. Scanned image of the «FIANIT »halo event ($E_0 \sim$ 400 PeV, S_{halo} = 1020 mm^2) (left); calculated halo formed by the primary proton (E_p = 420 PeV, S_{halo} = 2100 mm^2) (right).

The criteria for selection of the γ-ray families with halos in the XREC (Pamir) experiment are as follows [2,3]:

- $\Sigma E_\gamma \geq 500$ TeV;
- an area of halo $S_{D=0.5}$ bounded by isodense with optical density D = 0.5.
- $S_{D=0.5} \geq 4$ mm^2- for single-center halo;
- $\Sigma S_{D=0.5} \geq 4$ mm^2, if $S_{D=0.5} \geq 1$ mm^2 -for a multi-center halo.

Model calculations show that the γ-families with ΣE_γ = 100–400 TeV ($E_\gamma \geq 4$ TeV) are produced mainly by PCR particles (protons) with E_0 = 1–10 PeV, while the halo γ-families with $\Sigma E_\gamma \geq 500$ TeV are produced by PCR particles with $E_0 \geq 10$ PeV.

The statistics of the Pamir experiment includes 61 γ-ray families with halos at ST = 3000 m$^2\cdot$ year·sr.

Calculations have shown that γ-ray families with halo are created as a result of the overlapping of SDFs of EPhC initiated by $e^{\pm}$ and γ-rays. Therefore, the halo is not some exotic phenomenon but the central part of a high energy EAS.

The analysis of γ-ray families with halos allows us to apply these events for estimating the PCR mass composition because the halo phenomenon observed in XREC experiments is predominantly formed by protons and helium nuclei ($\geq$ 90%). Its origin is in the SDF overlapping of EPhC initiated by $e^{\pm}$ and γ-rays and is almost independent of the simulation model.

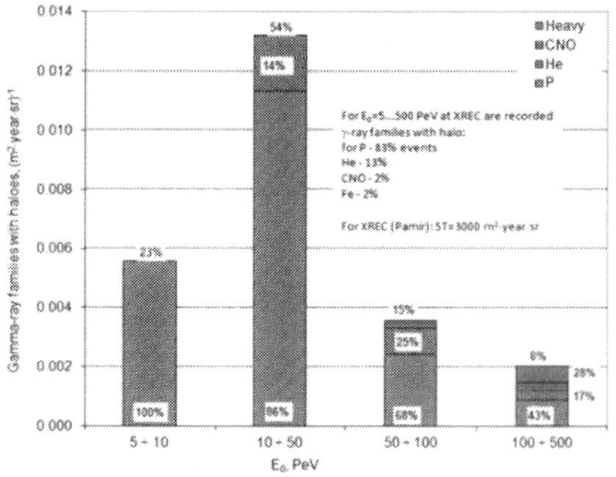

Figure 3. Dependence of the intensity of γ-ray families with halos and relative fractions of different PCR nuclei producing these families on E_0.

Table 1. The mass composition in the MC0 model.

E PeV	1	10	100
p%	33	26	20
He%	22	17	15

3. Dependence on E$_0$ of the fraction of γ-ray families with halos

The dominant part of the halo-containing γ-ray families is formed by protons (Fig. 3). Observation of a large number of such events indicates a notable presence of protons in the PCR spectrum up to $E_0 \sim 100$ PeV.

Model calculations show that when the PCR consists of protons only, the number of observed halos should be equal to 140 instead of 61 registered by the Pamir experiment, 34 and 5 in the cases for pure helium and heavy nucleus spectra, respectively. It follows from the above that the fractions of PCR protons and helium nuclei are higher than 15% and 24%, respectively. The summary p+He fraction is higher than 39% at E_0 = 10 PeV.

The mass composition in MC0 model is presented in Table 1. Calculations revealed that 83% of all events are produced by primary protons, 13% are produced by helium nuclei and not more than 4% are produced by heavier nuclei. This conclusion is almost independent of the model and thus provides a possibility to estimate the proton fraction in PCR in the range of E_0 = 1–100 PeV.

4. Fractions of γ-ray families with multicenter halos generated by different nuclei

In some cases the halo consists of several centers. The multicenter of a halo is a parameter which strongly depends on the type of primary nuclei (Table 2).

It is evident that recorded γ-ray families with one halo are almost entirely generated by primary protons with a possible small addition of He. This conclusion

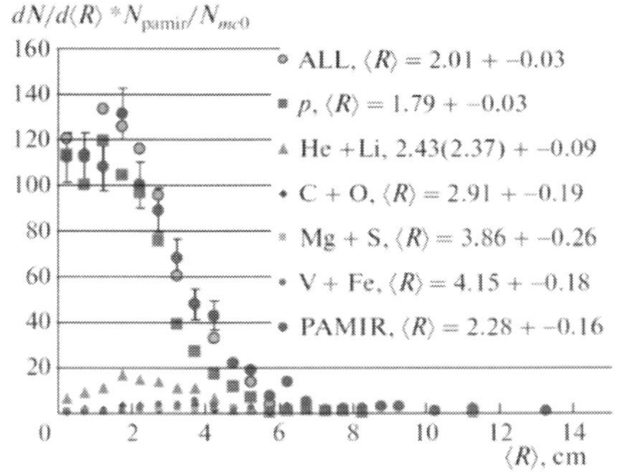

Figure 4. Distribution over average size of the γ-ray families, dN/d⟨R⟩. Simulated distributions for MC0 are normalized to the number of experimental γ-ray families observed in the XREC (Pamir) experiment.

Table 2. Fraction of γ-ray families with multicenter halos generated by different nuclei.

p	He	C	Fe	MC0	PAMIR
0.25 $\pm$0.03	0.45 $\pm$0.09	0.59 $\pm$0.11	0.70 $\pm$0.03	0.28 $\pm$0.03	0.23 $\pm$0.07

Table 3. The probabilities of halo appearance.

p	He	>He
1.76%	0.44%	0.13%

is confirmed by the probabilities of halo appearance (Table 3).

MC0-based calculations show that at $E_0 > 5$ PeV the efficiency of formation of the halo-containing γ-ray families by protons is four times higher than by He nuclei (Table 3). This explains the preferential registration with XREC of such γ-ray families produced by protons.

5. Average size of γ-ray families in the XREC experiment

According to calculations [2,3], the average size ⟨R⟩ of the γ-ray families formed by protons is indeed 1.8 cm; ⟨R⟩ = 4.5 cm for events formed by iron nuclei. Experimental and calculated ⟨R⟩ distributions are shown in Fig. 4. γ-ray families with energy $\Sigma E_\gamma = 100$–400 TeV, where reliable identification of all cascades is possible, were used to construct these distributions. At high family energies, a halo can occur in the central part of an event; single low-energy cascades, initiated by γ-rays in lead, can be lost inside these halos, and the determination of ⟨R⟩ can be complicated. It can be seen from Fig. 4 that the experimental ⟨R⟩ distribution is well described by the set of spatial distributions of events formed by primary protons. The average size of γ-ray families in the XREC of the Pamir experiment is $\langle R_{Pamir} \rangle = 2.28$ cm, which is somewhat higher than that predicted by the MC0 model, i.e., $\langle R_{MC0} \rangle = 2.01$ cm, and virtually coincides with the average size of events formed by He, $\langle R_{He} \rangle = 2.37$ cm (Fig. 4).

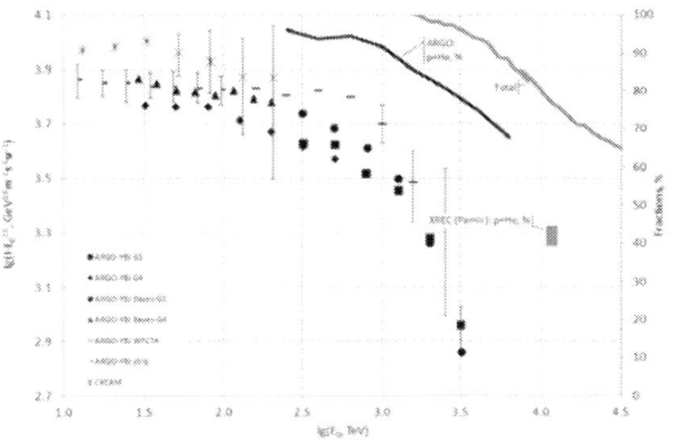

Figure 5. The p+He fraction according to the ARGO, CREAM and XREC (Pamir) experiments depending on E_0.

An analysis of XREC response shows that the spatial distribution of cascades in the γ-ray families as it passes through XREC increases by 15%. The overestimation of ⟨R⟩ in the experiment is associated with the registration due to fluctuations of a large number of families with energies close to the threshold energy ($\Sigma E_{\gamma thr} = 100$ TeV). Such families have a broader spatial distributions. The true size of experimental γ-ray families is $\langle R_{Pamir} \rangle = (1.94 \pm 0.06)$ cm, which proves the above made conclusion on the leading role of primary protons in the formation of γ-ray families registered in the XREC.

6. Proton+helium fraction at $E_0 = 1$–100 PeV according to the ARGO, CREAM and XREC (Pamir) experiments

The p+He fraction at $E_0 = 1$–100 PeV according to the ARGO, CREAM [4,5] and XREC (Pamir) experiments depending on E_0 is shown in Fig. 5.

The estimate of the p+He fraction found in the XREC (Pamir) experiment is based on the analysis of the flow of γ-ray families with halos and known probabilities of formation of halos by protons, helium and heavier nuclei.

The experimental number of the γ-ray families with halos (61) is compared with the expected number of halos created by protons, helium and heavier nuclei. It is shown, that if the PCR would consist of only protons, the number of halos must be equal to 140. In the case of dominance of helium or heavier nuclei, the number of halos must be equal to 34 or 5, respectively. Thus we conclude there is a significant fraction of protons in the PCR on the basis of the experimental number of γ-ray families with halo.

To explain the observed number of γ-ray families with halos in the XREC (Pamir) experiment, the p+He fraction should be higher than 39%.

Conclusions

1. Our calculation explains the nature of γ-ray families with halos in the framework of a traditional model of hadron interactions, which gives a rather good description of the XREC data.

2. According to the XREC (Pamir) and HADRON (Tian Shan) experiments the p+He component is noticeably present at $E_0 = 1\text{–}100$ PeV.
3. Fractions of protons and He nuclei in the PCR mass composition at $E_0 = 10$ PeV are as follows:
   - protons - higher than 15%
   - p+He - higher than 39%.

References

[1] G.F. Fedorova, R. A. Mukhamedshin. Bull. Soc. Sci. Lett. Lodz, Ser. Rech. Def., XVI, 137 (1994)

[2] V.S. Puchkov, A.S. Borisov, Z.M. Guseva, V.G. Denisova, E.A. Kanevskaya, M.G. Kogan, V.M. Maximenko, A.E. Morozov, R.A. Mukhamedshin, S.E. Pyatovsky, M.D. Smirnova. Bulletin of the Russian Academy of Sciences Physics, 2011, **75**(3), pp. 392–394

[3] V.S. Puchkov, A.S. Borisov, Z.M. Guseva, V.G. Denisova, E.A. Kanevskaya, M.G. Kogan, V.M. Maximenko, A.E. Morozov, R.A. Mukhamedshin, S.E. Pyatovsky, M.D. Smirnova. 32nd International Cosmic Ray Conference, Beijing 2011, HE1.2

[4] Zhang S.S., Cao Z. 34th International Cosmic Ray Conference, 30 July – 6 August – 2015 – Hague, Netherlands. PoS (ICRC2015) 261

[5] I. De Mitri on behalf of the ARGO-YBJ collaboration. EPJ Web of Conferences **99**, 08003 (2015). – doi: 10.1051/epjconf/20159908003

Study of the peculiarities of multiparticle production via event-by-event analysis in asymmetric nucleus-nucleus interactions

Anastasiya Fedosimova, Adigam Gaitinov, Ekaterina Grushevskaya, and Igor Lebedev[a]

Institute of Physics and Technology, Almaty, Kazakhstan

Abstract. In this work the study on the peculiarities of multiparticle production in interactions of asymmetric nuclei to search for unusual features of such interactions, is performed. A research of long-range and short-range multiparticle correlations in the pseudorapidity distribution of secondary particles on the basis of analysis of individual interactions of nuclei of ^{197}Au at energy 10.7 AGeV with photoemulsion nuclei, is carried out. Events with long-range multiparticle correlations (LC), short-range multiparticle correlations (SC) and mixed type (MT) in pseudorapidity distribution of secondary particles, are selected by the Hurst method in accordance with Hurst curve behavior. These types have significantly different characteristics. At first, they have different fragmentation parameters. Events of LC type are processes of full destruction of the projectile nucleus, in which multicharge fragments are absent. In events of mixed type several multicharge fragments of projectile nucleus are discovered. Secondly, these two types have significantly different multiplicity distribution. The mean multiplicity of LC type events is significantly more than in mixed type events. On the basis of research of the dependence of multiplicity versus target-nuclei fragments number for events of various types it is revealed, that the most considerable multiparticle correlations are observed in interactions of the mixed type, which correspond to the central collisions of gold nuclei and nuclei of CNO-group, i.e. nuclei with strongly asymmetric volume, nuclear mass, charge, etc. Such events are characterised by full destruction of the target-nucleus and the disintegration of the projectile-nucleus on several multi-charged fragments.

1. Introduction

Cosmic rays,which are measured at ground stations, are a natural source of high-energy particles. The primary cosmic particle enters the Earth's atmosphere, in which numerous nuclear and electromagnetic interactions take place giving rise to a cascade of secondary particles. To determine the primary cosmic particle parameters, a complex inverse task must be solved, because a lot of interactions happen before the observation level. The most correct way for a solution of the problem is by simulating the shower. For the simulation the characteristics of the elementary act (interaction cross section, inelasticity, multiplicity of the process, etc.) from data obtained at accelerators, are used. Therefore, to understand the features of the development of cascade processes, detailed information about the fluctuations of distributions of secondary particles in different types of nucleus-nucleus interactions in the elementary act, is required [1].

Anomalous events (high multiplicity, large transverse momenta, large multiparticle correlations in pseudorapidity distributions of secondary particles, etc.) can significantly change the dynamics of the cascade process. For the research of such peculiarities in multiparticle production, different approaches and methods are used [2–6].

[a] e-mail: `lebedev692007@yandex.ru; lebedev@sci.kz`

A detailed event-by-event analysis, taking into account possible kinematic distinctions, can provide more information on the dynamics of the process of nucleus-nucleus interactions [7,8]. Analyzing in detail the characteristics of each individual event, background fluctuations can be separated from experimental peculiarities which must be taking into account to describe the development of cascade processes on the basis of interaction models.

In this work the study on peculiarities of multiparticle production in interactions of asymmetric nuclei to search for unusual features of such interactions, is performed.

In addition to the analysis of standard distributions (multiplicity, pseudorapidity, etc.), the existence of correlated groups of secondary particles and their dislocation in pseudorapidity distributions was analyzed.

For analysis of the correlated groups, Hurst's method was used [9,10]. To study the pseudorapidity correlations we analyzed the behaviour of the Hurst curves, which are dependent on the normalized range, R/S (where R and S are the range and the standard deviation), versus the size of the pseudorapidity interval, $d\eta$, using the function $R/S = (a \cdot d\eta)^{\mathrm{h}}$, where a is a free parameter and h is the Hurst index. The analysis of the behaviour of the Hurst curve allows to estimate the "force" and "length" of multiparticle correlations in pseudorapidity distributions of secondary particles and to distinguish fluctuations of dynamically correlated distributions of secondary particles from stochastic, connected with statistical reasons [4].

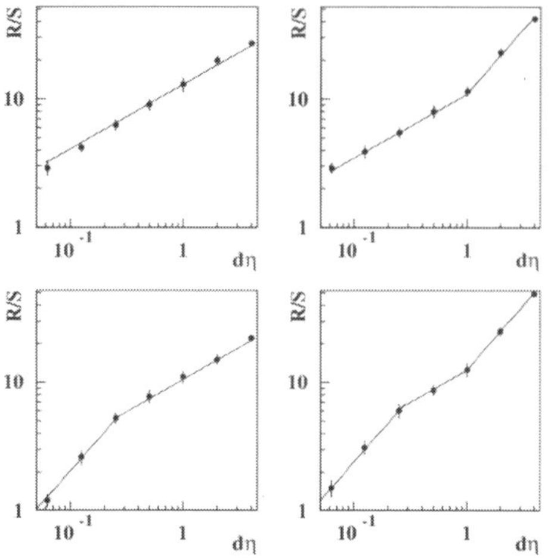

Figure 1. Behaviour of the Hurst curve for four typical events of Au+Em 10.7 AGeV.

In the present paper we analyzed experimental pseudorapidity distributions of secondary particles obtained in 10.7 A GeV ^{197} Au interactions with emulsion nuclei [11].

Nuclear photoemulsion experiments in comparison with other approaches used for investigating nuclei interactions, is one of the most informative. At first, it has high spatial resolution. Secondly nuclear photoemulsions allow us to observe an interaction in 4π geometry of the experiment. Most other methods have essentially dead zones in which secondary particles are not registered.

Besides, the method of nuclear emulsion allows to define the charges of fragments and provides a chance to register rather small excited nucleus-targets, and also has no energy threshold for the registration of the fragments of a nucleus-projectile.

Moreover, the emulsion technique allows us to investigate parameters of fragmentation and multiparticle processes in the same interaction.

2. Pseudorapidity correlations

Based on a detailed analysis of individual interactions all events are divided into four types. Characteristic events of various types are shown in Fig. 1.

The first type is characterized by the linear behavior of the Hurst curve with $h \sim 0.5$ (Fig. 1 left-upper). It corresponds to an uncorrelated pseudorapidity distribution of secondary particles. Events of such a type are well described on the basis of simulating with a stochastic pseudorapidity distribution.

Events of the second type are characterised by the Hurst index $h > 0.64$ in the field of small pseudorapidity intervals and $h \sim 0.5$ at other $d\eta$ (Fig. 1 left-lower). Such behaviour of a correlation curve can be initiated by short-range multiparticle correlations. It corresponds to processes of jet type.

Events with a change of the Hurst index to $h > 0.64$ in the field of large values of pseudorapidity intervals have been related to the third type (Fig. 1 right-upper). Such behaviour of the correlation curve corresponds to the essential display of long-range multiparticle correlations and such events are referred to processes of the explosive type.

Events with $h > 0.64$ at both small values and at large values of the pseudorapidity interval, but with $h \sim 0.5$ in the middle region of $d\eta$, have been referred to the fourth type (Fig. 1 right-lower). Such behaviour of the Hurst curve corresponds to events of the mixed type, including processes of the explosive and jet types.

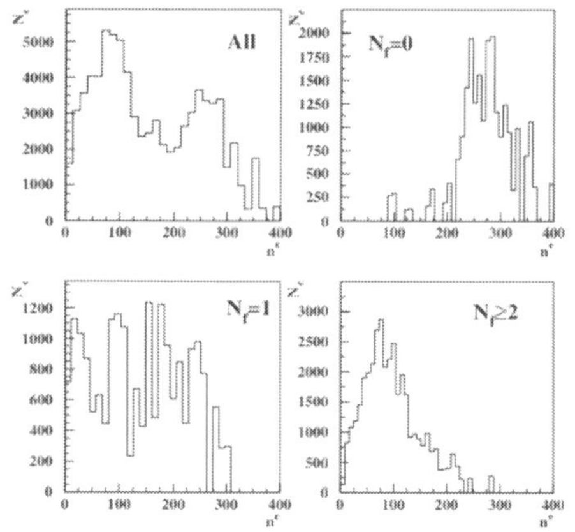

Figure 2. Multiplicative n^e-distributions for events with different numbers of multi-charge fragments in Au+Em 10.7 AGeV.

3. Fragmentation of the projectile nucleus

Detailed analysis has shown that events of different types had different fragmentation parameters.

Events of the explosive type are processes of full destruction of the projectile nucleus, in which multi-charge fragments are absent. Events with several multi-charge fragments of the projectile nucleus give Hurst curves corresponding processes of the mixed type. In uncorrelated events and in processes of the jet type only one multi-charge fragment is discovered.

In addition, interesting results follow from a joint analysis of the fragmentation parameters and multiplicity distributions, which are presented in Fig. 2.

In order to reveal a deposit of low-probability (but very important) events with large multiplicity, in Fig. 2 a multiplicative distribution is presented. In this approach a weight of each event is equal to the number of particles in the event, i.e. $N_c = \sum n^e$, where n^e is the multiplicity of an individual event.

As seen in Fig. 2, the multiplicity distribution has two clear humps (Fig. 2 left-upper).

Events of the explosive type, which are characterised by the complete absence of multi-charge fragments $N_f = 0$ (Fig. 2 right-upper), have large multiplicity. They form the hump with average multiplicity $< n^e > \sim 272$.

Mixed type events, which have several multi-charge fragments, give their main deposit to the hump with $< n^e > \sim 97$ (Fig. 2 right-lower).

Events with $N_f = 1$, which correspond to uncorrelated events and processes of the jet type, have different multiplicities in wide intervals without clearly expressed peculiarities (Fig. 2 left-lower).

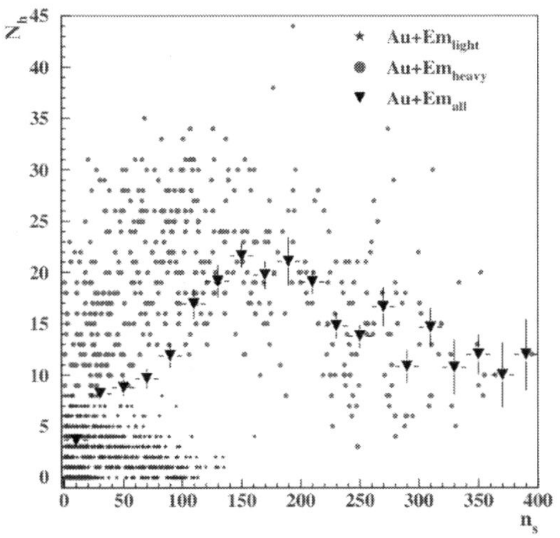

Figure 3. The distribution of events with various target-nucleus fragment number, N_h, versus multiplicity, n_s, for individual events of $Au + Em$ at 10.7 AGeV.

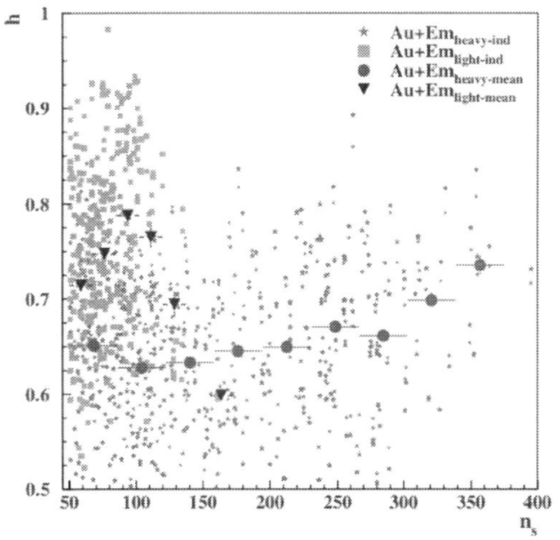

Figure 4. The distribution of events with various values of the Hurst index versus the multiplicity, n_s, for individual events of interactions of nuclei of ^{197}Au at 10.7 AGeV with heavy and light photo-emulsion nuclei.

4. Interactions with heavy and light nuclei

In the considered experiments, the EMU01-collaboration (^{197}Au + Em 10.7 AGeV) used standard nuclear emulsions of type BR-2. It includes hydrogen (39.2%), CNO-group nuclei (35.3%) and nuclei of AgBr (25.5%).

To separate events, which is connected with different asymmetries of interactions, we have analysed the dependence of the nucleus-target fragments number, N_h, versus multiplicity, n_s, for individual events of interactions of ^{197}Au with energy 10.7 AGeV with light and heavy nuclei of photoemulsion, which are shown in Fig. 3.

As follows from Fig. 3, events of interactions with light and heavy photoemulsion nuclei are separated well enough. For interactions with light nuclei the choice is limited by two conditions.

First, the maximum number of fragments of the target-nucleus cannot exceed 8, which corresponds to the largest charge of light photoemulsion nucleus - to oxygen nucleus.

Secondly, the maximum multiplicity, n_s, in interactions with light photoemulsion nuclei is much lower in comparison with interactions with heavy photoemulsion nuclei. Use of these facts allows us to separate heavy events, which have large multiplicities and numbers of target-nucleus fragments less than 8, from light events.

On the basis of the received separation, the dependence of the multiplicity, n_s, and the Hurst index, h, for individual events of interactions of ^{197}Au 10.7 AGeV with heavy and light photoemulsion nuclei, is shown in Fig. 4.

The mean value of the Hurst index for interactions of gold nuclei with light photoemulsion nuclei represents a peak behaviour with a maximum in area of $n_s = 100$, where events with the highest values of h are observed.

Hence, the most considerable multiparticle pseudorapidity correlations are discovered in the central interactions of heavy nuclei of gold and CNO-group nuclei, i.e. nuclei with strongly asymmetric volume, nuclear mass, charge, etc.

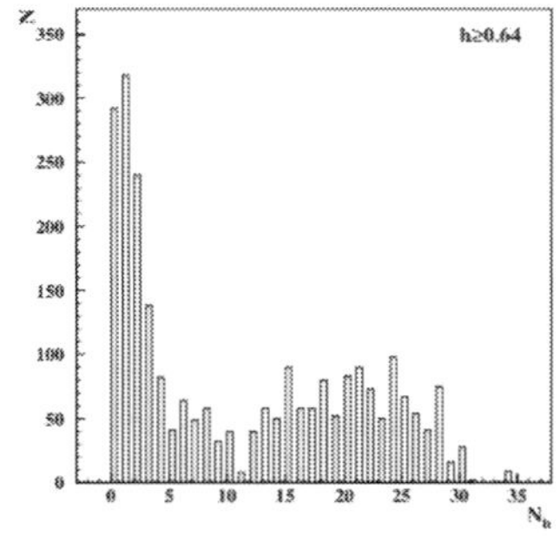

Figure 5. N_h-distribution of target-nuclei fragments for events of Au+Em 10.7 AGeV with Hurst index $h \geq 0.64$.

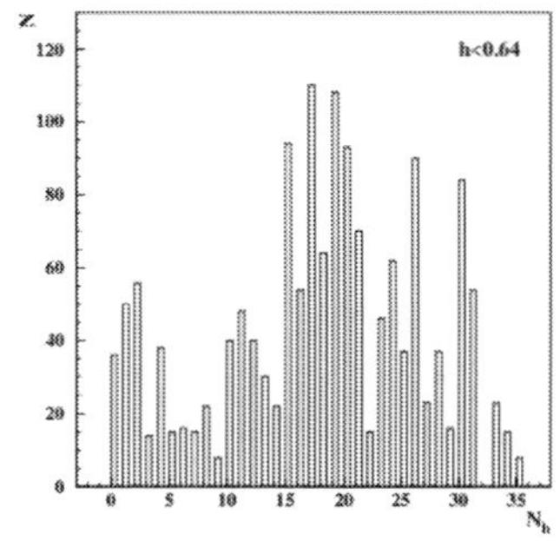

Figure 6. N_h-distribution of target-nuclei fragments for events of Au+Em 10.7 AGeV with Hurst index $h < 0.64$.

5. Fragments of the target nucleus

Essential differences for events with $h < 0.64$ and $h \geq 0.64$ are found in distributions of target nucleus fragments. The N_h distributions are shown in Figs. 5 and 6.

As seen in Fig. 5, the peak in the field of small values of N_h is observed for correlation events. For events with $h < 0.64$ the distribution maximum is located in average area of change of N_h (Fig. 6).

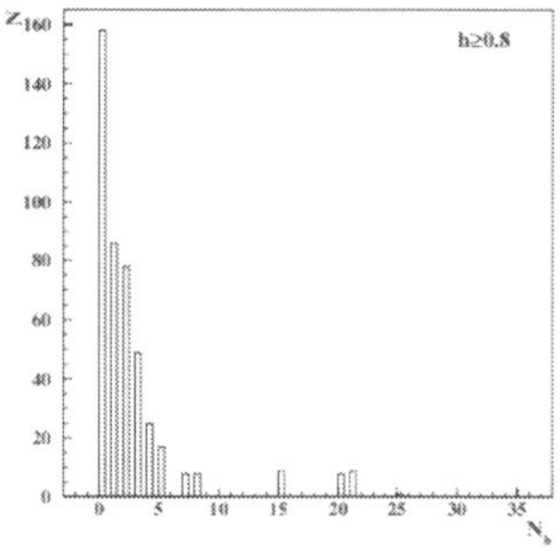

Figure 7. N_h-distribution of target-nuclei fragments for events of Au+Em 10.7 AGeV with Hurst index $h \geq 0.8$.

This effect is amplified if we consider more correlated events, i.e. by increasing the selection criterion.

The N_h-distribution for the events, selected by the criterion $h \geq 0.8$, is presented as a demonstration of the observation in Fig. 7.

As seen in Fig. 7, the majority of events with considerable multipartical correlations have in the final distribution full destruction of the target-nucleus.

6. Conclusion

Investigations of the density structure of pseudorapidity fluctuations in interactions of Au 10.7 AGeV nuclei with photo-emulsion nuclei by the Hurst method, have been carried out. As a result, events with long-range multiparticle correlations, short-range multiparticle correlations and mixed type in pseudorapidity distribution of secondary particles are discovered.

Detailed analysis showed, that events of different types had different fragmentation parameters and specific multiplicity distributions.

Based on the research of the dependence of multiplicity versus target-nuclei fragment numbers for events of various types it is revealed, that the most considerable multiparticle correlations are observed in interactions of the mixed type, which correspond to central collisions of gold nuclei and nuclei of CNO-group, i.e. nuclei with strongly asymmetric volume, nuclear mass, charge, etc. Such events are characterised by the full destruction of the target-nucleus and the disintegration of the projectile-nucleus into several multi-charged fragments.

This work was supported by grant N4824/GF4 of Ministry of Education and Science of Kazakhstan Republic.

References

[1] A. Haungs, Phys. Procedia **61**, 425–434 (2015)
[2] M.I. Adamovich et al., *Azimutal correlations of secondary particles in ^{32}S induced interactions with Ag(Br) nuclei at 4.5 GeV/c/nucleon* Part. and Nucl. Lett. **101**, 76 (2000)
[3] M.I. Adamovich et al., *Factorial Moments of ^{28}Si Induced Interactions with Ag(Br) Nuclei* APH N.S. Heavy Ion Phys. **113**, 213 (2001)
[4] I.A. Lebedev et al., J. Phys. G **23**, 637 (1997)
[5] T.N. Kvochkina et al., J. Phys. G **26**, 35 (2000)
[6] A.I. Fedosimova et al., J. Phys.: Conf. Ser. **668**, 012067 (2016)
[7] S. Bhattacharya, M. Haiduc, A.T. Neagub, E. Firub, Phys. Lett. B **726**, 194–205 (2013)
[8] T.K. Nayak, Int. J. Mod. Phys. E **16**, 3303–3322 (2008)
[9] H.E. Hurst, R.P. Black, Y.M. Simaika, *Long-Term Storage: An Experimental Study* (Constable, London, 1965)
[10] J. Feder *Fractals* (Plenum Press, New York, 1988)
[11] M.I. Adamovich et al. Eur. Phys. J. A **5**, 429–440 (1999)

Cherenkov telescope prototype

Lev Timofeev[a] and Anatoly Ivanov[b]

The Yu.G. Shafer Institute of Cosmophysical Research and Aeronomy, Yakutsk 677980, Russia

Abstract. This report presents a prototype Cherenkov telescope, the observation method of Cherenkov light from extensive air showers (EAS) generated by cosmic rays (CRs) above 10^{16} eV and preliminary observations. The interest in Cherenkov light differential detectors of EAS is caused by the possibility to measure the depth of the cascade maximum, X_{max}, and/or the shower age via angular and temporal distributions of the Cherenkov signal. In particular, it was shown, using EAS model simulations, that the pulse width measured at the periphery of the shower, r > 300 m, at sea level is pronouncedly connected with X_{max}. The Cherenkov detector is a wide-angle telescope working in coincidence with scintillation detectors, integral and differential Cherenkov detectors at the Yakutsk complex EAS.

1. Introduction

In conjunction with the Yakutsk Extensive Air Shower (EAS) modernization program we have developed a wide field of view Cherenkov telescope (Fig. 1), working in coincidence with scintillation detectors of the array. Currently, a prototype telescope with a photomultiplier (PMT) placed at the focus length of the spherical mirror is undergoing field tests. In this article the parameters and prototype telescope design are described.

The aim of the modernization program of the Yakutsk Extensive Air Shower array is to create an astrophysical instrument capable of measuring in detail the parameters of high energy galactic cosmic rays (CRs), their sources, the energy spectrum and the mass composition. The scheme of the Yakutsk EAS array can be found in Fig. 2. Another goal is to study the transition region between galactic and extragalactic CR components, in which some irregularities in the spectrum and composition can be identified [1]. A crucial role played by the Cherenkov telescope array in this program should be to observe the light emitted by the extensive air showers for measuring the angular and temporal signal structures associated with the cascade curve in showers with energies above $E = 10^{15}$ eV [2].

Our intention to develop a Cherenkov telescope functioning as a differential detector of EAS was motivated by the possibility to measure the depth of the cascade maximum and/or the shower age via the angular and temporal distributions of the Cherenkov signal [3]. Combining X_{max} and the shower age with other characteristics measured with the surface detectors of the EAS array, e.g. the energy and muon content, one is able to estimate the average mass composition of CRs. Experimental arguments in elucidating the origin of the knee and ankle in the CR spectrum will significantly strengthen due to the measurements of the angular and temporal distributions of the Cherenkov signal in the energy range above 10^{15} eV. Existing scenarios of CR acceleration in the sources are different in the expected model composition around the knee and in the transition region between galactic and extragalactic components [4]. So an accurate estimation of the average mass of the CR nuclei in addition to improved measurements of the sharpness of the knee and ankle should allow us to discriminate some scenarios.

[a] e-mail: timofeevlev@ikfia.ysn.ru
[b] e-mail: aivanov@ikfia.ysn.ru

2. Operation algorithm

The PMT in such a prototype telescope is used as the observing camera and placed at the focal length in front of a spherical mirror, see Fig. 1. There are two types of reading out the signal from position-sensitive PMTs: multi-anode and crossed wire anode [5]. Preferences for the former are the uniformity of the output signals and unique location on the anode area. Crossed wire anode signals are less uniform, and location of the signal is not unambiguous in the case of a smeared signal. Nevertheless, we have preferred Hamamatsu R2486 PMT with crossed wire anode for our telescope due to low price and small number of output channels needed, see Fig. 3. Indeed, in this case our data acquisition system (DAS) can have 32 independent channels instead of 256 in the multi-anode alternative. In order to reconstruct the signal distribution over the anode surface, one has a system of 32 equations

$$\sum_{j=1}^{16} V_{ij} = q_i^x, i = 1, .., 16, \quad (1)$$

$$\sum_{i=1}^{16} V_{ij} = q_j^y, j = 1, .., 16, \quad (2)$$

where V_{ij} is the unknown signal in a grid knot; q_i^x, q_j^y are the output signals of wires. The equations are dependent because of $\sum q_i^x = \sum q_i^y$. If the light source induces signals in more than one $x-$ and $y-$ wire, then there is an infinite number of solutions of the system. Only in the case of a symmetric signal with a single maximum is there

DOI: 10.1051/epjconf/201714519010

Figure 1. Wide field of view Cherenkov telescope prototype.

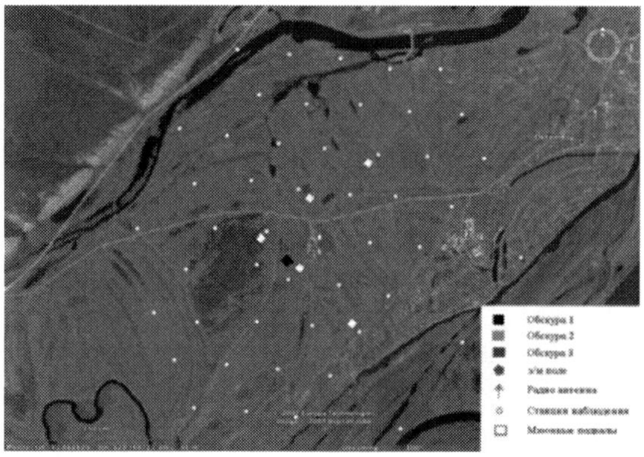

Figure 2. The Yakutsk Extensive Air Shower array scheme.

a possibility to locate the position of the maximum on the anode surface.

We composed a particular solution of the system as:

$$V_{ij} = q_i^x q_j^y / (\Sigma q_i^x \Sigma q_j^y) \qquad (3)$$

to locate the position of the signal maximum on the anode surface.

Since the radius of the photocathode R2486 equals 3 cm, the mirror diameter was chosen to be 26 cm with a radius of curvature of 22.5 cm. Mirror parameters have been chosen from considerations to ensure as wide a field of view as possible, but so that possible distortions do not exceed the pixel size of 3.8 mm. As a result, the width of the field of view is $-14° \leq \alpha \leq 14°$.

In the design chosen, the telescope provides an effective aperture $D_{eff}(0^0) = 21.8$ cm due to shadowing of the mirror by the PMT and the support. The telescope aperture is given in Fig. 4. We calculated the angular dependence as a ratio of the light intensity on the photocathode surface to the initial intensity falling into the actual aperture of the telescope, taking into account the reflectance of aluminum, 92.4%, in the PMT sensitivity interval $\lambda \in (300, 600)$ nm.

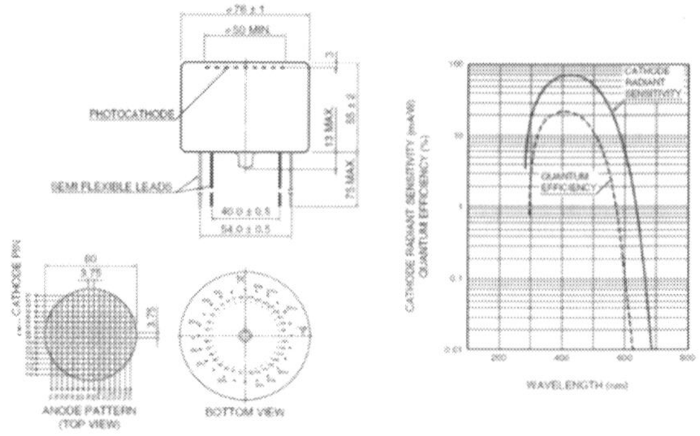

Figure 3. The photo multiplier Hamamatsu R2486 with coordinate-sensitive anode consisting of 16×16 crossed wires has been chosen for the Cherenkov telescope prototype.

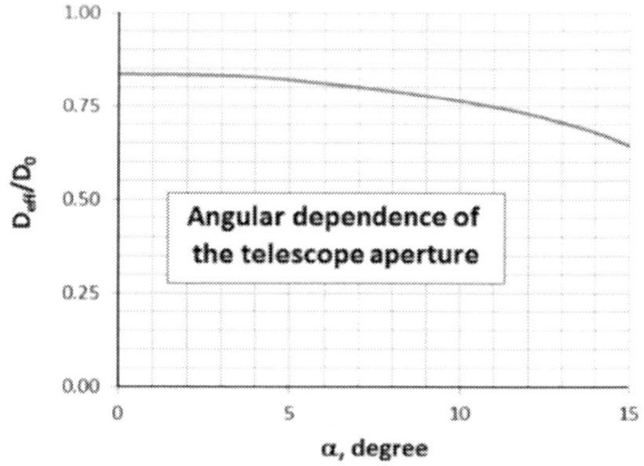

Figure 4. Shadowing of the mirror by the PMT and support + cables. The ratio of effective mirror diameter to the unshielded diameter as a function of incidence angle, α.

The signals from each channel of the multi-anode PMTs are continuously input to preamplifiers and onwards to the Analogue to Digital converters and stored in the buffer memory of a 32-channel industrial computer [6]. On the arrival of a trigger generated by the EAS array, an industrial computer saves the accumulated pre-history as a separate file labeled with the time mark and the assignment of a serial number.

We carried out the photomultiplier sensitivity check from plastic scintillator in the voltage range from 900 to 1300 volts and obtained a coordinate-dependent sensitivity of the Photo Multiplier Tube from which we obtained a radial non-uniformity of sensitivity and preamplifiers transition ratio as a function of input signal voltage, see Fig. 5. We also measured the angular dependence of the telescope, the energy spectrum of telescope channels, and found that the Photo Multiplier Tube dark current is minimal at a voltage less than 1300 volts. Currently, the telescope is installed at the Yakutsk EAS array at station obscura-3 at a distance 500 meters from the center of the array, where the field test is run.

An example of the output signals from the multi anode PMT is shown in Fig. 6. The average number of channels triggered in the event, where the signal definitely exceeds the noise level is 17 ± 6. So we get a good accuracy of the pulse duration by means of several independent measurement channels.

 DOI: 10.1051/epjconf/201714519010

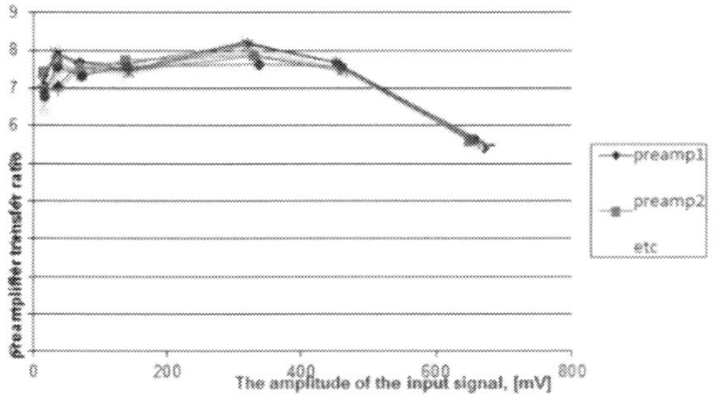

Figure 5. Preamplifiers transition ratio as a function of input signal voltage.

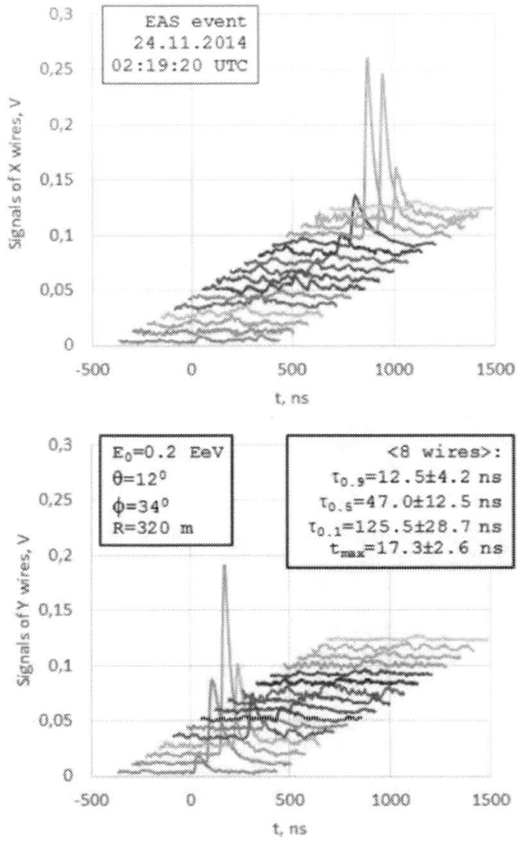

Figure 6. Signals of 32 wires from the multi-anode PMT at the focus of the telescope. The EAS parameters in a particular shower are estimated using data of the surface detectors. Pulse durations (τ) are measured at 0.1, 0.5, 0.9 levels of the pulse maximum V_{max}. The time difference (t_{max}) between V_{max} and 0.1 V_{max} of the rising signal is shown in the right panel.

3. First results of EAS measurements with the Cherenkov telescope working in coincidence with the surface detectors

During the field testing of the telescope in the period from 19.10.2012 to 11.04.2013 we had 604 hours of clear moonless nights that yielded 11124 EAS events detected by the scintillator subset of the array, from which 424 events resulted in a non-zero simultaneous signal

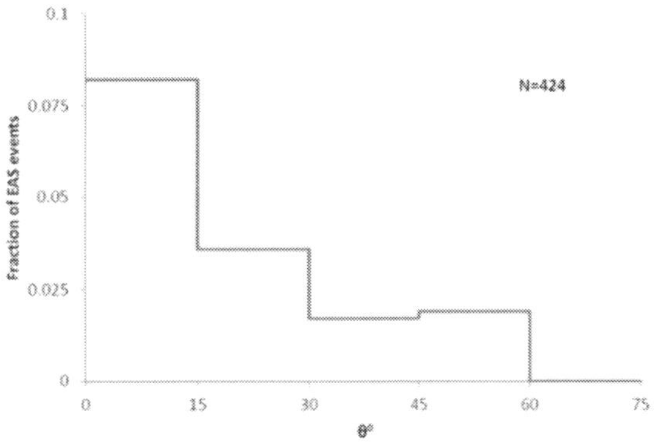

Figure 7. Zenith angle distribution of the shower fraction detected with the telescope.

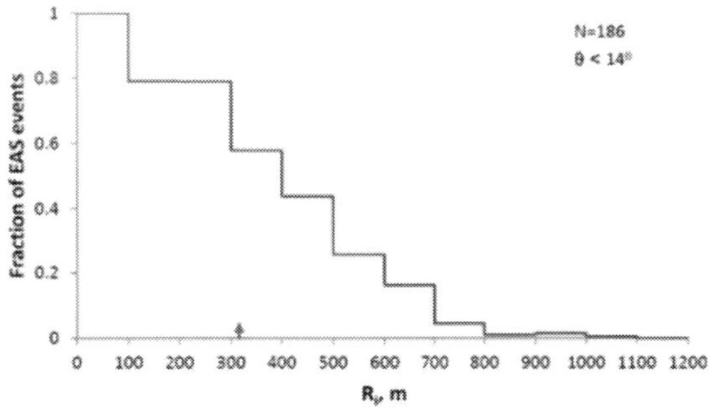

Figure 8. Axis distance distribution of the shower fraction detected with the telescope.

in the telescope. An additional 277 telescope signals were triggered by the Cherenkov detectors; these will be analyzed later and the results published in a forthcoming paper. From the numbers above (and the array area of 8.2 km^2) we determined the effective radius of the telescope acceptance $R_{eff} = 315$ m. This radius can be used in planning the grid of telescopes. The fraction of EAS events detected with a non-zero telescope signal is 3.8%. The dependence of the signal on the zenith angle and the shower axis distance is illustrated in Fig. 8; the primary energies were in the region $E_0 > 10^{16}$ eV.

While the telescope FOV is 308 sq. degrees ($\theta \in (0°, 14°)$), EAS events were detected with zenith angles up to $\theta = 60°$. This observation can be explained by the broad angular distribution of electrons emitted in the shower and photons scattering in the atmosphere. Another contribution is the angular uncertainty of the shower reconstruction procedure, which is considerably increased at the lower threshold energy of the array.

To avoid this uncertainty, we selected showers within $\theta < 14°$ to define a distribution of the shower axes (Fig. 7). The effective radius of the telescope detecting area, R_{eff}, is indicated by the arrow on the R_i axis as well.

We have measured the pulse shape parameters of the Cherenkov signal in individual showers. A typical set of signals from the EAS event detected on 12.11.2012, 13:20:11 UTC is shown in Figs. 9 and 10. A large part of the 16 μs buffer data is cut out where only noise is recorded, so the point $t = 0$ is arbitrary in this case. The signals of channels are not equalized here. The pulse shape

DOI: 10.1051/epjconf/201714519010

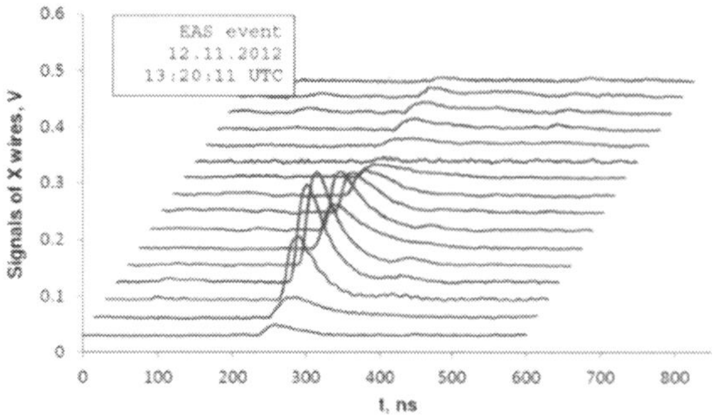

Figure 9. The Cherenkov pulse shape detected using 32 channels of the telescope DAS: 16 X wires.

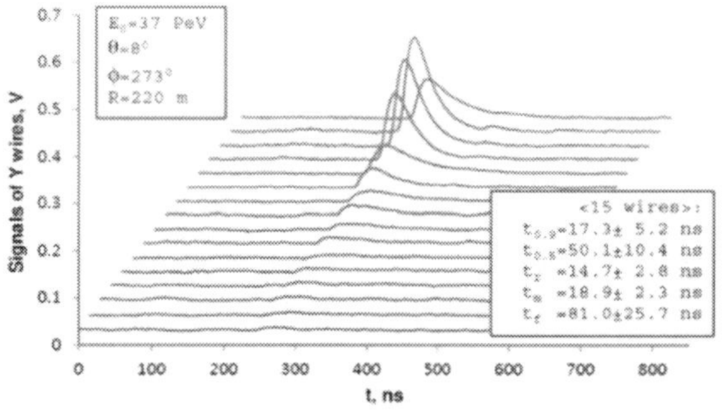

Figure 10. The Cherenkov pulse shape detected using 32 channels of the telescope DAS: 16 Y wires.

parameters are averaged over 15 wires for signals that are well above the noise. The parameters of EAS events are reconstructed from the data of the surface detectors of the Yakutsk array.

These data should be corrected for the response of the DAS to estimate the dispersion of the input signal. The measured average RMS deviation of signals in 15 wires in the EAS event under consideration is found to be $\sigma_{out} = 23.6$ ns. The variance of the output signal is the sum of variances of the input signal and the DAS. Consequently, the RMS deviation of the Cherenkov light signal is estimated as

$$\sigma_{Cher} = \sqrt{\sigma_{out}^2 - \sigma_{DAS}^2} = 9 \pm 5\,\text{ns}.$$

Note that the time constant of Hamamatsu R2486 PMT, $\tau_{PMT} = 5.5$ ns [5], is included in the DAS time constant.

When the detector input signal is a δ-function located at t_i, the DAS output signal is a function $g(t - t_i)$ with some time constant. An arbitrary input function can be represented as a sum of δ-functions:

$$f_{in}(t) = \lim_{N\to\infty} \sum_{i=1}^{N} f_{in}(t_i)\delta(t - t_i) = \int_{-\infty}^{\infty} f_{in}(\tau)\delta(t - \tau)d\tau.$$

The detector transforms every δ-function into $g(t - t_i)$

$$f_{out}(t) = \lim_{N\to\infty} \sum_{i=1}^{N} f_{in}(t_i)g(t - t_i) = \int_{-\infty}^{\infty} f_{in}(\tau)g(t - \tau)d\tau.$$

Hence, mathematically, our detector is equivalent to the linear integral operator with a difference kernel function that transforms f_{in} to f_{out}.

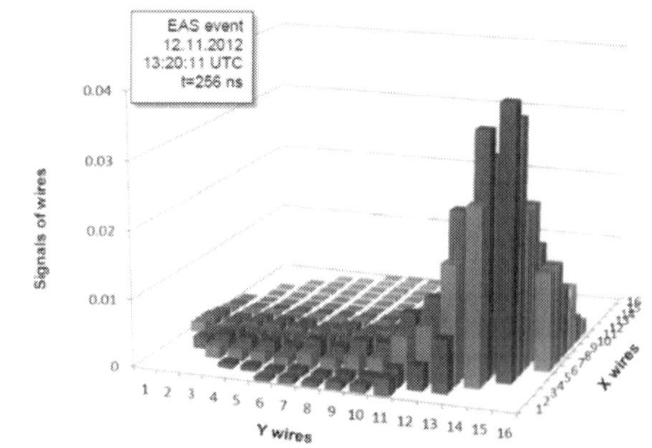

Figure 11. The particular distribution of the Cherenkov signal outputs on the surface of the photo-anode. A grid of wires has a pitch size of 3.75 mm. The distribution is taken at the time $t = 256$ ns after the event trigger when the signal amplitudes are approximately maximum.

The main properties of the convolution operator of interest to us are that the mean values and the variances of the functions are additive:

$$\overline{f_{out}} = \overline{f_{in}} + \overline{g}$$

$$\sigma_{f_{out}}^2 = \sigma_{f_{in}}^2 + \sigma_g^2.$$

The angular distribution of the Cherenkov signal is converted by the telescope into the spatial distribution of the signal on the PMT anode surface. Using the position-sensitive PMT, we are able to measure the spatial distribution of the signal. The spatial resolution of the PMT, σ_R, is connected with the angular resolution $\sigma_\theta \sim \sigma_R/F$. An example of the spatial distribution of the signal on the anode surface is given in Fig. 11 at the moment of the maximal amplitude. We used here a particular solution (Eq. (3)) considering the distribution of the signal on the anode as symmetric, with a single maximum.

The maximum of the light intensity points to the x_m^{Cher} position in the atmosphere. Combining it with EAS axis coordinates determined by the scintillation counters of the array, we found the height of the maximum to be $h_m \leq 1300$ m, corresponding to $x_m^{Cher} \geq 850\text{g/cm}^2$. In this particular shower, the lower limit to the depth is the result of the maximum of light intensity located at the edge of the telescope aperture.

In general, having a set of WFOV telescopes deployed at the EAS array, one is able to locate the shower maximum based on the angular distribution of the Cherenkov signals detected. Another method of the shower profile reconstruction is evident from the results of Cherenkov pulse shape measurements illustrated in Figs. 9 and 10. In this case, the position of the maximum in the temporal spread of signals, t_{max}, can be connected to the time of maximal radiation of Cherenkov photons [7].

4. Conclusions

Currently, the telescope is installed at the Yakutsk EAS array, near the station obscura-3 (500 m from the center of the array), where the telescope has completed field testing.

DOI: 10.1051/epjconf/201714519010

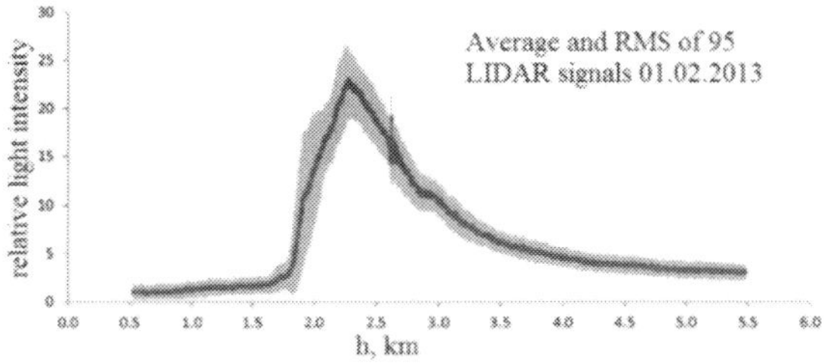

Figure 12. Average and RMS of 95 LIDAR signals.

Also, we were able to detect the LIght Detection And Ranging (LIDAR) signal, located at a distance from the Cherenkov telescope. The distance between the LIDAR and the Cherenkov telescope measured by GPS is equal to 490 ± 6 m.

The average LIDAR signal for 95 events shows that the signal amplitude dramatically increases when entering into the aperture of the telescope from 2 km height rather than reaching a maximum value and attenuation according to the square of the distance Fig. 12. The obtained data indirectly confirms the proper functioning of the Cherenkov telescope as a measuring device.

A Cherenkov telescope prototype obtained experimental data for the period from 19.10.2012 to 11.04.2016, on 203 clear moonless nights with about 1500 hours of observation, during the observation of the master load 30,000 times, of which the Cherenkov telescope gave a signal to the coincidence of about 1200 times. Measuring the width at half-maximum as a function of distance from the shower core, (R), will help determine the shower core in the array plane. Further modernization of the array will enable to calculate the height of the shower maximum by the time delay.

The work is supported in part by RFBR (grants 11-02-00158, 11-02-12193, 12-02-10005, 12-02-31550) and the Russian Ministry of Education and Science (contracts 02.740.11.0248, 16.518.11.7075). Grant NOFMU academic mobility and Head of Sakha Republic Grant for young scientist 17/16.

References

[1] A.A. Ivanov et al., J. Phys.: Conf. Ser. **012084**, 409 (2013) doi: 10.1088/1742-6596/409/1/012084
[2] A.A. Ivanov et al., Astrophys.Space Sci.Trans. **6**, 53 (2010) doi: 10.5194/astra-6-53-2010
[3] M.N. Dyakonov et al., Nucl. Instrum. Meth. A **248**, 224 (1986)
[4] E.G. Berezhko et al, Astropart. Phys. **36**, 31 (2012)
[5] `http://www.hamamatsu.com/`
[6] `http://www.rudshel.ru/`
[7] Ivanov, A.A et al., Nuclear Instruments and Methods in Physics Research A **772**, 34–42 (2015)

The search for primary particle tracks in nucleon-nucleus interactions with gamma ray energy $\Sigma E_\gamma \geq 3\,\text{TeV}$ registered in stratospheric X-ray emulsion chambers using data of the RUNJOB experiment

I.S. Zayarnaya[a] and T.A. Irkhina

P.N. Lebedev Physical Institute of RAS, 53 Leninskiy Prospekt, Moscow 119991, Russia

Abstract. We present here the result of a retreatment of data from the RUNJOB (RUssia-Nippon JOint Balloon) experiment of nucleon-nucleus interactions registered in stratospheric X-ray emulsion chambers (REC) using a new method for searching and tracing of galactic particles in nuclear emulsions. In about halfcof these interactions (~ 50) recorded in REC RUNJOB'96-3B, RUNJOB'97-6A and RUNJOB'99-11A,B with energy released in the electromagnetic component $\Sigma E_\gamma \geq 3\,\text{TeV}$ and $\Sigma E_\gamma \geq 5\,\text{TeV}$ respectively, single charged particle tracks are not found within the search area defined individually by the particle track location accuracy. The absence of primary proton tracks is consistent with the original treatment of the RUNJOB experimental data.

There is a difference in the zenith angular distribution for two groups of events in which a single charged particle track is observed or absent. The average penetration depth of the primary particles in REC to the interaction vertex in the zenith angle range from 60° to 79° differs by a factor two for these groups.

1. Introduction

The Russian-Japanese balloon experiments RUNJOB have been conducted from 1995 to 1999 for the purpose of studying the elemental composition and energy spectra of galactic cosmic rays in the energy range $\sim 10\text{–}10^3\,\text{TeV/particle}$ using stratospheric emulsion chambers exposed at an atmospheric depth $\sim 10\,\text{g/cm}^2$ in long duration flights (130–170 hr). The structure of REC RUNJOB'97 is shown in Fig. 1. As a result of the experimental data processing it was noted that the primary protons are identified in about half of the events related to the nucleon-nucleus interactions [1]. The single charged particle tracks are detected by nuclear emulsion with close to 100% efficiency (it is based on tracing the secondary $\pi^\pm$ mesons as well as primary protons through the emulsion layers). To explain a large number of unfound primary particles in the events related to nucleon-nucleus interactions, a detailed investigation of methodical reasons for this experimental fact was conducted [2].

As a result of this work we did not succeed in explaining why there are a lot of unfound single charged particle tracks by the method of searching and tracing of primary particles in the emulsion layers. For the analysis it was necessary to increase the statistics of events on account of low-energy nucleon-nucleus interactions registered in the REC and not included in the final primary proton spectrum in the RUNJOB experiment [3]. To be able to make a retreatment of these events we proposed a new method of searching and tracing of particles in the REC nuclear emulsions [4].

In this method as well as in the original one, the background nuclei are used as fiducials for locating the primary particle track. The results of the search for primary particle tracks in all nucleon-nucleus interactions registered in the RUNJOB'96-3B, RUNJOB'97-6A and RUNJOB'99-11A,B chambers with total electromagnetic energy $\Sigma E_\gamma \geq 3\,\text{TeV}$ and $\Sigma E_\gamma \geq 5\,\text{TeV}$ respectively by the two methods fully coincide, i.e., in about half of the events related to proton events the primary particle tracks are not found within the search area determined by the primary track location accuracy [5]. Figure 2 shows the deviation distribution of measured coordinates of the primary particle from the respective predicted ones for nucleon-nucleus and nucleus-nucleus interactions registered in the X-ray emulsion chambers using the new method. The measurement accuracy σ for the primary particle track location is near 31 microns.

2. The angular analysis of nucleon-nucleus interactions with found and absent primary particle tracks

In this work we analyzed 50 events registered in REC RUNJOB'96-3B, RUNJOB'97-6A and RUNJOB'99-11A,B with gamma ray energy $\Sigma E_\gamma \geq 3\,\text{TeV}$ and $\Sigma E_\gamma \geq 5\,\text{TeV}$ respectively at the zenith angle θ cut $\tan\theta \leq 5$ and related to the nucleon-nucleus interactions in the RUNJOB experiment. In each X-ray emulsion chamber the primary single charged particle track is not found in about half of the interactions. Figure 3 shows the zenith angle distribution for two groups of events in which the primary particle is observed or is not found. The graphs show that

[a] e-mail: zayar@sci.lebedev.ru

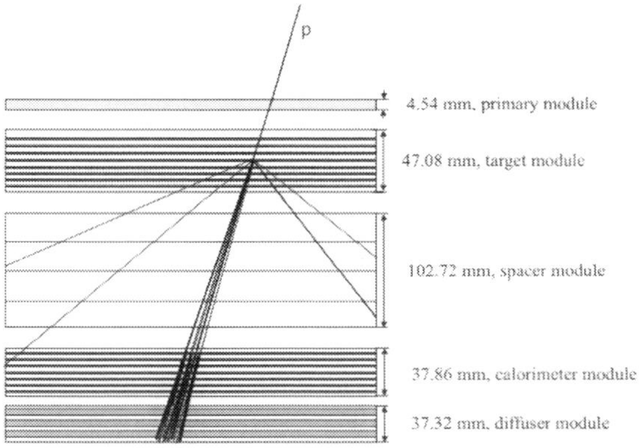

Figure 1. The structure of REC RUNJOB'97.

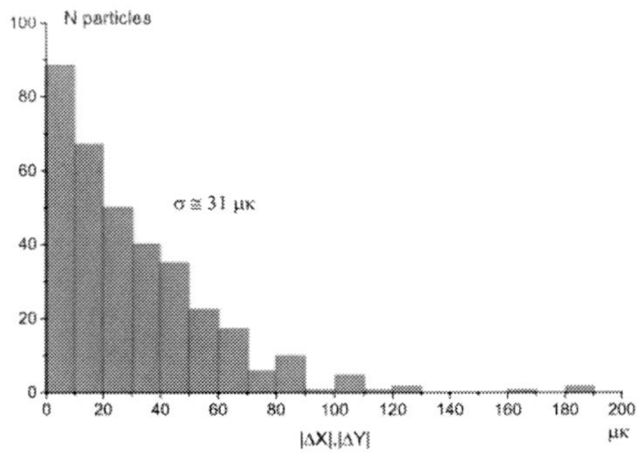

Figure 2. Deviation distribution of the measured co-ordinates of primary particles from the respective predicted ones.

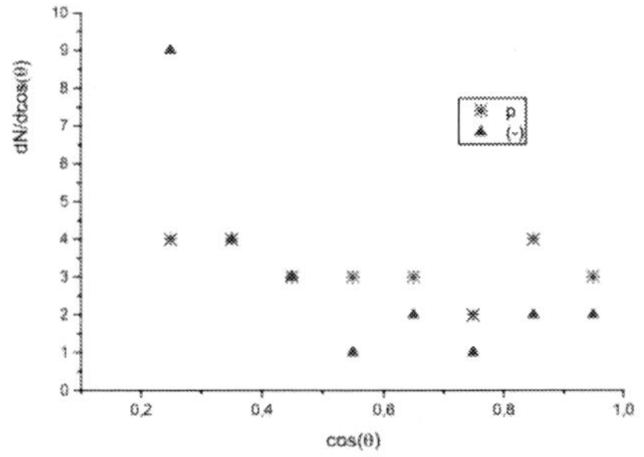

Figure 3. Zenith angle, θ, distribution for two groups of events: with observed primary proton track $(*)$ and unfound ones (Δ).

the zenith angle distribution for the group with observed proton track is nearly isotropic, that is not contrary to the expected zenith angle distribution of protons at the chamber exposure level ($\sim 10\,\mathrm{g/cm^2}$).

At the same time, an increased number of events with an unfound primary single charged particle track are observed in the zenith angle range $\theta > 60°$. It is interesting to note that the mean penetration depth, Λ, of primary particles within the REC from the first emulsion layer to the layer above the interaction vertex differs by a factor two for these groups in the zenith angle range from 60° to 79°. In units of mean free path length for a proton in the group with an observed proton track $< \Lambda(p) >= 0.18 \pm 0.12$, whereas for the group with an absent charged particle track $< \Lambda(-) >= 0.37 \pm 0.16$. Since the statistics of the studied interactions is low, we can only discuss about a possible indication of a distinction in the nature of the primary particles in the two groups of events.

3. Conclusion

According to the reprocessing of galactic particle interactions detected by REC RUNJOB'96-3B, RUNJOB'97-6A and RUNJOB'99-11A,B and related to the nucleon-nucleus interactions with gamma ray energy $\Sigma E_\gamma \geq 3$ TeV and $\Sigma E_\gamma \geq 5$ TeV respectively using a new method of searching and tracing of the primary particles in the REC nuclear emulsion plates the primary proton tracks are not observed in about half of the events. The absence of single charged particle tracks within the search area for the primary particle is consistent with the original processing of the RUNJOB experiment data. The nucleon flux incident on REC may include the secondary neutrons from the interaction of the primary cosmic rays with the residual atmosphere above the exposed installation ($\sim 10\,\mathrm{g/cm^2}$). According to calculations [6], their proportion in the flux of secondary particles is about 4% which is much smaller than the experimental relative number of events with absent proton track within the search area. To clarify the nature of the primary particle in the events related to nucleon-nucleus interactions in the RUNJOB experiment further processing and analysis of experimental data is required.

References

[1] A.V. Apanasenko et al. Astroparticle Phys., 2001, **16**, p. 13
[2] I.S. Zayarnaya, Physics of Atomic Nuclei, 2008, **71**, No. 2, p. 299–303
[3] I.S. Zayarnaya, Preprint FIAN, No. 1 (2010)
[4] M.Hareyama et al., Proton and Helium spectra obtained by RUNJOB experiment, Proc. 29th ICRC, **3**, OG1.1, p. 17 (2005)
[5] I.S. Zayarnaya., Bull. of the RAS, Ph., 2014, **78**, No.3, p. 221–224
[6] L.G. Sveshnikova et al., Preprint INP MSU 97-43/494 (1997)

Fast “swarm of detectors” and their application in cosmic rays

G.P. Shoziyoev[1,2,a] and Sh.P. Shoziyoev[3,4]

[1] P.N. Lebedev Physical Institute of the Russian Academy of Sciences, Russia
[2] Lomonosov Moscow State University, Dushanbe Branch, Tajikistan
[3] Khorog State University, Khorog, Tajikistan
[4] Faculty of Physics, M.V. Lomonosov Moscow State University, Russia

Abstract. New opportunities in science appeared with the latest technology of the 21st century. This paper points to creating a new architecture for detection systems of different characteristics in astrophysics and geophysics using the latest technologies related to multicopter cluster systems, alternative energy sources, cluster technologies, cloud computing and big data.

The idea of a quick-deployable scaleable dynamic system of a controlled drone with a small set of different detectors for detecting various components of extensive air showers in cosmic rays and in geophysics is very attractive. Development of this type of new system also allows to give a multiplier effect for the development of various sciences and research methods to observe natural phenomena.

1. Introduction

Traditionally, cosmic ray studies of extensive air showers (EAS) systems use different detectors with different geometry and sensitivity to the components of the EAS. Currently we know the most popular type of installation of YakuShAL [1], TAIGA [2,3], TA [4], Auger [5,6]. For example the size of the Auger installation is 3000 square kilometers and consists of 1600 water Cherenkov detectors. All these settings are collected in stages and initially deployed on a given grid on the earth's surface and detect the various components of the EAS with different types of detectors. Of course, for decades, these units were the best available at the time, and in the presence of an appropriate infrastructure with special conditions of service data units could produce great results in the field of ultra-high energies.

We have proposed the concept of a system with a rapidly deployable detection system architecture. This system can be used to measure various characteristics of the EAS Cherenkov radiation using the latest technology related to multicopter systems, alternative energy sources, cluster technologies, cloud computing and big data.

2. Cherenkov light from EAS

Cherenkov radiation plays an important role in the study of EAS. A large enough light output in the optical range and weak absorption in a pure, clear atmosphere can effectively help determine the total energy of the EAS, as well as the history of the shower development in the atmosphere.

The history of EAS Cherenkov radiation research began in the 1940s in Gttingen (1947), when photo-electron multipliers were first used to collect the Cherenkov light, and the Cherenkov light experiment conducted by Jelley (1951). The first measurements of Cherenkov light showers were made by Goldanskiy and Zhdanov [7] and independently by Galbraith [8]. Especially valuable at that time was the pioneering work of Chudakov [9] and the design work he made in conjunction with V.I. Zatsepin [10] for detecting Cherenkov light from EAS, made on the basis of the cascade theory of electron-photon showers primary gamma ray proton and two levels of observation.

At sea level, the Cherenkov angle is equal to 1.3° and decreases with height (Figure 1). The threshold energy for particles producing the Cherenkov light is, for example for electrons, protons and muons at sea level, 21 MeV, 4.4 GeV and 39 GeV respectively, and increases exponentially with height.

The pioneering work on Cherenkov light measurements by Chudakov also included the contribution of ionization and emission (fluorescent light) as the background to Cherenkov experiments. He later suggested to use this glow in experimental studies of EAS.

In our case we need to use a Cherenkov detector with small size and weight. A surface area of the Cherenkov detector of about 78.5 cm^2 is sufficient. It may also be used as a direct light Cherenkov detector or a Cherenkov aerogel detector [11]. Nanosecond timing collection of EAS Cherenkov light allows the determination of the arrival direction of showers and the characteristics of the shower front.

3. Drones or multicopter

The basic unit of the proposed system is a multicopter drone with a set of detectors on board. We called it a dedron (dedron - detecting drone) (Fig. 2).

Civil use of drones and multicopters [12] is quickly developing in the world.

[a] e-mail: gulmurod@mail.ru

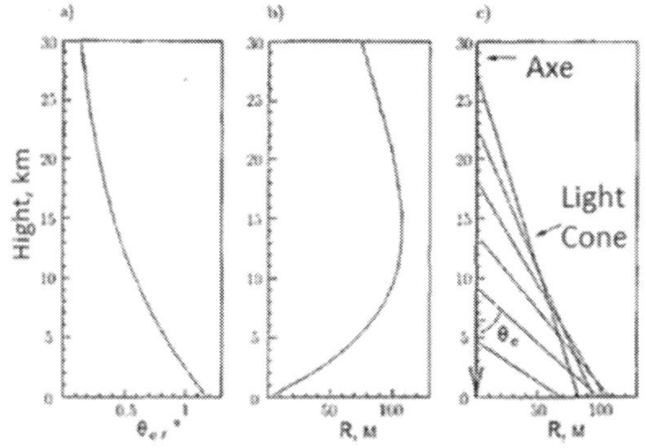

Figure 1. Cherenkov light height dependence from a) angle and b) circle radius. c) Shows the change of the Cherenkov light cone for different heights in the atmosphere.

Figure 2. Quadrocopter dedron.

There are also studies testing working management systems for drones [13]. Due to this, the purchase prices for them are falling rapidly.

The most effective power source for drones is a LiPo battery pack [14]. The voltage of the 4S LiPo battery assembly will be around 14.8 V. The discharge rate of the battery with a capacity of 1000 mAh will determine its maximum output current. It is necessary to choose the correct combination of power and discharge rate. It is also important to know that the battery cells are not taken into account for the maximum current consumption. The number of elements only define the voltage of the LiPo element needed for the hardware.

For example, the calculation of the rate of discharge shows that the A 2200 mAh / 30C LiPo can only provide 66A max. Therefore, the rate of discharge capacity is essential.

For example, 1A maximum current is required for the 4 motors of quadrocopters, and, for other equipment, 1A (detectors, lights, etc.), the required number of voltage 11.1V 8800 mAh and we have 134 minutes flying time and 8.8 Ampere given speed discharge.

A Quadrocopter with 1–2 kg weight drifted around 50 seconds over a distance of 1 km. It also depends on the mode of flight and other factors such as weight, wind resistance. The gyroscope, which allows quadrocopters to stay in the air also allows it to land vertically on the surface.

With the development of charging technology from alternative sources, time and efficiency will increase.

4. Communications technology, management, data collection and analysis

Technology at the beginning of the 2000s made great advances (4G [15], LTE [16] and WiMax [17]), and their use does not require special permission in everyday life. The Glonass module [18] and GPS [19] allow precise positioning of drones with detectors in space that allows the choice of positioning any network detectors structure. Controlling from a central control server connected with drone clusters will be optimal.

Deployment of the system will be possible taking into account the weather conditions and the transparency of the atmosphere on clear, moonless nights. Operation of the detection system starts after landing the chosen distribution of units on the surface. In fact, the position of dedrons is synchronized by GPS and wireless communication with a central base station. Dedrons land after flying to a predetermined coordinate, blades engines are turned off and the detection system switched on. After an overnight detection session the dedrons fly back to the central base station for safe charging from an alternative power source or solar cells. That also eliminates all sorts of negative scenarios with security detectors from both natural and man-made phenomena. It also prolongs the life of the detector system.

All signals detected by dedrons are taken simultaneously with their time stamps and stored in the server. The volume of incoming data for each event at each node can be evaluated in a few kilobytes of information. A server in online mode, following the data and the coincidence of signals on multiple nodes, classifies and stores the recorded events.

Subsequent data processing is performed in an offline mode by physics specialists with work skills on BIG DATA [20].

5. Summary

Due to the rapid emergence of new technologies we propose to develop a new direction in the detection and monitoring of various parameters of natural phenomena (EAS, storm effects, the consequences of natural and manmade disasters) by placing different types of sensors and detectors on each dedron.

A special interest is the study of Cherenkov light and the fluorescent light from EAS.

In this type of experiment, a study of the spectrum and mass composition of cosmic rays in the energy range 10^9–10^{18} eV can be carried out by a single method - registration of EAS Cherenkov radiation. By covering an area of 1–2 sq. km. and positioning 100 optical detectors at distances of 1 meter to 100 meters from each other, one can study the structure of the EAS front. The grid can be both regular or irregular. The nodes can be either positioned centrally symmetricly relative to the central base station or distributed uniformly in a polygonal structure.

In fact, this system is a telescope with a variable focal length. If this solution is technically implemented in practice, it will create a revolutionary "telescope of the XXI century".

Widespread use is possible for other types of detectors on this type of system for geophysical studies and real-time monitoring of deep structures of the Earth's surface.

References

[1] Ivanov, A.A., S.P. Knurenko, A.D. Krasilnikov, Z.E. Petrov, M.I. Pravdin, I.Ye. Sleptsov, L.V. Timofeev. Nuclear Instruments and Methods in Physics Research A **772** (2015) 34–42

[2] Antokhonov B et al., 2011, Nucl. Instrum. Meth. A **639**, 42

[3] Berezhnev S et al., 2012, Nucl. Instrum. Meth. A **692**, 98

[4] H. Kawai et al., Telescope Array Experiment, Nuclear Physics B – Proceedings Supplements, 10.1016 j.nuclphysbps.2007.11.002

[5] Louedec, Karim "Atmospheric Monitoring at the Pierre Auger Observatory – Status and Update" (2011)

[6] The Pierre Auger Collaboration, Astropart. Phys., 2010, **33**, 108–129

[7] Gol'danskii VI Zhdanov GB JETP Letters, 1954, **26**, s.405

[8] Galbraith W., Jelley J.V., J. Atmosph. Terr. Phys., 1955, **6**, 250, p. 304

[9] Chudakov AE, NM Nesterova et al., Proc. 6th Int. Conf. by CL. M.: Publishing House of the USSR Academy of Sciences, 1960, **2**, p. 47

[10] Zatsepin VI, AE Chudakov JETP Letters, 1969, **42**, c.1622

[11] Fields, D.E. et al., Nuclear Instruments and Methods 1994, **349**, 431–437

[12] Hoffman, G., Huang, H., Waslander, S.L., Tomlin, C.J., "Quadrotor Helicopter Flight Dynamics and Control: Theory and Experiment" (2007). Conference of the American Institute of Aeronautics and Astronautics. Hilton Head, South Carolina

[13] Martin Saska et al., Journal of Intelligent & Robotic Systems, 2016, p. 1–24, doi: 10.1007 / s10846-016-0338-z

[14] Zhang, Heng; Liu, Chengyong; Zheng, Liping (1 July 2014). Electrochimica Acta **133**, 529–538, doi: 10.1016 / j.electacta.2014.04.099

[15] International Mobile Telecommunications-2000 (IMT-2000), 1997, M.687–2 (02/97)

[16] Stefan Parkvall, Erik Dahlman, Anders Furuskär et al; Ericsson, Robert Syputa, Maravedis; Vehicular Technology Conference, 2008. VTC 2008-Fall. IEEE 68th 21-24 Sep. 2008, p. 1–5

[17] G.S.V. Radha Krishna Rao, G. Radhamani. WiMAX: A Wireless Technology Revolution. - 2007, ISBN: 0-8493-7059-0

[18] GLONASS: principles of construction and operation / Ed. AI Perov, VN Kharisova, 3rd ed., Revised. – M.: Radio Engineering, 2005, 688 p. - 1000 copies, - ISBN: 5-93108-076-7

[19] Navstar GPS user equipment introduction, september 1996, Public release version `http://www.navcen.uscg.gov/pubs/gps/gpsuser/gpsuser.pdf`

[20] Morrison, Alan and other Big Data:. How to extract information from them. PricewaterhouseCoopers, Technology Forecast. The quarterly magazine, Russian edition, 2010, Issue 3

The array for investigations of EAS neutron component

D.M. Gromushkin[1,a], N.V. Ampilogov[1], N.S. Barbashina[1], F.A. Bogdanov[1], R.P. Kokoulin[1], V.V. Ovchinnikov[1], A.A. Petrukhin[1], I.A. Shulzhenko[1], Yu.V. Stenkin[1,2], I.I. Yashin[1], and K.O. Yurin[1]

[1] National Research Nuclear University MEPhI (Moscow Engineering Physics Institute), 115409 Moscow, Russia
[2] Institute for Nuclear Research of RAS, Moscow 117312, Russia

Abstract. The neutron component of EAS has not been practically studied for many years. Creation of the neutron detector arrays PRISMA-32 and PRISMA-YBJ demonstrated the effectiveness of such investigations. It is important to underline that recording the EAS neutron component is carried out over the whole array area. Development of a new larger scale neutron detector array (URAN) in MEPhI will continue these studies. During the construction of the URAN array, a new design of en-detectors based on the scintillator with natural boron and new electronics was developed. The use of new electronics with a switchable sampling rate allows one to reach EAS fast timing accuracy up to 10 ns. The en-detectors of the URAN are located on the roofs of the buildings. The sensitive area of each detector is 0.36 m^2, while the array's starting area is $\sim 10^3$ m^2. Total number of the initial array detectors is 72. They are subdivided into 6 independent clusters each of 12 en-detectors, connected with the central data acquisition station.

1. Introduction

The neutron component of EAS is produced in interactions of the hadronic component with atmospheric nuclei or substance close to the detector. In its turn, the hadron component carries important information about the primary particle energy. Therefore, studying it is of a great interest. Investigations of the hadronic component are mainly performed using hadron calorimeters, but they are quite expensive and have a limited area. In 2001, a new method for the registration of EAS hadronic component using thermal neutrons was proposed [1]. The first results on the EAS neutrons were obtained at the PRISMA-32 setup [2] (MEPhI, Russia). The obtained data have shown the possibilities of studying the EAS neutrons using the proposed method [3–5]. The main feature of this method is the usage of en-detectors which register EAS electromagnetic and neutron components simultaneously.

In 2014–2015 in MEPhI within the frameworks of the Program for support of the Russian Unique Scientific Facilities the first stage of a new URAN (Upstairs Registration of Atmospheric Neutrons) array [6] was created. It is aimed at the registration of thermal neutrons produced by the EAS hadronic component over the whole area of the array. The URAN detectors are located on the roofs of MEPhI laboratory buildings. Such location allows to decrease the influence of the building floor structure.

2. The URAN array

The first stage of the URAN array includes 72 en-detectors. The detectors are located on the roofs of two laboratory buildings and are combined into cluster structures. On each roof, three clusters (36 en-detectors) are deployed. The total area of the array is about 10^3 m^2. Figure 1 shows the layout of the URAN array.

2.1. Detectors

The en-detector is a basic element of the URAN array. The construction of the en-detector is shown in Fig. 2. A scintillator with an area of 0.36 m^2 is mounted at the bottom of a 200-liter plastic tank [7]. In order to enhance the light collection, a cone with a diffusely reflecting surface is used. In the cone vertex an FEU-200 PMT (Novosibirsk, Russia) is installed. The PMT is attached to the tank lid with a diameter of 350 mm. For protection against environmental influences, detectors are installed inside the external housings made of 0.7 mm galvanized steel.

For the registration of thermal neutrons produced by the EAS high-energy hadrons, an inorganic scintillator based on ZnS with the addition of B_2O_3 is used. The content of ^{10}B in B_2O_3 is 18%. This scintillator has the following features:

- high light yield at the registration of heavy particles produced during neutron capture;
- sufficiently high effectiveness of thermal neutron registration due to the large cross-section of thermal neutron capture by ^{10}B nuclei;
- fast response (40 ns);
- low sensitivity to single charged particles.

Registration of thermal neutrons in the scintillator $ZnS(Ag)+B_2O_3$ occurs due to the following reactions:

$$^{10}B + n \to {}^{7}Li + \alpha + 2.792\,MeV$$
$$^{10}B + n \to {}^{7}Li^{*} + \alpha + 2.31\,MeV$$
$$\downarrow$$
$$^{7}Li^{*} \to {}^{7}Li + \gamma + 482\,keV.$$

[a] e-mail: DMGromushkin@mephi.ru

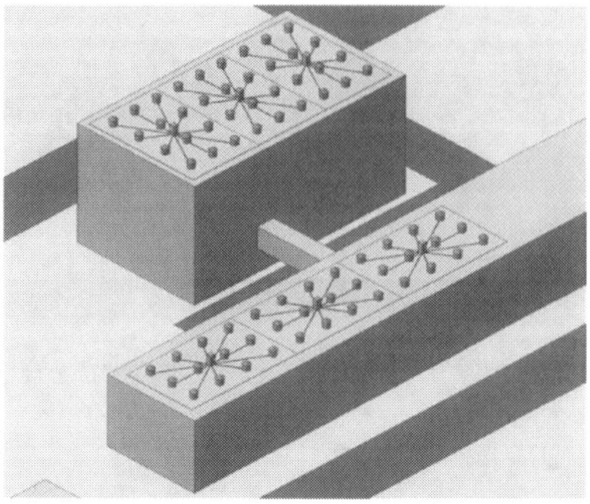

Figure 1. The layout of the URAN array.

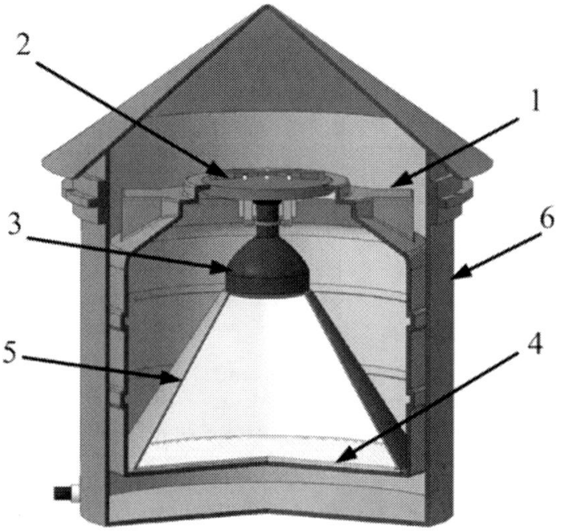

Figure 2. The en-detector design; 1 – light protecting housing; 2 – tank lid; 3 – FEU-200 PMT; 4 – scintillator; 5 – light reflecting cone, 6 – outer metal housing.

The formed nuclei cause luminosity of ZnS(Ag) in the visible wavelength range [8]. Due to the use of ZnS(Ag), the detectors also allow to register the EAS electron component at the group passage of charged particles.

Before the array deployment, all detectors were tested and adjusted. The response of en-detectors to the neutron registration was studied using the test facility allowing to separate neutron and charged particle signals according to the pulse shape. The pulse from the neutron capture products has a longer rise time (several μs) than a pulse from charged particles or PMT noise due to a more intensive luminosity of the scintillator slow components. In order to increase statistics and to decrease the influence of meteorological factors, a neutron source ^{252}Cf (with neutron flux of $\sim 1000\,\mathrm{s}^{-1}$) is installed under the detector. To obtain thermal-energy neutrons, the source is placed in a cube made of a hydrogen-containing substance (polyethylene with a width of 100 mm).

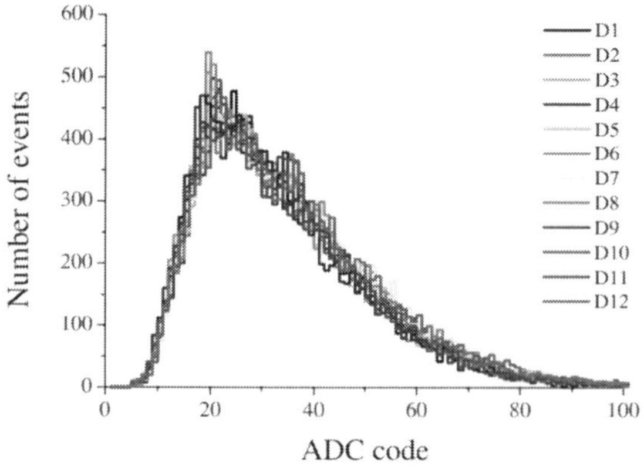

Figure 3. Neutron signal spectra for 12 detectors.

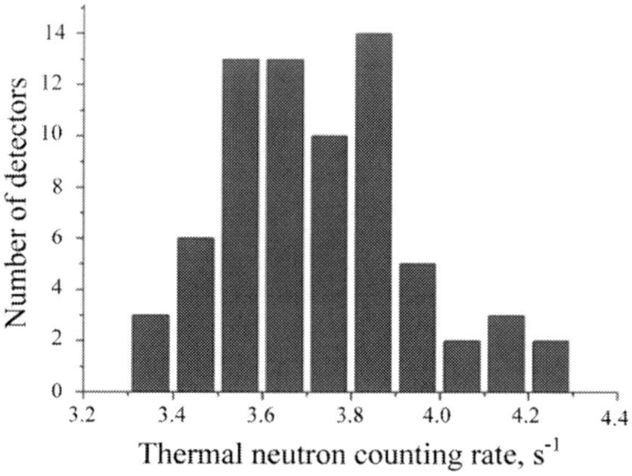

Figure 4. Distribution of 72 detectors in the thermal neutron counting rates.

En-detector testing included 1-hour measurements of the thermal neutron amplitude spectra. An example of the measured spectra for 12 detectors is shown in Fig. 3. For better separation, signals are integrated with an integration constant of 20 μs. The obtained amplitude spectra are similar. The maxima are located at about 25th code of ADC.

Figure 4 shows the distribution of 72 en-detectors in the thermal neutron counting rates. The average thermal neutron counting rate is 3.7 neutrons/s, maximal counting rate deviation for all tested en-detectors does not exceed 14%.

2.2. Electronics

Electronics for signal digitizing is installed in the Local Post (LP) of primary data processing which, like the detectors, is located on the building roof. Cluster electronics includes two 12-channel blocks of amplitude analysis designed according to the 6U Euromechanics standard. One block performs digitizing of analog signals from the 12th dynodes of the cluster PMTs, the second block digitizes analog signals from the 7th dynodes. Usage of a two-dynode readout system allows to increase the dynamic range for measuring the energy deposit of the EAS electromagnetic component. The structural

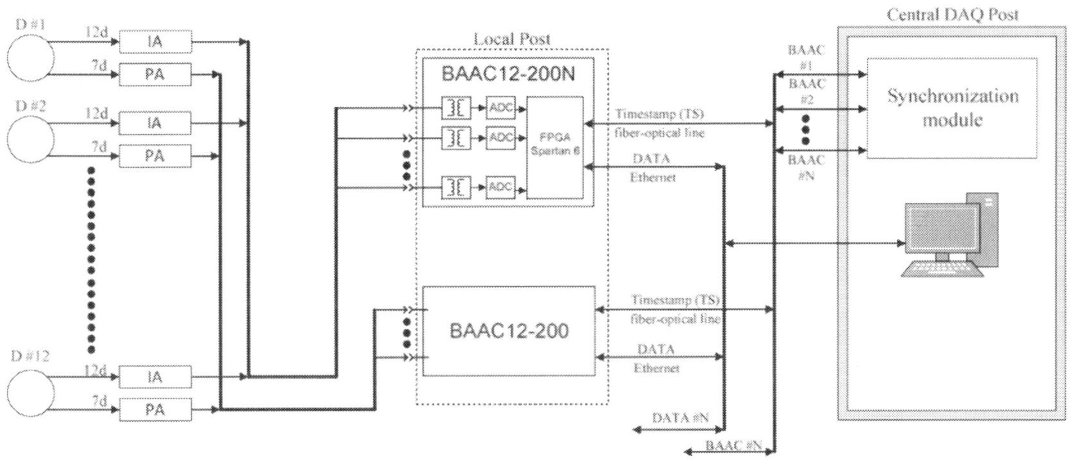

Figure 5. Structural scheme of the URAN array registering system.

scheme of the URAN array electronics is shown in Fig. 5. Signals from the detectors are fed to the inputs of the block of amplitude analysis via 25 m coaxial cables. Beside this, signals from the 12th dynodes are fed through the integrator-amplifier with an integration time of 1 μs, and signals from the 7th dynodes – through the preamplifier. Integrator-amplifier (IA) and preamplifier (PA) are installed inside the external housing of the detector. Both blocks have the same design, but differ in the software configuration. Block ADCs digitize analog signals from the en-detectors with a sampling frequency of 200 MHz in the amplitude range from -3.5 V to +3.5 V.

The main block digitizing signals from the 12th dynodes constantly records data into the 1024-count circular buffer. If the triggering conditions are fulfilled, the sampling frequency changes from 200 MHz to 1 MHz and 20 thousand counts are recorded. Due to such a digitizing scheme, the first 1024 points are recorded with a step of 5 ns. This allows to obtain information about the energy deposit of the electromagnetic component. The following 20000 points digitized with a step of 1 μs provide information about thermal neutrons accompanying the shower. The second block has no sampling frequency variation and only records 1024 points with a step of 5 ns (information on the energy deposit of the electron component comes from the 7th dynode of the PMT).

Blocks operate independently from each other. The data of separate clusters are associated according to the time of the cluster event registration. The common time for all blocks is set by the synchronization module which broadcasts the unified clock signal with a frequency of 100 MHz via a fiber-optical communication line and ensures simultaneous starting of internal clocks of all blocks. Time synchronization accuracy is 10 ns.

Such a registering system allows to receive and store data in the form of oscillograms of all detectors for each event (data are stored in binary files).

The experimental data processing is performed by special software in the off-line mode. This software allows to obtain information about registered EAS: the energy deposit of electromagnetic component and the number of neutrons detected during 20 ms after the EAS front arrival. The oscillogram of the registered thermal neutron is shown in Fig. 6.

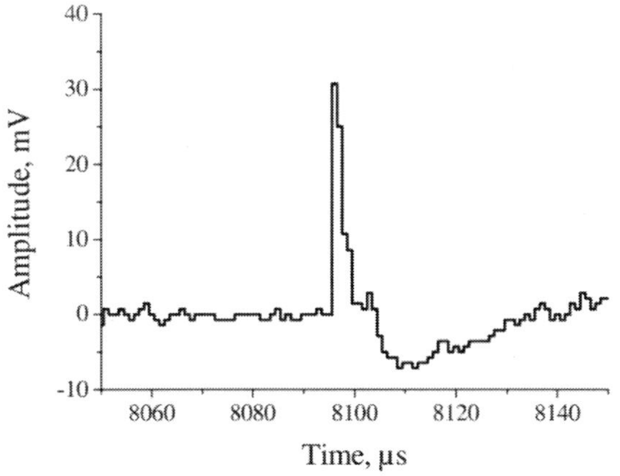

Figure 6. Example of the oscillogram of the neutron pulse integrated over 1 μs.

3. Test run

For studying the operation of en-detectors, a test run with the first three clusters of the URAN array was started. During the test run, all detectors and electronics blocks were examined, the registration thresholds for EAS events were adjusted. At the next step, a week-long exposure was started. The cluster triggering condition was a coincidence of not less than 2 en-detectors with signal amplitudes of at least 12 mV.

During the test run, the first EAS events were registered. Figure 7 shows an example of the event registered by 3 clusters of the URAN array. The responses of en-detectors to the EAS electromagnetic component A_e (in ADC codes) and the numbers of neutrons registered during 20 ms after the shower front arrival are indicated. In this event, 36 detectors registered 24 neutrons. Using the values of responses to the electromagnetic component, the EAS axis position can be estimated (marked by a cross in Fig. 7).

The events with high energy deposit of the EAS electromagnetic component have been associated with events registered by other setups of the NEVOD complex (PRISMA-32 [2], NEVOD-EAS [9,10] and

DOI: 10.1051/epjconf/201714519013

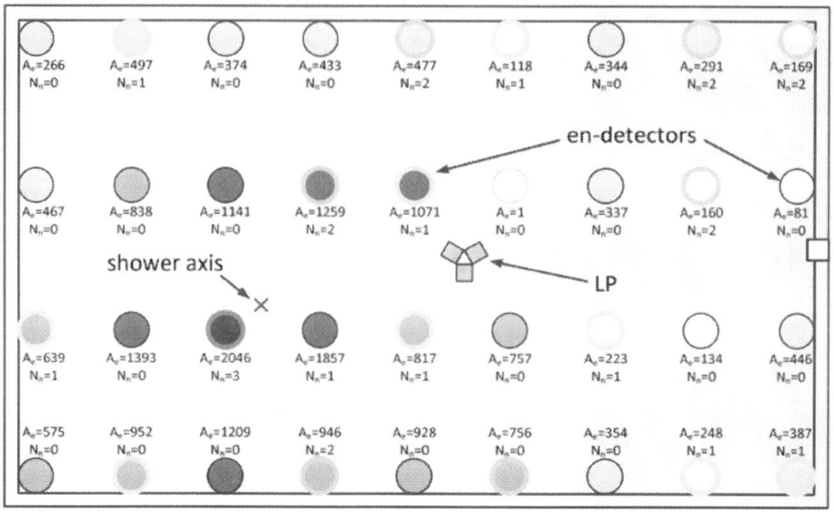

Figure 7. Example of the EAS registration with three clusters of the URAN array.

NEVOD-DECOR-CTS [11]). It has been revealed that the topologies of the events are in good agreement with each other. In the near future, a system of global synchronization with the NEVOD-EAS array will be started. It will allow to obtain quantitative estimations of the EAS registration.

4. Conclusion

The first stage of the setup for registration of EAS neutrons has been created. Test runs have been started. The first EAS events registered during test runs are in good agreement with the data of other installations. The URAN array is an inseparable part of already existing detectors of the Experimental Complex NEVOD and installations which are now under construction. The combination of setups using different methods of EAS registration into a single experimental structure will allow to carry out multicomponent studies of extensive air showers.

This work was performed at the Unique Scientific Facility "Experimental complex NEVOD" with the support of the Ministry of Education and Science of the Russian Federation (grant RFMEFI59114X0002, MEPhI Academic Excellence Project 02.a03.21.0005 of 27.08.2013) and the Russian Foundation for Basic Research (grant 16-29-13028-ofi-m).

References

[1] Yu.V. Stenkin and J.F. Valdes-Galicia, Modern Phys. Letters A **17**, 1745 (2002)
[2] D.M. Gromushkin et al., J. Instrumentation **9**, C08028 (2014)
[3] D.M. Gromushkin et al., Phys. Atomic Nuclei **78**, 349 (2015)
[4] D.M. Gromushkin et al., Astrophys. Space Sci. Trans. **7**, 115 (2011)
[5] D.M. Gromushkin et al., J. Phys.: Conf. Ser. **409**, 012044 (2013)
[6] D.M. Gromushkin, et al., J. Phys.: Conf. Ser. **675**, 032043 (2016)
[7] M.B. Amelchakov et al., PoS (ICRC2015) **651** (2015); `http: pos.sissa.it`
[8] W.D. Allen, Neutron detection. Harwell (1960)
[9] I.A. Shulzhenko et al., Bull. Russ. Acad. Sci. Phys. **79**, 389 (2015)
[10] I.I. Yashin et al., J. Phys.: Conf. Ser. **632**, 012029 (2015)
[11] V.V. Kindin et al., Physics Procedia **74**, 435 (2015)

Model-stable universality of the air shower electromagnetic component: An approach to solving the mass composition problem

R. Raikin[a], T. Serebryakova, A. Lagutin, and N. Volkov

Altai State University, 61 Lenin Ave., Barnaul 656049, Russia

Abstract. On the basis of the scaling approach and CORSIKA simulations data the radial scale factor of the lateral distribution of electrons in extensive air showers is confirmed as a potentially effective primary mass estimator, and its sensitivity to hadronic interaction model is investigated. It is shown that improved composition results both on average and event-by-event basis can be achieved taking into account the universality property of air shower development expressed by the relation between the radial scale factor and the longitudinal age parameter. The enhancements of such a theoretically motivated tool for an unbiased cosmic ray composition deduction in a wide primary energy range for current and future (multi-)hybrid air shower measurements are discussed.

1. Introduction

Uncertainties in hadronic physics and chemical composition are two basic obstacles for understanding the origin of very high energy cosmic rays. A physical interpretation of features of the cosmic ray energy spectrum in terms of sources and propagation properties relies on the assumed mass composition while its robust estimation from extensive air shower (EAS) observables interferes with their sensitivity to nuclei interaction models at energies not accessible with accelerators.

At present, numerous methods and techniques are implemented to infer the mass composition of cosmic rays (e.g. see [1]). They include the analysis of mean values, fluctuations, correlations and even particular features of distributions of different EAS observables such as depth of the shower maximum, muon production depths, total number of electrons and muons at the observation level and local densities far from the shower axis, as well as particle arrival time distributions and the spatial distribution of EAS radio signals. However, despite considerable efforts that have been made in recent years, the composition and its variation with energy remain quite uncertain in almost the entire primary energy range of the cosmic rays available for EAS studies [2–5].

In this paper we examine the efficiency of the formalism describing the lateral distribution function (LDF) of EAS electrons as a scale-invariant and its dependence on the shower longitudinal development stage, that reflects EAS universality properties, for reducing the uncertainties in the current analysis and for improving the estimation of the mean mass composition at a certain energy or the primary particle type in the case of individual showers.

In the following section we give a brief overview of the scaling approach for LDF, proposed in [6–9], and an outline for the following simulations data analysis. In Sect. 3 we present the results of CORSIKA simulations of EAS carried out with EPOS LHC and SIBYLL nuclei interaction models: distributions of depth of shower maximum $X_{\max}$ and radial scale factor R_0 for vertical showers initiated by protons and iron nuclei, correlations between R_0 and the longitudinal age parameter, s. Section 4 contains the discussion and final conclusions.

2. Radial scaling of the electron component and universality in EAS development

One of the key EAS quantities necessary for basic shower parameters reconstruction is the lateral distribution of charged particles at a fixed observation depth, X. The exact form of LDF has been debatable for decades. The majority of analytical parametrizations of LDF of different EAS components is traditionally based on the well known Nishimura-Kamata-Greizen (NKG) function [10] originally obtained for electromagnetic cascade showers:

$$\rho(r; E, s) = \frac{N(E,s)}{r_0^2}\frac{\Gamma(4.5-s)}{2\pi\Gamma(s)\Gamma(4.5-2s)} \times \left(\frac{r}{r_0}\right)^{s-2}\left(1+\frac{r}{r_0}\right)^{s-4.5}. \quad (1)$$

Here $\rho(r; E, s)$ is the local particle density at a radial distance, r, from the core position in the shower with primary energy, E, and the longitudinal age parameter, s, $N(E, s)$ – total number of particles at the observation depth (shower size), r_0 – shower scale radius, which does not depend on the primary particle type and energy (originally – the Molière unit r_M). Various modifications of the NKG function, such as introducing different fixed scale factors, lateral ($s_\perp$) or local ($s(r)$) age parameters and also

[a] e-mail: raikin@theory.asu.ru

generalizations of the function by using a third power-law term were suggested. A comprehensive review is beyond the scope of this paper (some discussions can be found in e.g., [11–13]).

A different theoretically motivated approach, so-called scaling formalism, was proposed in our papers [6–9] for the lateral distribution of electrons in both electromagnetic and hadronic showers:

$$\rho(r;E,X) = \frac{N(E,X)}{R_0^2(E,X)} F\left(\frac{r}{R_0(E,X)}\right). \quad (2)$$

Here the radial scale factor, R_0, in contrast to the commonly used r_M, depends on the primary particle type, the shower age and (in case of extensive air showers) properties of hadronic interactions. The function $F(x)$ is the common scaling part of LDF. According to our calculations based on a semi-analytical approach [6,7,9], the factor R_0 is equal to the root mean square radius of the electron component, R_{ms}, which is defined in a standard way as

$$R_{ms}(E,X) = \left(\frac{2\pi}{N(E,X)}\int_0^\infty r^2\rho(r;E,X)rdr\right)^{1/2}, \quad (3)$$

and the following expression suggested in [8] for $F(x)$ could be used

$$F(x) = Cx^{-\alpha}(1+x)^{-(\beta-\alpha)}(1+(x/10)^\gamma)^{-\delta}, \quad (4)$$

where $C = 0.28$, $\alpha = 1.2$, $\beta = 4.53$, $\gamma = 2.0$, $\delta = 0.6$.

On the basis of simulations using the simplified hadronic generator, it was also found [9,14,15] that the relation between the lateral shape of the electron distribution and the longitudinal shower age can be expressed by the $R_0(s)$ functional dependence, that is a consequence of shower universality properties [16].

3. Results

Monte-Carlo simulations of EAS initiated by protons and iron nuclei in the energy range $10^{15} \div 10^{19}$ eV were performed using CORSIKA v.7.4100 [20] with EPOS LHC v.3400 and SIBYLL v.2.1 (FLUKA 2011.2c.2) hadronic interactions models. In order to get reliable data on electron LDF at very large distances from the shower core the thinning level and particle weight limit were set as $\varepsilon_{th} = 10^{-8}$ and $\omega = 10^2$ respectively.

The analysis showed that for the averaged lateral distributions the whole set of calculated electron densities are reproduced by the scaling representation (Eq. (2)) within 20% in relative discrepancy over the interval of scaling variable $x = r/R_0 = (0.05 \div 25)$, corresponding to the region of radial distances from $r \sim 10$ m to $r \sim (2 \div 4)$ km. For smaller distances discrepancies become large. It does not allow to use the scaling approach for shower size evaluation from the experimentally measured local densities far from the axis. This limitation of the scaling approach was also noted in [17–19]. Nevertheless, scaling formalism provides precise description of the shape of lateral distribution of electrons at radial distances, where local densities can be measured by ground detectors of large air shower arrays. It should be particularly emphasized that the dependence of LDF on hadronic interaction models is completely governed by the variation of the single parameter, R_0.

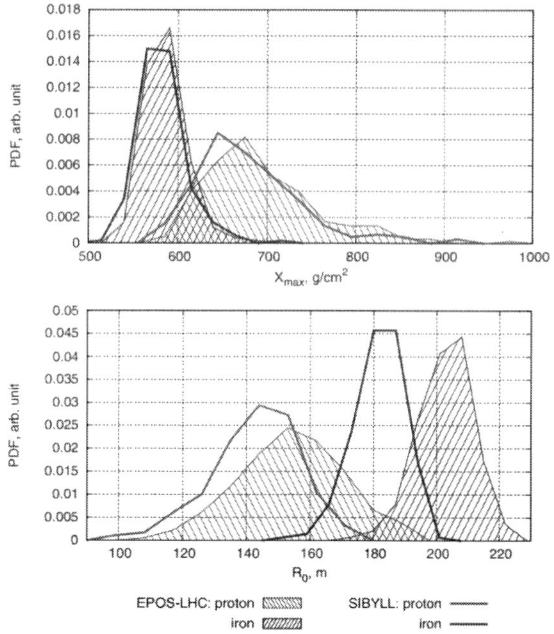

Figure 1. Distributions of depth of shower maximum X_{max} (top) and radial scale factor R_0 (bottom) at sea level for 10^{17} eV vertical showers initiated by protons (red) and iron nuclei (blue).

Radial scale factors were evaluated for both average and individual showers as parameters of simulated LDFs fitting. Note, that to obtain the unbiased estimates of R_0 an iterative procedure was implemented for discrimination of data at distances where the scaling formalism is not valid and also for using fixed distances ranges with respect to the scaling variable $x = r/R_0$ for showers of a specific energy and primary particle type. For $F(x)$, the expression of function (4) was used. We have checked out function (4) with a refined set of parameters α, β, γ, δ and other representations, including polynomial approximation for $x^2F(x)$ giving a considerably better overall fit of simulated data. It was found that the resulting R_0 values demonstrate only a weak dependence on the explicit form of $F(x)$ choosing for fitting.

The distributions of depth of the shower maximum, X_{max}, and radial scale factor, R_0, at sea level for 10^{17} eV vertical showers initiated by protons and iron nuclei are shown in Fig. 1. 200 simulated showers are included in each data set. Results for EPOS LHC and SIBYLL interaction models are shown with shaded and open distributions respectively. The figure clearly shows that R_0 estimated from simulations is more sensitive to both primary mass and interaction model in comparison with X_{max}.

Figure 2 shows the scale factor, R_0, for average EAS of fixed energy at sea level versus the longitudinal shower age, s, defined in terms of a classic cascade theory as $s(X, X_{max}) = 3/(1 + 2X_{max}/X)$. At the first stage electrons produced from the decay of low energy muons were eliminated from local particle densities. In this case electron LDF is a pure superposition of partial

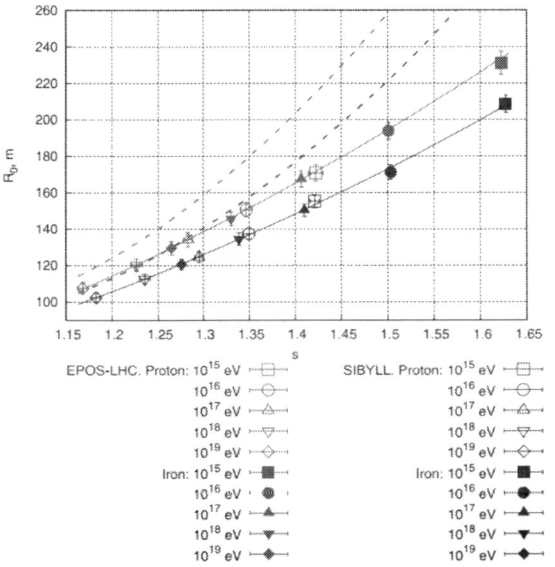

Figure 2. Radial scale factor, R_0, for average EAS of fixed energy as a function of shower age, s, (see text for details).

electromagnetic subshowers definitely possessing scaling and universality features. This data is shown by symbols and solid lines of red (EPOS LHC) and blue (SIBYLL) color. When the contribution of electrons from muon decays is taking into account the LDF become narrower at very large distances, which results in increased R_0. This effect wanes with energy. The corresponding data is presented by dashed lines in Fig. 2. It is important that in both cases there is a functional dependence $R_0(s)$, i.e., one-valued relation between parameters of the shower lateral (R_0) and the longitudinal (s) development stage representing the air shower universality [14, 15].

One can see from Fig. 2, that the rate of decrease of R_0 with energy is almost equal for the two interaction models used in the calculations. Thus the $\partial R_0/\partial \log E$ value can be suggested [9, 21–24] to be a model-independent measure of primary composition variations with energy similar to the widely used elongation rate approach.

Finally we investigated the possibility of using R_0 together with X_{max} for a two-component event-by-event analysis available in the case of hybrid measurements by the surface array and fluorescent telescopes. In Fig. 3 correlations and fluctuations of R_0 and s at sea level are shown for 10^{17} eV vertical proton- and iron-induced showers simulated with the EPOS LHC model. Strong anticorrelation useful for primary mass discrimination in individual showers is observed.

4. Discussion

Discrepancies between mass composition estimates, obtained by different methods from different air shower arrays data, are probably caused by the sum of instrumental and methodical systematic biases, as well as a strong model dependence of observables using as primary mass indicators (mostly in case of muon component characteristics), the inability of effective discrimination of electromagnetic and muon contributions to the ground

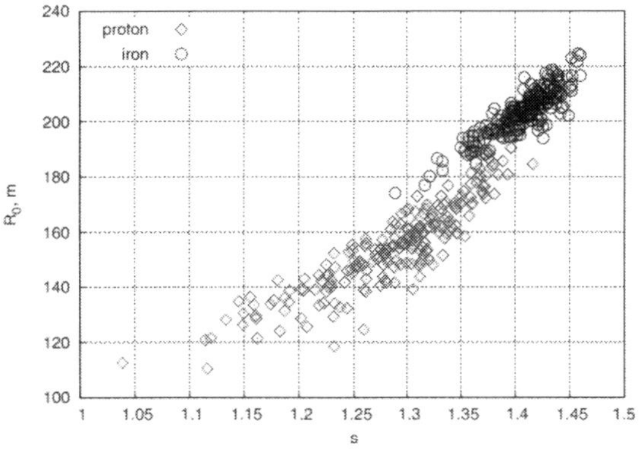

Figure 3. R_0 vs. s at sea level scatter plot. Data for 200 simulated vertical showers initiated by protons (red diamonds) and iron nuclei (blue circles) of 10^{17} eV (EPOS LHC model) is given.

EAS signal, the lack of understanding of meteorological effects and specific detectors properties etc.

A possible solution to the problem might be achieved with refined multi-hybrid measurements by existing EAS arrays taking into account the capabilities of their anticipated upgrades (e.g., separate detection of electromagnetic and muon contributions to the ground signal by "Auger Prime" [25]) together with generalizations of the analysis by revealing universal features, new parameters and functionals that exhibit weak sensitivity to the interaction model being good primary mass indicators.

The slope of the lateral distribution of EAS charged particles far from the shower core is known to be a primary mass discriminator. But it is rarely used in recent experimental works because precision measurements of the shape of the lateral distribution can not be realised by air shower arrays with large separation between ground stations. On the other hand the important advantage of the LDF as a source of information about mass composition is that ground detectors of charged particles provide an order of magnitude higher duty cycle in comparison with atmospheric telescopes. It is important to note here that using the characteristics of the muon component as basic composition estimates faces the "muon excess problem" [26–28], which requires extended investigations.

5. Conclusions

On the basis of the analysis of CORSIKA simulations we show that, when an adequate model-independent description of the lateral distribution of electrons in the framework of scaling formalism is adopted, the single integral parameter of the lateral distribution – scaling factor R_0 – is a potentially effective primary mass estimator. When combining with the data on longitudinal shower age, it provides improved composition results for both averaged shower measurements and on an event-by-event basis.

The proposed approach exhibits promises as a tool for both primary composition studies and hadronic interaction models tests in a wide primary energy range. It could be implemented for present and future (multi-)hybrid EAS observations, especially in realising the potential of Auger and Telescope Array observatories upgrades, as well as for re-analysis and cross-calibration of the data collected from different air shower arrays.

This work was supported by Russian Foundation for Basic Research (grant #16-02-01103 a).

References

[1] K.-H. Kampert, M. Unger, Astropart. Phys. **35**, 660–678 (2012)
[2] A. Haungs, Physics Procedia **61**, 425–434 (2015)
[3] A. Haungs, J. Phys.: Conference Series **632**, 012093 (2015)
[4] K.-H. Kampert, P. Tinyakov, Comptes Rendus Physique **15**, 318–324 (2014)
[5] S. Buitink et al., Nature **531**, 70–73 (2016)
[6] A.A. Lagutin et al., Proc. 25 ICRC, Durban **6**, 285–288 (1997)
[7] A.A. Lagutin et al., Nucl. Phys. B (Proc. Suppl.) **60B**, 161–164 (1998)
[8] A.A. Lagutin, R.I. Raikin, Nucl. Phys. B (Proc. Suppl.) **97B**, 274–277 (2001)
[9] A.A. Lagutin et al., J. Phys. G: Nucl. Part. Phys. **28**, 1259–1274 (2002)
[10] K. Greizen, Ann. Rev. Nucl. Sci. **10**, 63–108 (1960)
[11] W.D. Apel et al., Astroparticle Physics **24**, 467–483 (2006)
[12] R.K. Dey, A. Bhadra, J.N. Capdevielle, J. Phys. G: Nucl. Part. Phys. **39**, 085201 (2012)
[13] R. Dey, S. Dam, Eur. Phys. J. Plus **131**, 402 (2016)
[14] A.A. Lagutin, R.I. Raikin, T.A. Serebryakova, Bull. Russ. Acad. Sci. Phys. **77**(5), 623–625 (2013)
[15] A.A. Lagutin, R.I. Raikin, T.A. Serebryakova, J. Phys.: Conf. Ser. **409**, 012092 (2013)
[16] P. Lipari, Phys. Rev. D **79**, 063001 (2009)
[17] N.N. Kalmykov, G.V. Kulikov, V.P. Sulakov et al., Bull. Russ. Acad. Sci. Phys. **71**, 522–525 (2007)
[18] N.N. Kalmykov, G.V. Kulikov, V.P. Sulakov et al., Moscow Univ. Phys. **62**, 31–34 (2007)
[19] Yu.A. Fomin et al., Nucl. Phys. B (Proc. Suppl.) **175–176**, 334–337 (2008)
[20] D. Heck et al., **Report FZKA 6019** (Forschungszentrum Karlsruhe, 1998)
[21] R.I. Raikin et al., Nucl. Phys. B (Proc. Suppl.) **75–176**, 559–562 (2008)
[22] R.I. Raikin et al., Nucl. Phys. B (Proc. Suppl.) **196**, 383–386 (2009)
[23] R.I. Raikin, A.A. Lagutin, Bull. Russ. Acad. Sci. Phys. **75**(3), 305–308 (2011)
[24] R.I. Raikin, A.A. Lagutin, Proc. of 32 ICRC, Beijing, China **1**, 299–302 (2011)
[25] A. Aab et al. (Pierre Auger Collaboration), `arXiv:1604.03637` (2016)
[26] T. Abu-Zayyad et al. (HiRes-MIA Collaboration), Phys. Rev. Lett. **84**, 4276 (2000)
[27] A. Aab et al. (Pierre Auger Collaboration), Nucl. Instrum. Methods Phys. Res., Sect. A **798**, 172 (2015)
[28] A. Aab et al. (Pierre Auger Collaboration), Phys. Rev. Lett. **117**, 192001 (2016)

EPJ Web of Conferences **145**, 19015 (2017) DOI: 10.1051/epjconf/201714519015
ISVHECRI 2016

Search for dark matter by the Yakutsk array data

A.A. Mikhailov[a] and V.A. Kolosov

Yu.G. Shafer Institute of Cosmophysical Research and Aeronomy, 31 Lenin Ave., 677980 Yakutsk, Russia

Abstract. We analyzed the Yakutsk Extensive Air Shower array data registered from 1974 to 2011. 14 particles were found that come directly from the Galactic Center to Earth without any deflection in the Galactic magnetic field.We suppose that these are particles of dark matter.

1. Introduction

We analyzed the distribution of particles detected by the Yakutsk array for energies $E \geq 8.10^{18}$ eV from 1974 until 2011. We considered the arrival directions of extensive air showers (EAS) with zenith angles $\theta < 60°$ and whose cores were situated inside the perimeter of the array. The energy of the particles was determined by registering Cherenkov light with an accuracy of $\sim$30% and for angles $\sim 5° - 7°$.

2. The distribution of particles in right ascension

At first we analyzed the distribution of particles in right ascension by harmonic analyses for energies $E \geq 8.10^{18}$ eV. For these energies we found the amplitude and phase in right ascension of the 1st harmonic of Fourier analysis to be $r_1 = 0.14 \pm 0.04$, $\phi_1 = 0.6° \pm 19°$, $r_2 = 0.02 \pm 0.04$, $\phi_2 = 8° \pm 125°$. The total number of particles detected was 938 and the chance probability, P, having amplitudes $r_1 \geq 0.14$ is $\sim$0.01.

Earlier we suggested a new method to analyze data only along two coordinates, declination and right ascension, and we analysed the data by this method [1]. We did not find any statistical significance in the amplitude of arrival directions of particles by Fourier analyses, but phases of the 1st harmonic at energies $E > 10^{19}$ eV start to change at right ascension RA$\sim$ 180° or at high galactic latitudes, b (Fig. 1). Possibly this change of phase of the 1st harmonic most likely points to a change of sources of ultrahigh energy cosmic rays to high latitudes.

3. The distribution of particles with energies $E > 8.10^{18}$ eV in the Celestial sphere

In a previous paper [1] we concluded that cosmic rays up to energy $E < 4 \times 10^{19}$ eV are galactic and their sources are pulsars. We have found 6 regions, including data from the Telescope Array (TA), with many pulsars: inside the 1st region – 39 pulsars, the 2nd region – 16 pulsars, the 3rd region – >10 pulsars (the Galactic Center), the 4th region – 4. Inside these 6 regions there were 183, 11, 17, 23, 14, 11 particles respectively. The chance probabilities to find such a number of events from 938 particles are: $P_1(938, \geq 183) \sim 6.10^{-5}$, $P_2(938, \geq 11) \sim 0.6$, $P_3(938, \geq 17), \sim 2.10^{-3}$, $P_4(938, \geq 13) \sim 0.5$, $P_{4,5} \sim 0.1$.

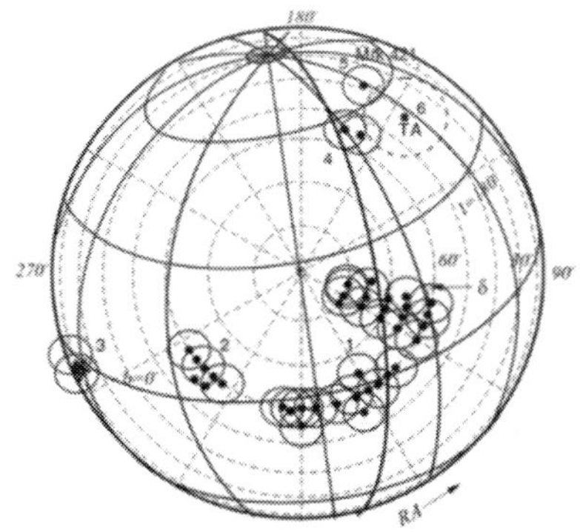

Figure 1. Yakutsk, $E \geq 8.10^{18}$ eV, distribution of particles in the celestial sphere • within a radius R < 6° – ≥16 particles were observed. 1, 2, 3, 4 – data of the Yakutsk array; 5 – data of TA around galaxy Mkr 421; 6 – TA data – R < 20°.

Note that the chance probabilities are only significant around the 1st and 3rd groups of pulsars at radius R < 6° and are 3σ more than expected in the case of isotropy. The maximum number of events $20 \leq n \leq 23$ at R < 6° are observed around the 1st groups of pulsars: PSR 0141+6009, 0146+6145, 0147+5922, 0157+6212, 0215+6218. These pulsars are in the galactic plane and have been considered earlier as possible sources of particles [1].

A flux of particles was observed from the side center of the Galaxy (Fig. 1, number 3). All 14 particles arrive at the Yakutsk array at large zenith angles, $60° \geq \theta \geq 38°$, and from only one direction as neutral particles of dark matter (the trajectories of these particles are not similar to other particles arriving directly from one direction only – the Galactic Center without any deviation). From 6 showers we have no information about the muon component data – these showers were registered in the summer (in summer muon detectors do not operate

[a] e-mail: `mikh@ikfia.ysn.ru`

DOI: 10.1051/epjconf/201714519015

because they are underground in permafrost and possible melting permafrost). In the autumn after repair the muon detectors begin to operate. Note that an excess of particles was found from the direction of the Galactic center mainly in autumn at energy $E \sim 10^{18}$ eV (according to theory maximum fluxes of particles of dark matter are expected in autumn [2]). We suppose that we registered particles of dark matter. The AGASA and Sydney arrays also observed a flux of particles from the same direction – the side center of the Galaxy [3,4].

4. Conclusion

We have found correlations between the arrival directions of ultrahigh energy particles with $E < 4 \times 10^{19}$ eV and the positions of pulsars and an anisotropy of the arrival directions of these particles from the side of the galactic plane for both the Yakutsk array data and the Telescope Array data at energy $E > 5.7 \times 10^{19}$ eV at high galactic latitudes. Our analysis of the data shows that the cosmic rays observed up to 4×10^{19} eV are galactic and above this energy most likely extragalactic.

References

[1] Mikhailov A.A., Analysis of the Arrival Directions of Ultrahigh Energy Particles. Discovery of Cosmic Rays, book, Denver, June 26–28, 2012, ed. J. Ormes
[2] Kuzmin V.A., Rubakov V.A., Yadernaya Fizika, 1998, **61**, 1122
[3] Anda R., Aguirre C., Tachi K., Proc.17-th ICRC, Paris, 1981, **2**, 164
[4] Bray A.D., Horton., McCusker C.B.A., Proc.17-th ICRC, 1981, **2**, 168

Liquid scintillator composition optimization for use in ultra-high energy cosmic ray detector systems

Dmitriy Beznosko[a], Ayan Batyrkhanov, Alexander Iakovlev, and Khalykbek Yelshibekov

Nazarbayev University, Physics Department, Astana, Kazakhstan

Abstract. The Horizon-T (HT) detector system and the currently under R&D HT-KZ detector system are designed for the detection of Extensive Air Showers (EAS) with energies above $\sim 10^{16}$ eV ($\sim 10^{17}$ eV for HT-KZ). The main challenges in both detector systems are the fast time resolutions needed for studying the temporary structure of EAS, and the extremely wide dynamic range needed to study the spatial distribution of charged particles in EAS disks. In order to detect the low-density of charged particles far from the EAS axis, a large-area detector is needed. Liquid scintillator with low cost would be a possible solution for such a detector, including the recently developed safe and low-cost water-based liquid scintillators. Liquid organic scintillators give a fast and high light yield (LY) for charged particle detection. It is similar to plastic scintillator in properties but is cost effective for large volumes. With liquid scintillator, one can create detection volumes that are symmetric and yet retain high LY detection. Different wavelength shifters affect the scintillation light by changing the output spectrum into the best detection region. Results of the latest studies of the components optimization in the liquid scintillator formulae are presented.

1. Introduction

In large-area distributed detectors for ultra-high energy cosmic rays, such as the Horizon-T detector system [1] and the HorizonT-KZ (HT-KZ) detector system [2] (currently under development), there is always a question about a medium for charged particle detection. One can choose water and detect Cherenkov light; it is a very fast method but produces low numbers of photons per passing particle. Another choice is an organic scintillator [3], that can be in the form of plastic or liquid scintillator (LS).

An important part of either the plastic or LS is the right choice of fluor and wavelength shifter, and their concentrations. The choice affects the spectral distribution of the light output and the detection efficiency as each photodetector has its own spectral sensitivity region. In this study, the 1,2,4-Trimethylbenzene, or Pseudocumene (PC) based LS was used together with PPO (2,5-diphenyloxazole) as a fluor and MSB (1,4-Bis(2-mehylstyryl)benzene) as a wavelength shifter. Both the total light yield (LY) and the spectral differences in the outputs with different amounts of components were measured. This study can be applied for plastic scintillators as well as for the liquid ones. For large-volume detectors planning to use LS with high particles density signals, LY and cost can be reduced by using the novel water-based liquid scintillator material [4].

2. Fluor concentration optimization

The schematics for the setup used to measure the LY at different PPO concentrations is shown in Fig. 1. The sample is viewed by two MELZ-FEU [5] PMT-115M connected to the CAEN [6] DT5743 ADC. The ADC internally handles the double coincidence between the PMT signals. ^{60}Co was used with each sample as the source of radiation. All measurements were conducted inside the black box. PC+PPO solution was used with this setup. A sample histogram of the light intensity for a measurement with 1 g/L PPO concentration in the sample is shown in Fig. 2.

The results of the analysis of light intensity dependence on PPO concentration are presented in Fig. 3. From the figure, we can see that that there is no optimal concentration for PC + PPO solution - the dependence is logarithmic and this function increases with the smoothly decreasing slope. The sample with a concentration of 2 g/L PPO was chosen for future tests.

After completion of the PPO tests, the wavelength shifter was added to observe how different concentrations will affect the LY. The addition of MSB to the sample caused a decrease in the total amount of light detected by the PMTs. This outcome could be only explained by the spectral sensitivity characteristics of the MELZ-FEU PMT. It has a very uniform sensitivity between $\sim$380 nm and $\sim$410 nm, and in the first approximation the MSB is shifting the emitted by PPO light (peak at $\sim$380 nm) within this range of wavelengths (to $\sim$410 nm) with some finite efficiency. This observation has warranted detailed studies of the spectral distributions of LS light to see the actual changes and access the MSB efficiency.

3. Spectroscopy and MSB contribution

Sample spectra for different concentrations of shifter in the solution were obtained. Measurements were done using an Agilent Cary Eclipse spectrophotometer [7]. In order to preserve the output spectra of the PPO as seen by the

[a] e-mail: dmitriy.beznosko@nu.edu.kz

DOI: 10.1051/epjconf/201714519016

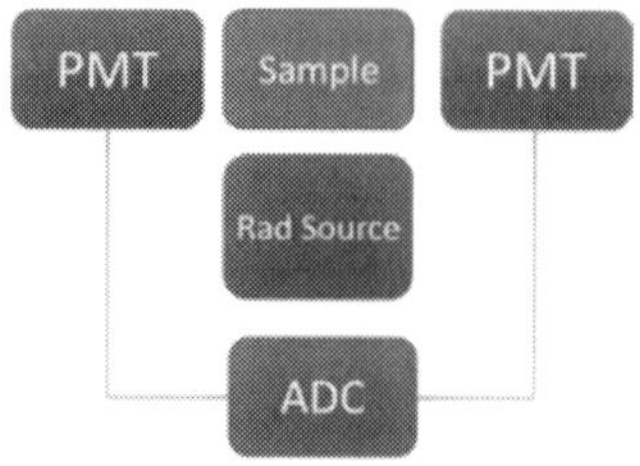

Figure 1. Experimental setup schematics for PPO concentration optimization.

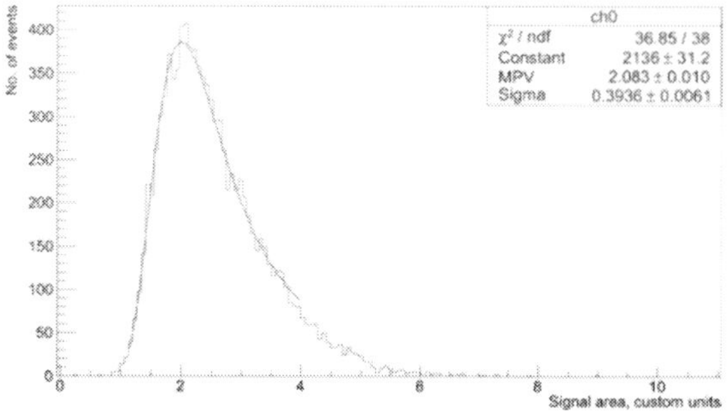

Figure 2. Sample data for 1 g/L PPO concentration.

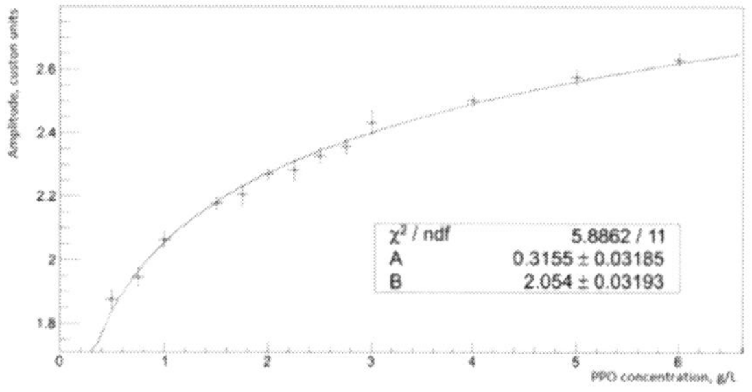

Figure 3. Light yield vs. PPO concentration.

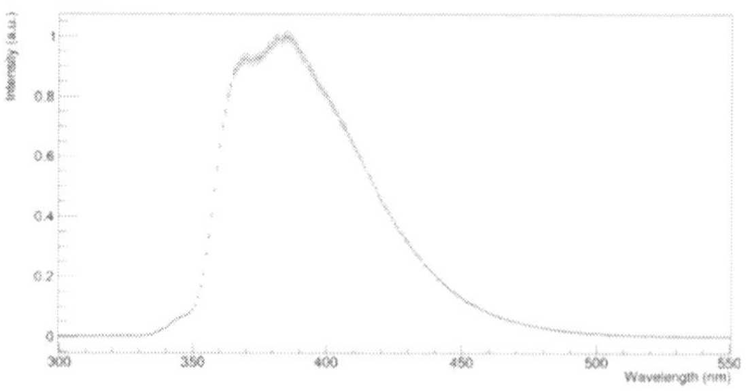

Figure 4. LS + PPO output spectrum.

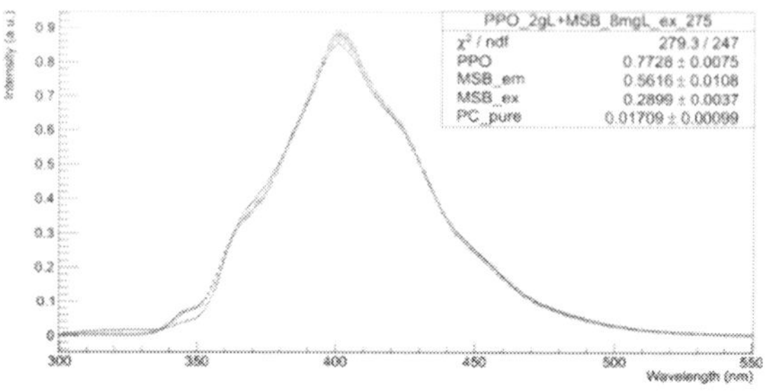

Figure 5. Output spectrum with MSB concentration of 8 mg/L.

shifter and minimally affect the added shifter directly, the wavelength of the excitation light was chosen at 275 nm that corresponds to MSB minimal absorption so that it will be absorbed by PPO only. The read-out range of the emitted light was chosen between 300–550 nm.

The output of the LS with only PPO added is presented in Fig. 4. Changes in the shape of the spectra with MSB concentrations of 8 mg/L can be seen in Fig. 5. Data is fitted by the empirical equation to determine the contribution of each component. PPO, MSB_{em}, MSB_{ex} and PC_{pure} are the parameters for light emitted by PPO, MSB absorption of PPO light, MSB re-emission, and PC/other background contributions. The parameters of most interest are MSB_{ex} and MSB_{em} that represent the light absorbed and emitted by MSB. The ratio of these two parameters shows that only about 50% of the light is re-emitted. Thus, integral LY is reduced.

4. WbLS testing

WbLS potentially combines both the Cerenkov ring and scintillation light capabilities. It can preserve the particle identification for particles above the Cerenkov threshold, and detect the charged particles below the threshold via scintillation light. In addition, WbLS features the lower cost than pure LS and it is safer to handle [4]. For high energy cosmic rays, WbLS can be a medium with a tunable LY. The ability to reach the desired LY can be checked using a mono-energetic proton beam with different WbLS concentrations. For the test, two different WbLS formulations (0.4% and 0.99%), pure water and pure LS samples were chosen. Three different proton beam energies were used with each sample. The choice of the energies comes from the following considerations:

- 2000 MeV protons behave as minimum ionizing particle (MIP)
- 475 MeV protons are just below the Cerenkov limit in water
- 210 MeV protons have ~same energy deposition as K^+ from the proton decay simulation.

The details of the experimental setup and measurements can be found in [4]. A surfactant (Linear AlkylBenzene Sulfonic Acid, LAS) is used to dissolve the PC, a common LS material also used for WbLS, in water. The PPO and MSB are used as a fluor and a wavelength shifter to downshift the ultraviolet light from the LS into the blue region of the spectrum where water has higher transparency.

The composition of the 4 samples tested in this experiment:

- Water (purified)
- WbLS-1: 0.4% PC + 0.4 g/L PPO + 3 mg/L MSB+surfactant in water
- WbLS-2: 0.99% PC + 1.36 g/L PPO + 7.48 mg/L MSB+surfactant in water
- LS: PC + 2 g/L PPO + 15 mg/L MSB.

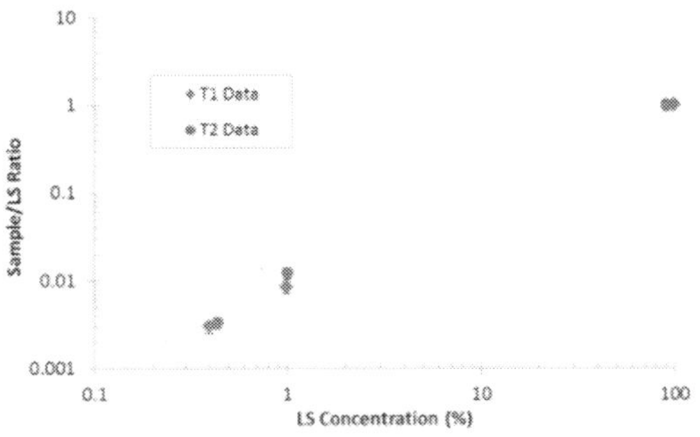

Figure 6. WbLS output ratio to LS for 475 MeV proton energy. Some points are shifted for visibility.

The water sample is used to control systematics and to subtract the Cherenkov light contribution to the scintillation light. The result for the two test volumes (T1 and T2) at 475 MeV proton beam energy is shown in Fig. 6 as WbLS LY ratio to pure LS. The results for other energies is similar.

The LY for pure LS and two formulations of the WbLS have been successfully measured. The 0.99% WbLS sample yields ~1% light of the pure LS; assuming that typical LY of LS is 10000 optical photons per MeV suggests 100 photons/MeV for this concentration. Therefore, the WbLS that satisfies the requirements for different detector needs can be fabricated. The result also illustrates that different LY can be easily achieved by adjusting the WbLS components concentration. Additional work is planned at Nazarbayev University to test the LS optimization with WbLS.

5. Conclusion

Work has been done on the optimization of LS composition. The results can also be applied to plastic scintillators as well. The concentration of each LS component can be chosen based on the needs from the fit parameters of the LY. Experiment showed that the amount of output light is reduced when the shifter is added to base + fluor solution as shifter efficiency is ~50%. The choice of the wavelength shifter depends on the sensitivity of the detector used for the experiment. This study plans to see the results applicability to WbLS and continue the optimization tests.

The project is financially supported by the grant of the Corporate Fund "Fund of Social Development"

References

[1] RU Beisembaev et al., `arXiv:1605.05179` (2016)
[2] A. Duspayev et al., in proceedings to *PhotoDet2015* conference, **PoS(PhotoDet2015)056**, (2015)
[3] A Baitenov et al., `arXiv:1601.00086` (2016)
[4] L. J. Bignell et al., Journal of Instrumentation, **10**, 12009 (2015)
[5] MELZ-FEU, 4922-y pr-d, 4c5, Zelenograd, g. Moskva, Russia, 124482 `http://www.meltz-feu.ru`
[6] CAEN S.p.A. Via della Vetraia, 11, 55049 Viareggio Lucca, Italy, `http://www.caen.it`
[7] Agilent INC. 5301 Stevens Creek Blvd, Santa Clara, CA 95051, US, `http://agilent.com`

Hadrons anti-hadrons production in reaction planes in relativistic heavy ion collisions on two common event generators: HIJING and HYDJET++

M.H.M. Soleiman[a], S.S. Abdel-Aziz, and M.S.E. Sobhi

Cairo University, Faculty of Science, Physics Department, Giza, Egypt

Abstract. The study of hadron production in heavy ion collisions provides the researcher with valuable techniques to investigate the properties of quark gluon plasma (QGP). The hadron and anti-hadron production in the reaction plane, at relativistic heavy ion collisions are studied for the energy per nucleon available at the RHIC-STAR experiment of $\sqrt{(s_{NN})} = 200\,\text{GeV}$. In order to test the models used for hadron and anti-hadron production, two common event generators (*HIJING* 1.35 and *HYDJET++* 2.0.2), established on Monte Carlo techniques, are discussed for hadron and anti-hadron production in the reaction plane for heavy ions $^{197}Au_{79} - {}^{197}Au_{79}$ collisions at $\sqrt{S_{NN}} = 200\,\text{GeV}$. We find that, *HYDJET++* 2.0.2 is more realistic in comparing the results of simulations with those of the experimental data published by the STAR-Collaboration.

1. Introduction

Relativistic heavy-ion collisions create suitable conditions for a phase transition from hadron to deconfined quark matter QCD and a test benchmark for lattice QCD calculations. Many evidences have demonstrated that the quark gluon plasma (QGP) matter has been produced mainly in the central $Au - Au$ collisions at RHIC energies [1]. In the process, large amounts of energy are deposited into a more extended volume than that achieved in elementary particle collisions. These nuclear interactions briefly produce hot and dense matter containing roughly equal numbers of quarks and antiquarks. Then the QGP expands rapidly and cools down. QGP undergoes a transition into unstable hadron gas, producing hadrons, anti-hadrons, and even complete nuclei as was reported by the RHIC-STAR experiment – USA in 2010 [2], where the discovery of anti-Helium 4 ($^4He_2$ or α) show the possibility of new dynamics in producing anti hadrons from QGP. Therefore the relativistic heavy-ion collision can not only provide an environment to study strong interacting phase transition and QCD matter but also an ideal venue to produce antimatter particles [3].

There are research proposals due to Kharzeev and Voloshin [4–6] about the rule of following the steps of chiral symmetry breaking: a- Rotation of charged quarks in QGP, b- Self-induced magnetic field, c- Generation of Faraday and Hall currents, d- Chiral flipping of left helicity to right helicity quarks. e- Charge separation due to self-induced magnetic field.

The global study of the ratio over all generated events is carried out in [7–9]. The event by event asymmetry calculation and then averaging is carried out in this investigation and assigned a convention of a local study.

The present work investigates the ability of the common models "*HIJING* 1.35", and "*HYDJET++* 2.0.2" used in studying QGP phenomenology with and without collective flow, respectively, by Monte Carlo techniques [3]; to reproduce phase space (i.e. spatial and momentum) distributions asymmetry in antihadron to hadron ratios. The spatial asymmetry distributions take the reaction plane as a spatial reference in estimating the asymmetries in antihadron to hadron ratios. We restrict our investigation to two baryons and two mesons. That are; proton and lambda ($p^{\pm}$, $\Lambda^{\pm}$) for baryons, and pi and kaon ($p^{\pm}, \kappa^{\pm}$) for mesons. Our results could be investigated at the relativistic heavy ion collider (RHIC), and applied on the data from the experiments of STAR, PHOBOS, PHENIX, BRAHMS in USA-BNL.

The paper is organized in order to show the reader the possibility to detect charge asymmetry in samples of events generated by each the event generators *HIJING* 1.35, and *HYDJET++* 2.0.2. The charge asymmetry appear and are detected without regarding the Kharzeev and Voloshin effects of the self-induced global chiral magnetic field and the subsequent charge separation of quarks and anti-quarks, since these effects are not included in any of the studied event generators. After the introduction, Sect. 2 will introduce basic definitions and methods used throughout the discussion. Section 3 gives the results and discussions. Finally Sect. 4 summarizes the conclusions and future work.

2. Methods and concepts

Hijing 1.35 is an event generator written in *Fortran* 77, based on QCD inspired models for multiple jet production, and designed to study jet and mini-jet production and associated particle production in pp, pA, and AA ultra-relativistic collisions. This

[a] e-mail: `mhm.soleiman@gmail.com; mhmsoleiman@cu.edu.eg`

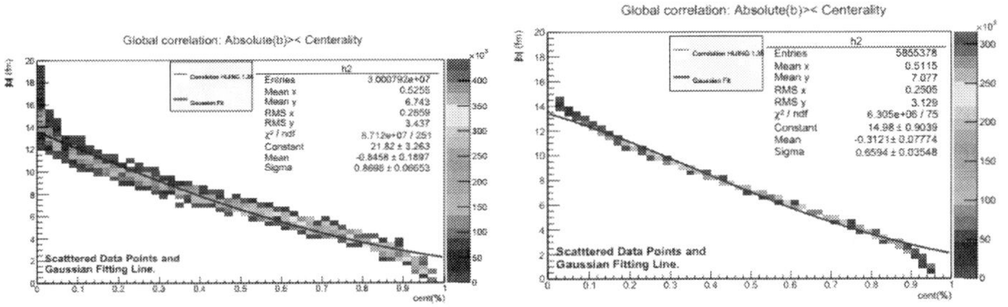

Figure 1. The correlation between the impact parameter |**b**| and the centrality. The solid line is the most probable Gaussian fit. The colors represent the densities of points. Graph (a) gives the results obtained on *HIJING* 1.35, and graph (b) is the results deduced on *HYDJET++* 2.0.2.

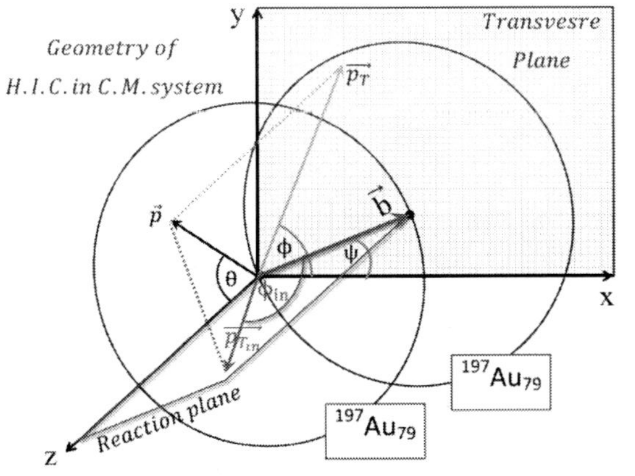

Figure 2. The geometry of heavy ion collision event. The reaction plane is represented in red. The transverse plane is represented as xy-plane.

model incorporates mechanisms such as multiple minijet production, soft excitation, nuclear shadowing of parton distribution functions and jet interactions in dense hadronic matter. Each generated event and the produced particles represents one probable state of a physical system. The flexibility of programming allows for testing several hypothesis and modifications on the dynamical models of QGP. *HYDJET++* 2.0.2. *HYDJET++* event generator [11] includes detailed treatment of soft hadro-production as well as hard multi-parton production, and takes into account medium-induced parton re-scattering and energy loss. The main program *HYDJET++* is written in the object-oriented *C ++* language under the *ROOT* environment [12, 13]. In our study *HIJING* 1.35 and *HYDJET++* 2.0.2 are tuned to $^{197}Au_{79} - {}^{197}Au_{79}$ relativistic heavy ion collisions at $\sqrt{s_{NN}} = 200\,\text{GeV}$, and 1000 collision events are generated.

The centrality of the heavy ion collision event is defined as $c = 1 - \frac{N_{participants}}{N_{total}}$, $N_{total} = A_{projectile} + A_{target} = 2 \times 197$ for $^{197}Au_{79}$. Figure 1 shows the scatter graphs of the absolute value of the impact parameter $|\overrightarrow{b}|$ in *Fermi units (fm)* and the centrality of $Au - Au$ collision as produced by *HIJING* 1.35 in graph (1-a) and *HYDJET++* 2.0.2 in graph (1-b). The monotonically decreasing strong correlation between c and $|\overrightarrow{b}|$ is confirmed by Gaussian fitting to the most probable values. The illustration in Fig. 2 is the geometry of a single collision event. It represents the reaction plane in red as the plane containing both the impact parameter vector $\overrightarrow{b}$ and the incidence direction of both heavy ions.

The present work studies the asymmetry in antihadron to hadron production ratios concerning four selected hadrons in the reaction plane. Two mesons: $\pi^{\pm}$ as a low energy scalar meson and $\kappa^{\pm}$ as a strange meson, and two baryons: $p^{\pm}$ the protons and $\Lambda^{\pm}$ the lambda strange baryon, are selected in our study. The investigation of their multiplicity distributions are commonly reported [9].

The asymmetry sum is defined as,

$$\zeta = \Sigma_{events} \frac{(No.\, Antiparticles - No.\, particles)}{(No.\, Antiparticles + No.\, particles)},$$
$$-N_{events} \leq \zeta \leq N_{events} \qquad (1)$$

3. Results and discussions

The momentum and space distributions of the hadron for the collision events produced by the *HIJIJNG* 1.35 event generator are obtained and presented in Fig. 3 for global antihadron to hadron ratios, and Fig. 4 for local asymmetry sum ζ. In general there is a fault in the *Hijing* 1.35 hadronization mechanism for producing lambda baryons. The results for lambda particles are not trusted quantitatively, but the qualitative behavior is concerned as it is consistent with the validation made by V. V. Uzhinsky [9].

The symmetry in antihadron to hadron productions is balanced, if the ratios are equal to one and the asymmetry sums, ζ, equals zero over all the space and momenta.

The distributions in Fig. 3 show, in general, larger values of antihadron to hadron ratios for $\pi^{\pm}$, than for $\kappa^{\pm}$ and finally for $p^{\pm}$ as p^{+} already exist in the colliding heavy ions, respectively. That is, the production of the anti-partner of the lightest meson is a candidate to test the *HIJING* model ability to produce the asymmetry in antihadron-hadron production [14]. Except for statistical fluctuations, the in reaction plane transverse momentum $P_{T_{in}}$ behaviors are semi–decreasing $P_{T_{in}}$ values. As shown in graph (3-Left) for *HIJING* 1.35 and graph (3-Right) for *HYDJET++* 2.0.2. Kharzeev models also predict the enhancement in the production of anti-baryons with rapidity y [4].

Figures (4-left) and (4-right) shows asymmetry ratios sum ζ distributions with $P_{T_{in}}$ for baryons $p^{\pm}$ and $\Lambda^{\pm}$. The mesons $\pi^{\pm}$ and $\kappa^{\pm}$ are symmetrical in the production of anti-hadrons. The anti-kaon production is enhanced in the

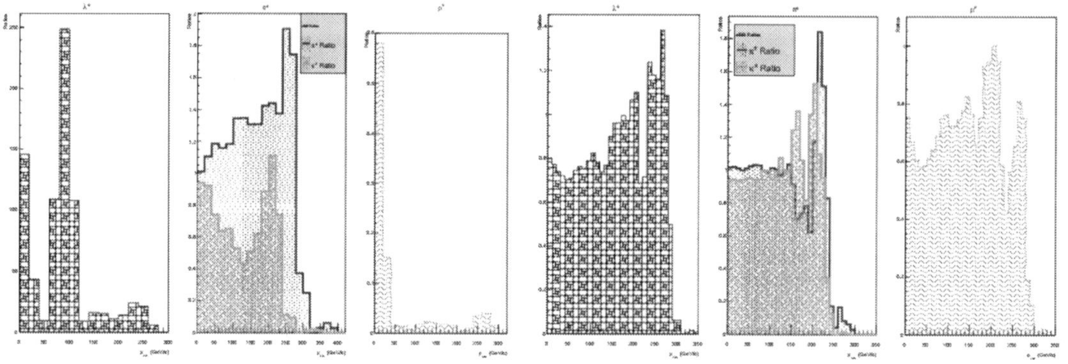

Figure 3. The ah/h ratios in relation with $p_{T_{in}}$ for mesons $\pi^{\pm}$ (blue), $\kappa^{\pm}$ (green)), and baryons ($P^{\pm}$ (pink), $\Lambda^{\pm}$ (black)); (Left) *HIJING* 1.35. (Right) *HYDJET++* 2.0.2.

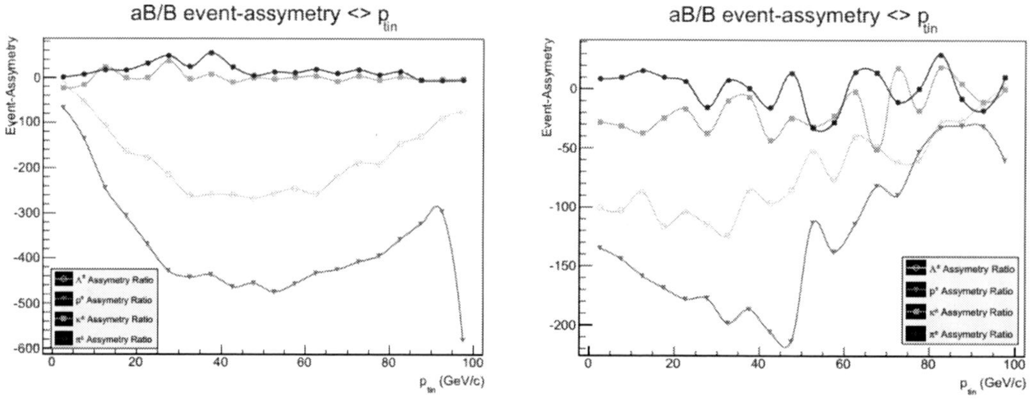

Figure 4. ah/h event asymmetry ratio sum with respect to in reaction plane momentum $p_{T_{in}}$ using (Left) *HIJING* 1.35 and (Right) *HYDJET++* 2.0.2 simulator.

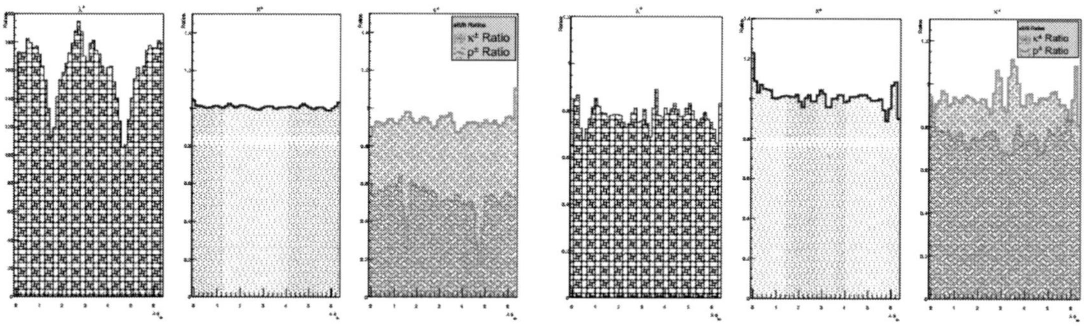

Figure 5. The ah/h ratios in relation with $\Delta\phi_{in}$ for mesons $\pi^{\pm}$ (blue), $\kappa^{\pm}$ (green)), and baryons ($P^{\pm}$ (pink), $\Lambda^{\pm}$ (black)); (Left) *HIJING* 1.35. (Right) *HYDJET++* 2.0.2.

backward hemisphere, but the anti-proton production is enhanced in the forward hemisphere.

Figures (5-left) and (5-right) are the ratio distributions for the four particles with the azimuthal angle referenced with respect to the reaction plane (= direction of the impact parameter vector $|\overrightarrow{b}|$,) $\Delta\phi = \phi - \psi$, and the in-reaction plane azimuthal angle, $\Delta\phi_{in}$. The distributions are symmetrical except for the proton $p^{\pm}$ ratio distribution for $\Delta\phi_{in}$. The production of anti-protons is equally probable except for the angles $\Delta\phi_{in} = \pi/2, 3\pi/2$ radians.

Figures (6-left) and (6-right) are the asymmetry ratio sum ζ for $\Delta\phi_{in} = \phi_{in}$. Irrespective of $p^{\pm}$ the production of particles and anti particles are uniform over the angles $\Delta\phi_{in} = \phi_{in}$, while for protons, $p^{\pm}$, the protons production exceeds the anti-proton production at the angles $\Delta\phi_{in} = \pi/2, 3\pi/2$ radians.

4. Conclusions

HIJING 1.35 could reproduce the asymmetry in baryons and anti-baryons productions, but not for meson and

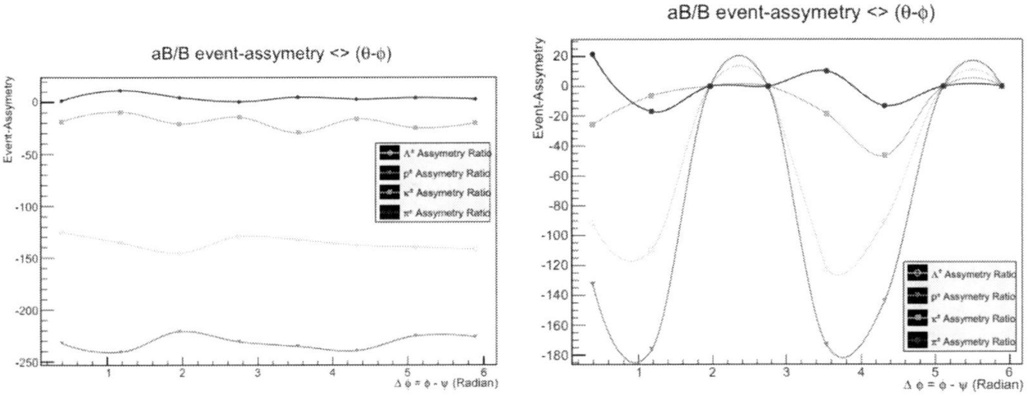

Figure 6. ah/h event asymmetry ratio sum with respect to in reaction plane azimuthal angle $\Delta\phi_{in}$ using (Left) *HIJING* 1.35 and (Right) *HYDJET++* 2.0.2 simulator.

anti-mesons. The new defined parameter of asymmetry sum is statistically stable, especially, for baryons (which have larger masses than mesons), and shows a general asymmetry for baryons but not for mesons. The proton and anti-proton production according to the *HIJING* model shows a deficiency in anti-proton to proton production in the directions normal to the reaction plane, *HYDJET++* 2.0.2. The asymmetry ratio ξ with $p_{T_{in}}$ is decaying for $p^{\pm}$ in *HIJING* 1.35 and semi decaying for $p^{\pm}$ and $\Lambda^{\pm}$ in *HYDJET++* 2.0.2. *HYDJET++* 2.0.2 predicts that the production of anti-baryons is enhanced with the $p_{T_{in}}$ value increasing. For the in-reaction plane azimuthal angle $\Delta\phi_{in}$, it is found that the asymmetry ratio sum ζ for the studied hadrons produced in events generated by *HYDJET++* 2.0.2 in the interval $2 \leq \Delta\phi_{in} \leq 4rad$ appears to decrease with some fluctuations. *HYDJET++* 2.0.2 has been shown to be more realistic than $HIJIN$ 1.35 in the study of dynamical parameters $p_{T_{in}}$. It could be considered that the behavior of the asymmetry ratio sum ζ is a good representation of what can be obtained from the experimental measurements. The asymmetry in the distributions of charged particles (i. e. hadrons) could be observed very effectively using the statistically stable parameter of the asymmetry ratio sum, ζ.

In future we intend to enter the effect of Kharzeev of the induced chiral magnetic field and the charge separation.

Thanks are due to Cairo University's research committee for partial funding of attendance at the symposium and encouraging the publishing of this work.

References

[1] STAR Collaboration, J. Adams, et al., Nucl. Phys. A **757**, 102 (2005)
[2] STAR Collaboration, B. I. Abelev, et al., Nature **473**, 353 (2011)
[3] Xin-Nian Wang, Miklos Gyulassy, Phys. Rev. D **44**, 3501 (1991), Comput. Phys. Commun. **83**, 307 (1994)
[4] Umut Gürsoy, Dmitri Kharzeev, Krishna Rajagopal, Nuclear Physics A **931**, 986–991(2014)
[5] D.E. Kharzeev, et al., Nuclear Physics A **803**, 227–253 (2008)
[6] S. A. Voloshin and A. M. Poskanzer, Phys. Lett. B **474**, 27–32 (2000)
[7] STAR Collaboration, B. I. Abelev, et al., Phys. Rev. C **76**, 024915 (2007)
[8] Jian-Hua Gao, Shou-Wan Chen, et al., Phys. Rev. C **77**, 044902 (2008)
[9] V. V. Uzhinsky, `arXiv:hep-Ph/0312089v2` (2003)
[10] I.P. Lokhtin, L.V. Malinina, S.V. Petrushanko, A.M. Snigirev, I. Arsene and K. Tywoniuk, Compu Phys. Commun. **180**(5), 779–799 (2009)
[11] I.P. Lokhtin, L.V. Malinina, S.V. Petrushanko, A.M. Snigirev, I. Arsene and K. Tywoniuk, Compu. Phys. Commun. **180**(5), 779–799 (2009)
[12] R. Brun and F. Rademakers, Nucl. Instrum. Meth. A **389**, 81 (1997), `http://root.cern.ch`
[13] Igor.Lokhtin, `http://cern.ch/lokhtin/hydjet++`, Last update: November 30 (2015) by `Igor.Lokhtin@cern.ch`
[14] AlLICE Collaboration, Abelev, and B. Bezverkhny, et al., Phys. Lett. B **728**, 25–38 (2014), `arXiv:nucl-ex/1307.6796v2`, CERN-PH-EP-2013-135

Average value of the cosmic ray injection exponent at Galactic sources

Anatoly Lagutin[a] and Nikolay Volkov[b]

Altai State University, Radiophysics and Theoretical Physics Department, 656049, 61 Lenin ave, Barnaul, Russia

Abstract. The energy spectrum of high energy cosmic rays emitted by sources change during propagation in the Galaxy. Using the spectrum observed at the Earth and two different transport models, based on anomalous and normal diffusion equations, we have retrieved an average value of the cosmic ray injection exponent at Galactic sources taking into account the inhomogeneity of the interstellar medium. We have shown that the average value of the injection index p, obtained in the framework of both transport models, equals to $p \sim (2.8 - 3.0)$.

1. Introduction

The spectrum of cosmic rays (CRs) measured at the Earth is different from their source spectrum. A key to understanding this difference is the determination of how cosmic ray (CR) particles propagate through the turbulent interstellar medium (ISM).

Assuming that the medium is quasi-homogeneous, the propagation process is described by Ginzburg-Syrovatsky's normal diffusion equation [1,2]

$$\frac{\partial N(\mathbf{r},t,E)}{\partial t} = D(E)\Delta N(\mathbf{r},t,E) + S(\mathbf{r},t,E), \quad (1)$$

where $N(\mathbf{r},t,E)$ is the density of particles with energy E at the location $\mathbf{r}$ and time t, $D(E) = D_0 E^{\delta}$ is the diffusion coefficient assumed to be spatially constant and $S(\mathbf{r},t,E)$ is the source term.

The steady state solution of (1), frequently used for interpretation of the CR phenomena, has the form

$$N(\mathbf{r},E) \sim E^{-p-\delta}, \quad (2)$$

where p is the injection spectrum exponent at the Galactic sources. The same result may be obtained using the entirely uniform Leaky-Box model.

The steady state energy spectrum (2) leads to a very simple rule to retrieve the exponent p from the spectrum $N(E) \propto E^{-\eta}$ observed at the Earth: $p = \eta - \delta$. Since $\eta \approx 2.7$, for $\delta \sim 0.3$ we find that the slope of the CR injection spectrum at the source is ~ 2.4.

However, recent results show that homogeneous diffusion models, based on (1), failed to reproduce observations (see, for example, [3]). Multiscale structures in the Galaxy, found during the last few decades, may be taken as a support to this conclusion.

Theory and observations show that the ISM is inhomogeneous (fractal-like) on the hundreds of parsecs scales [4–6]. Stars formation regions also demonstrate fractal features with spatial scales up to about a kpc [7–9].

The particles emitted by Galactic sources en route to the Solar system pass through regions of the Galaxy that have different properties. In such a inhomogeneous ISM, the normal diffusion model is certainly not kept valid.

The main goal of this paper is to retrieve the cosmic ray injection exponent at the Galactic sources from the spectrum observed at Earth taking into account the inhomogeneity of the turbulent interstellar medium. Two different models, based on anomalous and normal diffusion equations, are used to relate the spectrum injected by the source and that measured at Earth.

In the following section we give an overview of the anomalous diffusion model, discuss our injection exponent retrieval technique and the results obtained in this model. In Sect. 3 we present a new approach to describe cosmic ray propagation in the non-homogeneous ISM, based on the normal diffusion Eq. (1). Our findings and conclusions are given in Sect. 4.

2. Retrieval of the injection exponent using the anomalous diffusion model

The non-homogeneous character of matter distribution and associated magnetic field leads to the need to incorporate these ISM features into the cosmic ray diffusion model. A possible way to generalize the normal diffusion model is to replace the assumption about statistical homogeneity of inhomogeneities distribution by their fractal distribution. An important consequence of this generalization is the power-law distribution of free paths, r, in such a medium $p(\mathbf{r},E) \propto A(E,\alpha)r^{-\alpha-1}, r \to \infty, 0 < \alpha < 2$ — so-called Lévy flights. Since an intermittent magnetic field of the fractal-like ISM leads to nonzero probability of a long stay of particles in inhomogeneities, the presence of the Lévy trap can not be excluded. In the general case, the probability density function $q(t,E)$ of time, t, during which a particle is trapped in the inhomogeneity

[a] e-mail: lagutin@theory.asu.ru
[b] e-mail: volkov@theory.asu.ru

(Lévy trap), also has a power-law behavior: $q(t, E) \propto B(E, \beta)t^{-\beta-1}$, $t \to \infty$, $\beta < 1$.

A generalization of the homogeneous normal diffusion model to the case of inhomogeneous (fractal-like) ISM, has been made for the first time in our papers [10, 11]. Later, it was shown [12–17] that an anomalous cosmic ray diffusion model, developed by the authors, allows to describe the main features of nuclei, electron and positron spectra observed in the Solar system. Particularly, in the anomalous diffusion model the key feature of the all particle energy spectrum — the knee at $3 \cdot 10^{15}$ eV — appears naturally without additional assumptions.

The equation for the density of particles with energy, E, at the location, $\mathbf{r}$, and time, t, generated in a fractal-like medium by Galactic sources with a distribution density $S(\mathbf{r}, t, E)$ can be written as [12, 14]

$$\frac{\partial N(\mathbf{r}, t, E)}{\partial t} = -D(E, \alpha, \beta)\mathrm{D}_{0+}^{1-\beta}(-\Delta)^{\alpha/2}N(\mathbf{r}, t, E) + S(\mathbf{r}, t, E). \quad (3)$$

Here D_{0+}^{μ} denotes the Riemann-Liouville fractional derivative [18] and $(-\Delta)^{\alpha/2}$ is the fractional Laplacian ('Riesz operator') [18]. The anomalous diffusion coefficient $D(E, \alpha, \beta) \sim A(E, \alpha)/B(E, \beta) = D_0(\alpha, \beta)E^{\delta}$.

The solution of Eq. (3) for a point instantaneous source with a power-law injection spectrum $S(\mathbf{r}, t, E) = S_0 E^{-p}\delta(\mathbf{r})\delta(t)$ has the form [12, 14]

$$N(\mathbf{r}, t, E) = S_0 E^{-p}(D(E, \alpha, \beta)t^{\beta})^{-3/\alpha} \times \Psi_3^{(\alpha,\beta)}(|\mathbf{r}|(D(E, \alpha, \beta)t^{\beta})^{-1/\alpha}),$$

where $\Psi_3^{(\alpha,\beta)}(\rho)$ is the density of the fractional stable distribution [19, 20]

$$\Psi_3^{(\alpha,\beta)}(\rho) = \int_0^{\infty} g_3^{(\alpha)}(r\tau^{\beta})q_1^{(\beta,1)}(\tau)\tau^{3\beta/\alpha}d\tau.$$

Figure 1 demonstrates the behaviour of the function $\Psi_3^{(\alpha,\beta)}(\rho)$ for different parameters α and β.

The asymptotic behaviour of the scaling function $\Psi_3^{(\alpha,\beta)}(\rho)$ for $\alpha < 2$, $\beta < 1$,

$$\Psi_3^{(\alpha,\beta)}(\rho) \propto \rho^{-(3-\alpha)}, \quad \rho \to 0,$$

$$\Psi_3^{(\alpha,\beta)}(\rho) \propto \rho^{-3-\alpha}, \quad \rho \to \infty,$$

is shown on Fig. 2.

Note that in the energy range below the knee the cosmic ray density is

$$N(\mathbf{r}, t, E) \sim \frac{E^{-p+\delta}}{r^{3+\alpha}}. \quad (4)$$

Taking into account that $\rho \equiv |\mathbf{r}|(D(E, \alpha, \beta)t^{\beta})^{-1/\alpha}$, we find

$$N \sim E^{-\eta} = E^{-p+\delta}, \quad E \ll E_k$$

and

$$N \sim E^{-\eta} = E^{-p-\delta}, \quad E \gg E_k,$$

where E_k is the knee energy. Since $\Delta\eta = \eta|_{>E_k} - \eta|_{<E_k}$ is known from experimental data to equal $\Delta\eta \sim 0.6$, the last equations permit to, self-consistently, retrieve both spectral exponents p and δ:

$$\delta = \Delta\eta/2 \sim 0.3,$$

$$p = \eta|_{<E_k} + \delta = \eta|_{>E_k} - \delta \approx 2.8 \div 2.9.$$

At $E = E_k$ the spectral index $\eta(E_k)$ is equal to the injection exponent p at the Galactic source.

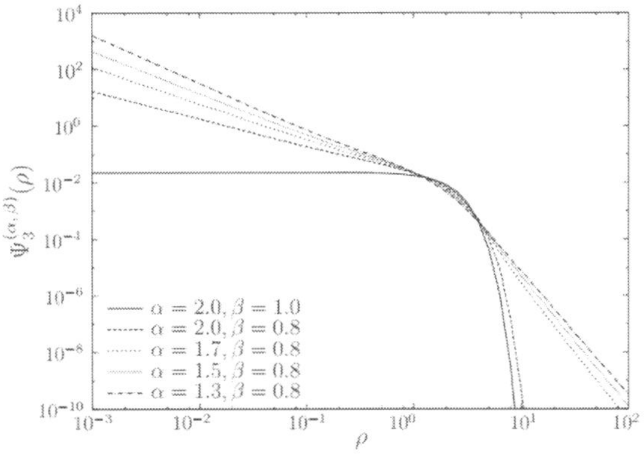

Figure 1. Probability density of fractional stable distribution $\Psi_3^{(\alpha,\beta)}(\rho)$ for different parameters α and β.

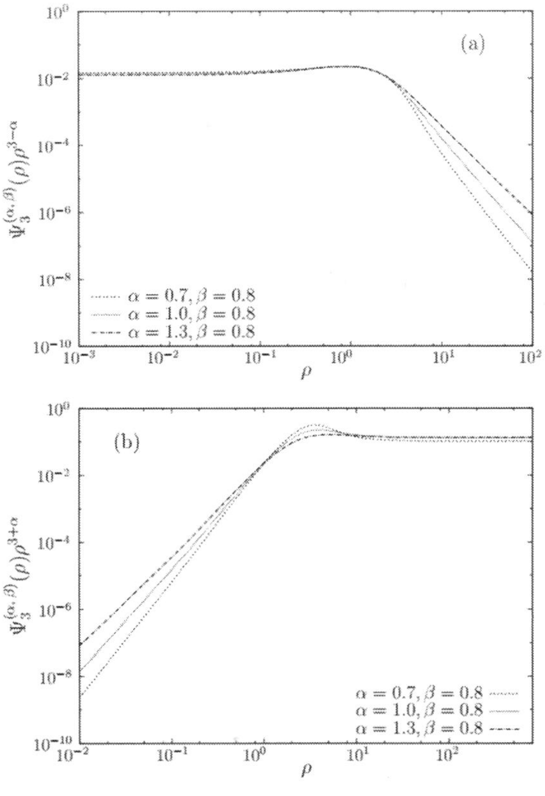

Figure 2. Asymptotic behavior of the function $\Psi_3^{(\alpha,\beta)}(\rho)$ for $\rho \to 0$ (a) and for $\rho \to \infty$ (b).

3. Normal diffusion of cosmic rays in a non-homogeneous ISM

The medium properties of an inhomogeneous ISM vary in space and time. The particles emitted by Galactic sources en route to the Solar system pass through regions of the Galaxy that have different diffusion coefficients.

Here we suppose that the particles with probability density $\varphi(D_0)$ propagate in the ISM with a diffusion coefficient D_0. Since for a homogeneous ISM, with the constant diffusion coefficient D_0, the solution of a normal diffusion Eq. (1) for a point instantaneous source with a power-law injection spectrum has the form

$$N(\mathbf{r},t,E)=\frac{S_0E^{-p}}{(4\pi D_0E^{\delta}t)^{3/2}}\exp\left(-\frac{r^2}{4D_0E^{\delta}t}\right), \quad (5)$$

we can find the density of particles visiting this part of the Galaxy,

$$dN_D(\mathbf{r},t,E)=\frac{S_0E^{-p}\varphi(D_0)dD_0}{(4\pi D_0E^{\delta}t)^{3/2}}\exp\left(-\frac{r^2}{4D_0E^{\delta}t}\right),$$

and then the observed density $N(\mathbf{r},t,E)$

$$N(\mathbf{r},t,E)=\int dN_D(\mathbf{r},t,E)=\frac{S_0E^{-p-3\delta/2}}{(4\pi t)^{3/2}} \times\int_0^{\infty}\frac{1}{D_0^{3/2}}\exp\left(-\frac{r^2}{4D_0E^{\delta}t}\right)\varphi(D_0)dD_0. \quad (6)$$

In order to have insight into possible types of diffusivity distribution we follow the approach presented in [21]. If r is the random jump length of a particle and τ is the time required to make this jump, the diffusion coefficient is given by equation

$$D_0=\frac{r^2}{2\tau}.$$

Since r/τ is the speed v, we have

$$D_0=\frac{v^2\tau}{2}.$$

Assuming that the speed of particle motion v along its trajectory is the same for all particles, for $\varphi(D_0)$ we find

$$\varphi(D_0)=\psi(\tau)\left|\frac{d\tau(D_0)}{dD_0}\right|=\frac{2}{v^2}\psi(\tau).$$

It is seen, for example, that the assumption that the time, τ, of a single step is distributed normally leads to a normal distribution for diffusivity. In the case of $\psi(\tau)\propto\tau^{-\beta-1}$, $\beta<1$ the power-law distribution for $\varphi(D)$ may be obtaned.

Similar distributions may be obtained if we assume that all particles during the random walk have the same value of r. In this case $\tau(D_0)=r^2/(2D_0)$ and

$$\varphi(D_0)=\psi(\tau)\left|\frac{d\tau(D_0)}{dD_0}\right|=\frac{r^2}{2{D_0}^2}\psi(\tau). \quad (7)$$

Using, for example, a normal distribution for τ

$$\psi(\tau)=\frac{1}{\sqrt{\pi}\Delta\tau}\exp\left[-\left(\frac{(\tau-\tau_0)}{\Delta\tau}\right)^2\right],$$

where τ_0 and $\Delta\tau$ are parameters ($\Delta\tau\ll\tau_0$), for large D_0 from (7) we find

$$\varphi(D_0)\sim {D_0}^{-2}.$$

The probability distribution function $\varphi(D_0)$ has a power-law form.

Here we consider the power-law distribution $\varphi(D_0)\sim D_0^{-\gamma}$ for $D_0\to\infty$, where γ is a positive parameter. Since $\varphi(D_0)$ has the meaning of a probability density, that is $\int_0^{\infty}\varphi(D_0)dD_0=1$, we must assume that $\gamma>1$.

In order to calculate the integral in Eq. (6), we need to make some additional assumptions about the properties of $\varphi(D_0)$ at small and also intermediate values of D_0. We adopt the following functional form

$$\varphi(D_0)=AD_0^{-\gamma}\exp\left(-\frac{\mathfrak{D}}{D_0}\right), \quad (8)$$

where $A=\dfrac{\mathfrak{D}^{\gamma-1}}{\Gamma(\gamma-1)}$ is the scaling factor, $\mathfrak{D}$ is the characteristic diffusivity for the Galactic medium, $\gamma>1$ is an auxiliary parameter, $D_0>0$. We additionally define $\varphi(0)=0$.

Having substituted equation (8) into equation (6) we find

$$N(\mathbf{r},t,E)=\frac{S_0AE^{-p-3\delta/2}}{(4\pi t)^{3/2}} \times\int_0^{\infty}\frac{dD_0}{D_0^{\gamma+3/2}}\exp\left[-\left(\mathfrak{D}+\frac{r^2}{4E^{\delta}t}\right)\frac{1}{D_0}\right]. \quad (9)$$

Using new variables $p=\mathfrak{D}+\dfrac{r^2}{4E^{\delta}t}$ and $D_0=\dfrac{p}{z}$, the integral in (9) is transformed as follows:

$$\int_0^{\infty}D_0^{-\gamma-3/2}\exp\left(-\frac{p}{D_0}\right)dD_0 =-p\int_{\infty}^{0}\frac{p^{-\gamma-3/2}}{z^{-\gamma-3/2}}\exp(-z)\frac{dz}{z^2} =p^{-\gamma-1/2}\int_0^{\infty}z^{\gamma-1/2}e^{-z}dz=p^{-\gamma-1/2}\Gamma\left(\gamma+\frac{1}{2}\right).$$

As a result, we obtain

$$N(\mathbf{r},t,E)=\frac{S_0AE^{-p-3\delta/2}}{(4\pi t)^{3/2}}\left(\mathfrak{D}+\frac{r^2}{4E^{\delta}t}\right)^{-\gamma-1/2} \times\Gamma\left(\gamma+\frac{1}{2}\right). \quad (10)$$

At any given moment of time, t, and for sufficiently large r, so that $r^2 \gg 4\mathfrak{D}E^{\delta}t$, the density of particles, $N(\mathbf{r}, t, E)$, has the form

$$N(\mathbf{r}, t, E) = \frac{S_0 \mathfrak{D}^{\gamma-1}}{\pi^{3/2}(4t)^{1-\gamma}} \frac{\Gamma\left(\gamma + \frac{1}{2}\right)}{\Gamma(\gamma - 1)} \frac{E^{-p-\delta+\delta\gamma}}{r^{2\gamma+1}}$$

or

$$N(\mathbf{r}, t, E) \sim \frac{E^{-p-\delta+\delta\gamma}}{r^{2\gamma+1}}. \qquad (11)$$

Thus, it can be seen, that the function $\varphi(D_0)$ given by equation (8) leads to an anomalous spatial distribution of the particle density. Instead of the Gaussian distribution of particle density (5), which is predicted by the homogeneous normal diffusion models, the large-distance asymptotical behavior, we have obtained, is described by a power law $\sim r^{2\gamma+1}$.

In the discussed model (10) the energy spectrum observed at Earth is $N(E) \propto E^{-\eta} = E^{-p-\delta+\delta\gamma}$. From the slope $\eta = p + \delta(1-\gamma)$ we can find the injection spectrum exponent, p, at the Galactic source of the cosmic rays: $p = \eta - \delta(1-\gamma)$. It differs from the steady state energy spectrum exponent $p = \eta - \delta$. Since $\gamma > 1$, we have $p > \eta$.

It should be noted that a similar anomalous spatial distribution of the particle density may be found in the case of $\varphi(D_0) \sim D_0^{\theta} \exp\left(-\frac{D_0}{\mathfrak{D}}\right)$, where $\mathfrak{D}$ is the characteristic diffusivity for the Galactic medium and $0 < \theta < 1$. This distribution describes a model with exponential decrease at $D_0 \to \infty$. For this scenario we have found

$$N(\mathbf{r}, t, E) \sim r^{\theta-1} \exp\left(-\frac{r}{\sqrt{\mathfrak{D}E^{\delta}t}}\right) E^{-p-\delta(\theta+1)/2}.$$

4. Conclusions

We have performed the retrieval of the cosmic ray injection exponent at the Galactic sources from the spectrum observed at Earth taking into account the inhomogeneity of the turbulent interstellar medium. Two different transport models, based on anomalous and normal diffusion equations, have been used to relate the spectrum injected by the source with that measured at Earth.

Using the normal diffusion model, we have shown that the variation of the ISM properties leads to an anomalous spatial distribution of the particle density. Its asymptotic behavior (11) for $1.5 < \gamma < 2$ practically coincides with our result (4) obtained in the framework of the anomalous diffusion model for $1 < \alpha < 2$. In other words, the cosmic rays observed at Earth collectively appear to display non-diffusive (superdiffusive) characteristics although an individual particle has moved in a diffusive manner.

We have shown that the average value of the injection index p, obtained in the framework of both transport models, equals $p \sim (2.8 \div 3.0)$.

We note that a similar steep spectrum of accelerated particles has been observed from supernova remnants W44 [22] and IC 443 [23] with the Fermi Large Area Telescope. The value $p = 3$ has been found in [24] for RX J1713.7-3946.

The authors acknowledge support from The Russian Foundation for Basic Research grant No. 16-02-01103.

References

[1] V.L. Ginzburg, S. I. Syrovatskii, *Origin of cosmic rays* (Pergamon Press, 1964)
[2] V.S. Berezinskii, S. V. Bulanov, V. A. Dogiel et al., *Astrophysics of Cosmic Rays* (Amsterdam, North-Holland, 1990)
[3] G. Jóhannesson, R. Ruiz de Austri, A. C. Vincent et al., ApJ **824** (2016)
[4] B.G. Elmegreen, J. Scalo, Annu. Rev. Astron. Astrophys. **42**, 211 (2004)
[5] E.A. Bergin, M. Tafalla, Annu. Rev. Astron. Astrophys. **45**, 339 (2007)
[6] N. Sánchez, E. J. Alfaro, ApJ. Suppl. Ser. **178**, 1 (2008)
[7] Y.N. Efremov, A. D. Chernin, Physics — Uspekhi, **46**(1), 1 (2003)
[8] R. de la Fuente Marcos, C. de la Fuente Marcos, ApJ **700**, 436 (2009)
[9] N. Sánchez, N. Añez, E. J. Alfaro, M. C. Odekon, ApJ **720**, 541 (2010)
[10] A.A. Lagutin, Y.A. Nikulin, V.V. Uchaikin, Preprint ASU (2000/4)
[11] A.A. Lagutin, Y.A. Nikulin, V.V. Uchaikin, Nucl. Phys. B (Proc. Suppl.) **97**, 267 (2001)
[12] A.A. Lagutin, V.V. Uchaikin, Nucl. Instrum. Meth. **B201**, 212 (2003)
[13] A.D. Erlykin, A.A. Lagutin, A.W. Wolfendale, Astropart. Phys. **19**, 351 (2003)
[14] A.A. Lagutin, A.G.Tyumentsev, Bulletin of Altai State University **5**, 4 (2004 in Russian)
[15] A.A. Lagutin, A.V. Yushkov, A.G. Tyumentsev, IJMP A **20**, 6834 (2005)
[16] A.A. Lagutin, N.V. Volkov, A.S. Kuzmin, A.G. Tyumentsev, Bull. of RAS: Physics **73**(5), 581 (2009)
[17] N.V. Volkov, A.A. Lagutin, A.G. Tyumentsev, J. Phys. Conf. Ser. **632**, 012027 (2015)
[18] S.G. Samco, A.A. Kilbas, O.I. Marichev, *Fractional integrals and derivatives — Theory and Applications* (Gordon and Breach, New York 1993)
[19] V.V. Uchaikin, V.M. Zolotarev, *Chance and Stability* (VSP, Netherlands, Utrecht, 1999)
[20] V.M. Zolotarev, V.V. Uchaikin, V.V. Saenko, JETP **88**(4), 780 (1999)
[21] S.V. Petrovskii, A.Y. Morozov, Am. Nat. **173**, 278 (2009)
[22] A.A. Abdo, M. Ackermann, M. Ajello et al., Science **327**, 1103 (2010)
[23] A.A. Abdo, M. Ackermann, M. Ajello et al., ApJ **712**, 459 (2010)
[24] T. Tanaka, Y. Uchiyama, F. Aharonian et al., ApJ **685**, 988 (2008)

DOI: 10.1051/epjconf/201714519019

The primary energy spectrum with the Linsley method correcting heavy primary effects in the LAAS experiments

Hiroki Matsumoto[1,a], Atsushi Iyono[2,b], Kazuhide Okei[1], Shuhei Tsuji[1], Nobuaki Ochi[3], Nobusuke Takahashi[4], and Saya Yamamoto[2]

[1] Kawasaki Medical School, Kurashiki 701-0192, Japan
[2] Dept. of Fundamental Science, Faculty of Science, Okayama University of Science, Okayama 700-0005, Japan
[3] Toyo University, Tokyo 112-8606, Japan
[4] Dept. of Physics, Hirosaki University, Hirosaki 036-8560, Japan

Abstract. The compact EAS Extensive Air Shower (EAS) array (200 m^2) and the Linsley method have been used with the aim to observe the primary energy spectrum. The spectral indexes $\alpha_{\rm proton}$ and $\alpha_{\rm iron}$ are obtained as -2.46 ± 0.17 and -3.08 ± 0.24, respectively, in the primary energy region of 10^{16} eV – $10^{18.5}$ eV in the single observation of the compact EAS array. Additionally, $\alpha_{\rm proton}$ and $\alpha_{\rm iron}$ are obtained as -3.12 ± 0.13 and -3.52 ± 0.17, respectively, in the same primary energy region in the observation by restricting the zenith angle of EAS events.

1. Introduction

The Large Area Air Shower (LAAS) experiments [1,2] have been observing cosmic rays with compact Extensive Air Shower (EAS) arrays, and have maintained EAS arrays at several institutes since 1996. The purposes of LAAS experiments are to study the primary energy spectrum, anisotropy of high energy cosmic rays, and simultaneous and parallel EAS events [3–7]. The group of Okayama University of Science has mainly observed cosmic rays in the primary energy range above 10^{15} eV and studied its primary energy spectrum.

In order to observe EASs caused by high energy cosmic rays on the ground, many detectors scattered over a large area are required such as the Pierre Auger Observatory [8] and the Telescope Array Observatory [9] due to the feature of primary energy spectrum. However, these methods have a high cost in installing and maintaining detectors.

To solve the problem and carry out low-cost observation, we adapted the Linsley method [10], which is the result of the EAS core distance, r, dependent on the dispersion, σ_t, of the arrival time of EAS particles to the compact EAS array in LAAS experiments. Bezboruah [11] has also reported on the results of the primary energy spectrum obtained by adapting the Linsley method to their mini-array.

In the LAAS experiments, the observation of the primary energy spectrum has been carried out by combining Linsley's method with the compact EAS array since 2006, and an apparatus for improving the determination accuracy of the primary energy was installed in 2008.

2. Apparatus

Okayama University of Science (OUS) is in Okayama, Japan and located at 34°42′(N) and 133°56′(E), at sea level. There are five EAS arrays (OUS1∼OUS5) and these arrays, except the OUS4 which is the apparatus for obtaining information of the EAS zenith angle, each consist of eight scintillation counters.

The arrangement of the OUS1 array is shown in Fig. 1. In this work, the obtained data of OUS1 and OUS4 are used. Each scintillation counter equips a plastic scintillator of size 50 cm × 50 cm × 5 cm, a PMT (HAMAMATSU H7195) and a stainless steel case of 5 mm thickness. The OUS1 array is installed on the rooftop of a building in Okayama University of Science, covers an area of about 200 m^2. The data acquisition system (DAQ) consists of a CAMAC TDC (Kaizuworks Model 3781), a CAMAC ADC (Lecroy Model 2249W), and a shift register (Mbeware Model PL-320E). In this work, we use EAS event data obtained by the shift register because the shift register allows storing more information of time and the number of EAS singles from detectors than the TDC system. The event trigger time and clock synchronization are maintained by a GPS module (Kaizuworks Model 3051A) of time accuracy 1 μs. An EAS event is triggered by a coincidence detection of more than three detectors within a 2.5 μs time window.

The shift register module is able to store the time series signal from each detector as digital data with ± 2.5 μs time window from an EAS trigger time. The time resolution of the shift register is 5 ns.

OUS4 is the apparatus for obtaining the zenith angle of EAS events and is shown in Fig. 1. This apparatus equips scintillation counters for the EAS event trigger on the top and bottom layer, respectively, and for anti-counting at the side panels. The scintillation counter, the size of which is 20 cm × 50 cm × 1 cm, for the event trigger, is equiped

[a] e-mail: `h.matsumoto@med.kawasaki-m.ac.jp`
[b] e-mail: `iyono@das.ous.ac.jp`

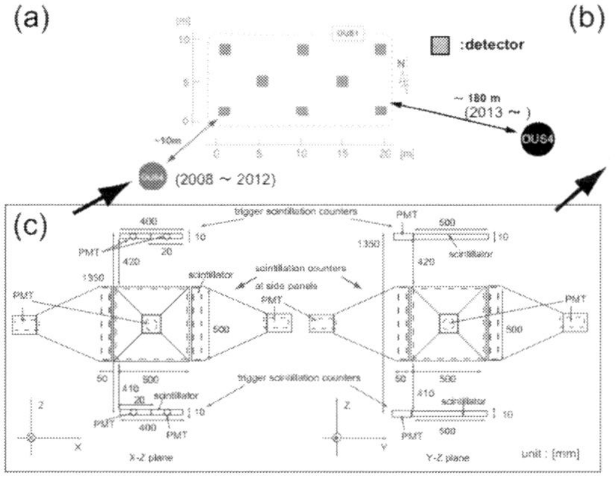

Figure 1. (a) The arrangement of OUS1 and OUS4. (b) Photograph of OUS4. (c) The schematic drawing of OUS4.

with a PMT and a black plastic cover. The total area of each of the bottom and top counters is 0.2 m^2 respectively. Anti-counters at the side panels are the same as those of OUS1. DAQ is also the same as the OUS1 and the GPS module is equipped for synchronizing the observation clock. The zenith angle, θ, of EAS events is distinguished by using anti-counters events of $\theta < 25.6°$ from all events. The OUS4 was located at about a distance of 10 m from the OUS1 in 2008 $\sim$ 2012. In 2013, we relocated OUS4 to a distance of 180 m from OUS1.

3. Method

3.1. EAS Core distance

The estimation of the core distance, r, of EAS is performed by using Linsley's method and the result of Okita [6]. Linsley showed that the relation between the average dispersion of arrival time distribution of EAS particles, $\langle\sigma_t\rangle$, and r follows the empirical formula

$$\langle\sigma_t\rangle = \sigma_{t0}\left(1 + \frac{r}{r_t}\right)^b, \qquad (1)$$

where

$$b = (2.08 \pm 0.08) - (0.4 \pm 0.06)\sec\theta + (0 \pm 0.06)\log\left(E/10^{17}\text{eV}\right), \qquad (2)$$

$\sigma_{t0} = 1.6$ ns, $r_t = 30$ m, θ is the zenith angle and E is the primary cosmic ray energy. We adapt b to the averaged value $\bar{b} = 1.65$ by the EAS zenith angle distribution expected by the observation instead of the original b, and ignore the contribution of the primary energy. In Fig. 2, $\langle\sigma_t\rangle$ is plotted as a function of r.

The EAS arrival distribution is assumed to be the gamma distribution by Linsley, and the best estimator, discussed in reference [6], of the dispersion of arrival time distribution of EAS particles, σ_t, is

$$\sigma_t = \frac{\sqrt{2}}{1.67}t_m, \qquad (3)$$

where t_m is the median of the time series data of the observed EAS events. The core distance, r, is obtained by substitution of Eq. (1) into Eq. (3), and rearrangement of the terms

$$r = 30\left((1.35t_m)^{\frac{1}{1.65}}\right). \qquad (4)$$

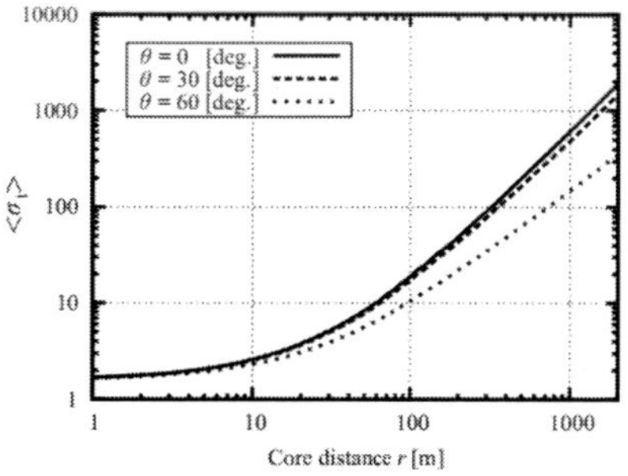

Figure 2. The relation between r and $\langle\sigma_t\rangle$. Each curve is Eq. (1) for $\theta = 0°, 30°, 60°$, respectively. The averaged θ is $\sim 23°$ in OUS1.

Note that the systematic error of r increases with increasing EAS zenith angle θ, because we use the averaged $\bar{b}$ instead of the original b.

3.2. Primary energy

In order to determine the primary energy of EAS events, we combine the database of lateral distributions of muons and electrons obtained by carrying out the EAS simulation program AIRES, and the observed data of the core distance and the particle density. In the EAS simulation, hadronic interaction models are assumed to be QGSJET II-3 and the Hillas Splitting model. The primary nuclei are assumed to be protons and irons, respectively. This simulation procedure is described in Ref. [7] in detail.

Figure 3 shows the calculated energy distribution assumed to be that the input primary energy E_0 is at 10^{17} eV in the primary nuclei of protons and irons in the simulation. The OUS1+4 in Fig. 3-(B) indicates the synchronized observation between OUS1 and OUS4. By restricting the zenith angle of EAS events by OUS4, the accuracy of the determination of the primary energy of OUS1+4 is able to improve more than that of the OUS1.

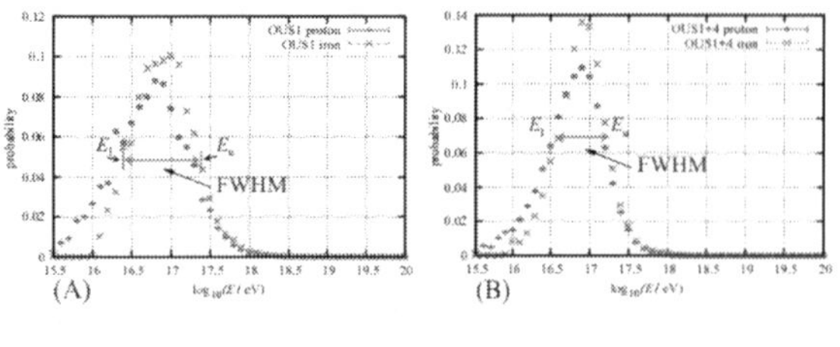

Figure 3. The simulated energy distributions in OUS1 (A) and OUS1+4 (B). Symbols of (+) and (∗) indicate proton primaries and iron ones, respectively.

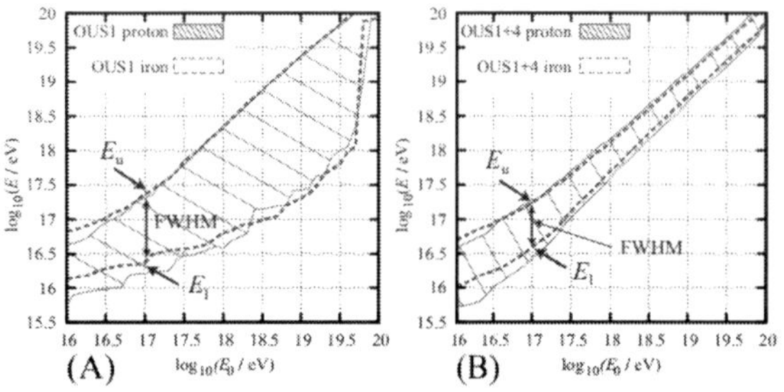

Figure 4. The energy resolution bands in the OUS1 (A) and the OUS1+4 (B). The shaded bands indicate the result of assumption of proton primaries. The dashed line bands indicate the result of assumption of iron primaries.

We define the energy resolution of detectors as the FWHM value of the calculated energy distribution. Figure 4 shows that the relation between E_0 and the FWHM. E_l and E_u in Fig. 4 correspond to the same letters in Fig. 3, respectively.

In order to determine the primary energy spectrum, the observed data spectrum corresponds to the spectrum calculated by the detector simulation. The true primary spectrum is assumed to be a single power-law E^{α}, where α is the spectral index value in the simulation. The α value is obtained as α' by being affected by the detector sensitivity and the fluctuation of the shower size. The calculation method of α from α' is described in Refs. [7] and [12].

4. Result

4.1. Data analysis

In this analysis, the data period is from 4/2006 to 3/2016 for OUS1, and from 8/2008 to 12/2012 in the synchronized observation between OUS1 and OUS4 (OUS1+4). The observation time, the total number of events and the number of events enabled to determine the primary energy are summarized in Table 1. The assumption of the primary nuclei are protons and irons, and N_{proton} and N_{iron} indicate the number of events enabled to determine the primary energy in each assumption, respectively.

The selection criteria for available events are the range of the observed core distance r_{obs}, $100\,\mathrm{m} < r_{\mathrm{obs}} < 2000\,\mathrm{m}$. Additionally, in OUS1+4, we used the coincidence events of the time difference between OUS1 and OUS4 within 10 μs, and the events of the EAS zenith angle θ within $\theta < 25.6°$.

Table 1. The data summary of the OUS1 and OUS4 observations.

	T_{obs} [day]	N_{obs}	N_{proton}	N_{iron}
OUS1	2171	1.5×10^7	8.6×10^4	1.8×10^5
OUS1+4	1336	2.0×10^5	5.5×10^2	1.9×10^3

T_{obs}: Total observation time
N_{obs}: Total number of triggered events
N_{proton}: Number of events enabled to determine primary energy ($\geq 10^{16}$ eV) assumed nuclei to be protons
N_{iron}: Number of events enabled to determine primary energy ($\geq 10^{16}$ eV) assumed nuclei to be iron.

Table 2. The spectral index α' and α.

		Primary energy region [eV]		
		(i) $10^{16} - 10^{19.5}$	(ii) $10^{16} - 10^{18.5}$	(iii) $10^{18} - 10^{19.5}$
OUS1	α'	-2.37 ± 0.11	-2.62 ± 0.15	-2.09 ± 0.18
Proton	α	-2.19 ± 0.11	-2.46 ± 0.17	-1.93 ± 0.17
OUS1	α'	-2.61 ± 0.13	-2.96 ± 0.15	-2.07 ± 0.07
Iron	α	-2.60 ± 0.18	-3.08 ± 0.24	-2.00 ± 0.07
OUS1+4	α'	-2.92 ± 0.15	-3.14 ± 0.25	–
Proton	α	-2.87 ± 0.16	-3.12 ± 0.13	–
OUS1+4	α'	-3.23 ± 0.17	-3.47 ± 0.15	–
Iron	α	-3.23 ± 0.20	-3.52 ± 0.17	–

4.2. Primary energy spectrum

The primary energy spectra obtained by this analysis are shown in Fig. 5. In order to calculate the spectral index value of the data spectrum, each spectrum is fitted by the least squares method to a single power-law spectrum, $E^{\alpha'}$. Because the spectral index seems to manifest a flattening around $10^{18.5}$ eV, the primary energy regions of the fitting are

(i) 10^{16} eV $\sim 10^{19.5}$ eV,
(ii) 10^{16} eV $\sim 10^{18.5}$ eV and
(iii) 10^{18} eV $\sim 10^{19.5}$ eV.

The OUS1 obtained spectral indexes $\alpha'_{\mathrm{proton}}$ and α'_{iron} -2.62 ± 0.15 and -2.96 ± 0.15, respectively, in the primary energy region (ii). Thus, α_{proton} and α_{iron} are equal to -2.46 ± 0.17 and -3.08 ± 0.24, respectively, by calculating from the relation between α and α'.

In OUS1+4, α is also calculated by using the same procedure. $\alpha'_{\mathrm{proton}}$ and α'_{iron} are -3.14 ± 0.25 and -3.47 ± 0.15, respectively, in the primary energy region (ii), and α_{proton} and α_{iron} are equal to -3.12 ± 0.13 and -3.52 ± 0.17, respectively. The obtained spectral indexes are summarized in Table 2.

5. Conclusion

In the LAAS experiments, we have observed cosmic rays and studied various phenomena attributed by those with compact EAS arrays since 1996. Especially, the primary energy spectrum has been observed by adapting Linsley's method to our compact EAS array in Okayama University of Science since 2008. The primary energy spectrum is updated in this work, α_{proton} and α_{iron} are equal to -2.46 ± 0.17 and -3.08 ± 0.24, respectively, in the primary energy region of 10^{16} eV – $10^{18.5}$ eV in the

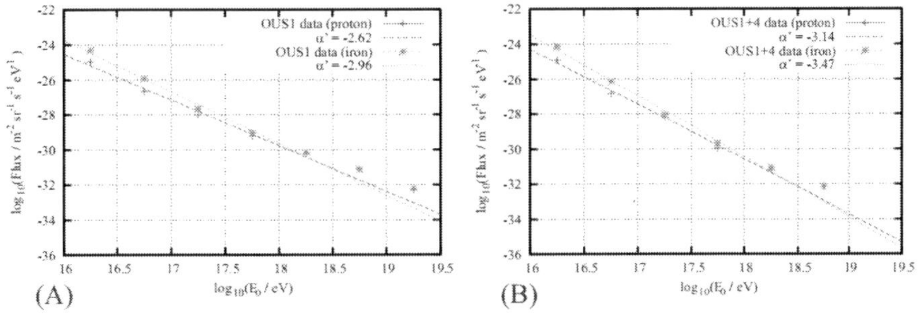

Figure 5. The data spectrum assumed to be proton and iron primaries, respectively, in the OUS1 (A), and in the OUS1+4 (B). Symbols of (+) and (∗) indicate observed data. Lines show the result of the least square fittings.

OUS1. also, α_{proton} and α_{iron} are equal to -3.12 ± 0.13 and -3.52 ± 0.17, respectively, in the same primary energy region in OUS1+4.

References

[1] T. Wada et al., Nucl. Phys. B (Proc. Suppl.) **75**, issue 1-2, 330–332 (1999)
[2] N. Ochi et al., J. Phys. G: Nucl. Part. Phys. **29**, 1169–1180 (2003)
[3] A. Iyono et al., Proceedings of the 30th ICRC Merida, 114 (2007)
[4] S. Ohara et al., Proceedings of the 32nd ICRC Beijing, 406 (2011)
[5] A. Iyono et al., Astrophysics and Space Sciences Transactions **7**, 327–333 (2011)
[6] M. Okita et al., Nucl. Phys. B (Proc. Suppl.) **175–176**, 322–325 (2008)
[7] H. Matsumoto et. al., Nucl. Instr. & Meth. A **614**, 475–482 (2010)
[8] I. Valino, Proceedings of the 34th ICRC Hague, 271 (2015)
[9] C. Jui, Proceedings of the 34th ICRC Hague, 35 (2015)
[10] J. Linsley, J. Phys. G: Nucl. Phys. **12**, 51–57 (1986)
[11] Bezboruah et al., Astropart. Phys. **11**, 395 (1996)
[12] H. Matsumoto et. al., Proceedings of the 34th ICRC Hague, 257 (2015)
[13] S. J. Sciutto, *AIRES* (1999) [astro-ph/9911331]
[14] S. Ostapchenko, Phys. Rev. D **74**, 014026 (2006)
[15] A. M. Hillas, Nucl. Phys. B (Proc. Suppl.) **52**, issue 3, 29–42 (1997)

Latest nuclear emulsion technology

Production, readout and interaction analysis

Hiroki Rokujo[a], Hiroaki Kawahara, Ryosuke Komatani, Misaki Morishita, Toshiyuki Nakano, Naoto Otsuka, and Masahiro Yoshimoto

Nagoya University, Furo-cho, Chikusa-ku, Nagoya 464-8602, Japan

Abstract. Nuclear emulsion is a extremely high-resolution 3D tracking detector. Since the discovery of the pion by C.F. Powell et al. in 1946, experiments with nuclear emulsions have contributed to the development of particle physics. (e.g. the OPERA collaboration reported the discovery of $\nu_\mu \rightarrow \nu_\tau$ oscillations in appearance mode in 2015) The technology of nuclear emulsion still keeps making progress. Since 2010, we have introduced a system of nuclear emulsion gel production to our laboratory in Nagoya University, and have started self-development of the new gel, instead of from the photographic film companies. Moreover, a faster automated emulsion scanning system is developed. Its scanning speed reaches 4000 cm^2/h, and the load for analyzing becomes more and more lighter. In this presentation, we report the status of nuclear emulsion technologies for cosmic ray experiments.

1. Introduction

Nuclear emulsion is an ultra high-precision position-sensitive detector. It consists of mostly gelatin and crystals of silver bromide (AgBr). The crystals penetrated by a charged particle grow up to silver grains (the diameter is about 0.8 μm) via a chemical development process, then we can observe the trajectory of the particle as a line of black points using optical microscopes. The intrinsic position accuracy is expected to be $200/\sqrt{12} = 56$ nm. In addition, although typical electronic trackers consist of readout channels in two or more independent projections for 3D reconstruction, nuclear emulsion directly gives 3D position information of the trajectory.

S. Kinoshita discovered that the radiation of alpha decay is a "track" of a charged particle by the observation of nuclear emulsion in 1915 [1]. C. F. Powell and his colleagues discovered the decay chain of $\pi \rightarrow \mu + \nu$ in nuclear emulsions exposed to cosmic rays [2].

Double-side poured nuclear emulsion plate realizes a precise measurement of incident angle. In the 1950's, a Japanese group developed a chamber, called Emulsion Cloud Chamber (ECC), that had a sandwich structure of nuclear emulsion and other material plates (carbon, lead, etc.). ECCs mounted on balloons or airplanes enabled to measure angles of jet particles from high-energy cosmic-ray interactions. A fire ball model of multiple meson production in nucleon-nucleon collisions [3] and the discovery of charmed meson in cosmic-ray interactions [4] were produced by the observation of ECCs in the era when no accelerator was in Japan.

Though the observation and the measurement of tracks in nuclear emulsion had been done by human eyes and microscopes, the realization of fully automated scanning systems enabled high resolution and large-scale experiments which use the advantages of nuclear emulsion. For example, the OPERA experiment discovered $\nu_\mu \rightarrow \nu_\tau$ oscillation in appearance mode on the basis of these technologies [5].

Here, we report the recent improvements of the nuclear emulsion technology since 2010.

2. Emulsion gel production

In 2010 we introduced a system of emulsion gel production to the laboratory in Nagoya University, and started to supply the new gel instead of relying on photographic film companies. Figure 1 shows a nuclear emulsion gel produced in Nagoya on the left, and an image of AgBr crystals captured by SEM after removing gelatin on the right.

Our self-production system enables us to develop ambitious nuclear emulsion gels. We succeeded in the development of a high-sensitive nuclear emulsion by increasing the occupancy of AgBr crystals from typically 30–35% to 55%. Figure 2 shows a comparison of microscopic images between two type of nuclear emulsion. The grain density (the average number per unit length) of the silver particles of this new type of gel is two times higher than that of a typical gel.

The size control of AgBr crystals is also studied. 200 nm-diameter crystal has been used for detection of minimum ionization particles for a long time. Recently, we produced various-sized crystals according to the purposes of experiments. The minimum-sized crystals (the diameter is 30–40 nm) are being developed for dark matter directional searches [6], and the maximum-sized crystals (~800 nm) for the long term observation requesting the stability of the latent image [7].

[a] e-mail: `rokujo@flab.phys.nagoya-u.ac.jp`

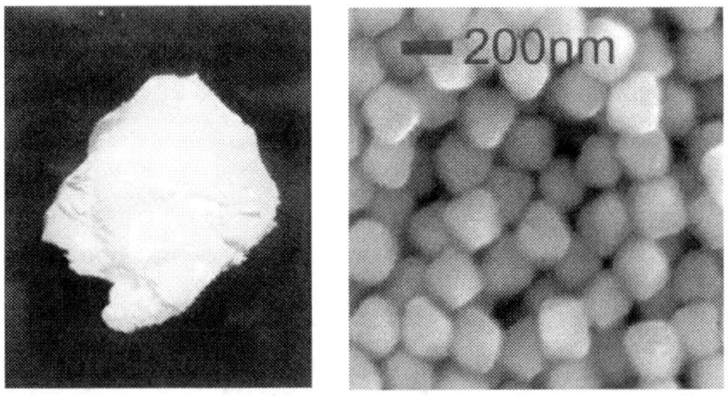

Figure 1. Nuclear emulsion gel produced in Nagoya on the right, and SEM image of AgBr crystals after removing gelatin on the left.

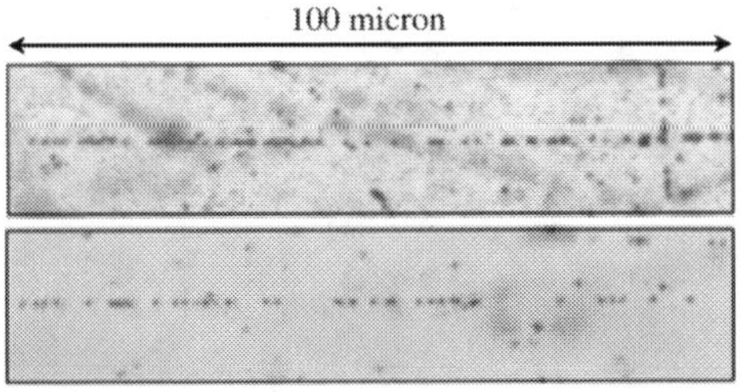

Figure 2. Microscopic view of highly sensitive film (upper) and typical film (lower).

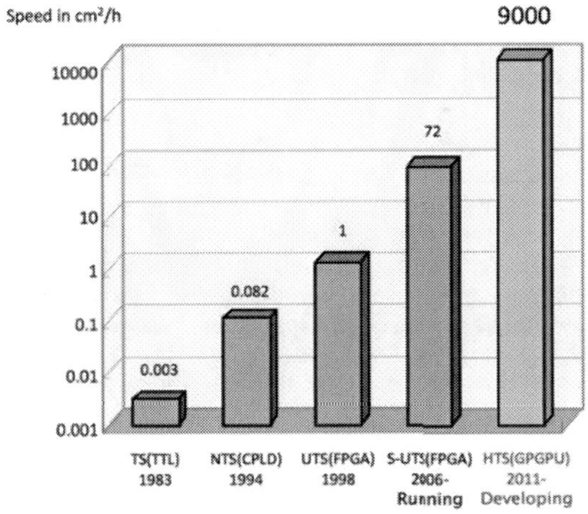

Figure 3. Evaluation of scanning speed by generations of Track Selector.

The environment of the mass production of films is constructed in the laboratory. We have already produced the nuclear emulsion film at a speed of $100\,m^2$/year and supplied several experiments. In a few years, the production speed will increase to about $1000\,m^2$/year for accelerator neutrino experiments, balloon-borne experiments, muon radiography projects, etc.

3. Emulsion readout system

The concept of fully automated emulsion readout system, Track Selector (TS), was invented by K. Niwa [8], and was put to practical use in the 1990's. The latest TS is called Hyper Track Selector (HTS). HTS has been developed in Nagoya University since 2011. Figure 3 shows scanning speed by generations of TS, and HTS is designed for $0.9\,m^2$/hour.

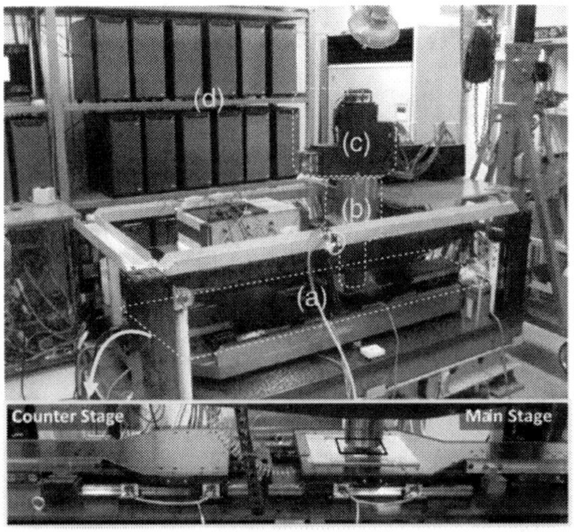

Figure 4. Picture of HTS. It consists of (a) precision stages which move to the next view precisely and quickly after capturing, (b) a huge objective lens that realizes a 5 mm × 5 mm-size FOV, (c) 72 mosaic CMOS sensors that cover the large FOV (each 2M pixels, 340 fps.), and (d) 36 PSs for realtime process of track-recognizing with 27 GB/s image data.

HTS consists of precision stages, a huge objective lens, 72 mosaic CMOS sensors, and 36 PSs. These are shown in Fig. 4. The most characteristic point of HTS is a very wide field of view (FOV), 5 mm × 5 mm. The FOV of HTS is 600 times larger than that of the previous system, S-UTS. To cover its FOV with 0.4 μm-pitch pixel, the image is divided into six imaging planes through prisms, and 12 CMOS sensors (each 2M pixels, 340 fps) capture it at each plane. The image data from 72 sensors are sent to 36 PCs, then the track recognition and the calculation (angles, positions, etc.) are processed immediately. A motor-driven XY axes stage moves to the next view precisely and quickly. At that time, the counter stage moves in the opposite direction from the main stage to cancel acceleration.

Figure 5 shows the pulse height distribution obtained by the high-sensitive film and HTS. TS recognizes a track from 16 images (32 images in a double-side poured film) captured over changing a focal plane, and the number of hit layers in a signal track is called the pulse height. By improvement of the grain density, clear signals were observed. Figure 6 shows the track-finding efficiency obtained by the high-sensitive film and HTS. At each incident angle from $\tan\theta = 0.0$ (a vertical angle) to $\tan\theta = 2.0$, more than 95% was evaluated [9].

The operation of HTS was started in 2015 at half of the designed speed. Nuclear emulsion films employed in the GRAINE-2015 balloon experiment were scanned as the first practical data taking of HTS, and it took three months to finish $41\,m^2$ in total [10].

Table 1. Candidates of previous emulsion experiments for the archive project.

Experiments	Original purpose	Total Area for Scanning
RUNJOB (balloon experiment)	primary cosmic-ray observation	$120\,m^2$
JACEE (balloon experiment)	primary cosmic-ray observation	$900\,m^2$
CHORUS (accelerator ν experiment)	search for $\nu_\mu \rightarrow \nu_\tau$ oscillations	$400\,m^2$
DONUT (accelerator ν experiment)	detection of ν_τ	$100\,m^2$

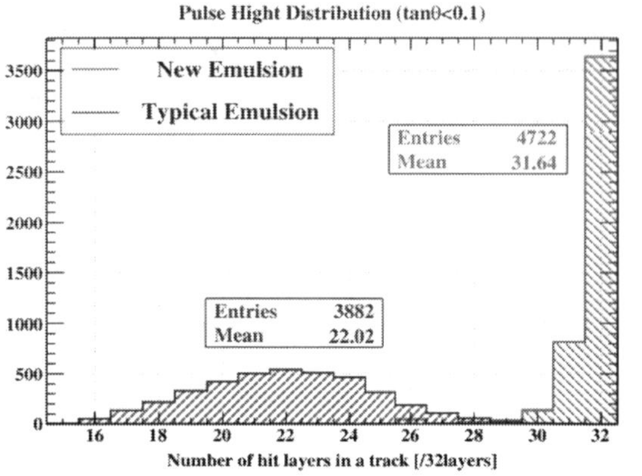

Figure 5. Pulse height distributions. Red histogram was given by high-sensitive emulsion film and HTS. Blue histogram was given by typical emulsion film (OPERA type).

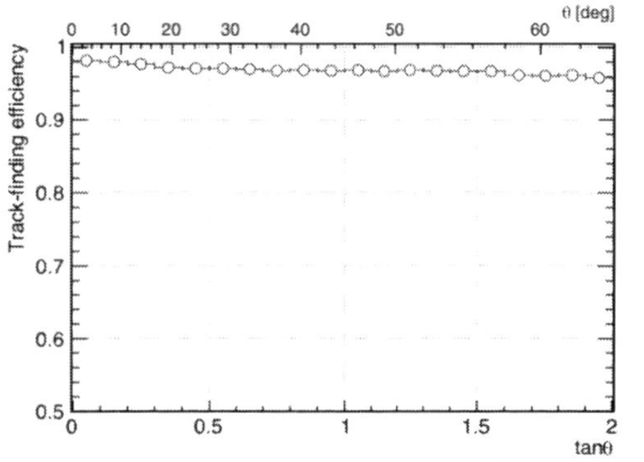

Figure 6. Track-finding efficiency as a function of incident angle obtained by the high-sensitive film and HTS.

4. Applications for cosmic-ray experiments

4.1. Balloon-borne gamma-ray telescope

GRAINE is a project aimed at a precise observation of gamma-ray sources (the energy region is sub-GeV/GeV) with a balloon-borne nuclear emulsion chamber. The gamma-ray telescope with the original timestamp technique [11, 12] has the following advantages: high angular resolution (0.1deg at 1 GeV); polarization sensitive; large aperture area($> 10\,m^2$). Recently, a small-scale observation was made to demonstrate the performance of the detector [13].

4.2. Emulsion archives project

We also plan to archive "digital" data of emulsion plates of past experiments and to make it public. In previous (accelerator and cosmic-ray) emulsion experiments, the analysis was performed by focusing on special interesting events. A re-scan data set of the whole of the detector volume provides for minimum-bias and no-dead-time analysis without an event trigger of another detector (electronic counters or X-ray films). We will start a test scan of old emulsion plates (see Table 1) and discuss about a standard data format etc, toward opening archive data to the public.

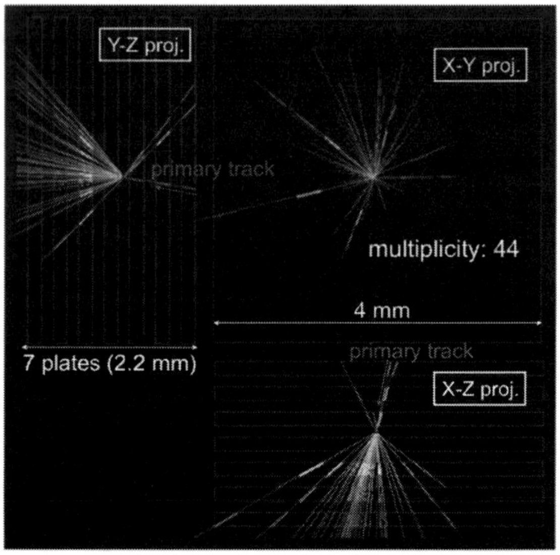

Figure 7. Cosmic-ray interaction event recorded in a balloon-borne chamber. It was detected by an offline analysis method under development.

4.3. Study for cosmic ray interactions

We are developing an offline analysis method to detect hadronic interactions recorded in the emulsion chamber. Figure 7 shows a sample event recorded in a balloon-borne chamber. If a systematic analysis flow is established, it is possible to search for short-lived particles from a large number of cosmic-ray interactions, and to measure the production rate of charmed particles at balloon altitudes. In accelerator experiments, the measurement of differential cross section of D_s mesons produced in proton interactions is starting at the CERN SPS. Nuclear emulsions can record all tracks emitted from the vertex with 4-π angular acceptance, and cover the wide range of rapidity in the laboratory frame. A precise comparison of the topological values between hadron interaction models will also be tried.

5. Conclusion

We report the status of nuclear emulsion technologies for cosmic ray experiments. The highest quality emulsion films became available by introducing self-gel production system to the laboratory level. The emulsion readout systems are getting increasingly faster and the next generation system is being planned. These innovations of nuclear emulsion technology will lead cosmic ray physics to the next stage.

This work was supported by JSPS KAKENHI Grant Number 26247039, 26105510, and 16K17691.

References

[1] S. Kinoshita and H. Ikeuti, Philosophical Magazine and Jounal of Science Ser. 6, **29**, 171, pp. 420-425 (1915)
[2] C.M.G. Lattes, G.P.S. Occhialini and C.F. Powell, Nature **160**, 453 (1947)
[3] K. Niu et al., Nuovo cimento **10**, p. 994, (1958)
[4] K. Niu et al., Progress of theoretical physics **46**(5), 1644–1646 (1971)
[5] N. Agafonova et al. Physical review letters **115**(12), 121802 (2015)
[6] T. Naka et al., Nuclear Instruments and Methods A **718**, 519–521 (2013)
[7] T. Ariga et al., Journal of Instrumentation **11**(03), P03003 (2016)
[8] K. Niwa et al., Proc. Int. Cosmic Ray Symp. High Energy Phenomena, Tokyo, p. 149 (1974)
[9] K. Ozaki et al., Journal of Instrumentation **10**(12), P12018 (2015)
[10] H. Rokujo et al., Proc. of ISVHECRI 2016, Moscow (2016)
[11] S. Takahashi et al., Nuclear Instruments and Methods A **620**, 192 (2010)
[12] H. Rokujo et al., Nuclear Instruments and Methods A **701**, 127 (2013)
[13] S. Takahashi et al., Prog. Theor. Exp. Phys. **2015**, 043H01 (2015)

Study of the EAS size spectrum in thermal neutrons measured by the PRISMA-YBJ array

Oleg Shchegolev[1,a], Victor Alekseenko[1], Anastasia Bagrova[1,2], Yayun He[3], Bingbing Li[3], Shuwang Cui[3], Xinhua Ma[4], Yuri Stenkin[1,2], and Vladimir Stepanov[1]

[1] Institute for Nuclear Research, Russian Academy of Sciences, Russia
[2] Nuclear Research National University MEPhI, Russia
[3] College of Physics and Information Engineering, Hebei Normal University, China
[4] Key Laboratory of Particle Astrophysics, Institute of High Energy Physics, Chinese Academy of Sciences, China

Abstract. An Extensive Air Shower array of a novel type has been constructed in Tibet at an altitude of 4300 m above sea level. It consists of specially designed scintillator en-detectors capable of measuring two main EAS components: hadrons (n) and electrons (e). Results of the experiment are presented. It is demonstrated that the EAS integral distribution in the number of thermal neutrons can be well fitted by a single power law function. Comparison with detailed Monte-Carlo simulations using CORSIKA6.9 and GEANT4.10 codes are also shown.

1. Introduction

The hadronic component is the main one in Extensive Air Showers (EAS), but is still not studied well due to the absence of inexpensive detectors capable of covering a large area. In 2001 a novel method was developed in INR RAS to solve this problem [1]. The main feature of this method is a new electron-neutron detector (en-detector) capable to simultaneously measure two components of EAS: the electromagnetic component and thermal neutrons produced by the EAS hadronic component. This method allows us to make a large area array of a novel type (PRISMA project). The prototype of this array called PRISMA-YBJ is already running in Tibet (Yangbajing, 4300 m above sea level) since January 2013. Details of the detector and array can be found elsewhere [2]. The EAS size in thermal neutrons was never studied before, and we present here results obtained after about 3 years of the PRISMA-YBJ continuous operation.

2. Method

The array is deployed inside the ARGO-YBJ [3] experimental hall just above RPC plates with a layout shown in Fig. 1. The efficiency of the scintillator for thermal neutron capture was found to be 20% but the efficiency of the en-detector for thermal neutron recording is 14% due to pulse height selection. The neutrons are recorded as delayed pulses within a time gate of 20 ms. It is more than enough to collect a great bulk of thermal neutrons produced by EAS hadrons [4]. An on-line program pre-analyzes the data and stores the energy deposit (above a threshold of 2.5 m.i.p.) and the number of recorded neutrons in each detector. A two-fold coincidence trigger condition is applied for the array to store the data. Additionally, during off-line data processing, 4-fold coincidence with energy deposit above 5 m.i.p. is used for the event selection due to the complicated task of core location.

We applied a traditional maximum-likelihood analysis of EAS, employing the Nishimura-Kamata-Greisen (NKG) function for the electromagnetic component to find the position of the EAS core, its age, EAS size (Ne), etc. As has been shown earlier [5], the total number of secondary neutrons (mostly evaporation ones) is proportional to the number of hadrons passing through the array area. Therefore, measuring the EAS size spectrum in thermal neutrons is equivalent to that in the number of high energy hadrons.

3. Simulation

We made a detailed GEANT4-based simulation of our array, including en-detectors and experimental hall. The models we used are QGSP, BIC and NeutronHP. This set is often used in GEANT4 simulations of experiments with neutrons, for example neutron monitors [6]. It allows us to calculate neutron production by EAS hadrons and the neutron thermalization process. The following results were obtained due to the simulations: energy deposit in the en-detector as a function of energy of EAS particles, dependence of neutron multiplicity on EAS hadrons energy and lateral distribution of recorded neutrons around the hadron trajectory [7]. The CORSIKA-based Monte-Carlo simulation uses dependences obtained with GEANT4 to simulate our experiment. The models used in CORSIKA are QGSJET-II.03 and GHEISHA-2002. Simulations were performed for primary protons and iron in the energy range 30 TeV (100 TeV for iron) 300 PeV. The integral primary energy spectra index was $= -1.7$ in the whole range. Simulated data are stored at the same format as the experimental ones.

[a] e-mail: shchegolev_oleg@mail.ru

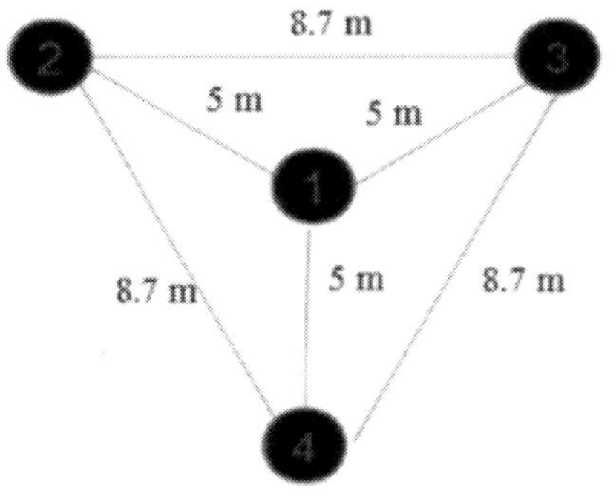

Figure 1. PRISMA-YBJ array layout.

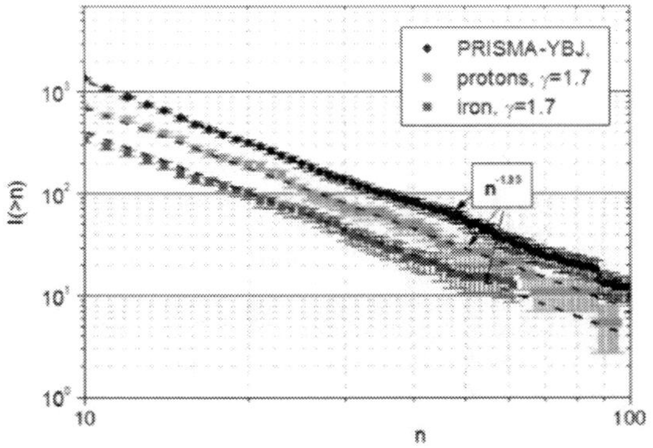

Figure 2. The integral neutron EAS size distribution. Black points experimental data from PRISMA-YBJ array. Red points simulation using CORSIKA and GEANT4 codes for primary protons. Green points the same for primary iron. Curves are shifted to prevent data overlapping

4. Results

The experimental integral distribution of showers on the number of recorded thermal neutrons as measured by the PRISMA-YBJ array with en-detectors is shown in Fig. 2 (black points). The spectrum follows a power law with an index of -1.95 ± 0.05 in the range from 10 to 100 neutrons per shower.

As simulations show (Fig. 2, green and red points), the integral EAS distribution on the number of thermal neutrons also follows a power law with index -1.95 ± 0.05. Results of the experiment and our simulations are in good agreement.

5. Conclusion

A prototype of a novel type of EAS experiment (PRISMA-YBJ) is running in Tibet for more than 3 years. It is noticeable that, for the first time, the EAS size spectrum in thermal neutrons at mountain level measured with the array shows a pure power law behavior in the truncated neutron multiplicity range corresponding to cosmic ray primary energies 1–10 PeV. This result is in excellent agreement with that obtained by us at sea level [8].

This study was performed at the Institute for Nuclear Research Russian Academy of Sciences. The Computational Cluster of the Theoretical Department of INR RAS was used for simulations. This work was supported in Russia by RFBR (grants 14-02-00996, 16-32-00054 and 16-29-13067_ofi_m), RAS Praesidium Program "Fundamental properties of matter and astrophysics", and in China by NSFC (No. 10975046, No. 11375052).

References

[1] Yu.V. Stenkin and J.F. Valdes-Galicia, Proc. of 27th ICRC, Hamburg **4**, 326 (2001)
[2] B. Bartoli et al., Astropart. Phys. **81**, 49–60 (2016)
[3] G. Staiti et al., Nucl. Instr. Meth. Phys. Res. A, **588**, 1-2, 7–13 (2008)
[4] Yu.V. Stenkin, Nucl. Phys. B (Proc. Suppl.) **196**, 293–296 (2009)
[5] Yu.V. Stenkin et al., Chinese Physics C, **37**, 015001 (2013)
[6] P. Paschalis et al., New Astronomy, **33**, 26–37 (2014)
[7] O.B. Shchegolev et al., Journ. of Phys.: Conf. Ser. **718**, 052038 (2016)
[8] A.A. Petrukhin et al., Proc. 34th ICRC, Hague, PoS(ICRC2015), 427 (2015)

AUTHOR INDEX

AUTHOR INDEX

AUTHOR INDEX

AUTHOR INDEX

AUTHOR INDEX